천수한

Digital Communications by Satellite

PRENTICE-HALL INFORMATION AND SYSTEM SCIENCES SERIES
Thomas Kailath, *Editor*

ANDERSON & MOORE	*Optimal Filtering*
ASTROM & WITTENMARK	*Computer Control Theory and Design*
GOODWIN & SIN	*Adaptive Filtering, Prediction and Control*
KAILATH	*Linear Systems*
MACOVSKI	*Medical Imaging Systems*
MELSA & SAGE	*An Introduction to Probability and Stochastic Processes*
SPILKER	*Digital Communications by Satellite*

Digital Communications by Satellite

J. J. SPILKER, JR. Ph.D.

Chairman, Stanford Telecommunications, Inc.

PRENTICE-HALL, INC., *Englewood Cliffs, New Jersey*

Library of Congress Cataloging in Publication Data

Spilker, J J

Digital communications by satellite.

(Prentice-Hall information and system sciences series)

Bibliography: p. 625

Includes index.

1. Artificial satellites in telecommunication.
2. Data transmission systems. I. Title.

TK5104.S64 621.38'0423 75–43878

ISBN 0–13–214155–8

10 9 8 7

Printed in the United States of America

PRENTICE-HALL INTERNATIONAL, INC., *London*
PRENTICE-HALL OF AUSTRALIA PTY. LIMITED, *Sydney*
PRENTICE-HALL OF CANADA, LTD., *Toronto*
PRENTICE-HALL OF INDIA PRIVATE LIMITED, *New Delhi*
PRENTICE-HALL OF JAPAN, INC., *Tokyo*
PRENTICE-HALL OF SOUTHEAST ASIA PTE. LTD., *Singapore*

To my Mother and Father

Contents

Preface

During the past few years, the importance of digital communications, particularly digital satellite communications, has increased rapidly. The accumulation of a vast body of engineering literature in the various technical journals has accompanied the design and development of digital coding and transmission systems, and the launch of an increasing number of communications satellites. At the same time, however, there has been no attempt to present and to unify this material in a single-volume textbook useful both to the student and to the communications systems designer.

Much of the material in this book was first presented in a graduate communications theory course at Stanford University in 1970. Its scope is appropriate for a two-quarter course of graduate instruction. While the level of the mathematics used has been purposely restrained to allow comprehension by a wide audience, it is assumed that the reader has already been exposed to a first course in random processes [Davenport, 1970; Davenport and Root, 1958; Cramer, 1945] and a first course in communications theory [Stein and Jones, 1967; Sunde, 1968; Lee, 1971; Schwartz, 1970; Wozencraft and Jacobs, 1965].

No book is created by a single author, and this one is no exception. The author has made extensive use of important material presented in the IEEE Transactions on Communication Systems, the Bell System Technical Journal, and papers presented at various conventions by members of the U. S. Army Satellite Communications Agency, the U. S. Air Force Space and Missile System Organization, the Defense Communications Agency, the National

Aeronautics and Space Administration, the California Institute of Technology Jet Propulsion Laboratory, and COMSAT Corporation. For example, substantial portions of Part 1 are based on excellent contributions by members of The Bell Telephone Laboratories. The author has attempted to give adequate credit to the authors of the cited material; however the sheer volume of published material makes that task almost impossible, and the author apologizes in advance to any author not adequately credited.

The author has had the valuable help of numerous individuals in reviewing the final versions of this book. In particular, the author expresses his appreciation to: Mr. Herman A. Bustamante, Dr. Charles R. Cahn, Dr. Horen Chang, Mr. C. L. Cuccia, Dr. Burton I. Edelson, Col. E. D. Frankhouser, Dr. Floyd M. Gardner, Mr. Walter J. Gill, Dr. Robert Y. Huang, Mr. Jay J. Jones, Dr. William C. Lindsey, Dr. D. Thomas Magill, Dr. Francis D. Natali, Dr. John F. Ohlsen, Col. Bradford W. Parkinson, USAF, Dr. Paul D. Shaft and Dr. Andrew J. Viterbi. Each of these colleagues has made important contributions to one or more chapters of the book. Furthermore, the author has drawn heavily from their published works in preparation of the manuscript. Finally, the author gratefully acknowledges the patience and hard work of Mrs. Deane Saltzman, Mrs. Ann Rowe and Ms. Joanna McClean who typed most of the manuscript, and the assistance and encouragement of Anna Marie Spilker during the final writing and editing process.

J. J. Spilker, Jr.
Palo Alto, California

Digital Communications by Satellite

CHAPTER 1

PRINCIPLES OF DIGITAL SATELLITE COMMUNICATIONS

1-1 INTERNATIONAL COMMUNICATIONS BY SATELLITE

The first communication satellite SCORE was placed in orbit in 1958. Communications by active satellite repeaters began in 1962 with Telstar, followed by the first synchronous satellite Syncom in 1963. The first INTELSAT satellite, INTELSAT I (Early Bird), was launched April 6, 1965, and numerous other INTELSAT satellites have been orbited since that initial launch. Satellite communications is now a major means for international as well as domestic communications over long or moderate distances. Various organizations and governments provide operational satellite communication serivice, including the INTELSAT Consortium through COMSAT Corporation, the U.S. Government, and the Canadian domestic satellite communications network. The use of satellites for communication continues to expand as existing networks are enlarged, and numerous other countries are implementing their own domestic satellite communication networks.

In almost all systems, the communications satellites are in orbits that are synchronous with the earth rotation, which allows considerable system simplification. Each earth terminal then operates continuously with the same communications satellite. The Initial Defense Communications Satellites (IDCS) were not synchronous, but instead they drifted slowly around the earth, relative to a fixed point on earth in approximately two weeks. They required each earth terminal to hand-over from satellite-to-satellite thereby causing time gaps in communications. Synchronous satellites avoid the hand-over problem but require additional fuel to make longitudinal and possibly latitudinal corrections to the orbit. This additional fuel is a small price to pay however for the simplicity in system operation and the avoidance of communication gaps.

1-2 ADVANTAGES OF DIGITAL TRANSMISSION

Communications by digital signaling is an increasingly important technique for radio communication by satellite relay and other means. Digital transmission has a number of advantages over other techniques. These include: (1) the ease and efficiency of multiplexing multiple signals or handling digital messages in "packets" for convenient switching; (2) the relative insensitivity of digital circuits to retransmission noise, commonly a problem with analog systems; (3) potential for extremely low error rates and high fidelity through error detection and correction; (4) communications privacy; and (5) the flexibility of digital hardware implementation, which permits the use of microprocessors and miniprocessors, digital switching, and the use of large scale integrated (LSI) circuits.

Digital transmission techniques are gaining increased usage for satellite communication, microwave relay, and cable or waveguide transmission. However, the original and final forms of the information transmitted by the digital link may be analog voice or video and therefore the analog/digital interface is an important element of the communications system.

Emphasis in this book is on transmissions at microwave frequencies and UHF. Most satellite communication is at microwave frequencies largely because the available bandwidth is substantial. However, transmission in the UHF frequency band has important application to relatively low data rate mobile users where near-omnidirectional antennas are employed.

1-3 SYSTEM CONFIGURATION

Inputs can arrive at the ground terminal in a variety of analog and digital forms. These original information sources, analog and digital, commonly include:

Analog	Digital
Voice	Teletype or multiplexed Teletype
Multiplexed voice	Computer input/output
Video	Digital television or imagery
Scanned mail or letters	Digitized voice

Several configurations of these signal sources are shown in Fig. 1-1. Analog-to-digital (A/D) conversion of the analog signals can occur at the

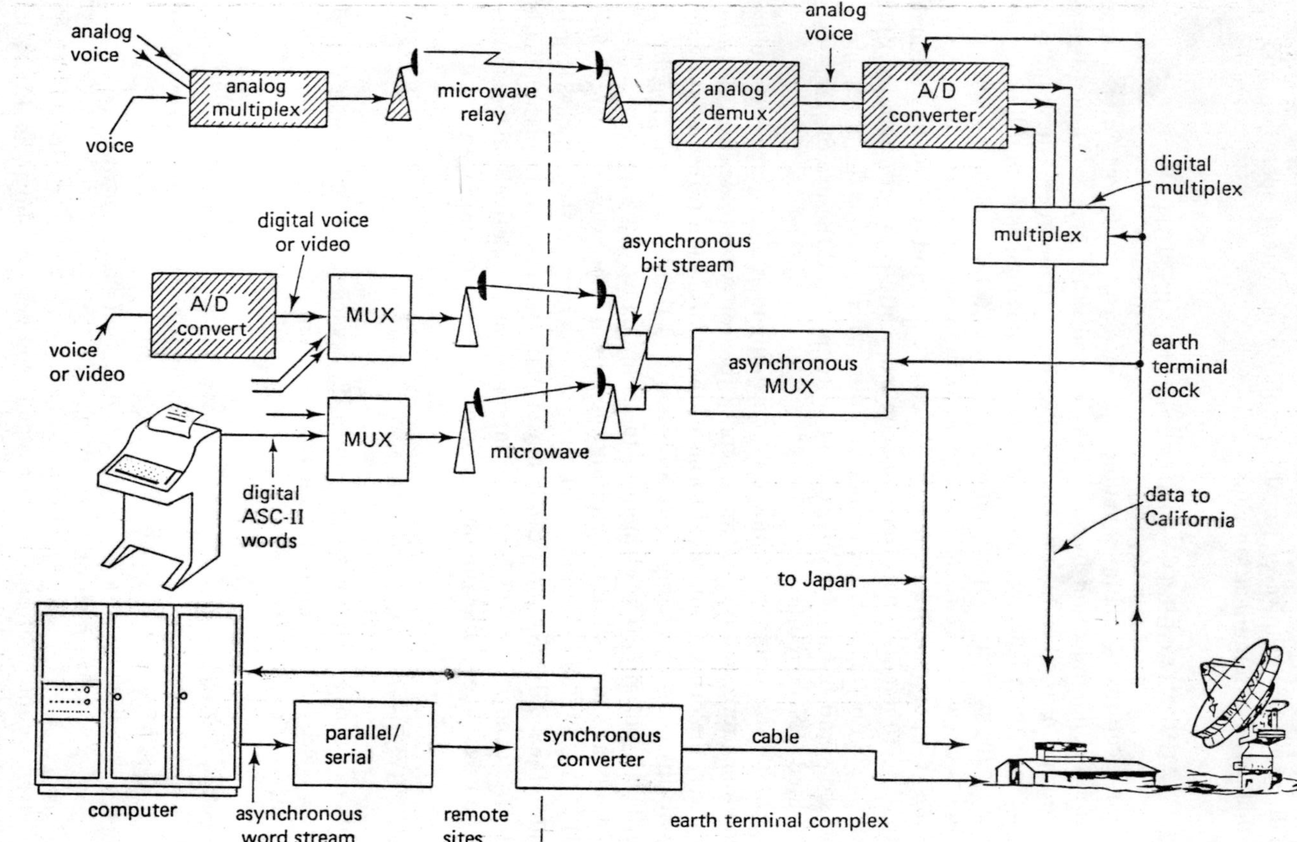

Fig. 1-1 Illustrative configuration of data inputs to an earth station. Analog communication system elements are shown shaded.

information source or at the earth terminal itself. If the analog quantities are quantized at their source, they arrive at the ground terminal in digital form each as a serial bit stream. These parallel bit streams from different sources generally are quasi-synchronous or plesiosynchrous with one another (not

Table 1-1 Seven-bit American Standard Code for Information Interchange (ASCII)

b_7				0	0	0	0	1	1	1	1
b_6				0	0	1	1	0	0	1	1
b_5				0	1	0	1	0	1	0	1
b_4	b_3	b_2	b_1								
0	0	0	0	NULL	DC_0	b	0	@	P	UNASSIGNED	UNASSIGNED
0	0	0	1	SOM	DC_1	!	1	A	Q		
0	0	1	0	EOA	DC_2	"	2	B	R		
0	0	1	1	EOM	DC_3	#	3	C	S		
0	1	0	0	EOT	DC_4 (STOP)	$	4	D	T		
0	1	0	1	WRU	ERR	%	5	E	U		
0	1	1	0	RU	SYNC	&	6	F	V		
0	1	1	1	BELL	LEM	' (APOS)	7	G	W		
1	0	0	0	FE_0	S_0	(	8	H	X		
1	0	0	1	HT SK	S_1	)	9	I	Y		
1	0	1	0	LF	S_2	*	:	J	Z		
1	0	1	1	VTAB	S_3	†	;	K	[		
1	1	0	0	FF	S_4	, (COMMA)	<	L	\		ACK
1	1	0	1	CR	S_5	—	=	M	]		①
1	1	1	0	SO	S_6	.	>	N	↑		ESC
1	1	1	1	SI	S_7	/	?	O	←		DEL

Standard 7-bit set code positional order and notation are shown with b_7 the high-order, and b_1 the low-order bit position. An 8th. bit can be added as a parity bit or a fixed bit to form an 8-bit byte.

Table 1-1 (Continued)

Example:

	b_7	b_6	b_5	b_4	b_3	b_2	b_1
The code for "R" is:	1	0	1	0	0	1	0

NULL	Null/Idle	DC_1–DC_3	Device control
SOM	Start of message	DC_4 (Stop)	Device control (stop)
EOA	End of address	ERR	Error
EOM	End of message	SYNC	Synchronous idle
EOT	End of transmission	LEM	Logical end of media
WRU	"Who are you?"	S_0–S_7	Separator (information)
RU	"Are you . . . ?"	b	Word separator (space, normally nonprinting)
BELL	Audible signal		
FE_0	Formal effector	$<$	Less than
HT	Horizontal tabulation	$>$	Greater than
SK	Skip (punched card)	↑	Up arrow (exponentiation)
LF	Line feed	←	Left arrow (implies/ replaced by)
V_{TAB}	Vertical tabulation		
FF	Form feed	\	Reverse slant
CR	Carriage return	ACK	Acknowledge
SO	Shift out	①	Unassigned control
SI	Shift in	ESC	Escape
DC_0	Device control reserved for data link escape	DEL	Delete/Idle

exactly at the same clock phase or frequency), and must be multiplexed and reclocked by pulse or word stuffing to convert them to a common stable earth terminal clock rate. Conversion to synchronous digital form with high clock-rate stability is desired in order to permit efficient demodulation at the receive earth station.

When the source generates a serial bit stream composed of ASCII words (Table 1-1), the signal arrives at the earth terminal already coded in digital form. This source format is typical of Teletype or computer-computer installations. On the other hand, some analog signals arrive at an earth terminal in multiplexed analog form. In this case, they can be demultiplexed, sampled, and quantized in synchronism with that earth terminal/clock, and the synchronous digital streams are then multiplexed according to their ultimate earth terminal destination. The resulting one or more parallel synchronous bit streams is then fed to a digital modem that modulates a carrier at some convenient IF frequency (usually 70 MHz or 700 MHz). Each modulated carrier is then up-converted to the appropriate RF frequency for transmission to the satellite. In some terminals the modulation of the carrier takes place directly at RF.

The modulated signals from several earth terminals then arrive at the synchronous satellite transponder where they are relayed to the appropriate destination earth terminal by means of either an earth-coverage or narrow-beam satellite antenna.

Some signals must be relayed through a second satellite (Fig. 1-2) to reach their final destination. This second relay is accomplished by demodulating and demultiplexing the signal destined for relay through the second satellite, multiplexing this signal with others destined for that satellite, and transmitting the signal through a second earth terminal. This second multiplexing operation may require asynchronous or quasi-synchronous multiplexing, since bit rates through the first and second satellites are not usually perfectly synchronous with one another.

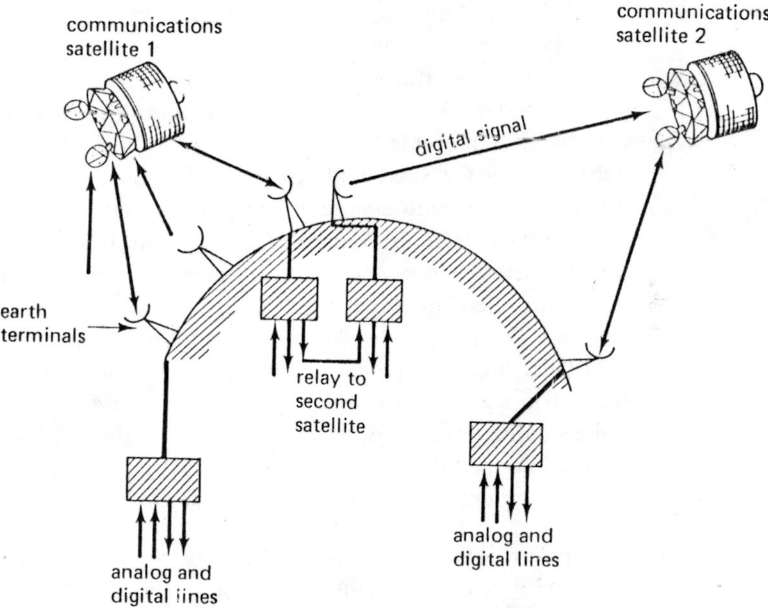

Fig. 1-2 Model of digital communications by satellite relay. The model illustrates analog information which is relayed to a second terminal by a digital communication channel and one or more ground relays.

1-4 THE TEXT: OBJECTIVES AND ORGANIZATION

Objectives

The basic objective of this book is to present an up-to-date version of digital communications in a context useful to the designer, analyst, and user. The principles of real, rather than idealized, digital communication channels are presented in terms of the operation of each element of a digital communication link. Key areas of design and analysis have been selected from the literature and from the author's own work.

Specific objectives of the book include:

1. To describe the basic principles and performance of synchronous satellite relay, orbits, propagation, transponders, and multiple-access techniques.
2. To provide a general review of digital communications in its broadest sense—from the original analog source or computer through sampling, quantizing, multiplexing, coding, modulation, transmission link, transponder relay, and back through the inverse operations to the final analog or digital output. Although the emphasis is on satellite communications, much of the material presented is applicable to any digital microwave, coaxial cable, or waveguide communication system.
3. To provide a mathematical analysis of the principles and performance of each element of a real digital communication system. Emphasis is on the nonlinear, dispersive, or other distorting elements that exist in any actual communication channel as opposed to the idealized additive Gaussian noise channels. For completeness, however, performance calculations for signal transmission in an ideal additive Gaussian noise channel are reviewed. The results for this additive noise channel are restated, but it is generally assumed that the reader has encountered these results previously.

The digital communication analyses concentrate on the effects of nonidealized channels exhibited by satellite transponder or ground terminal nonlinearities, oscillator phase-noise effects, and filter time dispersion and amplitude distortion. These effects are typical of real telecommunications links, and they often dominate or are at least comparable to the thermal noise effects treated in idealized signal detection calculations. Specific analyses include the effects of transponder amplitude nonlinearities, transponder AM/PM distortion, review of PSK and digital baseband modulation/detection, maximum likelihood coding/decoding, oscillator phase noise and carrier reconstruction, bit synchronization, and filter distortion effects.

Organization

SAMPLING AND MULTIPLEXING. Part 1 deals with the model and functions shown in Fig. 1-3. The concepts discussed apply to nearly all types of digital transmission links, whether or not they use satellites. Typically, at a given earth terminal a set of duplex analog (and digital) lines enters an earth

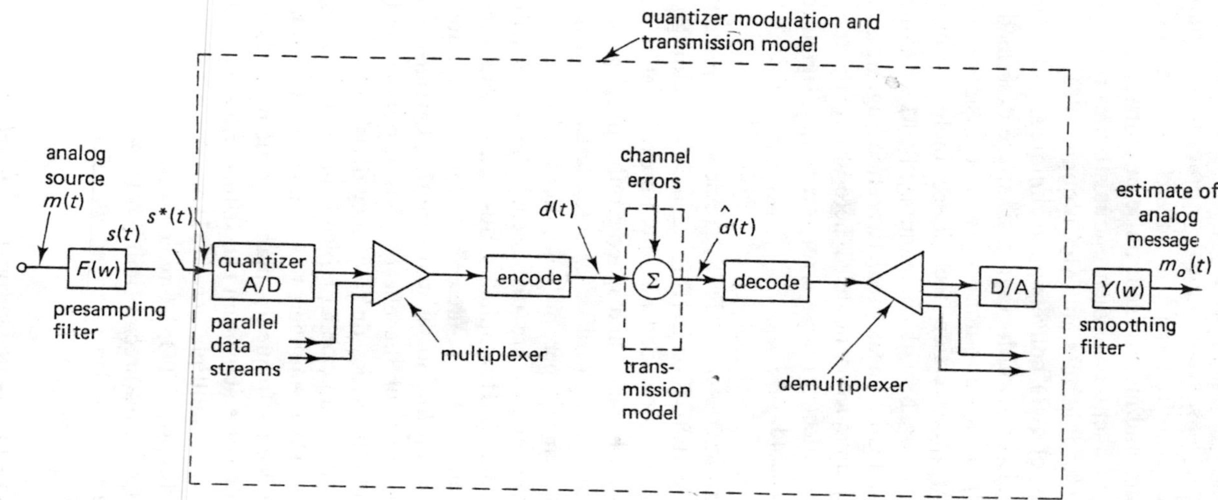

Fig. 1-3 Sampling and quantizing model

terminal complex (Fig. 1-1). These analog lines, which may be some distance from the terminal, are filtered, sampled, and quantized in an analog/digital (A/D) converter. Quasi-synchronous or plesiosynchronous parallel bit streams or word streams from the A/D converters and encoders are converted to serial bit streams and multiplexed; pulse or word stuffing typically is used to convert these parallel quasi-synchronous streams to a synchronous clock rate [Buchner, 1970; Johannes and McCulough, 1966].

Several types of A/D conversion techniques, both with and without feedback to provide bandwidth compression, are described and analyzed in terms of the output signal-to-noise ratio in the smoothed D/A converter output. These A/D conversion techniques include uniformly quantized PCM, optimal PCM, PCM with companding, PCM with dither, delta modulation, differential PCM, linear predictive encoding, and adaptive DPCM. The transmission channel is modeled here solely by the introduction of channel errors; the effects of these errors on output signal-to-noise ratio are analyzed and discussed.

SATELLITE COMMUNICATIONS. Part 2 reviews the principles of synchronous satellite relay at SHF or EHF frequencies, including such elements as satellite antennas and transponders. Satellite orbits, effects of orbit inclination, satellite station-keeping, and space-ground propagation effects are also considered. Calculations are made of satellite-earth terminal link budgets and the resulting energy per bit/noise density ratio for digital signals. Multichannel transponders are discussed. The key earth terminal communication subsystems—IF amplifiers, up- and downconverters, and power and low-noise amplifiers—are described, and calculations are made of satellite-earth terminal link losses and resulting carrier-to-noise ratios.

Analyses are given of the effects of transponder and earth terminal power amplifier nonlinearities on frequency-division multiple access signals (Fig. 1-4). The satellite transponder typically has multiple bandpass channels, some with separate TWT amplifiers to minimize intermodulation effects. The use of properly selected frequency plans and demand assignment multiple access (DAMA) or voice activation can reduce intermodulation interference further. Time-division multiple access (TDMA) system configurations and data formats are discussed. TDMA avoids some of the effects of transponder nonlinearities but requires precise timing between earth terminals. Satellite-switched multiple access is also described where satellite channels are switched in time sequence.

MODULATION AND CODING IN DISTORTED CHANNELS. Part 3 reviews the performance of various PSK, FSK, and MSK modulation/demodulation techniques, including phase-lock loop carrier recovery for synchronous detection, bit synschronization, and error-rate performance in a Gaussian noise

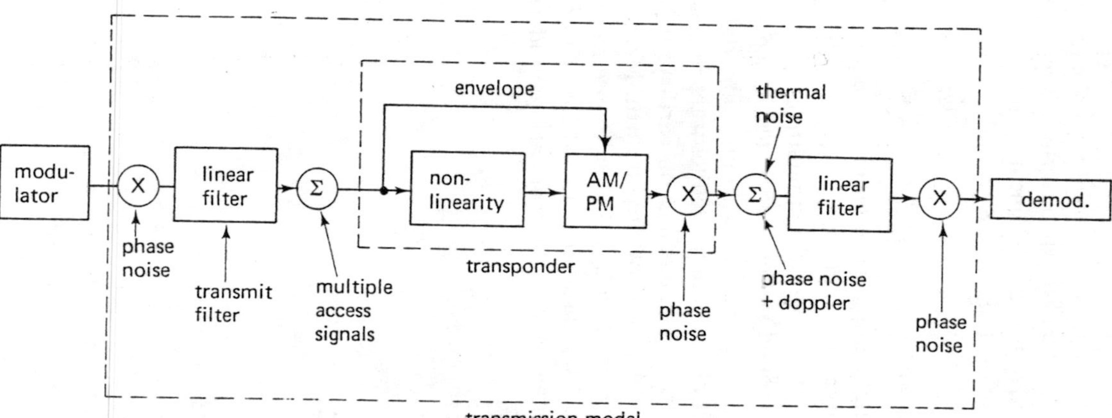

Fig. 1-4 Transmission channel model. The transmission channel includes as a minimum the multiplicative phase noise effects of oscillators used for up- or down-conversion, filter distortion in transmitters, transponders, and receivers, and transponder nonlinearities present in any power amplifier to some degree. In addition, of course, thermal noise enters at one or more points.

environment. The effects of transmission channel distortions, and noise (Fig. 1-4) are described and analyzed. These noise perturbations include the effects of multiplicative phase noise introduced by (1) the local oscillators and frequency synthesizers used for frequency translation in the satellite transponder, and (2) the earth terminal up- and downconverters. Sources of signal distortion and intersymbol interference are the linear filters and transmission lines in the transmit and receive portions of the ground terminal as well as in the transponder. Adaptive and fixed equalization techniques are described.

WORLDWIDE TIMING BY SATELLITE RELAY. Part 4 discusses accurate methods for synchronizing multiple earth terminals to a real or hypothetical clock in the satellite, or to a master clock in an earth station. Precise timing and ranging are important for time-division multiple-access satellite communications, in which each ground station is allocated its separate, nonoverlapping time slot. Precise timing has an even more important application to navigation satellites. A set of navigation satellites transmitting precisely timed signals can permit passive user equipment on earth to indicate accurately their position in three dimensions anywhere on earth. Methods are discussed for measuring time delay using phase coded pseudonoise sequences and delay-lock code-tracking loops. Code properties useful for analyzing the acquisition properties of the timing recovery loop are also considered.

PART I

Signal Quantizing and Multiplexing

A large class of information sources—speech, television, imagery, and thermal, vibration, and position measurements—are analog in nature and must be sampled and quantized for digital transmission. The resulting parallel bit streams addressed to a given destination are then multiplexed together and fed to the earth terminal modulator for transmission to the communications satellite or other transmission media. Signal quantization and multiplexing are described in Part 1 for several types of realistic signal statistics. Chapter 2 deals with optimal prefiltering and smoothing of sampled nonbandlimited signals. No real signal in nature is strictly bandlimited to a bandwidth B Hz. Hence the Nyquist $2B$ sampling rate commonly referred to is generally inadequate, for signals of nominal bandwidth B, and signals must be processed at a higher sampling rate f_s or be filtered more sharply bandlimited to $f_s/2$ Hz. Chapters 3 and 4 deal primarily with quantization of the sampled signal and the effects of channel errors on the fidelity of the analog output. Several types of quantizers are considered. These include:

- conventional Pulse Code Modulation (PCM) quantization with equal quantizing intervals,
- optimal quantization with a fixed number of intervals,
- quantizers coded for channel errors,
- effects of dither signals on quantizer statistics,
- delta modulators, and
- differential Pulse Code Modulation or ℓ-bit delta modulators.

Adaptive versions of some of these quantizers are also discussed. Mean-square error is the primary fidelity criterion. In several cases, however, the average magnitude of error is also used. Chapter 5 deals with techniques and data formats for multiplexing various quasi-synchronous bit streams together. Both pulse- and word-stuffing techniques are considered. The measure of performance here is the probability of loss in bit integrity—that is, the probability that the entire bit stream is shifted in time by an improperly added or deleted data bit.

CHAPTER 2

SAMPLING NONBANDLIMITED SIGNALS

2-1 INTRODUCTION

The basis for PCM signaling began with the principle that a bandlimited baseband signal can be transmitted by sampling at a rate f_s, which is two or more times the maximum message frequency f_m. The message can be completely recovered by passing the sampled message through an ideal low-pass filter with a cutoff at f_m [Oliver et al., 1948, 1324–1331].

In a practical communication system, the message spectrum is not strictly bandlimited; signal samples are quantized, and noise and channel errors are introduced in transmission. These effects, of course, prevent the signal from being recovered with zero error. It is possible, however, to recover the signal with small error by filtering the message prior to appropriate sampling and quantizing, and by carefully choosing the smoothing filter at the receiver.

Since Wiener's original publication in 1949, many significant contributions have been made to the optimal filtering problem for sampled-data communications. Franklin [1955] and Lloyd and McMillan [1955, 241–247] were among the first to study optimal filtering of sampled stationary random time functions. Later Stewart [1956, 253–258] and Spilker [1960, 335–341] studied the sampling of stationary nonbandlimited random signals. Spilker [1960], and Brown [1961, 269–270] investigated optimal presampling and smoothing problems and showed results in terms of percentage error in the smoothed output for nonbandlimited signals. These results are reviewed in the following sections.

These same results also apply to the sampling of bandpass signals having a symmetrical spectrum with center frequency f_o, and bandwidth $2f_m$. In this bandpass signal example, the signal is first down-converted to form two baseband signals by mixing it with an oscillator at two phases, sin $\omega_o t$ and cos $\omega_o t$. The resulting inphase and quadrature signals each have bandwidth f_m (one-sided) then and can be treated as separate baseband signals each to be sampled at rate f_s for a total sampling rate $2f_s$.

2-2 SAMPLED-DATA COMMUNICATION SYSTEM

Figure 2-1 shows the main elements of a sampled-data communication system for filtering and sampling an analog message. The message $m(t)$ (noise free) enters the system from the left, passes through the presampling filter $F(j\omega)$ to form $s(t)$, and is sampled. The transmission system passes the sampled message but perturbs it with additive "white" noise. The relation between the added noise $n(t)$ and the noise at the input to the receiver (inside the transmission system) depends on the type of modulation used. The smoothing filter $Y(j\omega)$ converts its input $s^*(t) + n(t)$ into an output $m_o(t)$, which closely matches the original message delayed a time τ, $m(t - \tau)$.

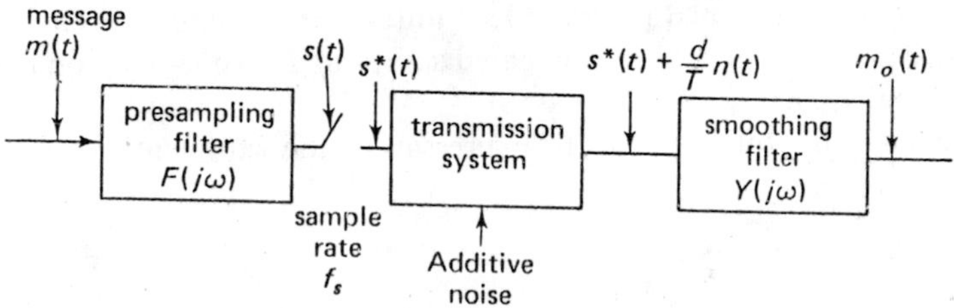

Fig. 2-1 Filtering and sampling of an analog message in a sampled data communication system. The additive noise is normalized by d/T.

The PCM quantizer is assumed to be part of the transmission system of Fig. 2-1. Quantization is assumed to be sufficiently accurate to avoid signal suppression and to permit representation of the quantizer noise by the independent additive noise $n(t)$. Further, it is assumed that (1) the analog message input $m(t)$ is wide-sense stationary and has power spectral density $G_m(\omega)$, which decreases monotonically with ω; and (2) the periodic sampling function $r(t)$ has finite width and fixed period T. (Landau [1967, 1701–1706] has shown that sampling at a nonuniform rate cannot be done at an average rate lower than $2B$.)

The sampled signal $s^*(t)$ is simply the product $s(t)r(t)$. The periodic finite-width sampling function (Fig. 2-2) has sampling width d and amplitude

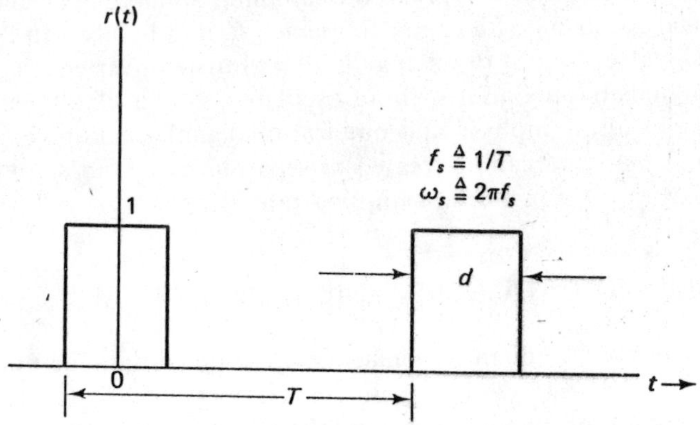

Fig. 2-2 Periodic finite sampling of width d and rate $f_s = 1/T$

unity. Note that at each sampling interval the sampling function $r(t)$ takes a finite width segment of the input waveform. If $d/T = 1$, the sampler output is identical to the input. For application to quantized systems, the ratio d/T is set arbitrarily close to zero (zero-width sample) since only one ℓ-bit word is generally assumed per sample. Finite-width sampling is encountered again later in connection with time-gated sampling for time-division multiple-access signals (Chaps. 10, 12, and 18).

The sampling function can be expressed as the exponential series

$$r(t) = \frac{d}{T} \sum_{\nu=-\infty}^{\infty} C_\nu e^{j\nu\omega_s t}$$

where $C_\nu = \text{sinc}\ \pi\nu d/T \triangleq (\sin \pi\nu d/T)/(\pi\nu d/T), C_o = 1$ and $\omega_s \triangleq 2\pi/T$. We define the sampled signal $s^*(t)$, its Fourier transform $S^*(j\omega)$, and power spectral density $G_s^*(\omega)$ as

$$s^*(t) \triangleq s(t)r(t) \qquad S^*(j\omega) \triangleq \frac{d}{T} \sum_\nu C_\nu S(j\omega - j\omega_s\nu)$$

$$S(j\omega) \triangleq F(j\omega)M(j\omega) \qquad G_s^*(\omega) \triangleq \left(\frac{d}{T}\right)^2 \sum_\nu C_\nu^2 G_s(\omega - \nu\omega_s)$$

where $M(j\omega)$ is the Fourier transform of $m(t)$, and $\omega = 2\pi f$ the power spectral density of the prefiltered signal $s(t)$ is $G_s(\omega) = |F(j\omega)^2| G_m(\omega)$, where $F(j\omega)$ is the prefilter transfer function.

Typical spectra of the prefiltered signal $s(t)$ and the sampled signal $s^*(t)$ are shown in Fig. 2-3. Note that the sampled signal spectrum centered at $2\omega_s$ can fold back into the baseband spectrum if ω_s is not sufficiently high.

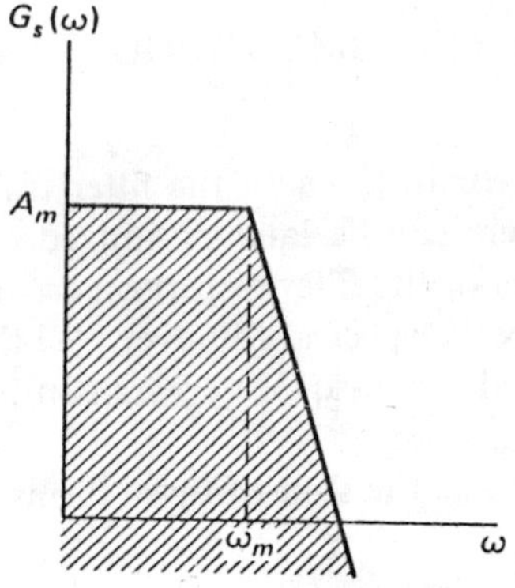

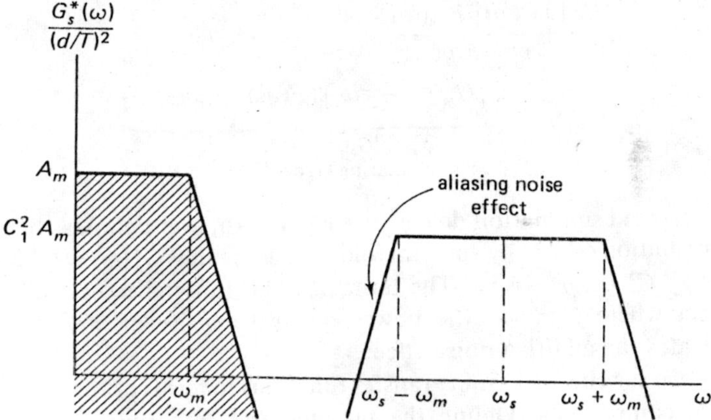

Fig. 2-3 Typical power spectra of the filtered signal $s(t)$ and the sampled signal $s^*(t)$

This interference phenomenon is often termed *aliasing*. The error $e(t)$, the smoothed output, $m_o(t)$, and its transform $M_o(j\omega)$ are defined by

$$e(t) = m_o(t) - \frac{d}{T} m(t - \tau)$$

$$M_o(j\omega) = Y(j\omega)[S^*(j\omega) + \frac{d}{T} N(j\omega)]$$

The error transform is

$$E(j\omega) = Y(j\omega)\left[S^*(j\omega) + \frac{d}{T} N(j\omega)\right] - \frac{d}{T} M(j\omega)e^{-j\omega\tau}$$

2-3 OPTIMUM PREFILTERING AND SMOOTHING FILTERS

We now evaluate the mean-square error in terms of the filter transfer functions $Y(j\omega)$, $F(j\omega)$ so that these filters can be later optimized. The mean-square error can be written as the sum of the filter distortion error, the noise error, and the aliasing error caused by the spectral foldover. All these effects are uncorrelated. Hence the individual mean-square errors can be summed to obtain the total mean-square error.

The error spectrum can be written as the sum of three terms:

$$\frac{G_e(\omega)}{(d/T)^2} = G_m(\omega)\underbrace{|Y(j\omega)F(j\omega) - e^{-j\omega\tau}|^2}_{\text{filter distortion}} + \underbrace{|Y(j\omega)|^2[G_n(\omega)}_{\text{noise error}} + \underbrace{\sum_{\nu=-\infty}^{\infty}{}' C_\nu^2 G_m(\omega - \nu\omega_s)|F(j\omega - \nu j\omega_s)|^2]}_{\substack{\sum' C_\nu^2 G_s(\omega - \nu\omega_s) \\ \text{aliasing error}}} \tag{2-1}$$

where the primed summation deletes the $\nu = 0$ term and denotes the aliasing error contribution caused by spectral foldover. Note that $G_s^*(\omega) = (d/T)^2 G_s(\omega) + (d/T)^2 \sum' C_\nu^2 \; G_s(\omega - \nu\omega_s)$. The term $(d/T)^2 \, G_n(\omega)$ indicates the additive noise effect where $G_n(\omega)$ is the power spectral density of $n(t)$. The $G_n(\omega)$ term indicates the additive noise effect.

Next, we rewrite both filter transfer functions in terms of their magnitude and phase components. Define the prefilter and smoothing filter transfer functions:*

$$\begin{aligned} F(j\omega) &\triangleq F(\omega)e^{j\phi(\omega)} && \text{prefilter} \\ Y(j\omega) &\triangleq Y(\omega)e^{j\theta(\omega)} && \text{smoothing filter} \end{aligned} \tag{2-2}$$

We optimize the smoothing filter phase $\theta(\omega)$ in (2-1) to minimize

$$\overline{e^2} = \frac{1}{2\pi}\int \frac{G_e(\omega)}{(d/T)^2}\, d\omega$$

The only terms involving θ in $G_e(\omega)$ are filter distortion effects

$$|Y(j\omega)F(j\omega) - e^{-j\omega\tau}|^2 = |(Y(\omega)F(\omega)\, e^{j(\phi+\theta)} - e^{-j\omega\tau}|^2 \tag{2-3}$$

*No constraints are placed on these filter transfer functions because of realizability, since it is assumed that the delay τ in the smoothed output is noncritical and any filter function can be approximated arbitrarily closely, given enough delay.

For any YF, the mean-square error is minimized by setting the two phasors in (2-3) at the same angle, namely,

$$\phi(\omega) + \theta(\omega) = -\omega\tau$$

or
$$\theta(\omega) = -[\omega\tau + \phi(\omega)]$$

Then the error spectrum of (2-1) can be rewritten as

$$\frac{G_e(\omega)}{(d/T)^2} = G_m[YF - 1]^2 + Y^2[G_n + \sum{}' C_v^2 G_s(\omega - v\omega_s)] \tag{2-4}$$

The optimal $Y(\omega)$ is found using Euler's equation [Courant and Hilbert, 1953]:

$$\begin{aligned}\frac{\partial G_e}{\partial Y} &= 2G_m F[YF - 1] + 2Y[G_n + \sum{}' C_v^2 G_s(\omega - v\omega_s)] \\ &= 0\end{aligned} \tag{2-5}$$

By solving for Y, we see that the optimal smoothing filter is the ratio

$$\begin{aligned}Y &= \frac{G_m(\omega)F(\omega)}{\underbrace{G_m(\omega)F^2(\omega)}_{G_s(\omega)} + G_n(\omega) + \sum{}' C_v^2 G_s(\omega - v\omega_s)} \\ &= \frac{G_m(\omega)F(\omega)}{G_n(\omega) + \sum C_v^2 G_m(\omega - v\omega_s)F^2(\omega - v\omega_s)} \\ &= \frac{G_s(\omega)}{G_n(\omega) + G_s^*(\omega)}\frac{1}{F(\omega)} = \frac{G_s(\omega)}{G_T(\omega)F(\omega)}\end{aligned} \tag{2-6}$$

where $G_T(\omega) \triangleq G_s^*(\omega) + G_n(\omega)$.
It is easily determined that the series filter combination $Y(j\omega)F(j\omega)$ is the optimum smoothing filter for a message spectrum $G_s(\omega)$ and addition noise spectrum $G_n(\omega)$. Substitute the optimal filter of (2-6) into the error spectrum equation (2-4) and use $G_s(\omega) = F^2(\omega)G_m(\omega)$ to obtain

$$\begin{aligned}\frac{G_e}{(d/T)^2} &= G_m\left(\frac{G_s}{G_T} - 1\right)^2 + \frac{G_s^2}{G_T^2}\frac{1}{F^2}(G_T - G_s) \\ &= G_m + \frac{G_m G_s}{G_T}\left(-2 + \frac{G_s}{G_T} + 1 - \frac{G_s}{G_T}\right) \\ &= G_m - \frac{G_m G_s}{G_T} = G_m - G_T Y\end{aligned}$$

where we have omitted the functional dependence on ω for simplicity.

Integrate this expression for the error spectrum to obtain

$$\overline{e^2} = \frac{1}{2\pi}\int \frac{G_e(\omega)}{(d/T)^2}\,d\omega$$

$$= \frac{1}{2\pi}\int G_m(\omega) - [G_n(\omega) + G_s^*(\omega)]Y^2(\omega)d\omega \geq 0$$

$$= \frac{1}{2\pi}\int_{-\infty}^{\infty} \{G_m(\omega) - [(G_n(\omega) + G_s^*(\omega)]Y(\omega)^2\}\,d\omega \geq 0 \quad (2\text{-}7)$$

for $\quad Y(j\omega) = Y(j\omega)|_{\text{optimum}}$

Next we adjust the presampling filter $F(j\omega)$ to minimize $\overline{e^2}$. Note that Y and G_s^* both depend on F, which is not yet determined. To minimize $\overline{e^2}$, we first define the second term in (2-7):

$$I \triangleq \frac{1}{2\pi}\int_{-\infty}^{\infty} [G_n(\omega) + G_s^*(\omega)]Y^2(\omega)\,d\omega$$

$$I = \frac{1}{2\pi}\int_{-\infty}^{\infty} [G_n + \sum C_\nu^2 G_m F^2]Y^2\,d\omega$$

which can be written as

$$I = \frac{1}{2\pi}\int_{-\infty}^{\infty} \frac{G_m^2(\omega)F^2(\omega)\,d\omega}{G_r(\omega) + \sum C_\nu^2 G_m(\omega - \nu\omega_s)F^2(\omega - \nu\omega_s)}$$

$$I < \frac{1}{2\pi}\int_{-\infty}^{\infty} G_m(\omega)\,d\omega \triangleq P_m \quad (2\text{-}8)$$

Since $\overline{e^2} = P_m - I$, P_m is the message power, and $I < P_m$, the mean-square error is minimized by maximizing $I = P_m - \overline{e^2}$.

If we let $d/T = 0$—that is, the samples are of zero width—then $C_\nu = 1$ for all ν. Assume white noise, $G_n(\omega) = N_o/2$, where $N_o/2$ is the 2-sided noise density or equivalently N_o is the one-sided noise density. The denominator is then periodic with period $\omega_s = 2\pi/T$, and the integral I is equal to the finite integral of the infinite summations

$$I = \int_{-\omega_s/2}^{\omega_s/2} \frac{\sum G_m^2(\omega - \nu\omega_s)F^2(\omega - \nu\omega_s)}{N_o + \sum G_m(\omega - \nu\omega_s)F^2(\omega - \nu\omega_s)}\,d\omega$$

$$= \int_{-\omega_s/2}^{\omega_s/2} \frac{\sum G_m(\omega - \nu\omega_s)G_s(\omega - \nu\omega_s)}{N_o + \sum G_s(\omega - \nu\omega_s)}\,d\omega \triangleq \int_{-\omega_s/2}^{\omega_s/2} R(\omega)\,d\omega \quad (2\text{-}9)$$

where the summation is added to the numerator to give the same result as the infinite integral. Use $G_m F^2 \triangleq G_s(\omega)$; then maximize the ratio $R(\omega)$ for all values of ω.

$$R(\omega) = \frac{\sum G_m(\omega - \nu\omega_s) G_s(\omega - \nu\omega_s)}{N_o + \sum G_s(\omega - \nu\omega_s)} = \frac{\mathbf{A} \cdot \boldsymbol{\theta}(\omega)}{N_o + \sum \theta_\nu(\omega)} \tag{2-10}$$

where we define the spectrum vector **A**

$$\mathbf{A} = [G_m(\omega), G_m(\omega - \omega_s), \ldots] \tag{2-11}$$

for a given value of ω; and the vector $\boldsymbol{\theta}$

$$\boldsymbol{\theta} = [G_s(\omega), G_s(\omega - \omega_s), \ldots], \theta_\nu \triangleq G_s(\omega - \nu\omega_s) \tag{2-12}$$

For any ω, R is maximized by maximizing the dot product $\mathbf{A} \cdot \boldsymbol{\theta}$ with a fixed constraint on the summation $\sum \theta_\nu(\omega) + N_o = C$. This ratio is maximized for each ω by maximizing the projection of $\boldsymbol{\theta}$ on **A** as shown in the two-dimensional example of Fig. 2-4.

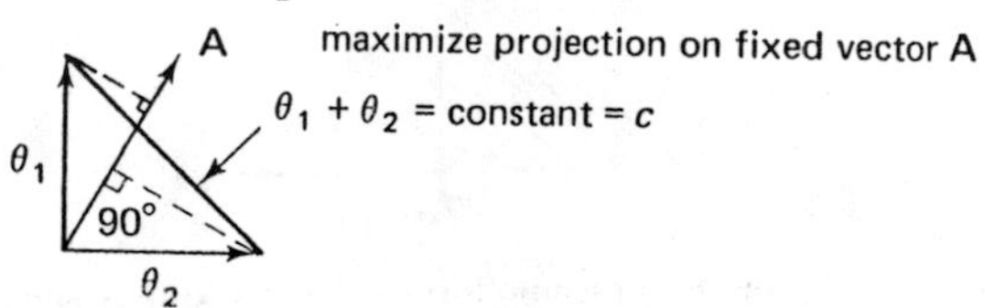

Fig. 2-4 The projection $\mathbf{A} \cdot \boldsymbol{\theta}$ is maximized for fixed $\sum \theta_\nu$ by putting $\theta_i = C$ and all others $\theta_j = 0$ where the largest component of **A** is the ith component. The largest projection of θ_ν on **A** is by θ_i for this example.

Consider first a value of $\omega = \omega_1 < \omega_s/2$. Let $G_m(\omega)$ be monotone decreasing. Then $G_m(\omega_1) > G_m(\omega_1 - \omega_s), G_m(\omega_1 - 2\omega_s), \ldots$ for all $|\omega_1| < \omega_s/2$, and the largest component of **A** is the $\nu = 0$ component, namely, $G_m(\omega_1)$, for this frequency range. We maximize $R(\omega_1)$ by setting all $\theta_\nu = 0$ except for θ_o since it is along the axis where **A** has its largest component. This result is accomplished by setting all $F(\omega - \nu\omega_s) = 0$ except for the value of $\nu = 0$ where $G_m(\omega - \nu\omega_s)$ is largest (Fig. 2-5). The value of $F(\omega_1)$, $\omega_1 < \omega_s/2$ selected for the nonzero component is not important, but we set it equal to 1 for simplicity. Similarly, the values of $F(\omega)$ for $\omega > \omega_s/2$ are set to zero. Hence, the optimum prefilter is a constant (which can be unity) for all $\omega < \omega_s/2$ and is zero elsewhere as shown in Fig. 2-6. The presampling filter thus simply removes any spectral components above $\omega_s/2$. This filter can be approximated arbitrarily closely with a sufficiently large delay τ. Performance

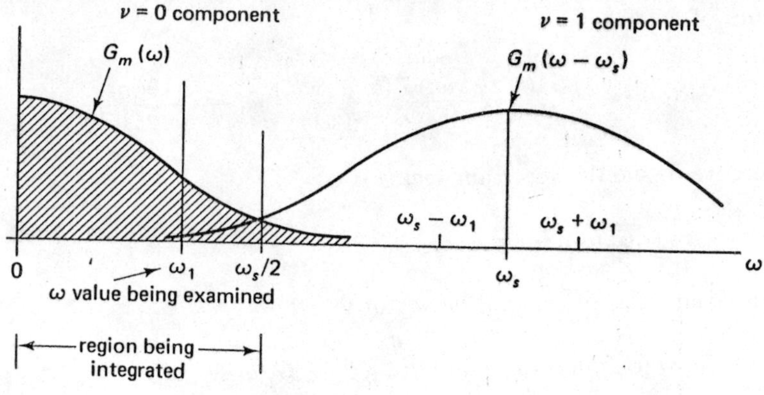

Fig. 2-5 Spectral components of the sampled message

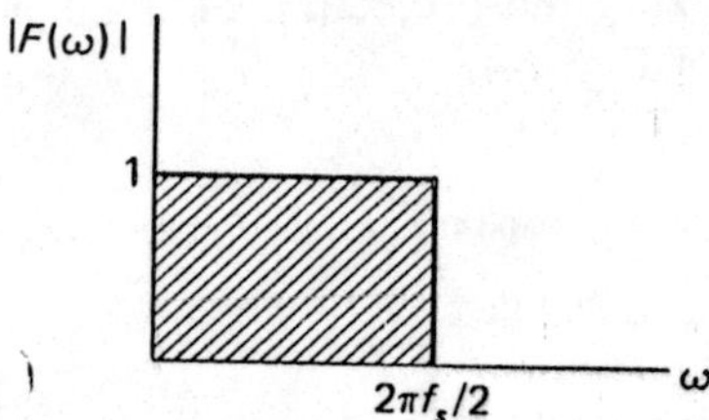

Fig. 2-6 Optimum presampling filter for additive white noise

obtained with more realistic presampling and smoothing filters, such as Butterworth filters, has been studied by Davis and McRae [1969, 65–73].

2-4 SAMPLED-DATA SYSTEM PERFORMANCE

Clearly, the minimum smoothing error for the optimum *F*, *Y* filters is simply the integral of the message spectrum above half the sampling rate

$$\overline{e^2} = 2\int_{f_s/2}^{\infty} G_m(f)\, df \tag{2-13}$$

As an example, assume an *RC* input message spectrum with unit input power

$$G_m(f) = \left(\frac{1}{\pi f_c}\right)\frac{1}{1 + (f/f_c)^2} \tag{2-14}$$

Thus, the smoothing error is

$$\overline{e^2} = \frac{2}{\pi f_c}\int_{f_s/2}^{\infty} \frac{1}{1+(f/f_c)^2}\,df = \frac{2}{\pi}\tan^{-1}\left(\frac{f}{f_c}\right)\Big|_{f_s/2}^{\infty}$$

$$= 1 - \frac{2}{\pi}\tan^{-1}\frac{f_s}{2f_c} \tag{2-15}$$

which is plotted in Fig. 2-7. Note that $\overline{e^2} = 1/10$ corresponds to only a 10-dB output signal-to-noise ratio (SNR).

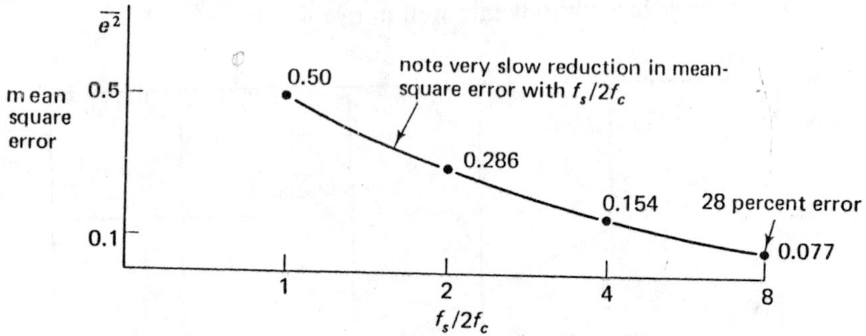

Fig. 2-7 Mean square error $\overline{e^2}$ versus normalized sampling rate $f_s/2f_c$ for an RC input spectrum

Because the *RC* spectrum rolls off rather slowly with frequency, the mean square error (2-15) decreases very slowly with increases in sampling rate. Normally, the signal spectrum of real interest to the user falls off more rapidly than the *RC* spectrum, and the required sampling rates are not as high.

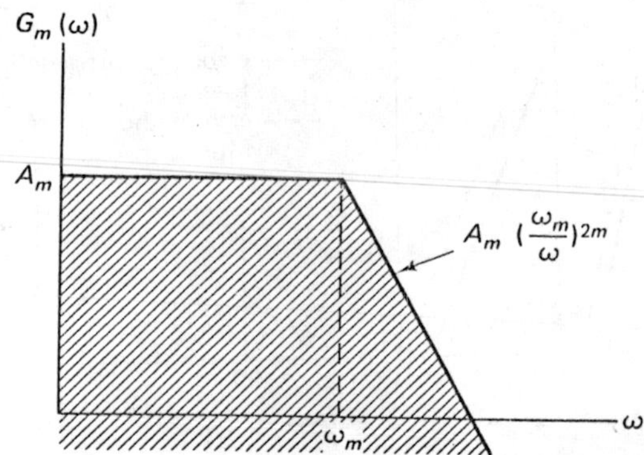

Fig. 2-8 Example signal-power spectral density of message $m(t)$ (log-log scale)

Figure 2-8 shows example signal-power spectral densities with a cutoff frequency f_m and a cutoff rate of 6 m dB/octave. Figure 2-9 shows the output percentage error for the smoothed inputs with and without presampling filters. Note that for 1 percent rms error (40 dB SNR), the sampling rate required for a signal spectrum $2m = 6$ is $f_s/f_m = 10$ without presampling filters. The use of a presampling filter permits a decrease in sampling rate by approximately 10 percent for the same percent error. Note also that in some cases it is somewhat arbitrary as to what the message really is. In speech, for example, one often considers the message to be limited to 4 kHz since that bandwidth is clearly adequate for intelligible speech even though the actual spectrum of speech has components well above 4 kHz.

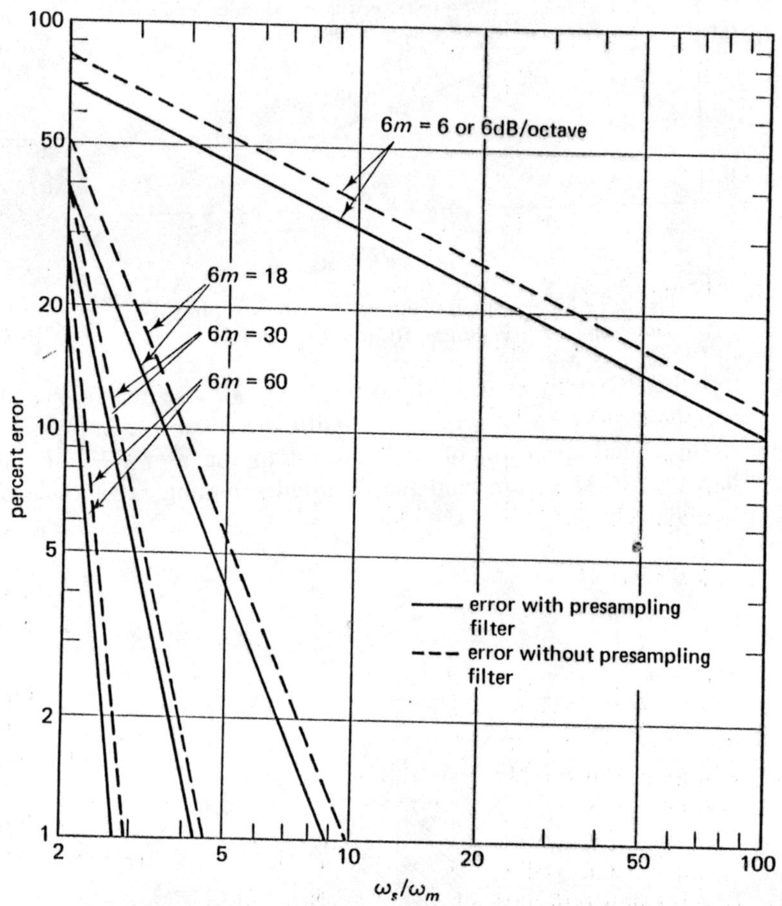

Fig. 2-9 Output error percentages with and without presampling filter. The input signal spectrum cuts off at 6m dB/octave.

CHAPTER 3

QUANTIZING BY PULSE CODE MODULATION

3-1 INTRODUCTION

The sampling operations of Chap. 2 are only the first steps in Pulse Code Modulation (PCM) quantizing. In this chapter, the signals are assumed to be bandlimited and are sampled and quantized by an instantaneous quantizer. The sampling rate f_s is high enough to avoid the aliasing errors discussed in Chap. 2.

The chapter begins with a brief discussion of the fundamentals of quantizing and the measures of quantizer performance—for example, mean-square error in recovered analog output. The discussion continues with a simplified summary of expected performance through a digital communication channel that introduces errors in transmission. Bounds are obtained for performance both with and without coding delay.

In general, the amplitude probability density of the analog input is nonuniform, and a uniform quantizer can be surpassed in performance by various nonuniform quantizers. Both optimal and quasi-optimal quantizers are discussed, including several companders and quantizers used in operational systems. Adaptive PCM is also discussed as a means for matching the PCM quantizer step sizes to the signal dynamic range—an important consideration for nonstationary signals, such as speech.

The quantizer output must be formatted and coded in a serial bit stream or other form suitable for transmission over a communication channel.

Several signal formats are discussed, and the possible benefits of coding the quantizer output are examined briefly.

For certain types of signal sources, including voice and video, the mean-square-error criterion is only a partial measure of performance. In such cases, the use of dither signals can randomize the quantizing error statistics and remove certain disturbing "contouring" effects or other signal-dependent error patterns.

3-2 FUNDAMENTALS OF AMPLITUDE QUANTIZATION

We first examine uniform quantizing of the sampled signal and investigate the accuracy of the reconstructed sampled signal [Viterbi, 1966, 214]. The quantization noise produced by this operation is a component of the transmission system noise discussed in Chapter 2. Figure 3-1 depicts a simplified PCM link. Assume first that a stationary, uniformly distributed sampled analog signal is to be quantized and transmitted in a channel with errors (binary symmetric channel (BSC)). The objective of this analysis is to determine the effect of

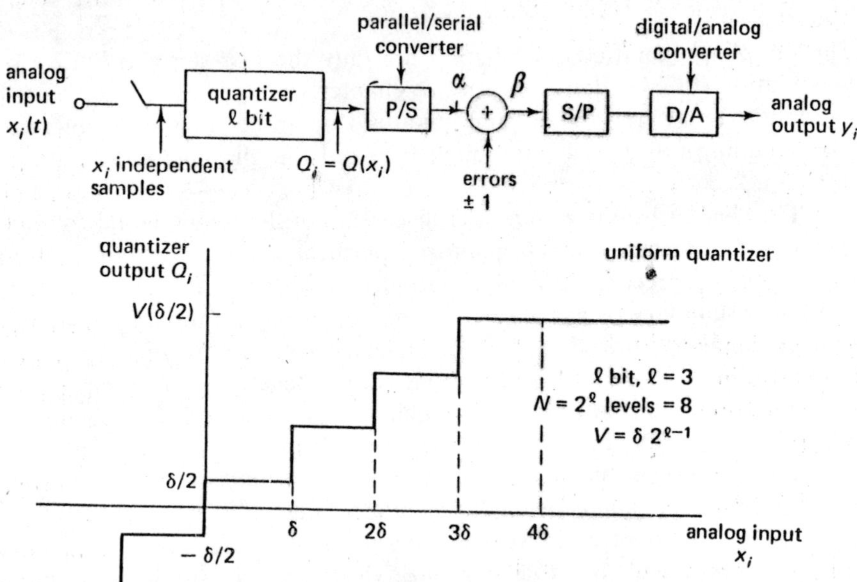

Fig. 3-1 Quantizing with an ℓ-bit uniform quantizer. The α and β represent the serial bit stream before and after introduction of channel errors, respectively.

independent channel errors on SNR in the sampled analog output. The samples are quantized with an ℓ-bit uniform quantizer as shown.

Quantizer Output

The multilevel output of the quantizer at time $t = iT$ is $Q(S_i)$ and is expressed in terms of the ℓ-bit words $(\alpha_{i1}, \alpha_{i2} \ldots, \alpha_{i\ell})$. Thus, the digital/analog (D/A) converter output sample at $t = iT$ is

$$Q(x_i) = V \sum_{j=1}^{\ell} \alpha_{ij} 2^{-j} \tag{3-1}$$

where $\alpha_{ij} = \pm 1$ are the serial quantizer output bits for the ith sample before the addition of the channel errors. The maximum quantizer output magnitude is

$$Q_{\max} = \frac{V}{2}\left(1 + \frac{1}{2} + \cdots + \frac{1}{2^{\ell-1}}\right) = \frac{V}{2}\frac{1-(1/2^{\ell})}{1-(1/2)} = V - \frac{\delta}{2} \tag{3-2}$$

where δ is the quantizer step size and $V = \delta 2^{\ell-1}$. Clearly, the output signal magnitude cannot exceed this value, and any input signal above $Q_{\max}$ causes quantizer overload.

Reconstructed Sample

The reconstructed sample is derived from the received bits β_{ij} as

$$y_i = V \sum_{j=1}^{\ell} \beta_{ij} 2^{-j} \tag{3-3}$$

where y_i is the estimate of x_i and the β_{ij} may have had errors. Assume that the errors in β_{ij} are independent of one another. The mean-square error is then the sum of the uncorrelated quantization error and channel error effects:

$$\mathbf{E}[(x_i - y_i)^2] = \mathbf{E}\{[x_i - Q(x_i)]^2\} + \mathbf{E}\{[Q(x_i) - y_i]^2\} \tag{3-4}$$

where $\mathbf{E}(\;)$ represents the ensemble average, the first term represents quantization error, and the second is the channel error effect. We define the quantization error $e \triangleq x_i - Q(x_i)$, and next assume that the input is uniformly distributed over the interval $(-V, V)$. Then the quantizing error is also uniform, and the mean-square quantization error is

$$\overline{e^2} = \int_{-\delta/2}^{\delta/2} e^2 p(e)\, de = 2\int_0^{\delta/2} e^2 \frac{1}{\delta}\, de = \frac{1}{3}\left(\frac{\delta}{2}\right)^2 = \frac{\delta^2}{12} = \frac{V^2}{3N^2} \tag{3-5}$$

Since $2V/\delta = N$ and $p(e)$ is the probability of the quantization error, we have used $p(e) = 1/\delta$ for $e \in (-\delta/2, \delta/2)$ and zero elsewhere.

Channel Noise Error

The effects of the channel noise error can be computed using the independence property of different bits in each quantizer code word and the independence of different code words from one sample to the next. This channel error effect can be evaluated using (3-1), (3-3) as

$$\begin{aligned}
\mathrm{E}\{[y_i - Q(x_i)]^2\} &= V^2\mathrm{E}[\textstyle\sum (\beta_{ij} - \alpha_{ij})2^{-j}]^2 \\
&= V^2 \sum_j \sum_k [\delta_{jk} + \delta_{jk} - 2\delta_{jk}\mathrm{E}(\alpha_{ij}\beta_{ij})]2^{-j-k} \\
&= V^2 \textstyle\sum 2^{-2j}[2 - 2\,\mathrm{E}(\alpha_{ij}\beta_{ij})] \\
&= \textstyle\sum V^2 2^{-2j}[2 - 2(1 - 2\mathcal{P}_B)]
\end{aligned}$$

where $\mathcal{P}_B$ is bit error probability and δ_{jk} is the Kronecker delta function; $\delta_{jk} = 1$ if $j = k$ and is zero otherwise; $\mathrm{E}(\beta_{ij}\beta_{ik}) = \delta_{jk}$, etc. because of the independence in bit errors. Since different bit positions are independent and $\mathrm{E}(\alpha_{ij}\beta_{ij}) = 1 - 2\mathcal{P}_B$, can be rewritten as

$$\begin{aligned}
\mathrm{E}\{[y_i - Q(x_i)]^2\} &= 4V^2\mathcal{P}_B \sum_{j=1}^{\ell} 2^{-2j} = 4V^2\mathcal{P}_B \frac{1}{4}\left(\frac{1 - 1/4^{\ell}}{1 - 1/4}\right) \\
&= 4V^2\mathcal{P}_B\left[\frac{2^{2\ell} - 1}{3(2^{2\ell})}\right] = \frac{4V^2}{3}\frac{(N^2 - 1)}{N^2}\mathcal{P}_B \\
&= \frac{\delta^2}{3}(2^{2\ell} - 1)\mathcal{P}_B
\end{aligned} \tag{3-6}$$

The useful signal power can be evaluated as

$$\mathrm{E}(y_i^2) = \int_{-V}^{V} y_i^2 p(y_i)\, dx_i = \frac{1}{2V}\int_{-V}^{V} x_i^2\, dx_i = \frac{V^2}{3} \tag{3-7}$$

where $p(y_i)$ is the probability density of y_i. Hence, the output SNR is

$$\text{SNR} = \frac{V^2/3}{4V^2[(N^2 - 1)/3N^2]\mathcal{P}_B + (V^2/3N^2)} = \frac{N^2}{1 + 4\mathcal{P}_B(N^2 - 1)} \tag{3-8}$$

and is bounded by SNR $< N^2$. For large $4\mathcal{P}_B(N^2 - 1) \gg 1$—that is, a sufficiently high bit error probability and $N \gg 1$—we have

$$\text{SNR} = \frac{1}{4\mathcal{P}_B}\left(\frac{N^2}{N^2 - 1}\right) \cong \frac{1}{4\mathcal{P}_B}$$

and SNR is approximately independent of N for $N^2 \gg 1$. Figure 3-2 is a

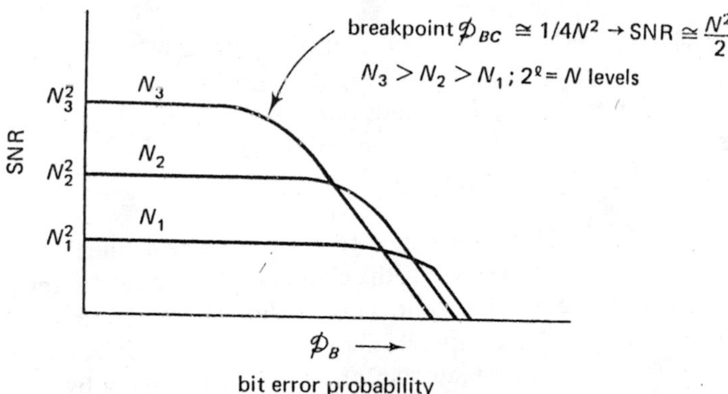

Fig. 3-2 General behavior of output SNR (3-8) versus bit error probability for different numbers of quantizer levels N_i

plot of the general behavior of SNR for various N. The maximum output SNR $= N^2$ for $\mathcal{P}_B = 0$, and the SNR decreases as $1/(4\mathcal{P}_B)$ for larger $\mathcal{P}_B$ and $N \gg 1$.

If the input signal is not uniformly distributed but has zero mean and an rms value σ_s, then a useful "rule of thumb" (4σ-loading) for minimal quantizer overload is to set the quantizer step size at $\delta = 8\sigma_s/N$. The output SNR using (3-5) for no channel errors is then

$$\text{SNR} = \frac{\sigma_s^2}{\delta^2/12} = \frac{\sigma_s^2}{(8\sigma_s)^2/12N^2} = \frac{12}{64}N^2 = \frac{3}{16}N^2 = \frac{3}{16}2^{2\ell} \tag{3-9}$$

as compared to N^2 for a uniform input. Similarly if 3σ-loading is used and $\delta = 6\sigma_s/N^2$, the output signal-to-noise ratio becomes

$$\text{SNR} = \frac{\sigma_s^2}{\delta^2/12} = \frac{\sigma_s^2}{36\sigma_s^2/12N^2} = \frac{N^2}{3} \tag{3-10}$$

Effects of Channel Bandwidth and Noise

System signal and noise parameters are defined as follows:

$T = 1/f_s$ = sampling period
$R_b = \ell f_s/2f_m$ = normalized bit rate
$\mathcal{R}$ = information bit rate $= \ell f_s = 1/\tau$
W = transmission channel bandwidth
N_0 = one-sided noise density—white noise channel
P_S = signal power
$N = 2^\ell$ = number of quantizing levels
f_s = sampling rate of input analog signal
$B = f_m = f_s/2$ = input information bandwidth
$\mathcal{R}_s$ = symbol rate for quadriphase PSK
$E_b = P_S\tau = P_S/\mathcal{R}$ signal energy per bit

To relate the PCM performance to transmission channel parameters, it is necessary to draw on results for the channel bit error rate in terms of the energy-per-bit-to-noise-density ratio. For coherent detection of biphase-phase-shift-keyed (BPSK) or quadriphase-shift-keyed (QPSK) signals, the bit error rate $\mathcal{P}_B$ for a white Gaussian noise channel is given by

$$\mathcal{P}_B = \frac{1}{2}\text{erfc}\sqrt{\frac{E_b}{N_o}} < \exp\frac{-P_S/N_oB}{2\ell} \tag{3-11}$$

where

$$\text{erfc}\, x \triangleq 2\int_x^\infty \frac{e^{-y^2}}{\sqrt{\pi}}\,dy \leq 2 \tag{3-12}$$

is the complementary or co-error function. This probability of bit error for coherent biphase PSK (correlation between 0 and 1 waveforms $\rho = -1$) is shown in Fig. 3-3; Fig. 3-4 is a plot of erfc'(x). Note that another definition of the carrier function is sometimes used:

$$\text{erfc}'(x) = \frac{1}{\sqrt{2\pi}}\int_x^\infty e^{-y^2/2}\,dy \leq 1$$

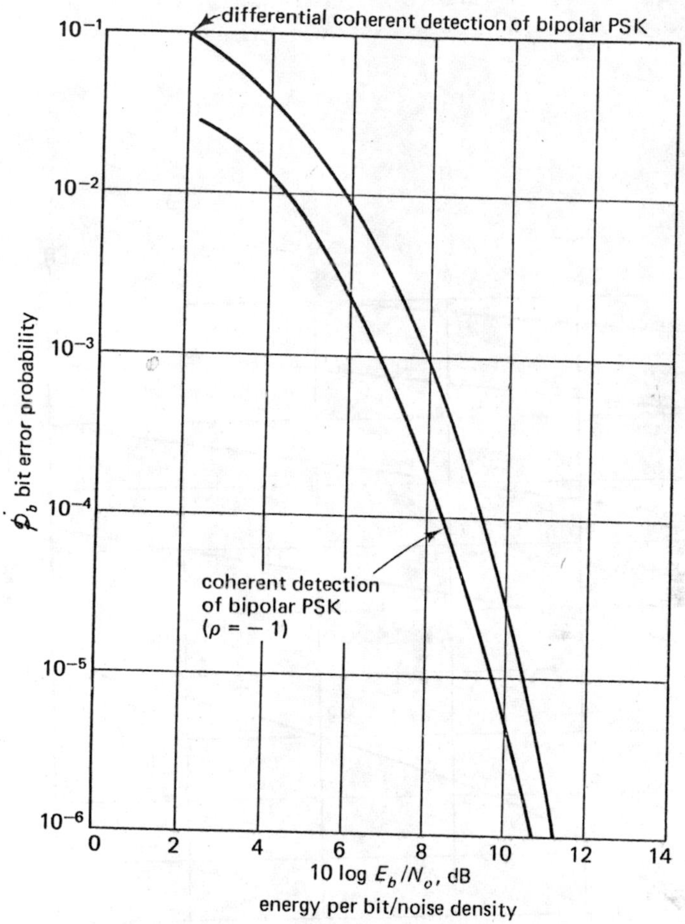

Fig. 3-3 Error rate performance for coherent detection of biphase PSK and differentially coherent detection of bipolar PSK

Figure 3-3 also shows the probability of bit error $\mathcal{P}_b$ for differentially coherent detection of PSK. The expression for $\mathcal{P}_b$ for this signal is [Viterbi, 1966, 214]

$$\mathcal{P}_b = \frac{1}{2} \exp -\frac{E_b}{N_o} \tag{3-13}$$

A more extensive discussion of error rates for MPSK and other signals is given in Part 3.

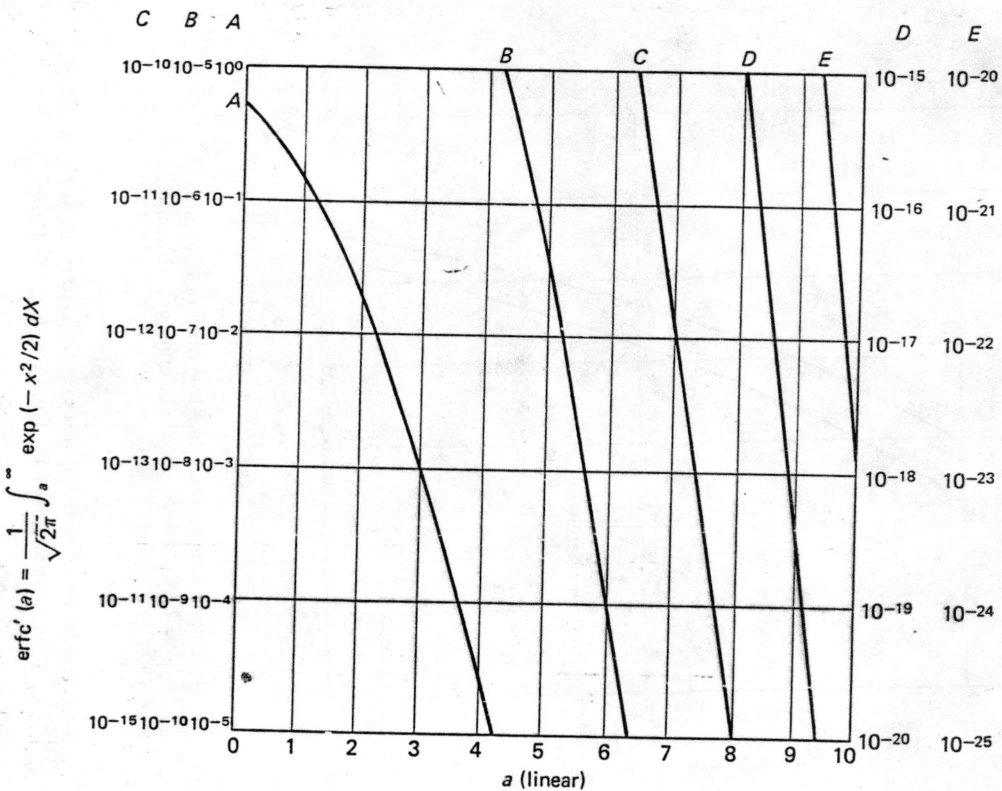

Fig. 3-4 Plot of $\text{erfc}'\, a = (1/\sqrt{2\pi}) \int_a^\infty \exp(-x^2/2)\, dx$. The different vertical scales correspond to the indicated curves labeled A, B, C, D, E.

The resulting output SNR for coherent PSK and various values of N is shown in Fig. 3-5. If the optimum value of N is used for each SNR, one can establish a lower bound graphically by tracing the envelope of the breakpoints in the SNR curve; this lower bound on SNR performance without coding is indicated by the dashed curve in Fig. 3-5 and is given by

$$\max_N \text{SNR} > \frac{1}{5} \exp \sqrt{\frac{P_S \log 2}{N_o B}} \triangleq \text{SNR}_{\text{LB}} \tag{3-14}$$

where log is the natural logarithm.

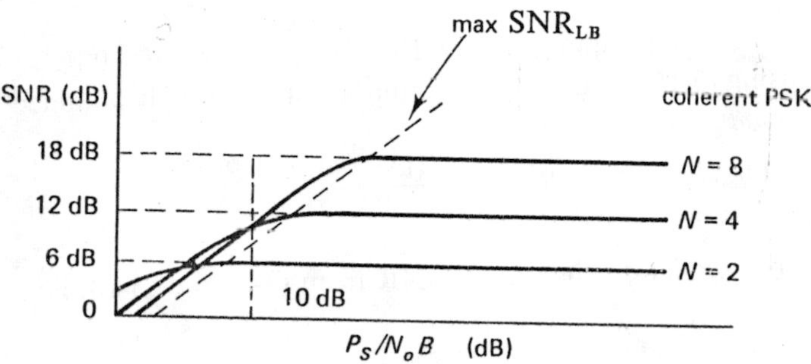

Fig. 3-5 Plot of general behavior of output SNR versus P_S/N_0B for coherent PSK and N-level quantization of the analog input

The output SNR can increase exponentially with P_S/N_oB, as shown, by increasing the number of quantizing levels in an optimum manner. Of course, a practical system generally has a fixed N and can be optimum only at one value of SNR.

3-3 BOUND ON CODED SYSTEM WITH DELAY AND INFINITE BANDWIDTH

Performance can be improved if an infinite bandwidth and delay is permitted in the channel and several ℓ-bit words representing several signal samples are grouped in a larger block prior to transmission. If we block n sample words of ℓ bits each into one code block of $n \log_2 N = n\ell$ bits where $N \triangleq 2^\ell$, zero error rate could be produced and the output SNR $= N^2$ for all rates less than the channel capacity and n is arbitrarily large.

For transmission of f_s ℓ-bit samples/sec the transmission rate is defined as

$$\mathcal{R} = f_s \log_2 N = 2B \log_2 N \qquad \text{bits/sec} \tag{3-15}$$

where $f_s = 2B$. We let the block size n become arbitrarily large. The rate $\mathcal{R}$ must be bounded by channel capacity $\mathcal{C}$ for a white Gaussian noise channel with unlimited bandwidth [Lindsey and Simon, 1973]:

$$\mathcal{R} = 2B \log_2 N < \frac{P_s}{N_o \log 2} = \mathcal{C} \qquad \text{bits/sec} \tag{3-16}$$

In theory, the energy per bit/noise-density ratio can then decrease to

$$\frac{E_b}{N_0} = \frac{P_S}{\mathcal{R}N_o} \geq \log 2 = 0.693 \qquad \text{or } -1.59 \text{ dB} \tag{3-17}$$

as compared to a requirement of 10.8 dB for uncoded BPSK at $\mathcal{P}_b = 10^{-6}$. Thus, using (3-16) and (3-17), the number of levels N is bounded by

$$\log N = (\log 2) \log_2 N = \frac{(\log 2)\mathcal{R}}{2B} < \frac{P_S}{2BN_o} \tag{3-18}$$

Hence, the bound on the output SNR from (3-8) and (3-18) is

$$\max \text{SNR} < N^2 < \exp \frac{P_S}{N_o B} \qquad \text{or} \qquad N < \exp \frac{P_S}{2BN_o} \tag{3-19}$$

for an infinite transmission bandwidth channel. Thus, as seen by comparing (3-14) and (3-19), a large increase in SNR is permitted by the addition of coding delay and an arbitrarily large bandwidth.

Capacity Constraint versus Bandwidth (Infinite Delay)

The performance with infinite delay is further bounded by channel capacity for a channel of maximum bandwidth W. Here the rate $\mathcal{R}$ must decrease because of capacity constraints to

$$2B \log_2 N = \mathcal{R} < W \log_2 \left(1 + \frac{P_S}{N_o W}\right) = \mathcal{C} \tag{3-20}$$

By solving for N, we obtain the bound

$$\log_2 N < \log_2 \left(1 + \frac{P_S}{N_o W}\right)^{W/2B} \tag{3-21}$$

where W/B is the bandwidth expansion factor, the ratio of channel bandwidth to input signal bandwidth. Thus the number of quantizing levels must be less than

$$N < \left(1 + \frac{P_S}{N_o W}\right)^{W/2B} \tag{3-22}$$

Hence, the output SNR is bounded by

$$\text{SNR} = N^2 < \left(1 + \frac{P_S}{N_o W}\right)^{W/B} = \left[1 + \frac{P_S}{N_o B}\left(\frac{B}{W}\right)\right]^{W/B} \tag{3-23}$$

The above result bounds the output SNR attainable with a given transmission bandwidth W and data bandwidth B. As shown in Fig. 3-6, no improvement is possible without an increase in bandwidth—that is, $W/B > 1$. As W/B increases, the output SNR gradually approaches the exponential relationship (3-19) since $(1 + x/n)^n \longrightarrow \exp x$ as $n \longrightarrow \infty$.

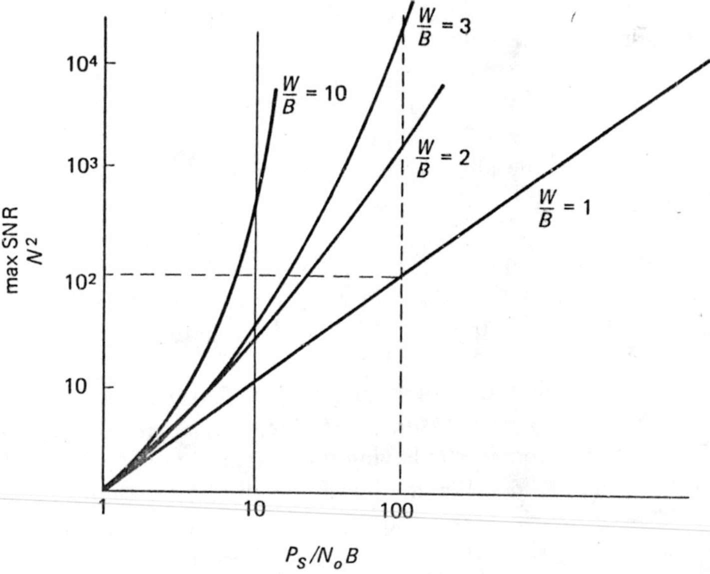

Fig. 3-6 Bound on output SNR versus input SNR for various bandwidth expansion ratios W/B; coding with ∞ delay is assumed.

An example of an efficient coding technique is discussed in Chap. 15 for rate 1/2 coding where two coded bits are transmitted for each data bit. This technique is capable of permitting a 5-dB reduction in the signal energy per bit, $E_b = P_s/\mathfrak{R}$, for a given noise spectral density N_0 while providing a given performance in the output (at a 10^{-5} output error rate). This performance is close to the improvement shown for $W/B = 2$ in Fig. 3-6.

3-4 OPTIMUM QUANTIZATION

We have reviewed the performance of uniform PCM quantizers in the presence of channel errors. Since the input signal was uniformly distributed, the uniform quantizer was optimum. In this section, the input signal statistics are not generally uniform; the quantizer levels are allowed to be nonuniform and are selected to optimize performance for a given input statistic. The objectives of the next sections are:

- To determine the optimum quantizer (or compandor) characteristic for minimum mean-square error [Max, 1960, 7–12; Wood, 1969, 248–252]
- To show the small performance difference between optimum and equal-interval quantizing
- To examine the large dynamic range improvement by logarithmic companding [Smith, 1957, 653–710] as compared to optimal quantizing for a fixed input variance
- To discuss the advantages of adaptive PCM for inputs of time-varying signal levels.

In general, the optimum quantizer is impractical to implement. However, its performance serves as a useful bound on performance and can be closely approximated at high output SNR—that is, where fine quantizing is employed.

We first determine the optimal quantizing characteristic for a given input probability density $p(x)$. Assume that independent samples are taken of a "white" stationary process $x(t)$ having a differentiable probability density $p(x)$; that is, $|p'(x)| < \infty$. We minimize the distortion $\mathfrak{D}$ by defining a differentiable error metric $f(e)$ with $|f'(e)| < \infty$. The distortion metric can be written in terms of the output levels y_i and defined as

$$\begin{aligned}\mathfrak{D} &\triangleq \mathbf{E}[f(x - y_i)] \\ &= \sum_{i=1}^{N} \int_{x_i}^{x_{i+1}} - f[x - y_i(x)]p(x)\,dx \qquad (3\text{-}24)\end{aligned}$$

where $x_1 = -\infty$, $x_{N+1} = \infty$, and y_i is the output produced for an input $x_i < x \leq x_{i+1}$ as shown in Fig. 3-7.

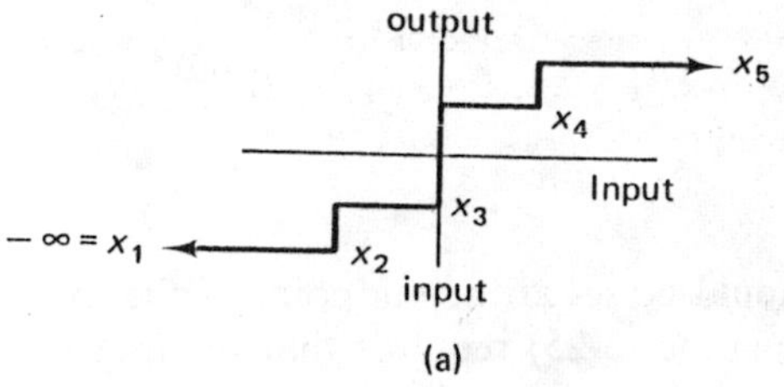

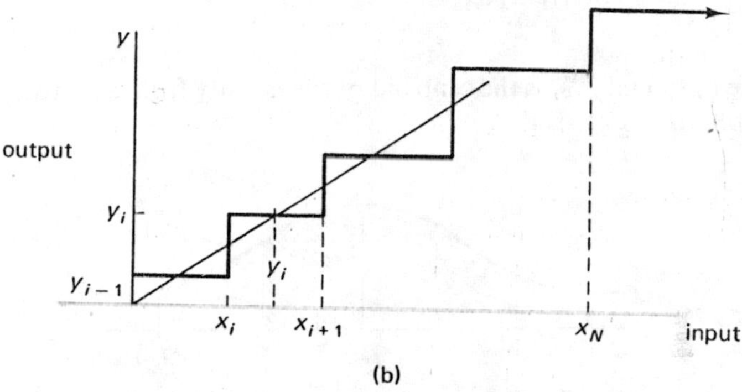

Fig. 3-7 Arbitrary quantizer input-output characteristic: (a) a four-level symmetrical quantizer shown as an example, and (b) a generalized quantizer

Next we minimize $\mathfrak{D}$ by differentiating $\mathfrak{D}$ with x_i, y_i to obtain the following sets of equations for the optimum x_i, y_i:

$$\frac{\partial \mathfrak{D}}{\partial y_j} = \int_{x_j}^{x_{j+1}} f'(x - y_j)\, p(x)\, dx = 0$$

$$\frac{\partial \mathfrak{D}}{\partial x_j} = f(x_j - y_{j-1})p(x_j) - f(x_j - y_j)p(x_j) = 0 \qquad j = 2, \ldots, N \tag{3-25}$$

or $$f(x_j - y_{j-1}) = f(x_j - y_j) \tag{3-26}$$

is required to satisfy (3-25).

Minimum Mean-Square Error

Using the mean-square-error criterion $f(e) = e^2$, we use (3-25), (3-26) to obtain the constraint equations for x_i, y_i. Equation (3-26) then yields

$$(x_j - y_{j-1})^2 = (x_j - y_j)^2 \qquad (x_j - y_{j-1}) = \pm(x_j - y_j) \tag{3-27}$$

and

$$x_j = \frac{y_j + y_{j-1}}{2} \tag{3-28}$$

That is, the points x_j should be set at the midpoint of the interval (y_i, y_{i-1}) to satisfy Eq. (3-26). Equation (3-25) requires that the integral

$$\int_{x_{j-1}}^{x_j} (x - y_{j-1})p(x)\,dx = 0 \tag{3-29}$$

Thus, the optimum y_{j-1} is the centroid of the density from x_{j-1} to x_j as shown in Fig. 3-8.

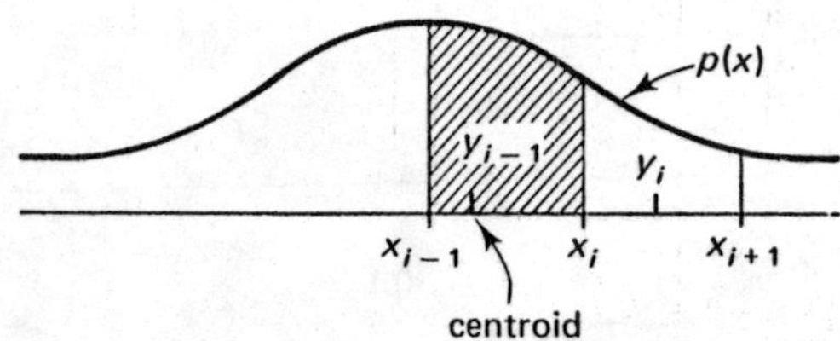

Fig. 3-8 Example showing x_j at the midpoint of the y_{i+1} and y_i, and y_j at the centroid between x_{i-1}, x_i

As an illustration of the parameters and optimality requirements, consider the uniform signal and quantizer characteristics shown in Fig. 3-9. In general, we must solve (3-25) and (3-26) by computer because of the complexity of the equations for nonuniform probability densities. Results for optimal quantization for a Gaussian input are shown in Table 3-1 for 4-, 16-, and 32-level quantizers. Also shown is the result for the gamma density representation of speech [Paez and Glissom, 1972]. Note that the gamma density gives a larger MSE than does either the Gaussian or uniform density.

Notice that for Gaussian inputs, the mean-square error 2.5×10^{-3} for the optimum quantizer ($N = 32$) is only slightly less relative to the MSE (3.5×10^{-3}) for the equispaced quantizer with optimum spacing. For uniform inputs, of course, the optimum quantizer and the equispaced quantizer are identical. As we would expect, the performance for a uniform input is improved relative to that for the Gaussian input.

For the error measure $f(x - y) = |x - y(x)|^r$ Panter and Dite [1951, 44–48] have defined an approximately optimum quantizer as the quantizer for which each of the N quantizing intervals makes an equal contribution to the $|x - y(x)|^r$. The distortion metric approximation for large N and a

Table 3-1 Quantizing Levels and Corresponding Mean-Square Error for Uniform and Optimal Instantaneous Quantizers for Various Values of the Number of Quantizing Levels N. The Optimum Quantizer Result Is also Shown for the Gamma Density $p(x) = \frac{1}{2}\sqrt{k/\pi}\,[\exp - k|x|]/\sqrt{|x|}$, $k = 0.866$, $\sigma = 1$.

		Gaussian Input $\sigma = 1$						Uniform Input Equispaced	Gamma Density Optimum
		Optimum*			Equispaced				
N	i	x_i	y_i	MSE	δ Interval	$(N/2)\delta$	MSE	$MSE = 1/N^2$	MSE
$N = 4$	3	0.0	0.4528						
	4	0.9816	1.510	0.1125	0.9957	1.9914	0.1188	0.0625	0.2326
$N = 16$	9	0.0	0.1284						
	10	0.2582	0.3851						
	...	...	...						
	15	1.844	2.064						
	16	2.401	2.733	9.5×10^{-3}	0.3352	2.6816	1.15×10^{-2}	3.9×10^{-3}	1.96×10^{-2}
$N = 32$	17	0.0	0.0659						
	18	0.132	0.1981						
	...	...	...						
	32	2.977	3.263	2.5×10^{-3}	0.1881	3.0096	3.5×10^{-3}	9.75×10^{-4}	5.2×10^{-3}

*One can use the approximation $\mathfrak{D} = \text{MSE} \cong 2.21N^{-1.96}$ for $N > 32$.

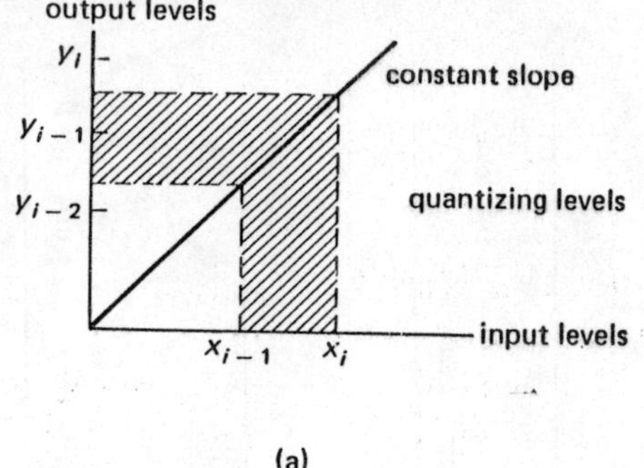

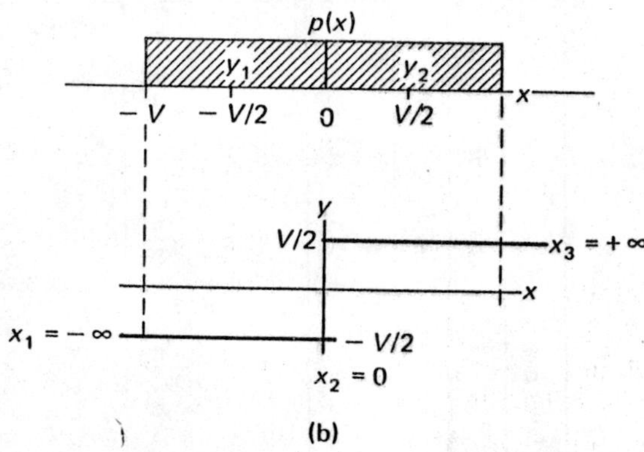

Fig. 3-9 Example of the optimal quantizer for a uniformly distributed input $x(t)$; (a) the constant slope quantizer for N levels, and (b) the two-level quantizer along with the probability density function

smooth probability density $p(x)$ is

$$\mathfrak{D} = \overline{|x - y|^r} \cong \frac{1}{2^r(1 + r)N^r}\left[\int_0^1 p(x)^{1/(1+r)}\, dx\right]^{1+r} \tag{3-30}$$

Optimum End Points for Uniform Quantizing

Because the optimal quantizer is generally difficult to build, an alternate uniform quantizer can be considered and its end points optimized. Results for this quantizer can then be compared with the performance of the optimal quantizer.

Assume a symmetric input signal $p(x) = p(-x)$, and solve for the distortion $\mathfrak{D}$. For uniform quantizing of step size δ,

$$\mathfrak{D} = 2 \sum_{i=1}^{(N/2)-1} \int_{(i-1)\delta}^{i\delta} f\left[x - \frac{(2i-1)\delta}{2}\right] p(x)\, dx + 2 \int_{[(N/2)-1]\delta}^{\infty} f\left[x - \left(\frac{N-1}{2}\right)\delta\right] p(x)\, dx \qquad (3\text{-}31)$$

for an even number of output levels N and quantization intervals δ. The quantizer end points are then $(N - 1)\delta$ apart. Next, minimize $\mathfrak{D}$ and solve for the optimal step size δ using Euler's equation.

$$\frac{\partial \mathfrak{D}}{\partial \delta} = - \sum_{i=1}^{(N/2)-1} (2i - 1) \int_{(i-1)\delta}^{i\delta} f'\left[x - \delta\left(\frac{2i-1}{2}\right)\right] p(x)\, dx - (2N - 1) \int_{(N-1)\delta/2}^{\infty} f'\left[x - \left(\frac{2i-1}{2}\right)\delta\right] p(x)\, dx = 0 \qquad (3\text{-}32)$$

Equation (3-32) can be solved by computer for the optimal δ to obtain results for Gaussian inputs. These results are shown in Table 3-1 in the column labeled equispaced and are compared with optimum quantizing. Note the relatively small improvement in MSE between equispaced and optimal quantizing for any value of N. Results for the equispaced quantizer (optimal) and a uniform input density are also shown in Table 3-1 for comparison since this result represents the lower bound on mean-square error.

3-5 QUASI-OPTIMAL QUANTIZER (COMPANDOR) FOR LARGE N

If a large number of nonuniform quantizer steps are used, a quasi-optimal quantizer [Smith, 1957, 653–710] can be solved without use of a computer. Furthermore, practical realizations of the quantizer often can be obtained. Assume sufficiently fine quantization and a relatively smooth probability density function $p(x)$ so that

$$p(x_i) \cong p(x_{i+1}) \qquad (3\text{-}33)$$

That is, there is little change in the probability density between quantizing intervals. Then the optimal output level is simply the midpoint of the input levels; namely,

$$y_i = \frac{x_i + x_{i+1}}{2} \triangleq \frac{2x_i + \Delta_i}{2} \qquad (3\text{-}34)$$

where $x_{j+1} \triangleq x_j + \Delta_j$ (3-35)

The mean-square error for inputs in the jth interval is then

$$\sigma_j^2 \cong \int_{x_j}^{x_{j+1}} (x - y_j)^2 p(x_j)\, dx = p(x_j)\Delta_j\left(\frac{\Delta_j^2}{12}\right) \tag{3-36}$$

Total mean-square error can then be written, using (3-36),

$$\sigma^2 = \sum \sigma_j^2 = \tfrac{1}{12} \sum p_j \Delta_j^2$$

where we define the probability that the signal is in the jth level as

$$p_j \triangleq p(x_j)\, \Delta_j \cong \int_{x_j}^{x_{j+1}} p(x)\, dx \tag{3-37}$$

Hence, the mean-square error can be written as the expected value

$$\sigma^2 = \tfrac{1}{12} \mathrm{E}(\Delta_j^2) \tag{3-38}$$

Compandor Characteristics

Any quantizer characteristic can be obtained by the series combination of an instantaneous nonlinearity, compressor $v(x)$, followed by a uniform quantizer as shown in Fig. 3-10. At the other end of a digital channel the

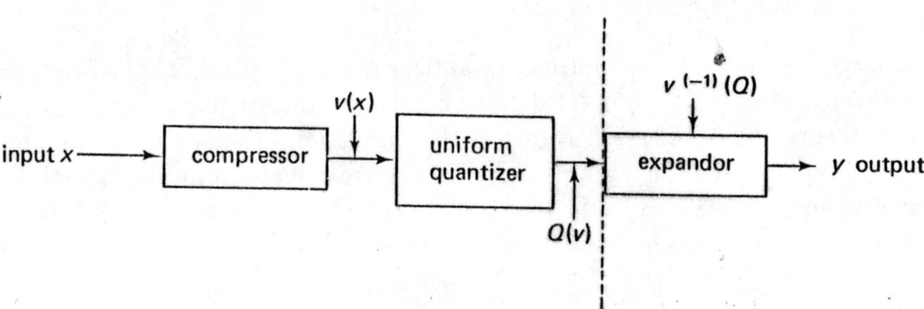

Fig. 3-10 Configuration of a compandor

quantizer output is fed to an expandor $v^{(-1)}(x)$. The compressor and expandor can represent continuous nonlinearities, or the same effect can be obtained by use of nonuniform quantizer Analog/Digital (A/D) and Digital/Analog (D/A) converters; however, the use of a uniform quantizer may be more economically sound. The expandor has the inverse characteristic of $v(x)$;

thus, the series combination of $v(x)$ and $v^{(-1)}(x)$ is a unit gain transfer function, that is, $v^{(-1)}[v(x)] = x$.

Assume symmetry of the input probability density and that the $N/2$ positive quantizer input and output intervals are each of increment $\delta = 2V/N$ which matches the increments in the compressor output. Define the compressor slope $u(x) \triangleq dv/dx$. The compressor input increments are then of width

$$\Delta_j \simeq \delta \frac{1}{dv/dx} = \delta/u \tag{3-39}$$

Hence, the mean-square error (3-38) is

$$\sigma^2 = \frac{1}{12}\mathbf{E}(\Delta_j^2) = \frac{\delta^2}{12}\mathbf{E}\left[\left(\frac{dx}{dv}\right)^2\right] = \frac{\delta^2}{12}\int_{-V}^{V}\left(\frac{dv}{dx}\right)^{-2} p(x)\,dx \tag{3-40}$$

Assume a symmetrical input $p(x) = p(-x)$. The compressor characteristic is then also symmetric, and we need show only the positive portion, omitting the negative portion $v(-x) = -v(x)$, as in Fig. 3-11.

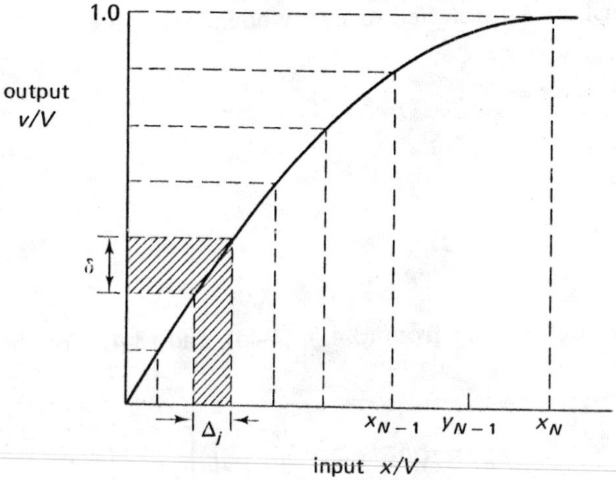

Fig. 3-11 Instantaneous compression characteristic $v(x)$

Next, we obtain the optimum compandor $v(x)$ function. Minimize the mean-square error σ^2 using Euler's equation [Courant and Hilbert, 1953, 184–206]. We first rewrite σ^2 of (3-40) in terms of $u(x)$:

$$\sigma^2 = \frac{2\delta^2}{12}\int_{0}^{V} u(x)^{-2}\, p(x)\,dx \qquad \delta = 2V/N \tag{3-41}$$

with a constraint

$$\int_0^V u(x)\,dx = V \tag{3-42}$$

and define the integral

$$I = \sigma^2 + \lambda v = \int_0^V \left[\frac{2\delta^2}{12}\frac{p(x)}{u(x)^2} + \lambda u(x)\right] dx \tag{3-43}$$

where λ is the Lagrangian multiplier. Euler's equation, simply, is

$$\frac{\partial I}{\partial u} = \frac{2\delta^2}{12} p(x) \frac{-2}{u^3(x)} + \lambda = 0 \tag{3-44}$$

or the optimum compressor slope is

$$\frac{dv}{dx} = u(x) = [3\lambda p(x)]^{1/3} = Kp^{1/3}(x) \tag{3-45}$$

where the constant K is selected to satisfy boundary value (3-42)

$$\int_0^V Kp^{1/3}\,dx = V \tag{3-46}$$

That is,

$$K = \frac{V}{\int_0^V p^{1/3}(x)\,dx} \tag{3-47}$$

Thus, the minimum σ^2 from (3-41), (3-45), and (3-47) for the optimum compandor is

$$\begin{aligned}\sigma^2 &= \frac{2\delta^2}{12}\int_0^V \frac{1}{K^2 p^{2/3}(x)} p(x)\,dx = \frac{2\delta^2}{12}\frac{1}{V^2}\left[\int_0^V p^{1/3}(x)\,dx\right]^3 \\ &= \frac{2}{3N^2}\left[\int_0^V p^{1/3}(x)\,dx\right]^3\end{aligned} \tag{3-48}$$

which for uniform $p(x)$ yields $\sigma^2 = V^2/3N^2$.

Note that we have selected the input and output intervals to give

$$\delta \triangleq \int_{v_i}^{v_{i+1}} dv = \int_{x_i}^{x_{i+1}} \left(\frac{dv}{dx}\right) dx = \int_{x_i}^{x_{i+1}} Kp^{1/3}(x)\,dx \tag{3-49}$$

where $v_i \triangleq v(x_i)$, and we then have equal-area increments under each $p^{1/3}(x)$ curve segment from x_i to x_{i+1}. Hence the area under the $p^{1/3}(x)$ curve from x_1 to x_{i+1} is

$$\int_{x_1}^{x_{i+1}} Kp^{1/3}(x)\,dx = \delta i \tag{3-50}$$

Thus, large differences in $p(x)$ from one region to the other are somewhat de-emphasized by the 1/3 power effect.

In a more general sense, Elias [1970, 172–184] has shown that for an input x which has a probability density that is constant over each quantizing interval and has an absolutely continuous probability distribution, the quantizer that is optimized to minimize the error measure $|x - y(x)|^r$ still sets $y_i = (x_i + x_{i-1})/2$. Then the expected value of the error measure is bounded (for large N) by

$$\mathrm{E}[|x - y(x)|^r] \lesssim \frac{1}{2^r(1+r)}\left[\int_0^1 p^m(x)\right]^{r/n} \tag{3-51}$$

where $\quad n = \dfrac{r}{1+r} \qquad m = \dfrac{1}{1+r} \qquad m + n = 1$

Example: Speech Represented by Laplace Distribution Let the probability density of a speech waveform be represented by the Laplace density*

$$p(x) = G \exp - \frac{|x|}{x_o} \tag{3-52}$$

where $G \triangleq 1/2x_o$ and $\overline{x^2} = \sigma_x^2 = 2x_o^2$ is the mean-square value of the speech. Using Eqs. (3-45) and (3-47), the optimum compressor characteristic can be written as

$$\begin{aligned} v(x) &= \int_0^v dv = \int_0^x \left(\frac{dv}{dx}\right) dx = \int_0^x Kp^{1/3}\,dx \\ &= \int_0^x KG^{1/3}e^{-x/3x_o}\,dx = -KG^{1/3}3x_o e^{-x/3x_o}\Big|_0^x \\ &= KG^{1/3}3x_o(1 - e^{-x/3x_o}) = V\frac{1 - e^{-x/3x_o}}{1 - e^{-V/3x_o}} \end{aligned} \tag{3-53}$$

Define the normalized overload level C as the ratio of quantizer range to

---

*The gamma density $p(x) = \sqrt{k/4\pi|x|}\exp(-k|x|)$ is a more complex but better approximation to speech [Paez and Glissom, 1972, 225–230].

speaker volume $C = V/\sigma_x$. For negligible overload effect, we set $C \gg 1$. Small speaker overload corresponds to a large value of C.

Quantization Error-Power Ratio Variation with Speaker Volume

We next generate a family of compressor characteristics of the same general form as the optimal characteristic of (3-53), but having a fixed set of parameters independent of σ_x.

$$\frac{v(x)}{V} = \frac{1 - e^{-mx/V}}{1 - e^{-m}} \tag{3-54}$$

One value of m is optimum for a given speaker volume level σ_x^2. We can now determine how compandor performance varies with σ_x^2, when the compandor has been optimized for one value of σ_x^2.

Define the distortion ratio as

$$\mathfrak{D}^2 \triangleq \frac{1}{\text{SNR}} = \frac{\sigma^2}{\sigma_x^2} \tag{3-55}$$

where (3-40) gives

$$\sigma^2 = \frac{2}{3}\frac{V^2}{N^2}\int_0^V \frac{1}{(dv/dx)^2}p(x)\,dx \tag{3-56}$$

The compressor slope characteristic corresponding to (3-54) is

$$\frac{1}{V}\frac{dv}{dx} = \frac{1}{1 - e^{-m}}\frac{m}{V}e^{-mx/V} \tag{3-57}$$

Inserting (3-57) in (3-56), we then obtain a mean-square error which varies with $\sigma_x^2 = 2x_0^2$ for the Laplace density of (3-52) as

$$\begin{aligned}
\frac{\sigma^2}{V^2} &= \frac{2}{3}\left(\frac{V}{N}\right)^2 \int_0^V \left[\frac{m/e^{-mx/V}}{1 - e^{-m}}\right]^{-2} Ge^{-x/x_0}\,dx \\
&= \frac{1}{2x_o}\frac{2}{3}\left(\frac{1}{N}\right)^2\left(\frac{1}{m}\right)^2(1 - e^{-m})^2 \int_0^V e^{x[(2m/V)-(1/x_o)]}\,dx \\
&= \frac{-1}{2x_o}\frac{1}{m^2N^2}\frac{2}{3}\left\{\frac{1 - e^{V[(2m/V)-(1/x_o)]}}{(2m/V) - (1/x_o)}\right\}(1 - e^{-m})^2
\end{aligned} \tag{3-58}$$

Define $M \triangleq 1 - e^{-m}$ and use $C = V/\sqrt{2}\,x_o$, where C is the normalized dynamic range, to simplify (3-58). The distortion ratio can then be written:

$$\mathfrak{D}^2 = \frac{\sigma^2}{x^2} = \frac{1}{3x_o}\frac{V^2M^2}{m^2N^2}\frac{-1+e^{2m-(V/x_o)}}{2x_o^2[(2m/V)-(1/x_o)]}\frac{1}{3}\frac{V^2M^2}{m^2N^2}$$

$$\cdot\frac{e^{2m-\sqrt{2}C}-1}{(2m-\sqrt{2}C)x_o^3 2V}$$

$$= \frac{1}{3}\left(\frac{M^2}{Nm}\right)^2\left(\frac{V}{x_o}\right)^3\frac{1}{2}\frac{e^{2m-\sqrt{2}C}-1}{2m-\sqrt{2}C}$$

$$= \frac{\sqrt{2}C^3M^2}{3N^2m^2}\frac{e^{2m-\sqrt{2}C}-1}{2m-\sqrt{2}C} \qquad \text{(3-59)}$$

Equation (3-59) shows the sensitivity of performance to the speaker volume term $C = V/\sigma_x$ and is plotted for one example in Fig. 3-12. The compandor

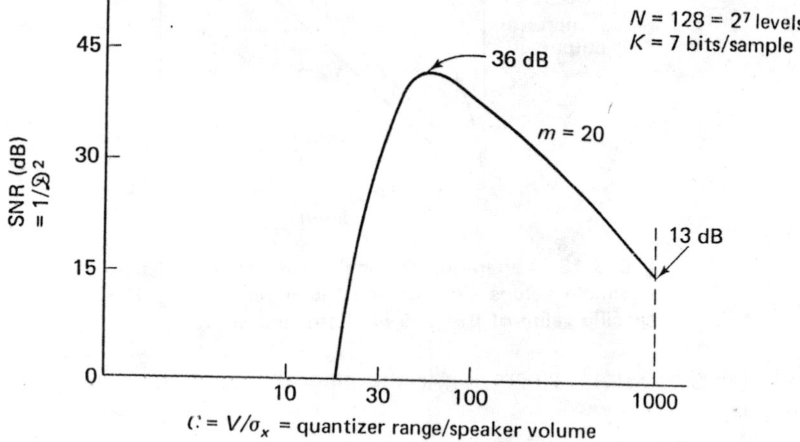

Fig. 3-12 Variation of SNR with rms input speaker volume σ_x; optimum performance occurs at $C = (3/\sqrt{2})m = 42.4$.

illustrated is of limited value for PCM voice because of the limited dynamic range over which SNR > 30 dB, even though it is optimum for one particular rms amplitude level. For small values of C, speaker overload occurs, and for large C the quantizer/compandor is too coarse.

Logarithmic Companding (μ Law Quantizers)

Logarithmic companding of speech is nonoptimal as compared to the compandor of (3-57) for any given speaker volume. However, it yields an improvement in SNR relative to a uniform quantizer for a wider range of volume levels than the optimal compandor and has important practical use. For appropriate parameter values, logarithmic companding can provide toll quality speech with 7 bits or 128 levels/sample, while a uniform quantizer needs about 11 bits/sample for similar performance [Smith, 1957; Jayant, 1974].

This logarithmic compressor characteristic, shown in Fig. 3-13, is expressed by

$$\frac{v(x)}{V} = \frac{\log(1 + \mu x/V)}{\log(1 + \mu)} \qquad 0 \leq x \leq V \tag{3-60}$$

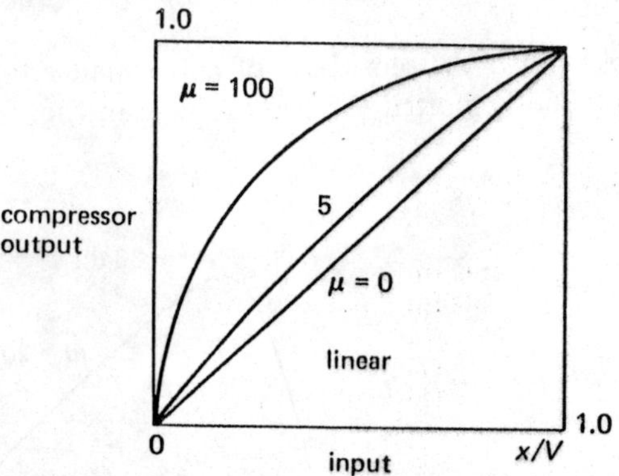

Fig. 3-13 Logarithmic compressor characteristics. Desirable values of μ are on the order of 100. The specific value of $\mu = 255$ is often used in practice.

and $v(-x) = -v(x)$, where log is the natural logarithm. The compressor slope is

$$u(x) \triangleq \frac{dv}{dx} = \frac{\mu}{\log(1 + \mu)} \frac{1}{1 + \mu|x|/V} \tag{3-61}$$

Note that $\mu = 0$ corresponds to no companding. The effective input step size Δx is given by

$$\Delta x \cong \frac{2V}{N} \frac{dx}{dv} \qquad N \gg 1 \tag{3-62}$$

and is approximated as

$$\begin{aligned}
\frac{\Delta x}{|x|} &\cong \frac{2V/N}{|x|} \frac{\log(1 + \mu)(1 + \mu|x|/V)}{\mu} = \frac{2 \log(1 + \mu)}{N} \frac{V + \mu|x|}{\mu|x|} \\
&\cong \frac{2}{N} \log(1 + \mu) = \text{constant} \qquad \left|\frac{\mu x}{V}\right| \gg 1 \\
&\cong \frac{2}{N} \log(1 + \mu) \frac{V}{\mu|x|} \qquad \mu|x| \ll 1
\end{aligned} \tag{3-63}$$

Thus, the effective step size is approximated by

$$\Delta x \cong \frac{2V}{N\mu} \log(1 + \mu) \triangleq \alpha V \tag{3-64}$$

for small $|x|$, and Δx is very small compared to the uniform quantizing step size $2V/N$ for large $\mu \gg 1$. Assume a symmetrical input $p(x) = p(-x)$. The mean-square quantizing error is obtained by using

$$\sigma^2 = \frac{\overline{\Delta x^2}}{12} = \frac{\alpha^2}{12}\overline{(V + \mu|x|)^2}$$

$$= \frac{\alpha^2}{12}(V^2 + 2V\mu\overline{|x|} + \mu^2\overline{x^2}) \quad \text{where } \overline{|x|} = 2\int_0^V xp(x)\,dx \tag{3-65}$$

Evaluate the SNR and distortion ratio for the logarithmic compandor:

$$\text{SNR} = \frac{1}{\mathfrak{D}^2} = \frac{\text{mean square input}}{\text{mean square error}} = \frac{\overline{x^2}}{\sigma^2} \equiv \frac{\sigma_x^2}{\sigma^2}$$

$$= \frac{12}{\alpha^2}\frac{\overline{x^2}}{(V^2 + 2V\mu\overline{|x|} + \mu^2\overline{x^2})} \tag{3-66}$$

Thus, the output SNR from (3-64), (3-66) is

$$\text{SNR} = \frac{12N^2\mu^2}{4\log^2(1 + \mu)}\frac{1}{(\mu^2 + 2\mu AC + C^2)}$$

$$= \frac{3N^2}{\log^2(1 + \mu)}\frac{1}{(1 + 2AC/\mu + C^2/\mu^2)} \tag{3-67}$$

where $A \triangleq \overline{|x|}/\sqrt{\overline{x^2}}$ = average absolute input/rms input

$C \triangleq V/\sigma_x$ = compressor overload voltage/rms input

Example: Exponential Input Density

Again consider the exponential density model of speech and calculate the performance of the logarithmic compandor. For this density

$$p(x) = \frac{\lambda}{2}e^{-\lambda|x|} \tag{3-68}$$

we have the constants

$$A = \frac{1}{\sqrt{2}}$$

$$C = \frac{V}{\sigma_x} = \frac{V}{\sqrt{2}/\lambda}, \quad C^2 = V^2/\overline{x^2}$$

$$\overline{x^2} = \frac{2}{\lambda^2}$$

Thus, the SNR of (3-67) for logarithmic companding varies with compressor overload C as

$$\text{SNR}_q = \frac{3N^2}{\log^2(1+\mu)} \frac{1}{1 + \sqrt{2}\,C/\mu + C^2/\mu^2} \tag{3-69}$$

By comparison, using (3-5), (3-68), standard PCM ($\mu = 0$) yields a SNR $= \overline{x^2}/(V^2/3N^2) = 3N^2/C^2$ for $C \geq 3$. These results are plotted in Fig. 3-14 for both standard PCM and PCM with companding ($\mu = 100$). Note that the companded output exceeds performance with no companding for $C > 5$ and has good performance over a wide dynamic range of inputs ($\simeq 40$ dB).

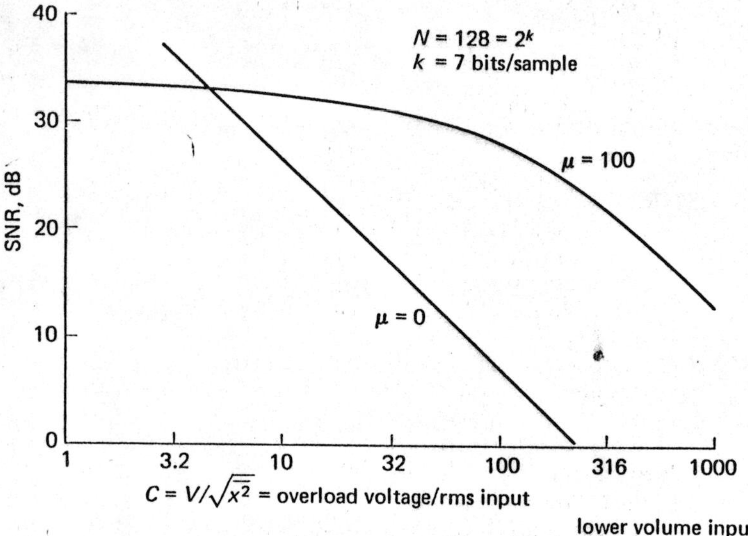

Fig. 3-14 Variation of SNR with PCM with and without logarithmic companding, $\mu = 100$, and $\mu = 0$, respectively

At $C = 100$. corresponding to a low volume input, companding produces an improvement of > 20 dB. Thus, this compandor gives an improvement in output SNR over a much wider range of input variance than did the quantizer of (3-54) optimized for a specific input variance.

D2 Channel Bank PCM Coding Technique

The widely used D2 channel bank, used in conjunction with the T1 carrier system, employs a nonlinear (μ law) PCM coder for quantizing voice channels in the Bell System [Damman et al., 1972, 1675–1700]. These voice channels are sampled at an 8-kHz rate, compressed using a nonlinear 15-segment approximation to the $\mu = 255$ law, and quantized with 8 bits per sample. Actually, 8 bits are used 5/6 of the time; the remaining 1/6 of the time, 7-bit quantizing is used; the remaining bit is used for signaling. Thus, compression is an approximation to the logarithmic compression discussed above and is designed to cope with voice and message inputs that are expected to have a dynamic range over 40 dB.

The compressor input/output characteristic has the segments listed in Table 3-2. The resulting input/output characteristic curve is shown in Fig. 3-15. Note that the step sizes are all in powers of 2; hence, the compressor has a so-called "digitally linearizable" compression law.

Table 3-2 CODER SEGMENTS FOR 15-SEGMENT, $\mu = 255$ COMPRESSOR CHARACTERISTIC; ONLY THE POSITIVE SEGMENTS ARE SHOWN.

Coder Input Level	*Segment Code*	*Step Size*
0	111	2
31	110	4
95	101	8
223	100	16
479	011	32
991	010	64
2015	001	128
4063	000	256
8159		

The code format at the output of the quantizer is selected to be an inverted folded-binary code. The first bit (not shown in Table 3-2) indicates the sign, and the following bits indicate the magnitude. For input magnitudes between 0 and 31, the inverted binary coding of the magnitude indicated in Table 3-2 gives X111, a sign bit, and three 1s. Hence, for all except the highest magnitudes, 1s are present in the coder output. This code has been selected because of its (1) lower sensitivity to channel errors, and (2) high density of 1s to provide good synchronization properties when later transmitted in bipolar format. (See Chapter 16.)

The relatively low sensitivity to channel errors results because speech signals have a high probability of zero input (pauses in speech), and an error

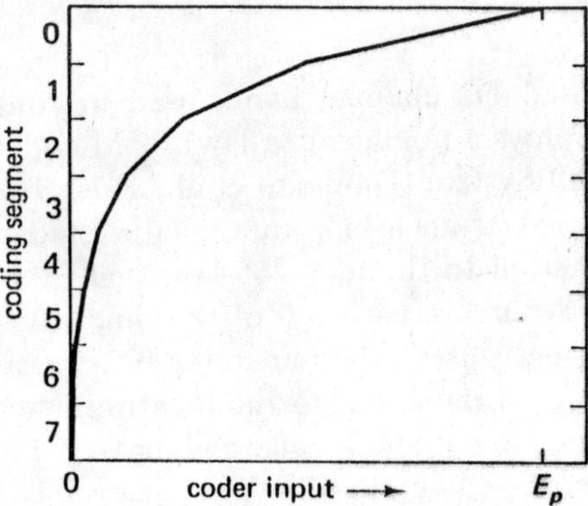

Fig. 3-15 The asymmetric compression characteristic for the 15-segment $\mu = 255$ compressor; only the positive input segment is shown. [From Damman, 1972]

in the sign bit causes an amplitude error proportional to the signal level at that instant. For ordinary binary codings, on the other hand, a bit error can cause an error of half the total amplitude range.

Another alternative, the Gray code (Table 3-3) also has a sign bit, but it does not exhibit as high a density of 1s for small signal levels.

Table 3-3 THE 4-BIT GRAY CODE FOR 16-LEVEL INPUTS ($k = 4$)

Level n	*Input Interval m*	*Gray Code Output $G_k(m)$*	
		Sign Bit	
16	+8	1000	
15	+7	1001	
14	+6	1011	
13	+5	1010	
12	+4	1110	
11	+3	1111	
10	+2	1101	
9	+1	1100	Mirror image
	0		except for
8	−1	0100	sign bit
7	−2	0101	
6	−3	0111	
5	−4	0110	
4	−5	0010	
3	−6	0011	
2	−7	0001	
1	−8	0000	

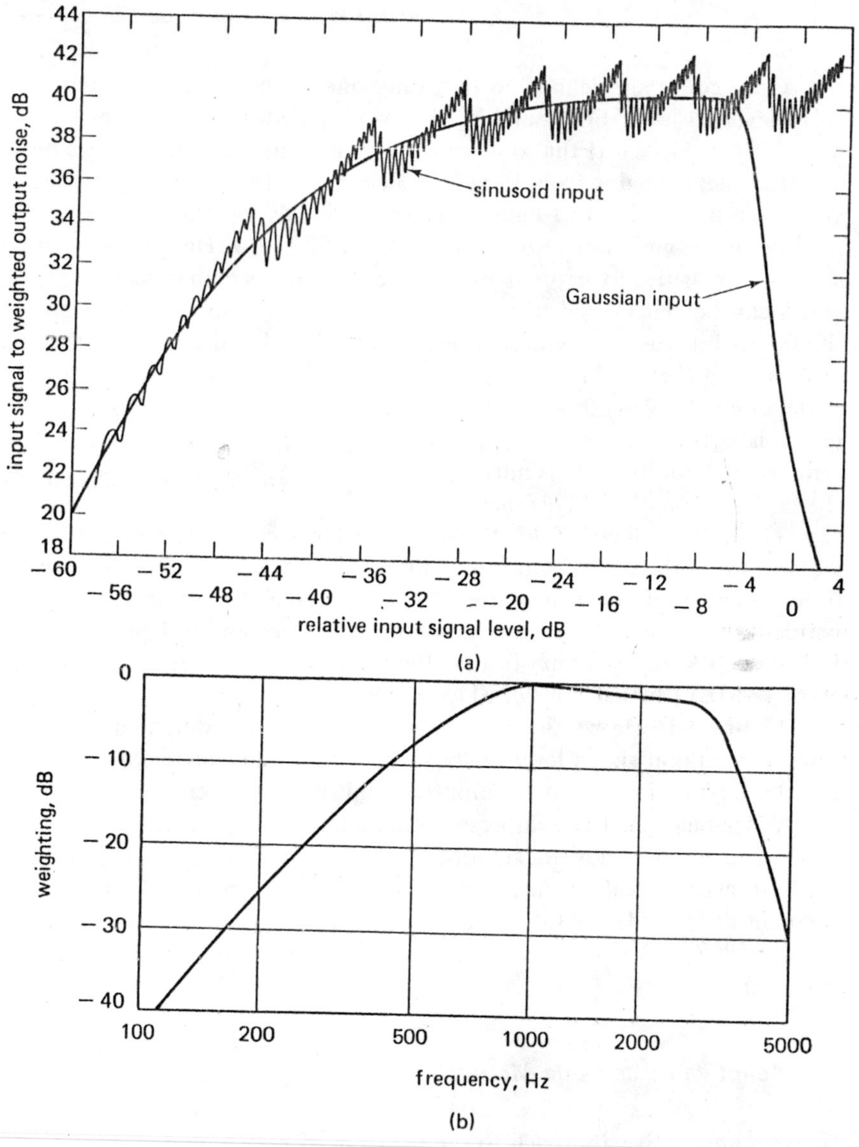

Fig. 3-16 (a) Output SNR as a function of relative input signal level for both Gaussian and sinusoidal inputs to the 8-bit, 15-segment, μ-law PCM quantizer [Damman et al., 1972] for C-message weighting of the noise. (b) The frequency response of C-message weighting includes the frequency characteristics of the Bell type 100 telephone as well as the hearing of the average subscriber. [Bell Telephone Laboratories Staff, 1971, 31–34]

Gray codes are defined so that only one bit changes with each change from level to level. For example, the two-bit codes for the levels 1, 2, 3, 4 are 00, 01, 11, 10. Note that only one bit changes as the level changes by one: for example, changing from level 2 to 3 changes 01 to 11. However, the converse is not true, and a single bit error obviously can cause an error from maximum to minimum—for example, from 00 to 10. Hence, some bits are still more sensitive to errors than others even for this coding technique. Gray codes can be generalized to more levels simply by adding a zero sign bit to the four 2-bit words to represent levels 1 through 4 and a one to 5 through 8 and making the original two bits the mirror image for levels 5 through 8. If the binary representation is $n = \sum_{i=1}^{k} a_i 2^{i-k}$, then the Gray code representation is $G_k(n) = (a_k, a_k \oplus a_{k-1}, a_{k-1} \oplus a_{k-2}, \ldots a_2 \oplus a_1)$ where $\oplus$ represents mod 2 addition. This mirror image property, shown in Table 3-3, gives the code the name *reflected binary*.

The natural binary, folded-binary, and Gray codes are all Harper codes, which exhibit the same average magnitude of error (AME) when errors occur independently of the transmitted bits and adjacent errors are also independent. For sufficiently small error probabilities and independent errors, Harper codes are optimum in that they minimize the average magnitude of error (AME) [Bucher, 1969*b*, 3113–3130].

Figure 3-16 shows the output SNR (*C*-message weighted noise) as a function of input signal level for both Gaussian and sinusoidal inputs. Note that the results for sinusoidal inputs exhibit a cyclic oscillation corresponding to segments in the compressor characteristic. The Gaussian input, on the other hand, tends to smooth out this fine structure. In addition, it exhibits a high peak factor, which overloads the compressor in the outer segment at lower levels than does the sinusoidal input. The Gaussian input is somewhat more representative of actual speech inputs than the single sinusoid.

Adaptive Pulse Code Modulation

An alternative approach to the problem of matching quantizer step size to input signal variance is to adapt the step sizes based on memory of previous quantizer output samples. In this approach, one can utilize a simple quantizer configuration: for example, use uniform step sizes. The step size δ_i at time iT is then adapted to larger or smaller values depending on whether the previous samples (or some weighted summation of previous quantizer outputs) are close to the saturation level or at the minimum output magnitude.

This approach bears some resemblance to automatic gain control (AGC) in the amplifier prior to the quantizer but has several important distinctions.

The adaptive PCM technique (APCM) discussed here is based solely on the quantizer output rather than analog inputs; hence, correct step sizes can be derived at the receiver by observations of received quantized samples alone.

Jayant [1974] describes one approach where an ℓ-bit uniform step-size quantizer has step sizes based on the previous step size and previous quantizer output code word magnitude. Let the output of the adaptive ℓ-bit quantizer for an analog input sample x_i be [refer to (3-1)]

$$Q_i(x_i) = 2^{\ell-1} \sum_{j=1}^{\ell} \alpha_{ij} 2^{-j} \delta_i = 2^{\ell} \frac{\mathcal{A}_i \delta_i}{2} \tag{3-70}$$

where $\mathcal{A}_i = \sum_{j=1}^{\ell} \alpha_{ij}\, 2^{-j+1} \qquad \alpha_{ij} = \pm 1 \qquad -(2^{-\ell} - 1) \leq \mathcal{A}_i \leq 2^{\ell} - 1$

Thus the magnitude of the quantizer code word is bounded by

$$|\mathcal{A}_i| \leq 2^{\ell} - 1 = N - 1$$

for an N-level quantizer. The quantizer step size δ_i is chosen to be scaled to the previous step size δ_{i-1} by a time invariant adaption function of the previous quantizer code word magnitude

$$\delta_i = \delta_{i-1} M(|\mathcal{A}_{i-1}|) \tag{3-71}$$

The adaption scale factor $M(\ )$ is selected so as to respond quickly to requirements for step-size increases, thereby lessening the effects of quantizer overload when δ_i is too small. For example, a step size increase is specified if the previous value of A_{i-1} exceeded half of the overload value. These step-size increases cannot be made at too large a ratio, however, or the response can become unstable and never settle down in response to a step input. Step-size decreases are to be made less rapidly. The advantage sought for adaption is to increase the dynamic range of the quantizer rather than to increase the peak value of output SNR at a particular signal level. At the same time, overload errors in general can be more harmful (as large as the peak signal level) than the granular quantizing error which is limited to the step size itself. Overload errors can be as large as the peak signal.

Table 3-4 gives an example set of parameters for an $N = 16$-level quantizer. Thus the value of M during step-size increases is selected to be as large as 2.4, whereas the value for step-size decreases is only slightly less than unity.

Table 3-4 ADAPTION SCALE FACTOR FOR PCM AS A FUNCTION OF THE QUANTIZER OUTPUT CODE WORD MAGNITUDE $\mathfrak{a}$ FOR AN $N = 16$ LEVEL QUANTIZER

Quantizer Output Code Magnitude $\lvert a \rvert$ $\lvert a \rvert + \frac{1}{2}$	*Adaption Step Size Scale Factor* M
1	0.80
2	0.80
3	0.85
4	1.00
5	1.20
6	1.60
7	2.00
8	2.40

Note that over a substantial range, $3 \leq |\mathfrak{a}| + \frac{1}{2} \leq 5$ the scale factor M is approximately unity. Step size decreases at a maximum rate of 0.80; on the other hand, step size can increase by a factor of 2.4 with each new sample, which overloads the quantizer.

Application of this adaption algorithm can theoretically lead to an infinite ratio between δ_{max} and δ_{min}, the maximum and minimum step sizes. In practice this ratio $\delta_{max}/\delta_{min}$ is held to a maximum for simplicity of implementation. Results for a value of $\delta_{max}/\delta_{min}$ of 100 have been described by Jayant [1974], and lead to quasi-optimal performance over a dynamic range of more than 20 dB.

3-6 QUANTIZER CODING FOR TRANSMISSION ERRORS

The digital outputs of quantizers are converted to serial bit streams for transmission. Then multiple bit streams are often multiplexed together in a single higher rate stream for transmission over a trunk or satellite link. Errors in the received bit stream can have widely differing effects, depending on which bit is in error. If coding is performed after a number of PCM channels have been multiplexed, the significance of different bits may change as different bit streams are accommodated, and it is usual to ignore this difference. Here, however, we examine a coding concept operating prior to multiplexing and directly tied to the quantizer. In this coding not all bits are treated equally. For example, the most significant bits are coded while the others are left uncoded [Buchner, 1969*a*, 1219–1248]. We will define the structure of significant-bit codes, and determine their effect on the average magnitude of output error for a channel with errors.

As mentioned above not all bits of a quantizer output are of equal importance, and hence some bits can be provided with coding protection while others are left unprotected. Even with Gray coding a single bit error can produce an error from maximum to minimum (see Table 3-3). The system model for examining there error effects is shown in Fig. 3-17 where binary-symmetric-channel (BSC) errors are inserted in the serial bit stream.

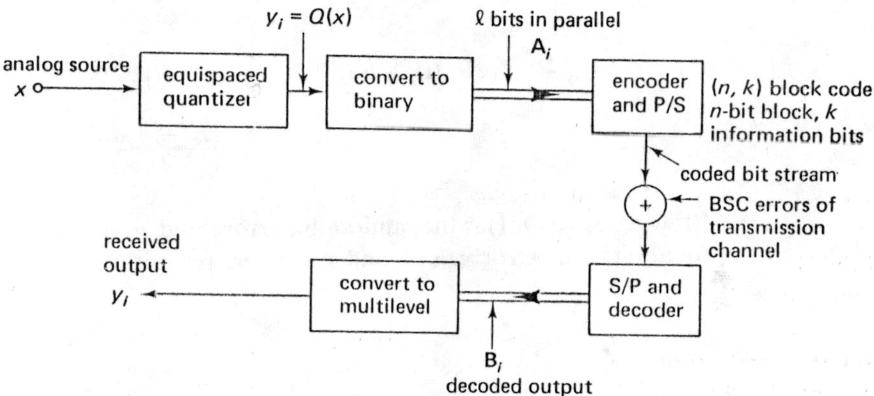

Fig. 3-17 Block diagram of a quantized channel with coding and BSC errors

Assume that the sampled analog source, $x(iT)$ is quantized with an equispaced ℓ-bit quantizer with input and output levels x_j, y_j:

$$\begin{array}{ccccccccc} x_1 & \delta & x_2 & & x_3 = 0 & & x_4 & & x_5 \\ & y_1 & & y_2 & & y_3 & & y_4 & \end{array} \quad \begin{array}{l}\text{input levels, } x_j;\\ \text{spacing } \delta \\ \text{output levels, } y_j\end{array}$$

$$\longleftarrow (2^\ell - 1)\delta \longrightarrow$$

The transmitted quantizer output levels y_i are related to the quantizing bits α_{ij} or a_{ij} by

$$\begin{aligned} y_i &= V \sum_{j=1}^{\ell} 2^{-j}\alpha_j(i) = V\left[\frac{\alpha_1(i)}{2} + \frac{\alpha_2(i)}{2^2} + \cdots + \frac{\alpha_\ell(i)}{2^\ell}\right] \\ &= \frac{V}{2^{\ell-1}}\left[\sum_{j=1}^{\ell} 2^{j-1} a_j(i)\right] - \frac{(2V - \delta)}{2} \\ &= \delta[a_1(i) + a_2(i)2 + \cdots + a_\ell(i)2^{\ell-1}] - \frac{(2^\ell - 1)\delta}{2} \end{aligned} \tag{3-72}$$

where $a_j(i) = \pm 1$ and $\alpha_j(i) = (0, 1)$.

The quantizer output at time iT can also be written

$$y(iT) = \delta A(i) - \frac{(2^\ell - 1)\delta}{2} \qquad 0 \leq A(i) \leq 2^\ell - 1 = N - 1 \quad (3\text{-}73)$$

where $A(i)$ is the quantizer code level.

Note that the previously defined quantizer level $a(i) = A(i) - (2^\ell - 1)/2$. The final decoded output similarly can be defined in terms of the received bits $b_j(i)$:

$$\hat{y}_{(i)} = \delta B(i) - \frac{(2^\ell - 1)\delta}{2} = \delta[b_1(i) + b_2(i)2 + \cdots + b_\ell(i)2^{\ell-1}] - \frac{(N-1)\delta}{2} \quad (3\text{-}74)$$

where $b_j(i) = a_j(i) \oplus e_j$; $e_j = (0, 1)$ is the random bit error; and $p = 1 - q$, and q are the probabilities of error $e_j = 1$ and no error, respectively.

System Performance

AVERAGE MAGNITUDE OF ERROR (AME) In order to compute the Average Magnitude of Error (AME) for this system, we assume a uniform distributed input

$$\begin{aligned} p(x) &= \frac{1}{2V} \qquad |x| \leq V \\ &= 0 \qquad |x| > V \end{aligned} \quad (3\text{-}75)$$

Write the generalized average system error as the sum of channel and quantizing error components:

$$\mathbf{E}[f(\epsilon)] \triangleq \mathbf{E}[|\hat{y}_i - x|^\gamma] = \mathbf{E}[|\underbrace{(\hat{y}_i - y_i)}_{\text{channel error effect}} + \underbrace{(y_i - x)}_{\text{quantizing error}}|^\gamma] \quad (3\text{-}76)$$

where $\mathbf{E}$ represents the expected value operation. $f(\epsilon) = |\epsilon|^\gamma$ is the error metric, and the error $\epsilon \triangleq \hat{y} - y$. For $\gamma = \mathbf{E}[f(\epsilon)] = \mathbf{E}[\epsilon^2]$ the average system error is the mean-square error examined previously.

Now consider $f(\epsilon) = |\epsilon|$, the magnitude of error. The AME can be evaluated as using

$$\mathbf{E}|\epsilon| = \mathbf{E}\left|\delta(B - A) + \left[A\delta - (2^{\ell} - 1)\frac{\delta}{2} - x\right]\right|$$

$$= \frac{1}{N\delta}\delta \sum_{A=1}^{N-1} \sum_{B=1}^{N-1} \int_{\delta(A-1/2)}^{\delta(A+1/2)} dx \left|(B - A) + \left(A - \frac{x}{\delta}\right)\right| p(B|A)$$

error probability (pointing to $p(B|A)$)

$$= \frac{1}{N}\left[\sum_A \int_{\delta(A-1/2)}^{\delta(A+1/2)} \left|A - \frac{x}{\delta}\right| dx\, p(0|0)\right.$$

probability of no error $p(A|A) = p(0|0)$

$$\left. + \sum_A \sum_{\substack{B \\ A \neq B}} \int_{\delta(A-1/2)}^{\delta(A+1/2)} \left|B - A + A - \frac{x}{\delta}\right| p(B|A)\, dx\right]$$

$$= \frac{\delta}{N}\left[\sum_A \int_{-1/2}^{1/2} |\Delta| p(0|0)\, d\Delta\right.$$

$$\left. + \sum_A \sum_{\substack{B \\ A \neq B}} \int_{-1/2}^{1/2} |B - A + \Delta| p(B|A)\, d\Delta\right] \qquad \Delta \triangleq \frac{A - x}{\delta}$$

for any $A \neq B$ the integral is $|B - A|$

(3-77)

Hence, the AME is the sum of quantizing and channel error effects

$$\mathbf{E}|\epsilon| = \frac{\delta}{N}\left[\frac{N}{4}p(0|0) + \sum_A \sum_B |B - A| p(B|A)\right. \tag{3-78}$$

quantizing error (pointing to $\frac{N}{4}p(0|0)$); channel error effect (pointing to $\sum_A \sum_B |B - A| p(B|A)$)

Since $p(0|0)$ is simply q^{ℓ}, the probability of no error in ℓ bits, and the second term is the expected value of $|B - A|$, the AME can be rewritten as

$$\mathbf{E}|\epsilon| = \frac{\delta}{4}q^{\ell} + \delta\mathbf{E}|B - A| \tag{3-79}$$

Channel Error Effect Next we evaluate the effect on the AME of channel errors; the second term in (3-78) is

$$\frac{\delta}{N}\sum_A \sum_B |B - A| p(B|A) \tag{3-80}$$

where $A = \sum_{j=1}^{\ell} 2^{j-1}a_j$

$$B = \sum_{j=1}^{\ell} 2^{j-1}b_j = \sum 2^{j-1}(a_j \oplus e_j)$$

where $\oplus$ represents mod 2 addition. Rewrite $|B - A|$ in terms of a_j, e_j:

$$|B - A| = |\sum 2^{j-1}(a_j \oplus e_j - a_j)| \tag{3-81}$$

The value of $|B - A|$ for a given error vector $\mathbf{e} = (e_1, e_2, \ldots, e_\ell)$ depends on A. Define a set of error vectors $\{\mathbf{e}^{(m)}\}$ with members $\mathbf{e}^{(m)}$ where $\mathbf{e}^{(m)} = (xx, \ldots, xx\,100, \ldots, 00)$* has a 1 in the mth position. These error vectors cause an average error magnitude $|B - A|$ equal to 2^{m-1} when averaged over all values of A_m. That is, the average over all $\mathbf{e}$ in this set is

$$\underset{\mathbf{e}}{\mathbf{E}}\left(\frac{1}{2^\ell}\sum_{A=1}^{2^\ell}|B - A|\right) = 2^{l-1} \qquad 2^{l-1} \le \sum e_j^{2^{j-1}} \le 2^l - 1 \tag{3-82}$$

where the expectation $\mathbf{E}$ is taken over all error vectors $\mathbf{e}$.

To show that (3-83) is valid, consider any given input vector $\mathbf{a} = (a_1, a_2, \ldots, a_\ell)$ and error $\mathbf{e} = (x, \ldots, x\,10, \ldots, 0)$ where the first m-1 components are x. Since the block of xs go through all 2^{m-1} equally likely values, there are 2^{m-1}equally likely values for $\mathbf{e}$ of this form. The resulting $\mathbf{b}$ is then of the form $\mathbf{b} = (xx, \ldots, x\bar{a}_m\, a_{m+1}, \ldots, a_\ell)$ where $\bar{a}_m = a_m \oplus 1$, and $\mathbf{b} = \mathbf{a} + \mathbf{e}$. Thus, the bits in $\mathbf{a}$ above the mth bit are not in error for this $\mathbf{e}^{(l)}$ and can be ignored.

Define the integers that represent the first n bits of the quantizer output as

$$A^{(m)} = \sum_{j=1}^{m} a_j 2^{j-1} \qquad \text{and} \qquad B^{(m)} = \sum_{j=1}^{m} b_j 2^{j-1} \tag{3-83}$$

If $a_m = 1$, then $B^{(m)}$ lies somewhere between 2^m and 0, where $2^m \geq A^{(m)} \geq 2^{m+1} - 1$ and $0 \leq 2^{m+1} - 1$. This region for $B^{(m)}$ is shown in the quantizer output scale for an error $\mathbf{e}^{(m)}$

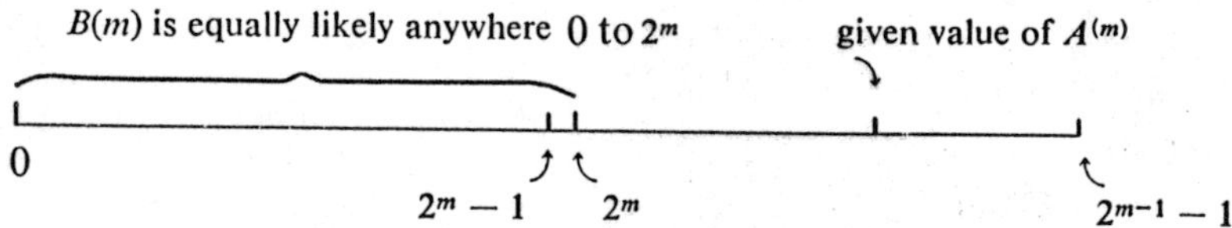

Clearly, the picture would simply be reversed if $a_l = 0$. Thus, $|A - B| =$

---

*The xs represent "don't care" bits. Note that $(00, \ldots, 0100, \ldots, 00)$ with a 1 in the mth position gives an amplitude $\sum e_j 2^{j-1} = 2^{m-1}$ and $(11, \ldots, 110, \ldots, 00)$ represents $2^m - 1$ where the most significant 1 is in the mth position.

$|[A^{(m)} - 2^m] + [B^{(m)} + 1]|$. Averaging over all **e** in the set [i.e., all $B^{(m)}$] gives $\mathbf{E}|A - B| = [A^{(m)} - 2^m] + (2^{m-1}/2)|$. Averaging over all A and $\mathbf{e} \in \{e^{(m)}\}$ gives $\mathbf{E}|A - B| = 2^{m-1}$ for $\mathbf{e} \in \{e^{(m)}\}$.

The probability that $\mathbf{e} \in \{\mathbf{e}^{(m)}\}$ is simply $pq^{\ell-1}$ where $p = 1 - q$. Hence, the expected value of $|A - B|$ is

$$\mathbf{E}|A - B| = \sum_{i=1}^{\ell} 2^{i-1} pq^{\ell-1} = p2^{\ell-1} \sum_{i=1}^{\ell} \left(\frac{q}{2}\right)^{\ell-1} = p2^{\ell-1} \frac{(1 - (q/2)^{\ell})}{(1 - q/2)} \tag{3-84}$$

and the average magnitude of error (AME) for a channel with bit error probability p is

$$\mathbf{E}|\epsilon| = \frac{\delta}{4} q^{\ell} + \delta p 2^{\ell} \frac{[1 - (q/2)^{\ell}]}{(2 - q)} \tag{3-85}$$

and

$$\frac{\mathbf{E}|\epsilon|}{2V} = \frac{1}{4}\left(\frac{q}{2}\right)^{\ell} + p\frac{[1 - (q/2)^{\ell}]}{2 - q} \tag{3-86}$$

where $\delta = 2V/2^{\ell}$. For small bit error probability $p \ll 1$, the AME is approximately equal to

$$\begin{aligned} \frac{\mathbf{E}|\epsilon|}{\delta} &\cong \frac{1}{4}(1 - p)^{\ell} + \frac{p2^{\ell}}{1 + p}\left[1 - \left(\frac{1 - p}{2}\right)^{\ell}\right] \cong \frac{1}{4}(1 - \ell p) \\ &\quad + \frac{p2^{\ell}}{1 + p}\left[1 - \frac{1}{2^{\ell}}(1 - \ell p)\right] \\ &\cong \frac{1}{4} + p\left(2^{\ell} - 1 - \frac{\ell}{4}\right) \\ &= \frac{1}{4} \qquad p = 0 \end{aligned} \tag{3-87}$$

Thus, channel errors dominate when the bit error probability is sufficiently large that $p[(2^{\ell} - 1) - \ell/4] \geq 1/4$. The critical error probability p_c, where channel and quantizing errors are equally large in (3-87), becomes

$$p_c \cong \frac{1}{4[(2^{\ell} - 1) - \ell/4]} \cong \frac{1}{2^{\ell+2}} \qquad 2^{\ell} \gg 1 \tag{3-88}$$

for 2^{ℓ} quantizing levels. Comparison of Eq. (3-88) with Fig. 3-2 confirms that the breakpoint is essentially the same whether mean-square error or AME is used as the performance measure. The AME is plotted in Fig. 3-18.

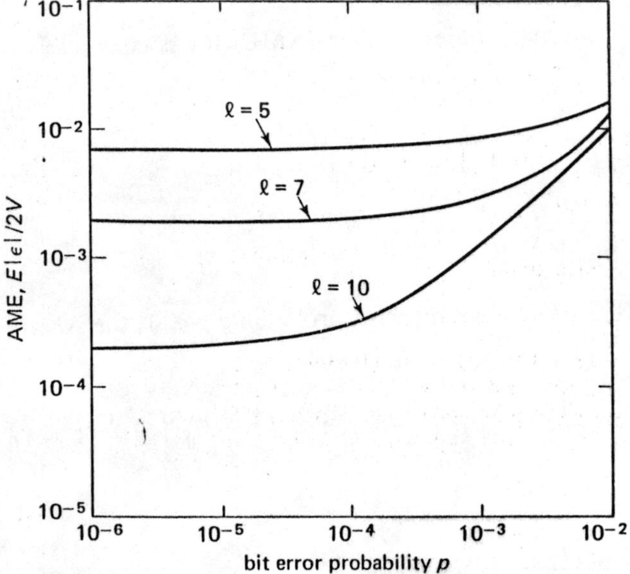

Fig. 3-18 AME versus bit error probability for no coding. The quantizer is uniform with ℓ-bits/sample.

Coded Performance

Quantize so that there are ℓ information bits/sample. Then group α of these ℓ bit samples together into one code block. Error-correcting codes can be used either to provide protection to all $\alpha\ell$ information bits, or to protect only the σ most significant bits of each sample. The two general formats of packed codes are shown in Fig. 3-19. Refer to Fig. 3-19b. The total number of bits transmitted per frame is

$$N_f = \alpha(\ell - \sigma) + \alpha\sigma + (n - k) = \alpha\ell + n - \alpha\sigma = \alpha(\ell - \sigma) + n$$

for a block code of n bits and k information bits. Thus the average number of

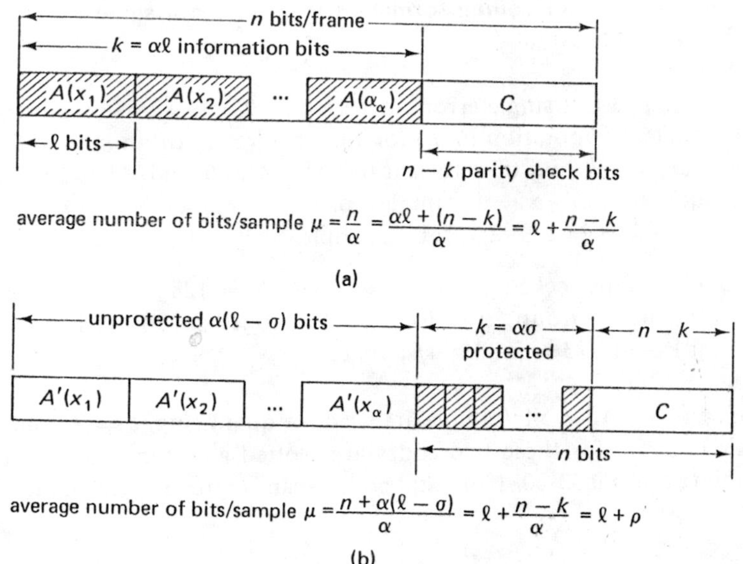

Fig. 3-19 Two possible formats for packed code vectors; (a) protect all bits, and (b) protect only the σ most significant bits of each sample. The protected bits are indicated by crosshatch.

information bits per sample increases with coding from ℓ to $\mu = N_f/\alpha = \ell + (n - k)/\alpha$ where $(n - k)$ is the number of parity check bits/frame.

Suppose that the σ most important positions are protected and $\ell - \sigma$ positions are left unprotected. Assume that sufficient protection is provided so that the protected bits have an error probability much less than p for $p < p_0$ and that their error contribution is negligible. Then the AME can be obtained from (3-79) and (3-84) as

$$\mathbf{E}|\epsilon| = \delta\left(\frac{q^\ell}{4} + p\sum_{i=1}^{\ell-\sigma} 2^{i-1} q^{\ell-\sigma-i}\right) \qquad \ell > \sigma \tag{3-89}$$

where the first term is the quantization error for the ℓ-bit quantizer and the second is the effect of errors in the $\ell - \sigma$ unprotected bits (not valid if $\ell \cong \sigma$). For small probability of error, these results can be approximated as

$$\frac{\mathbf{E}|\epsilon|}{2V} \cong \frac{1}{2^\ell}\left[\frac{1}{4} + p\left(2^{\ell-\sigma} - 1 - \frac{\ell - \sigma}{4}\right)\right] \qquad p \ll 1 \tag{3-90}$$

To examine the effects of coding on system performance, consider the following examples of coding formats, for which we assume an average of $\mu = 7$ bits/sample.

Code 1 (3, 1) perfect single error correction (SEC), or simple majority vote, code. The notation (n, k) for the code refers to an n-bit code word carrying k bits of information. Use $\ell = 5$ and $32 = 2^5$ levels of quantization, $\alpha = 1$ samples packed together to yield $\mu = 5 + (n - k)/\alpha = 5 + 2 = 7$ bits/sample.

Code 2 (31, 26) perfect SEC code; $\ell = 7$ and $N = 128 = 2^7$ levels of quantization. Group $\alpha = 13$ samples together to yield $\mu = 7 + (31 - 26)/13 = 7.38$ bits/sample.

Uncoded form: All 7 bits/sample are used for quantizing, $\ell = 7$. The resulting AME curves for these two codes are plotted as a function of bit error probability p in Fig. 3-20. Note that coding can improve performance for bit

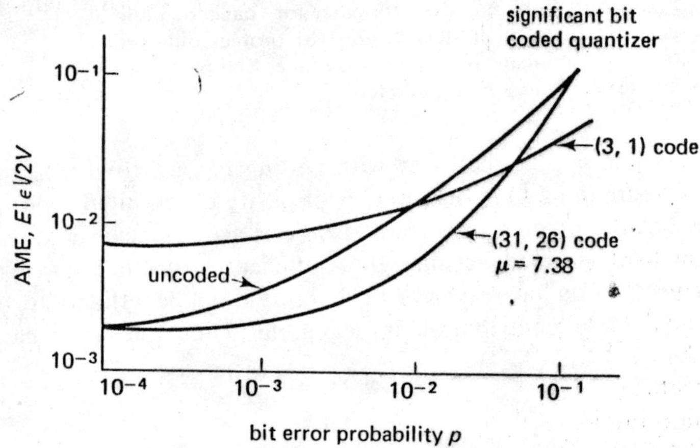

Fig. 3-20 AME versus bit error probability for coded and uncoded channels; $\mu \cong 7$ bits/sample

error probabilities on the order of 10^{-2} to 10^{-3}; for inefficient majority codes, however, performance can be degraded for values of p below this range. The (31, 26) code provides improvement over a wide range of bit error probabilities. Clearly the (3, 1) code has a larger minimum AME because it uses only $\ell = 5$ bit quantization.

3-7 USE OF DITHER SIGNALS WITH A UNIFORM QUANTIZER

In the preceding sections, we have evaluated quantizer performance in terms of average magnitude of error or mean-square error. For television signals or speech, however, other effects are important to the observer or listener. For example, a ramp change in gray scale after uniform quantizing produces a staircase effect and picture "contours." Contouring results from step changes in intensity across the picture, analogous to the terracing of a hillside, as opposed to a continuous more gradual change; contouring of facial features is particularly disturbing. The eye performs a spatial averaging function that smooths very fine grain gray scale fluctuations but does not eliminate the coarse contouring effects. Thus pictures with contouring are often more objectionable than pictures with the same mean-square error but no contouring.

Here we show that objectionable contouring effects of conventional PCM can be avoided through the use of pseudorandom *dither signals* [Roberts, 1962, 145–154; Schuchman, 1964, 162–165]. Similar applications of dither to speech signals have been described by Jayant and Rabiner [1972]. A statistical linearization procedure using these techniques is described below. These techniques are most useful for quantizers with less than 6-bits per sample where signal-dependent errors are most apparent.

System Concept

With a conventional quantizer, a video ramp input produces the staircase (contoured) output of Fig. 3-21. Constant gray levels are produced within each quantizer interval. If we add a pseudorandom dither signal d_i to the input before quantizing and subtract it from the received output (Fig. 3-22), the dither signal in effect randomizes the quantizing noise and decreases contouring. Note that because the dither is pseudo-random it is possible with appropriate synchronization to regenerate it at the receiver.

The pseudo-random dither d_i is selected to be uniformly distributed from $-\delta/2$ to $+\delta/2$. The effective quantizer characteristic $Q(x + d)$, plotted as the input x, changes form depending on the value of d as shown in Fig. 3-23. The signal estimate is related to the quantizer output by $y = Q(x + d) - d$. If $x = 0$, then $y = Q(d) - d$, and y is simply quantizing noise. The system output is plotted versus input x in Fig. 3-24. Since y is a random function, only the envelope of the boundaries is shown along with a plot of y for $d = 0$. For a given value of x, different values of d produce a value of y within the boundaries shown. Thus, $p(y \mid x)$ is uniformly distributed over this interval if d is uniformly distributed.

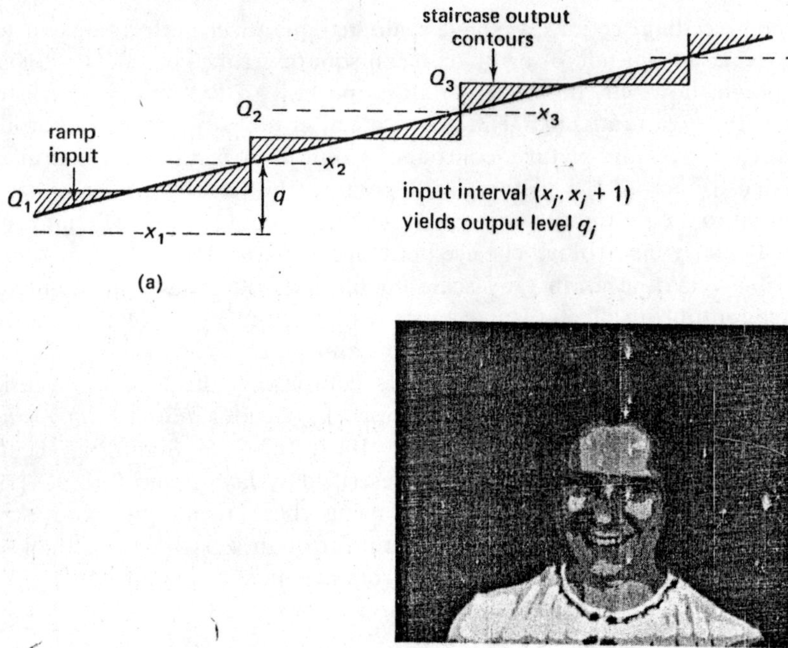

Fig. 3-21 Contouring effects (a) Staircase (contoured) output resulting from a ramp video input. (b) Quantized image (8 gray levels) showing contouring. (Courtesy NASA Ames Research Center)

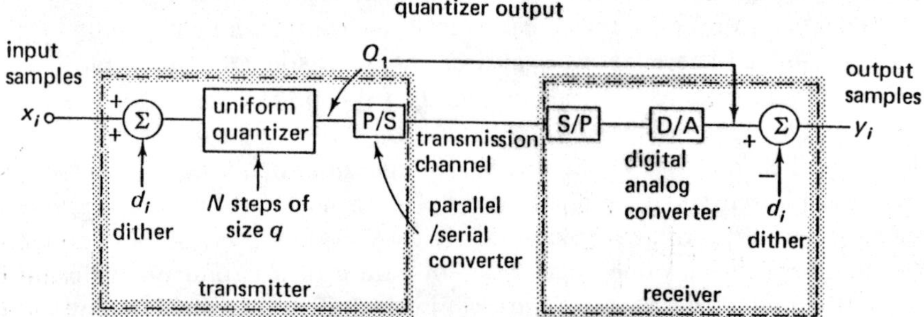

Fig. 3-22 Use of dither signal with a uniform quantizer

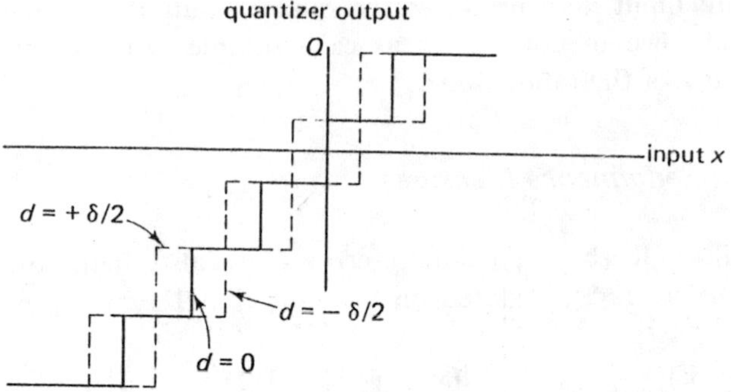

Fig. 3-23 Quantizer output Q for various values of dither d_i

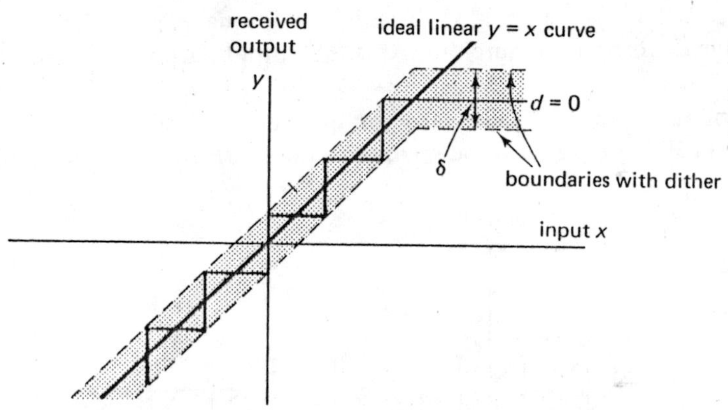

Fig. 3-24 System output versus input with dither

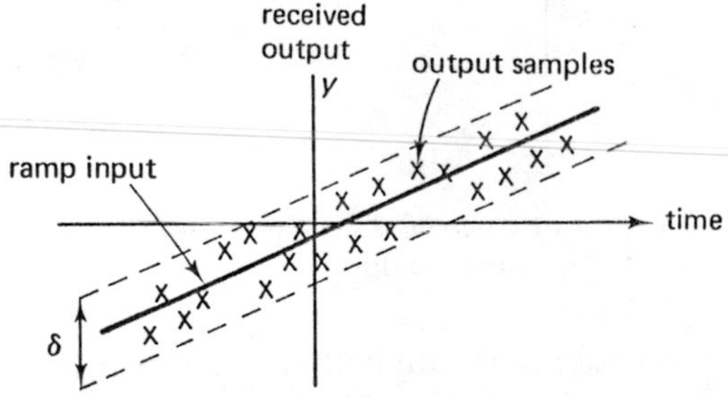

Fig. 3-25 System output versus time for a ramp input

A ramp input now produces the random output shown in Fig. 3-25. Note the absence of contouring as each sample is uniformly distributed between $\pm\delta/2$ of the input value.

System Performance Measures

In addition to the mean-square error ϵ, we also define the conditional mean-square errors $\mathfrak{D}$, $\mathcal{V}$ based on the conditional expectations:

$$\begin{aligned}\epsilon &= \mathbf{E}[(x-y)^2] = \mathbf{E}[(x-\bar{y}_x)^2] + \mathbf{E}[(y-\bar{y}_x)^2] \triangleq \mathfrak{D} + \mathcal{V} \\ \mathfrak{D} &= \mathbf{E}[(x-\bar{y}_x)^2] \geq 0 \\ \mathcal{V} &= \mathbf{E}[(y-\bar{y}_x)^2] \geq 0 \end{aligned} \tag{3-91}$$

where $\bar{y}_x = \mathbf{E}(y|x) = \int yp(y|x)\,dx$ is the conditional expectation of y given x, $\mathfrak{D}$ is the distortion measure, and $\mathcal{V}$ is the output variance.

Conditional Density   The variation in y versus d for a given value of x is shown in Fig. 3-26. As can be seen, the value of the input x biases up and

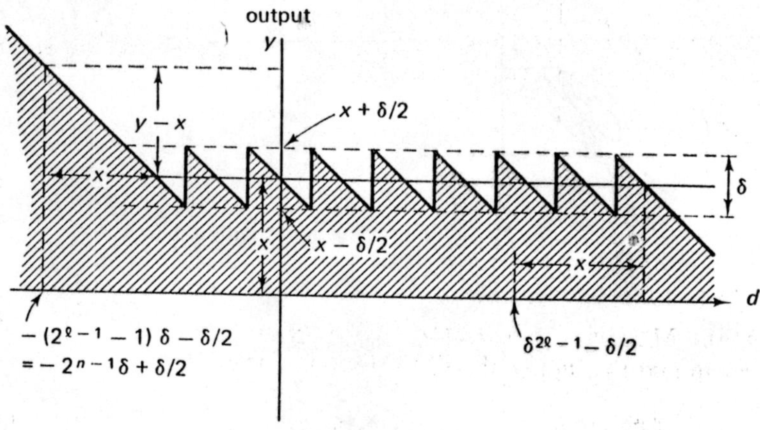

Fig. 3-26 Plot of y versus d for a given input x for a quantizer of 2ℓ levels of width δ

down the mean of this entire sawtooth pattern. Assuming 2^ℓ quantizer levels of interval δ, we can write the conditional density of the quantizer output y, dependent on $p(d)$, as

$$p(y \mid x) = \begin{cases} p(d = -y - 2^{\ell-1}\delta + \delta/2) & \text{for } x + \delta/2 \leq y < \infty \\ & \text{saturation} \\ \sum_{k=0}^{2n-1} p(d = -y - 2^{\ell-1}\delta + \delta/2 + k\delta) & \\ & \text{for } x - \delta/2 < y < x + \delta/2 \\ p(d = -y - \delta/2 + 2^{\ell-1}\delta) & \text{for } -\infty < y \leq x - \delta/2 \\ & \text{saturation} \end{cases} \tag{3-92}$$

Conditional Mean of y The conditional mean $\bar{y}_x$ can be computed for a given dither probability density. Assume that $p(d) = 1/\delta$ for $|d| \leq \delta/2$, and zero otherwise. Then the conditional mean $\mathbf{E}(y \mid x) \triangleq \bar{y}_x$ is equal to x except when saturated as shown in Fig. 3-27. Saturation recurs for inputs x which exceed $(\delta/2)(N - 1)$.

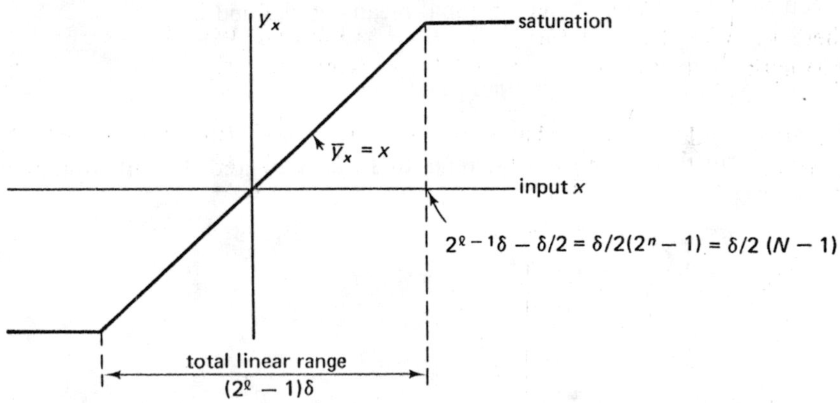

Fig. 3-27 Variation in output conditional mean with x for a finite quantizer with a uniformly distributed dither

Distortion Measure The mean square difference between the conditional mean and the input—the distortion measure $\mathfrak{D}$—is defined as

$$\mathfrak{D} = \iint (x - \bar{y}_x)^2 p(x) p(y \mid x)\, dx\, dy \tag{3-93}$$

For uniform dither $p(d) = 1/\delta$ for $|d| < \delta/2$, and $p(d) = 0$ for $|d| > \delta/2$, the distortion measure $\mathfrak{D}$ can be computed, using (3-92), as

$$\mathfrak{D} = \int_{x \in X_{\text{sat}}} \left[x - \frac{\delta}{2}(2^{\ell} - 1) \right]^2 p(x)\, dx \qquad |x| > \frac{\delta}{2}(2^{\ell} - 1) \tag{3-94}$$

where x is integrated over the saturation region $|x| > (\delta/2)(N-1)$. The other regions $|x| < (\delta/2)(N-1)$ have no saturation, and $\bar{y}_x = x$. If x is bounded and uniform for $|x| < 2^{\ell}\delta/2$, then the distortion (resulting from saturation) is from (3-94)

$$\mathfrak{D} = \frac{2}{2^{\ell}\delta}\int_0^{\delta/2} x^2\,dx = \frac{1}{2^{\ell-1}}\frac{1}{3}\frac{\delta^3}{8} = \frac{\delta^2}{12}2^{-\ell} \tag{3-95}$$

for ℓ-bit quantizing. Compare this result with $\mathfrak{D} = \delta^2/12$ for a quantizer with no dither. Clearly, as ℓ increases, the distortion $\mathfrak{D}$ approaches zero for uniform dither. The disappearance of contouring corresponds to this decrease in $\mathfrak{D}$.

VARIANCE The output error variance—the mean-square difference between the output and the conditional mean—is defined as

$$\mathcal{V} \triangleq \iint (y - y_x)p(x)p(y\,|\,x)\,dx\,dy \tag{3-96}$$

For uniform dither and any x, refer to Figs. 3-24 and 3-27 to obtain the variance

$$\begin{aligned}\mathcal{V} &= \int (y - \bar{y}_x)^2 p(y - \bar{y}_x)\,d(y - \bar{y}_x) \\ &= \frac{\delta^2}{12}\end{aligned} \tag{3-97}$$

and is the same for no dither. Thus, the mean-square error does not improve with dither, but contouring disappears.

ERROR PROBABILITY DENSITY We define the analog error as $e \triangleq y - x$. Then $p(e)$ is computed from

$$p(e) = \int p(e|\,x)p(x)\,dx \tag{3-98}$$

$$p(e) = \frac{1}{\delta}\int_{\delta/2(2^{\ell}-1)-e}^{\delta/2(2^{\ell}-1)-e} p(x)\,dx \qquad |e| < \frac{\delta}{2} \tag{3-99}$$

$$p(e) = \frac{1}{\delta}\int_{\delta/2(2^{\ell}-1)-e}^{-\delta/2(2^{\ell})-e+\delta} p(x)\,dx \qquad e > \frac{\delta}{2} \tag{3-100}$$

These results are plotted for both uniform and Gaussian inputs in Fig. 3-28.

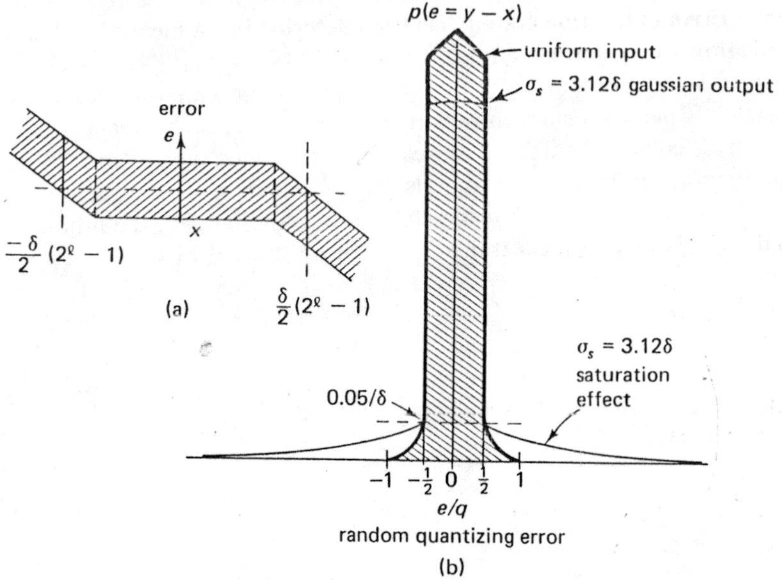

Fig. 3-28 Quantizing error probability density for uniform and Gaussian types of dither; (a) shows the uniform distribution of error versus x, and (b) shows the resultant density of error e.

Clearly, e is essentially uniformly distributed $|e| < \delta/2$ when saturation effects are small.

3-8 SUMMARY

This chapter has discussed the performance of quantizers with respect to three primary performance measures, mean-square error, average magnitude of error (AME), and a distortion measure (the difference between the conditional mean of the output and the input). We have examined simple uniform quantizers, optimal quantizers, and approximately optimal quantizers employing compandors. Compandors can improve the performance of the quantizer substantially and can increase the quantizer dynamic range significantly. Performance calculations for these quantizers have assumed bandlimited inputs which are sampled at twice the maximum input frequency.

Channel error effects on performance have been examined both with respect to the bounds in performance established from channel capacity and bandwidth constraints, as well as the effect of errors on realizable transmission systems. Various coding techniques used to convert the parallel words at the quantizer output to a serial bit stream suitable for transmission have been examined. Error correction coding has been described as a means for providing protection to the more significant bits.

In image transmission or TV, coarse quantizing can produce a contouring effect whereby a smooth variation in gray scale, such as in a person's face, is converted to a disturbing staircase of gray scales such as in a TV test pattern. Pseudo-random dither signals have been described as a possible means for reducing these contouring effects without changing the number of quantizing levels or requiring transmission of additional bits.

CHAPTER 4

DELTA MODULATION AND DIFFERENTIAL PCM

4-1 INTRODUCTION

When the input data are bandlimited but have a nonuniform frequency response, perhaps heavily weighted at the lower portion of the frequency band, performance can be improved by means of feedback around the quantizer. In essence, one attempts to predict the next sample value and quantizes only the difference between the predicted and actual value. In this way, the variance of the quantizer input can be reduced below the value of the system input.

Several forms of predictive quantizing are analyzed here. The simplest is *delta modulation* [Goodman, 1969, 1197–1218; Van de Weg, 1953, 367–385], in which only one-bit quantization is used and the sampling rate is raised above the Nyquist rate, $2W$. Two types of quantizing noise appear here: *granular quantizing noise*, analogous to the PCM quantizing noise of Chap. 3; and *slope overload noise*, which is peculiar to quantizers with delta modulation or differential PCM (DPCM). Slope overload noise in DPCM is somewhat analogous to quantizer amplitude saturation in PCM.

The one-bit delta modulation (ΔM) concept can be generalized to the use of ℓ-bit quantizers, nonuniform quantizers, and more sophisticated predictive filtering. Several quantizer types in this family of differential PCM (DPCM) systems are discussed, including adaptive DPCM. In the more complex adaptive predictors for DPCM, the prediction coefficients vary with time, and these optimized coefficients are transmitted along with the quantized difference between the predicted signal and the actual input. This input signal may be nonstationary, such as speech.

Examples are given for both speech and television quantizing. In the transmission of television or imagery, the input might be the output of a Hadamard, Haar, or Fourier transform of the picture as well as the analog television signal itself.

4-2 DELTA MODULATION

The quantizers discussed in Chap. 3 operate instantaneously on the sampled signals. This chapter deals with *feedback*, or *predictive*, quantizers, in which the operations of sampling and quantizing are combined in a single feedback loop and are no longer separable as they were with PCM. These DPCM systems take advantage of the correlation between successive samples of the input data in order to reduce quantizing noise.

The delta modulator (ΔM) is one of the most easily implemented types of A/D converters for communications (Fig. 4-1) and is the simplest form of predictive quantizer, or differential PCM. It provides one-bit quantization of the difference in actual versus predicted input values and thus forms a serial bit stream that can be fed to a PSK modulator or other digital transmission device.

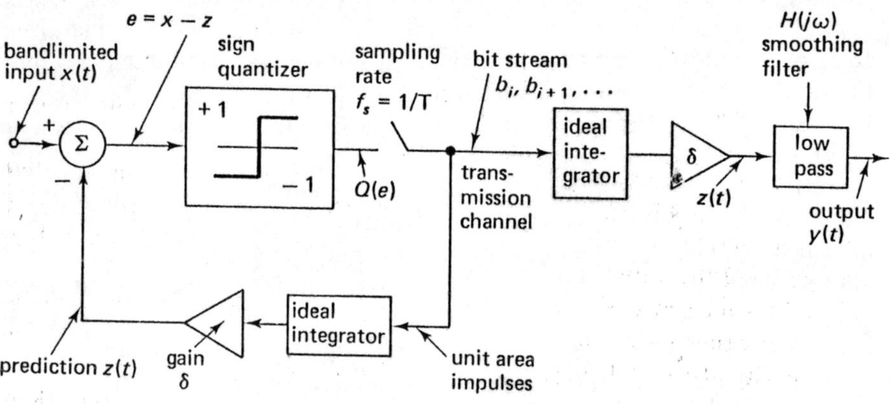

Fig. 4-1 One-bit delta modulator system configuration

As shown in Fig. 4-1, the input predicted value $z(t)$ is subtracted from the input $x(t)$, and the sign of the error signal is sampled at a periodic sampling rate. This binary output bit stream is then transmitted. If the data are received without error, the output of the filter/amplifier is identical to $z(t)$ since the same type of filter (an integrating device) is used at the transmit and receive sites. The output $z(t)$ is then smoothed to obtain an estimate $y(t)$ of the input signal.

In practice, the ideal integrator (Fig. 4-1) is periodically reset or replaced by a finite memory integrator (RC filter, or second-order filter) to prevent a single transmission error from persisting in the output for an infinite time.

A typical input waveform produces an output $z(t)$ (before smoothing) as shown in Fig. 4-2. The ΔM step size is δ. Notice that in regions where the slope of the input exceeds the maximum slope of the ΔM, a slope overload condition occurs in which the input can depart substantially from the output. At other times, when slope overload does not occur, the quantizing error has a more granular behavior.

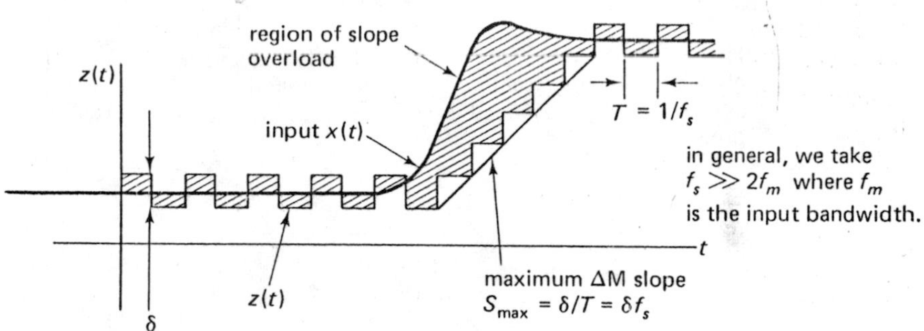

Fig. 4-2 Typical delta modulator output. The shaded area depicts the error between the input and predicted value.

If z is even at $t = 0$, note that $z(t) = 2j\delta$, an even-level output, at even sampling instants $t = 2i\text{T}$, and that $z(t) = (2j - 1)\delta$, an odd-level output, at odd sampling instants $t = (2i - 1)\text{T}$. Since z changes by $\pm\delta$ each sampling instant, z changes by 0 or ± 2 between 2 sampling instants. The roles of odd and even outputs would be reversed if the output at $t = 0$ were odd. It is assumed that the initial condition is equally likely to be odd or even.

With sufficiently slowly varying inputs (no slope overload), the samples $z(i\text{T})$ at any time instant match the output of a uniform quantizer Q_e of even-order parity at $t = 2i\text{T}$ and of odd-order parity Q_o at odd time instants, or vice versa if the quantizer started at $t = 0$ at an odd level. Thus, one alternates back and forth in time between the even- and odd-order quantizers shown in Fig. 4-3, and hence acts as if the effective step size is 2δ for a fixed uniform quantizer at any given instant of time.

One expects to see an output SNR of the ΔM varying with step size with a general behavior similar to that shown in Fig. 4-4. That is, with step sizes too small, slope overload dominates because of an inability to track. For too large a step size, granular quantizing noise dominates.

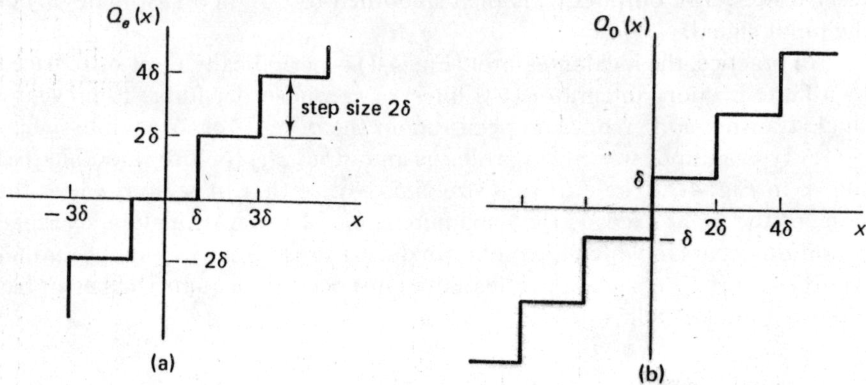

Fig. 4-3 (a) Even- and (b) odd-order quantizer characteristics representing the delta modulator in the absence of slope overload at even and odd instants of time

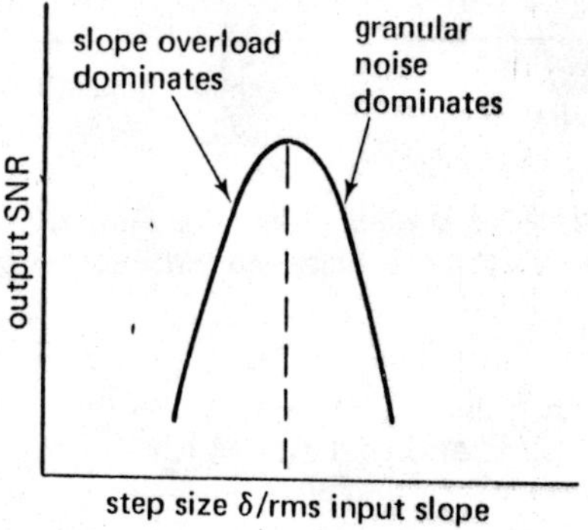

Fig. 4-4 General behavior of output SNR with step size for a fixed sampling rate

Granular Quantizing Noise with No Output Smoothing

Consider first the simplest measure of performance, the quantizing noise with no slope overload and no output smoothing. Since the quantizer step size is 2δ and there is no output filtering, the output granular quantizing noise [see (3-5)] for PCM with the same 2δ step size is

$$\sigma_g^2 = \frac{(2\delta)^2}{12} = \frac{\delta^2}{3} \tag{4-1}$$

for uniform distributed inputs and no slope overload. Thus, the same result is achieved as for a uniform instantaneous quantizer with quantizing interval (2δ). The ΔM, however, has an arbitrarily large dynamic range so long as slope overload is avoided, whereas a conventional quantizer is limited in amplitude range to $\pm 2^{\ell-1}\delta$ for an ℓ-bit quantizer.

Output SNR with No Output Smoothing

Performance analyses require definition of the following system parameters:

input power $\sigma_x^2 = \overline{x^2} = \int G_{xx}(f)\,df$,

where $G_{xx}(f)$ is power spectral density of the input x

input spectrum bandlimited to $|f| < f_m$

$$\text{effective input bandwidth } f_e = \left[\frac{\int_{-f_m}^{f_m} f^2 G_{xx}(f)\,df}{\int_{-f_m}^{f_m} G_{xx}(f)\,df}\right]^{1/2}$$

rms slope of input $\sigma_{\dot{x}} = 2\pi\sigma_x f_e$

where the superbar denotes the time average. We also define an approximate maximum input signal slope $S_{max} \triangleq 4\sigma_{\dot{x}} = 8\pi\sigma_x f_e$, or 4σ loading value, and set the maximum ΔM slope equal to S_{max}—that is, $\delta f_s = 8\pi\sigma_x f_e$. Thus,

$$\delta = \frac{8\pi\sigma_x f_e}{f_s} \tag{4-2}$$

is the step size necessary to prevent slope overload. Since the output signal power is σ_x^2, the maximum output SNR from (4-1), (4-2) for no slope overload is

$$\text{SNR}_0 = \frac{\sigma_x^2}{\sigma_g^2} = \frac{3\sigma_x^2}{\delta^2} = \frac{3\sigma_x^2}{(8\pi\sigma_x f_e/f_s)^2} = \frac{3}{(8\pi)^2}\left(\frac{f_s}{f_e}\right)^2 \tag{4-3}$$

If the input $x(t)$ has a rectangular spectrum, then the effective bandwidth f_e and output SNR are

$$f_e = \sqrt{\frac{1}{f_m}\int_0^{f_m} f^2\,df} = \frac{f_m}{\sqrt{3}}$$

and $$\text{SNR}_0 = \frac{9}{(8\pi)^2}\left(\frac{f_s}{f_m}\right)^2 \qquad \text{SNR}_0 = \frac{9}{(4\pi)^2}R_b^2 \tag{4-4}$$

where $R_b = \ell f_s/2f_m$ is the normalized bit rate, ($\ell = 1$ for 1 bit ΔM). Thus, this unsmoothed output SNR increases with sampling rate at 6 dB/octave. Performance can be improved, however, by smoothing the output samples since normally $f_s > 2f_m$ and the output samples are correlated.

Differential PCM with ℓ-Bit Quantization and No Smoothing

In a variation of the delta modulation quantizer configuration, the sign operation of Fig. 4-1 can be replaced by an ℓ-bit quantizer as in Fig. 4-5. Note that the equivalent quantizer $Q(e)$ has increments of 2δ as consistent with the 1 bit ΔM of Fig. 4-1. This approach is known as differential PCM or DPCM.

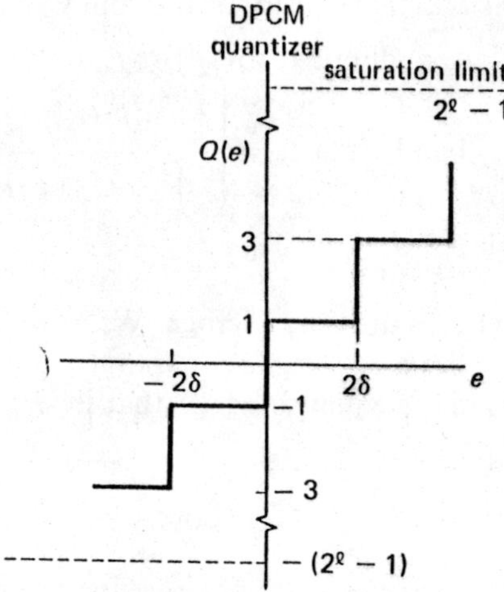

Fig. 4-5 Quantizing of the DPCM error signal with a multi-level quantizer, where $e = x - z$, the difference between the input x and the predicted value z. The increment between quantizing levels is 2δ.

However, in this paragraph we restrict our attention to one version of DPCM where the feedback filter is a perfect integrator as in Fig. 4-1. More general versions are discussed in Sec. 4-6. The maximum positve slope for DPCM with an equispaced 2^ℓ-level (rather than a two-level) quantizer is

$$S_{max} = (2^\ell - 1)\,\delta f_s \tag{4-5}$$

Set the maximum slope of the DPCM equal to the maximum input slope, $S_{max} = 8\pi\sigma_x f_e$ = maximum slope of x to obtain a quantization interval requirement (4σ-loading)

$$\delta = \frac{8\pi\sigma_x f_e}{(2^\ell - 1)f_s} \tag{4-6}$$

For no slope overload and uniform inputs, the SNR then becomes

$$\text{SNR}_0 = \frac{\sigma_x^2}{\sigma_g^2} = \frac{\sigma_x^2}{(2\delta)^2/12} = \frac{3}{(8\pi)^2}(2^\ell - 1)^2\left(\frac{f_s}{f_e}\right)^2 \tag{4-7}$$

or an increase of $(2^\ell - 1)^2$ relative to the two-level quantizer. However, ℓ bits per sample are transmitted rather than one. If the input spectrum is rectangular, the equivalent input bandwidth is $f_e = f_m/\sqrt{3}$, and the output SNR before filtering (4-7) is

$$\text{SNR}_0 = \frac{9}{(8\pi)^2}(2^\ell - 1)^2\left(\frac{f_s}{f_m}\right)^2 = \left(\frac{3}{4\pi}\right)^2\left(\frac{2^\ell - 1}{\ell}\right)^2(R_b)^2 \tag{4-8}$$

where $R_b = \ell f_s/2f_m$ = the normalized transmitted bit rate. The improvement factor F obtained by increasing the number of bits in the quantizer while maintaining constant bit rate is $F = (2\ell - 1)^2/\ell^2$ for no smoothing. The values of F for several values of ℓ are

ℓ bits/sample	*Improvement F*
1	1
2	2.25
3	5.4
4	14.1

It can be shown that the quantizing noise bandwidth goes approximately as $1/\ell$ so that changing ℓ from 1 to 2 with uniform quantizing produces essentially no improvement when output smoothing is used. Use of larger values of ℓ does produce substantial improvement, however.

4-3 GRANULAR QUANTIZING NOISE IN DELTA MODULATION WITH OUTPUT SMOOTHING

We have computed the output SNR with no output smoothing. We now compute the output SNR after a smoothing filter has reduced the noise effects. When the sampling rate is large compared to the input bandwidth, output smoothing can greatly enhance the output SNR because of the correlation between samples. We will compute the output SNR for binary delta modulation for both the optimum smoothing filter and an ideal low-pass smoothing filter. A slope overload constraint is placed on the ΔM parameters. We also make the following assumptions:

1. Draw the input $x(t)$ from a stationary bandlimited (to a maximum frequency $f_m < f_s/2$) random process with autocorrelation $\sigma_x^2\rho(\mu)$ having specific values $\rho(n\mathrm{T}) = \rho_n$, where $T \triangleq 1/f_s$.
2. Assume the absence of slope overload so that the output samples z_n at any time instant match the output of an instantaneous quantizer with an even-order parity Q_e or an odd-order parity Q_o with step size 2δ.
3. Assume random initial phase of the ΔM clock so that the probability (ensemble) of being in either of the quantizer states at $t = 0$ is 1/2; that is, $p(Q_e) = p(Q_o) = 1/2$. This assumption assures that z_n is wide-sense stationary.

Performance with an Output Smoothing Filter

Because the sampling rate with a one-bit ΔM is often much greater than the input bandwidth and hence samples are correlated, substantial reductions in quantizing noise variance can be obtained by smoothing (see Fig. 4-1). To determine the system performance with smoothing, we first evaluate the quantizing error autocorrelation function $R_e(\mu T)$

$$
\begin{aligned}
R_e(\mu T) &= \mathrm{E}(x_{n\mathrm{T}} - z_{n\mathrm{T}})[x_{(n+\mu)\mathrm{T}} - z_{(n+\mu)\mathrm{T}}] \\
&= \mathrm{E}[x_{n\mathrm{T}}x_{(n+\mu)\mathrm{T}} - z_{n\mathrm{T}}x_{(n+\mu)\mathrm{T}} - x_{n\mathrm{T}}z_{(n+\mu)\mathrm{T}} + z_{n\mathrm{T}}z_{(n+\mu)\mathrm{T}}] \\
&= (\sigma_x^2\rho_\mu - \phi_{-\mu} - \phi_\mu + r_\mu) \qquad (4\text{-}9)
\end{aligned}
$$

where $\mathrm{E}[x_{n\mathrm{T}}x_{(n+\mu)\mathrm{T}}] \triangleq \sigma_x^2\rho_\mu$ is the input autocorrelation

$\mathrm{E}[z_{n\mathrm{T}}x_{(n+\mu)\mathrm{T}}] \triangleq \phi_\mu, \ldots, \mathrm{E}[z_{nT}\, z_{(n+\mu)T}] \triangleq r\mu$

where ρ_μ is the normalized input autocorrelation and ϕ_μ is the crosscovariance. It has been shown [Goodman, 1969, 1197–1218] that $\phi_\mu = \phi_{-\mu} = c\sigma_x^2\rho_\mu$ where c is a constant to be evaluated later. Thus, we have an error autocorrelation

$$
R_e(\mu T) = \sigma_x^2\rho_\mu - 2\phi_\mu + r_\mu = \sigma_x^2\rho_\mu(1 - 2c) + r_\mu \qquad (4\text{-}10)
$$

Each of these newly defined parameters is evaluated in the next paragraph and Appendix A.

Spectra of the Sampled Components

Since all signal elements are sampled at f_s, the spectrum of the sampled input $x^*(t)$ is periodic with period f_s and can be evaluated as

$$G_x^*(f) \triangleq \frac{1}{T} \sum_{\nu=-\infty}^{\infty} G_x(f - \nu f_s) \qquad f_s = \frac{1}{T} \tag{4-11}$$

and hence is represented by the cosine (Fourier) series coefficients

$$\begin{aligned} G_n &= 2T \int_0^{1/T} G_x^*(f) \cos 2\pi n f T \, df = \int_{-\infty}^{\infty} G_x(f) \cos 2\pi n f T \, df \\ &= R_x(nT) = \sigma_x^2 \rho_n \end{aligned} \tag{4-12}$$

since it is an even function of ω, namely

$$G_x^*(f) = \sum G_n \cos \frac{2\pi n f}{f_s} \tag{4-13}$$

The asterisk superscript in G_x^* denotes the power spectral density of the sampled signal $x^*(t)$. Similar sampling notation is used for the spectral density of the sampled $z(t)$ and $e(t)$; see Sec. 2-2. Similarly, the spectrum of the error samples is

$$G_e^*(f) = G_z^*(f) + G_x^*(f) - 2G_{zx}^*(f) \tag{4-14}$$

where $G_{zx}(f) = F[\phi_\mu]$ is a function of ϕ_μ.

The mean square error in the smoothed output $y(t)$ is then

$$\sigma^2 = \mathbf{E}[(y - x)^2 = \mathbf{E}\left[\int h(\sigma)\, z^*(t - \sigma)\, d\sigma - x(t)\right]^2$$

where $h(t)$ is the impulse response of the smoothing filter having a transfer function $H(j\omega)$. Notice that $y - x$ is a special case of the more general form

$$s(t) = \int \sum h_n(\sigma) s_n(t - \sigma)\, d\sigma \tag{4-15a}$$

which has been shown (Davenport and Root, 1958, 183–184) to have a power spectral density

$$G_s(f) = \sum_n \sum_m H_n(j\omega) \overline{H_m}(j\omega) G_{nm}(f) \tag{4-15b}$$

where $\bar{H}$ is the complex conjugate of H. Thus the spectrum of $e = y - x$ can be obtained by setting

$$h_1(\sigma) = h(\sigma), \quad h_2(\sigma) = \delta(\sigma), \quad \text{other } h_i = 0$$
$$s_1(t) = z^*(t), \quad s_2(t) = -x(t), \quad \text{other } s_i = 0$$

in (4-15a), and the resulting output power in the frequency range of interest $(0, f_s/2)$ from (4-15b) is

$$\sigma^2 = 2T \int_0^{f_s/2} \{G_x^*(f) - 2Re[H(j\omega)G_{xz}^*(f)] + |H(j\omega)|^2 G_z^*(f)\} \, df \tag{4-16}$$

The optimum (nonrealizable) Wiener filter to minimize σ^2 can be computed as

$$H_{\text{opt}}(j\omega) = \frac{G_{zx}^*(f)}{G_z^*(f)} \qquad |f| \leq \frac{f_s}{2}$$
$$= 0 \qquad |f| > \frac{f_s}{2} \tag{4-17}$$

Thus, minimum mean-square error (with this optimal filter) is obtained by inserting (4-17) into (4-16)

$$\sigma_{\min}^2 = 2T \int_0^{f_s/2} \left[G_x^*(f) - \frac{[G_{zx}^*(f)]^2}{G_z^*(f)}\right] df \tag{4-18}$$

The result for white Gaussian input is expressed in terms of output SNR in Fig. 4-6 [Goodman, 1969].

Rectangular Low-Pass Smoothing Filter

If an ideal rectangular low-pass filter of bandwidth f_m is used as the smoothing filter, the mean-square output error is

$$\sigma_{\text{LP}}^2 = 2T \int_0^{f_m} G_e^*(f)\, df \triangleq 2T \int_0^{f_m} \sum G_{en} \cos n2\pi\left(\frac{f}{f_s}\right) df$$
$$= 2T \sum G_{en}\left[\frac{\sin 2\pi n(f/f_s)}{2\pi n/f_s}\right]_0^{f_m}$$
$$= 2Tf_m \sum G_{en} \operatorname{sinc} 2\pi n\frac{f_m}{f_s} = \frac{1}{R_s} \sum_{n=-\infty}^{\infty} G_{en} \operatorname{sinc}\left(\frac{\pi n}{R_s}\right)$$
$$= \frac{1}{R_s} \sum R_e(nT) \operatorname{sinc} \frac{\pi n}{R_s} \tag{4-19}$$

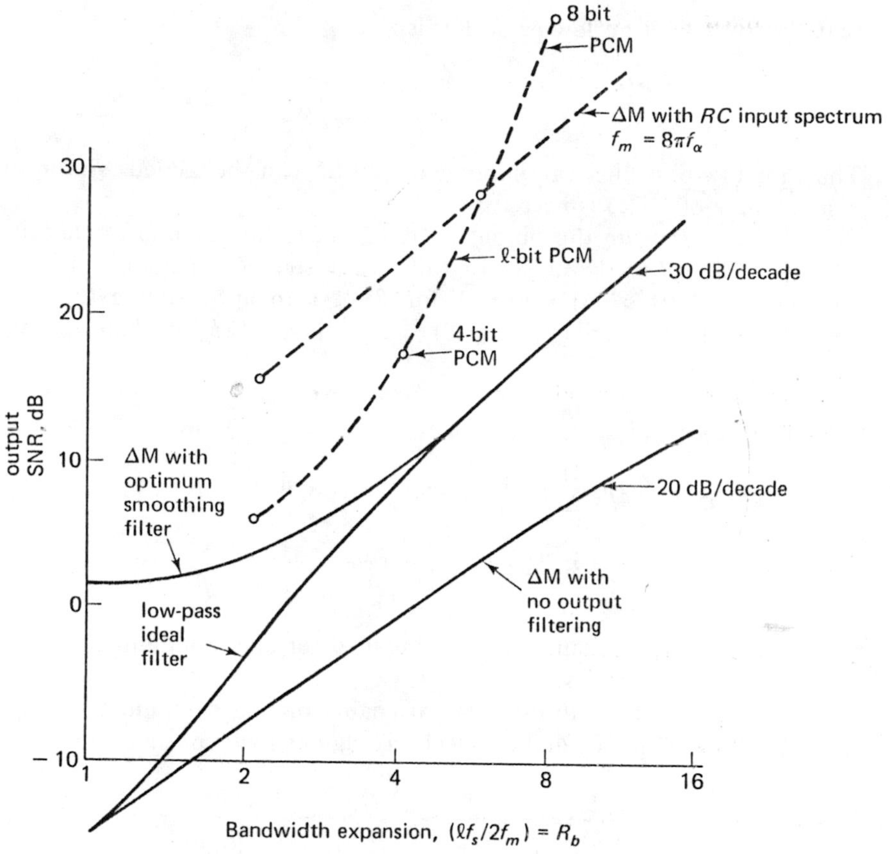

Fig. 4-6 Output SNR for PCM and one-bit ΔM and "white" or RC input spectra versus the normalized bit rate $R_b = \ell f_s/2f_m$

where $R_s \triangleq f_s/2f_m$, the normalized sampling rate. Since $R_e(nT) = R_e(-nT) = \sigma_x^2 \rho_n - 2\phi_n + r_n$, we can write the mean-square error for the ideal low-pass filter as

$$\sigma_{\text{LP}}^2 = \frac{1}{R_s}\left[R_e(0) + 2\sum_{n=1}^{\infty} R_e(nT)\,\text{sinc}\,\frac{\pi n}{R_s}\right] \tag{4-20}$$

The output SNR is then the ratio

$$(\text{SNR})_{\text{LP}} = \frac{\sigma_x^2}{\sigma_{\text{LP}}^2} = \frac{R_s \sigma_x^2}{R_e(0) + 2\sum_{n=1}^{\infty} R_e(nT)\,\text{sinc}\,(\pi n/R_s)} \tag{4-21}$$

The error covariance coefficients $R_e(nT)$ are evaluated in Appendix A. Note that the SNR with no low-pass filter is

$$\text{SNR} = \frac{\sigma_x^2}{R_e(0)} \tag{4-22}$$

Thus, the low-pass filter can improve output SNR if the second term in the denominator of (4-21) is negative or $R_s > 1$.

We next evaluate the output SNR of (4-21) for some example input spectra. Define β as the normalized quantizer step size $\beta \triangleq \delta/\sigma_x$. Use the covariance results of Appendix A for $\beta^2 \ll 1$ to approximate the error covariance. The covariance $R_e(\mu T) = \sigma_x^2 \rho_\mu + r_\mu - 2\phi_\mu$ is then approximately equal to

$$R_e(0)\sigma_x^2 \cong \frac{\beta^2}{3} \tag{4-23}$$

$$R_e(\mu)\sigma_x^2 \cong 4\beta^2 \sum_{n=1}^{\infty} \frac{(-1)^{\mu n}}{n^2} e^{-n^2/\rho^2} \rho_\mu{}^2 \sinh \frac{n^2 \rho_\mu}{\beta^2} \qquad \beta^2 \ll 1 \tag{4-24}$$

$$\cong 2\beta^2 \sum_{n=1}^{\infty} \frac{(-1)^{\mu n}}{n^2} e^{-n^2(1-\rho_\mu)/\beta^2} \qquad \frac{\rho_\mu}{\beta^2} \gg 1 \tag{4-25}$$

The output error covariance can be computed for any input autocovariance using these infinite series.

Expressions (4-23) through (4-25) enable one to compute the output SNR for "white" inputs and an ideal rectangular low-pass smoothing filter

$$(\text{SNR})_{\text{LP}} = \frac{\sigma_x^2}{\sigma_{\text{LP}}^2} = \frac{R_s}{R_e(0) + 2\sum_{\mu=1}^{\infty} R_e(\mu T)\operatorname{sinc}\left(\frac{\pi\mu}{R}\right)} \triangleq \frac{R_s}{D(\beta)} \tag{4-26}$$

where $R_s = f_s/2f_m$ is the normalized sampling rate. To avoid slope overload with uniform quantized ℓ-bit ΔM, β must be at least as large as

$$\beta \triangleq \frac{\delta}{\sigma_x} = \frac{8\pi\sigma_x f_e/f_s}{\sigma_x(2^\ell - 1)} = \frac{8\pi}{(2^\ell - 1)} \frac{f_m}{\sqrt{3} f_s} = \frac{4\pi}{(2^\ell - 1)\sqrt{3} R_s} \tag{4-27}$$

The denominator of the SNR expression can thus be written

$$\begin{aligned} D(\beta) &\triangleq R_e(0) + 2\sum_{\mu=1}^{\infty} R_e(\mu T)\operatorname{sinc}\frac{\pi\mu}{R_s} \\ &\cong \frac{\beta^2}{3} + 4\beta^2 \sum_n \sum_\mu \frac{(-1)^{\mu n}}{n^2} \exp\left[\frac{-n^2}{\beta^2}\left(1 - \operatorname{sinc}\frac{\pi\mu}{R_s}\right)\right] \operatorname{sinc}\frac{\pi\mu}{R_s} \\ &\cong \frac{\beta^2}{3} + 4\beta^2 \sum_n \sum_\mu \frac{(-1)^{\mu n}}{n^2} \\ &\quad \times \exp\left[\frac{-n^2 - 3R_s^2}{(4\pi)^2}(2^\ell - 1)^2 \frac{(\pi\mu/R_s^2)^2}{(2\cdot 3)}\right] \operatorname{sinc}\frac{\pi\mu}{R_s} \end{aligned} \tag{4-28}$$

If $R_s \gg 1$, then (4-28) simplifies to

$$D(\beta) \cong \frac{\beta^2}{3} + 4\beta^2 \sum_n \sum_\mu \frac{(-1)^{\mu n}}{n^2} \exp\left[\frac{-n^2\mu^2(2^\ell - 1)^2}{32}\right] \frac{\operatorname{sinc} \pi\mu}{R_s} \tag{4-29}$$

Examine the series for ℓ large; the denominator term is then approximated by

$$D(\beta) \cong \frac{\beta^2}{3} \qquad (2^\ell - 1)^2 \gg 32 \quad \text{or} \quad \ell \geq 3 \tag{4-30}$$

For a one-bit ΔM ($k = 1$), the negative correlation in the $R_e(\mu T)$ terms in (4-28) is not negligible, and the approximate expression for SNR is

$$(\text{SNR})_{\text{LP}} \cong 2\left(\frac{3}{4\pi}\right)^2 R_b^3 \tag{4-31}$$

where the normalized bit rate $R_b = R_s$ for one-bit delta modulation. That is, it varies in proportion to R_b^3 rather than simply as R_b^2 obtained for the unsmoothed result of (4-4).

This behavior of SNR $\sim R_b^3$ is consistent with the result [DeJager, 1952] for the output SNR with an input sinusoid of frequency f_o, an output bandwidth of f_m, and a bit rate f_s. This output SNR for the single integrator delta modulator is

$$(\text{SNR})_{\text{LP}} = \frac{f_s^3}{25 f_o^2 f_m} = \frac{8R_b^3}{25}\left(\frac{f_m}{f_o}\right)^2 \qquad R_b \triangleq \frac{f_s}{f_m}, \qquad \ell = 1 \text{ bit/sample}$$

Clearly this result can exceed the result for a white noise input for low frequency input sinusoids at $f_o \ll f_m$.

The approximate SNR for high SNR and $\ell \geq 3$ and normalized sampling rate R_s is obtained using (4-26) and (4-30)

$$(\text{SNR})_{\text{LP}} \cong \frac{3R_s}{\beta^2} = \frac{9R_s^3(2^\ell - 1)^2}{(4\pi)^2} = \left(\frac{3}{4\pi}\right)^2 R_s^3(2^\ell - 1)^2 \tag{4-32}$$

The normalized transmitted bit rate is $R_b = \ell R_s$. Then the SNR can be expressed as

$$(\text{SNR})_{\text{LP}} \cong \left(\frac{3}{4\pi}\right)^2 R_b^3 \frac{(2^\ell - 1)^2}{\ell^3} \qquad \ell \geq 3 \tag{4-33}$$

a result first derived by Van de Weg [1955, 367–385]. Notice that the lowpass filter again gives an output SNR proportional to R_b^3 rather than to R_b^2

as obtained in (4-8). Proportionality to R_b^2 occurs with no output smoothing. The improvement factor $(2^\ell - 1)^2/\ell^3$, $\ell \geq 3$, for white inputs is

Bits/Sample ℓ	*Improvement Factor* $(2^\ell - 1)^2/\ell^3$
1	1.0
2	1.125
3	1.82
4	3.52

Note that for white inputs, little improvement is obtained by going from one-bit ΔM to two-bit DPCM.

4-4 COMPARISON OF DELTA MODULATION AND PCM QUANTIZING NOISE

Figure 4-6 shows the output SNR for a variety of quantizers, including a one-bit ΔM with and without output smoothing, and ℓ-bit PCM instantaneous uniform quantizers. A Gaussian input with an ideal rectangular spectral density of bandwidth f_m is assumed. The output SNR is obtained for a white input with autocovariance $\rho_\mu = \text{sinc } 2\pi f_m \mu T$. The results are plotted from Eqs. (4-4) for no output smoothing, and (4-21) for a rectangular low-pass smoothing filter for a white input with autocovariance $\rho_\mu = \text{sinc } 2\pi f_m T$. The results are plotted versus the normalized bit rate $R_b = f_s/2f_m$. Note that there is essentially no difference in ΔM between the ideal low-pass filter and optimum smoothing (4-18) for SNR above 9 dB. In this region, the output SNR is

$$(\text{SNR})_{\text{LP}} \cong 2\left(\frac{3}{4\pi}\right)^2 R_b^3 \qquad \text{if SNR}_{\text{LP}} > 10 \text{ dB} \tag{4-34}$$

There is a substantial improvement for output smoothing relative to performance with no output filtering obtained using either type of filter for $R_b \geq 4$.

For comparison, the result for ℓ-bit PCM sampled at a rate $f_s = 2f_m$ is also shown (no output filtering). The PCM SNR [see (3-9)] for a Gaussian input is approximately (neglects saturation)

$$(\text{SNR})_{\text{PCM}} \cong \frac{\sigma_x}{\delta^2/12} = \frac{12}{8^2}2^{2\ell} = 3(2^{2(\ell-2)}) \tag{4-35}$$

where $2^\ell\delta = 8\sigma_x$, called 4σ loading, $\ell f_s = 2\ell f_m$, and $R_b = \ell f_s/2f_m$. Note that PCM gives a higher output SNR than ΔM for white inputs; for example,

PCM gives an SNR 18 dB higher for $R_b = 8$. On the other hand, delta modulation can show an improvement for most real inputs where the spectrum is not white and higher correlation exists between samples.

RC-Shaped Gaussian Input

An *RC*-shaped spectral density bandlimited at f_m is an example of a nonwhite input spectrum and is a useful representation of a TV signal. This low-pass input spectrum can be written as

$$\frac{G_x(f)}{\sigma_x^2} = \left[\frac{f_\alpha}{\tan^{-1}(f_m/f_\alpha)}\frac{1}{(f^2 + f_\alpha^2)}\right] \qquad 0 \leq f \leq f_m \tag{4-36}$$

$$= 0 \qquad f > f_m \tag{4-37}$$

where f_α is the 3-dB bandwidth. Let the cutoff frequency be $f_m = 8\pi f_\alpha$ or $f_\alpha = f_m/8\pi$. The autocorrelation for this input is then [O'Neal, 1966, 117–141]

$$\frac{R_x(T)}{\sigma_x^2} = \frac{(\pi/2)e^{-\omega_\alpha|T|} - \omega_\alpha|T|\operatorname{cosc}\omega_m|T| + \operatorname{Si}(\omega_m|T| - \pi/2)}{\pi/2 - f_\alpha/f_m} \qquad f_\alpha \ll f_m \tag{4-38}$$

where $\omega_\alpha = 2\pi f_\alpha$, $\omega_m = 2\pi f_m$, Si(x) is the sine integral function, and cosc $x \triangleq \cos x/x$. The resulting performance for a one-bit ΔM, plotted with a smoothed output, is shown in Fig. 4-6. Note that the correlation between samples for the low-pass filtered *RC* spectrum gives about a 14-dB improvement for $R_b > 2$ over the performance obtained with a white Gaussian input of the same bandwidth. (This result corresponds to the peak SNR when both quantizing and slope overload noise effects are included and the step size is optimized.) For this input spectrum, simple one-bit ΔM provides better performance than PCM for $R_b < 6$.

Comparison of ℓ-Bit PCM with ℓ-Bit DPCM with White Inputs

A comparison can be made between ℓ-bit PCM and ℓ-bit DPCM with output smoothing for the Gaussian white noise inputs using the results of Van de Weg [1955, 367–385] and Bennett [1948, 446–472]. Results are shown in Fig. 4-7. The PCM SNR shown for a fixed number of bits/sample, increases with bit rate (sampling rate) at approximately 2 to 3 dB/octave because of decreasing correlation of the quantization samples and subsequent low-pass filtering. Note that the PCM curves of Fig. 4-7 are for a fixed

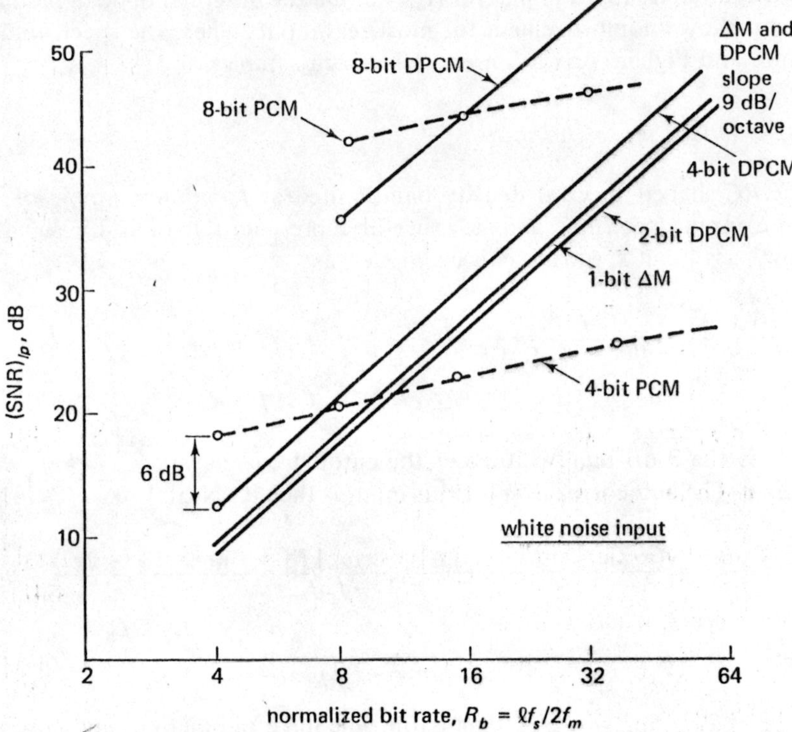

Fig. 4-7 Comparison of ℓ-bit ΔM and DPCM with ℓ-bit PCM as a function of normalized bit rate; outputs of both are smoothed with a low-pass filter [After Van de Weg, 1955, Fig. 4]

number of bits/sample, ℓ, and variable sampling rate, whereas those in Fig. 4-6 are for a fixed sampling rate with uncorrelated samples and a variable number of bits/sample, ℓ. Clearly, when sampling is done at the minimum rate with uncorrelated samples ($f_s = 2f_m$), PCM has a 6-dB advantage over ΔM. As the sampling rate increases, however, DPCM and ΔM can give the superior performance.

4-5 SLOPE OVERLOAD NOISE IN DELTA MODULATION

Thus far we have assumed that the signal slope does not exceed the maximum ΔM slope, and that granular quantizing noise predominates. Here we calculate the noise contribution caused by slope overload transients [O'Neal, 1966, 117–141].

Consider a one-bit ΔM having a sampling rate f_s and step size δ. Assume a Gaussian input $x(t)$ having a power spectral density $G(f)$. Slope overload can occur whenever the slope exceeds the maximum ΔM slope:

$$\frac{dx}{dt} \triangleq x' \geq \delta f_s \tag{4-39}$$

Define the beginning of the slope overload interval as the time t_0 where the input slope is equal to the maximum ΔM slope. $x'(t_0) = x'_0 \triangleq \delta f_s$ and $z_0 = x_0 = x(t_0)$; that is, the input x and the output z are equal at t_0 but are separating. We neglect any errors due to granular quantizing noise and consider time in a continuous sense. At the end of the slope overload interval, at time t_i the input and output are again equal:

$$x_1 = z_1 = x_0 + (t_1 - t_0)\,\delta f_s = x_0 + (t_1 - t_0)x'_0 \tag{4-40}$$

The burst noise in the overload interval is

$$n(t) \triangleq x(t) - z(t) = x(t) - [x_0 - (t_1 - t_0)x'_0] \triangleq x(t) - [x_0 + \Delta x'_0] \tag{4-41}$$

for $\quad t_0 \leq t \leq t_1$

where $t_1 - t_0 \triangleq \Delta$, the time duration of this particular slope overload event (Fig. 4-8). The noise energy in a given burst event is then

$$\mathcal{E}_n \triangleq \int_{t_0}^{t_1} n^2(t)\,dt \tag{4-42}$$

During the overload interval we expand $x(t)$ as a power series about x_0 for t:

$$x(t) = x_0 + (t - t_0)x'_0 + \frac{(t - t_0)^2 x''_0}{2!} + \frac{(t - t_0)^3 x'''_0}{3!} + \cdots \tag{4-43}$$

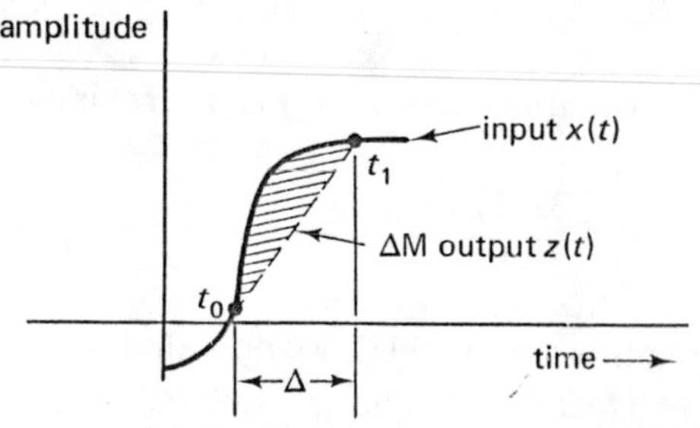

Fig. 4-8 A ΔM slope overload event of duration $\Delta = t_1 - t_0$ sec

Since $x_1 = x_0 + \Delta x'_0$ at $t = t_1$, the power series (4-43) is equal to

$$x_1 = x_0 + \Delta x'_0 = x_0 + \Delta x'_0 + \frac{\Delta^2 x''_0}{2!} + \frac{\Delta^3 x'''_0}{3!} + \cdots \tag{4-44}$$

For $\Delta \ll 1$, neglect terms of order $\Delta^4\ x''''_0/4'$ and higher, and solve (4-44) for the overload duration which is approximated by

$$\Delta \cong -\frac{3x''_0}{x'''_0} \tag{4-45}$$

Statistics of the Gaussian Input

Define the derivatives $\mu = x'(t)$, $\eta = x''(t)$, $\xi = x'''(t)$ for Gaussian inputs. Then the joint density of these variables can be shown to be [Rice, 1944, 282–332]

$$P(\mu, \eta, \xi) = \frac{1}{(2\pi)^{3/2}\sqrt{B_2 B}} \exp\left[-\frac{\mu^2}{2B_0} - \frac{\eta^2}{2B_2} - \frac{(\xi + B_2 B_0^{-1}\mu)^2}{2BB_0^{-1}}\right] \tag{4-46}$$

where $B_n \triangleq \int_0^\infty (2\pi f)^{n+2} G_x(f)\, df$

and $B \triangleq B_0 B_4 - B_2^2$

That is, the B_n are bandwidth measures. Note that if x has unit power, $B_2 = (2\pi f_e)^2$; f_e is the equivalent bandwidth used in Eq. (4-4). The average value of x'''_0 for a given x'_0 is $\mathbf{E}(x'''_0 \mid x'_0) = -b_2 x'_0/B_0$, and x'''_0 has standard deviation $\sqrt{B/B_0}$. If the overload x'_0 is large so that $\mathbf{E}(x'''_0/x'_0)$ is large compared to the standard deviation, then x'''_0 can be approximated by its conditional mean

$$x'''_0 \cong -\frac{B_2 x'_0}{B_0} \tag{4-47}$$

Thus, the overload duration from (4-45) is approximately equal to

$$t_1 - t_0 = \Delta = \frac{3x''_0 B_0}{B_2 x'_0} \tag{4-48}$$

and, by multiplying (4-47) and (4-48), we note that the approximation checks with (4-45)

$$\Delta x'''_0 = -3x''_0$$

Noise Burst Energy

The noise burst energy for a given slope overload event can now be evaluated, using (4-44), as

$$\begin{aligned}\mathcal{E}_n &\triangleq \int_{t_0}^{t_1} n^2(t)\,dt = \int_{t_0}^{t_1} [z(t) - x(t)^2\,dt] \\ &\cong \int_{t_0}^{t_1} \{x(t) - [x_0 + (t - t_0)x_0']\}^2\,dt \\ &= \int_0^1 \Delta\left[\frac{\Delta^2\tau^2 x_0''}{2!} + \frac{\Delta^3\tau^3 x_0'''}{3!}\right]^2 dz \end{aligned} \tag{4-49}$$

where $t - t_0 \triangleq \Delta\tau$, $dt = \Delta d\tau$. Hence, the energy in a burst is

$$\begin{aligned}\mathcal{E}_n &\cong \Delta \int_0^1 \left(\frac{\Delta^2}{2!}\right)^2\left[\tau^4(x_0'')^2 + \frac{2\Delta}{3}x_0''x_0'''\tau^5 + \frac{\Delta^2}{9}(x_0''')^2\tau^6\right]d\tau \\ &\cong \frac{\Delta^5}{4}\left[\frac{(x_0'')^2}{5} + \frac{2}{3}\frac{\Delta}{6}x_0''x_0'' + \frac{\Delta^2}{7\cdot 9}(x_0''')^2\right] \\ &\cong \frac{1}{4}\left(\frac{3x_0''B_0}{B_2x_0'}\right)^5\left[\frac{(x_0'')^2}{5} + \frac{2}{3\cdot 6}(-3x_0'')x_0'' + \frac{1}{9\cdot 7}(x_0'')^2\right]\end{aligned} \tag{4-50}$$

where we have used (4-48) $\Delta x_0''' \cong 3x_0''$. The energy then is approximated by

$$\mathcal{E}_n \cong \frac{3}{4}\left(\frac{B_0}{B_2x_0'}\right)^5 (x_0'')^7 \underbrace{\left[\frac{1}{5} - \frac{1}{3} + \frac{1}{7}\right]}_{1/105} = \frac{81}{140}(x_0'')^7\left(\frac{B_0}{B_2x_0'}\right)^5 \tag{4-51}$$

and can be evaluated in terms of the input bandwidth measures and variance, and the ΔM overload slope $x_0' = \delta f_s$. Thus, the energy per burst is proportional to $(x_0'')^7$.

Average Noise Power

The quantity of real interest in evaluating ΔM performance is the average power in the slope overload noise bursts. This average power is the expected noise energy/burst times the expected number of bursts/sec. we begin by evaluating the expected value of $\mathcal{E}_n$ in (4-51). The joint probability of a slope overload $x_t' = x_0' \triangleq \delta f_s$ and $x_t'' = (x'', x'' + dx'')$ during the interval $(t, t + dt)$ is obtained noting $x''dt = dx'$ and is written using (4-46)

$$\begin{aligned}p(x_0', x'')\,dx'\,dx'' &= p(x_0', x'')x''\,dt\,dx'' \\ &= \frac{x''}{2\pi\sqrt{B_0B_2}}\exp\left[-\frac{\mu^2}{2B_0} - \frac{\eta^2}{2B_2}\right]dx''\,dt\end{aligned} \tag{4-52}$$

From a large ensemble of signal sources of size M, we find that the expected number of members having critical slopes $x'_t = x'_0$ during $(t, t + dt)$ is

$$M_0 = M\,dt \int_0^\infty p(x'_0, x'')\,dx'' = M\,dt\sqrt{\frac{B_2}{B_0}}e^{-(x_0')^2/2B_0} \tag{4-53}$$

When (4-51) is used, the average noise energy in an overload burst during $(t, t + dt)$ is proportional to

$$\mathbf{E}[(x''_0)^7] = \frac{1}{M_0}\int_0^\infty (x'')^7 \underbrace{[x'' p(x'_0, x'')\,dx''\,dt]}_{\substack{\text{probability of}\\ \text{overload from}\\ \text{(4-52)}}} = (2B_2)^{7/2}\Gamma\left(\frac{9}{2}\right) \tag{4-54}$$

where we have integrated using (4-53) and we are averaging only over the M_0 members of the ensemble in overload during $(t, t + dt)$ and $\Gamma(x)$ is the gamma function

$$\Gamma(x) \triangleq \int_0^\infty u^{v-1}\,e^{-u}\,du \qquad v > 0$$

Hence, the expected noise energy per burst from (4-51), (4-54) is

$$\mathbf{E}(\mathcal{E}_n) \triangleq \mathbf{E}\left[\int_{t_0}^{t_1} n^2(t)\,dt\right] \cong \frac{81}{140}\left(\frac{B_0}{B_2 x'_0}\right)^5 \mathbf{E}(x''_0)^7$$

$$= \frac{81}{140}\left(\frac{B_0}{B_2 x''_0}\right)^5 (2B_2)^{7/2}\Gamma\left(\frac{9}{2}\right)$$

$$\mathbf{E}(\mathcal{E}_n) \cong \frac{\sqrt{2\pi}}{8}\left(\frac{3B_0}{x'_0}\right)^5 B_2^{-3/2} \tag{4-55}$$

We next must compute the expected number of these overload events/sec. The final assumption is that a noise burst occurs every time $x'(t)$ increases through x'_0 or decreases through $-x'_0$. (In reality, one or more crossings can occur in the middle of a noise burst in addition to the beginning; hence, the number of crossings is slightly larger than the true number of noise bursts.) Rice [1945, 45–156] has computed the expected number of these crossings per second as

$$r_b \cong 2\left(\frac{1}{2\pi}\right)\sqrt{\frac{B_2}{B_0}}e^{-(x_0')^2/2B_0} \qquad \text{bursts/sec} \tag{4-56}$$

Hence, the expected overload noise power is the product of the expected number of slope overloads per second times the average energy in a slope overload event:

$$r_b \mathrm{E}(\mathcal{E}_n) \cong \frac{1}{4\sqrt{2\pi}}\left(\frac{B_0}{B_2}\right)^2\left(\frac{3B_0^{1/2}}{x_0'}\right)^5 e^{-(x_0')^2/2B_0} \tag{4-57}$$

where $x_0' \triangleq \delta f_s$. The output signal-to-burst-noise ratio for input signals with autocovariance $R_x(0) = \sigma_x^2 = 1$ is then

$$(\mathrm{SNR})_{\mathrm{burst}} = \frac{\sigma_x^2}{r_b \mathrm{E}(\mathcal{E}_n)} = \frac{1}{r_b \mathrm{E}(\mathcal{E}_n)}$$
$$\cong \frac{4\sqrt{2\pi}}{3^5}\left(\frac{B_2}{B_0^2}\right)\left(\frac{\sqrt{B_0}}{f_s\delta}\right)^5 e^{(f_s\delta)^2/2B_0} \sim \frac{\exp\left[(f_s\delta)^2/2B_0\right]}{(f_s\delta)^5} \tag{4-58}$$

Thus, in the region of slope overload (small $f_s\delta$) where this burst noise dominates, the SNR increases exponentially with $f_s\delta$.

For white signals bandlimited at $(0, f_m)$, the expressions for B_0, B_2 from (4-46) are

$$B_0 = \frac{(2\pi f_m)^2}{3} \qquad B_2 = \frac{(2\pi f_m)^4}{5} \tag{4-59}$$

where $\quad R_x(\tau) = \operatorname{sinc} 2\pi f_m \tau \qquad (4\text{-}60)$

is the signal autocorrelation function for the unity power signal.

These expressions allow the computation of the output SNR for (4-58) and (4-59) for white inputs given below.

$$(\mathrm{SNR})_{\mathrm{burst}} \cong \frac{4\sqrt{2\pi}}{3^5}\frac{(2\pi f_m)^4}{5}\frac{3^2}{(2\pi f_m)^4}\frac{(2\pi f_m)^5}{3^{5/2}}\frac{1}{(f_s\delta)^5}e^{(f_s\delta)^2/2[(2\pi f_m)^2/3]}$$
$$\cong \frac{4}{5}\sqrt{\frac{2\pi}{3^{11}}}\frac{e^{(3/2)R^2}}{R^5} \tag{4-61}$$

where $R \triangleq f_s\delta/2\pi f_m$ is a normalized slope of the ΔM.

Bandlimited RC Input

RC inputs are of even greater interest because they correspond more closely to video and speech. Assume a unit-power bandlimited *RC*-filtered input spectrum of the form

$$G_x(f) = \frac{f_\alpha}{\tan^{-1}(f_m/f_\alpha)}\left(\frac{1}{f^2 + f_\alpha^2}\right) \qquad 0 \le f \le f_m$$
$$= 0 \qquad f_m < f \tag{4-62}$$

where f_α is the 3-dB bandwidth. The values of B_0 and B_2, defined in (4-46), have been found to be

$$\frac{B_0}{(2\pi)^2} = \frac{f_m f_\alpha}{\tan^{-1}(f_m/f_\alpha)} - f_\alpha^2 \qquad \frac{B_2}{(2\pi)^4} = \frac{f_m^3 f_\alpha - 3 f_\alpha^3 f_m}{3 \tan^{-1}(f_m/f_\alpha)} + f_\alpha^4 \tag{4-63}$$

As an example, let the 3-dB bandwidth $f_\alpha = f_m/8\pi$, that is, $1/8\pi$ the band-limiting frequency f_m. The resulting SNR, including both slope overload and granular quantizing noise, is given in Fig. 4-9. Computer simulations have given good argument with these approximate results. Note that where $f_s\delta$ is small and slope overload dominates, the SNR increases exponentially with f_s until granular quantizing noise dominates.

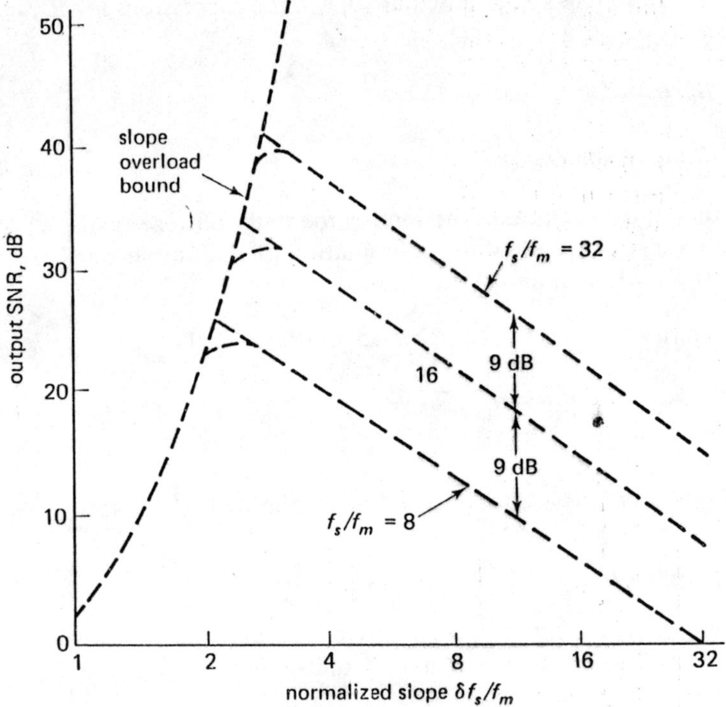

Fig. 4-9 Output SNR versus one-bit ΔM normalized slope for various sampling rates and an *RC* input $f_\alpha = f_m/8$, where f_α is the *RC* filter 3-dB point and the signal is bandlimited to f_m

4-6 DIFFERENTIAL PCM—LINEAR PREDICTIVE QUANTIZATION

As discussed in Sec. 4-4, delta modulation can be extended by using ℓ bits per sample quantization with either a uniform or a nonuniform quantizer. Whenever more than one-bit quantization is employed in a feedback quantizer, we shall use the terminology differential PCM (DPCM) or linear predictive encoding [O'Neal, 1966, 689–721]. Furthermore, the loop filter or predictor can be optimized as opposed to using the simple integrator of conventional ΔM. In this section, we will optimize the performance of linear predictive quantizers, determine the SNR performance with quasi-optimal quantization and predictive filters, examine the effects of channel errors, and investigate possible methods of adapting the quantizer step size or predictor coefficients to improve performance. Speech and television both exhibit substantial correlation between samples when sampled at the Nyquist rate. Thus, quantization of the difference between actual and predicted sampler can be expected to produce improved performance.

Linear Predictive Quantizer

Figure 4-10 is a block diagram of the linear predictive quantizer, including the reconstruction filter F. The quantizer input samples are x_i and the predicted sample values are z_i. The predictive filter contains delay elements, $D, 2D. \ldots$, and weight coefficients, $a, a_1, \ldots$. Note that the predicted

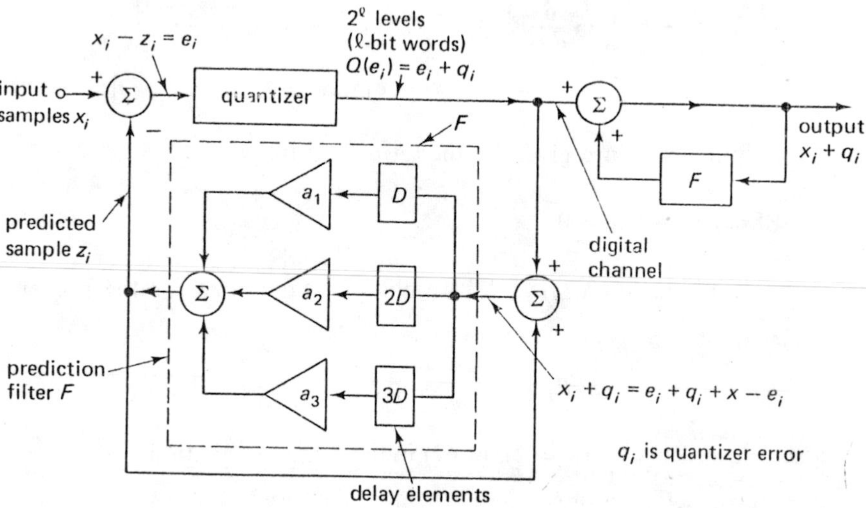

Fig. 4-10 Block diagram of the linear predictive quantizer and reconstruction filter

samples z_i are based solely on the quantizer output, available at the receiver and not on the analog input.

We assume stationary input samples x_i having covariance $\mathbf{E}(x_i x_{i+j}) = R_j$, and optimize prediction in the absence of quantizing errors by selection of the filter weights a_i. If there were no quantizing error and $q = 0$, we would have

$$z_i = a_1 x_{i-1} + a_2 x_{i-2} + \cdots = \sum_{i=1}^{n} a_j x_{i-j} \tag{4-64}$$

and the mean-square error in this predicted sample value would be

$$\sigma_e^2 = \mathbf{E}[(x_i - z_i)^2] = \mathbf{E}[(x_i - a_1 x_{i-1} - a_2 x_{i-2} - \cdots)^2] \tag{4-65}$$

The smaller σ_e^2 becomes for a given input variance, the finer the quantization and the smaller the output quantizing error q_i for a given number of bits/sample.

With some analog signals—for example, speech in a two-way conversation—the signal statistics vary with time in a substantial manner, and significant performance improvements can be obtained by adjusting the predictive filter coefficients a_i with time [Dunn, 1971, 1021–1032]—in accord with the speech formants. However, assume for the moment that the input statics are stationary.

We minimize the mean-square error with respect to a fixed set of a_j by setting

$$\begin{aligned} \frac{\partial}{\partial a_j}\mathbf{E}\Big[(x_i - z_i)^2\Big] &= \frac{\partial}{\partial a_j}\mathbf{E}\Big[(x_i - a_1 x_{i-1} - a_2 x_{i-2} - \cdots)^2\Big] \\ &= -2\mathbf{E}[x_{i-j}(x_i - a_1 x_{i-1} - a_2 x_{i-2} - \cdots)] \end{aligned} \tag{4-66}$$

Thus, the optimum a_j are given by the set of equations

$$\mathbf{E}[x_{i-j}(x_i - z_i)] = 0 \tag{4-67}$$

Using the definition $\mathbf{E}(x_i x_{i-j}) \triangleq R_j$, we obtain n equations for the n coefficients a_j:

$$R_j = a_1 R_{j-1} + a_2 R_{j-2} + \cdots + a_n R_{j-n} \qquad j = 1, 2, \ldots, n \tag{4-68}$$

For a second-order filter ($n = 2$), (4-67) reduces to the equations

$$\begin{Bmatrix} R_2 = a_1 R_1 + a_2 R_0 \\ R_1 = a_1 R_0 + a_2 R_1 \end{Bmatrix} \longrightarrow \begin{Bmatrix} R_1 R_2 - R_0 R_1 = a_1(R_1^2 - R_0^2) \\ R_0 R_2 - R_1^2 = a_2(R_0^2 - R_1^2) \end{Bmatrix} \tag{4-69}$$

and the optimum coefficients are

$$a_1 = \frac{R_1(R_0 - R_2)}{R_0^2 - R_1^2} \qquad a_2 = \frac{R_0 R_2 - R_1^2}{R_0^2 - R_1^2} \tag{4-70}$$

In general, the optimum coefficients can be written in terms of the input covariance matrix [Jayant, 1974]. Define the vector **A** of predictor coefficients and the covariance matrix $\mathbf{\Gamma}$ and vector $\mathbf{\Sigma}$ as

$$\mathbf{A} = \begin{bmatrix} a_1 \\ a_2 \\ \cdots \\ a_n \end{bmatrix} \qquad \mathbf{\Sigma} = \begin{bmatrix} R_1 \\ R_2 \\ \cdots \\ R_n \end{bmatrix} \tag{4-71}$$

$$\mathbf{\Gamma} = \begin{bmatrix} R_0 & R_1 & R_2 & \cdots & & R_{n-1} \\ R_1 & R_0 & R_1 & \cdots & & \\ R_2 & R_1 & & & & \cdots \\ \cdots & \cdots & & & & R_1 \\ R_{n-1} & & & \cdots & R_1 & R_0 \end{bmatrix} \tag{4-72}$$

Then the optimum set of prediction coefficients is

$$\mathbf{A}_{\text{opt}} = \mathbf{\Gamma}^{-1}\mathbf{\Sigma} \tag{4-73}$$

Mean-Square Value of Error Signal

The mean-square error in the predictor output for an optimum set of filter coefficients becomes

$$\sigma_e^2 = R_0 - (a_1 R_1 + a_2 R_2 + \cdots + a_n R_n) \qquad \sigma^2 = R_0 \tag{4-74}$$

For a second-order predictor, this expression reduces to

$$\begin{aligned} \sigma_e^2 &= \sigma_x^2 \left[\frac{R_1^2(R_1 - R_2) + R_2(R_0 R_2 - R_1^2)}{R_0^2 - R_1^2} \right] \\ &= \sigma_x^2 - \frac{R_0(R_1^2 + R_2^2) - 2R_1^2 R_2}{R_0^2 - R_1^2} \end{aligned} \tag{4-75}$$

For a first-order predictor (optimum for $R_i = \sigma_x^2 e^{-i\alpha}$), the error variance is

$$\sigma_e^2 = \sigma_x^2 \left(1 - \frac{R_1^2}{\sigma^2}\right) \tag{4-76}$$

When the number of quantization intervals is large, $2^\ell = N > 8$, the effect of the quantizer error on the value of σ_e is relatively small, and Eqs. (4-70), (4-71), and (4-72) computed for a linear system are quite accurate.

Television Picture Covariance

Consider a typical TV sampled line scan with covariance values as shown in the raster:

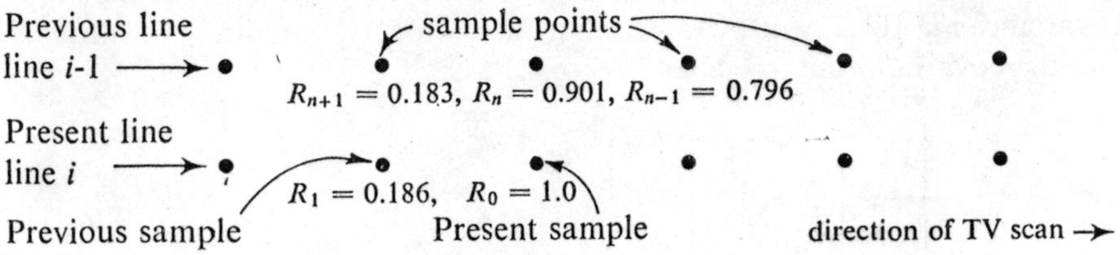

Previous line — sample points
line i-1 ⟶ • • • • •

$R_{n+1} = 0.183$, $R_n = 0.901$, $R_{n-1} = 0.796$

Present line
line i ⟶ • • • • • •

$R_1 = 0.186$, $R_0 = 1.0$

Previous sample — Present sample — direction of TV scan ⟶

For a two-sample predictor using previous sample and previous line feedback, the optimum filter coefficients are:

$$a_1 = 0.270 \qquad a_n = 0.686 \tag{4-77}$$

and the reduction in variance produced by prediction is

$$\frac{\sigma_e^2}{\sigma_x^2} = 1 - \frac{R_0(R_1^2 + R_n^2) - 2R_1^2R_n}{R_0^2(R_0^2 - R_1^2)} = (0.402)^2 = 0.1616 = \frac{1}{2.49} \tag{4-78}$$

Thus, the rms input to the quantizer has been reduced by a factor of 2.49 or 7.92 dB. Thus the linear predictor permits a finer quantization and results in a higher SNR at the output.

Speech Signal Autocorrelation

Noll [1972; see also Jayant, 1974] has shown typical autocorrelation functions for speech signals sampled at an 8-kHz sampling rate. Table 4-1 gives the range of the autocorrelation coefficients for a typical set of speakers for input speech in the frequency band 300–3400 Hz.

Table 4-1 TYPICAL RANGE OF AUTOCORRELATION COEFFICIENTS FOR BANDPASS SPEECH (300–3400 Hz) SAMPLED AT 8 kHz [NOLL, 1972]

	Autocorrelation R_i	
i	*Max*	*Min*
0	1.0	1.0
1	0.9	0.8
2	0.6	0.1
3	0.2	−0.1
4	0	−0.2
5	−0.1	−0.3

The wide range in the values in Table 4-1 indicate a strong dependence on the speaker. There is an equally strong dependence on the speech material not indicated in the table. Furthermore, there is a wide difference between voiced and unvoiced sounds. Voiced sounds generally have $R_1 > 0.5$, whereas unvoiced speech segments with noise-like fricatives generally have $R_1 \approx 0$. Because of the rapid decrease in the autocorrelation values of speech, there is little benefit in increasing the order of the predictor beyond the second order. The improvement in performance demonstrated by Noll for the second-order predictor ranged from 5 to 8 dB relative to PCM.

Optimal Nonuniform Quantization of the Prediction Errors

Experimental tests on DPCM have shown that the probability density of the error signal is approximately Laplacian for Gaussian inputs $x(t)$:

$$p(e) = \frac{1}{\sqrt{2}\,\sigma_e} e^{-\sqrt{2}e/\sigma_e} \tag{4-79}$$

The error statistic (4-79) is now assumed to be the input to a compandor/quantizer, and the quantizer can be optimized as described in Sec. 3-5.

The minimum mean-square error for the N-level optimum quantizer has been shown [Eq. (3-48)] to be approximately equal to

$$\begin{aligned}\sigma_q^2 &= \frac{2}{3N^2}\left(\int_0^V p(e)^{1/3}\,de\right)^3 = \frac{2}{3N^2}\left(\frac{1}{\sqrt{2}\,\sigma_e}\right)\left(\int_0^V e^{-\sqrt{2}V/3\sigma_e}\,de\right)^3 \\ &= \frac{\sqrt{2}\,(3\sigma_e/\sqrt{2})^3}{3N^2\sigma_e}(1 - e^{-\sqrt{2}V/3\sigma_e}) \cong \frac{1}{2}\left(\frac{3\sigma_e}{N}\right) \qquad V \gg 3\sigma_e/\sqrt{2}\end{aligned} \tag{4-80}$$

that is, for negligible quantizer overload. Thus, the output SNR for optimized predictive quantization or DPCM is

$$\text{SNR} = \frac{\sigma^2}{\sigma_q^2} = \frac{2N^2}{9}\left(\frac{\sigma}{\sigma_e}\right)^2 \quad \text{for } R_o = \sigma^2 \tag{4-81}$$

Compare this SNR to standard PCM (uniform quantizing) with a peak-to-peak quantizer range of $2V = 2[(N/2)\delta] = 2(3\sigma)$ required for a Gaussian input (3σ loading). The output SNR for an $N = 2^\ell$-level PCM quantizer (3-5) is

$$\text{SNR} \cong \frac{\sigma^2}{\delta^2/12} = \frac{12\sigma^2}{(6/N)^2\sigma^2} = \frac{1}{3}(N^2) \tag{4-82}$$

Clearly, the larger that $(\sigma/\sigma_e)^2$ becomes (the greater the input signal redundancy), the greater the improvement in SNR produced by DPCM, using nonuniform quantizing.

EXAMPLE If $\sigma/\sigma_e = 2.5$, as described for the two-sample predictor in the TV example in (4-78), the performance comparison between quasi-optimal DPCM and PCM is as shown in Fig. 4-11. The DPCM error is obtained from (4-81) and the PCM result is (3-10). Note the 6.2-dB improvement in output SNR for the two-sample predictor.

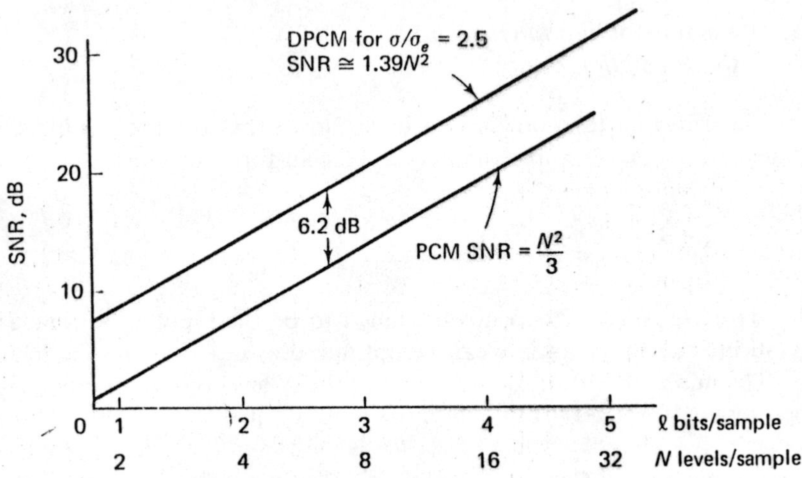

Fig. 4-11 Performance comparison between PCM and DPCM with Gaussian input signals and no output smoothing filters

Effects of Channel Errors on DPCM

Channel errors affect both DPCM and PCM (Fig. 4-12). A single channel error for a 2^ℓ-level quantizer of step size δ produces a mean-square error [Wolf, 1966, 2–7] in the quantizer output at point B:

$$\sigma_n^2 = \frac{\delta^2}{3}(2^{2\ell} - 1)\mathcal{P}_b \tag{4-83}$$

where $\mathcal{P}_b$ is the bit error probability (Chap. 3). Any time a bit error and a D/A converter error pulse appear, the error effect begins and continues in the output, producing an error "streaking" effect in imagery transmission. The effect of the error remains approximately for a memory time constant. Thus, the lower redundancy in the DPCM transmission can sometimes make it

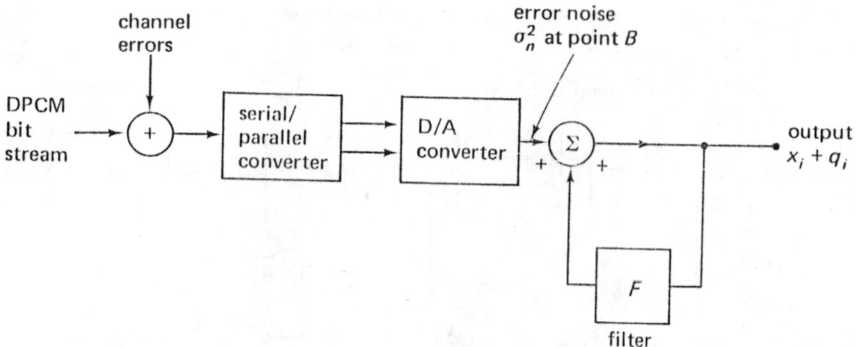

Fig. 4-12 Receive unit of DPCM system; the closed-loop filter (integrator) has an effective memory time constant N samples

somewhat more susceptible to errors. On the other hand, errors in PCM coding of speech are more disturbing than the same errors in DPCM. An error in PCM can cause a large error spike of peak-peak magnitude corresponding to the quantizer dynamic range. In DPCM the same error causes a smaller error spike because of the smaller quantizer dynamic range within the feedback loop. Thus, the perceptual effects of DPCM errors on speech have been reported to be less than for PCM [Yan and Donaldson, 1972].

The quantitative effects of channel errors on the total mean-square output error in a DPCM system can be calculated in much the same manner as for a PCM system (Chap. 2)—namely, by separating the errors into sampling errors, quantizing errors, and channel error effects. Here we calculate the errors for a first-order Markov input process and a first-order linear predictor as shown in Fig. 4-13, where the compressor characteristic $v(x)$ acts to convert the uniform quantizer ($v(x) = x$ and $q = q'$, $d = d'$) into a nonuniform quantizer.

Consider first the DPCM system with uniform quantization—that is, a unit slope compandor $v(x) = x$, and $q' = q$, $d' = d$. For simplicity in notation, we define the kth sample of the input process $x(k)$ where the samples are taken T sec apart. Define the number of quantizer bits per sample as ℓ. Hence, the transmission data bit rate is $\mathcal{R} \triangleq \ell/T = \ell f_s$ bits/sec. Define further the quantizing noise $n_q(k)$ and the channel error noise $n_c(k)$ as

$$n_q(k) \triangleq q(k) - e(k) \qquad n_c(k) \triangleq d(k) - q(k) \tag{4-84}$$

The channel error noise depends on the channel bit error rate, which, in turn, depends on the transmission channel signal-power/noise-density ratio C/N_0, the transmission rate $\mathcal{R}$, and the modulation method used. Channel errors are assumed to be independent of one another.

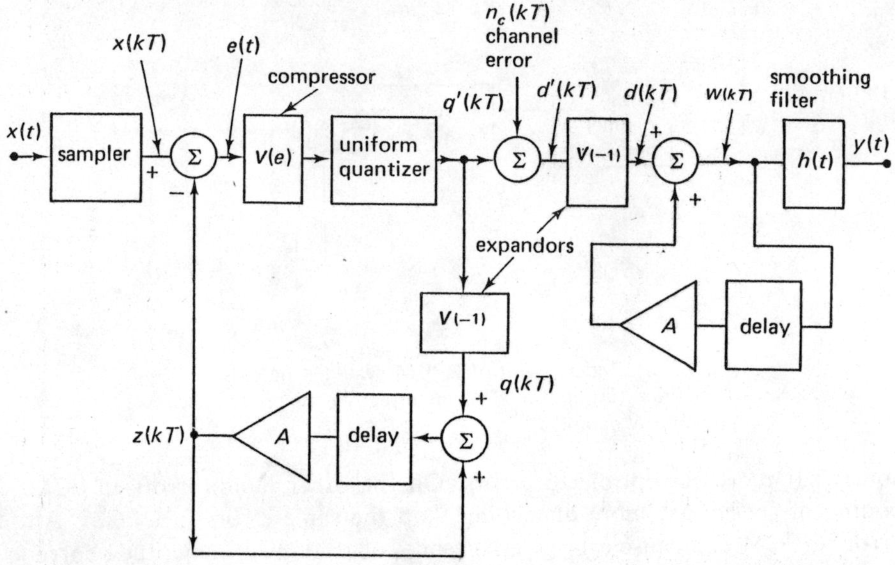

Fig. 4-13 Channel error effects in a DPCM system

The predicted input value $z(k)$ and the input difference $e(k)$ to be quantized are related to the input waveform $x(k)$ by

$$z(k) = A[q(k-1) + z(k-1)] = \sum_{i=1}^{\infty} A^i q(k-i) \tag{4-85}$$

$$e(k) = x(k) - z(k) \tag{4-86}$$

where A is the prediction filter gain constant. The input $w(k)$ to the smoothing reconstruction filter $h(t)$ can be expressed in terms of the input $x(k)$ and the noise terms by using (4-84), (4-85), (4-86)

$$\begin{aligned} w(k) &= d(k) + Aw(k-1) = d(k) + \sum_{i=1}^{\infty} A^i d(k-i) \\ &= e(k) + n_q(k) + n_c(k) + \sum_{i=1}^{\infty} A^i[q(k-i) + n_c(k-i)] \\ &= [x(k) - \sum_{i=1}^{\infty} A^i q(k-i)] + n_q(k) + n_c(k) \\ &\quad + \sum_{i=1}^{\infty} A^i[q(k-i) + n_c(k-i)] \end{aligned} \tag{4-87}$$

Thus, the estimate $w(k)$ is

$$w(k) = x(k) + n_q(k) + \sum_{i=0}^{\infty} A^i n_c(k-i) \tag{4-88}$$

and is the sum of the input samples plus quantizing noise plus a weighted sum of present and past channel error noise effects. This dependence on past channel errors, of course, causes the error correlation or streaking effect referred to earlier.

The smoothed estimate of $x(k)$ at the output of the reconstruction filter is then

$$\begin{aligned} y(k) &= \sum_{j=0}^{\infty} h(k-j)w(j) \\ &= \underbrace{\sum_{j=0}^{\infty} h(k-j)x(j)}_{x(k)+\epsilon_s(k)} + \underbrace{\sum_{j=0}^{\infty} h(k-j)\,n_q(j)}_{\epsilon_q(k)} \\ &\qquad + \underbrace{\sum_{j=0}^{\infty}\sum_{i=0}^{\infty} h(k-j)A^i n_c(j-i)}_{\epsilon_c(k)} \\ &\triangleq x(k) + \epsilon_s(k) + \epsilon_q(k) + \epsilon_c(k) \end{aligned} \tag{4-89}$$

that is, the sum of a sampling error effect $\epsilon_s(k)$, quantizing error $\epsilon_q(k)$, and a channel error effect $\epsilon_c(k)$.

It can be shown that when the quantizing interval is small, the three error components (4-89) are independent [Widrow, 1956, 266–276; Totty and Clark, 1967, 336; Katzenelson, 1962, 58–68]. Furthermore, because the input process is Markovian, adjacent samples are uncorrelated in e, d, q. [Papoulis, 1965]. Thus, the autocorrelation function and the cross-correlation function at the quantizer input are equal to

$$\phi_{ee}(k) \triangleq \mathbf{E}[e(i)e(i+k)] \begin{cases} = \sigma_e^2 & k = 0 \\ = 0 & k \neq 0 \end{cases} \tag{4-90}$$

$$\phi_{dd}(k) = \phi_{qq}(k) = \phi_{qd}(k) = \phi_{dq}(k) = 0 \qquad k \neq 0 \tag{4-91}$$

Assume that the filter $h(k-j)$ is absent, $h(k-j) = \delta_{kj}$, and the autocorrelation function of the output error noise, using (4-84) and (4-89), ϵ_c is then

$$\begin{aligned} \phi_{\epsilon_c\epsilon_c}(k) &= \sum_{i=0}^{\infty}\sum_{j=0}^{\infty} A^{i+j}[\phi_{qq}(k-i+j) - \phi_{qd}(k-i+j) \\ &\qquad - \phi_{dq}(k-i+j) + \phi_{dd}(k-i+j)] \\ &\simeq \sum_{i=0}^{\ell} 2A^{2i+|k|}[\phi_{qq}(0) - \phi_{qd}(0)] \end{aligned} \tag{4-92}$$

where $\phi_{dd}(0) \simeq \phi_{qq}(0)$, $\phi_{dq}(0) = \phi_{qd}(0)$. Thus, as expected, the channel error samples are clearly correlated.

For a uniform quantizer (ℓ bit) the quantizer output levels over $\pm V$ are

$$q(k) = V\sum_{j=1}^{\ell} \alpha_{jk}2^{-j} \qquad \alpha_{jk} = \pm 1 \tag{4-93}$$

$$d(k) = V \sum_{j=1}^{\ell} \beta_{jk} 2^{-j} \qquad \beta_{jk} = \pm 1 \tag{4-94}$$

where β_{jk} are the received data bits and α_{jk} are the transmitted bits. The channel error autocorrelation can be calculated for the quantizer as

$$\phi_{e_c e_c}(k) = \sum_{j=0}^{\infty} A^{2j+|k|} \frac{4V^2}{3} \mathcal{P}(1 - 2^{-2\ell}) \tag{4-95}$$

where $\mathcal{P}$ is the received bit error probability; bit errors are assumed independent of one another. The corresponding channel error spectral density is

$$G_{e_c}(\omega) = \frac{(1/T)4V^2\mathcal{P}(1 - 2^{-2\ell})}{[(1 + A^2) - 2A \cos \omega T]} \tag{4-96}$$

which shows that the spectral density increases at low frequency because the denominator is a minimum at $\omega = 0$.

As noted earlier in (4-79), the difference signal e at the input to the DPCM quantizer in Fig. 4-13 is approximately Laplacian, and the optimal quantizer is exponential [see (3-53)]. The error probability density is then

$$p(e) = \frac{1}{2\sigma_e} \exp\left(\frac{-\sqrt{2}\,|x|}{\sigma_e}\right)$$

The optimal compressor for a quantizer with saturation level V is then

$$v(e) = V\left[\frac{1 - \exp(-\sqrt{2}\,|e|/3\sigma_e)}{1 - \exp(-\sqrt{2}\,V/3\sigma_e)}\right]$$

The variance of this difference signal $e(k)$ is $\sigma_e^2 = (1 - R_1^2)\sigma_x^2$ for a first-order predictor. We assume a Markovian input process with correlation R_1 between adjacent samples where x has an input variance equal to σ_x^2. The optimal expandor characteristic $v^{(-1)}(\quad)$ gives

$$q(k) = v^{-1}[q'(k)] = -\frac{V}{v} \log\left\{\frac{1 - q'(k)[1 - \exp(-v)]}{V}\right\} \qquad q'(k) \geqq 0 \tag{4-97}$$

$$= \frac{V}{v} \log\left\{\frac{1 + q'(k)[1 - \exp(-v)]}{V}\right\} \qquad q'(k) < 0 \tag{4-98}$$

Note that the series combination of compression $v(x)$ and expandor $v^{(-1)}(x)$ is $v^{(-1)}[v(x)] = x$.

where $v \triangleq \sqrt{2}\, V/3\sigma_e$ and σ_e is the rms value of the difference signal $e(k)$. The autocorrelation function for this quantizer for $\ell \geqq 3$ bits/sample is [Essman and Wintz, 1973, 867–877]

$$\phi_{\epsilon_c\epsilon_c}(k) = \frac{A^k}{1-A^2}\frac{63}{6}\mathcal{P}\sigma_e^2 \tag{4-99}$$

where A is the prediction filter coefficient.

Consider now a zero-order-hold (ZOH) output smoothing filter $h(t)$ used with the DPCM exponential quantizer. The input signal $x(k)$ is Markovian with autocorrelation function $\phi_{xx}(k) = \exp(-b|kT|)$ and $R_1 = \phi_{xx}(1) = e^{-bT}$. The mean-square error $\epsilon \triangleq x - y$ in the smoothed output for DPCM is then [Essman and Wintz, 1973]

$$\sigma_\epsilon^2 = 2\left\{1 - \frac{[1-\exp(-bT)]}{bT}\right\} + \frac{9\sigma_e^2}{2^{2\ell+1}} + \left(\frac{63}{6}\right)\frac{\sigma_e^2}{(1-A^2)}\mathcal{P} \tag{4-100}$$

where $\sigma_e^2 = (1 - e^{-2bT})$ and the predictor coefficient $A = e^{-bT}$. This result can be compared to the mean-square error for PCM using a uniform quantizer

$$\sigma_\epsilon^2|_{\text{PCM}} = 2\left\{1 - \frac{[1-\exp(-bT)]}{bT}\right\} + \frac{V^2}{3(2^{2\ell})} + \frac{4V^2\mathcal{P}}{3T}(1 - 2^{-2\ell}) \tag{4-101}$$

Figure 4-14 compares mean-square error for three-bit DPCM and six-bit PCM for fixed values of P_s/N_0 and $A = \exp(-bT)$. PSK transmission is employed, and the bit error rate is

$$\mathcal{P} = \frac{1}{2}\left[1 - \operatorname{erf}\sqrt{\frac{P_s/R}{N_0}}\right] = \frac{1}{2}\operatorname{erfc}\sqrt{E_b/N_0} \tag{4-102}$$

where P_s is the carrier power $E_b = P_s/R$, N_0 is the one-sided noise density, and R is the bit rate ℓ/T. Note that the DPCM has a considerable advantage ($\approx$3 dB) over PCM at the optimum bit rates. As expected, the channel error noise effect begins to dominate as the bit rate increases above its optimal value.

For smaller values of the P_s/N_0, where the error rate is significant the optimum value of the prediction coefficient [Essman and Winzt, 1973, 867–877] is less than its optimum value $A = \exp(-bT)$ for high P_s/N_0. As the correlation between samples $\exp(-bT)$ gets closer to 1.0, the optimum value of A increases as the mean-square error decreases until the channel error noise effect dominates and performance no longer improves. For example, for $P_s/N_0 = 1000$, the maximum value of A which should be used is 0.95.

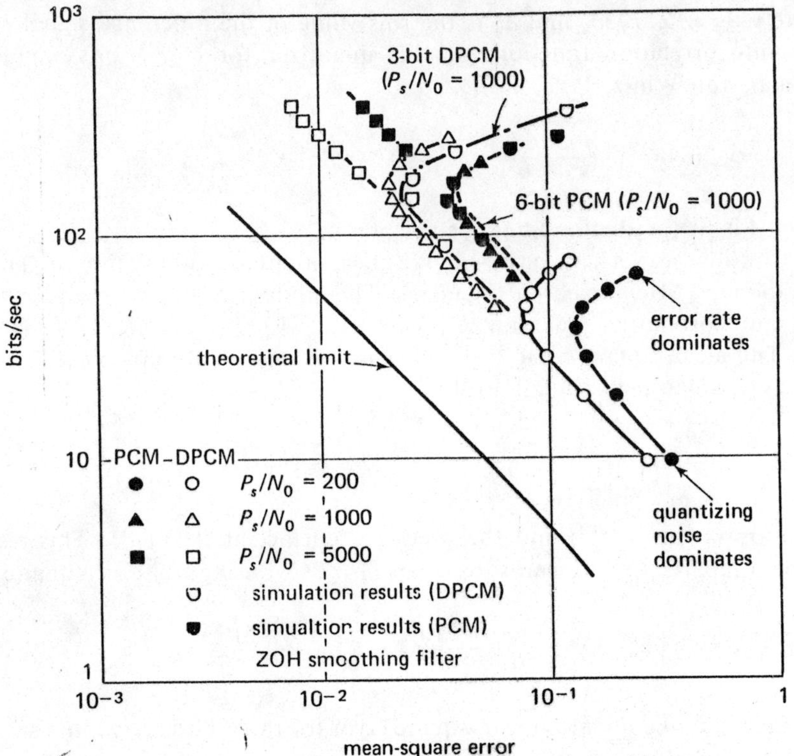

Fig. 4-14 Mean-square error vs bit rate for DPCM and PCM for several values of carrier to noise density P_s/N_0. Uniform quantization is used for the PCM and exponential quantizing for DPCM [From Essman and Wintz, 1973]

4-7 VARIABLE SLOPE DELTA MODULATOR (VSDM)

It is possible to improve the performance of a one-bit ΔM by changing the ΔM step size, depending on the number of consecutive positive or negative steps [Bosworth and Candy, 1969, 1459–1480]. This approach produces greater improvement in SNR for typical video signals of varying dynamic range than for a Gaussian input and yet is simpler to implement than the use of an ℓ-bit quantizer inside the feedback loop of DPCM. Figure 4-15 is a block diagram of a VSDM in which the sequence detector determines the step size or amplitude weight of the plus (+) or minus (—) step. The sequence detector increases step size when a succession of + or — steps occurs accord-

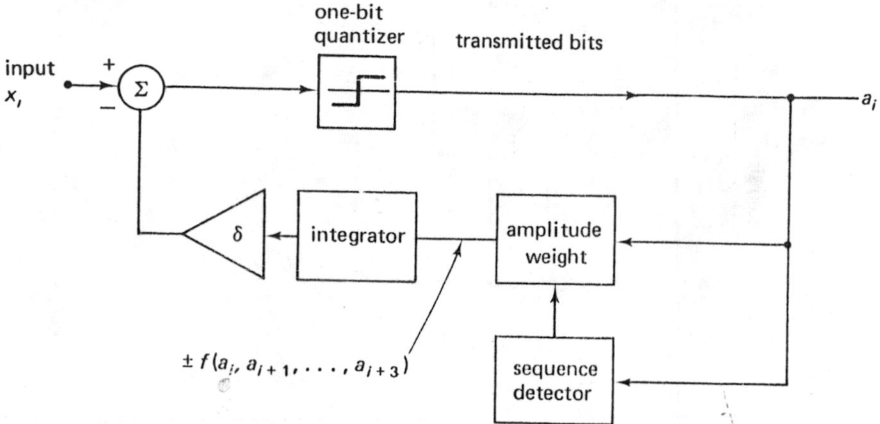

Fig. 4-15 Block diagram of the Variable Slope Delta Modulator (VSDM)

ing to the relationships

	$f(\mathbf{a})$	a_i	a_{i+1}	a_{i+2}	a_{i+3}
		a			
f_1	$+\delta$	x	x	0	1
f_2	$+\delta$	x	0	1	1
f_3	$+2\delta$	0	1	1	1
f_4	$+4\delta$	1	1	1	1

in which the transmitted bits are a_i, the weighted amplitude steps are $\pm f_i$, and x indicates "don't care," either a 1 or 0 bit. Many sequences of step sizes other than this one can be selected, but the steps cannot increase too rapidly, as will be seen, or the VSDM will not be "stable."

Step Response

With this particular sequence of step sizes, the response to a large positive step input produces a consecutive sequence of +1 output bits and the VSDM response corresponding to the sequence of Fibonacci numbers. Thus, consecutive positive steps are of increments 1, 1, 2, 4, 4, 4, for this truncated Fibonacci response. The step response is shown in Fig. 4-16.

Transient Response-Stability

To minimize quantizing noise effects for a fixed level input, the system output must always converge to the *minimum* step size δ from its maximum

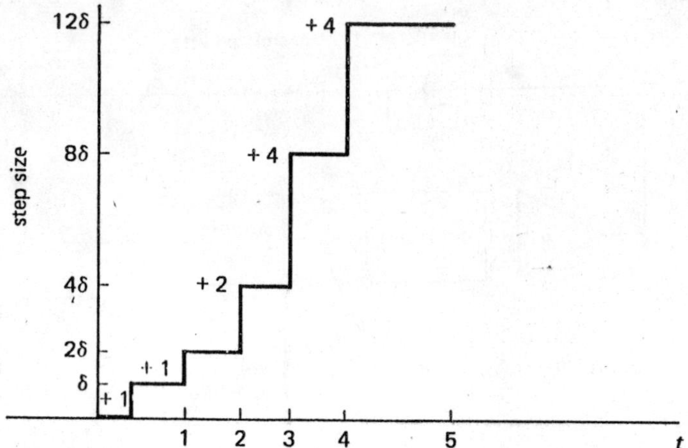

Fig. 4-16 Fibonacci step response of a VSDM with a maximum step size of 4δ

step. It can be shown that this stability constraint requires that

$$f_j \leq \sum_{i=1}^{j-1} f_i \qquad f_1 = 1 \tag{4-103}$$

for any j. Thus, the Fibonacci response $f_j = \sum_{i=1}^{j-1} f_i$ is the most rapid stable response. A higher rate of increase in step size such as the sequence of step sizes 1, 2, 8, 64, for example, will not generally converge to the smallest increment.

EXAMPLE Examine the transient response and overshoot to a large step input as shown in Fig. 4-17 for a Fibonacci sequence of steps. Note that Fibo-

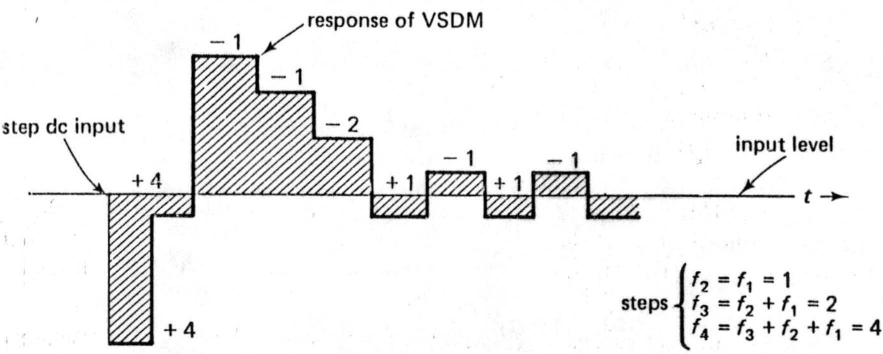

Fig. 4-17 Step response and peak overshoot of a Variable Slope Delta Modulator (VSDM)

nacci response eventually settles to the smallest step sizes and decays in a "stable" manner to its quiescent value.

Adaptive Delta Modulation

Jayant [1974] has described another step-size adaption algorithm where the step size δ_i at time i is related to the previous step size by

$$\delta_i = \delta_{i-1}\alpha^{a_i a_{i-1}} \tag{4-104}$$

where for speech the optimum value of α is $1 < \alpha_{\text{opt}} < 2$, and a_i and a_{i-1} are the present and previous bits, respectively. A value of $\alpha = 1.5$ used at a 60-kHz sampling rate for speech bandlimited to 3.3 kHz showed a 10-dB increase in output SNR as compared to conventional ΔM. Furthermore, the dynamic range obtained was on the order of 30 to 40 dB.

Notice that this adaption rule for $\alpha = \sqrt{2}$ is similar to the VSDM approach. If a succession of +1 samples appears, the step sizes increase as shown below.

Number of Successive +1	*Step Size/δ*	
	Adaptive ΔM	*VSDM*
1	1	1
2	1.41	1
3	2.0	2
4	2.83	4
5	4.0	4
6	5.64	4

Continuously Variable Slope Delta Modulation of Speech

Continuously variable slope delta modulation (CVSD) [Greefkes and DeJager, 1968] is another variation of ΔM that is useful for speech quantization. In this approach, a composite signal consisting of bandpass speech (300 to 3200 Hz) plus a speech envelope signal is fed to a one-bit variable delta modulator (Fig. 4-18). The speech envelope signal is low-pass filtered to approximately 100 Hz so that it does not interfere with the speech waveform itself.

The low-pass envelope waveform is used to adapt the step size at a slow rate with a time constant of approximately 10 msec. The adaption occurs in a manner similar to that in the sequence detector just described for VSDM

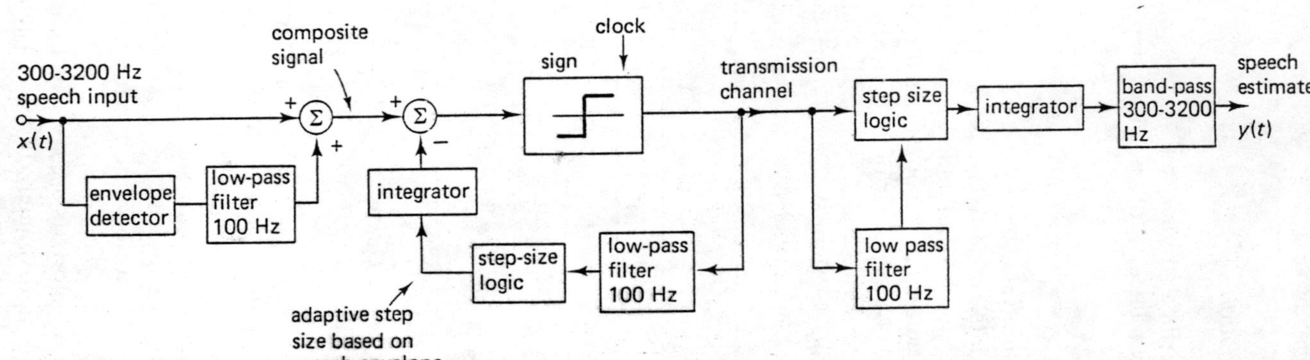

Fig. 4-18 Continuously Variable Slope Delta modulation (CVSD) of speech [After Greefkes and DeJager, 1968]

(Fig. 4-15) and adaptive ΔM. However, rather than adapting the step size on each successive sample based on the number of consecutive ones or zeros in a row, which is an estimate of the rate of change in the input signal, the step size is adapted slowly on the basis of the mean number of ones within the 10-msec averaging time. The mean number of ones in a row is now based on the 100-Hz speech envelope waveform, which of course is a direct measure of the input signal level. This CVSD approach has proven particularly useful for speech quantization at bit rates of 25 kbps and below.

An alternative approach is to base the step size on the magnitude of the integral of $a_i a_{i-1}$, where the integral has a finite memory. The step size for this CVSD is thus developed in a manner somewhat analogous to that of (4-104), except that the step size is continuously variable.

Adaptive DPCM and Linear Predictive Coding

DPCM can be adapted in two different manners—the quantizer step size can be varied with input signal dynamic range, and the predictor coefficients can be slowly updated with changes in the input statistics. The quantizer step-size adaption can be performed in a manner similar to that already discussed for step-size adaption, and is relatively simple to implement. Adaption of the predictor coefficients, on the other hand, requires storage of short segments of the input waveform, calculation of the autocorrelation coefficients for the stored section, optimizing the predictor coefficients, and transmitting the slowly varying predictor coefficients to the receiver.

Adaption of the quantizer step size can be accomplished in a manner essentially identical to that discussed for adaptive PCM in Sec. 3-5. Thus, the step size δ_i at time i is related to the step size at time $i - 1$ by the algorithm

$$\delta_i = \delta_{i-1} m(|\mathfrak{a}_{i-1}|) \tag{4-105}$$

where $|\mathfrak{a}_{i-1}|$ is the magnitude of the quantizer step size code word at the previous sample instant. Again, only a minimal one-word memory is required to store $|a_{i-1}|$. The step size adaption algorithm m is essentially the same as that shown in Table 3-4 for adaptive PCM, except that the rate of decrease of step size is made even less rapid as shown in Table 4-2. With DPCM, there is even less correlation between adjacent samples than for PCM when both are sampled at the Nyquist rate.

Adaption of the DPCM prediction coefficients can be accomplished nearly independently of the step-size adaption. One approach described by Cummiskey [1973] for speech processing is to store 32 sample speech segments (4 msec) and to perform a steepest descent gradient search to minimize the magnitude error at each new sample. Referring to Fig. 4-10 and Eq. (4-64),

Table 4-2 ADAPTION SCALE FACTOR OF DPCM AS A FUNCTION OF THE QUANTIZER OUTPUT CODE WORD $\mathcal{Q}$ FOR $N = 8$, 16-LEVEL QUANTIZERS [AFTER JAYANT, 1974] * $N = 2^\ell$

$\lvert\mathcal{Q}\rvert + \frac{1}{2}$	$m(\lvert\mathcal{Q}_{i-1}\rvert)$	
	$N = 8$	$N = 16$
1	0.90	0.90
2	0.90	0.90
3	1.25	0.90
4	1.75	0.90
5		1.20
6		1.60
7		2.00
8		2.40

*Reprinted with permission from *The Bell System Technical Journal*, Copyright 1972, The American Telephone and Telegraph Company.

the magnitude of error at time i is

$$e_i = |x_i - z_i| = e_i \text{ sgn } e_i$$

where $z_i = \sum_{j=1}^{n} a_j x_{i-j}$. The predictor coefficients are now slowly adapted at each new time sample by a new value

$$a_{ji} = a_{ji-1} + \Delta_{ji} \tag{4-106}$$

where $$\Delta_{ji} = \frac{\beta \text{ sgn } (e_i)\, x_{i-j}}{\sum_{\ell=1}^{n} |x_{i-\ell}|} \tag{4-107}$$

The adaption speed should be relatively slow, and typically $\beta < 0.1$. The predictor coefficients can then be transmitted to the receiver. For DPCM with ℓ bits/sample, transmission of these coefficient changes can be avoided by computing them both at the transmitter and the receiver based on quantized values of x_i, namely, the $Q(e_i)$ shown in Fig. 4-9. Additional discussion of linear predictive coding is beyond the scope of this book.

4-8 SUMMARY

This chapter has discussed a variety of feedback type quantizers that attempt to take advantage of correlation between adjacent samples of the input waveform. These feedback quantizers include one-bit delta modulation, ΔM, and ℓ-bit differential PCM, DPCM. There are a variety of quantizer variations, ranging from the use of fixed ℓ-bit quantizers, as discussed in Chap. 3, to variable slope ΔM or DPCM where the step size is adapted, depending on the number of times a step of the same sign has occurred in succession, the closeness to saturation of the previous quantized sample, or the amplitude of the input envelope. One-bit delta modulation is one of the simplest quantizers to implement and is capable of operating at relatively high bit rates ($\approx 10^9$ bits/sec.).

The essential feature in ΔM and DPCM is the use of a prediction filter that attempts to predict the value of the next input sample by using a linearly weighted summation of past input estimates. This linear predictive encoding then reduces the quantizing problem to one of quantizing the difference between the actual input and the predicted value of that input. For nonstationary inputs, such as speech, the slowly varying predictor coefficients can vary with time, and the value of the coefficients can be transmitted along with the quantizer output.

The performance of ΔM and DPCM quantizers is analyzed by considering two types of quantizing noise—granular quantizing noise and slope overload noise. Because the ΔM and DPCM quantizers are feedback systems, they can in principle operate over a wide dynamic range of input amplitudes so long as the difference between the present and the previous sample does not change too markedly and cause slope overload. In the absence of slope overload, the feedback quantizer of step size δ acts as a PCM quantizer of step size 2δ and unlimited amplitude range. The noise in this condition is granular quantizing noise.

For any realistic input, the step size and sampling rate must combine so that the maximum DPCM slope $(2^{\ell} - 1)\delta f_s$ matches that of the maximum input slope. If the DPCM slope is too small, momentary overloads will occur and cause burst noise effects. These burst noise effects can dominate the granular quantizing noise.

In general, feedback quantizers can out-perform PCM quantizers for inputs where the power spectral density is nonwhite in a significant manner (for example, for speech or video). PCM quantizers, on the other hand, can provide better performance for white or nearly white inputs or inputs where the spectral density varies widely with time and the filter coefficients for DPCM are not adapted.

CHAPTER 5

TIME DIVISION MULTIPLEXING

5-1 INTRODUCTION

Digital bit streams arriving in parallel at a satellite communications terminal or intermediate switching center originate at different sources and some of these may have the same destination. It is usually convenient to time-division-multiplex (time interleave) bit streams that have the same ultimate destination into one single serial bit stream for transmission on a single RF carrier. However, the bit streams to be multiplexed may have slightly different bit rates. Even if these bit streams are nominally at the same bit rate, they have imperfect clock stabilities and hence are not exactly synchronous with one another or with the earth terminal clock. Furthermore, some of these bit streams may have arrived via a satellite link in a multihop transmission where the satellite–earth station propagation time is changing because of satellite motion. Many of these bit streams have the same nominal frequency and are sometimes called pleisiosynchronous or quasi-synchronous. Given enough time the difference in the number of input bits in the two streams can be arbitrarily large. In order to time-multiplex these bit streams together, they must be converted to the same rates or multiples of the same rates. If the bit streams are pleisiosychronous, one or all of the bit streams must be buffered and have "stuff" bits added to synchronize them.

A simpler type of multiplexing problem occurs where the two bit streams are frequency-locked—that is, have the same average frequency and the number of bits of clock phase offset between the two bit streams is bounded by some maximum. In this case, an "elastic" buffer of sufficiently large size can smooth out short-term fluctuations in the data rates without adding "stuff" bits. The problem here is one of determining the required size of the buffer to prevent overflow or underflow by acting as a storage reservoir. This requirement in turn depends on the overall system timing (Chap. 17).

In this chapter, we consider a set of n data sources at approximately equal but fluctuating bit rates in the range $f_a < f_i < f_a + \Delta f$. Each data bit stream is to be fed into a time division multiplexer (MUX), which combines them to produce a single fixed stable output bit rate $f_0 > n(f_a + \Delta f)$, as shown in Fig. 5-1. In general, input bit rates may differ widely from one another, and several multiplexers may be used in a "tree" of multiplexers

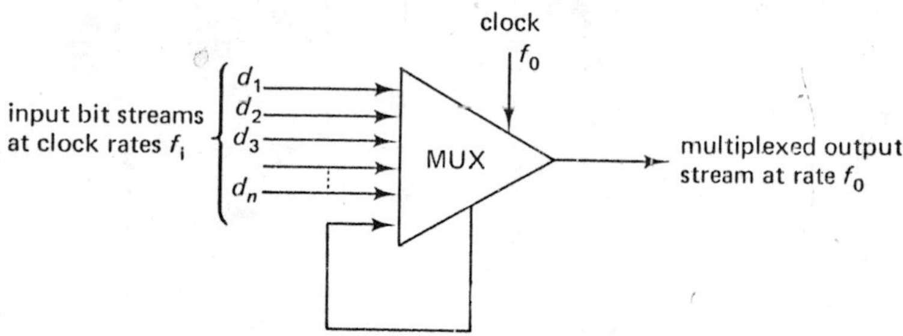

Fig. 5-1 Configuration of a time division multiplexer having inputs of nominal bit rate f_a

to permit all inputs to each multiplexer to have approximately the same frequency. For example, in Fig. 5-2, the second multiplexer has inputs at a nominal input rate f_0 and is fed by other multiplexers or other data streams. In this manner, bit streams of widely different rates can be accommodated.

Closer examination of the internal functions of a typical pleisiosynchronous multiplexer reveals the configuration shown in Fig. 5-3. Each input bit

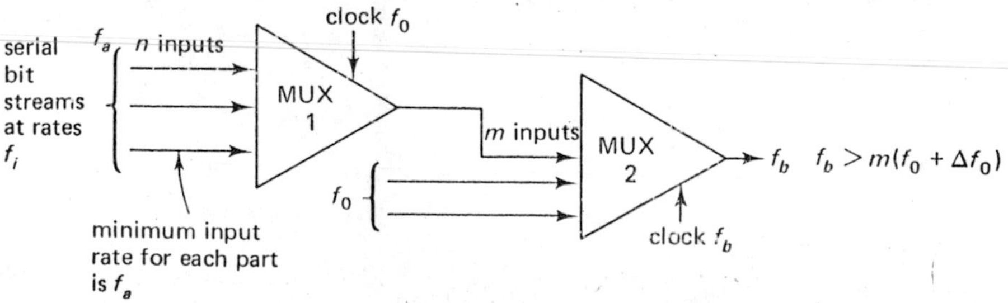

Fig. 5-2 Typical "tree" configuration of multiplexers; the second multiplexer has inputs at a nominal rate f_0, where $f_0 > n(f_a + \Delta f)$.

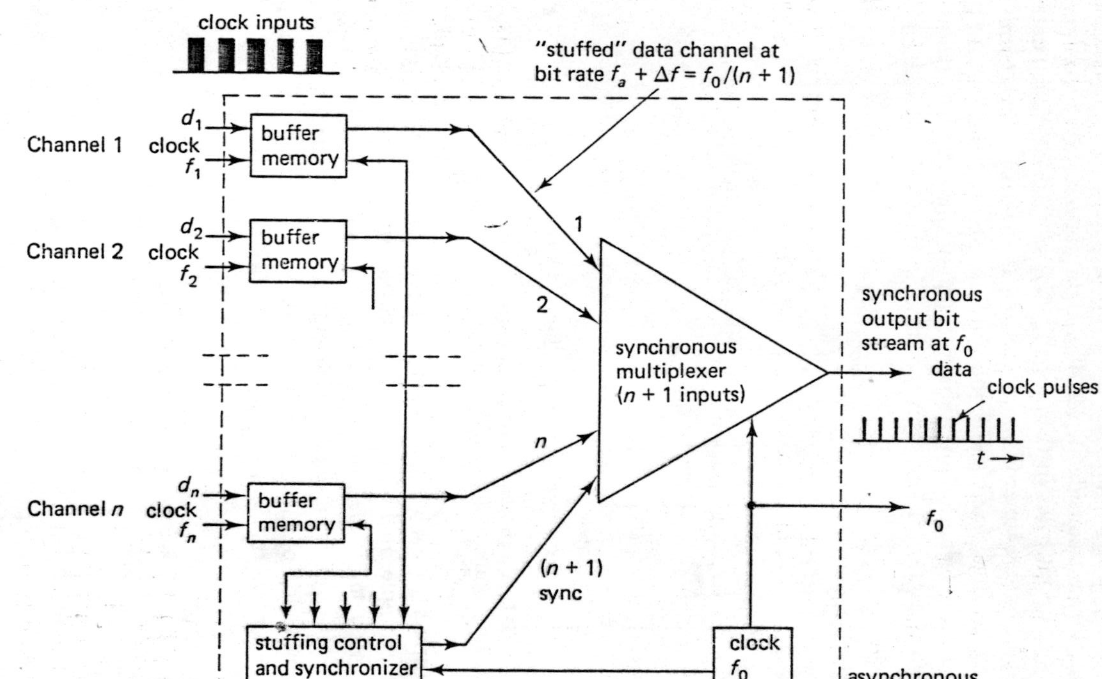

Fig. 5-3 Simplified internal configuration of a time division multiplexer; the minimum bit rate for all inputs is f_a.

stream is at clock rate $f_i \leq f_a + \Delta f$. The MUX accepts the pleisiosynchronous input bit streams and associated clock pulses, buffers them, and emits a synchronous bit stream at f_0 or a word stream at the submultiple of the MUX output clock rate f_0. In order for the buffer output channel to be synchronous, additional pulses or words are stuffed into the output bit stream. At least one additional channel $(n + 1)$ carries the information as to which bits or words are stuffed and where frame start occurs. Sometimes two or more sync channels are used—for example, channels $(n + 1)$, $(n + 2)$, $(n + 3)$, could be used if necessary.

In Fig. 5-3 the $(n + 1)$ buffer output bit streams are all clocked at rate $f_a + \Delta f$, and are then multiplexed together synchronously to form a single serial bit stream at rate $f_0 = (n + 1)(f_a + \Delta f)$.

There are two commonly used methods for stuffing the output channel to a fixed, synchronous clock rate, namely, pulse stuffing and word stuffing. Each of these techniques is discussed in the next section and has several variations which can produce different output bit rates f_0 for the same inputs.

5-2 PULSE STUFFING

We begin with a simple example of a multiplexer with a single variable rate input bit stream. Pulse stuffing is an operation by which identifiable extra pulses are added to the data stream to convert a varying input bit rate to a slightly higher but fixed output bit rate as in Fig. 5-4. The identifiable stuff bits are inserted into the transmitted bit stream in Fig. 5-4(b). The recovered destuffed bit stream in the demultiplexer output is also shown after the output bit rate has been smoothed in Fig. 5-4(*d*). Each pulse stuffing buffer operates

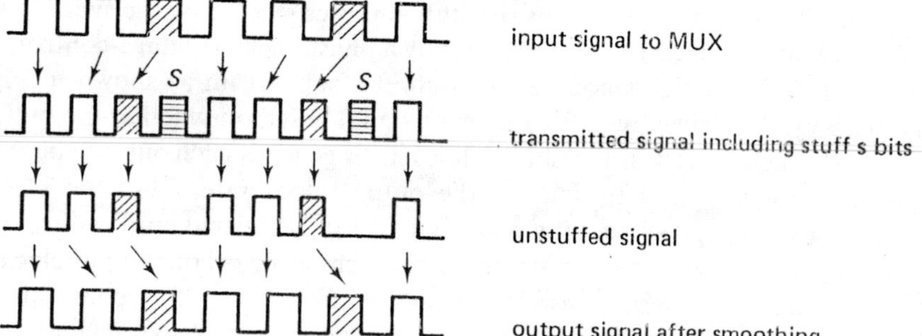

Fig. 5-4 Buffer memory inputs and outputs for multiplexed transmitted signal and demultiplexed output shown for one channel of the pleisiosynchronous multiplexer

so that every $T = 1/\Delta f$ sec a pulse can be stuffed in the bit stream from the pulse stuffer/synchronizer if the buffer memory of K bits memory capacity is not sufficiently full. The buffer acts in a manner analogous to a water storage reservoir which has to provide a smoothly changing output flow rate even though the input flow rate stored may change in abrupt random increments. The fullness of the buffer can be measured by an up-down counter which counts the difference between the number of output clock pulses and the input clock pulses. As shown in Fig. 5-5, each time the buffer memory drops below a given fullness threshold, an artificial pulse is stuffed into the buffer memory; the time at which this stuffing occurs is signaled by the synchronizer so that the stuffed pulse can be removed at the demultiplexer buffer. Note that the stuff demands in Fig. 5-5 occur before a stuff bit is actually inserted.

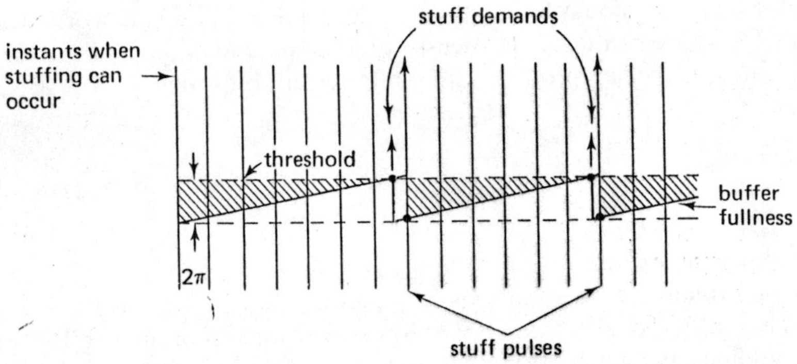

Fig. 5-5 Variation in input buffer fullness with time for a pulse stuffing synchronizer

At the demultiplexer in Fig. 5-6, the received bit stream is synchronously demultiplexed into parallel synchronization and data channels. Each data channel is fed into a buffer minus the stuffed pulses deleted by the synchronization receiver, and the buffer is read out by a phase-locked voltage-controlled oscillator (VCO) which smoothes the output clock stream as shown in Fig. 5-6. The VCO is driven by a filtered error signal based on whether the buffer is above or below a given threshold fill level. In general, each output channel requires its own VCO to smooth the output clock rate. Thus the VCOs smooth out the gaps in the bit stream caused by the deleted stuff pulses and produce output clocks that closely resemble each of the original input clocks. The dynamics of these phase-locked oscillators, hence the phase jitter characteristics on the output bit streams, depend on several parameters: (1) multiplexer and demultiplexer buffer storage in bits, (2) input bit stream dynamics, (3) framing periods and synchronization format, and (4) the closed-loop response of the phase-locked loop.

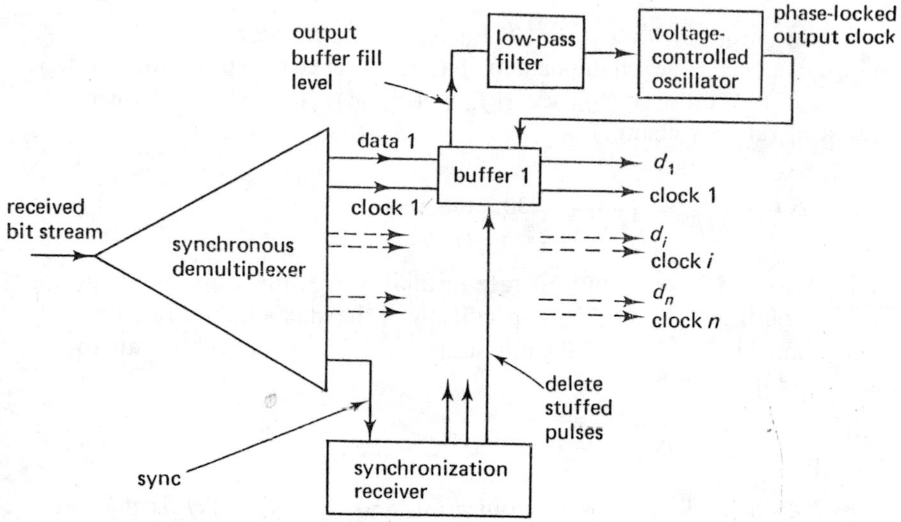

Fig. 5-6 Simplified configuration of the time division demultiplexer

5-3 PULSE STUFFING SYNCHRONIZATION SIGNALING

The format of the output bit stream is typically configured to have an N-bit frame and an M frame superframe. Thus we have a $M \times N$ bit matrix of bits a_{ij}

$$\begin{pmatrix} \mathbf{a}_1 \\ \mathbf{a}_2 \\ \cdot \\ \cdot \\ \cdot \\ \mathbf{a}_M \end{pmatrix} = \begin{pmatrix} (a_{11} & \cdots & a_{1N}) \\ (a_{21} & \cdots & a_{2N}) \\ & \cdot & \\ & \cdot & \\ & \cdot & \\ (a_{M1} & \cdots & a_{MN}) \end{pmatrix}$$

where $\mathbf{a}_i$ represents the ith frame in the superframe. Within each frame are n data channel bits and at least one frame marker bit and a synchronization bit, each of which can change in pattern from frame to frame. Thus, there are M sync bits ($M > n$) in a superframe to indicate in a coded form which particular data bits are stuff bits, etc. An example, $N = 145$ bit frame [Witt, 1965, 1843–1886], is shown in Fig. 5-7; in this example, the frame sync bit simply alternates between 0 and 1 from one frame to the next. There are $n = 143$ data channels plus two overhead bits in each frame; that is, $N = n + 2$.

At an output bit rate $f_0 = 10^6$ bits/sec and a frame rate $R_f \triangleq f_0/N$. A stuff bit can occur in each data channel at a rate once per superframe of MN bits or one stuff bit in MN/f_0 sec. If $f_0 = 10^6$, $N = 100$, $M = 326$ the maximum stuff rate per channel is

$$\Delta f \leqq \frac{f_0}{MN} = \frac{10^6}{32{,}600} = 30.7 \text{ bits/sec} \tag{5-1}$$

Channels with a larger input bit rate instability Δf, therefore, must have a smaller superframe period. More specifically, if the maximum bit rate for each input channel is $f_a + \Delta f$ and the minimum rate is f_a, the output bit rate for an n-channel input can be set at

$$f_0 = (f_a + \Delta f)N = R_f N \tag{5-2}$$

where $N > n$ to allow for synchronization and $R_f = f_a + \Delta f$ is the frame rate. The M-bit sync word contains c redundant synchronization bits for each data channel plus word sync of w bits in a superframe ($M = cn + w$). Thus, using (5-1), (5-2), the maximum allowable peak-to-peak frequency uncertainty Δf is

$$\Delta f \leq \frac{f_0}{MN} = \frac{(f_a + \Delta f)N}{(cn + w)N} = \frac{f_a + \Delta f}{cn + w} = \frac{R_f}{cn + w} \tag{5-3}$$

The maximum relative frequency uncertainty ϵ allowed in each input bit stream is then

$$\epsilon \triangleq \frac{\Delta f}{(f_a + \Delta f)} \leq \frac{1}{cn + w} = \frac{1}{M} \tag{5-4}$$

If $\Delta f/f_a$ increases above a given value, n must be made smaller, and the superframe must shorten.

The throughput efficiency η in this channel is the ratio of the total input bit rate to the output bit rate, or

$$\eta = \frac{n(f_a + \Delta f)}{f_0} = \frac{n}{N} = \frac{n}{n + 2} \qquad N = n + 2 \tag{5-5}$$

since two bits have been added in each frame, one for frame synchronization and one for stuffing synchronization (Fig. 5-7).

The synchronization signal must be transmitted with sufficient redundancy to tolerate channel errors without losing bit integrity—that is, without leaving in a stuffed bit or removing a valid data bit, thus shifting the entire succeeding output bit stream one bit ahead or behind in time. This loss of bit

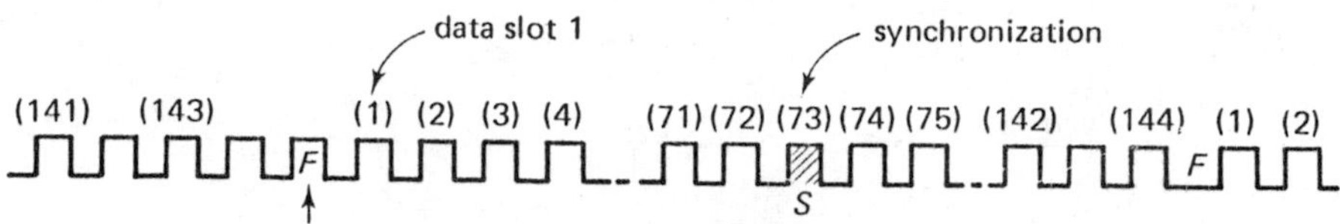

Fig. 5-7 Example of an N = 145-bit frame with the synchronization bit in slot number 73

integrity would at least cause the remaining bits in the frame to be read out erroneously, probably to the incorrect demultiplexer output port and hence is a much more serious degradation than a data bit error. User equipment receiving this bit stream is likely to be thrown out of sync and the data lost for a significant time interval.

The synchronization channel, multiplexed in the data bit stream once every N-bit frame, contains redundantly coded synchronization words, one for each of the $n = N - 2$ input data channels, plus additional words used for identifying the beginning of the superframe and rapid reframing. One possible synchronization channel superframe format for $n = 143$ is

$$\cdots C_{143} \Big| \overbrace{W\, C_1 C_2 C_3 C_4 C_5 \cdots C_{143}}^{M\text{-bits}} \Big| W\, C_1 \cdots \quad \text{sync channel bit stream}$$

marker word (0110101001011010. . .) — points to W

channel stuff word for channel 5 — points to C_5
(000) not stuffed
(111) stuffed
c-bits

where the synchronization channel (observed after demultiplexing each frame) has 3-to-1 redundancy ($c = 3$). A simple majority decision reduces the stuff error probability to $3\mathcal{P}^2$ where $\mathcal{P}$ is the probability of a bit error. For channels with a moderate error rate such as $\mathcal{P} = 10^{-3}$, this redundancy level and simple majority coding may not be sufficient.

With the format shown above, the superframe duration in bits is

$$MN = (cn + w)N = (3n + w)N \text{ bits} \qquad \text{if} \qquad c = 3 \tag{5-6}$$

where c is the redundancy used for each of the n data channels. For example, where $n = 98$ input data channels, $N = 100$, and $w = 32$, the superframe period is

$$MN = [3(98) + 32](100) = 326(100) = 32{,}600 \text{ bits} \tag{5-7}$$

Efficiency η can also be expressed in terms of allowable relative frequency uncertainty of the input bit rate ϵ using (5-4), (5-5). We can also express n in terms of ϵ

$$\eta = \frac{n}{n+2} \quad \text{and} \quad \epsilon = \frac{1}{cn + w} \longrightarrow n = \frac{1}{c\epsilon} - \frac{w}{c} \tag{5-8}$$

for the format of Fig. 5-7. The pulse stuffing efficiency for this format is obtained from (5-8)

$$\eta = \frac{(1/c\epsilon) - (w/c)}{(1/c\epsilon) - (w/c) + 2} = \frac{(1 - w\epsilon)}{(1 - w\epsilon) + 2c\epsilon} \tag{5-9}$$

For majority decisions and c bits (c odd) per stuff word, $c/2 + 1/2$ errors must occur to cause a stuff error. If the bit error probability is $\mathcal{P}$, the probability of a stuff error is

$$\mathcal{P}_s = \sum_{i=c/2+1/2}^{c} \binom{c}{i} \mathcal{P}^i (1 - \mathcal{P})^{c-i} \tag{5-10}$$

The stuff error rates for a range of input error rates and stuff word redundancy is shown in Table 5-1. For $c = 7$ and $\mathcal{P} = 10^{-4}$, the stuff error probability is 3.5×10^{-15}.

Table 5-1 STUFF ERROR RATE $\mathcal{P}_s$ FOR VARIOUS BIT ERROR RATES

Redundancy	*Bit Error Rate* $\mathcal{P}$					
c	10^{-1}	10^{-2}	10^{-3}	10^{-4}	10^{-5}	10^{-6}
3	2.8×10^{-2}	2.98×10^{-4}	3×10^{-6}	3×10^{-8}	3×10^{-10}	3×10^{-12}
5	8.56×10^{-3}	9.93×10^{-6}	9.98×10^{-9}	1×10^{-11}	1×10^{-14}	
7	2.7×10^{-3}	3.4×10^{-7}	3.5×10^{-11}	3.5×10^{-15}	3.5×10^{-19}	
9	8.9×10^{-4}	1.22×10^{-8}	1.26×10^{-13}	1.26×10^{-18}		
11	2.96×10^{-4}	4.42×10^{-10}	4.6×10^{-16}			
13	9.93×10^{-5}	1.63×10^{-11}	1.7×10^{-18}			
15	3.22×10^{-5}	6×10^{-13}	6.4×10^{-21}			

The expected time τ until loss in bit integrity for an individual channel stuffed once every M frame superframe of duration $T = \frac{MN}{f_0} = \frac{MN}{(f_a + \Delta f)N} = \frac{M}{f_a + \Delta f} \cong \frac{M}{f_a}$ is related to the stuff error probability $\mathcal{P}_s$ by

$$\tau = \frac{T}{\mathcal{P}_s} = \frac{M/(f_a + \Delta f)}{\mathcal{P}_s} \tag{5-11}$$

As an example, if $M = 3 \times 10^4$, $\mathcal{P}_b = 10^{-3}$, $c = 11$, $f_a = 10^4$ bits/sec, $\Delta f =$ 1 bit/sec from Table 5-1 we have $\mathcal{P}_s = 4.6 \times 10^{-16}$ and the expected time till loss of bit integrity using (5-11) is

$$\tau = \frac{3 \times 10^4}{10^4} \frac{1}{4.6 \times 10^{-16}} = 6.5 \times 10^{15} \text{ sec} \tag{5-12}$$

5-4 WORD STUFFING

Word stuffing is a variant of the pulse stuffing technique and is somewhat more tolerant of bit rate changes. Assume as in Fig. 5-1 that there are n input data channels at a data rate $f_i \geq f_a$ and $f_i \leq f_b$ bits/sec where $f_b = f_a + \Delta f$ is the maximum bit rate. Each of the n input data channels is subdivided into k-bit words at a word rate per channel of $R = f_b/k$. These k-bit words are then coded into m-bit words where $m > k$. Each data channel word is transmitted once per frame, the frame rate R_f is also equal to R.

$$\text{Frame Rate} = R_f = f_b/k \quad \text{frames/sec} \tag{5-13}$$

Each m-bit word either determines k information bits or is a control word to determine whether the entire m-bit word is a stuff word or a valid data word.

A possible format for word stuffing, which includes a single frame sync bit, is shown in Fig. 5-8. The frame is $nm + 1 = N$ bits in duration. Word

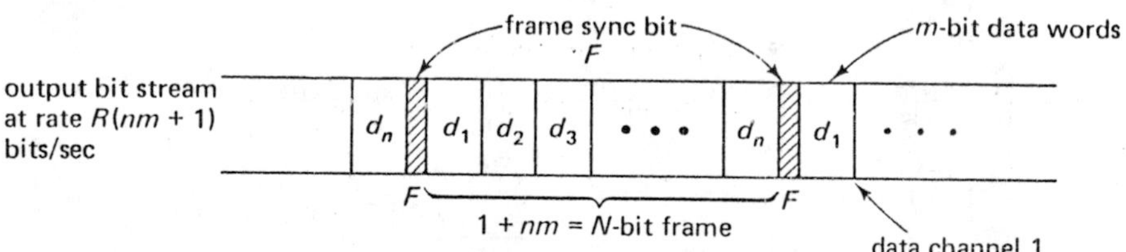

Fig. 5-8 Possible format for word stuffing

stuffing allows k bits to be stuffed per data channel every frame. Thus, the maximum bit suffing rate is one stuff word per frame per data channel and the maximum tolerable peak-to-peak clock frequency variation is

$$\Delta f = R_f k = f_b \text{ bits/sec} \tag{5-14}$$

and in principle word stuffing can accommodate very high bit rate variations (0 to f_b bits/sec) compared to the pulse stuffing technique.

The total set of 2^m m-bit words is subdivided into three types of code word sets [Buchner, 1970, 379–398]: a single stuff word **S** (for example, the all-zero word), the set **D** consisting of 2^k-data words, and the set **U** consisting of the $2^m - 2^k - 1$ unused words. The set **D** can be selected by maximizing the minimum distance d_m between **S** and any member of **D**. Thus, the probability of a data word being confused with a stuff word is minimized, and stuff word errors are minimized in the presence of channel errors. Because of a need for high transmission efficiency η,

$$\eta = \frac{nk}{nm+1} \cong \frac{k}{m} \qquad \text{if} \qquad n \gg 1 \tag{5-15}$$

only values of $m = k + 1$, or $k + 2$ are considered here.

The minimum distance d_m between the stuff word and any of the 2^k data words can be computed for $m = k + 1$ by the following analysis. After optimal selection of **D**, the unused set **U** contains the following numbers of words at each of the listed distances from **S**:

$$\left.\begin{array}{ll} \binom{m}{1} & m\text{tuples of distance 1 from } \mathbf{S} \\ \binom{m}{2} & m\text{tuples of distance 2 from } \mathbf{S} \\ \cdots & \\ \binom{m}{d_m - 1} & m\text{tuples of distance } d_m - 1 \text{ from } S \end{array}\right\} \begin{array}{l}\text{Set } \mathbf{U} \\ \text{contains} \\ M_d \\ \text{words}\end{array} \tag{5-16}$$

and r_m additional $m$tuples of distance d_m from **S** such that

$$\binom{m}{d_m} > r_m = \underset{\text{total}}{\underset{\downarrow}{2^m}} - \underset{\text{set } \mathbf{D}}{\underset{\downarrow}{2^k}} - \underset{\mathbf{S}}{\underset{\downarrow}{1}} - \underbrace{\sum_{j=1}^{d_m-1} \binom{m}{j}}_{M_d} \geq 0 \tag{5-17}$$

The total number of unused words is thus easily bounded, using (5-17), by

$$M_d = \sum_{j=1}^{d_m-1} \binom{m}{j} \leq \underbrace{2^m - 2^k - 1}_{\substack{\text{total number of} \\ \text{unused words } M}} < \sum_{j=1}^{d_m} \binom{m}{j} \tag{5-18}$$

where the inequality on the right hand of (5-18) exists because some of the $\binom{m}{d_m}$ words of distance d_m are by definition data words and not unused words.

The bound (5-18) can now be used to determine d_m. If $m = k + 1$, then

$M = 2^m - 2^k - 1 = 2^{k+1} - 2^k - 1 = 2^k - 1$, and the bound becomes

$$\sum_{j=1}^{d_m-1} \binom{k+1}{j} \leq 2^k - 1 < \sum_{j=1}^{d_m} \binom{k+1}{j} \tag{5-19}$$

Note that from the binomial expansion

$$2^{k+1} = \sum_{j=0}^{k+1} \binom{k+1}{j} = \sum_{j=1}^{k} \binom{k+1}{j} + 2 \tag{5-20}$$

or $$2^{k+1} - 2 = 2(2^k - 1) = \sum_{j=1}^{k} \binom{k+1}{j}$$

and thus $$M = (2^k - 1) = \frac{1}{2} \sum_{j=1}^{k} \binom{k+1}{j} \tag{5-21}$$

For k even, the total number of unused words $2^k - 1$ in (5-19) is equal to

$$M = 2^k - 1 = \frac{1}{2} \sum_{j=1}^{k} \binom{k+1}{j} = \sum_{j=1}^{k/2} \binom{k+1}{j} \tag{5-22}$$

since the binomial coefficients for $j = 1$ to $k/2$ are the same as those from k to $k/2 + 1$ in (5-22). For k odd, there is an additional contribution from the middle or $[k/2] + 1$ term*

$$M = 2^k - 1 = \sum_{j=1}^{[k/2]} \binom{k+1}{j} + \frac{1}{2} \binom{k+1}{\left[\frac{k}{2}\right] + 1} \tag{5-23}$$

Hence, for k even, the bound (5-19) can be rewritten, using (5-22):

$$\sum_{j=1}^{d_m-1} \binom{k+1}{j} \leq \sum_{j=1}^{k/2} \binom{k+1}{j} = 2^k - 1 \tag{5-24}$$

Thus, the largest d_m satisfying (5-24) corresponds to $d_m - 1 \leqq k/2$ or $d_m \leqq k/2 + 1$, where equality occurs only if $r_m = 0$. Similarly, for k odd the bound becomes, using (5-23),

$$\sum_{j=1}^{d_m-1} \binom{k+1}{j} \leq \sum_{j=1}^{[k/2]} \binom{k+1}{j} + \frac{1}{2} \binom{k+1}{\left[\frac{k}{2}\right] + 1} \tag{5-25}$$

and the general result for the minimum distance d_m is

$$d_m \leqq \left[\frac{k}{2}\right] + 1 \tag{5-26}$$

*The notation $[k/2]$ represents the largest integer $\leqq k/2$.

It takes $[d_m/2] + 1$ errors to convert a data word to a sync word decision error. The upper bound on the minimum distance d_m is plotted for $m = k + 1$ and $m = k + 2$ in Fig. 5-9.

Note again that the distance d_m is not the minimum distance between data words but rather the minimum distance between the single (but arbitrary) stuff word and any data word. Thus, the emphasis here is on using code *redundancy* to reduce the probability of loss of bit integrity and not on reducing the data error probability. In fact, as shown in the next example, the data bit error rate can be slightly increased by this coding.

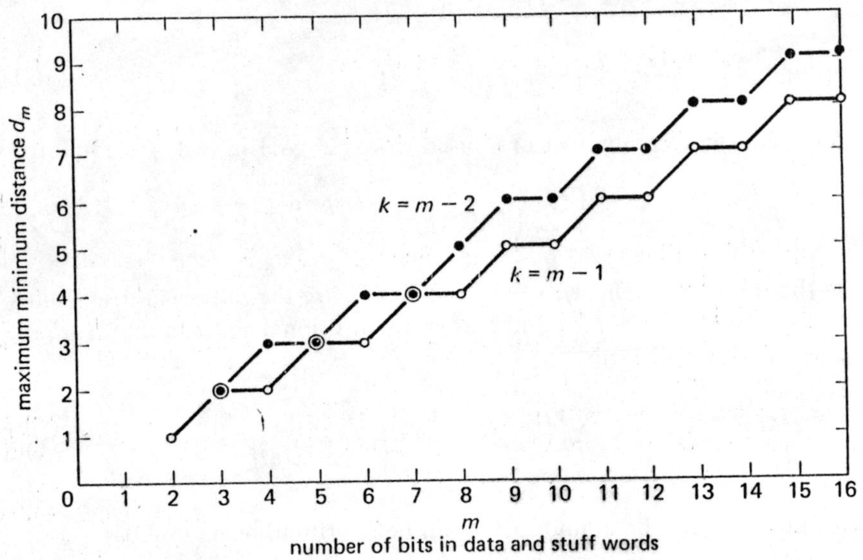

Fig. 5-9 Upper bound on minimum distance d_m between stuff word and any data word for various word sizes m and $k = m - 2, m - 1$

Example Codes for Word Stuffing

Consider, as an example, word length $m = k + 1$ where **S** is the all-zero $m$tuple [Buchner, 1970]. A k-bit input data word **A** with elements $(a_1, a_2, a_3, \ldots, a_k)$ is encoded into an m-bit $(m = k + 1)$ output word $\mathbf{B} = (b_1, b_2, b_3, \ldots, b_k, b_{k+1})$ where b_{k+1} is selected according to the algorithm

$$\left.\begin{aligned} &\text{If } w(\mathbf{A}) \geq \left[\frac{k}{2}\right] + 1 \qquad \text{then} \quad b_{k+1} = 0 \\ &\quad\;\; w(\mathbf{A}) \leq \left[\frac{k}{2}\right] \qquad\quad\;\; \text{then} \quad b_{k+1} = 1 \end{aligned}\right\} \tag{5-27}$$

and the other b_i are transmitted as $b_i = b_{k+1} \oplus a_i$, where $w(A)$ is the weight of **A** [the number of 1s in $\mathbf{A} = (a_1, a_2, \ldots, a_k)$]. To facilitate decoding, b_{k+1} is transmitted first and b_1 last. Decoding is performed to give output data words $\mathbf{A}' = (a'_1, a'_2, a'_3, \ldots, a'_k)$ by operating on the received code bits $\mathbf{B}' = (b'_1, b'_2, b'_3, \ldots, b'_k, b'_{k+1})$ to form

$$a'_i = b'_{k+1} \oplus b'_i \tag{5-28}$$

Examples for $m = 9$-bit words $\mathbf{B}_i$ with an input of $k = 8$-bit words $\mathbf{A}_i$ are

$$\begin{aligned} &\qquad\qquad\qquad\qquad\qquad b_{k+1} = b_9 \\ &\mathbf{A}_i = (00111111) \longrightarrow \mathbf{B}_i = (001111110) \qquad w(\mathbf{A}_i) = 6 > \frac{k}{2} + 1 \\ &\mathbf{A}_j = (00000110) \longrightarrow \mathbf{B}_j = (111110011) \qquad w(\mathbf{A}_j) = 2 \leq \frac{k}{2} = 4 \end{aligned} \tag{5-29}$$

Note that this coding algorithm causes an error multiplication effect in that if b_{k+1} is received in error, eight errors are caused in the output rather than just one.* These data words, however, have minimum distance $d_m = 5$ from the stuff word, and three or more channel errors are required to cause a stuff error if only words of $w(\mathbf{B}) \leq 2$ are considered stuff words.

The expected time until loss of bit integrity is the expected time until a triple error or worse event occurs. If the data rate is 80,000 bits/sec and $k = 8, m = 9$, then $R = f_b/k = 10^4$, nine-bit words are transmitted per second. For a bit error probability $\mathcal{P} = 10^{-5}$, the probability $\mathcal{P}_s$ of a stuff word being converted to a data word is approximately $\mathcal{P}_s \cong \binom{9}{3}\mathcal{P}^3 = (9!/6!3!)\mathcal{P}^3 = 84\mathcal{P}^3$, and the expected time until a stuff-to-data error for a stuff word rate R_s is

$$\begin{aligned} T_{s,d} &\cong \frac{1}{R_s \mathcal{P}_s} = \frac{1}{R_s(84)\mathcal{P}^3} \text{ sec} = \frac{1.19 \times 10^{13}}{R_s} \text{ sec} \\ &= \frac{3.3 \times 10^9}{R_s} \text{ hr} \qquad \mathcal{P} = 10^{-5} \end{aligned} \tag{5-30}$$

If the data-bit-rate stability is *moderately* high so that the ratio of stuff word rate to data word rate $R_s/R_d = \Delta = 10^{-3}$ and $R_s = 10$, $R_d = 10^4$, then (5-30) can be written as

$$T_{s,d} \cong \frac{1}{84\Delta R_d \mathcal{P}^3} = \frac{3.3 \times 10^9}{10} = 3.3 \times 10^8 \text{ hr} \tag{5-31}$$

*This coding can cause the data error rate to increase slightly. However, through the use of an appropriate code of smaller d_m, the error multiplication factor can be decreased.

The probability of a data word being converted to a stuff word depends on the number of data words of minimum distance to the stuff word, which, in turn, depends on the particular code in use. The effects of more distant data words can usually be neglected. In this example, where $k = 8$ is even, the distance $d_m = 5$, and there are exactly

$$\binom{8}{5} = \frac{8!}{3!5!} = 56 \text{ words} \tag{5-32}$$

of distance 5 from S out of a total of $2^8 = 256$. The probability of a data word being converted to the stuff word by one of these triple errors $[d_m/2] + 1$ is then

$$\mathcal{P}_{d/s} \cong \binom{5}{3} \mathcal{P}^3 \frac{56}{256} = 2.18\, \mathcal{P}^3 \tag{5-33}$$

and the expected time until a data word is converted to a stuff word is approximately

$$\begin{aligned} T_{d/s} &\cong \frac{1}{R_d \mathcal{P}_{d/s}} = \frac{1}{10^4(\mathcal{P}^3)2.18} = \frac{10^{11}}{2.18} \text{ sec} \\ &= 1.27 \times 10^7 \text{ hr} \qquad \mathcal{P} = 10^{-5} \end{aligned} \tag{5-34}$$

Thus both $T_{s/d}$ and $T_{d/s}$ are exceedingly long in this example.

Note that if the bit-rate-clock stability is *extremely* high so that the stuff rate is very small compared to the rate at which weight $[k/2] + 1 = 5$ words occur, the threshold decision level between stuff and data decision can be weighted more heavily in favor of data words. For example, only received words of $w(\mathbf{B}) \leq 1$ are decoded into stuff words, and stuff word errors can be made even less likely.

PART 2

Satellite Communications

Major elements in a satellite communications system include the satellite, a network of earth terminals, and multiple-access communications equipment by which many earth terminals can operate through a single satellite. Each of these system elements has many subsystems, is a complex subject in itself, and could individually be the subject of an entire book. In addition, the performance of this satellite communications system is influenced greatly by the satellite orbital configuration and the earth-space propagation channel. In the chapters that follow, only those subsystems of the satellite and earth terminals and system parameters relating directly to the communications system performance are given detailed attention, and many of the satellite and earth terminal subsystems less directly related are omitted entirely. Emphasis is on synchronous satellite systems and microwave frequency transmissions. The specific areas of satellite communications systems covered in Part 2 include:

- Satellite orbits and antenna coverage patterns
- Earth space propagation effects and link calculations
- Satellite transponders
- Earth terminal frequency converters and amplifiers
- Frequency-division multiple access (FDMA)
- Time division multiple access (TDMA)
- Effects of instantaneous nonlinearities on multiple-access signals

Instantaneous nonlinearities occur in any amplifier but are particularly important in the satellite transponder and earth terminal amplifiers. These amplifiers must be efficient and often operate with multiple carriers, and thereby produce intermodulation products in frequency-division multiple access operation. Other important satellite transponder and earth terminal effects relate more closely to the modulation type effects of cochannel interference, such as filter distortion and oscillator phase noise and are discussed in Part 3.

The transponder exhibits both the effects of an instantaneous amplitude nonlinearity and AM/PM conversion, whereby any input envelope variation (AM) causes a phase modulation (PM) of each input signal. These instantaneous nonlinearities produce several effects:

Signal suppression—suppression of weak signals by one or more strong signals

Intermodulation distortion—creation of cross-product distortion terms by two or more input signals

Cross-product spreading—bandwidth spreading of the intermodulation products relative to the input signals.

Cross-talk—modulation induced on one carrier by one or more other carriers.

Transponders used in satellite communications are usually channelized into separate frequency channels. Each has a peak-power-limited amplifier which operates in a quasi-linear or limiting mode and has a frequency translating repeater. Each channel operates with multiple carriers from separate earth terminals. Bandpass filters are used to separate carrier channels and to provide isolation between high-level satellite power output and low-level inputs.

The effects of instantaneous nonlinearities on FDMA are analyzed in Part 2 because they are relatively insensitive to the details of modulation type. Filter distortion effects and oscillator phase noise effects also occur in satellite transponders. However, these latter effects are substantially more dependent on data rate and modulation type and hence are discussed in Part 3, Modulation and Coding.

Earth terminals for satellite communications often are designed for multicarrier operation. As in satellite transponders, there must be adequate filtering to isolate the low-noise receiver from the high-power transmitter, the frequency converters must be designed to minimize spurious mixer cross-products, and the power amplifier may have to be backed off from full power level to reduce transmitted intermodulation products. Thus many of the effects of nonlinearities discussed above for satellite transponders apply equally well to earth terminal design.

Link calculations are given for a typical satellite communication channel.

These results provide the received carrier-to-noise density (C/N_o) ratio as a function of the earth terminal transmitted power, satellite radiated power, ground-terminal-gain/noise-temperature ratio ($G/\mathfrak{T}$), and propagation loss. The received carrier-to-noise density ratio can then be converted to the energy-per-bit/noise density ratio, E_b/N_o, and can be related to the digital signalling performance.

Because emphasis is on microwave communication, the multipath effects of considerable importance for VHF, UHF traffic to mobile users have not been included except for the very special example of cochannel interference on a nonfading signal discussed in Part 3.

Frequency-division multiple access (FDMA) and time-division multiple-access (TDMA) techniques, including several variations, are described. Single channel per access (SCPC-FDMA) is discussed as a variation of FDMA. Typical frame formats, system performance, and frame-rate considerations are discussed for TDMA. Satellite-switched TDMA (SW-TDMA) is described as a high-efficiency variation of TDMA that employs a special satellite transponder and multibeam satellite antennas.

CHAPTER 6

SYNCHRONOUS SATELLITE COMMUNICATIONS

6-1 INTRODUCTION

This chapter presents the principles of a satellite communications system that employs satellites in synchronous orbit. We begin with a brief discussion of the satellite orbits and consider the satellite visibility and geometry as viewed from earth stations at different points on earth. The possible advantages of multiple-spot beam antennas over a single earth-coverage satellite antenna are discussed as they affect the communications system performance.

Satellite/earth terminal mutual interference effects must also be considered for a system of satellites and earth terminals. These effects include the interference caused by sidelobes if an antenna interferes with an adjacent satellite or earth terminal equipment. Other sources of interference include ground microwave relay links, sun transit effects, and intermodulation products generated in the transponder or earth terminal.

The selection of satellite operating frequency depends on many factors, including the size and gain of the antennas, bandwidth allocations, the effects of rainfall, atmospheric attenuation, and ionospheric scintillation on the communications link; and the effects of various sources of noise.

The chapter concludes with both single- and multiple-access calculations of the link signal-to-noise performance. Link calculations are given to show the fraction of satellite power necessary to establish a given digital communication link.

6-2 RELAY BY SYNCHRONOUS SATELLITES

The number and type of satellites to be used in a satellite relay network depends on the network coverage desired. Although three satellites in geostationary orbits (Fig. 6-1) can provide global coverage (360° at the equator), the desire to satisfy increased communications demand, and to provide on-orbit redundancy and greater coverage at the higher latitudes makes four or more satellites and closer spacing an obvious consideration for a global

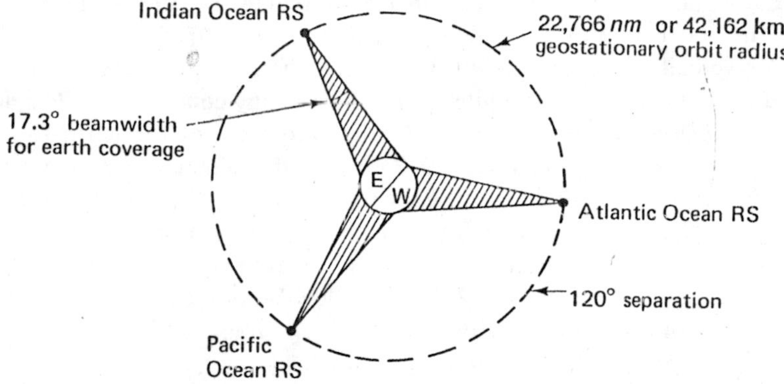

Fig. 6-1 Configuration of three relay satellites RS for global coverage. The orbit shown is calculated from the earth's center; that is, earth radius ($r_e \cong 3444$ nm) plus the distance from the earth's surface to the orbiting satellites (19,322 nm). The radius of the earth is approximately 3444 nm, or 6378.28 km. More precisely, the equatorial radius is 6378.165 km and the polar radius is 6356.785 km for a 21.4-km bulge at the equator; 1 nm = 6076.1 ft = 1.852 km.

system. Smaller numbers of satellites are required for domestic systems which must provide regional coverage of a nation or group of nations. However, these domestic links often must provide high satellite-effective-radiated-power (ERP) in order to permit the use of low-cost earth terminals and to provide high-density traffic.

There is also the possibility of using a larger number of moderately complex satellites each with narrow-beam antennas rather than a smaller number of more complex satellites. Consideration of initial cost, redundancy, replenishment, and on-orbit lifetime can influence this selection, and the earth coverage region of a given satellite configuration does not always determine the number of satellites in a given system.

Satellite Orbits

Before proceeding with the discussion of satellite geometry as related to the communication system performance, it is important to review the fundamentals of satellite orbits. The period of a satellite in an elliptic or circular orbit around the earth is

$$T = \frac{2\pi}{\sqrt{\mu}} a^{3/2} \quad \text{sec mean solar time} \tag{6-1}$$

where a is the semimajor axis (radius for a circular orbit) (see Fig. 6-1) and μ is the earth's gravitational parameter ($\mu = 1.407654 \times 10^{16}$ ft^3/sec^2). Satellite velocity for a circular orbit is $\sqrt{\mu/r}$ where r is the orbit radius. The period of a synchronous satellite is one sidereal day consisting of 24 sidereal hours. A *sidereal day* is defined as the time required for the earth to rotate once on its axis relative to the stars. This period is slightly less than a mean solar day (ordinary time) because in one solar day the earth makes one revolution plus an additional fraction of a revolution since it also travels 1/365th of its way in revolving about the sun. A sidereal day is thus less than a 24-hr solar day and consists of $23^h\ 56^m\ 04.009054^s$ of mean solar time [Bates, 1971]. Thus for a synchronous orbit, the value of a using (6-1) is as shown in Fig. 6-1,

$$a = \left(\frac{\sqrt{\mu}T}{2\pi}\right)^{2/3} = 22{,}766 \text{ nm} \quad \text{or} \quad 42{,}162 \text{ km}$$

Station Keeping

To maintain this synchronous orbit, the satellite must periodically make east-west position corrections or it will drift in longitude. North-south position corrections are also useful to prevent drift of the orbit inclination. If both north-south and east-west position corrections are made using satellite thrusters, then the earth terminal antenna potentially can be made less costly by avoiding the requirements for an automatically tracking (autotrack) antenna subsystem. These satellite position corrections are typically made by ground station commands to the satellite.

The causes of drift in orbital inclination of the satellite are primarily the gravitational attraction of the moon and the sun. The effect of the moon exceeds that of the sun by a factor of approximately three. These forces induce a daily oscillation of the orbit radius along with a more significant cumulative variation of the inclination plane. The mean rate of change of the inclination in the 1970–1980 time frame is 0.85°/year [Isley and Duck, 1974].

and if left uncorrected would build up to a maximum of 14.67° from an initial 0° inclination in 26.6 years. The inclination angle would then decrease back to 0° in a similar time period. The exact rates of change depend on the inclination of the lunar orbit with respect to the earth's equatorial plane.

The satellite can be injected into orbit with a small initial inclination to minimize the effects of inclination drift. If the ascending line-of-nodes (the line formed by the intersection of the orbit plane with the equatorial plane) is properly selected, the orbit inclination can be made to drift to zero in the first half of the satellite lifetime and to increase to its initial value during the second half. This approach is satisfactory for 5-year missions where the inclination angle tolerance is 2° to 3°.

If more accurate tolerance on inclination angle is required, for example, $<2°$, then some form of north-south station keeping is required. This station keeping is typically done with periodic thrusting by small gas jets. Thus the cost of station keeping is largely the cost of launching the extra weight of the fuel needed during the satellite operating lifetime. For a 3,000-lb satellite the total impulse required would exceed 15,000 lb-sec/year. Figure 6-2 shows the typical drift in inclination angle with and without north-south station keeping.

East-west station keeping is required to keep the satellite's longitude within prescribed bounds. Without this station keeping there is a mean daily drift in longitude and eccentricity. These accelerations are created by the J_{22} harmonic in the earth's gravitational field. Figure 6-3 shows the required

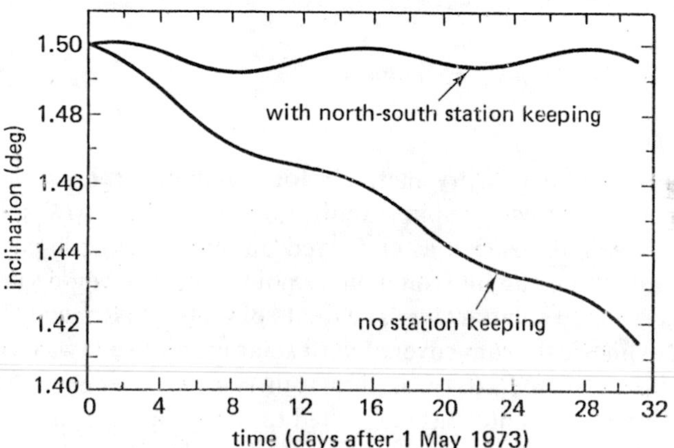

Fig. 6-2 Geosynchronous satellite monthly inclination trend with and without north-south station keeping. The small oscillation is caused by the varying solar/lunar attractions. [Isley and Duck article, reprinted from *Communications Satellite Technology*, P. Bargellini, ed., by permission of the M. I. T. Press, Cambridge, MA, © 1974. All rights reserved.]

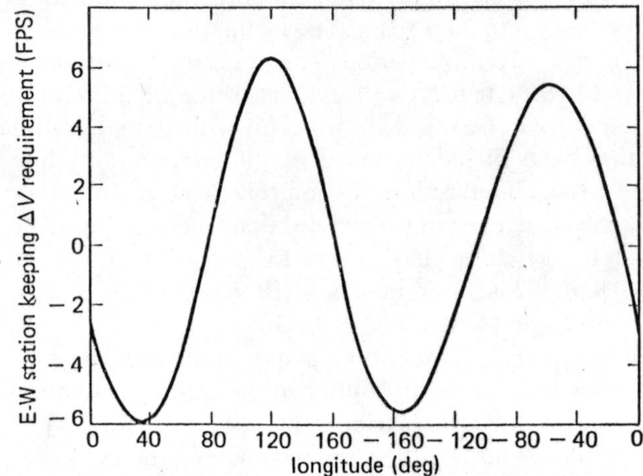

Fig. 6-3 Geosynchronous satellite east-west velocity corrections in feet per second [Isley and Duck article, reprinted from *Communications Satellite Technology*, P. Bargellini ed., by permission of the M. I. T. Press, Cambridge, MA, © 1974. All rights reserved.]

Δv (velocity change) required for east-west station keeping as a function of satellite longitude.

Satellite Attitude Stabilization and Power Generation

There are two primary methods for stabilizing the spacecraft attitude with respect to the earth: spin stabilization as used on dual-spin spacecraft, and three-axis stabilization as employed on body-oriented spacecraft. These methods allow the satellite antenna to point to earth while gathering in the sun's energy in the solar arrays. The dual-spin satellite design (Fig. 6-4) uses a spinning cylindrical drum covered with solar cells. A cutaway view of a dual-spin satellite, the Intelsat IV, is shown in Fig. 6-5. A despin platform contains earth-pointing antennas connected to the spinning section by a mechanical despin mechanism, which provides the relative motion and transfers signals and power. If the solar arrays are mounted only on the sides of the cylinder, output power decreases relative to its maximum by 8.3 percent at the summer and winter solstices when the sun angle is 23.5°.

The three-axis configuration, on the other hand, stabilizes the entire spacecraft and can have all solar cells mounted on planar or nearly planar paddles that are oriented to the sun (Fig. 6-6). The maximum sun power intercepted by this type of three-axis satellite with two planar solar arrays,

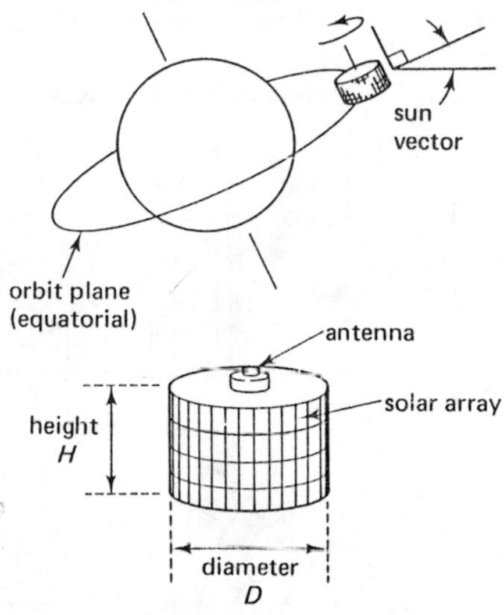

Fig. 6-4 Orbital configuration for a cylindrical spinning satellite [After Berks and Luft 1971, 265]

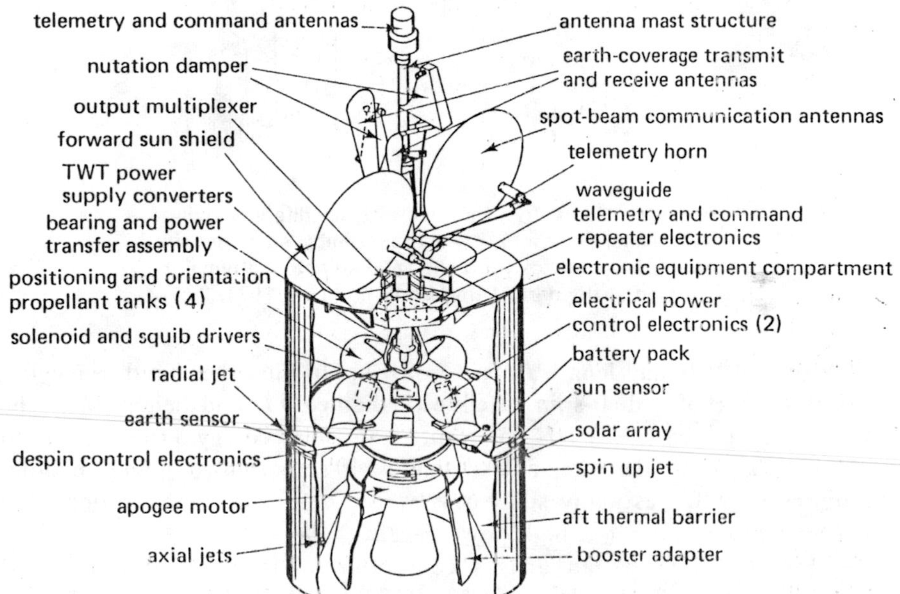

Fig. 6-5 Cutaway view of the INTELSAT IV communications satellite. The satellite has an on-station weight of 1600 lb. The two spot beam antennas are 30 in. in diameter (4.5° beam width at C-band) (Courtesy of COMSAT Corp.)

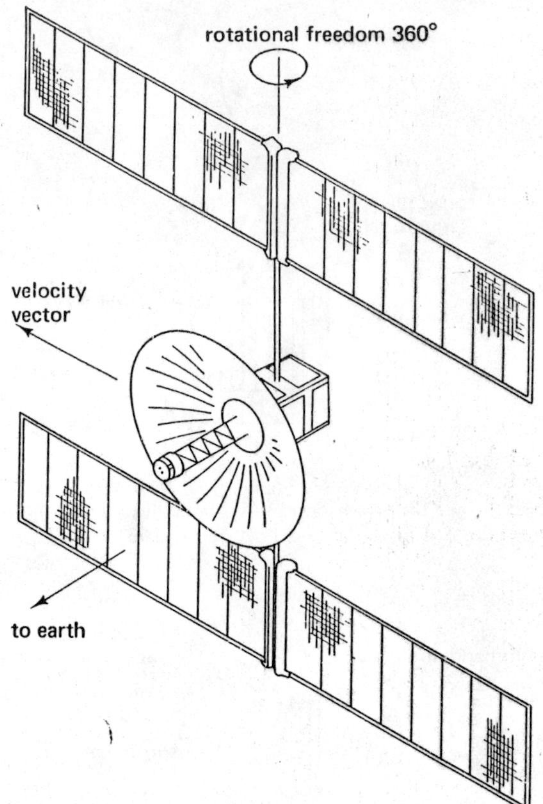

Fig. 6-6 Geometry of a three-axis satellite employing two planar solar arrays which are gimballed to remain pointed at the sun while the satellite antenna remains pointed at the earth [After Berks and Luft, 1971, 270]

each of width W and length L, is $A = 2WL$. On the other hand, the cylindrical surface of a dual-spin satellite of diameter D and height H of the same surface area $A = \pi DH$ intercepts only solar energy proportional to $DH = A/\pi$. Thus the three-axis design with sun-oriented planar paddles requires only $1/\pi$ as many solar cells as the dual-spin configuration, and hence is more efficient at high power levels.

Figure 6-7 shows an artist's conception of the three-axis stabilized Communications Technology Satellite (CTS). Each of the solar paddles, or sails, is 21 ft $\times$ 4 ft in dimension. The solar array delivers a power of over 1000 watts. Figure 6-8 shows the yaw, pitch, and roll axes of a body-oriented satellite such as the CTS. Each axis must have its separate control system and thrusters or momentum wheels.

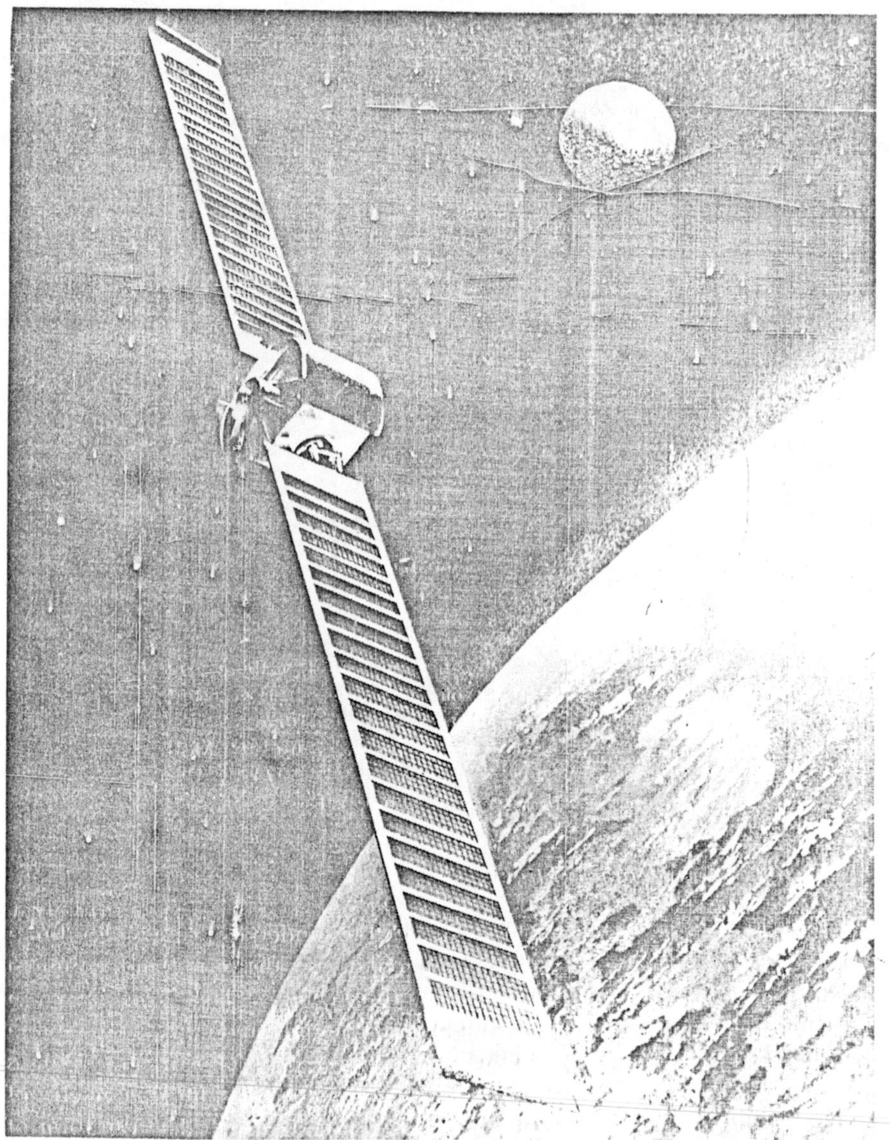

Fig. 6-7 Three-axis Stabilized Communications Technology Satellite (CTS) employing planar solar arrays (Reproduced through courtesy of National Aeronautics and Space Administration)

Power capabilities of the solar arrays for these spacecraft range from a maximum of approximately 1 kw for spinning satellites to more than 10 kw for three-axis stabilized satellites with sun-oriented solar arrays.

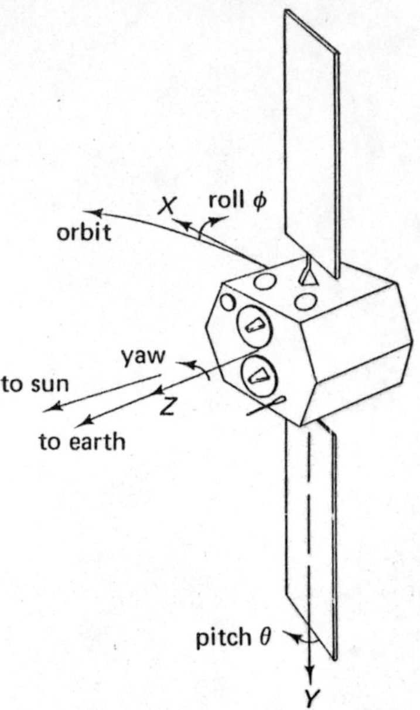

Fig. 6-8 Attitude and orbit control of a three-axis (roll, pitch, and yaw) stabilized spacecraft [Reprinted from Kaplan article in *Communications Satellite Technology*, P. L. Bargellini, ed., by permission of the M. I. T. Press, Cambridge, MA. Copyright © 1974 by The Massachusetts Institute of Technology]

Satellite-Sun Outage Effects

SATELLITE ECLIPSE All synchronous satellites undergo spring and autumn eclipses part of each day for a 46-day interval during the vernal and autumnal equinoxes. Batteries supply power during these eclipses. The eclipse varies in duration from approximately 10 min at the beginning and end of eclipse to a maximum of approximately 72 min at the equinox. The eclipse begins 23 days prior to equinox and ends 23 days after equinox. The simplified geometry of the geostationary satellite at equinox is shown in Fig. 6-9. The duration of eclipse per day versus the day of the year is shown in Fig. 6-10.

SUN TRANSIT OUTAGE A more serious but briefer interruption is the *sun transit outage*, which occurs when the pointing angles from a given earth station to the satellite and to the sun are so near coincidence that both are within the earth terminal antenna beamwidth (Fig. 6-11). The shadow of the satellite is then falling near the earth terminal. An earth station perceives the sun as a disc of extreme thermal noise 29 minutes or 0.48° in diameter with a minimum noise temperature for a mean quiet sun of 25,000°K for a

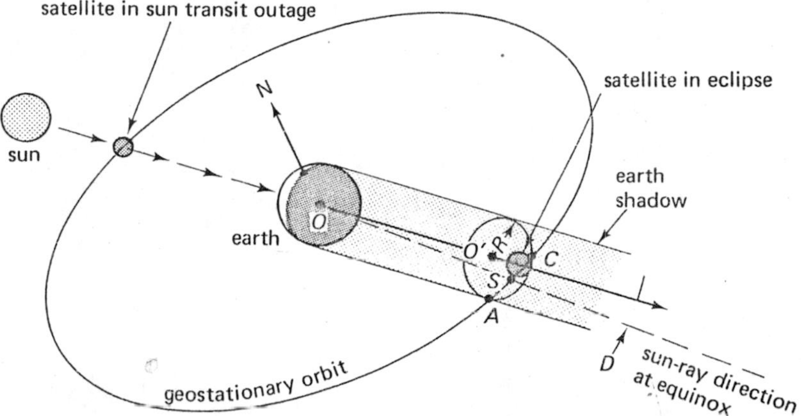

Fig. 6-9 Geometry of satellite eclipses and sun transit outage for geostationary orbit at equinox

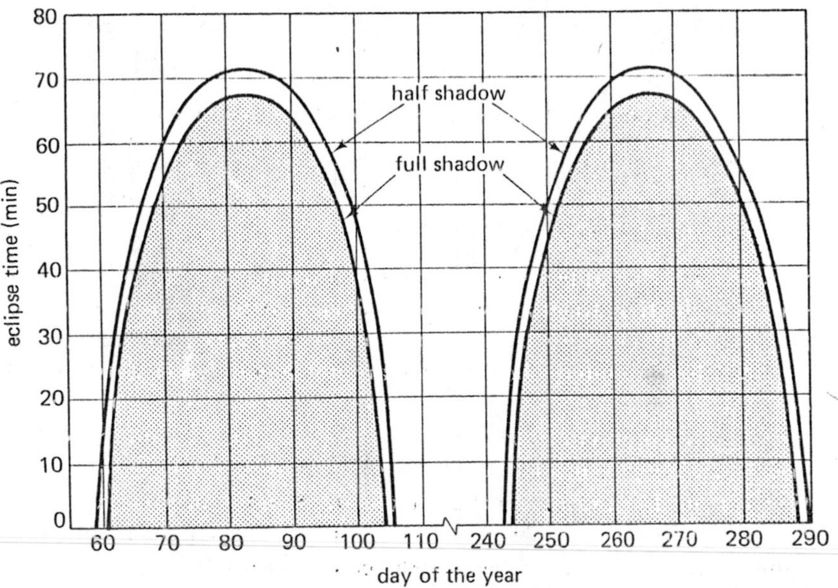

Fig. 6-10 Satellite eclipse time as a function of the current day of the year

single polarization [Hogg, 1968]. For large antennas of diameter 30 m at 4GHz the minimum acceptable angular displacement α of an earth terminal boresight from the sun center is approximately equal to 0.6° and is equal to the angle subtended by the sun. The sun thus corresponds to a noise disc of effective diameter d at synchronous satellite range where $d = 2S$

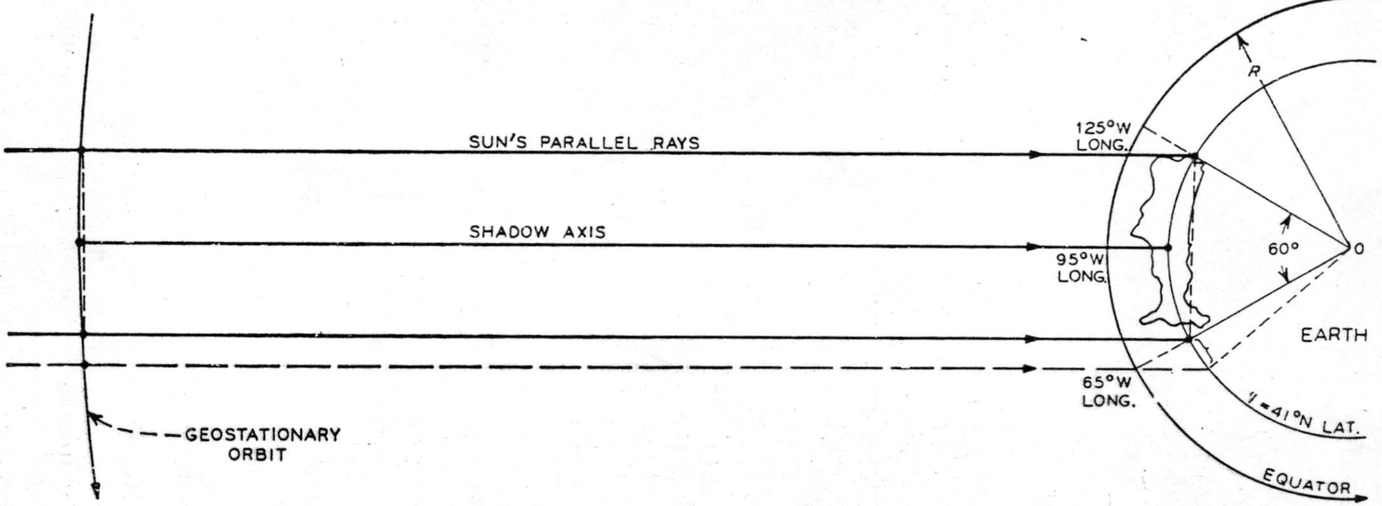

Fig. 6-11 Geometry describing motion of the sun transit outage region across a model of the contiguous United States [From Lundgren, 1970]

sin α = 775 km. The quantity $S = 37 \times 10^3$ km is the slant range to the satellite. For smaller diameter antennas the sun transit effect occurs over a wider region because the antenna beamwidth is wider, and the sun disc is then within the beamwidth and major sidelobes longer. For example, if the antenna diameter is 10 m, the required angular offset* is $\alpha = 1°$ at 4 GHz.

The peak outage time T_0 is the time for the satellite shadow to traverse the earth terminal or for the earth and synchronous satellite to move 2α (Fig. 6-10). Thus, if $2\alpha = 2°$, the peak outage time is

$$T_0 = \frac{2\alpha}{360°} 24 \times 60 = 8 \text{ min} \qquad 2\alpha = 2° \tag{6-2}$$

This outage occurs for approximately six days twice yearly at apparent noon at the satellite longitude. Lundgren [1970, 1943–1972] has described diversity arrangements of phased and slightly inclined orbit satellites, which avoid simultaneous outages by using a pair of satellites. However, these diversity satellites require earth terminal antenna or feed switching and satellite handover to avoid the outage. (It is possible to use two feeds on the same antenna for closely spaced satellites to permit two-satellite operation with a single earth terminal.)

Figure 6-11 illustrates the geometry of sun transit outage where the apparent declination of the sun is 6° 38′ relative to the equator. The north latitude of the satellite's shadow at the time of sun transit is 41° for this particular day in March. In general, the declination of the sun D and the north latitude of sun shadow at the time of sun transit φ are related by

$$\tan D = \frac{-\sin \varphi}{(1 - \cos \varphi) + h/r_e} \tag{6-3}$$

where h is the satellite altitude $h = 35{,}784$ km and r_e is the earth radius $r_e = 6378$ km.

The sun transit shadow moves off the earth at the largest declination angle (the latitude of the Tropic of Cancer), which is 23.5°. However, the meaningful sun transit outage cannot, of course, exceed the highest latitude of satellite visibility or 81°.

Figure 6-12 illustrates the path of the elliptical region of the sun transit outage (narrow-beam antennas) for several consecutive days in March 1970.

Orbit Plane Inclination Effects

Geostationary orbits are advantageous because they simplify the ground-station tracking requirements and avoid the handover problem—that is, transferral of link relay from satellite-to-satellite as one satellite goes out of

*Based on a 3 dB increase in noise temperature relative to 200°K.

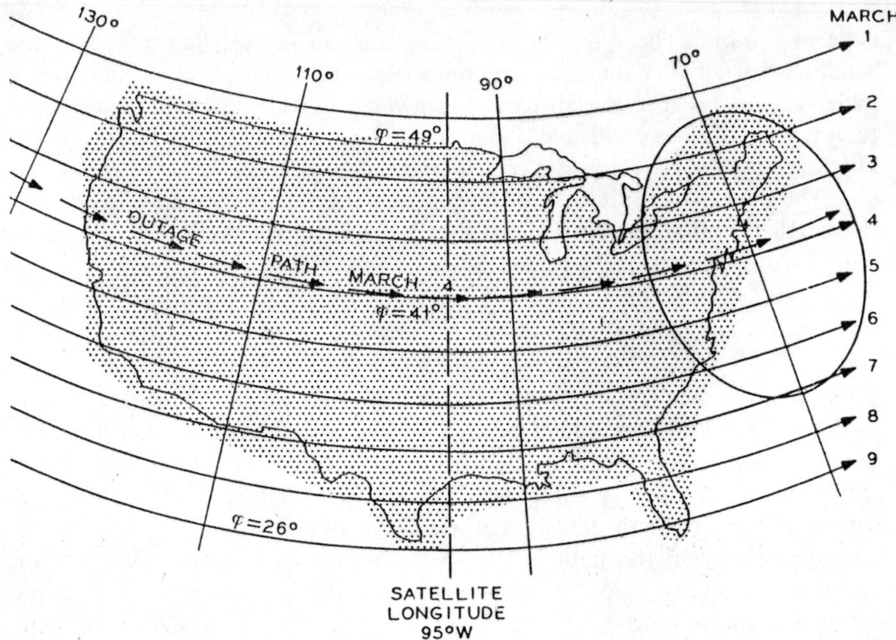

Fig. 6-12 Approximate paths of sun-transit outages for a geostationary satellite [From Lundgren. Reprinted with permission from *The Bell System* Technical Journal. Copyright 1970, The American Telephone and Telegraph Company.]

view and another appears in view. They also eliminate the possibility of two satellites drifting together within a single earth terminal beamwidth. A further advantage is a nearly constant range and very small Doppler shift.

As already stated, the satellites are not precisely geostationary. Unless corrected, the orbital inclination plane drifts ($\approx 0.86°$/year) as a result of lunar and solar gravitational attraction, and the orbits do not remain perfectly circular. We now consider how these orbital effects relate to the communication system. For example, orbital inclination and nonzero eccentricity cause the round-trip group delay to change slightly from its nominal 0.25-sec value which, in turn, can degrade the communication link.

An inclination of the satellite orbit causes the subsatellite point (directly below the satellite) to move in a figure **8** pattern. The traces of the subsatellite point* for 24-hr circular orbits are shown in Fig. 6-13 for various orbits from a stationary zero inclination orbit where the locus is simply a point, to slightly inclined orbits where the trace is a figure **8** with a 24-hr period. Note that the dimension of the figure **8** pattern increases with the inclination. For an inclination angle $\mathscr{I}$ rad, the width of the figure **8** is $\mathscr{I}^2/4$, and the peak-peak latitude variation is $2\mathscr{I}$ for circular orbits and $\mathscr{I} \ll 1$. The expression for the

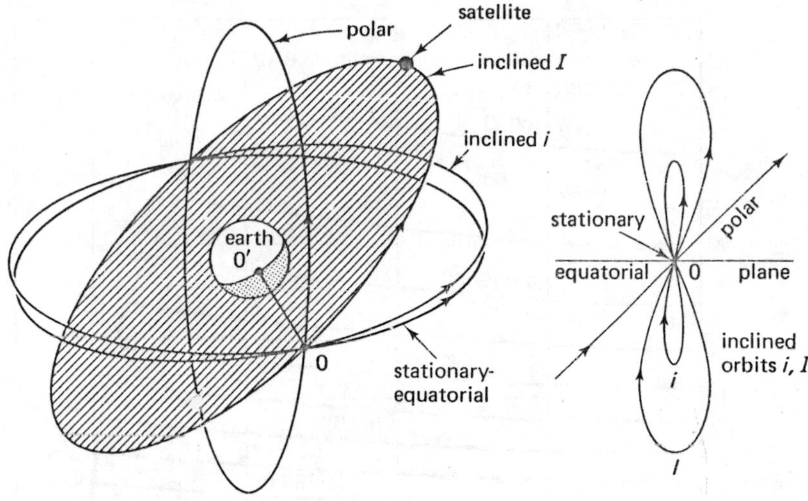

Fig. 6-13 Earth synchronous orbits and the figure 8 patterns of the subsatellite point

figure **8** in terms of the peak latitude and longitude excursions of the sub-satellite point, and normalized time of day, $\theta \triangleq 2\pi t/24$, (sidereal) is

$$\phi_{\text{lat}} = \sin^{-1}(\sin \mathscr{I} \sin \theta) \qquad \phi_{\text{long}} = \sin^{-1}\frac{\cos \mathscr{I} \sin \theta}{\sqrt{1 - \sin^2 \mathscr{I} \sin^2 \theta}} - \theta \tag{6-4}$$

for an inclination angle $\mathscr{I}$ and t in hours. Note that the orbit can be purposely inclined to give satellite visibility to the north and south poles or simply to provide greater coverage to higher latitudes. Three or more satellites in different inclination planes are then employed in different phases of the figure **8**.

Figure 6-14 shows the peak variation in range and range rate as a function of the earth station latitude and the relative longitude between the earth station and the mean satellite longitude. The example shown is for a satellite inclination of $\mathscr{I} = 1°$. The variation with time of the satellite range is

$$r = r_{\text{mean}} + A_i \sin \frac{2\pi t}{T} \tag{6-5}$$

where r_{mean} is the mean range, t is in hours, and T is one sidereal day. As one expects, the peak delay variation occurs at the maximum latitude of visibility $\mathscr{I} + 81°$ and at zero offset in longitude. At this point the peak variation is 111.3 km or 371 μsec when expressed in terms of peak variation in group delay.

The eccentricity, e, of an equatorial elliptic orbit causes the distance to

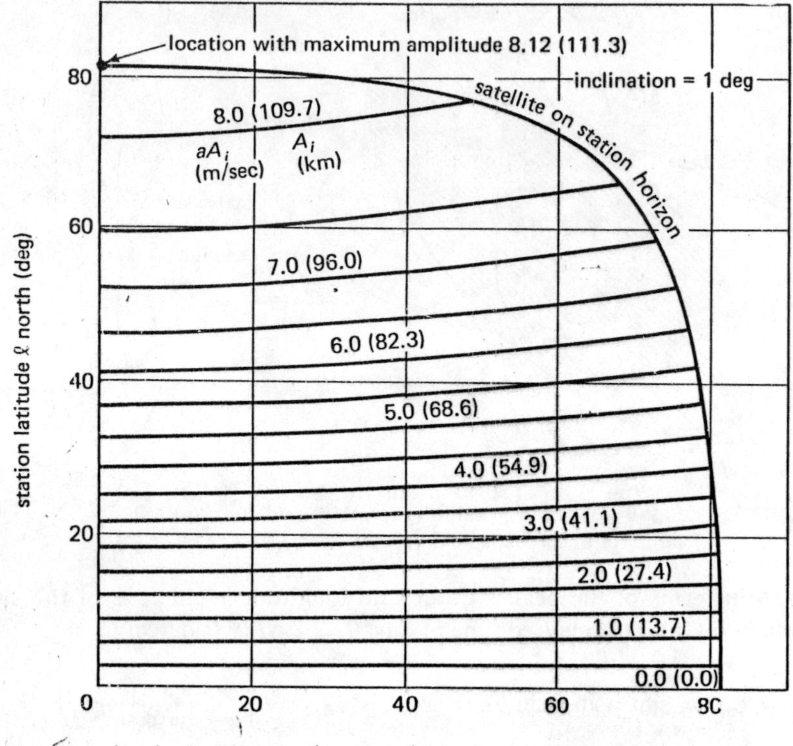

Fig. 6-14 Amplitudes for inclination part of variations, aA_i for range-rate and A_i for range in (6-5) [from Siabinski, 1974]. For south latitudes, take the negative of the quantity for the corresponding north latitude. For other inclinations $\mathcal{I}$, multiply quantity from graph by $\mathcal{I}$ in degrees.

an observer on the equator to vary by

$$\pm e(r_e + h) = e\ (22{,}766\ \text{nm})$$

The importance of this delay variation is examined in Chap. 10 when clock synchronization, buffers, and timing systems are examined for use in time-division multiple access (TDMA).

Inclined elliptic orbits, like that of the Russian Molniya satellite, can provide coverage to the higher latitude region of the earth at a higher earth-station elevation angle. The orbit can be designed so that the satellite is over a given region for a relatively long fraction of its period to minimize the handover problem. The Molniya satellite has a 65° inclination, 21,400-nm apogee, 270-nm perigee, and a subsynchronous 12-hr orbit period.

6-3 SATELLITE COMMUNICATION SUBSYSTEM

Earth-coverage satellite antennas from synchronous altitudes have a 17.3° beamwidth. The satellite is in view of ground stations on the equator which are offset ±81° in longitude relative to the subsatellite point. The satellite is similarly in view of ground stations offset by ±81° in latitude if they are directly north or south of the subsatellite point. Typical coverage patterns for various elevation angles are shown in Figs. 6-15 and 6-16. As satellite

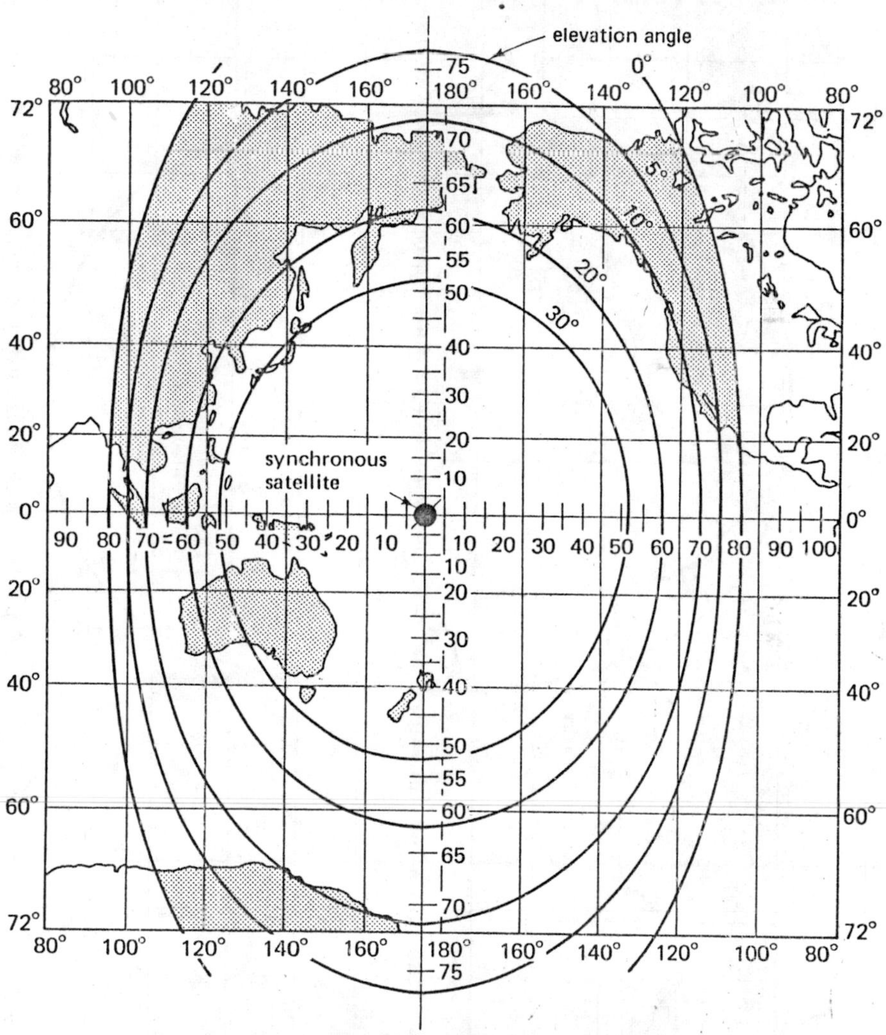

Fig. 6-15 Synchronous-satellite coverage pattern versus ground-station elevation angle

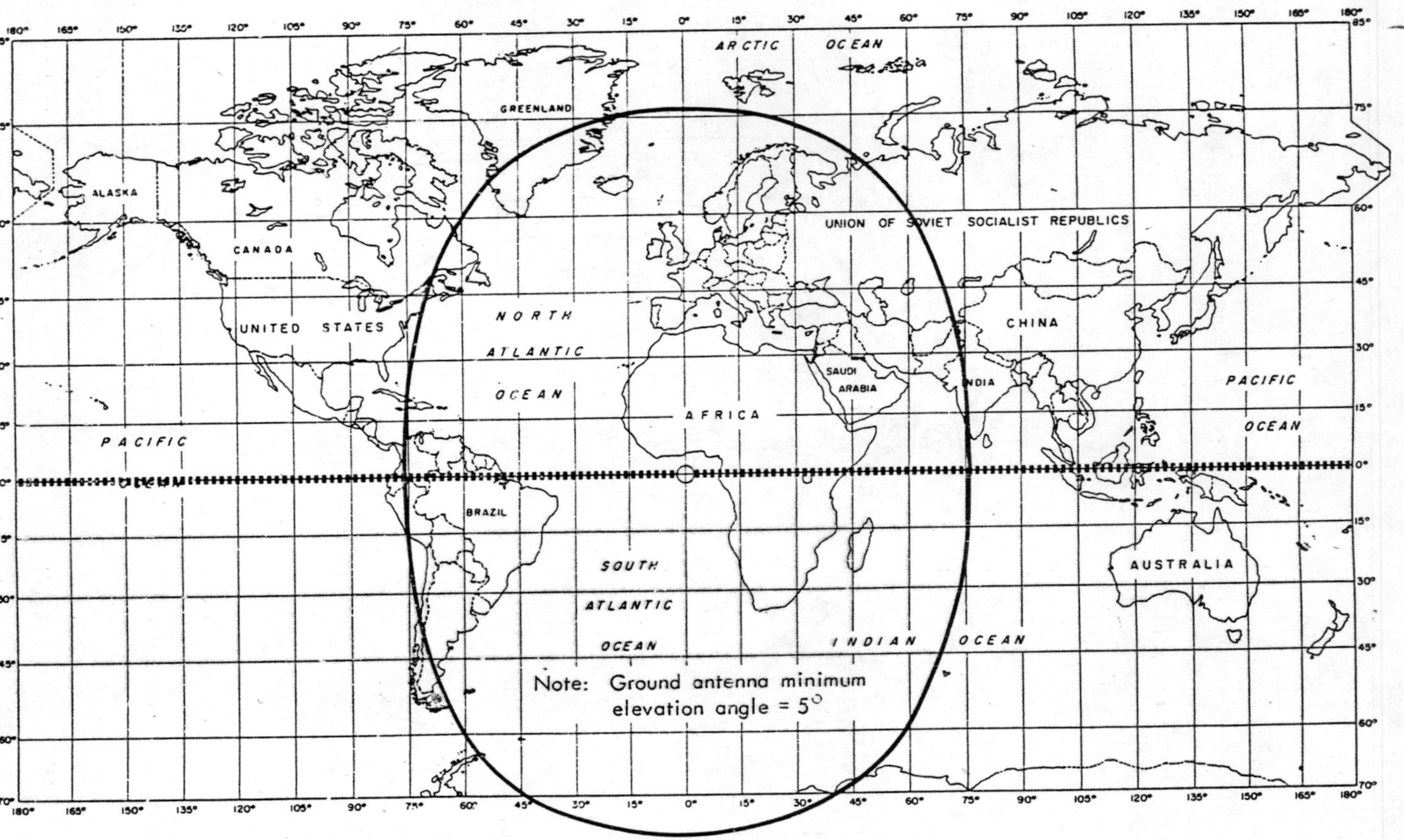

Fig. 6-16 Synchronous-satellite coverage pattern. A 5° elevation angle from the 19,310 nm synchronous altitude covers 25.8% of the earth's surface.

spacing decreases below 120°, higher elevation angles can be used on the equator and a higher minimum latitude coverage can be achieved.

Figure 6-17 shows a photograph of the earth as viewed from synchronous altitude. The photograph was taken by the NASA Synchronous Meteorological Satellite.

Fig. 6-17 View of the earth as observed from a synchronous satellite (Synchronous Meteorological Satellite). The subsatellite point is just west of South America. (Courtesy NASA Goddard Space Flight Center)

A useful nomogram giving earth-station elevation angles versus earth station latitude and relative ground-site longitude (Fig. 6-18) has been developed by Smith [1972, 394]. The latitude of the ground site may be either

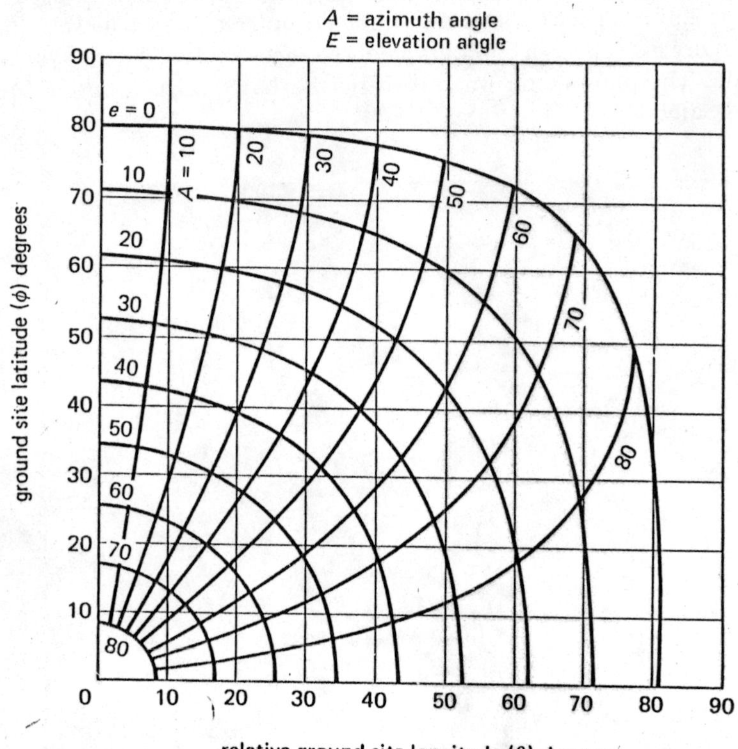

Fig. 6-18 Azimuth and elevation look angles to geostationary satellites [From F. L. Smith, 1972]

north or south, and the longitude may be either east or west of the subsatellite point. In Fig. 6-18, the azimuth angle is $A = 0°$ along the vertical axis, and $A = 90°$ along the horizontal axis. The satellite is at 0° elevation angle at 81° latitude for 0° relative longitude. The range (Fig. 6-19) to the satellite varies from 19,322 nm at 90° elevation angle and 20,400 nm at 45° elevation to 22,600 nm at 0° elevation angle [Yeh, 1972, 252], and path loss is easily computed versus elevation angle (see 17-3). The distance (slant range) r to the satellite is given by

$$r = [(h + r_e)^2 + r_e^2 - 2r_e(h + r_e)\cos\theta]^{1/2} \tag{6-6}$$

where r_e = the earth's radius of 3444 nm (6378 km); h = satellite altitude 19,322 nm (35,784 km); and $r_e + h$ = 22,766 nm (42,162 km).

Antennas with higher gain and narrower beamwidth than earth-coverage provide a larger satellite radiated power (ERP) for the same satellite power for limited coverage applications. These spot beams are especially useful for

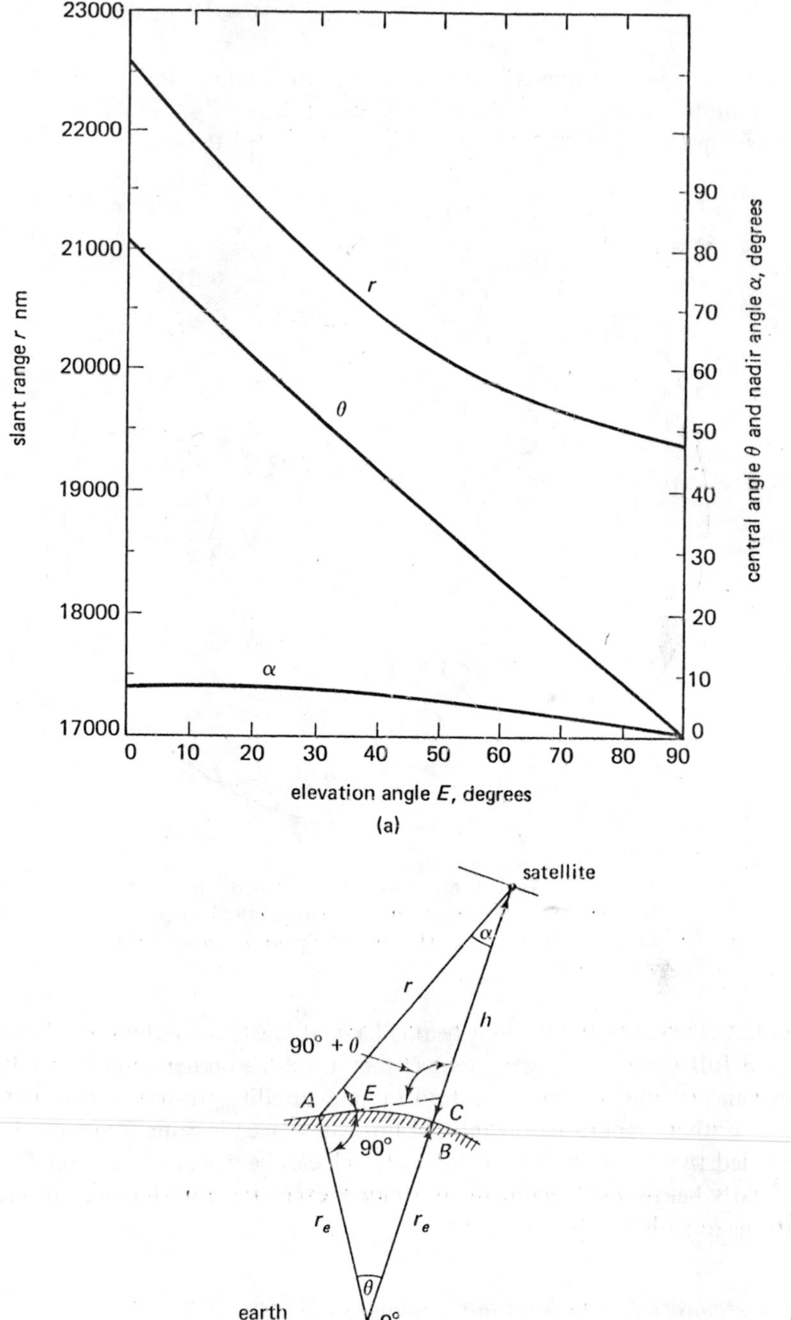

Fig. 6-19 Slant range, central, nadir, and elevation angles (geostationary orbit) [From Yeh, 1972]

r_e = earth's radius: 3444 nm (6378 km) h = satellite height 19,322 nm (35,784 km)
$r_e + h$ = 22,766 nm (42,162 km) r = slant range E = elevation angle
θ = central angle α = nadir angle

"trunking" applications where heavy communications loads exist between major nodes. Figure 6-20 shows a typical coverage pattern for three spot beams from a single multibeam antenna [Dion and Ricardi, 1971, 252–262].

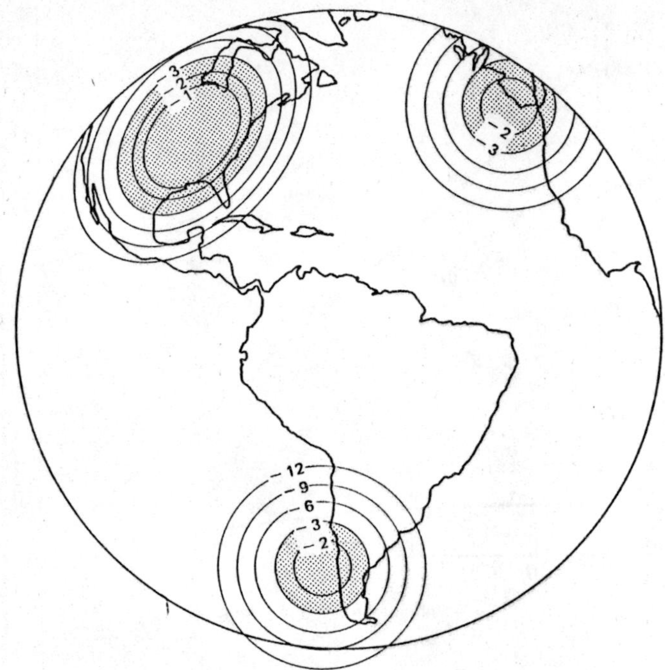

Fig. 6-20 Typical coverage patterns in dB for three-beam multibeam antenna having a minimum half-power beamwidth of ≈ 3° [From Dion and Ricardi, 1971, 252–262]

Limited-coverage satellite spot beams have at least two potential advantages over a full earth-coverage antenna: they produce higher satellite ERP, and they can provide spatial separation of one satellite/ground station network from another. Separate downlink beams can share the same frequency bands, provided that the antenna sidelobes, which can be designed to range from 20 to 30 dB below peak gain, do not cause excessive interference at ground stations outside the beam.

Satellite Transponders and Frequency Bands

Bandwidth availability for an individual satellite is limited to 500 MHz in the commercial C-band frequency region (5.925–6.425 GHz uplink and 3.7–4.2 GHz downlink) and 500 MHz for the military X-band (7.9–8.4 GHz uplink and 7.250–7.750 GHz downlink) satellite. Larger bandwidths are potentially available in the K-band and higher millimeter wave-frequency regions. Table 6-1 lists the principal satellite microwave frequency allocations. In addition, there is UHF traffic band in a narrower frequency band (225–400 MHz).

Table 6-1 PRINCIPAL COMMUNICATION SATELLITE RF BANDS

Band	*Bandwidth*	*Transmit (Downlink)*		*Receive (Uplink)*
6 and 4 GHz	500 MHz	3.7 to 4.2 GHz		5.925 to 6.425 GHz
14 and 12 GHz	250 MHz to 500 MHz	DOMESTIC	11.7 to 12.2 GHz	14.0 to 14.5 GHz
		INTERNATIONAL	10.95 to 11.2 GHz and 11.45 to 11.7 GHz	
29 and 19 GHz	2.5 GHz to 3.5 GHz	17.7 to 21.2 GHz		27.5 to 31.0 GHz
2.5 GHz BROADCAST	35 MHz	2500 to 2535 MHz		2655 to 2690 MHz
1.5GHz AERONAUTICAL	15 MHz	1543.5 to 1558.5		1645 to 1660 MHz
7 and 8 GHz MILITARY	500 MHz (50 MHz exclusive)	7.250 to 7.750 GHz		7.90 to 8.40 GHz
1.5 GHz MARITIME	7.5 MHz	1535 to 1542.5 MHz		1635 to 1644 MHz

Limited-frequency reuse is possible by transmitting orthogonal polarizations. For example, one can transmit right-hand circular and left-hand circular polarizations in the same frequency bands. At least 20-30 dB of polarization discrimination is achievable. This discrimination allows relatively low cross-channel interference for the two polarizations and is quite adequate for many digital modulation techniques. However, atmospheric and ionospheric effects can create cross polarization disturbances to reduce this isolation. [See Chap. 7.]

Figure 6-21 shows a typical satellite transponder configuration employing a frequency-channelized transponder and two downlink beams. Redundant receivers accept dual polarization uplink signals in frequency band f_u. Six different frequency channels in this uplink band are separated by means of frequency-selective filters in each of the input frequency multiplexers transmitted to the appropriate output frequency band, and are then fed to different

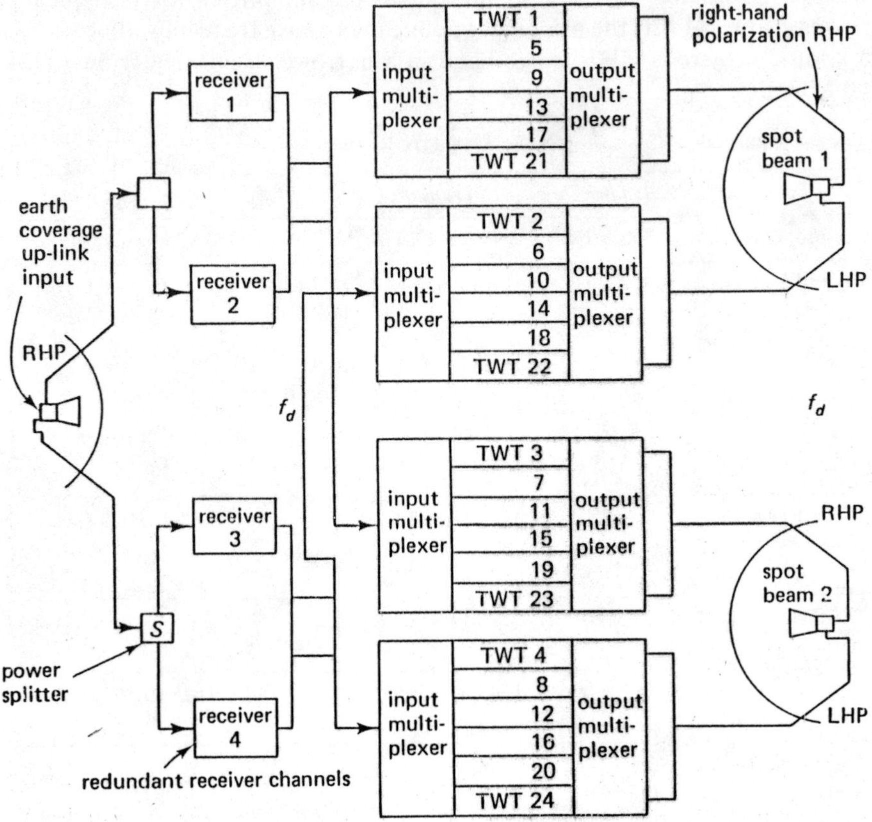

Fig. 6-21 Multichannel, multibeam satellite transponder employing frequency reuse. Orthogonal right-hand and left-hand polarizations are used on downlink.

TWT amplifiers. The outputs of different TWT amplifiers are combined in turn by the output multiplexer and are fed to the downlink spot-beam antennas at frequency band f_d. The input multiplexers shown consist of an array of bandpass filters and frequency converters, which both subdivide the input frequency band into the desired frequency channels and translate these channels to the desired output frequencies.

6-4 SATELLITE/EARTH TERMINAL MUTUAL INTERFERENCE

A network of ground stations and satellites operating in the same frequency band must be carefully coordinated to minimize interference effects. The most obvious example is a system of two satellites operating in the beamwidth or major sidelobes of a single ground station. Figure 6-22 shows a

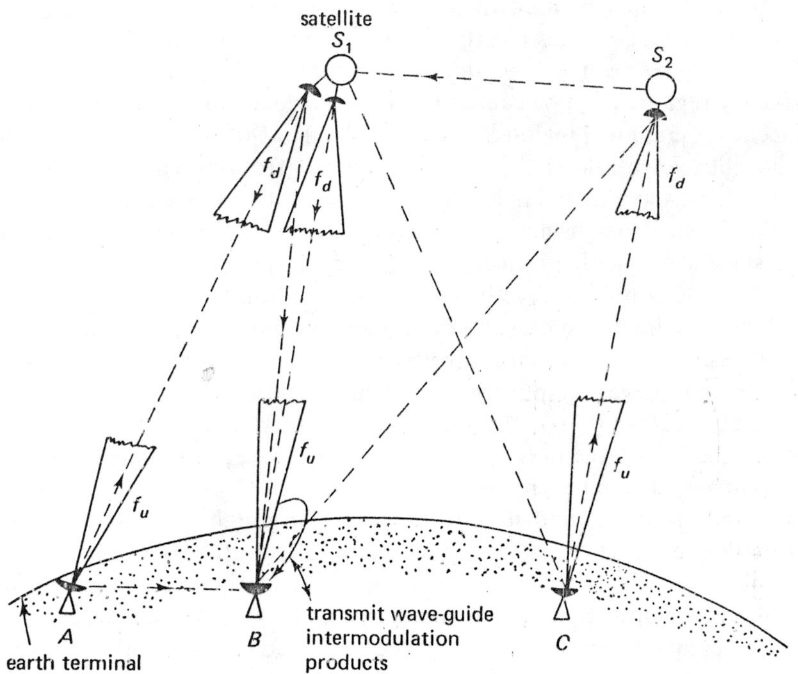

Fig. 6-22 Typical causes of potential satellite/earth station interferences

more complete interference model for satellites that employ spot-beam downlink antennas and earth-coverage uplink antennas. Multiple ground stations operate with narrow-beam antennas. As before, the uplink frequency band is centered at f_u and the downlink frequency at f_d. Interference to terrestrial radio-relay systems from the satellite downlink and vice versa is also of concern [May and Pagones, 1971, 81–102]. The International Radio Consultive Committee (CCIR) has recommended a limit on power flux density of +12 dBw Effective Isotropic Radiated Power (EIRP) in any 4-kHz band in the 3700–4200-MHz downlink band. Frequency dispersal waveforms are often used to spread the energy over a frequency band much wider than 4 kHz even when all data modulation is removed [Golding and Ball, 1973, 24–31].

Interference to the earth-station–satellite link B–S_1 in the uplink frequency region f_u can come from earth station A operating with the same satellite or earth station C operating with a nearby satellite. The adjacent satellite itself is a potential source of interference. In the frequency plan of Fig. 6-21, this interference is out-of-band and hence is usually completely negligible. (In principle, a frequency plan could have the downlink of one satellite in the same band as the uplink of another, in which case low sidelobe antennas must be used to minimize satellite-satellite interference. The spacing between satellites clearly has a strong effect on interference.

Interference to the satellite–earth-station link S_1–B in the downlink frequency region f_d can be caused by (1) spot-beam transmissions directed to adjacent geographical region S_1–A or S_2–C, (2) out-of-band interference from the satellite downlink earth coverage antennas from either S_1 or S_2, or (3) ground microwave or radar Radio Frequency Interference (RFI).

One must also consider the possibility of a direct beam from a microwave relay station on earth producing a grazing ray [Lundgren and May, 1969, 338–342], which points directly at a communication satellite.

Potential interference sources cannot be ignored merely because the main transmission from that source is out of band. Intermodulation products generated in power amplifiers can create strong interference out of the transmit band. One potential source of strong interference is the intermodulation product generated in transmit waveguide or feed, or passive receive-filter nonlinearities. These intermodulation products enter directly into the receiver of the very same ground-station receivers. Although the waveguide intermodulation products* may be extremely small, -100 dB compared to the transmit signal, the received signal is often sufficiently small that these intermodulation components may be significant and possibly even larger than the received signal if the terminal waveguides and filters are not properly designed.

6-5 ANTENNA SIDELOBE INTERFERENCE

Power from the sidelobes of an antenna aimed at one satellite can cause interference power to enter the receiver of an adjacent satellite. The interference level depends strongly on the sidelobe patterns of the antennas, the power levels, and the frequency bands utilized. The sidelobes depend in turn on the shape of the antenna, the illumination pattern on the reflector, and the rms surface error ϵ and surface error correlation distance C. For a parabolic reflector with a uniform illumination pattern, the amplitude gain G varies with off-axis angle θ as

$$G(\theta) \cong e^{-\delta^2} \frac{J_1(\pi D \sin\theta)}{\pi D \sin\theta} + e^{-\delta^2} \frac{2\pi C}{\lambda} \exp\left[-\left(\frac{2\pi c}{\lambda}\right)^2 \sin^2\theta\right] \qquad (6\text{-}7)$$

for an antenna diameter D, $\delta \triangleq 4\pi\epsilon/\lambda \ll 1$, and a signal wavelength λ [Zucker, 1968, 1637–1651; Ruze, 1966, 633–640; Hunt and Reinhardt, 1971,

*A waveguide or antenna feed is generally considered a linear device. This assumption is not necessarily valid, however, because metal/oxide layers or other metal-metal joints in waveguide can act in a nonlinear manner.

118–128]. The antenna gain in the absence of surface deviations is $J_1(x)/x$. The 3-dB beamwidth is 1.02 $\lambda/D = 58.4°/(D/\lambda)$. The second term in (6-7) represents scattered energy. Thus, for small surface deviations, the sidelobe levels fall off as $J_1(\alpha\theta)/\alpha\theta \sim 1/\theta\sqrt{\theta}$, or the power falls off as $1/\theta^3$. Typically, the sidelobes fall till their envelope reaches a point roughly 8 to 13 dB below isotropic.

Figure 6-23 shows the field gain pattern for a 64-ft dish operating at 7.5 GHz with first-order sidelobes at −17 dB. In practice, the antenna blockage and imperfect illumination often build up the first-order sidelobes to −13 dB.

For ground microwave application, parabolic antennas have been

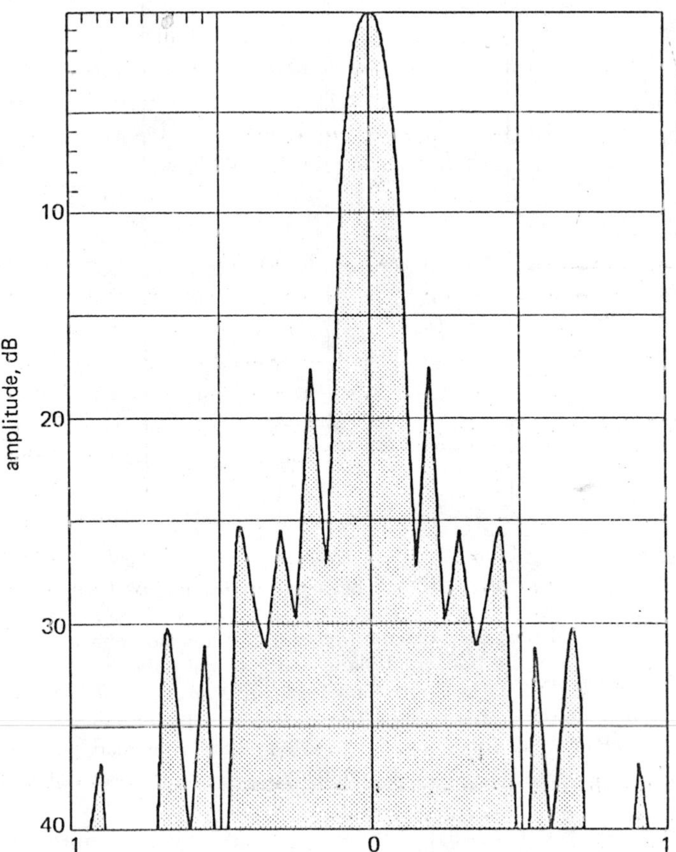

Fig. 6-23 Calculated radiation pattern for a 64-ft. dish at 7.5 GHz with a uniform illumination pattern and an rms surface deviation of 0.016 in. The peak gain for 70% efficiency is 62 dB. The first-order sidelobes are down approximately 17 dB with respect to peak gain.

developed with small enclosures that reduce substantially the sidelobe levels [Crawford and Turrin, 1969, 1605–1622]. Figure 6-24 shows the measured antenna pattern for an antenna operating at 11.2 GHz using a 30-in.-diameter parabolic reflector. Note that the maximum sidelobes here are at 90° or more off the beam axis in azimuth and are more than 34 dB below isotropic. A vertical plane measurement of the same antenna shows backlobes at angles just above the antenna $0 \simeq 90°$ are more than 10 dB below isotropic.

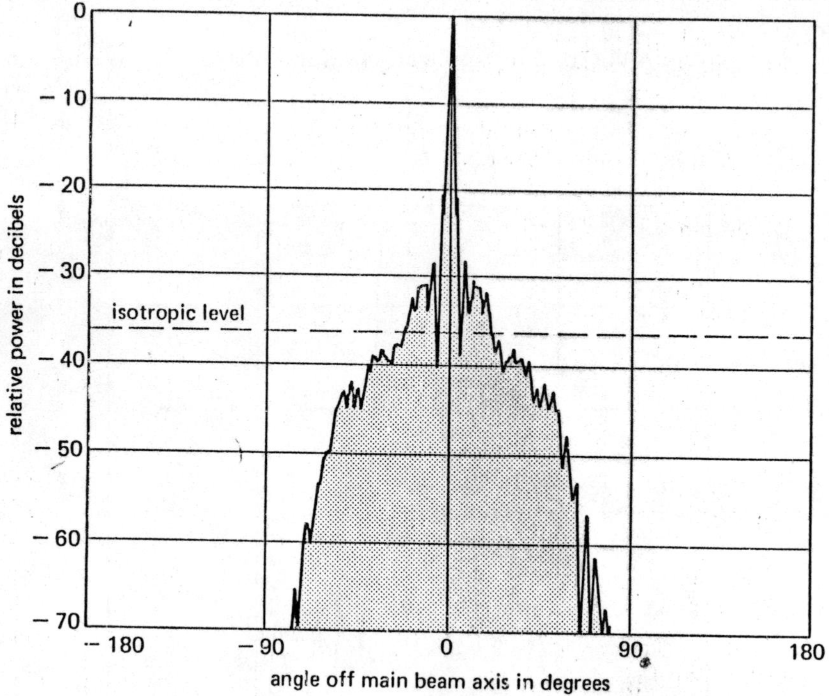

Fig. 6-24 Far-field radiation pattern of a low sidelobe antenna in the horizontal plane with vertical polarization at 11.2 GHz [From Crawford and Turrin article. Reprinted by permission from The Bell Telephone Laboratories, Inc. Copyright 1969, The American Telephone and Telegraph Company.]

The sidelobes must be carefully considered when planning satellite spacing and selecting antenna characteristics. Antenna illumination patterns or fewer antennas can usually be selected to give lower sidelobes at the expense of gain. The relative significance of these interference levels can be computed using the results of Chap. 11, where cochannel interference effects on PSK signals are discussed.

6-6 FREQUENCY-DEPENDENT ATTENUATION AND NOISE EFFECTS

There are a number of frequency-dependent effects on the propagation channel that increase the total transmission loss above the space loss of the satellite-earth terminal path. These propagation losses are primarily caused by rainfall, ionospheric scintillation, and atmospheric constituents which have molecular resonances in the microwave frequency band. The atmosphere and ionosphere also cause other effects, such as ray bending and group delay change, which are discussed in Part 4.

In addition, there are frequency-dependent and white noise effects, caused by the propagation losses as well as noise sources outside the earth, such as the white noise caused by sun transit effects already discussed and the cosmic noise discussed below.

Each of these effects must be considered by the satellite communications system designer in his choice of frequency as well as modulation bandwidth.

The use of frequencies below 1 GHz causes an antenna temperature increase because of *cosmic* (galactic) *noise*. The noise temperature at UHF

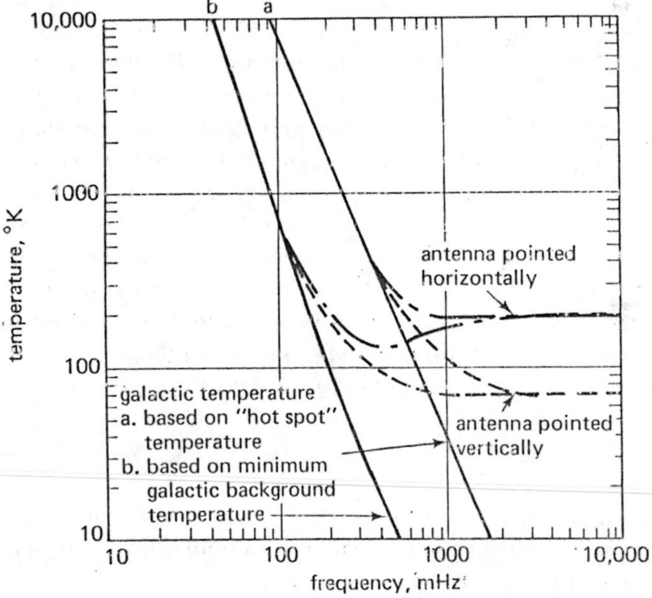

Fig. 6-25 Equivalent antenna temperatures including galactic background, atmospheric attentuation, and antenna sidelobes. (Assumes an aperture efficiency of 0.5, and that half the sidelobes' energy is received from the ground, which is assumed to be a blackbody at 280°K.)

varies roughly as $\lambda^{2.6}$, where λ is the RF wavelength. Figure 6-25 shows the typical variation of temperature with frequency for a parabolic reflector. (The noise effects of sidelobes and backlobes are calculated for this figure by assuming an aperture efficiency of 0.5, and that half the energy of the sidelobes is received from the ground, which is considered to be a blackbody at 280°K.)

The importance of this cosmic or galactic noise effect depends on the relative noise level of the low-noise receiver in the earth terminal and, of course, on the relative levels of the received signal from the satellite. At microwave frequencies above 1 GHz the cosmic noise effect is relatively small. However, at UHF frequencies 225–400 MHz the effects can be significant.

Rainfall Attenuation

Rainfall attenuation effects can have a severe effect on signal strength at frequencies above 10 GHz [Crane, 1971]. Figure 6-26 shows the attenuation coefficient in decibels/kilometer as a function of frequency for heavy rainfall (101.6 mm/hr). Experiments with Application Technology Satellite ATS V indicated that the use of a range $\approx$4.8 km gives a good approximation to the rainfall attenuation at zenith [Ippolito, 1971; Benoit, 1968, 73–80]. For example, at 20 GHz, a signal can encounter an attenuation of $4.8 \times 10 = 48$ dB in heavy rain for a 4.8-km path length. Fortunately, this level of rainfall occurs infrequently (0.1 percent of the time) and is very localized.

The rainfall attenuation more generally depends on the integrated rain content over the path of the ray and on the shape and sizes of the drops. Crane [1971, 181] has shown good measured correlation of the attenuation coefficient to $Z^{0.8}$ where $Z \triangleq \sum nd^6$ and n is the number of drops per unit volume with diameter d. The integrated rain, in turn, depends on the vertical and horizontal distribution of rain and the antenna elevation angles. The vertical distribution of precipitation rate can be approximated by

$$R(h) = R_s e^{-0.2h^2}$$

where R_s is the rain rate at the surface and h is the height in kilometers [Rice et al., 1967]. Empirical expressions for the attenuation coefficient α in decibels/kilometer in the 10–30-GHz region give

$$\gamma\alpha = KR^{\gamma}$$

where R is the rain rate, $1.0 \leq \gamma \leq 1.15$,

and $$K = [3(f-2)^2 - 2(f-2)] \times 10^{-4} \tag{6-8}$$

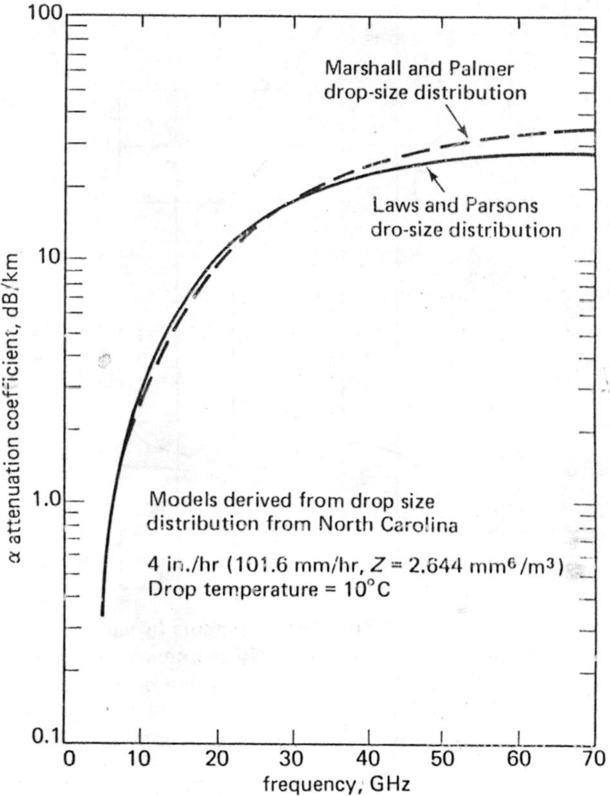

Fig. 6-26 Model estimates of the rain attentuation coefficient versus frequency. $Z = \sum nd^6$, where d is the drop diameter and n is the number of drops per unit volume with diameter d [From Crane, 1971, 178]

where f is in gigahertz. Thus, attenuation coefficient variation is nearly linear with rain rate in this frequency region.

For moderate and small rain rates the attenuation varies approximately as the cosecant of the elevation angle for angles above 10°. Very heavy rain rate cells, occurring in cloudbursts, can produce rain rate $\geq$ 100 mm/hr; however, these cells are usually very localized, less than 5 km in diameter (Fig. 6-27). Thus, at low elevation angles, the RF signal path can traverse a wide range of rain-rate distributions and the cosecant approximation is no longer valid. The rain cell height for intense rain can be $\approx$8 km, thus exceeding the 4.8-km rule of thumb discussed above. Hence, the horizontal and vertical distributions combine to produce an attenuation level which is almost independent of elevation angle at 4 km for high rain densities. Figure 6-28 shows the equivalent path length versus elevation angle for various rainfall

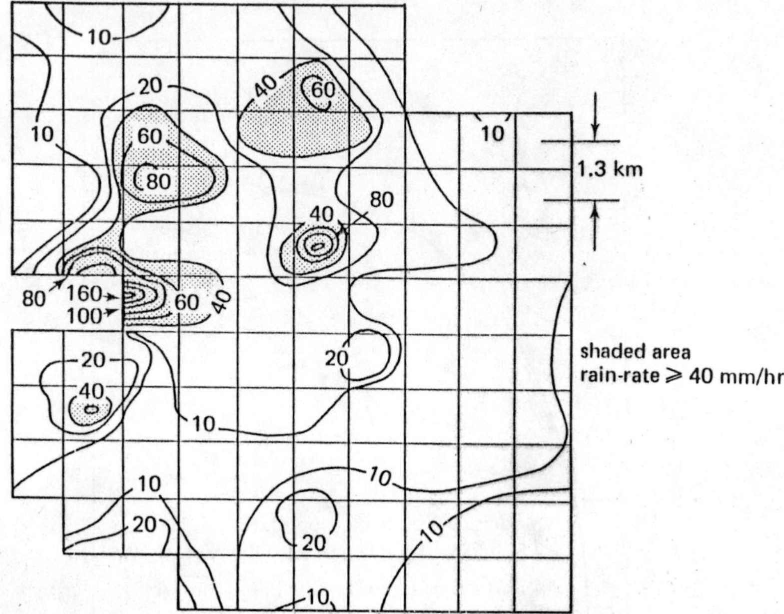

Fig. 6-27 Plot of rainfall-rate contours in millimeters/hour, showing several rain cells in an intense rain storm. Notice the highly localized configuration of the intense rain cells [From Semplak and Keller, 1969. Reprinted with permission from Bell Telephone Laboratories, Inc.]

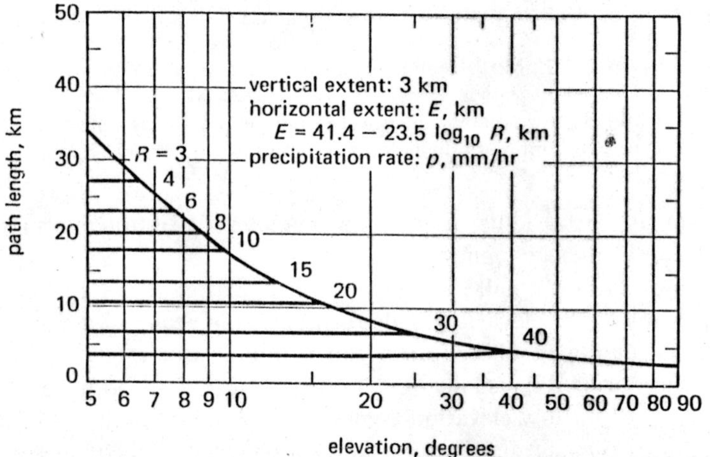

Fig. 6-28 Effective path length through rain versus elevation angle and rain rate R in millimeters/hour. At very heavy rain rates, $R > 40$ mm/hr, the effective path length is nearly independent of elevation angle. [From Benoit, 1968]

rates [Benoit, 1968, 73–80; Holzar, 1965, 119–125]. This result is based on the empirical relationship for the horizontal extent of rain

$$H = 41.4 - 23.5 \log_{10} R \text{ km}$$

where R is the rain rate in millimeters/hour.

Seltzer [1970, 1873–1892] has given computed values of rainfall attenuation for microwave frequencies using Mie extinction properties and rain drop sizes corresponding to the Laws and Parson's distribution. His results for the Mie extinction coefficient β_{ext} versus rain rate at various wavelengths are given in Fig. 6-29. The attenuation ratio in dB/km α of the power of the transmitted beam relative to the incident beam is related to the extinction coefficient by:

$$\alpha = e^{-2\beta_{\text{ext}}(\lambda)l} \text{ dB/km} \tag{6-9}$$

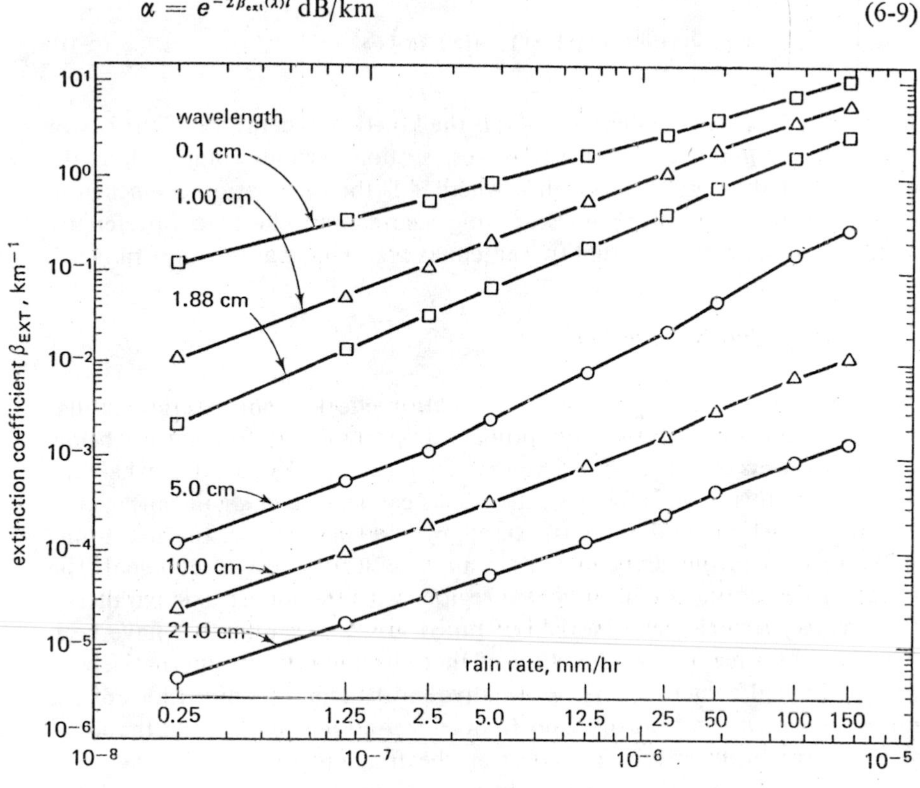

Fig. 6-29 Rainfall water content and rain rate versus extinction coefficient β_{ext} [From Seltzer article. Reprinted with permission from The Bell Telephone Laboratories, Inc. Copyright 1970, The American Telephone and Telegraph Company.]

where l is the length of the propagation path in kilometers. These results indicate that the variation in attenuation with frequency is approximately as f^2 for 10 GHz to 60 GHz and rain rates < 15 mm/hr, and is consistent with the expression for K in (6-8). The statistical distribution of the attenuation coefficient α in decibels is approximately log normal distributed for earth-space links during rainfall [Lin, 1973, 557–582]. Thus the quantity $2\beta_{\text{ext}}\ell = \log \alpha$ is normally distributed with zero mean, and the distribution of signal amplitude $V(t)$ is approximately log-log normal.

In addition the rain cells produce an increase in antenna temperature, which may be 100°K relative to a clear sky temperature [Crane, 1971]. The antenna temperature can be computed for the lower microwave frequencies ($f < 10$ GHz) from the integral along the path s,

$$\mathfrak{J}_A \cong \eta \int_0^\infty \mathfrak{J}(s)\beta(s) \exp\left(-\int_0^s \beta(x)\, dx\right) ds \tag{6-10}$$

where η is the antenna efficiency, $\mathfrak{J}(s)$ is the kinetic temperature of the gas or rain cell, and $\beta(s)$ is the attenuation cross section per unit volume along the direction of the ray; $\beta(s) = A/4.34$ where A is the attenuation coefficient in decibels per unit length. Rain scattering also can produce an interference effect in the unlikely event that two antennas are pointed at the same rain cell.

Ionospheric Scintillation

Another frequency-dependent attenuation effect is ionospheric scintillation. Ionospheric effects are of primary importance at frequencies below 1 GHz. However, even at the microwave frequencies above 1 GHz, where the attention of this book is focused, the effects can still be of significance.

Ionospheric scintillation is caused by irregularities in the night-time F-layer ranging from 200 to 600 km in altitude [Kent and Koster, 1966]. The irregularities appear to be elongated regions with the longer axis parallel to the earth's magnetic field lines. Axial ratios greater than 60 to 1 have been measured [Koster, 1966]. The effect of these irregularities is alternatively to produce signal enhancement and negative fades. The refractive index of the ionosphere is a function of radio frequency, and irregularities in the ionosphere have progressively less effect as the frequency increases. The exact nature of the frequency dependence appears to depend somewhat on the ionospheric conditions, but the absolute value of the scintillation attenuation seems to vary approximately as wavelength squared.

Two areas of the earth have the greatest incidence of scintillation fades, the subauroral-to-polar latitudes and a belt surrounding the geomagnetic equator. The effects on the equator are most dominant in the period of time within an hour or two after sunset to a time not usually beyond 1 : 00 to 2 : 00

AM local time. Figure 6-30 depicts the typical regions of fades for an earth longitude for a region where the magnetic equator is on the equator. The region of fade extends $\pm 15°$ relative to the magnetic equator, and the magnetic equator deviates from the equator in the Pacific Ocean area by as much as 7°.

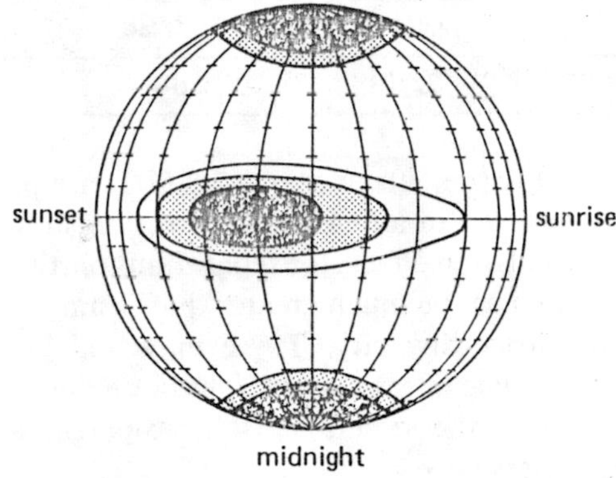

Fig. 6-30 Ionospheric irregularity structure at night. The density of the hatched area represents the occurrence of fading caused by ionospheric scintillation. [After Aarons, 1971]

For a statistical analysis that involves a long period of time, a scintillation index $\mathcal{S}$ can be defined [Whitney, Aarons, Malik, 1969]:

$$\mathcal{S} = \frac{\sqrt{\overline{P^2} - (\overline{P})^2}}{\overline{P}} = \frac{\sigma_p}{\overline{P}} \tag{6-11}$$

where P is the received power amplitude, σ_p is the standard deviation of received power, and $\overline{\mathrm{P}}$ is the average received power amplitude. Thus S is the normalized rms power amplitude. One theory of scintillation is restricted to the "weak scattering" approximation where the random phase deviation of the medium is required to be small compared to one radian. Briggs and Perkin [1963] have shown that the scintillation index can be expressed approximately by

$$\mathcal{S} \sim \lambda (\sec i)^{1/2} \left\{1 + \frac{\pi^2 L_c^4}{4\lambda^2 Z^2}\right\}^{-1/2} \tag{6-12}$$

where Z is the slant distance to the ionospheric irregularity 400 km $< Z <$ 1400 km, L_c is the irregularity autocorrelation distance in the medium, i is the zenith angle at the ionospheric intersection point, and λ is the wavelength. Typical values of these parameters are $Z \cong 600$ km, $L_c \cong 1$ km.

For short wavelengths ($f > 600$ MHz), we expect $L_c^2/\lambda Z \gg 1$, the far zone approximation holds, and $\mathcal{S}$ in (6-12) is approximately proportional to λ^2.

Measured data for a 95 percent confidence of fading less than a given value yield results generally consistent with the following:

f	250 MHz	2.3 GHz	7.3 GHz
Attenuation, dB	22 dB	2 dB	<0.5 dB

For the most part, scintillation effects above 7.3 GHz are practically absent.

Earth stations separated in an east-west direction show essentially zero correlation in their scintillation effects for separations on the order of 0.6 nm. North-south separation must be much greater than 1 nm to provide substantial decorrelation for space diversity. There may well be a correlation at separated stations for a time offset, however, because the ionospheric irregularity drifts with time over the second station. Apparent drift velocities of 280 m/sec have been observed.

The bandwidth of the scintillation amplitude variation is on the order of 0.2 Hz at the 3-dB point. Thus the scintillation fades are relatively slow and correspond roughly to the drift velocity of the irregularity and the width of the irregularity.

The bandwidth of the frequency region affected by the fade is very wide even at UHF frequencies. The 3-dB correlation bandwidth exceeds 100 MHz. Hence, the use of frequency diversity for scintillation would require a frequency separation considerably greater than 100 MHz in order to be effective and is generally impractical to implement.

Atmospheric Attenuation

At frequencies above 10 GHz, the atmospheric attenuation due to water vapor and oxygen can have significant effects on the communication link. These effects are even greater at some frequencies than the effects of rainfall. Curves for both 0 and 100 percent humidity are shown in Fig. 6-31. Water vapor causes the peaks at 22.2 GHz, and 183.3 GHz, and oxygen causes a family of absorption lines [Zimmerer and Mizushima, 1961, 152–155] at 60 GHz (56–65 GHz) and an isolated line at 118.8 GHz, as shown. The curves show the minimum attenuation values that occur between the oxygen absorption lines. Hence, the individual lines are not visible at 60 MHz in Fig. 6-31. The very high attenuation at the oxygen absorption frequencies renders these frequencies unusable for earth-satellite links. Over 100 dB of additional attenuation exists at these frequencies.

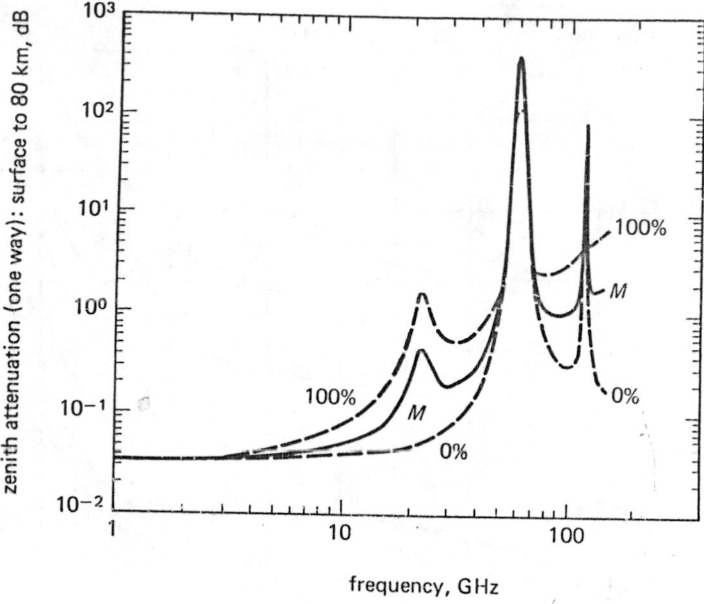

Fig. 6-31 Zenith attentuation versus frequency for various percent humidity [From Crane, 1971, 174]

Attenuation for other elevation angles e can be computed using the effective path length r as related to the vertical extent of the troposphere $r_v = 10$ km, and the effective earth radius r_0:

$$r = r_0^2 \sin^2 e + r_v(2r_0 + r_v)^{1/2} - r_0 \sin e \tag{6-13}$$

The total attenuation A dB for a given effective path length r relative to the vertical attenuation A_v dB is given by $A/A_v = r/r_v$. Oxygen and water vapor attenuation as a function of elevation angle is plotted in Fig. 6-32 using data for standard atmospheric conditions [Koelle, 1961; Fillipowski and Muehldorf, 1965], and the value of r calculated above.

6-7 COMMUNICATIONS LINK CALCULATIONS

Satellite Link

A satellite link, utilizing one channel of a multichannel transponder, is limited in transmission capability by satellite downlink power, ground-station uplink power, satellite and earth terminal noise levels, and channel bandwidth. One of these constraints usually dominates the others; most often

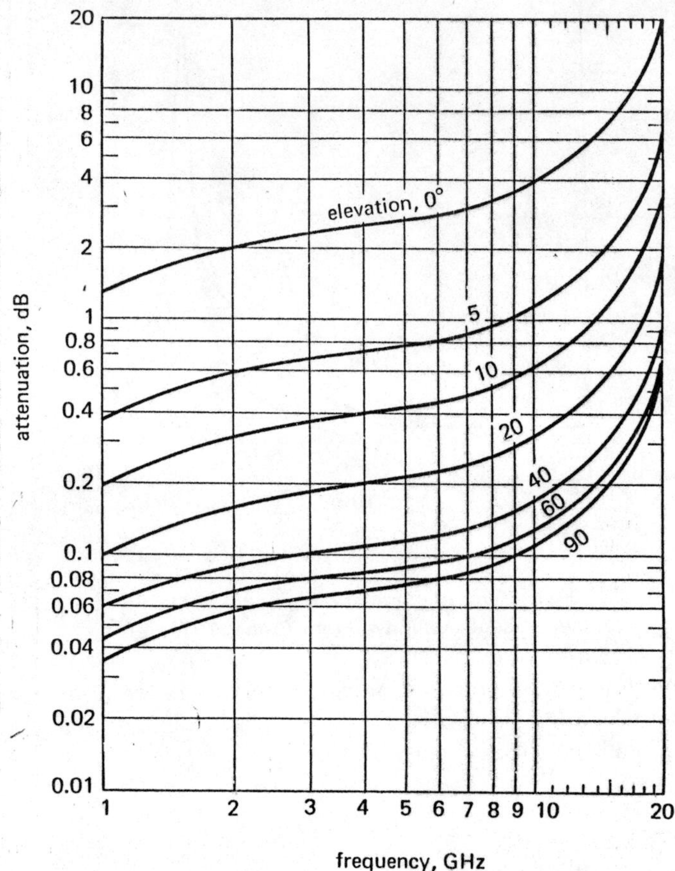

Fig. 6-32 Atmospheric attenuation due to oxygen and water vapor [From Benoit, 1968, 73–80]

downlink SNR or channel bandwidth is the major constraint. Figure 6-33 shows the simplified configuration of a linear satellite transponder channel where the satellite downlink power P_{sat} is shared in proportion to the power level in the received signals from an earth terminal relative to total received signal plus noise power.

The transmission begins from a ground station, which is transmitting a signal of bandwidth B_1 at an effective radiated power P_1 to a satellite transponder of bandwidth W. Other signals make up a total ERP of $\Sigma' P_i$ transmitted to this transponder. After the path losses (which can differ from one ground station to the next) and satellite receive antenna gains are taken into account, a total signal power $P_T \triangleq \Sigma A_i P_i$ is received along with received noise power of $N_s W$ in the transponder bandwidth W. For a linear trans-

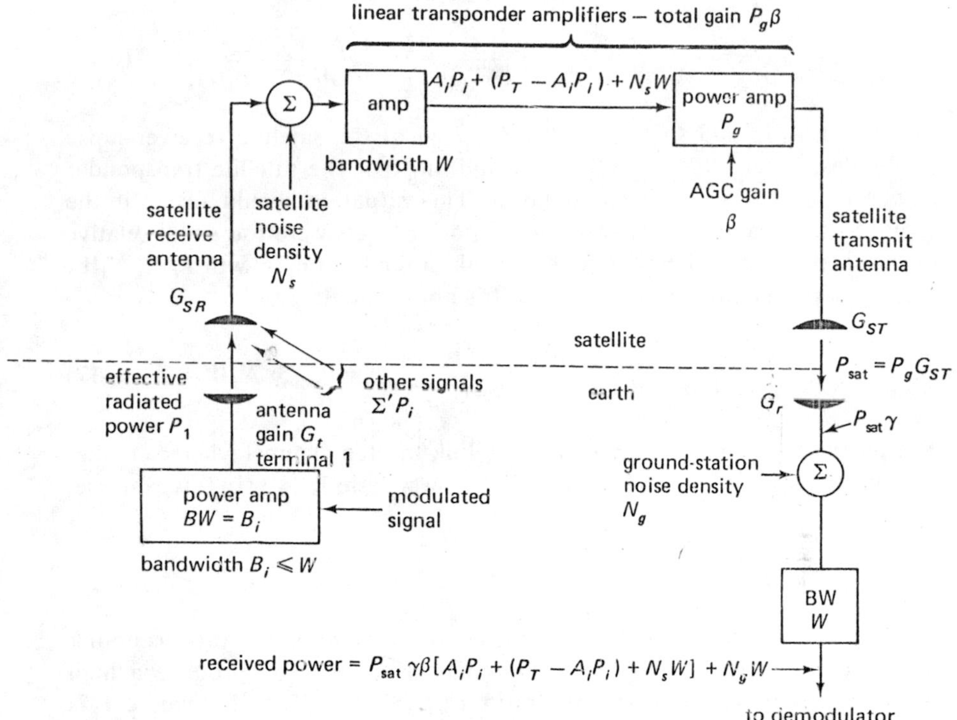

Fig. 6-33 Simplified satellite communication link

ponder, the total power P_{sat} available for this transponder channel is shared proportionately by the signal and noise according to their respective powers.

The transmitted satellite power P_{sat} depends, in turn, on the satellite transponder gain $G = P_{sat}/(P_T + N_sW)$ and the total input power level $P_T + N_sW$. Thus the transmitted effective radiated power (ERP) is composed of

$$P_{sat} = G(P_T + N_s W) = P_{sat}\beta[A_1P_1 + (P_T - A_1P_1) + N_sW] \quad (6\text{-}14)$$

where $\beta = 1/(P_T + N_sW)$ is the denominator of the power-sharing factor between signal and noise. The signal is finally received along with ground-station receiver noise N_gW for a total received power

$$P_{sat}\gamma_i\beta[A_1P_1 + (P_T - A_1P_1) + N_sW] + N_gW \quad (6\text{-}15)$$

where γ_i accounts for the downlink path losses and the receive antenna gain for the ith earth terminal. The effective received carrier power-to-noise density ratio C/N_0 for signal S_1 corresponding to terminal 1 is

$$\left(\frac{C}{N_0}\right)_{1i} = \frac{A_1\beta\gamma_i P_1 P_{\text{sat}}}{P_{\text{sat}}\gamma_i\beta N_s + N_g} = \frac{[A_1P_1/(P_T + N_sW)](P_{\text{sat}}\gamma_i)}{[\gamma_i P_{\text{sat}}/(P_T + N_sW)N_s + N_g]} \tag{6-16}$$

The link is said to be *uplink limited* when the satellite receiver noise dominates, giving $\mathbf{R} \triangleq P_T/N_sW \ll 1$, and most of the satellite transponder power is wasted as transmitted noise. This situation would occur if the ground-station antennas or transmitted power levels were too small relative to the satellite noise power. For the uplink limited example with $P_T \ll N_sW$, the signal-power-to-noise-density result is approximately

$$\left(\frac{C}{N_0}\right)_{1i} \simeq \frac{\gamma_i P_{\text{sat}}(A_1P_1/N_sW)}{(\gamma_i P_{\text{sat}}/W) + N_g} \simeq \frac{A_1P_1}{N_s} \qquad \text{if } \gamma_i P_{\text{sat}} \gg N_gW \tag{6-17}$$

The more common situation is the downlink-limited channel where satellite-transmitted ERP is limited and the uplink power ratio $\mathbf{R} \gg 1$. In this example,

$$\left(\frac{C}{N_0}\right)_{1i} \cong P_{\text{sat}}\frac{(\gamma_i P_1/P_T)}{N_g} \tag{6-18}$$

The power of the transponder is then shared only among the various uplink transmitted signals. This satellite communication link can produce a high signal-energy-per-bit/noise-density ratio $E_b/N_0 = TC/N_0 \gg 10$, where T is the duration of one bit.

In practice, of course, nonlinearities in the transponder and other distortion effects can have a significant impact on performance and change the power-sharing relationship. These quasi-linear, or limiting, effects are discussed in detail in Chap. 9.

For a particular modulation type, such as biphase PSK, the channel bandwidth, rather than downlink power, may be the limiting factor. This situation exists now in some links operating at X-band and C-band frequencies where efficient components are available and large antenna gains are utilized, but is less likely at K-band where efficiencies are not as great, bandwidths are wider, and rain losses are larger.

Earth Terminal Link

The downlink power budget for a given digitally modulated carrier is generally one of the key elements in the overall satellite communications system design and analysis. Uplink power is generally not as critical because the cost of ground terminal power is far less than that of satellite power (even after accounting for the lower sensitivity of the satellite RF receiver). Furthermore, the ground-terminal antenna size, and hence the transmit antenna gain, must generally be relatively large compared to the satellite antenna because of the downlink power constraints. Hence, it typically takes relatively little power from each of many earth stations to drive the satellite transponder to the desired output power level. It is assumed below that any uplink power required is available, and only the downlink power budget is calculated. However the more general results are available from the preceding discussion, using (6-16).

The critical parameter to be calculated is the ratio of received carrier power-to-noise density C/N_0. The minimum allowable value of this parameter is easily related to the required energy-per-bit-to-noise density ratio E_b/N_0 calculated in Chap. 11.

The value of C/N_0 in dB-Hz at the earth terminal IF output is given by

$$\left(\frac{C}{N_0}\right)_{1j} = P_{\max} - B_0 + \beta A_1 P_1 + \frac{G}{\mathfrak{T}} - \kappa - L_{Tj} \quad \text{dB-Hz} \qquad (6\text{-}19)$$

where

all terms are expressed in dB or dBm and $N_g = \kappa\mathfrak{T}$, $L_{Tj} = 1/\gamma_j$

$P_{\max}$ is the maximum (peak) ERP, in dBm, of the satellite transponder channel of interest (decibels above one milliwatt) $P_{\max} \geq P_{\text{sat}}$

B_0 is the amount in dB the total average satellite power is backed off from its peak value $P_{\max}$. Thus, the total average satellite power is $P_{\text{sat}} = P_{\max} - B_0$

$\beta A_1 P_1$ is the percentage (in decibels) of the available average satellite power (P_{sat}) to be used for the carrier of interest (10 percent gives $\beta A_1 P_1 = A_1 P_1/P_T = 1/10$ or -10 dB

$G/\mathfrak{T}$ is the ratio of effective receive antenna gain (dB) to received system noise temperature (°K, dB) at the received carrier frequency*

G is the antenna gain in dB $10 \log [(4\pi/\lambda^2)\mathfrak{A}\eta_a]$, where $\mathfrak{A}$ is the reflector area, λ is the carrier wavelength, and the aperture efficiencies and component losses are accounted for by η_a [Silver, 1949]

κ is Boltzmann's constant, −198.6 dBm/°K-Hz

L_{Tj} is the total loss, including free-space loss, antenna pointing loss, rainfall and radome loss, etc. to terminal j.

The system noise temperature $\mathfrak{T}$ is expressed by [*IEEE*, 1968; Blackwell and Kotzebue, 1961, Chap. 2] [F is the noise figure (not in dB)]

$$\mathfrak{T} = \frac{\mathfrak{T}_a}{L} + \frac{(L-1)}{L}\mathfrak{T}_0 + \mathfrak{T}_R \qquad {}^\circ\text{K}, \; \mathfrak{T}_R = (F-1)\mathfrak{T}_0 \qquad (6\text{-}20)$$

---

*As an example, standard earth stations complying with ICSC-45-13, nominally a 30-meter antenna, produce a $G/\mathfrak{T} = 40.7$ dB/°K at C-band.

where $\mathfrak{T}_a$ is the antenna noise temperature due to sources external to the antenna, for example, ground, rain, atmosphere, and sun; L is the loss in the antenna, feed, and waveguide components; $\mathfrak{T}_0$ is the ambient temperature; and $\mathfrak{T}_R$ is the noise temperature of the low-noise receiver and properly weighted losses. The antenna temperature is a weighted version of the sky temperature, which for an infinitely sharp beam is that given by Fig. 6-34. The free-space transmission loss included in L_{Tj} is

$$L_f = 20 \log_{10}\left(\frac{4\pi r}{\lambda}\right) = 92.446 + 20 \log_{10} f + 20 \log_{10} r \text{ dB} \quad (6\text{-}21)$$

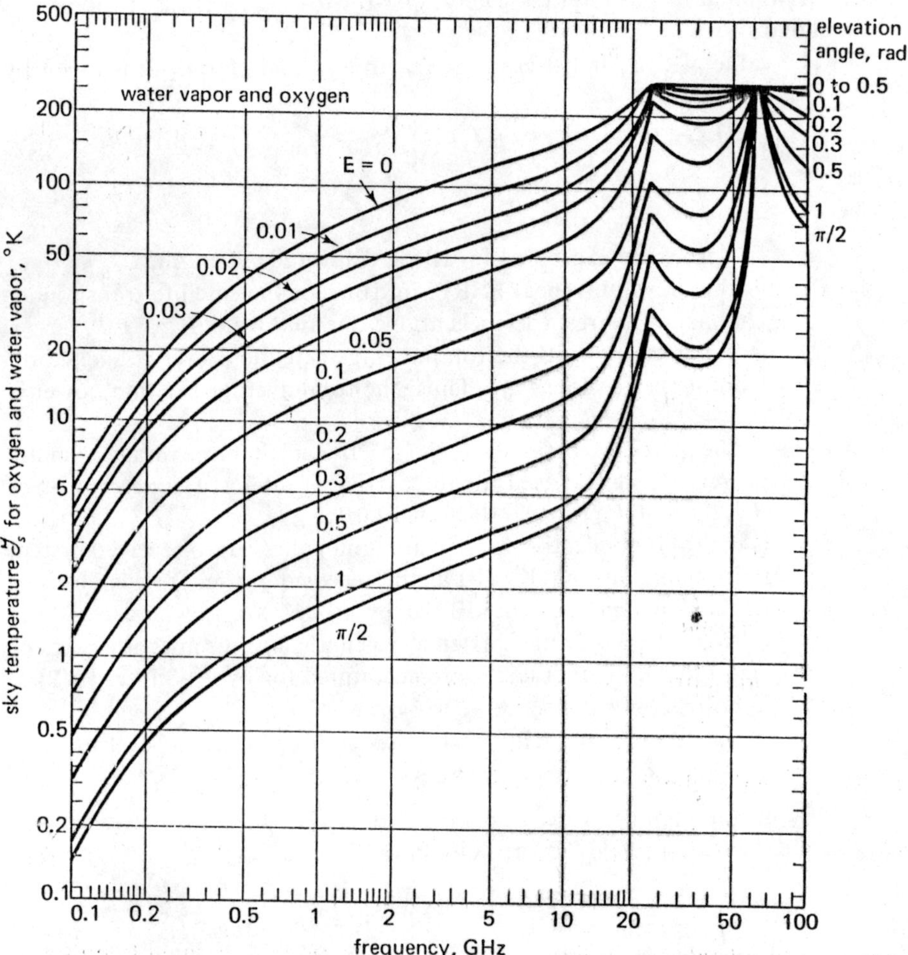

Fig. 6-34 Sky-noise temperature due to reradiation by oxygen and water vapor. The quantity E is the elevation angle in radians. [After D. L. Rice et al., 1967]

where frequency f is in gigaHertz (GHz) and the range r is in kilometers (Fig. 6-18), or

$$L_f = 97.796 + 20 \log_{10} f + 20 \log_{10} r' \tag{6-22}$$

where r' is in nautical miles. For synchronous altitude and elevation angle E the value of L_f is

E	L_f, dB
90°	$183.51 + 20 \log_{10} f$
45°	$183.95 + 20 \log_{10} f$
0°	$184.84 + 20 \log_{10} f$

The required C/N_0 for a given bit rate $\mathcal{R}$ is related to the E_b/N_0 required for the specified bit error rate (Part 3), and desired link margin M to allow for excess rain losses and other power degradations, by the expression

$$\left(\frac{C}{N_0}\right)_{req} = +\frac{E_b}{N_0} + \mathcal{R} + M \tag{6-23}$$

Typically, M is selected as 6 dB for X-band and 4 dB for C-band, and larger numbers are used for the higher K-band frequencies because of the higher rain losses at K-band noted earlier.

For an E_b/N_0 of 6 dB, these equations can be combined to relate the maximum permitted bit rate to the ground terminal $G/\mathfrak{T}$ and fractional satellite power $F = A_1 P_1 \beta$ required for the following example system parameters at X-band:

P_{max} = 58 dBm
κ = −198.6 dBm/°K-Hz
L_T = 203.6 dB (includes 2 dB miscellaneous losses)
B_0 = 3-dB power backoff or $P_{sat} = P_{max} - 3 \text{ dB} = 55$ dBm

Thus, the fractional power (using (6-19), (6-23), $M = E_b/N_o = 6$ dB) is:

$$F \triangleq A_1 \beta P_1 = \frac{E_b}{N_o} + \mathcal{R} + M - P_{max} + B_o - \frac{G}{\mathfrak{T}} + K + L_t = \mathcal{R} - 38 - \frac{G}{\mathfrak{T}} \tag{6-24}$$

The allowed bit rate versus earth terminal $G/\mathfrak{T}$ is shown for satellite power percentages of 1, 2, and 4 percent in Fig. 6-35. As an example, a 60-ft antenna [IEEE Staff, 1972] with a cooled paramp can provide an X-band receive $G/\mathfrak{T}$ of approximately 40 dB/°K, a 35-ft parabola with a cooled paramp

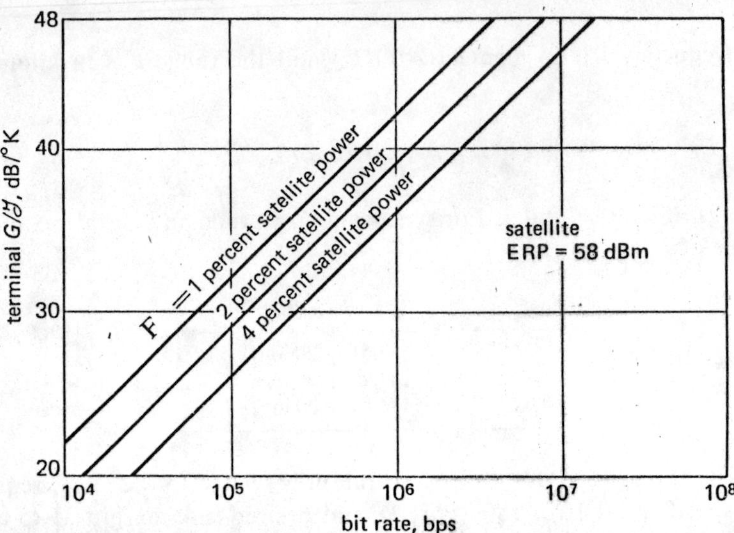

Fig. 6-35 Variations in required earth terminal G/𝒯 with bit rate for various percentages of total power usage for a given data link.

gives $G/\mathcal{T} = 35$ dB/°K. Thus, the 60-ft terminal can provide a bit rate of approximately 200 kbps using only 1 percent of the satellite power.

Dynamic network control of each earth terminal's uplink power can decrease the required operating margin M in (6-23) by responding precisely and promptly to changes in rain loss or other sources of attenuation or system degradation. A network control terminal can continually monitor the performance of the operating links and reallocate reserve satellite power as required.

CHAPTER 7

SATELLITE TRANSPONDERS

7-1 INTRODUCTION

The satellite transponder and associated antennas form the primary portion of the communications subsystem on a communication satellite. This transponder differs from some conventional microwave line-of-sight (LOS) repeaters in that many separate ground terminals access the satellite simultaneously at nearly the same instant from widely different points on earth. Thus, multiple carriers arrive at, and must be relayed by, the satellite. This chapter briefly describes some of the major elements and types of communication transponders. The discussion covers multichannel transponders, some of the advantages of transponder channelization, typical frequency plans and antenna assignments, and potential advantages of processing transponders. The detailed effects of the transponder on different types of multiple-access techniques are described in later chapters on frequency-division and time-division multiple access (Chaps. 9 and 10). Phase noise introduced by frequency translation is analyzed in Chap. 12, and filter distortion effects caused by channelization filters are covered in Chap. 13.

7-2 A TRANSPONDER MODEL

Most communication satellites contain several (four or more) parallel transponders, often with several narrow beam antennas to aid in the multiple-access problem, particularly where the received signal levels differ widely for different classes of users. A single channel of a typical transponder is modeled in Fig. 7-1. Only the most basic elements are shown: the channel separation bandpass filter, the frequency converter, the various amplifiers, and a possible limiter amplifier. Multiple input sinusoids enter the transponder in frequency band f_u and exit in band f_d. The frequency bands are separated sufficiently far to prevent "ringaround" oscillations in the transponder itself. This transponder uses a single frequency translation operation which converts the receive RF frequency directly to the transmit RF frequency. Other configurations first down-convert to a convenient IF frequency, for example, 150 MHz, and then up-convert to the transmit frequency.

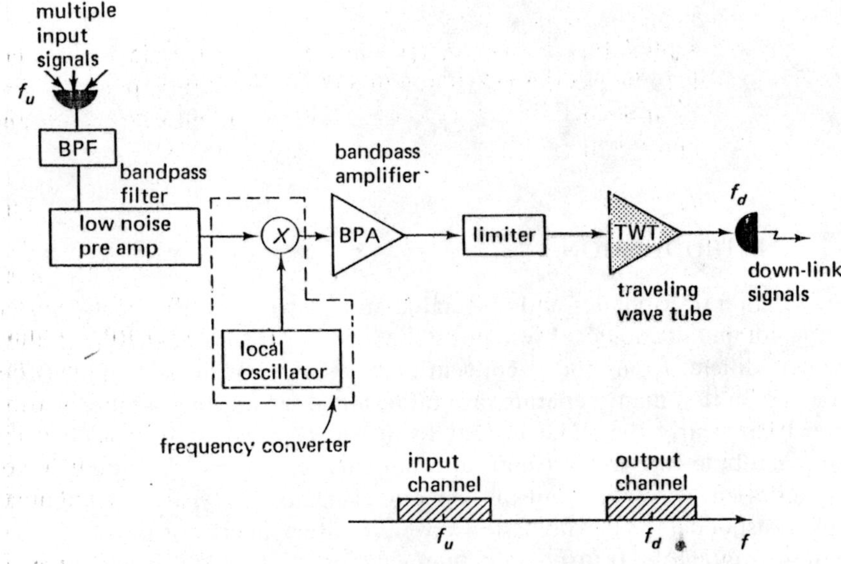

Fig. 7-1 Simplified transponder block diagram showing a single channel transponder

7-3 TRANSPONDER CHANNELIZATION

Satellite transponders used for multiple access are often *multichannel* in configuration. That is, the transponder is channelized by frequency-selective filters to allow different frequency bands to be handled by separate amplifiers and antennas. Transponder channelization has a variety of applications, including:

1. The total downlink power can be increased by using parallel power amplifiers (usually TWTs) in the satellite.

2. Uplink frequency selection in the earth terminal can control which polarization or downlink satellite antenna (for example, a spot beam) is to be used. Thus, this approach can use a change in uplink frequency to switch the signal to different geographical areas and hence to different receive ground stations.
3. The number of signals handled by a single TWT can be decreased and hence intermodulation (IM) product effects can be reduced.
4. Different classes of earth terminals can be isolated from one another by separate satellite channels—so that small mobile terminals of limited uplink ERP do not compete for satellite power with large fixed ground stations.
5. Uplink frequency selection controls which downlink frequency is to be used. Note that with a channelized transponder, there need not be a fixed offset between the uplink frequency band and the individual downlink frequency channels.

It is often convenient to set all transponder bandwidths equal. This bandwidth equality permits complete flexibility in the choice of transponder channel; a network of earth terminals occupying one transponder can be transferred to another transponder with minimal impact other than a frequency offset. Furthermore a spare unused transponder can provide a compatible redundant system element in the event of a transponder failure, thereby improving system reliability. Some of these potential uses of frequency channelization are illustrated in Fig. 7-2. In the multichannel transponder shown, frequency channels 1 and 2 are separately filtered and amplified for each of the orthogonal polarizations RHP and LHP. Either earth-coverage or spot-beam antennas can be employed on the downlink by mechanical switching (SW).

The received signals in Fig. 7-2 enter the earth-coverage antenna/diplexer, where they are bandpass filtered to isolate the receive preamp from the transmitted downlink signals. The isolation provided by the product of transmit and receive filters, and diplexer isolation must, of course, exceed the gain of the transponder by a significant amount to prevent ringaround oscillations in the transponder which can occur for a loop gain of unity or more. After RF preamplification, the received carriers in channels 1 and 2 are separated by the channel filters. Additional AGC amplification or limiting can be used to bring the signals in each channel to the proper level. However, it is common to have the gain adjustment of the transponder controlled by ground command. This approach is usually preferable to the use of AGC in the satellite which would provide an additional element of uncertainty in the earth terminal transmit power control problem. Thus, signal levels can

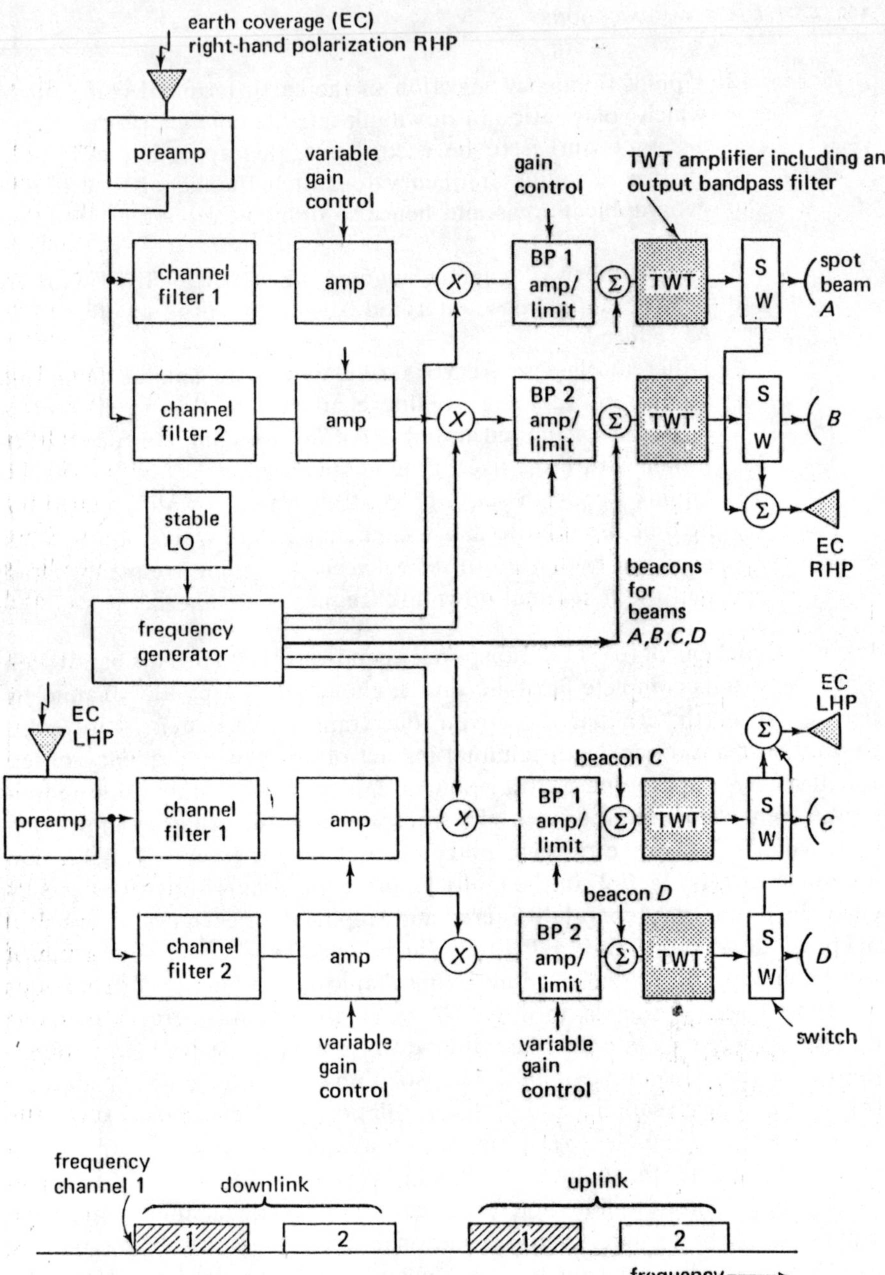

Fig. 7-2 Example of multichannel satellite transponder. The downlink has spot beams *A, B, C, D* available by switching.

be adjusted through the use of accurate uplink power control rather than hard limiting in the transponder.

The microwave carriers next are frequency converted to the transmit frequency channel, perhaps 750 MHz away, by a frequency generator containing a phase-locked frequency multiplier locked to a stable satellite frequency standard. A TWT or other amplifier then amplifies the output of one or more combined channels. The multiple carriers that may appear at the TWT input can cause significant intermodulation products in the output channel bandwidth unless the power level is backed off (Chap. 9). Other intermodulation products can spread all the way to the receive channel of the same transponder. This effect is particularly important if the same antenna is used for both transmit and receive. For this reason, the TWT amplifier output must be heavily filtered to prevent cross products from falling in the receive band and saturating the preamp. The filter also must be designed for low loss in the transmit channel to provide efficient power utilization. Care must be given also in this filter design to minimize intermodulation products that are generated in the passive (but slightly nonlinear) filter structure and enter the receive passband.

An earth terminal can transmit uplink to the earth-coverage satellite receive antenna in either polarization and in either frequency channel, thereby selecting the downlink spot beam to be used. The TWT power-amplifier output is then fed to the spot-beam antenna or to dual-polarized earth-coverage antennas, depending on the setting of a mechanical switch. The mechanical switch is controlled by ground command. Figure 7-3 shows an example satellite antenna beam pattern for a satellite operating over the Atlantic Ocean. A ground station anywhere in the earth-coverage uplink pattern can direct its signals to any one of the four downlink spot-beam regions (3.5° beamwidth) of the transponder shown. Frequency reuse is provided here by the use of separate carriers of LHP and RHP or cross-polarized linear polarizations in the same frequency band. Cross-polarization isolation ratios as high as 40 dB have been demonstrated for cross-polarized linear polarization antennas [Wilkinson, 1973, 27–62].

Satellite beacon frequencies are inserted in each output channel usually outside of the communication channel. The beacon frequencies provide an easily identifiable and permanent signal for autotrack receivers in the ground terminals to aid in satellite tracking. In addition, beacon frequency is often coherent with the stable local oscillator on board the satellite and can be used by the earth terminal to track long term drifts in the satellite translation frequency. It is possible to use an on-board atomic standard to improve this transfer oscillator stability (See Chap. 17). Finally, the beacon may be modulated to carry a satellite identification word and low-rate telemetry data on the satellite status. This direct satellite telemetry to a communications

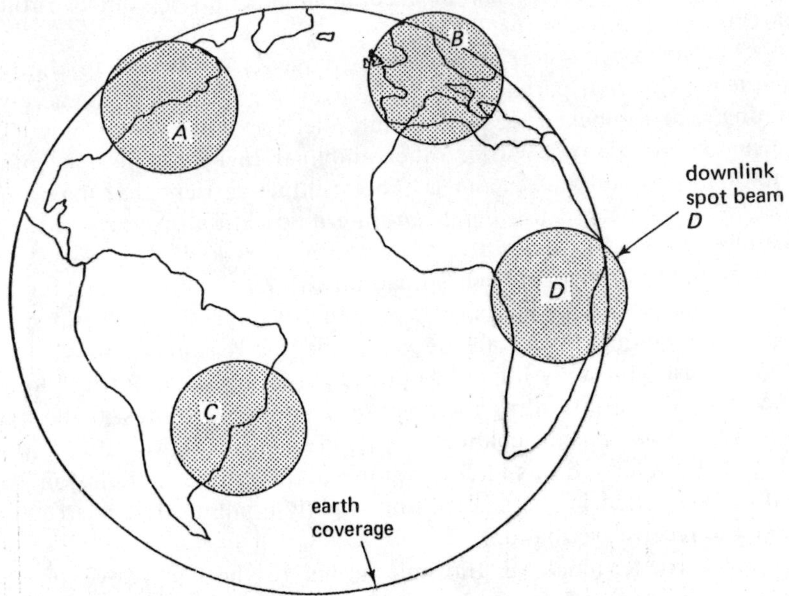

Fig. 7-3 A coverage pattern for an Atlantic satellite showing the four narrow beams and the earth coverage antenna beams

control earth station is sometimes used as an input to provide real time control of the satellite and earth terminal network.

The filters in the various bandpass amplifiers in the multiplexers are equalized in gain and group delay to reduce signal distortion. Each channel typically has its own set of equalizers. Both gain flatness and gain slope are often compensated to reduce signal distortion effects.

Figure 7-4 gives a simplified block diagram of the 12-channel Intelsat IV transponder. Each of the four receivers consists of a 6-GHz tunnel diode amplifier (TDA) low-noise preamplifier followed by a 4-GHz notch filter to provide out-of-band attenuation. The 500-MHz bandwidth receive signal band is then frequency translated to 4 GHz.

The even and odd input multiplexers contain six bandpass filters, each of bandwidth 36 MHz, as shown in Fig. 7-4. Since adjacent channels are separated in center frequency by 40 MHz, there is then a 4-MHz guard band between adjacent channels.

Each of the input multiplex filters separates the signals into even and odd channels and is separately equalized. The channel filters in the input multiplexers are followed by redundant power TWT amplifiers, harmonic filters, and switches to direct the signals to the appropriate output antenna.

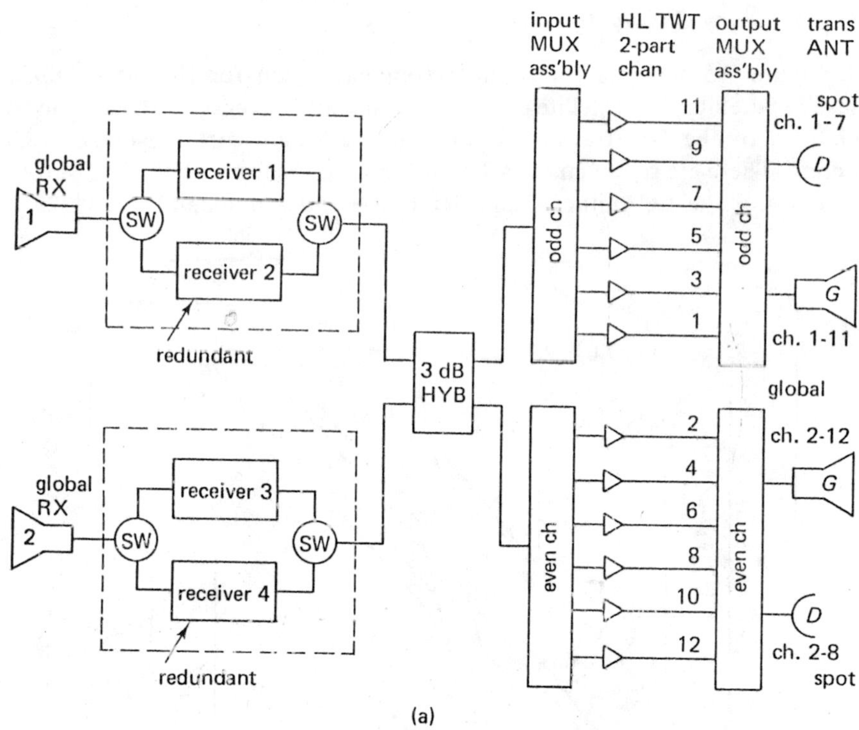

(a)

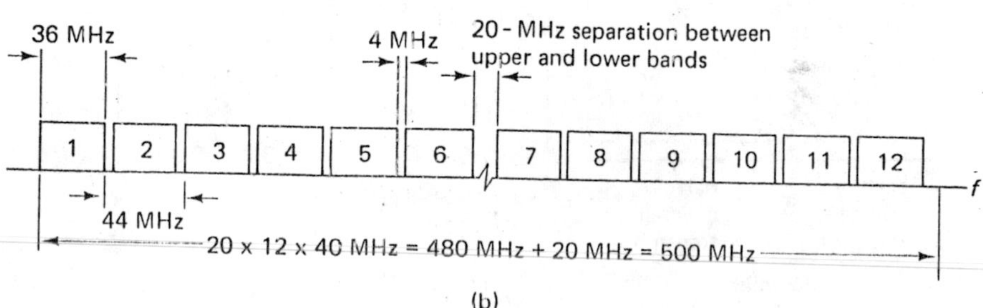

(b)

Fig. 7-4 Simplified configuration of the INTELSAT IV satellite transponder [After Bennett and Dostis, 1974, 375] (a) transponder block diagram, (b) transponder frequency plan

In practice, most elements of the transponder are made redundant to improve the reliability of the transponder. The redundant elements are not all shown in Fig. 7-4 for purposes of simplicity.

7-4 FREQUENCY PLANS

Frequency Channelization

Figure 7-5 shows a transponder frequency plan for the USA DSCS Phase II Satellite. The satellite employs two narrow-coverage (NC) antennas, which share the NC transmit power, and a single earth-coverage (EC) antenna. There are redundant TWT power amplifiers for both the NC output channel and the EC output channel [Huang and Hooten, 1971, 238–251].

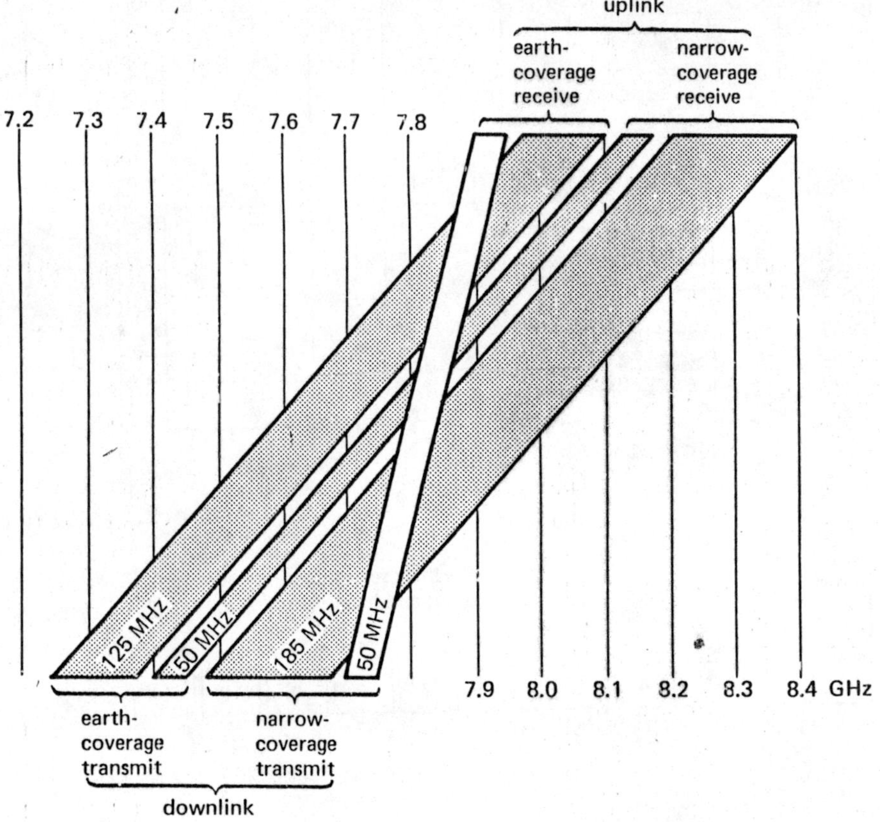

Fig. 7-5 DSCS Phase II satellite frequency plan for the uplink and downlink

In this transponder, uplink power from a user in the narrow-coverage beam ($\cong 2.5°$) of the satellite can be transmitted by either the earth-coverage or narrow-beam downlink antennas. The power is split to each of two narrow-beam antennas. Similarly, carriers in the earth-coverage uplink channel are directed to either downlink narrow coverage or earth coverage antenna, depending on the frequency of the uplink carriers. Thus an earth terminal situated in the beamwidth ($\approx$1000 nm diameter) of the NC antenna can transmit up to the satellite in either the NC or EC uplink bands and by proper frequency selection can transmit down in either NC or EC channels. Notice that the two 50-MHz channels in Fig. 7-5 are cross-strap channels, which convert an EC uplink signal to a NC downlink signal and vice versa.

The earth-coverage transmit channels (7250 to 7450 MHz) for a 200-MHz band are separated from the earth-coverage receive channels (7900–8100 MHz) by 450 MHz. Therefore, if there were not adequate transmit filtering in the TWT outputs, the seventh-order (4, 3) cross-product of two uplink carriers could fall as high as $7450 + 3(200) = 8050$ MHz and into the earth-coverage receive channel. The third- and fifth-order cross-products, however, cannot fall into a receive channel from the earth coverage transmit channel.

Frequency Reuse

Frequency reuse is the technique for transmission of two separate signals in the same frequency band by use of two separate types of antenna beams. The technique of particular importance here is the use of two coincident antenna beams of orthogonal polarizations, that is, vertical and horizontal polarization or right- and left-hand circular polarization. Figure 7-6 shows an artist's conception of a satellite employing vertical and horizontal polarizations, and employing polarizers in front of the antennas.

It appears feasible to obtain polarization isolation on the order of 30 dB. The polarization isolation of circularly polarized antennas depends on the axial ratios of the wave incident to the earth terminal receive antenna and on the receiving antenna axial ratio itself. For coincident elliptically polarized waves incident on the dual beam receive antenna, the coupling to the orthogonal port is given by [Duncan, Hamada, Wong, 1974]

$$\mathfrak{F} = \frac{(1 + \mathbf{r}_1^2)(1 + \mathbf{r}_2^2) - 4\mathbf{r}_1\mathbf{r}_2 \pm (1 - \mathbf{r}_1^2)(1 - \mathbf{r}_2^2)}{2(1 + \mathbf{r}_1^2)(1 + \mathbf{r}_2^2)} \tag{7-1}$$

where $\mathbf{r}_1$ is the axial ratio of the incident wave and $\mathbf{r}_2$ is the axial ratio of the orthogonal port antenna beam.

The plus sign in (7-1) yields the minimum coupling loss (coincident ellipse axes), and the minus sign yields the maximum coupling loss (orthogonal ellipse axes). Figure 7-7 illustrates the coupling loss for an incident wave axial ratio of 0.5 dB. If the receive antenna axial ratio is 0.5 dB, the minimum coupling loss is $\mathfrak{F} = 25$ dB.

Rainfall not only affects the attenuation of the received wave, as described in Chap. 6, but also depolarizes the incident wave. This rainfall can

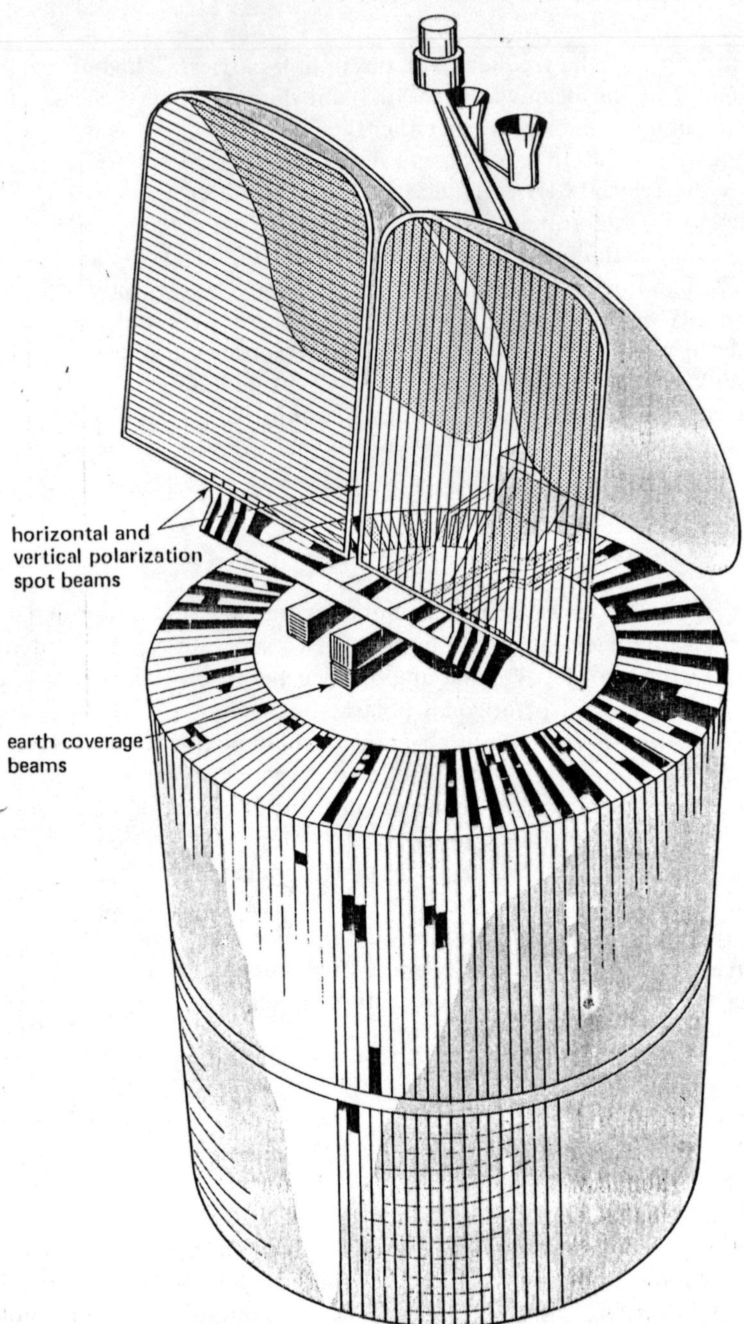

Fig. 7-6 Artist's conception of a satellite employing frequency reuse through transmission of vertical and horizontal polarizations (Courtesy of COMSAT General Corp.)

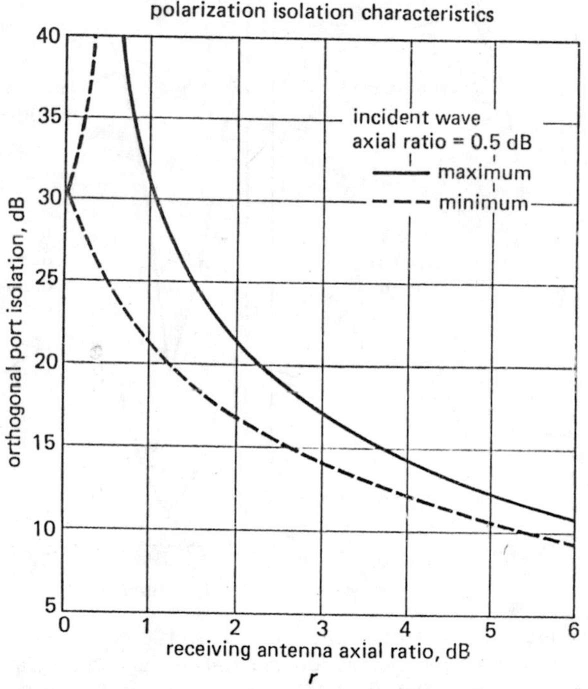

Fig. 7-7 Orthogonal port isolation versus antenna axial ratio [Reprinted from Duncan et al. article in *Communications Satellite Technology*, P. L. Bargellini, ed., by permission of the M. I. T. Press, Cambridge, MA. Copyright © 1974 by the Massachusetts Institute of Technology.]

significantly reduce the cross-polarization isolation needed for frequency reuse, particularly at frequencies above 10 GHz. Measured data illustrating this effect taken by Semplak [1974] show the effect of rain on 18-GHz signals of both vertical and horizontal polarization over a line-of-sight microwave link of 2.6 km. As can be seen in Fig. 7-8 the cross-polarization isolation for little or no rainfall was approximately 30 dB for both linearly and circularly polarized signals. At periods of high rainfall, the difference between the desired and depolarized component was only 8 dB for circularly polarized signals, but was about 18.5 dB for linear polarization. Thus linear polarization would appear to be the better choice with respect to rainfall depolarization effects.

Multibeam Satellite Antennas

Multibeam satellite antennas can increase both the power and bandwidth efficiency of the satellite link. The multiple narrow-beams provide

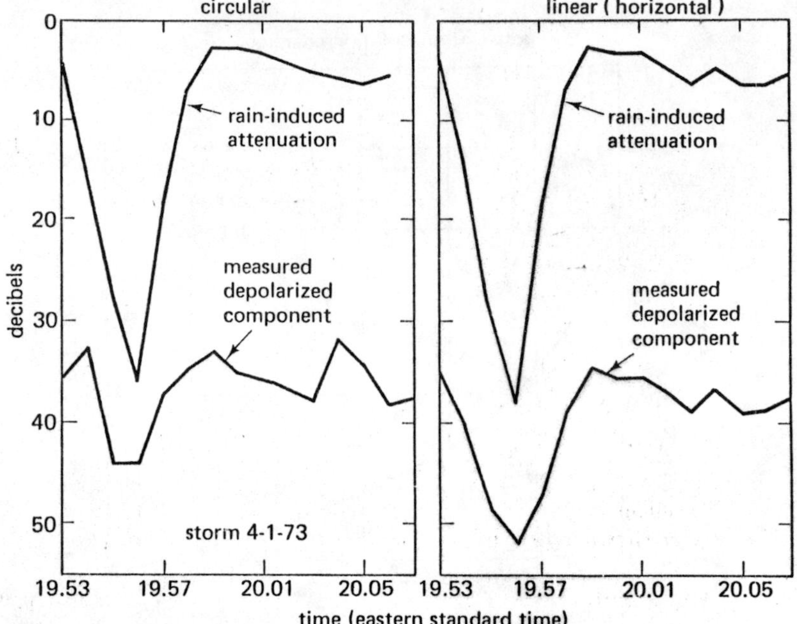

Fig. 7-8 Rain shower of April 1973 [Semplak article. Reprinted with permission from Bell Telephone Laboratories, Inc., Copyright 1974, The American Telephone and Telegraph Company]. Upper curves are rain-induced attenuation. Lower curves show the level of the depolarized component.

higher gain to localized areas of high density traffic thereby producing higher channel capacity for the same satellite RF power, relative to a 17.3° earth coverage antenna. Furthermore, with a multibeam antenna (MBA) the transponder channel can be connected with a multipole–multithrow switch to any one of a number of geographical regions on earth. High-gain multibeam antennas can be employed to reduce ground station cost at nodal communication points centered in the narrow beam.

The frequency reuse concept employing orthogonal polarizations can be generalized to provide many simultaneous uses of the same downlink frequency band by using a multibeam antenna [Shroeder, 1974]. (See Fig. 7-9.) If the frequency band is subdivided into two frequency subbands 1 and 2, and right-R and left-L hand circular polarizations are used for alternate beams, then contiguous beam coverage of earth's surface is provided with these four types of beams. Each type of beam is separated by one complete beamwidth from the same beam type appearing elsewhere. (Fig. 7-9(b)).

As described in the previous paragraph on frequency reuse, the isolation between individual narrow beams can provide additional frequency efficiency by using smaller guard bands between transponder channels. This ability

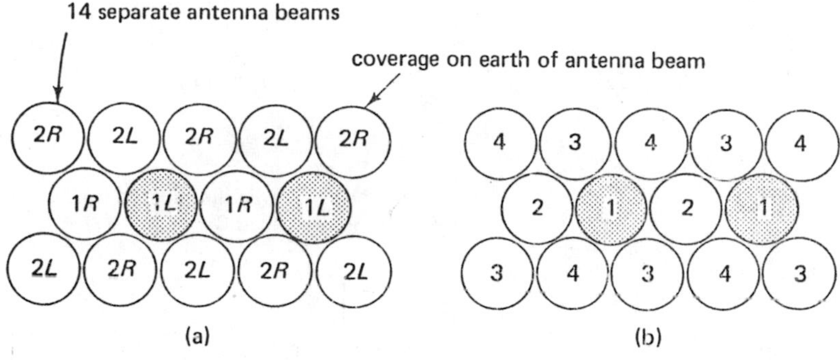

Fig. 7-9 Frequency reuse with a multibeam antenna. Frequency reuse provides contiguous coverage of the earth using frequency and polarization separation right and left hand (R & L) of the antenna beams in (a). This antenna provides four separate types of antenna beams as shown at the right in (b).

to improve bandwidth utilization becomes more important as satellite power levels increase.

This chapter is not intended as a discourse on antennas; however, some examples of multibeam antennas are appropriate in order to discuss some of their parameters affecting communication performance. These multibeam antennas require only a single reflector or lens, thus can be rather easily deployed on the satellite.

Figure 7-10 depicts one of the simplest forms of multibeam antennas to visualize. The antenna shown is a multibeam spherical reflector with multiple feeds to illuminate somewhat different sections of the reflector. It is desirable to be able to scan $\pm 8.6°$ relative to the center horn pointed at the subsatellite point, that is, to be able to scan over any section of the earth in view. From the geometry of the global scanning, the spherical reflector must have diameter L relative to the effective diameter of the reflector

$$L \cong 1.4D \tag{7-2}$$

For a focal length-to-diameter ratio $F/D = 0.7$ the physical surface area of the reflector is

$$A = 0.837L^2 \cong 1.64D^2 \tag{7-3}$$

The antenna gain is

$$G = \frac{\eta_a \pi^2 D^2}{\lambda^2} = 10\eta_a f^2 D^2 \tag{7-4}$$

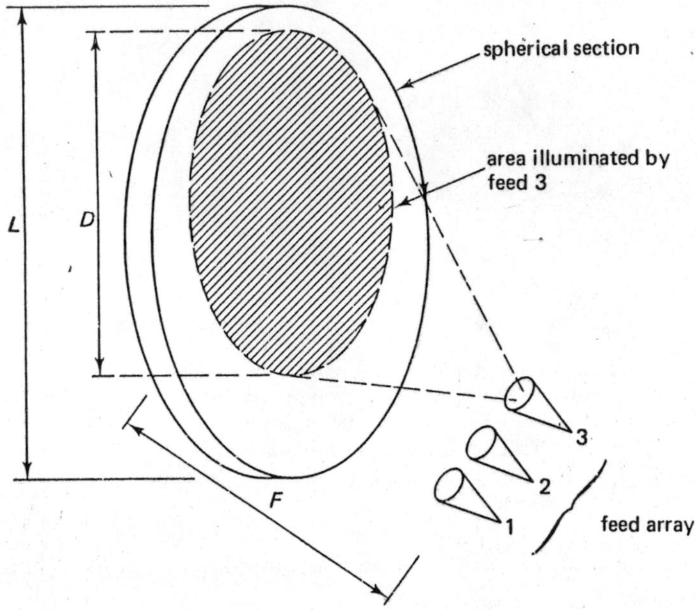

Fig. 7-10 Multibeam spherical reflector antenna [From Kiesling et al., 1972]

where η_a is the antenna efficiency, λ is the wavelength in the same units as D, and f is in GHz if D is in feet. The antenna 3-dB beamwidth in degrees is

$$\theta_{3\ \mathrm{dB}} = \frac{69^\circ}{fD} \tag{7-5}$$

Spherical reflectors have the advantage of being insensitive to beam steering because the illuminated region is always a section of a sphere. However, the sidelobe level is typically rather high next to the main beam, and the antenna aperture efficiency η_a is rather low—on the order of 25 percent. At frequencies in the millimeter waveband, there may be sufficient excess gain to permit use of this type of antenna.

The multibeam lens antenna has been described [Dion and Ricardo, 1971] as a means for achieving reasonably low sidelobe levels ≤ -20 dB. A photograph of an experimental model lens antenna is shown in Fig. 7-11. The lens is basically a double concave lens composed of waveguide segments that have been stepped (or zoned) to limit the width of the lens as shown in Fig. 7-12. The lens is illuminated by a cluster of feeds, one of which is located at point S in the figure. The antenna beam then points in direction α, the line from the feed to the center of the lens.

For earth coverage at X-band the diameter of the lens is 20 in., and the cluster of feedhorns is as shown in Fig. 7-12(b). When all nineteen of these

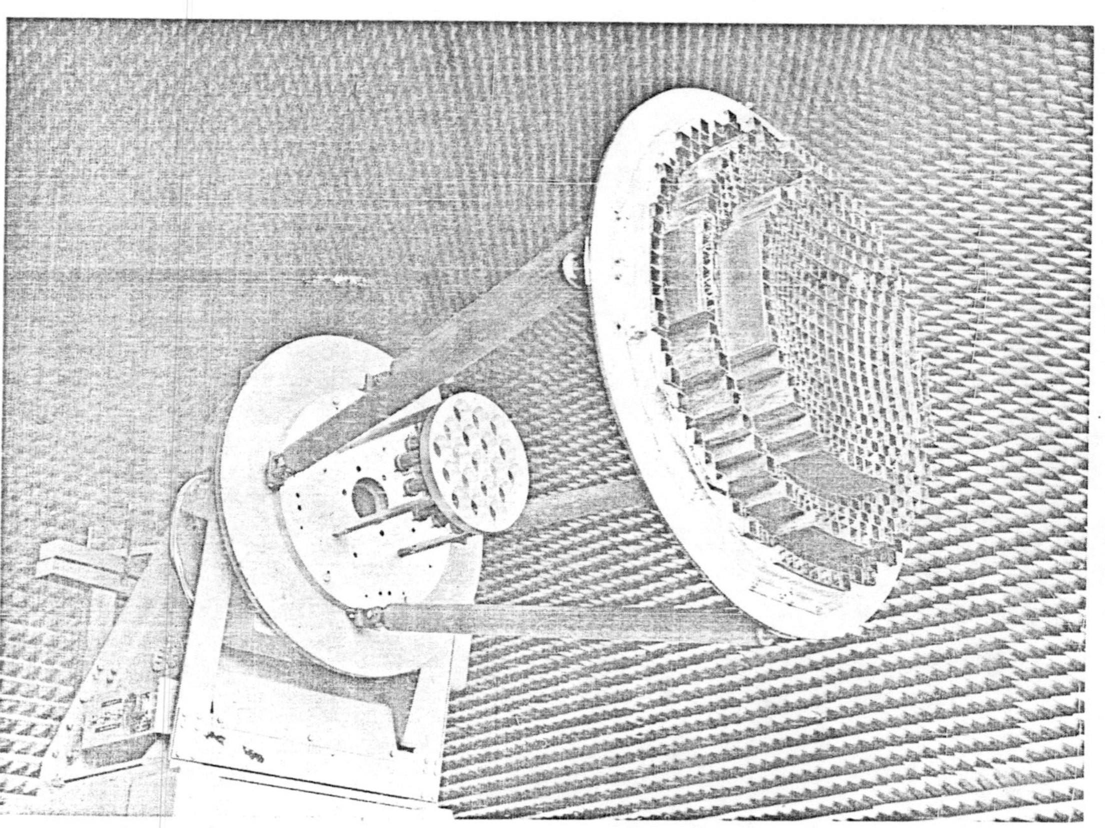

Fig. 7-11 Experimental lens antenna [Courtesy MIT Lincoln Laboratory, Dion and Ricardi, 1971] The stepped lens is at the right.

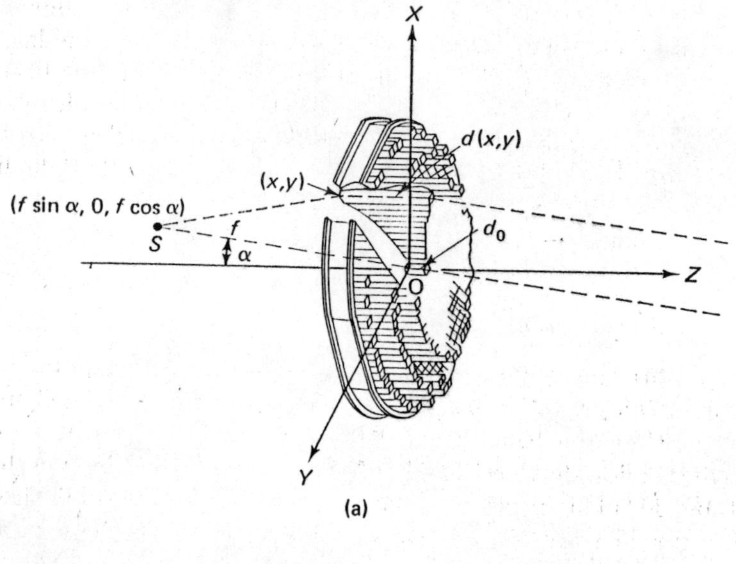

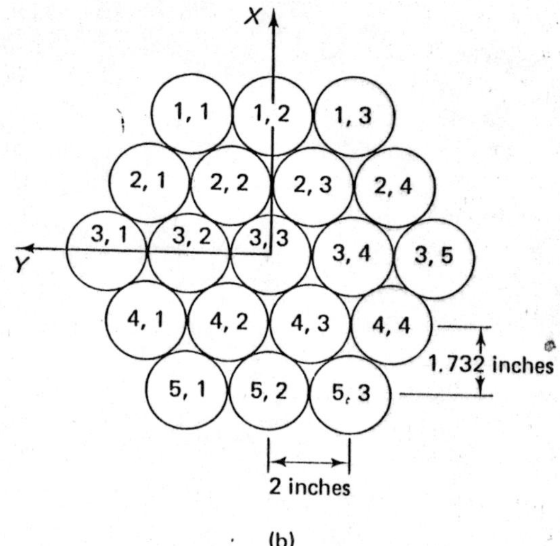

Fig. 7-12 Lens geometry of a multibeam antenna [Courtesy MIT Lincoln Laboratory, Dion and Ricardi, 1971] (a) Lens geometry, (b) feed cluster for the multibeam lens. Each of the feeds is driven by a separate variable power divider.

feedhorns are excited by RF power, the composite of the nineteen 3° beams produces a circular beam of 17.3° beamwidth, that is, earth coverage. The antenna gain is approximately 20 dB with less than 2 dB peak ripple. Excitation of a single feedhorn produces a gain of approximately 30 dB with a beamwidth of 3°. A feedhorn at point S offset by an angle α as shown in Fig. 7-12 produces a beam aligned at an angle α. The lengths of the lens waveguides properly phase all the contributing components for that beam.

The waveguide elements of the lenses are limited in bandwidth, however. The effects of waveguide dispersion based on a maximum phase error of $\pi/4$ radian is $B/f_0 = 25/(2 + k)$ percent, where k is one more than the number of steps in the lens. For this lens $k = 3$, and the bandwidth is then related to the center frequency by

$$\frac{B}{f_0} = \left(\frac{25}{2+k}\right) 5 \text{ percent} \tag{7-6}$$

However, this bandwidth is rather conservative and the use of a 0.5-dB decrease in efficiency leads to a 10 percent bandwidth as related to center frequency. Thus, at X-band the lens has sufficient bandwidth to accommodate the 500-MHz downlink channel, but the same lens cannot be used for both uplink and downlink frequency bands. A second effect occurs as the frequency is offset: the sidelobes of the antenna pattern increase. This increase in sidelobes must also be considered in the communication design.

Variable power dividers are connected to each feed to adjust the effective radiated power (ERP), P_i, in each of the M spot beams to give the desired MBA antenna patterns. The gain, and hence ERP, is subject to the constraint

$$\sum P_i = P_{\max} \tag{7-7}$$

If all beams are weighted equally, then each gain corresponds to that of an earth coverage antenna minus any power divider loss. (Power divider loss is approximately 0.3 dB per divider.)

In general, the number of beams in a nested set of circles in a 6-sided cluster is

$$M = \sum_{i=1}^{n} 6(i - 1) + 1,$$

where there are n layers in the nest. Thus M takes on values 1, 7, 19, 37, 61, ...

7-5 PROCESSING TRANSPONDERS

Onboard satellite processing can take a number of forms. Among these processing functions are: (1) active switching to distribute various uplink signals to the appropriate downlink amplifier and antenna; and (2) detection of the digital signals on the uplink and their regeneration for the downlink. See Fig. 7-13. The use of switching includes a "switchboard in the sky" concept, wherein different transponder input channels are switched by ground command to the appropriate downlink channel. An alternative switching

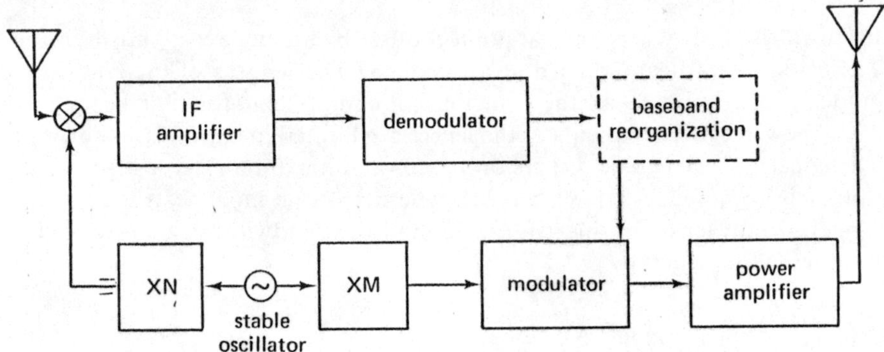

Fig. 7-13 Simplified block diagram of a demodulating-remodulating processing transponder [From D. P. Sullivan, 1970]

concept employs a preprogrammed switching sequence to provide satellite-switched time-division-multiple-access (SS-TDMA). The use of active time-division switching in a satellite transponder (Sec. 10-9) offers improved bandwidth and power efficiency compared with, for example, an FDMA technique.

Onboard demodulation of the uplink signals can improve the link performance. For example where up- and downlink SNRs are equal, this regeneration provides almost 2.6-dB improvement in performance relative to a linear transponder; while the error rate at the output of the ground terminal remains the same. Hence, if the SNR is the same at the regenerative satellite as at the receiving earth terminal, the error rates at the satellite and earth terminals are identical. Since these errors are independent, the total error rate at the earth terminal output includes those errors generated by the satellite as well as those generated by earth terminal demodulation. Since these error rates are equal, the total error rate is double that of the satellite itself. This tandem error effect corresponds to <0.5-dB loss in signal power. On the other hand, a 3-dB performance degradation occurs in a conventional linear transponder operating at the same power level when the earth terminal noise is added, and the error rate is thereby increased by approximately three orders of magnitude at low error rates.

Under many circumstances, however, the uplink SNR is relatively high, and there is little advantage to onboard regeneration. An exception occurs if either uplink interference is present or it is desired to demultiplex and remultiplex an uplink data channel in the satellite. The processing transponder constrains the type of signal that can be used to the particular modulation format built into the transponder. Thus the potential advantages of the regenerative transponder must be weighed against the constraints on signal modulation formats and the resulting lack of flexibility in changing modulation after the satellite is launched. In spite of these limitations, the potential for onboard processing, switching, and multiplexing of signals remains high.

CHAPTER 8

MULTIPLE-ACCESS EARTH TERMINALS

8-1 INTRODUCTION

Microwave earth terminals must be carefully designed to make most efficient use of the satellite power without undue cost to the total communications system. Clearly, the larger the earth terminal $G/\mathfrak{T}$, the more efficient is the use made of a given satellite power. However, earth terminal cost increases as the $G/\mathfrak{T}$ increases because of the added cost in the size of the structure and reflector and the costs of the low-noise receiver. Hence, in a total system, the costs of earth terminals and satellites must be weighed against each other. Although the gain and noise temperature are critical to the link performance, any detailed discussion of them in terms of implementation is more appropriately covered in books on antennas [Jasik, 1961; Hansen, 1966] and low-noise receivers [Blackwell and Kotzebue, 1961].

This chapter covers those elements of the earth terminal that relate more directly to the communications signal bandwidths and multiple-access transmission and reception. Specific areas covered include

1. Earth terminal configuration
2. Transmit/receive channel isolation—out-of-band interference and in-band (receive) intermodulation interference
3. Frequency conversion spurious effects—avoidance of spurious mixer outputs, control of oscillator phase noise and incidental

FM, reduction of power amplifier intermodulation products in the transmit band

4. IF/RF filter distortion effects on digital signals

The link performance calculations have already been discussed in Chap. 6 and are not repeated here. Other key elements of the earth terminal include the antenna and feed subsystem, antenna pedestal and drive system, automatic satellite tracking and servo system, power generation, and control subsystems. They are not, however, peculiar to the types of signals being utilized and are not dealt with here.

8-2 EARTH TERMINAL CONFIGURATION

Figure 8-1 is a simplified block diagram of a typical earth terminal communication system. Digital- or analog-modulated carriers from user modems enter the terminal at the IF frequency, often 70 or 700 MHz. The IF patch and switching panel enables a user carrier to access any one of a number of frequency converters and to switch automatically to a redundant frequency converter in the event of a unit failure. Figure 8-2 shows a photograph of the 60-ft diameter X-band earth terminal, the AN/FSC-78. This earth terminal has a $G/\mathfrak{T}$ of 39 dB/°K and is capable of transmitting signals over a 500-MHz frequency range.

Each frequency converter (Fig. 8-1) up-converts the IF signal in either one or two steps (single or double conversion) to the microwave band for further amplification and transmission. The tunable frequency synthesizers used in the converters are all phase locked to a frequency standard (usually cesium) and multiplied to the desired local oscillator (LO) frequency, perhaps by using a phase-locked multiplier. In simpler terminals, frequency conversion to the RF transmission frequency is often performed by a fixed frequency local oscillator rather than by a more complex microwave frequency synthesizer. Tunability, if any, can then be provided in the IF frequency range. The IF filters in the frequency converters are often equalized in amplitude and group delay versus frequency. In a general purpose earth terminal, the IF bandwidth is selected to equal or to exceed the widest expected transponder bandwidth. In simpler dedicated earth terminals, the IF bandwidth can be selected to match the transmitted signal bandwidth plus any tunability required for FDMA. Each up-converter also has an independent power control to allow adjustment of the transmitted carrier level.

Amplification at the microwave transmission frequency then takes place in the intermediate power amplifier (IPA) and high power amplifier (HPA), both of which are commonly wide bandwidth amplifiers (often TWTs). The output power of the HPA must be precisely controlled (perhaps to within

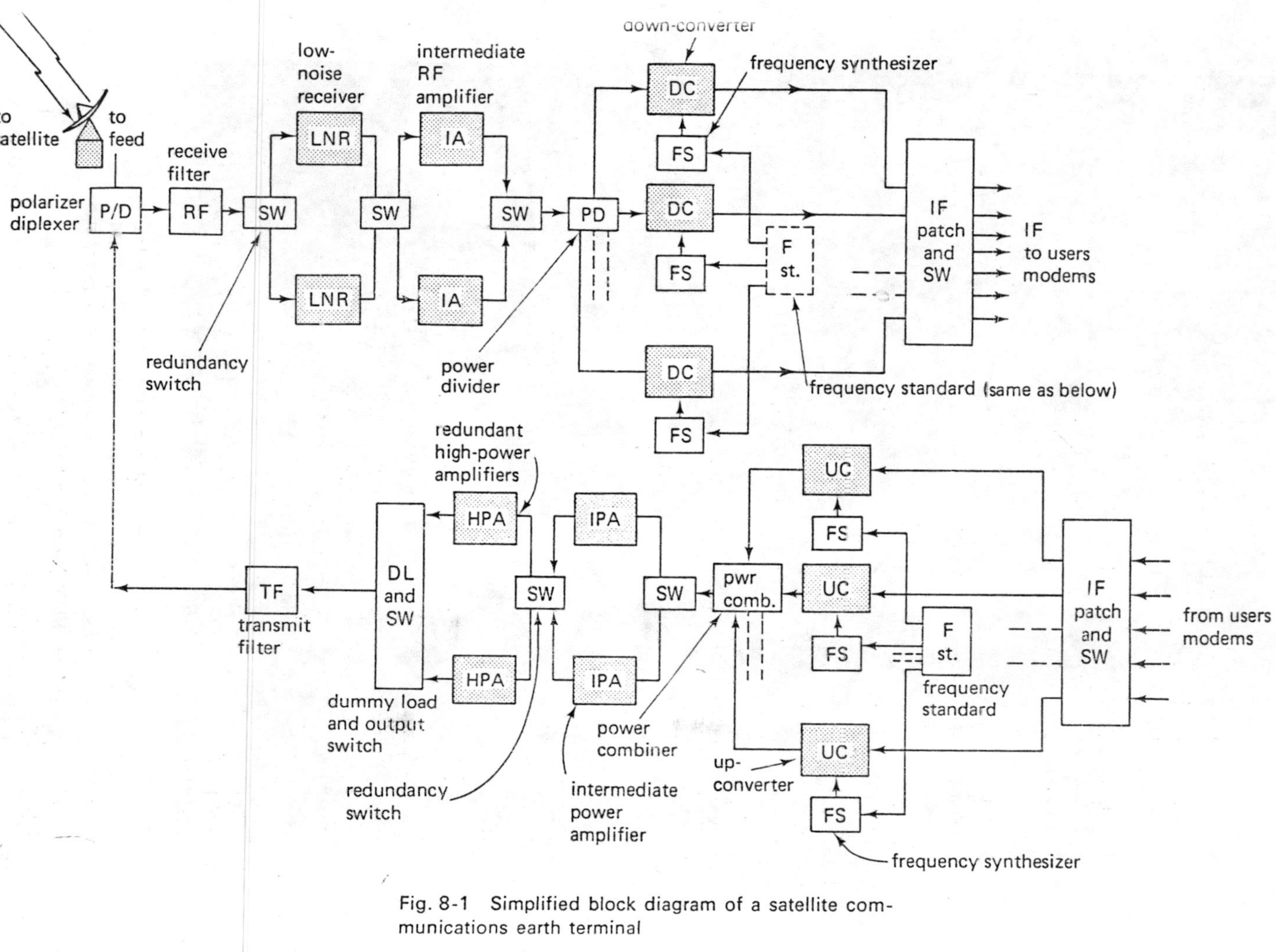

Fig. 8-1 Simplified block diagram of a satellite communications earth terminal

Fig. 8-2 Photograph of the 60-ft X-band satellite communications earth terminal AN/FSC-78 (Courtesy of the U.S. Army Satellite Communications Agency and Aeroneutronic-Ford Corporation).

less than 0.5-dB accuracy) over a wide range of power levels (100 w to 1 kw or more) in order that only the allocated portion of the satellite power is actually used by these carriers. As discussed later in Chap. 9, the HPA should have a peak power capability of 10 dB in excess of the total average carrier power for multiple FDM carriers to minimize intermodulation products. Furthermore, the power must be controlled by adjusting the HPA drive levels to compensate for the time-varying attenuation effects of rainfall on propagation and radome losses, and the relatively slow satellite-gain changes. The rainfall attenuation of the uplink can be estimated by measuring the attenuation of the satellite beacon and by compensating for the difference between transmit and receive attenuation levels caused by the frequency differential.

The output signal power in the transmit band is then heavily filtered to reduce undesired intermodulation and spurious components in the receive band, which otherwise could saturate the receive amplifiers for multicarrier transmission. Signal power is then passed through waveguide to the feed diplexer and emitted from the feed with appropriate polarization to illuminate the reflector itself.

At the receive end of the terminal, the satellite ERP has been attenuated by space losses of approximately 201.6 dB at X-band. Hence, there must be an extremely high level of isolation between transmit and receive signals provided by the diplexer and transmit and receive filters. This large difference in levels often requires precautions to avoid cross products generated by multiple transmitted carriers in passive elements such as waveguide and feed flanges, and filter tuning screws where a metal-oxide junction or other nonlinear impedance effects can exist. Even though these passive component intermodulation products may be more than 100 dB down from the transmitted signal level, they can fall in the receive band after transmit filtering and cause interference with a received signal.

The received signals pass through the diplexer and receive filter and are amplified by a wide-bandwidth, low-noise receiver (LNR)—for example, a paramp or tunnel-diode amplifier (TDA)—and an intermediate level amplifier (ILA). The microwave output of these (ILA) amplifiers is then passed to a power divider, which distributes the signals to the set of down-converters.

Down-conversion then usually translates the microwave output to the same IF frequency used in the up-converter. Commonalty in transmit and receive IF frequencies then allows the identical transmit and receive frequencies for the user modems, thus permitting self-testing or loop testing. The down-converter frequency synthesizers typically have tunability in 1-kHz steps or larger, thus allowing one to tune to any receive carrier and to compensate for drifts in satellite and ground station LO frequencies. This frequency synthesizer is phase locked to the same frequency standard used for the transmit up-converter—but obviously it is not the same standard as used at another terminal transmitting the carrier.

Several observations can be made regarding the effects of these elements on the digital modulation and the effects of the choice of multiple-access method on earth station design:

1. The use of wide-band digital modulation, particularly with wide-band TDMA, places restrictions on the various filter phase linearity (or group-delay distortions) and to a lesser extent on filter gain flatness (Chap. 13).
2. The use of lower rate digital modulation, such as that in single-channel-per-carrier Demand-Assignment-Multiple-Access (DAMA) or even lower rate systems (for example, 150 bps/carrier) places special requirements on the frequency convertor oscillator phase noise and incidental FM generated by oscillators and power amplifiers. The use of coding permits operation at lower E_b/N_o, hence often requires the use of even lower carrier recovery loop bandwidths. These coded signals can thus be more sensitive to oscillator phase noise in the earth terminal. (Chap. 12).
3. FDMA generally requires a multiplicity of frequency converters with full tunability over the satellite bandwidth ($\approx$ 500 MHz) or at least over the allocated portion of the transponder bandwidth. More specialized terminals can be designed to operate in a single 40-MHz bandwidth channel covered by one down-converter and to obtain tunability through the use of tunable digital modems.
4. Exclusive use of TDMA simplifies the terminal and allows the use of a single redundant up- and down-converter, since all terminals are addressed at the same center frequency. The wide bandwidth, perhaps $\approx$ 60 MHz, of this channel may make a higher IF center frequency than 70 MHz desirable. In general, this wide bandwidth channel must be carefully phase equalized to provide sufficient phase linearity. Thus the reduction in the number of frequency converters permitted by TDMA must be considered with the more stringent demands for frequency converter phase linearity as compared to FDMA.

8-3 TRANSMIT/RECEIVE CHANNEL ISOLATION

The large difference between earth terminal transmitted and received signal levels is caused by the very high path loss ($\approx$203 dB for an X-band satellite). Hence the combined isolation of the diplexer and transmit and receive filters must be correspondingly large (see Fig. 8-3). As the selectivity of these filters

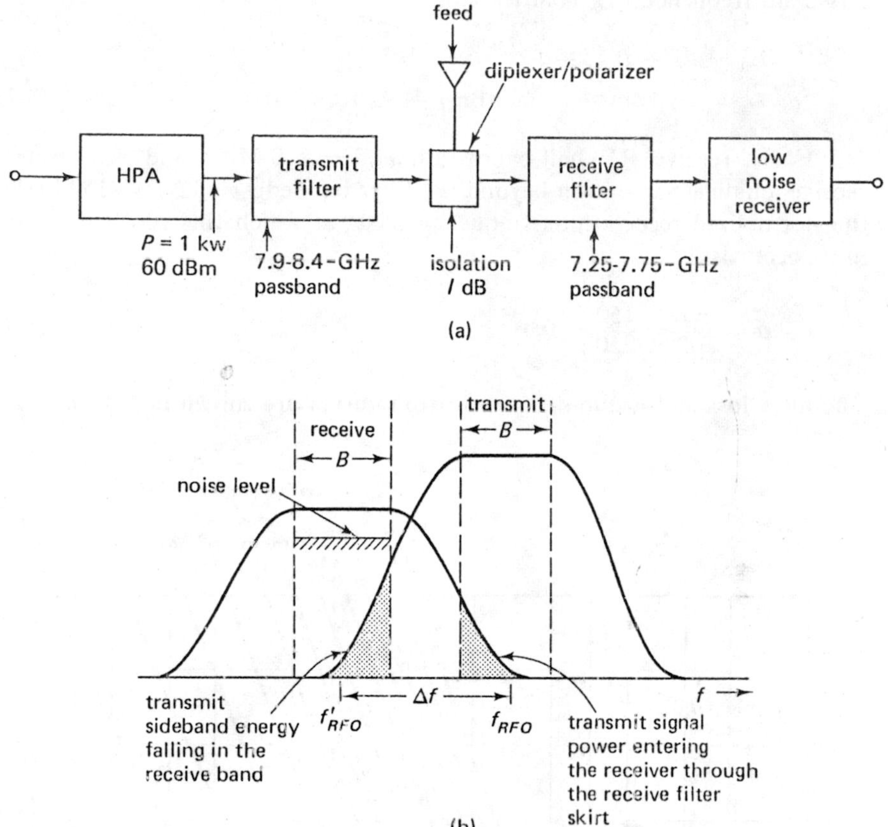

Fig. 8-3 Transmit/receive isolation filter for a typical 500-MHz satellite system bandwidth and $\Delta f = 150$ MHz separation between transmit and receive channels, (a) block diagram, (b) frequency relationships of earth terminal interference entering the receive band filter

increases, there is an increased filter group-delay distortion to signals near the bandedge.

An example of the signal levels and transmit/receive system isolation filter is shown in Fig. 8-3. Transmitted signal power outside this receive passband can still saturate the broadband LNR if this power is not sufficiently attenuated by the receive filter selectivity. (See Fig. 8-3b.) If the diplexer and feed polarizer provides $I = 50$ dB isolation between transmit signals and receive channels, the maximum transmitted power is $P = 1$ kw or 60 dBm, and the maximum out-of-band signal permitted at the paramp input is $P_{\max} = -60$ dBm, then the required receive filter loss to transmit passband frequencies L_R is at least

$$\begin{aligned} L_R &= P_T - P_{max} - I \\ &= 60\ \text{dBm} - (-60\ \text{dBm}) - 50\ \text{dB} = 70\ \text{dB} \end{aligned} \tag{8-1}$$

For a receive RF half-bandwidth $B/2 = 250$ MHz and a transmit-receive channel separation beyond the filter bandedges of $\Delta f = 150$ MHz, the normalized receive filter frequency offset at which this 70-dB isolation must occur is

$$\delta = \frac{\Delta f}{B/2} = \frac{150}{250} = 0.6 \tag{8-2}$$

The filter loss and group-delay distortion curves are shown in Figs. 8-4 and

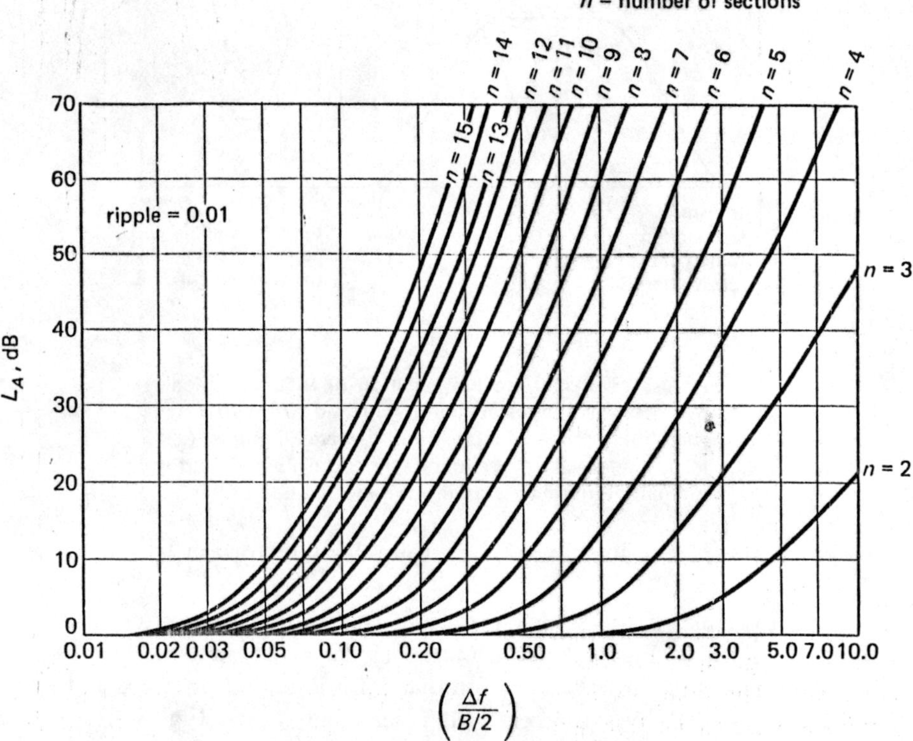

Fig. 8-4 Filter loss versus frequency offset Δf from bandedge for a 0.01-dB ripple Chebyshev filter of bandwidth B. The number of sections in the filter is n.

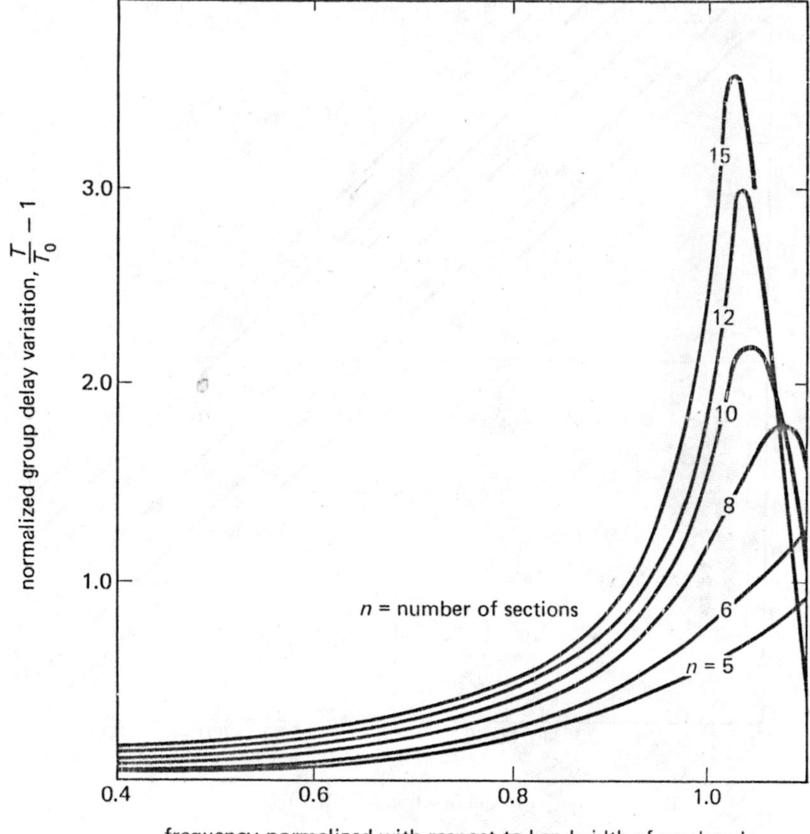

Fig. 8-5 Normalized group delay versus passband frequency for 0.01-dB Chebyshev filters, where f_{RFO} is the filter center frequency. The group delay is normalized to the midband group delay T_0. [From Lim and Scanlon, 1964]

8-5 for a 0.01-dB ripple Chebyshev filter. (See also Appendix B.) For an offset ratio $\delta = 0.6$ and 70-dB isolation, the filter requires 11 sections and produces a group-delay fluctuation at bandedge of approximately 75 nsec (Fig. 8-5). The relationships of typical group-delay distortions and phase nonlinearities versus frequencies to digital signaling losses are shown in Chap. 13.

The sidebands of the transmitted signal must be attenuated by the transmit filter to prevent them from directly entering the receive pass band and

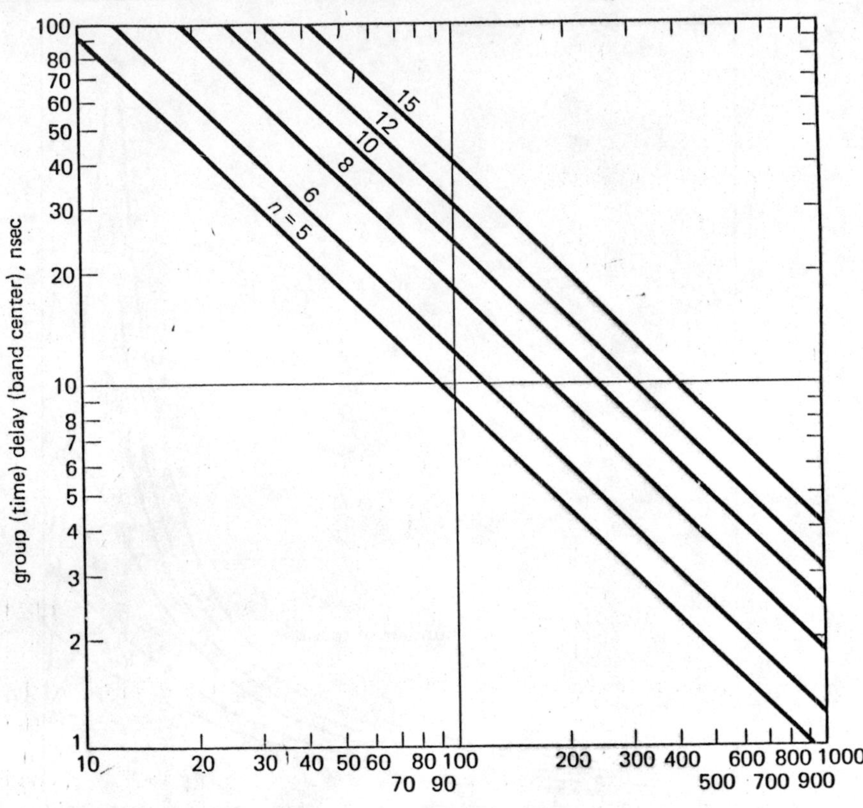

Fig. 8-6 Band-center group delay T_0 versus bandwidth and number of sections for 0.01-dB Chebyshev filters

exceeding the receive noise power spectral density over any significant frequency interval.

The allowable in-band noise and interference in the received signal band which is generated by the transmit channel can be selected so that in any minimum usable signal segment, say, 1 kHz, the interference level is 20 dB below the received noise power P_n in this frequency segment. Assume that intermodulation components are allowed to be no larger than P_T — 20 dB $= P_I$. Then the transmit filter loss L_T to the components in the receive band must be at least

$$\begin{aligned} L_T &= P_I + 20\text{ dB} - P_n - l \\ &= (P_T - 20) + 20 - (\kappa + \mathfrak{I}_s + B_{n1}) - l\text{ dB} \end{aligned} \tag{8-3}$$

where κ is Boltzmann's constant, -198.6 dBm/°K-Hz, $\mathfrak{I}_s$ is the system noise temperature, I is the diplexer isolation and B_m is the minimum signal bandwidth in hertz. For $P_T = 60$ dBm, $\mathfrak{T}_s = 200°\text{K} = 23$ dB, $B_m = 1$ kHz $= 30$ dB, we have a loss requirement of

$$L_T = 60 - (-198.6 + 23 + 30) - 50 = 155.6 \text{ dB} \tag{8-4}$$

A high-pass waveguide transmit filter can attenuate the lower frequencies in the transmitted signal spectrum if the waveguide cutoff is above the receive signal frequencies. Thus the transmitted signal power below the transmit pass band can be heavily attenuated by the waveguide below the cut off frequency.

A high-pass waveguide transmit filter with the receive band near or below cutoff can provide this degree of isolation (155.6 dB) to the lower receive frequencies. In a waveguide at a length 1 meter just above the cutoff frequency $f_c = 1/\lambda_c$, the theoretical attenuation L for a waveguide-intrinsic impedance η and surface resistivity R_s and dimension a, b is

$$L = \frac{8.686\, R_s}{b\eta\sqrt{1 - (f_c/f)^2}}\left[1 + \frac{2b}{a}\left(\frac{f_c}{f}\right)^2\right] \quad \text{dB/m} \tag{8-5}$$

for TE_{mo} waves where $\eta \triangleq \sqrt{\mu/\epsilon}$ and μ is the permeability and ϵ is the dielectric constant [Ramo and Whinnery, 1953, 368]. For free space the intrinsic impedance is $\eta_0 \cong 377$ ohms. The square root term in the denominator of (8-5) has a sharp frequency dependence which highly attenuates the lower frequency transmitted signals side bands. For high-power operation, this transmit filter may have to be cooled because the signal losses and intermodulation power absorbed by the filter can be substantial.

8-4 FREQUENCY CONVERSION AND SPURIOUS EFFECTS

The frequency converter design (for the up- and down-converters) and associated frequency synthesizers for an earth terminal depends heavily on the signaling requirements, including:

1. Total RF spectral range of tunability
2. IF bandwidth requirements
3. Constraints to use standard IF frequencies and modems
4. Avoidance of spurious mixer products in the mixer output
5. Filter distortion effects and requirements for amplitude/phase equalization
6. Oscillator phase noise and incidental FM

The IF bandwidth is relatively wide, and a wide range of tunability is often required; therefore, the frequency converter can be significantly more complex than that for a typical microwave LOS channel. In a general-purpose earth terminal, the IF bandwidth usually is in the range 5–130 MHz. In most cases the signals have significantly lower bandwidths than the IF. However, low-cost domestic terminals for special applications sometimes do not have nearly the tunability or bandwidth requirements of the general-purpose international communications terminal.

Up-Conversion

The up-converter translates an IF signal at a center frequency f_{IF} and bandwidth B_{IF} to an RF center frequency f_{RFO} in the range $f_{RFO} - B/2$ to $f_{RFO} + B/2$. (See Fig. 8-7.) Thus, a single-conversion frequency converter,

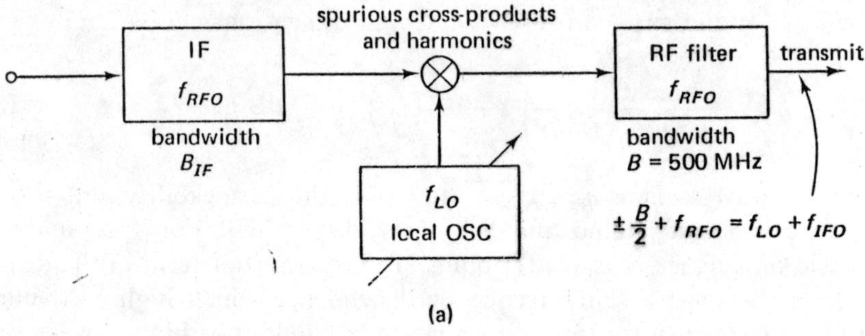

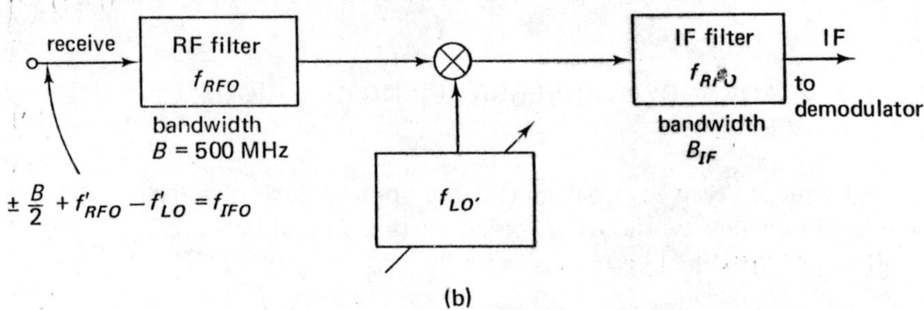

Fig. 8-7 Block diagram of the up-converter (a) and the down-converter (b). Note that the RF center frequencies $f_{RFO} \neq f'_{RFO}$ are not equal. Thus f_{LO} and f'_{LO} must vary in order to tune the transmitter and receiver over the entire RF bandwidth that is from $f_{RFO} - B/2$ to $f_{RFO} + B/2$, and $f'_{RFO} - B/2$ to $f'_{RFO} + B/2$, respectively.

or the second stage of a double-conversion technique, operates with a local oscillator (LO) at frequency f_{LO} where $f_{LO} = f_{RFO} - f_{IFO}$. This LO frequency selection avoids spectrum inversion (reversal of high and low frequencies) of the signal. Spectrum inversion is usually undesirable because of difficulties in running communication loop tests. Some multiphase signals reverse their coding under spectrum inversion. The LO synthesizer tunes over a range of B Hz, often in steps of 1 kHz.

The mixer used in this process is sometimes doubly balanced to minimize spurious frequency components. However, lower level cross products and harmonics are still generated at frequencies

$$m\frac{f_{IFO} \pm B_{IF}}{2} + nf_{LO} \tag{8-6}$$

The most significant spurious components are usually at frequencies corresponding to the third and fifth-order cross products.

$$\begin{aligned} &2f_{LO} \pm f_{IFO} \pm \frac{B_{IF}}{2}, \qquad f_{LO} \pm 2\left(f_{IFO} \pm \frac{B_{IF}}{2}\right) \\ &3f_{LO} \pm 2\left(f_{IFO} \pm \frac{B_{IF}}{2}\right), \qquad 2f_{LO} \pm 3\left(f_{IFO} \pm \frac{B_{IF}}{2}\right) \end{aligned} \tag{8-7}$$

and at harmonics of the LO and IF frequencies

$$2f_{LO}, \qquad 3f_{LO}, \qquad 2\left(f_{IFO} \pm \frac{B_{IF}}{2}\right), \qquad 3\left(f_{IFO} \pm \frac{B_{IF}}{2}\right) \tag{8-8}$$

where the $\pm$ signs indicate the edges of the cross-product spectrum.

The lowest order up-converter cross-product of significance is the (1, 2) cross-product at frequency f_{12}:

$$f_{12} = f_{LO} + 2\left(f_{IFO} \pm \frac{B_{IF}}{2}\right) = \left(f_{RFO} \pm \frac{B}{2} - f_{IFO}\right) + 2\left(f_{IFO} \pm \frac{B_{IF}}{2}\right) \tag{8-9}$$

where f_{LO} ranges from $f_{RFO} + B/2 \quad f_{IFO}$ to $f_{RFO} - B/2 - f_{IFO}$. We must constrain this cross-product to avoid the output RF frequency bands

$$f_{12} \neq f_{RFO} \pm \frac{B}{2} \qquad \text{constraint} \tag{8-10}$$

Thus, by combining (8-9) and (8-10) we obtain the constraints

$$f_{12} = \left(f_{RFO} \pm \frac{B}{2}\right) + f_{IFO} \pm B_{IF} \neq f_{RFO} \pm \frac{B}{2} \tag{8-11}$$

or

$$f_{IFO} \neq \pm \mathbf{B} \pm B_{IF} \tag{8-12}$$

where each $\pm$ can be selected independently, and the inequality denotes the frequency *region* that must be avoided. That is, the cross-product can fall anywhere within the region defined by the $\pm$ signs. Thus, f_{IFO} must exceed $\mathbf{B} + B_{IF} = 600$ MHz for $B_{IF} = 100$ MHz, and $f_{RFO} = 8.15$ GHz for $\mathbf{B} = 500$ MHz; and the IF frequency cannot be anywhere in the regions bounded by $\pm \mathbf{B} \pm B_{IF}$, respectively.

The second constraint is the (2, 3) cross-product at frequency f_{23}:

$$\begin{aligned} f_{23} &= 2f_{LO} - 3\left(f_{IFO} \pm \frac{B_{IF}}{2}\right) \\ &= 2\left(f_{RFO} \pm \frac{\mathbf{B}}{2} - f_{IFO}\right) - 3\left(f_{IFO} \pm \frac{B_{IF}}{2}\right) \end{aligned} \tag{8-13}$$

$$f_{23} \neq f_{RFO} \pm \frac{\mathbf{B}}{2} \qquad \text{constraint} \tag{8-14}$$

Thus we have the constraint on f_{IFO} from (8-13) and (8-14)

$$f_{23} = 2f_{RFO} \pm \mathbf{B} - 5f_{IFO} \pm 3\frac{B_{IF}}{2} \neq f_{RFO} \pm \frac{\mathbf{B}}{2} \tag{8-15}$$

Thus, the constraint on f_{IFO} is to avoid the frequency region defined by

$$5f_{IFO} \neq f_{RFO} \pm (\tfrac{3}{2})(B_{IF} + \mathbf{B}) \tag{8-16}$$

and one must avoid the IF center frequency region for $\mathbf{B} = 500$ MHz, $B_{IF} = 100$ MHz

$$f_{IFO} \neq 1630 \pm 180 \text{ MHz} = 1450 \text{ to } 1810 \text{ MHz} \tag{8-17}$$

The frequency constraints imposed to prevent these spurious components from falling in the output RF frequency band are depicted in Fig. 8-8 for an RF frequency band 7.9–8.4 GHz and an IF center frequency limited by $f_{IFO} < 2$ GHz. Thus, $f_{LO} > 7.9 - 2.0$ GHz $= 5.9$ GHz, and LO and IF harmonics are of little significance. Hence, the (1, 2) and (2-3) cross-products are the dominant constraints (8-11), (8-15).

Down-Conversion

Figure 8-7(b) illustrates the single conversion down-converter analogous to the up-converter of Fig. 8-7(a). For the same mixer cross-products, one must now avoid IF spurious outputs in the band $f_{IFO} \pm B_{IF}/2$. The lowest order cross product (2, 1) has components in the region

$$f_{21} = -f'_{RFO} \pm \frac{B}{2} + 2f'_{LO} = -f_{RFO} \pm \frac{B}{2} + 2\left(f'_{RFO} \pm \frac{B}{2} - f'_{RFO}\right) \tag{8-18}$$

This cross product must avoid falling in the IF frequency channel.

$$f_{21} \neq f_{IFO} \pm \frac{B_{IF}}{2} \qquad \text{constraint} \tag{8-19}$$

Thus we have the constraint based on (8-18), (8-19)

$$f_{21} = f'_{RFO} \pm 3\frac{B}{2} + 2f_{IFO} \neq f_{IFO} \pm \frac{B_{IF}}{2} \tag{8-20}$$

where the RF center frequency for receive is $f'_{RFO} = 7.5$ GHz. Thus, the constraint on f_{IFO} imposed by the (2, 1) cross product is

$$3f_{IF} \neq f'_{RFO} \pm 3\frac{B}{2} \pm \frac{B_{IF}}{2}$$

$$f_{IFO} \neq \frac{f'_{RFO}}{3} \pm \frac{B}{2} \pm \frac{B_{IF}}{6} = 2500 \pm 267 \text{ MHz} \tag{8-21}$$

and the IF center frequency must avoid the 2233–2767-MHz region.

The next-order down-conversion cross product (2, 3) then is in the range

$$\begin{aligned} f_{23} &= -2\left(f'_{RFO} \pm \frac{B}{2}\right) + 3f_{LO} \\ &= -2f'_{RFO} \pm B + 3\left(f'_{RFO} \pm \frac{B}{2} - f_{IFO}\right) \end{aligned} \tag{8-22}$$

Again, this cross product must avoid the IF output frequency band

$$f_{23} \neq f_{IFO} \pm \frac{B_{IF}}{2} \tag{8-23}$$

Thus, the constraint on the IF center frequency is obtained by combining the equations given as (8-22) and (8-23)

$$f_{23} = f'_{RFO} \pm 5\frac{B}{2} - 3f_{IFO} \neq f_{IFO} + B_{IF}/2 \tag{8-24}$$

$$4f_{IFO} \neq f_{RFO} \pm 5\frac{B}{2} \pm \frac{B_{IF}}{2} \tag{8-25}$$

$$f_{IFO} = \frac{f_{RFO}}{4} \pm 5\frac{B}{8} \pm \frac{B_{IF}}{8} = 1875 \pm 325 \text{ MHz} \tag{8-26}$$

and the receive IF must avoid the frequency 1550–2200-MHz region.

Since a common transmit and receive IF band is generally a requirement, the allowable IF center frequencies below 2.767 GHz are then in the frequency region 600 MHz $< f_{IFO} <$ 1550 MHz. In an actual design, of course, one may have to include additional cross-product constraints, or some of these already considered may be deleted because they have insignificant amplitudes. The significance of the components, of course, depends on the amplitudes of spurious products in the mixers being utilized.

The down-conversion constraints are also shown in Fig. 8-8.

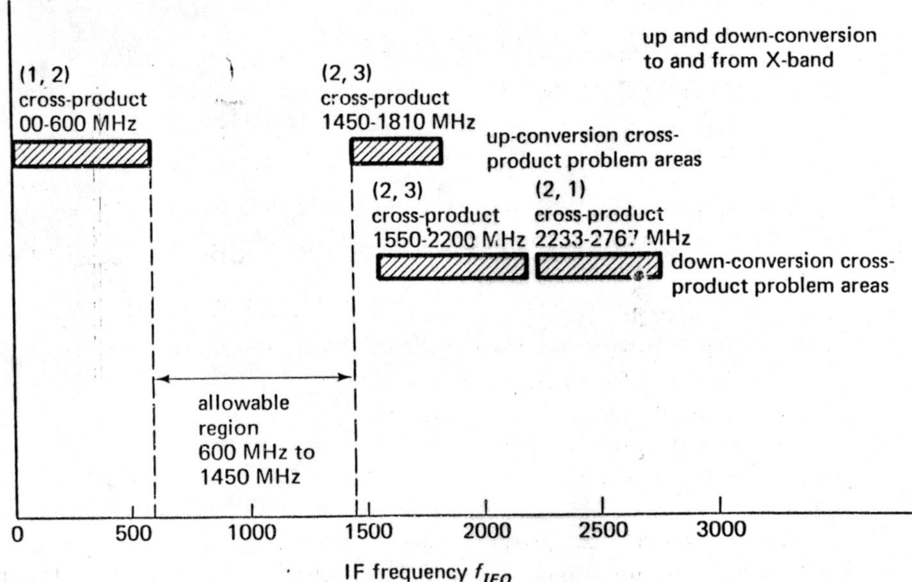

Fig. 8-8 Allowable regions for the IF center frequency f_{IFO}. Frequency regions with cross-hatch have cross-product interference.

Frequency Converter Filter Distortion

The 3-dB bandwidth of the earth terminal IF filters is often substantially broader than the frequently specified 1-dB bandwidth. This increase in bandwidth is desired in order to minimize distortion effects for broad bandwidth digital signals. Furthermore, for a significant number of examples even the 1-dB bandwidth is large compared to the bandwidth of a typical carrier. More restrictive bandpass filters to reduce noise and adjacent channel interference are usually part of the modem rather than the frequency converter. However, special dedicated terminals can be designed where the bandwidth is constrained to be the same as the modem signal. The frequency converter can be built as part of the modem for this type of terminal.

Nevertheless, the frequency converter and entire transmit/receive chain are usually specified with a restriction on amplitude flatness and phase linearity or group-delay distortion. The requirements on these parameters placed by high-performance demodulation of digital signals are calculated in Chap. 13 for some useful examples. These requirements are sometimes sufficiently severe that amplitude and group-delay equalizers are required in the IF or in the modem (perhaps using in-phase and quadrature channel transversal filters).

8-5 INTERMODULATION EFFECTS IN POWER AMPLIFICATION

An earth terminal operating at microwave frequencies typically uses klystron or TWT amplifiers, which are peak-power-limited devices and therefore become nonlinear as they reach saturation. Thus, for multicarrier transmission—for example, two carriers transmitted at frequencies f_1, f_2—there are intermodulation products generated at frequencies $nf_1 - mf_2$ where $|n - m| = 1$. Semiconductor amplifier devices may be used for the lower power applications, and they are already important at the lower microwave frequencies, L-Band, for example. These solid state devices have a similar peak power constraint.

Figure 8-9 shows a typical constraint on intermodulation products for a multicarrier environment consisting of 10 large carriers of equal power P_L and 10 small carriers of power $P_S = P_L/40$ each. The total desired average power for these 20-CW carriers is $410P_S$. For the cross products to remain 20 dB below the minimum carrier level P_S, they must be 46 dB below the total average output power. The total average output power of the TWT must then be backed off by some amount B_0 dB, which is calculated in Chap. 9 along with the signal suppression and various models for the nonlinearity.

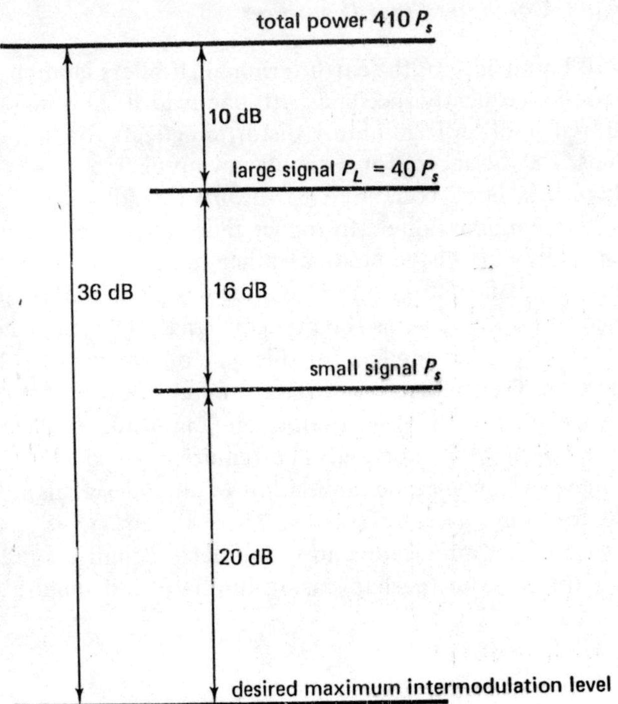

Fig. 8-9 Relative multicarrier and intermodulation product level for an example of 10 large and 10 small carriers. The large signal has 40 times the power of the small signal.

Methods of computing the allowable cross-product level are discussed in Sec. 11-6.

For TDMA operation, only a single signal passes through the HPA at one time, and therefore no power backoff is required. Furthermore, for very low-duty factors (1–0.1 percent), certain types of pulsed signal tubes, such as those used for radar, can be operated in an average power mode, provided that the thermal and power supply time constants are long compared to the TDMA pulse duration and that the HPA bandwidth is sufficiently large. If the TDMA pulse is too long, the TDMA pulse envelope will begin to sag, and there can also be significant phase modulation versus time introduced by the pulsed amplifier.

CHAPTER 9

FREQUENCY DIVISION MULTIPLE ACCESS AND SYSTEM NONLINEARITIES

9-1 INTRODUCTION

This chapter describes the fundamental properties of frequency division multiple access (FDMA) whereby multiple signals from the same or different earth terminals are transmitted on carriers at different RF center frequencies.

The FDMA class of signals includes many variations in the number and bandwidth of carriers transmitted by a given earth station. For example, we might transmit only one carrier per earth station, where the data to all receive terminals are multiplexed on that single carrier. Alternatively, separate carriers might be transmitted by each terminal for each receive earth terminal being addressed. This latter approach has the advantage that it requires the receive earth terminal to demodulate only the data intended for it, but this technique may not have any power or efficiency advantage. Finally, one can provide a separate carrier for each voice channel. This single-channel per carrier (SCPC) system has the advantage that it can be used in a demand-assigned mode and can thereby improve the system efficiency. These SCPC carriers can also be voice-activated such that carrier power is turned on only during time intervals when the voice envelope exceeds a threshold level.

The transponder and other amplifier nonlinearities have a significant effect on the FDMA signals, causing signal suppression and intermodulation products. An analysis is given of both of these effects. Examples are given later in the chapter for several different signal types. Both the effects of amplitude nonlinearities and AM/PM conversion effects are discussed. Methods of improving the signal-to-intermodulation ratio by proper choice of frequency plans and satellite power backoff are also described.

9-2 PRINCIPLES OF FREQUENCY DIVISION MULTIPLE ACCESS

Here we address the simplest form of multiple access wherein each carrier is transmitted at a different frequency. In FDMA, each signal is assigned a separate nonoverlapping frequency channel, and power amplifier intermodulation products are either accepted or minimized by appropriate frequency selection and/or reduction of input power levels to permit quasi-linear operation (Fig. 9-1). Attention is focused on the satellite transponder effects since

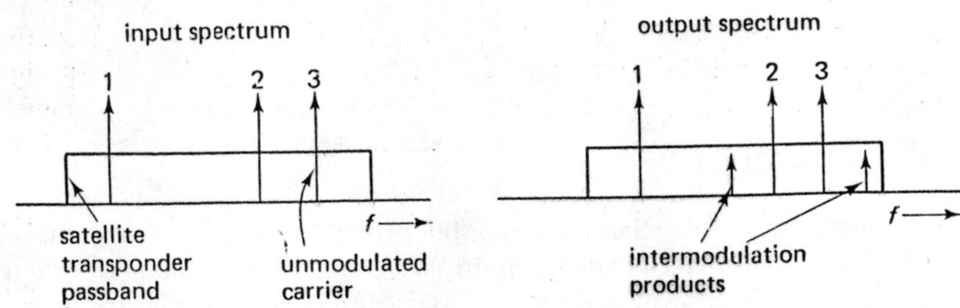

Fig. 9-1 Typical input and output spectra for FDMA. Intermodulation products are shown corresponding to nonlinear repeaters and unmodulated carriers.

this power is more critical and costly than earth terminal power. Typically, one might reduce the satellite average output power by 50 percent or more to reduce IM products to an acceptable level with a high density of input signals. Oscillators with good long-term stabilization are employed to keep the signals properly centered in nonoverlapping frequency bands.

FDMA Channel Formats

The format of the frequency channel utilized for FDMA depends on signal distortion, adjacent channel interference, and intermodulation effects caused by the satellite transponder nonlinearities. Figure 9-2 shows a simplified FDMA format for a single channel of a satellite transponder. Each FDMA carrier can either carry a multiplexed set of user data streams, or it can carry only a single user's bit stream as in the SCPC system described

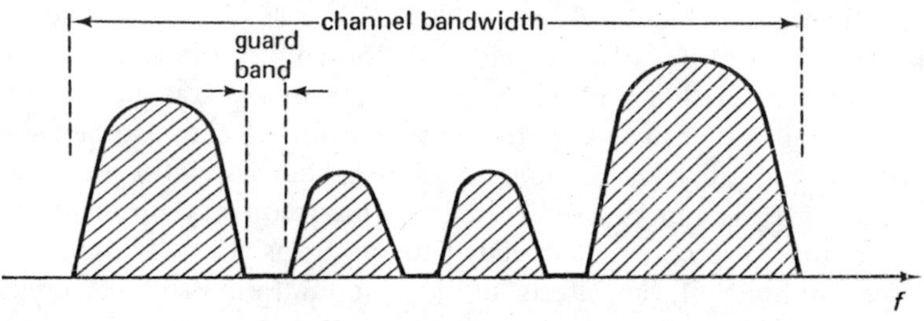

Fig. 9-2 Simplified format for FDMA signals in a single satellite channel

in Sec. 9-3. The carriers can either be destination oriented or a single carrier can carry data destined for several receive earth stations.

Guard bands must be used between adjacent frequency channels to minimize adjacent channel interference and these, of course, reduce the frequency utilization efficiency of the transponder channel. The required size of the guard band depends in part on the residual sidebands in each transmitted signal. Figure 9-3 shows the power spectral density of a QPSK signal at 1M symbol/sec (2M bps). Transmission filters can be employed to cut off the signal spectrum at IF bandwidths between 1 and 2 MHz. The smaller bandwidths must utilize some form of equalization. However, these sidebands can build back up when the signal is fed through a nonlinearity [Robinson, Shimbo, Fang, 1973] and envelope fluctuations produced by filtering are

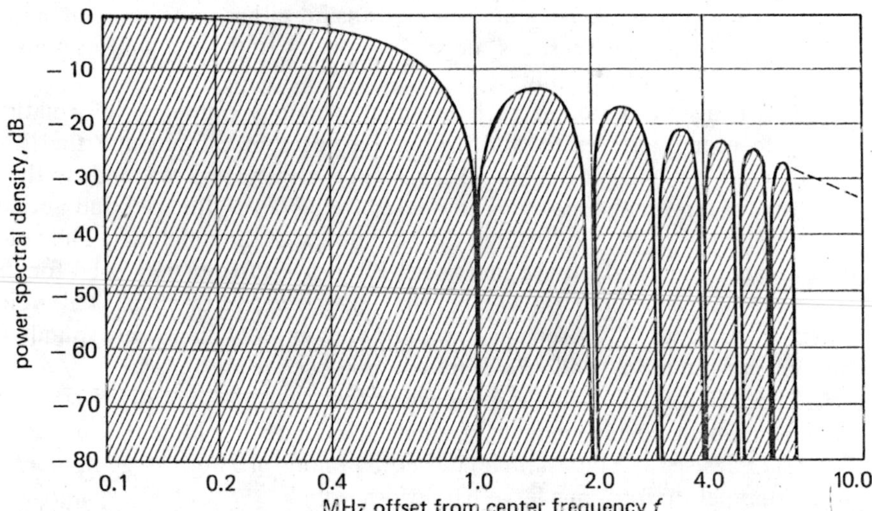

Fig. 9-3 Power spectrum of QPSK at 1.0M symbols/sec shown on a logarithmic frequency scale

reduced. The guard band between adjacent frequencies must also account for the frequency drifts of the oscillators controlling the signal center frequencies at the satellite and earth station frequency translators. Doppler shifts of satellites that are not perfectly synchronized can also be significant for very low data rate transmissions. Satellite beacons used for antenna tracking or pilot signals can be used to reduce this frequency uncertainty if the beacon frequency is coherently related to the translation frequency.

Computations of the effects of intermodulation products created by satellite and earth-station nonlinearities must account for changes in the relative signal strength received at the satellite. These signal levels can change because of ground-station transmit ERP fluctuation, localized rain losses that can significantly attenuate the signal received from one earth station and not another, and antenna pointing losses at both the ground station and the satellite.

9-3 SINGLE-CHANNEL DEMAND-ASSIGNMENT TECHNIQUE

An important variation of the FDMA transmission technique is Single-Channel Per Carrier Demand-Assigned Multiple-Access (SCPC-DAMA) [Puente et al., 1972, 218–229]. The terminology refers to the fact that each carrier is modulated by a bit stream representing a single user's voice channel (Fig. 9-4). One form of SCPC is the SPADE* system devised by COMSAT Corporation. A given earth station may transmit either none, one, or many of these single-channel carriers, depending on its traffic demand at a given time and equipment capability.

The transmit center at each terminal provides signaling information from the user interfaces to SPADE for purposes of establishing the link. The demand-assigned switching subsystem at each terminal responds to these requests for service and allocates an unused frequency to the user and notifies other terminals of its use through the common signaling channel. The voice detector gates on and off the individual PSK or QPSK carrier used for the voice channel, depending on whether the talker is active or silent. This voice detector can operate by sensing the harmonic content in voiced sounds as opposed to the more random nature of some background noise.

SPADE utilizes a single RF carrier for each 64-kbps digitized voice channel (which corresponds to a single 4-kHz voice channel sampled at 8000 samples/sec at 8 bits/sample) rather than multiplexing a large number of voice channels to form one large-bandwidth signal. QPSK at 32k symbols/sec

---

*Single-channel per carrier Pulse-code modulation multiple-Access Demand assigned Equipment (SPADE).

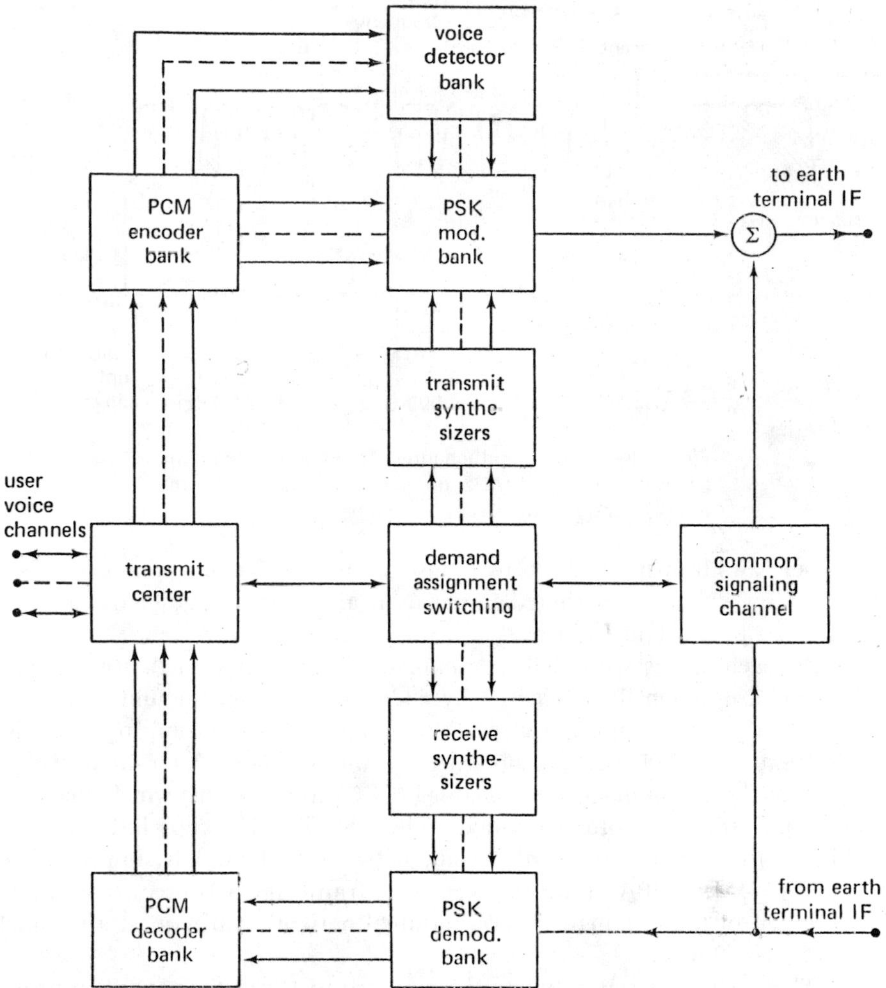

Fig. 9-4 Simplified diagram of SPADE demand-assigned multiple-access system

is used with a signal bandwidth of 38 kHz. A channel spacing of 45 kHz allows for a 7-kHz guard band between channels. The channel spacing allows for possible drifts in the satellite transponder frequency translators; however, pilot tones generated in an earth terminal provide a frequency reference that can be used by another terminal's receiver in an AFC loop to compensate for these drifts.

Figure 9-5 depicts the SPADE frequency spectrum for one of the Intelsat IV 36-MHz transponders. The figure shows a total of 800 channels (400

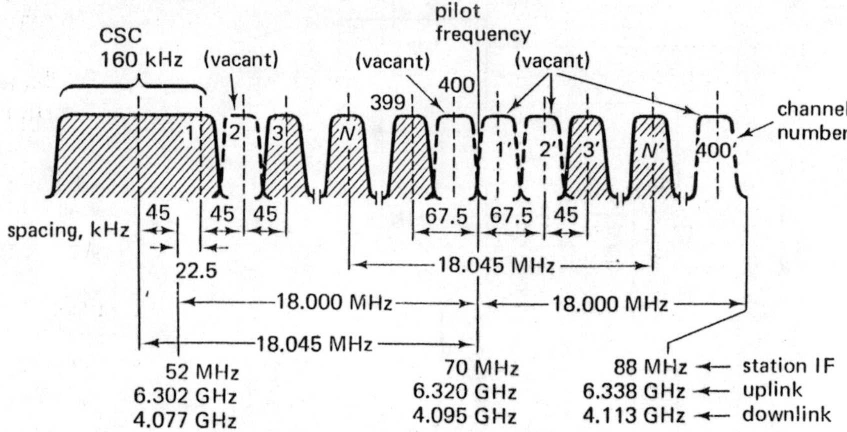

Fig. 9-5 SPADE multichannel frequency allocation spectrum for a 36 MHz transponder channel [From Edelson and Werth, 1972]

channel pairs transmit and receive) plus a common signaling channel (CSC) at 128 kbps which is time-shared among users and is used to assign frequencies and to request channel usage.

As each call is requested or completed by the user earth station and channels are assigned or released by a demand-assignment unit using the common signaling channel, each earth station updates a current log of available frequencies.* This log is updated continually by each terminal, which monitors the common signaling channel (CSC) to determine which channels are being utilized by other terminals. The CSC is a 128 kbps PSK channel which is time-shared among all terminals by using Time Division Multiple Access (TDMA). SPADE utilizes a 50-msec frame and a 1-msec access time. Thus each of the terminals in a 50-terminal network can request a channel once every 50 msec.

The earth terminal initialing the request must wait the roundtrip propagation time of 0.24 sec before the channel request is received by the destination terminal. The initiating terminal has requested at random a frequency pair of those remaining unassigned in the frequency log. If during the propagation time of the channel request another terminal requests the same frequency, the channel is considered busy and the transmitting terminal must initiate another frequency request. The random selection of a frequency channel makes it unlikely that two terminals will simultaneously request

*A net control terminal (NCT) can thereby simplify equipment at the remote terminals and control the assignments of each channel equally well or better than the separate assignment at each local terminal. Since the more complex control equipment is then located only at the NCT, the remote terminals can be simplified.

the same channel twice in a row unless there are very few frequencies remaining unassigned.

The use of SPADE can provide 800 voice channels in a 36-MHz band. This capability compares favorably with the use of FDM/FM analog transmission for multiple carriers as shown in Table 9-1 [Edelson, Werth, 1972].

Table 9-1 COMPARISON OF SPADE WITH FDM/FM VOICE CHANNEL TRANSMISSION FOR THE INTELSAT IV 36-MHz TRANSPONDER

Multiple Access	*RF Bandwidth/Carrier*	*Voice Channels/ Carrier*	*Total Accesses/ Transponder*	*Total Voice Channels/ Transponder*
FDM/FM	2.5 MHz	24	14	336
FDM/FM	5 MHz	60	7	420
SPADE	0.045 MHz	1	800	800

In addition, each carrier is gated on and off by voice activity on each channel [Miedema and Schachtman, 1962, 1455–1476; Farriello, 1972, 55–60]. Thus, the carrier is off while one speaker in a duplex link listens to the other speaker, or during pauses between words. This gating is accomplished by a separate voice-activity detector. For a large number of simultaneous conversations, fewer than 40 percent of the voice channels are active at any one instant. Hence, one achieves at least a 4-dB average power advantage. The time activation by each voice signal also time gates the intermodulation products in this random manner and significantly reduces their effect. McClure [1970] has shown that the worst intermodulation noise is reduced by 3 dB.

To accommodate the voice activation of the carrier, each PSK or QPSK demodulator must reacquire rapidly at the beginning of each speech segment or a large initial portion of the segment will be lost. Hence, the carrier recovery must be relatively rapid, that is, within the first 10 bits. Carrier acquisition and phase noise effects are analyzed in Chap. 12.

9-4 FDMA SIGNAL SUPPRESSION BY ARBITRARY BANDPASS NONLINEARITIES

Satellite transponder or earth terminal power amplifier amplitude nonlinearities generate intermodulation products and can cause suppression of the input signals and disproportionate power sharing of the satellite power. We first examine the simplest effect to calculate—namely, suppression of a desired signal in a multicarrier environment. We calculate signal suppression effects for two-signal and n signal inputs with arbitrary amplitude nonlinearities and examine specific examples of some transponder models. We then

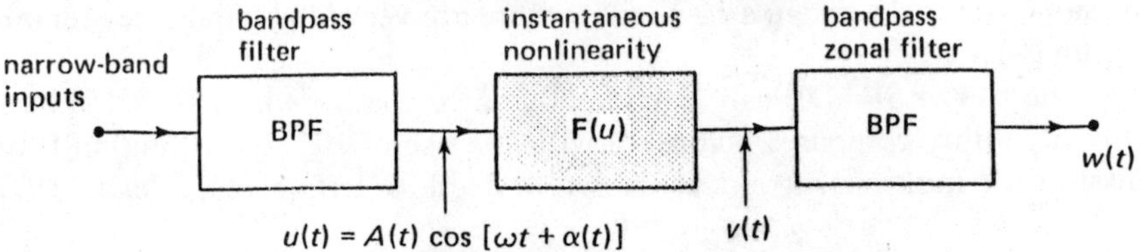

Fig. 9-6 Transponder model with bandpass nonlinearities. The function $F(u)$ is the instantaneous amplitude nonlinearity.

consider the generalized transponder model in Fig. 9-6 which contains an instantaneous amplitude nonlinearity. AM/PM conversion and combined nonlinearities are analyzed in Secs. 9-9, 9-10 respectively. The bandpass input to the transponder amplitude nonlinearity is defined as

$$u(t) = A(t) \cos [\omega t + \alpha(t)] \triangleq A(t) \cos \theta(t) \tag{9-1}$$

where $\theta = \omega t + \alpha(t)$. This signal representation can correspond to either one or a multiplicity of input signals, that is, an arbitrary bandpass input. The output of the instantaneous nonlinear element is defined as

$$v(t) = \mathbf{F}[u(t)] = \mathbf{F}(A \cos \theta) \tag{9-2}$$

Since v is a periodic function of θ, we can express v as a Fourier series in θ:

$$v = \tfrac{1}{2}g_0 + g_1 \cos \theta + g_2 \cos 2\theta + \cdots \tag{9-3}$$

The zonal bandpass filter passes only those frequency components in the vicinity of θ and neglects those at $0, 2\theta, 3\theta, \ldots$. The filter is assumed to be distortionless (linear phase and constant amplitude) within the fundamental zone. Thus, the signal component at the transponder output in the fundamental frequency zone* at ωt is

$$w(t) = g_1(A) \cos \theta \tag{9-4}$$

where

$$g_1(A) \triangleq g(A) = \frac{1}{\pi} \int_0^{2\pi} \mathbf{F}(A \cos \theta) \cos \theta \, d\theta \tag{9-5}$$

is the Chebyshev transform of $\mathbf{F}(x)$ of order 1, and gives the envelope of the signal output in the fundamental zone [Blachman, 1971, 398–404]. First-order

*It is clear that even though the output has the same angular frequency as the input (that is, zero crossings are unaffected), the output generally contains distortion components.

Table 9-2 FIRST-ORDER CHEBYSHEV TRANSFORMS

Amplitude Nonlinearity $F(u)$		Envelope in Fundamental Zone Chebyshev Transform $g_1(A) = \frac{1}{\pi}\int_0^{2\pi} F(A\cos\theta)\cos\theta\, d\theta$
u^n	n-odd $n > 0$	$2\binom{n}{\frac{n-1}{2}}\left(\frac{A}{2}\right)^n$
$\sin u$		$2J_1(A)$
u		$2I_1(A)$
$\sinh u$		$2I_1(A)$
$\sqrt{\frac{2}{\pi}}\int_0^u [e^{-x^2/2}]\, dx$		$\sqrt{\frac{2}{\pi}}\, Ae^{-A^2/4}\left[I_0\left(\frac{A^2}{4}\right) + I_2\left(\frac{A^2}{4}\right)\right]$
$F(Cu)$		$g_1(CA)$

Chebyshev transforms are listed in Table 9-2. Note that the effective gain of the nonlinearity is $g(A)/A$. The inverse transform of $g(A)$ is given by

$$F(A) = \frac{1}{2}\int_0^{2\pi} [g_1(u\cos\phi) + ug_1'(u\cos\phi)\cos\phi]\, d\phi \tag{9-6}$$

and gives the instantaneous amplitude nonlinearity.

Two Sinusoidal Inputs

The simplest example of signal suppression is where two sinusoids are applied to the nonlinearity [Blachman, 1964]. Add a weak signal $B\cos[\omega t + \beta(t)]$ to the signal $A\cos[\omega t + \alpha(t)]$ where $B \ll A$. The sum of these two signals can be represented as the sum of the two phasors, as shown; that is,

$$\begin{aligned} u &= A(t)\cos\theta(t) = Ae^{j\alpha} + Be^{j\beta} \\ &= [A + B\cos(\beta - \alpha)]e^{j\alpha} + jB\sin(\beta - \alpha)e^{j\alpha} \end{aligned} \tag{9-7}$$

Figure 9-7 illustrates this phasor representation of the input. Assume that $g(A)$ is sufficiently well behaved to be represented by a Taylor's series about A. Thus, the envelope output $g(A)$ of the nonlinearity becomes

$$g[A + B\cos(\beta - \alpha)] \cong g(A) + Bg'(A)\cos(\beta - \alpha) \quad B \ll A \tag{9-8}$$

Because the nonlinearity is instantaneous, the fundamental zone input and

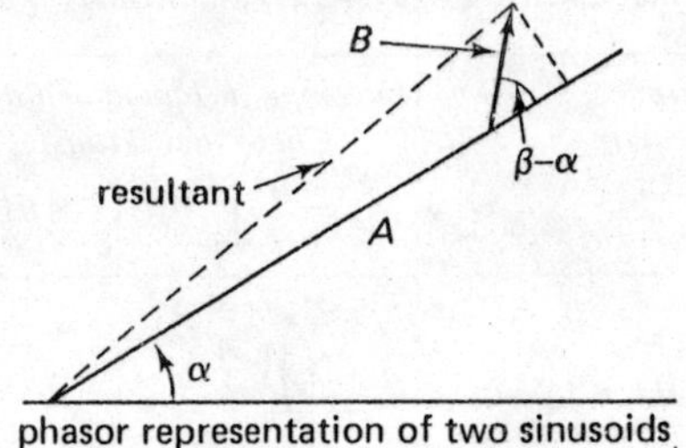

Fig. 9-7 Phasor representations of two sinusoids

output phase are identical and approximately equal to

$$\theta = \alpha + \frac{B}{A} \sin (\beta - \alpha) \tag{9-9}$$

This resultant output phase offset$(B/A) \sin(\beta - \alpha)$ introduces a quadrature component with respect to $e^{j\alpha}$. Hence, the output w can be represented as the sum of inphase and quadrature phasors

$$\begin{aligned} w &\cong [g(A) + Bg'(A) \cos (\beta - \alpha)]e^{j\alpha} \\ &\quad + j\left[g(A) + Bg'(A) \cos (\beta - \alpha)\right]\left[\frac{B}{A} \sin (\beta - \alpha)\right]e^{j\alpha} \\ &\cong [g(A) + Bg'(A) \cos (\beta - \alpha)]e^{j\alpha} \\ &\quad + j\frac{B}{A} g(A) \sin (\beta - \alpha)e^{j\alpha} \quad \text{for} \quad B \ll A \end{aligned} \tag{9-10}$$

Since $\cos \phi = (e^{j\phi} + e^{-j\phi})/2$, the fundamental zone output w in (9-10) can be written as

$$\begin{aligned} w &\cong g(A)e^{j\alpha} + \frac{e^{j\beta}}{2}\left[Bg'(A) + \frac{B}{A}g(A)\right] \\ &\quad + \frac{e^{j(2\alpha-\beta)}}{2}\left[Bg'(A) - \frac{B}{A}g(A)\right] \end{aligned} \tag{9-11}$$

Thus, the output contains components at angular frequencies

$$\omega + \dot{\alpha} \qquad \omega + \dot{\beta} \qquad \omega + 2\dot{\alpha} - \dot{\beta} \tag{9-12}$$

where $\dot{\alpha} \triangleq d\alpha/dt$, $\dot{\beta} \triangleq d\beta/dt$. Clearly, the $\omega + 2\dot{\alpha} - \dot{\beta}$ term is an intermodulation component (Fig. 9-8).

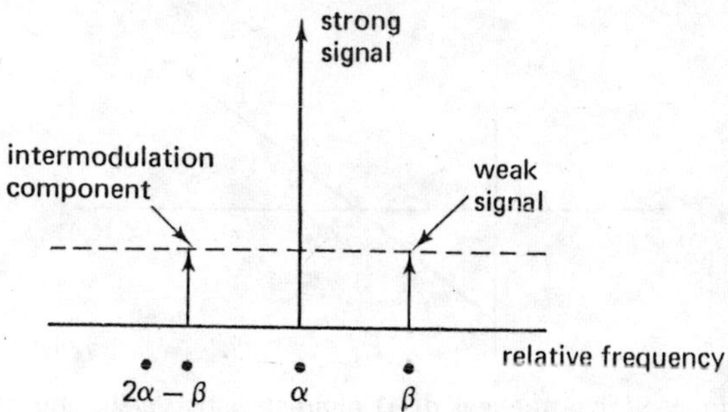

Fig. 9-8 Output spectrum with two input frequencies

Weak Signal Gain

The effective gain of the nonlinear element for the weak signal B is

$$G_B = \frac{1}{2}\frac{[Bg'(A) + g(A)B/A]}{B} = \frac{1}{2A}\frac{d[Ag(A)]}{dA} \tag{9-13}$$

and is termed the weak signal gain. Note that there is ∞ suppression of the weaker signal if the nonlinearity results in a $g(A) = 1/A$. The gain of the large interfering signal from (9-11) is $G_A = g(A)/A$. Similarly, the gain of the intermodulation term relative to the weak signal input is

$$G_I = \frac{1}{2}\frac{[Bg'(A) - g(A)B/A]}{B} = \frac{1}{2}\left[g'(A) - \frac{g(A)}{A}\right]$$

$$= \frac{A}{2}\frac{d[g(A)/A]}{dA} \tag{9-14}$$

Note that the difference between (9-13) and (9-14) is the sign change in the numerator. Obviously, the intermodulation product disappears if $g(A) = A$, as in a linear system. Note from (9-13), (9-14) that $G_I = G_B - G_A$.

A hypothetical $g(A)$ characteristic is shown in Fig. 9-9, where $A = A_0$ we obtain

$$g(A_0) = 0 \qquad \left.\frac{d[g(A)/A]}{dA}\right|_{A=A_0} = 0 \tag{9-15}$$

Thus for this one particular value of $A = A_0$, the stronger signal is completely suppressed and the weaker signal is emitted by itself. The instantaneous nonlinearity corresponding to this gain function can be computed using the inverse Chebyshev transform. However this $g(A)$ characteristic is

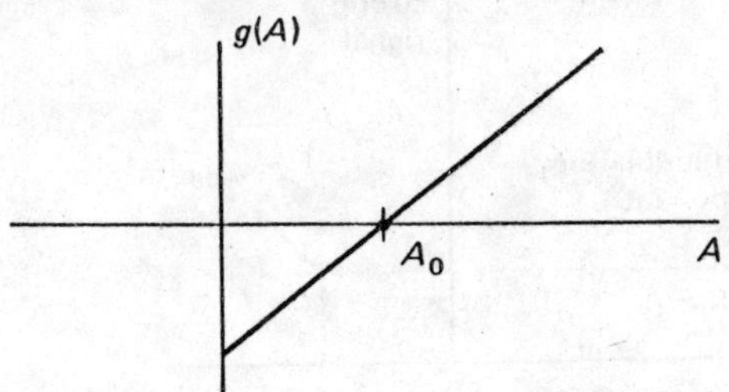

Fig. 9-9 Hypothetical $g(A)$ characteristic versus input amplitude A

probably of academic interest only, because this suppresion behavior is valid just for one value of A.

Signal Plus Gaussian Input

Consider a fixed envelope sinusoid of amplitude B, which is received in the presence of a large Gaussian interference as in (9-7). This Gaussian term is time-varying and can represent a large number of other sinusoidal signals. The probability density of the envelope of the interference A is Rayleigh

$$p(A) = \frac{A}{\sigma^2} e^{-A^2/2\sigma^2}, \quad A \geq 0, \quad \text{and} \quad 2\sigma^2 = \int_0^\infty A^2 p(A)\, dA \tag{9-16}$$

The expected value of the signal output over the ensemble of Rayleigh envelope inputs is

$$\mathrm{E}[BG_B] = B\mathrm{E}\left[\frac{1}{2A}\frac{d(Ag)}{dA}\right] = \frac{B}{2\sigma^2}\int_0^\infty \frac{d(Ag)}{dA} e^{-A^2/2\sigma^2}\, dA \tag{9-17}$$

where $B \ll \sigma^2$ is assumed. By integrating by parts, an average weak signal envelope output is computed as

$$\mathrm{E}[BG_B] = \frac{B}{2\sigma^2}\left[\overbrace{Age^{-A^2/2\sigma^2}\Big|_0^\infty}^{0} - \int_0^\infty Ag\left(-\frac{2A}{2\sigma^2}\right) e^{-A^2/2\sigma^2}\, dA\right]$$

$$= \frac{B}{2\sigma^2}\int_0^\infty Ap(A)g(A)\, dA \tag{9-18}$$

Thus, the output signal power at frequency $\omega + \dot{\beta}$ is

$$P_{B0} = \frac{1}{2}\left[\frac{B}{2\sigma^2}\int_0^\infty Ag(A)p(A)\, dA\right]^2 = \frac{1}{2}\left(\frac{B}{2\sigma^2}\right)^2 [\mathrm{E}(Ag)]^2 \tag{9-19}$$

and the total output interference power at center frequency $\omega + \dot{\alpha}$ in the first zone is approximately

$$P_{A0} = \frac{1}{2}\int_0^\infty g^2(A)p(A)\,dA = \frac{1}{2}\mathrm{E}(g^2) \tag{9-20}$$

Hence, the output signal-to-interference ratio is

$$\frac{P_{B0}}{P_{A0}} \triangleq \frac{B^2}{2\sigma^2}R \triangleq \frac{B^2}{2\sigma^2}\frac{\left[\int_0^\infty Ag(A)p(A)\,dA\right]^2}{\left[\int_0^\infty g^2(A)p(A)\,dA\right]\left[\int_0^\infty A^2p(A)\,dA\right]} \leq \frac{B^2}{2\sigma^2}$$

where $\dfrac{B^2}{2\sigma^2} = (\mathrm{SNR})_{\mathrm{in}}$ (9-21)

R is the signal-suppression ratio. The inequality in (9-21) is obtained using the Schwartz inequality $(f, g)^2 \leq (f, f)(g, g)$. No matter what $g(A)$ is used, the signal-suppression ratio cannot exceed unity for a strong Gaussian interference—that is, there can be no signal enhancement no matter what the nonlinearity. Note that P_{A0} represents the total interference in the first zone since it is large compared to the power at frequency $\omega + 2\dot{\alpha} - \dot{\beta}$. Several examples are given below for typical nonlinear transfer functions and different input signal classes.

9-5 FDMA SIGNAL SUPRESSION BY A BANDPASS HARD LIMITER

Sinusoidal Inputs

A symmetrical hard limiter is represented by $\mathrm{F}(u) = a =$ constant for $u > 0$ and $\mathrm{F}(u) = -a$ for $u < 0$ and $\mathrm{F}(0) = 0$. See Fig. 9-10. The output of the fundamental zone is then $4a/\pi \cos\theta(t)$; this result is simply the fundamental component of a square wave of amplitude a with transitions at $|\theta| = n\pi$. For sinusoidal inputs with envelopes $A(t) = A$, $B(t) = B$, we have a strong signal gain

$$G_A = \frac{g(A)}{A} = \frac{C}{A} \tag{9-22}$$

where $C \triangleq a(4/\pi)$. The weak signal gain is, from (9-13),

$$G_B \cong \frac{1}{2}\left[\frac{g(A)}{A} + \overbrace{g'(A)}^{0}\right] = \frac{1}{2}G_A \tag{9-23}$$

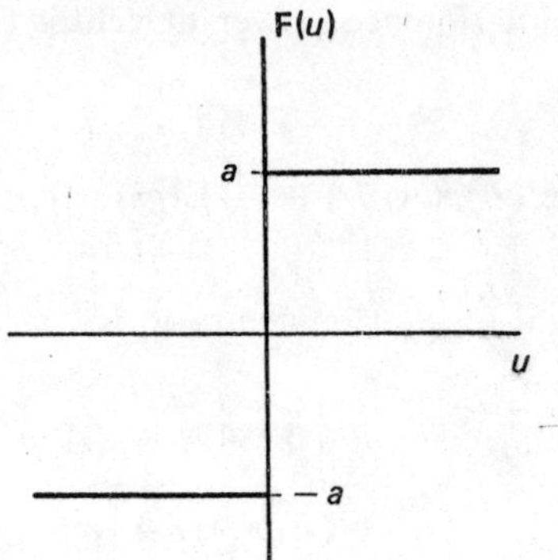

Fig. 9-10 Hard-limiter transfer function

Hence, small signal term B is suppressed by a factor of 2 in amplitude or —6 dB ($R = \frac{1}{4}$) relative to the strong signal for two sinusoidal inputs. It can be shown that the same 6-dB suppression results even if the weak signal term $B(t)$ is Rayleigh and represents the sum of a large number of small sinusoids.

Signal Suppression by a Strong Gaussian Term

If $A(t)$ has a Rayleigh density corresponding to a strong Gaussian interference (Fig. 9-11), then the output signal suppression ratio of (9-21) for the weak signal $B(t)$ in hard limiting is

$$R = \frac{C^2\left(\int_0^\infty Ap(A)\,dA\right)^2}{C^2\int_0^\infty A^2p(A)\,dA} = \frac{\overbrace{\left(\frac{1}{\sqrt{2\pi}\sigma}\int_0^\infty A^2e^{-A^2/2\sigma^2}\,dA\right)^2}^{(\sigma^2/2)^2}\left(\frac{2\pi}{\sigma^2}\right)}{2\sigma^2} \tag{9-24}$$

$$R = \frac{(\sigma^4/4)}{2\sigma^2}\left(\frac{2\pi}{\sigma^2}\right) = \frac{\pi}{4} = 0.7854 \quad \text{or} \quad 1.05\text{-dB suppression} \tag{9-25}$$

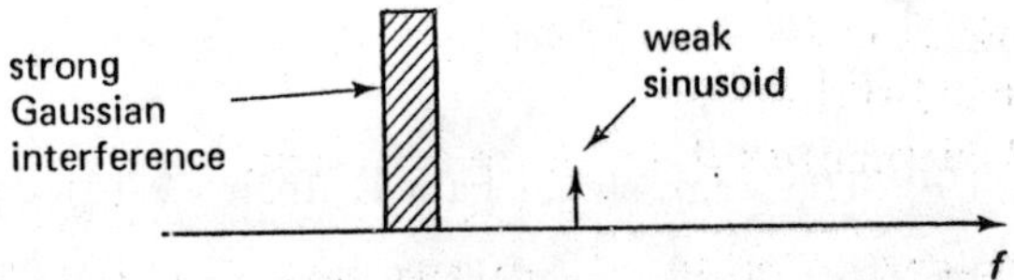

Fig. 9-11 Spectrum of a weak sinusoid plus a strong Gaussian interference

Thus, R gives a 1.05-dB suppression of the weaker signal when the strong signal is Gaussian.

Consider that the strong Gaussian signal may really be a composite signal made up of a large number of randomly phased sinusoids. Then each and every individual small sinusoid can play the role of the weak signal and is suppressed with respect to total output power by the same $\pi/4$ or 1.05 dB. Since the total output power is $(C^2/2)$, the total useful output power for all sinusoids is $(\pi/4)(C^2/2)$. The ratio of total useful output to total interference power in the fundamental zone is then

$$\frac{\pi/4}{1-\pi/4} = \frac{\pi}{4-\pi} = 3.66 \qquad \text{or} \qquad 5.63 \text{ dB} \tag{9-26}$$

Thus if all the interference power entered the receiver, then the input SNR = 5.63 dB. If a bandlimited white Gaussian input of bandwidth B is applied to the input of a bandpass hard limiter, the output spectrum is as shown in Fig. 9-12. The total power output of the limiter is P; the total signal output power

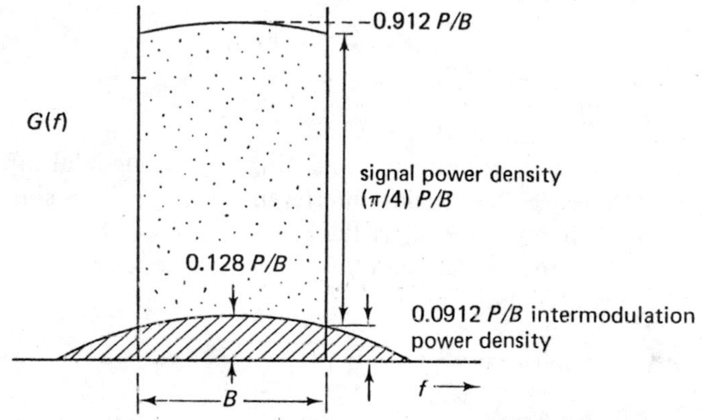

Fig. 9-12 Output spectrum of a bandpass limiter with a Gaussian input [From Price, 1955, 9-14a]

is $\pi P/4 = 0.785P$; the total intermodulation power within bandwidth B is $0.115P$; and the remaining output power in the tails has power $0.1P$. The third-order IM product contributes 57.4 percent of the intermodulation power at midband [Price, 1955].

Table 9-3 summarizes the major power contributors to the output noise spectrum and their relative levels. The total power in the fundamental zone has been set at $C^2/2 = (8/\pi^2)a^2 \triangleq P$. The power spectral density has been normalized to the bandwidth B. Note that within the bandwidth B there are

Table 9-3 SUMMARY OF OUTPUT INTERMODULATION AND SIGNAL COMPONENTS FOR A HARD LIMITER WITH A GAUSSIAN INPUT SIGNAL AND RECTANGULAR INPUT SPECTRUM; THE EDGE OF THE CHANNEL REFERS TO THE EDGE OF THE CHANNEL OF BANDWIDTH B [FROM BOND AND MEYER, 1970]

Component	*Watts*	*Decibels*
Total fundamental zone power output	P	0
Total signal power output	$0.785P$	-1.06
Signal power density output	$0.785P/B$	
Total intermodulation (IM) power	$0.215P$	-6.7
Ratio of total signal/total intermodulation		5.64
Total IM power in tails	$0.1P$	-10
Total IM Power in input band B	$0.115P$	-9.4
Center channel IM power density	$0.128P/B$	
Edge channel IM power density	$0.0912P/B$	
Total signal-to-inband-IM-power ratio		8.34
Center-channel-signal-to-center-IM-power ratio		7.8
Edge-channel-signal-to-edge-IM-power ratio		9.35

both signal and IM components; beyond $B/2$ from midband there are only distortion components.

Rician Interference

Consider an input environment consisting of a sinusoidal input signal plus Rician interference. The Rician interference consists of a sinusoid plus a Gaussian term where $r = P_c/P_n$ is the power ratio of the fixed envelope interference sinusoid to the Gaussian noise. The envelope of the interference has the Rician probability density

$$p(A_n) = A_n e^{-(A_n^2+2r)/2} I_0(\sqrt{2r}A_n) \tag{9-27}$$

The effective signal suppression factor for a weak sinusoid in the presence of this strong interference for a hard-limiting bandpass nonlinearity then can be computed as

$$R = \frac{\left[\int_0^\infty Ap(A)\,dA\right]^2}{\int_0^\infty A^2 p(A)\,dA} = \frac{\pi}{4}(1+r)\left[e^{-r/2}\cdot I_0\left(\frac{r}{2}\right)\right]^2 \tag{9-28}$$

which is plotted in Fig. 9-13. Note that in the limiting cases of large and small r, the results approach the suppression values for sinusoidal and Gaussian interference, (9-23), (9-25) respectively.

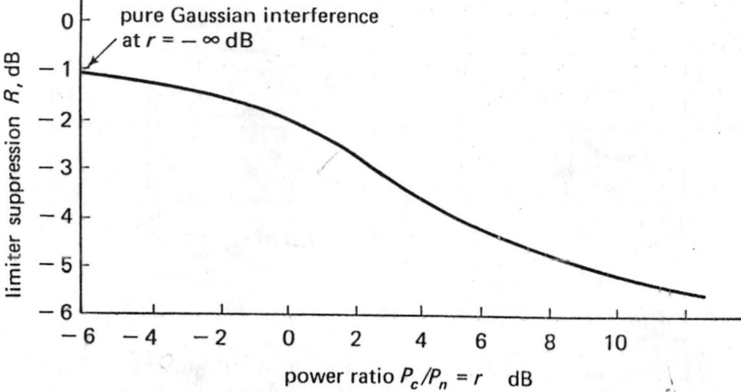

Fig. 9-13 Limiter suppression versus power ratio for a mixture of Gaussian and sinusoidal interference. The suppression approaches 6 dB as the interference becomes a pure sinusoid, $r = \infty$ dB.

9-6 DISTORTION CAUSED BY AMPLITUDE NONLINEARITIES

Signal suppression is only one of the effects caused by amplifier nonlinearities. One of the most important performance measures is the signal-to-distortion ratio—the ratio of individual signal power to the distortion power in the signal passband. We first determine the output signal-to-distortion spectral-density ratio versus the drive level. These results determine how much the transponder output power—power backoff—must be reduced to achieve a desired signal-to-distortion ratio for both sinusoidal and Gaussian inputs. As an intermediate step toward this calculation, the total signal power and total distortion power are calculated.

Piecewise Linear Bandpass Limiter

The transponder nonlinearity to be used here is the piecewise linear limiter, or clipping amplifier, shown in Fig. 9-14. The zonal filter passes only those frequency components in the fundamental frequency zone corresponding to the bandpass input. This model of a saturating amplifier is used in place of a hard limiter because it allows us to calculate the improvement resulting from power backoff. It is the simplest model of a saturating amplifier. Results for a hard limiter are obtained by letting the output be $y(t)/c$ and $c \longrightarrow 0$.

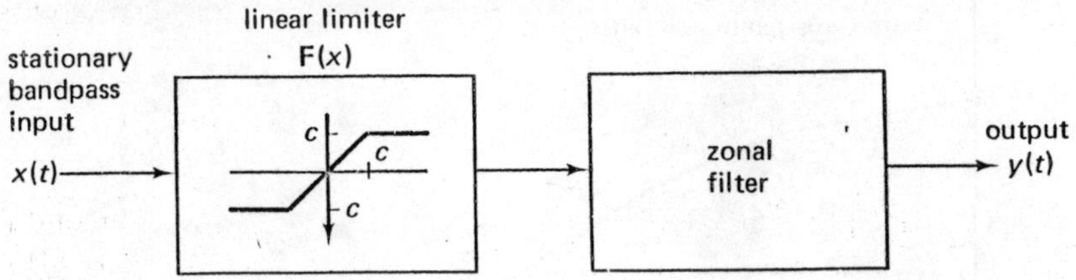

Fig. 9-14 Bandpass piecewise linear limiter model

Sinusoidal Input

As a first step, compute the variation in output power versus input power for the piecewise linear limiter. Let the input be a single sinusoid

$$x(t) = A\cos(\omega_0 t + \theta) = A\cos\phi \tag{9-29}$$

where $p(\phi) = \dfrac{1}{2\pi} \qquad |\phi| < \pi$

The output sinusoid amplitude in the fundamental zone is then

$$\begin{aligned} g(A) \triangleq B_1 &= \frac{1}{\pi}\int_0^{2\pi} f(A\cos\phi)\cos\phi\, d\phi \\ &= \frac{4}{\pi}\int_0^{\pi/2} f(A\cos\phi)\cos\phi\, d\phi \end{aligned} \tag{9-30}$$

For a linear limiter and a small amplitude input which produces no clipping, $A \leq c$, the system is linear and $B_1 = A$. For larger inputs $A \geq c$, where saturation occurs at least on the peaks of the sinusoid. The output is the sum of contributions from the limiting and linear regions:

$$\begin{aligned} B_1 &= \frac{4}{\pi}\Bigg[\underbrace{c\int_0^{\cos^{-1} c/A} \cos\phi\, d\phi}_{\text{limiting region}} + \underbrace{\int_{\cos^{-1} c/A}^{\pi/2} A\cos^2\phi\, d\phi}_{\text{linear region}}\Bigg] \\ &= \frac{4c}{\pi}\left\{\sin\phi\Big|_0^{\cos^{-1} c/A} + \frac{A}{2c}\left[1 + \frac{\sin 2\phi}{2}\right]_{\cos^{-1} c/A}^{\pi/2}\right\} \\ &= \frac{4c}{\pi}\Bigg[\sqrt{1-\left(\frac{c}{A}\right)^2} + \frac{A}{2c}\sin^{-1}\frac{c}{A} + \underbrace{\frac{A}{2c}\sin\phi\cos\phi\Big|_{\cos^{-1} c/A}^{\pi/2}}_{\frac{-A}{2c}\sqrt{1-\left(\frac{c}{A}\right)^2}\,\frac{c}{A}}\Bigg\} \end{aligned} \tag{9-31}$$

Thus, the output signal amplitude for this more general range of inputs is

$$B_1 = \frac{2c}{\pi}\left[\sqrt{1-\left(\frac{c}{A}\right)^2} + \frac{A}{c}\sin^{-1}\frac{c}{A}\right] \triangleq \gamma A \tag{9-32}$$

where γ is the equivalent gain. The power output at the fundamental frequency is $P_1 = B_1^2/2 \leq 8c^2/\pi^2$. The output power is plotted in Fig. 9-15 as a function of drive level $(A/c)^2$.

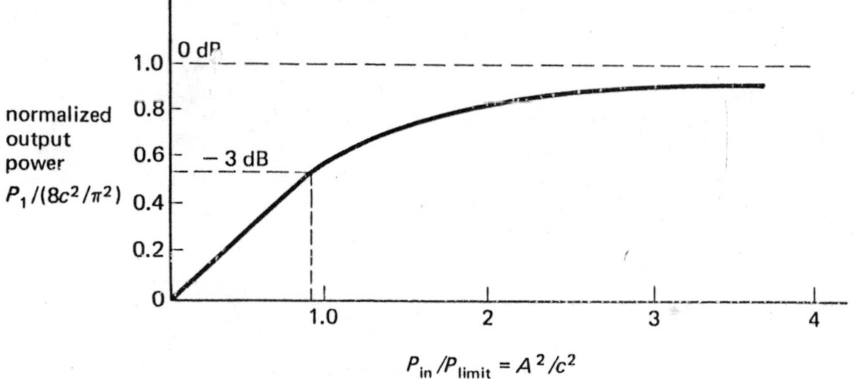

Fig. 9-15 Output versus input power for a piecewise linear limiter with a single sinusoidal input. The output power is normalized with respect to the maximum output power of the piecewise linear limiter.

An example of TWT transfer function (a "soft" limiter) is shown in Fig. 9-16. Both the output power and the intermodulation products are shown for two equal amplitude sine wave inputs.

Narrow-Band Gaussian Inputs

Consider a Gaussian input and compute the linear signal term and the intermodulation distortion components. Assume a bandpass input spectrum for the Gaussian input, as shown in Fig. 9-17. This spectrum can represent the sum of many frequency-multiplexed channels. If one channel is deleted, a notch would result in the input spectrum as shown. Intermodulation effects, caused by a nonlinear transponder amplifier or limiter, however, would partially fill in the notch in the amplifier output spectrum. The ratio of the intermodulation power in the notch to the signal power ordinarily in the notch gives the noise-power-ratio (NPR) or inverse SNR of that channel [Cahn, 1960, 53–59; Sunde, 1970, Chap. 8].

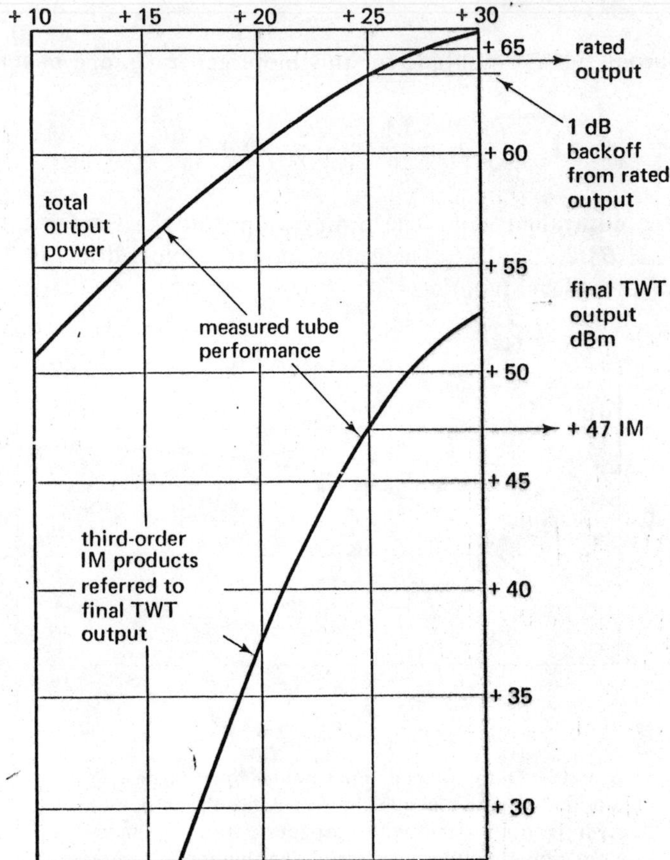

Fig. 9-16 Typical measured TWT output power and intermodulation products versus input drive levels with two equal amplitude input sinusoids, at 1-dB backoff from rated (not peak) output. The IM products are 14 dB below output power of one sinusoid and 17 dB below total output.

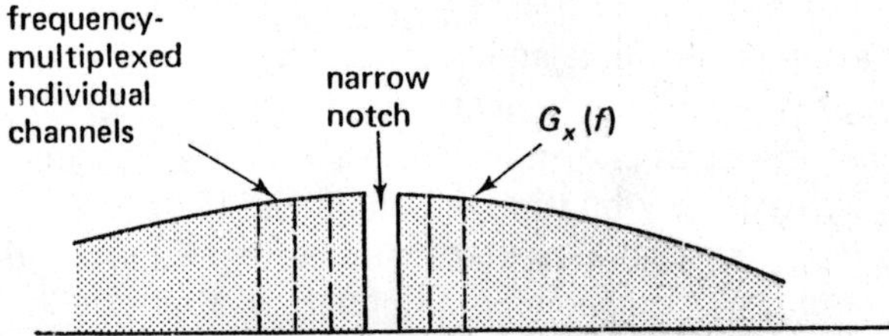

Fig. 9-17 Input spectrum representing the sum of contiguous narrow-band input channels, where one channel is missing—hence, the notch in the spectrum

Transform Method for Nonlinear Devices

The transform method of analyzing nonlinear devices with Gaussian inputs is discussed in detail in other references [Davenport and Root, 1958, Chap. 13]. The method is only briefly reviewed here as a step to obtain the distortion spectrum. The bilateral Laplace transform of the nonlinearity $f(x)$ can be written as

$$F(w) = \int_{-\infty}^{0} f(x)e^{-wx}\,dx + \int_{0}^{\infty} f(x)e^{-wx}\,dx \triangleq F_+(w) + F_-(w) \qquad \text{real } (w) > 0 \tag{9-33}$$

Thus, the nonlinear transfer function can be expressed in terms of its transform

$$f(x) = \frac{1}{2\pi}\int_{-\infty}^{\infty} F(jv)e^{jvx}\,dv \tag{9-34}$$

where $F(w) = F_+(w) + F_-(w)$ and v is the imaginary product of w. For the example of a piecewise linear limiter, this transform becomes

$$F_+(w) = \int_{0}^{c} xe^{-wx}\,dx + c\int_{c}^{\infty} e^{-wx}\,dx \tag{9-35}$$

Set $y \triangleq wx$; then the integral is simply

$$\begin{aligned} F_+(w) &= -\frac{1}{w^2}\Big[e^{-y}(y+1)\Big|_0^{cw} + \frac{c}{w}e^{-cw} \\ &= \frac{1}{w^2}[1 - e^{-cw}(cw+1)] + \frac{c}{w}e^{-cw} = \frac{1-e^{-cw}}{w^2} \end{aligned} \tag{9-36}$$

Since $f(x)$ is an odd function of x, the transform components are related by $F_-(w) = -F_+(-w)$, and

$$F(w) = F_+(w) - F_+(-w) = \frac{e^{-cw} - e^{+cw}}{w^2} \tag{9-37}$$

and thus for $w = jv$, that is, on the complex axis, the transfer function for the piecewise linear limiter is

$$F(jv) = 2j\frac{[e^{jcv} - e^{-jcv}]}{2jv^2} = 2j\frac{\sin cv}{v^2} \tag{9-38}$$

For a hard limiter of unity magnitude output, $f(x) = \text{sgn } x$, then

$$F(jv) = \lim_{c \to 0} \frac{2j}{v^2} \frac{\sin cv}{c} \cong 2j/v \qquad \text{for } v \ll 1/c$$

Output Autocorrelation Function

The autocorrelation function of the output of the nonlinearity $z(t) = f[x(t)]$ is expressible in terms of these transforms as

$$\begin{aligned} R_z(\tau) &= \mathbf{E}[f(x_t)f(x_{t+\tau})] \\ &= \left(\frac{1}{2\pi}\right)^2 \iint F(jv_1)F(jv_2)\, dv_1\, dv_2\, \mathbf{E}[e^{jv_1x_t + jv_2x_{t+\tau}}] \\ &= \left(\frac{1}{2\pi}\right) \iint F(jv_1)F(jv_2)\mathbf{C}(jv_1, jv_2)\, dv_1\, dv_2 \end{aligned} \tag{9-39}$$

where $\mathbf{C}(jv_1, jv_2)$ is the joint characteristic function of the input and $x_t \triangleq x(t)$. For stationary Gaussian inputs of variance σ^2, the characteristic function can be developed as

$$\mathbf{C}(jv_1, jv_2) = \exp\left[-\frac{\sigma^2}{2}(v_1^2 + v_2^2) - R(\tau)v_1v_2\right] \tag{9-40}$$

where $R(\tau) = \mathbf{E}[x_t x_{t+\tau}]$ is the autocorrelation function of the input. This exponential can be expanded as

$$\mathbf{C}(jv_1, jv_2) = e^{(-\sigma^2/2)(v_1^2 + v_2^2)} \sum_{k=0}^{\infty} \frac{R^k(\tau)(-1)^k(v_1v_2)^k}{k!} \tag{9-41}$$

From (9-39) and (9-41) one can show that the autocorrelation function for odd $f(x)$ contains only odd values of k and can be expressed as

$$R_z(\tau) = \sum_{\substack{k=1 \\ \text{odd}}}^{\infty} h_k^2 \frac{R^k(\tau)}{k!} \tag{9-42}$$

$$\text{where} \quad h_k \triangleq \frac{1}{2\pi} \int_{-\infty}^{\infty} F(jv_1)(jv)^k e^{-\sigma^2 v^2/2}\, dv \tag{9-43}$$

Even functions $f(x)$ yield even values for k. However, these terms are not of interest since they produce no components in the fundamental zone.

For the piecewise linear limiter, the series coefficients for k odd are

$$h_k = \frac{1}{2\pi}\int_{-\infty}^{\infty} 2j\frac{\sin cv}{v^2}(jv)^k e^{\sigma^2 v^2/2}\,dv$$

$$= 2\pi\int_{-\infty}^{\infty} 2(-1)^{(k+1)/2} v^{k-2} e^{-\sigma^2 v^2/2}\sin cv\,dv$$

$$= \frac{1}{2\pi}\int_{-\infty}^{\infty} 2(-1)^{(k+1)/2}\left(\frac{\sqrt{2}\,t}{\sigma}\right)^{k-2} e^{-t^2}\sin\frac{c\sqrt{2}\,t}{\sigma}\,dt\,\frac{\sqrt{2}}{\sigma}$$

$$= \frac{\sqrt{2}}{\pi\sigma}\int_{-\infty}^{\infty}(-1)^{(k-3)/2}\left(\frac{\sqrt{2}\,t}{\sigma}\right)^{k-2} e^{-t^2}\sin\left(\frac{c\sqrt{2}\,t}{\sigma}\right)dt \qquad (9\text{-}44)$$

where $t = \sigma v/\sqrt{2}$. The Hermite polynomial $\mathcal{H}_n(x)$ can be defined as the solution to the integrals (for n odd) [Abramowitz and Stegun, 1964]:

$$\mathcal{H}_n(x) \triangleq e^{x^2}\frac{2^{n+1}}{\sqrt{\pi}}\int_0^{\infty} e^{-t^2} t^n \cos\left(2xt - \frac{n}{2}\pi\right)dt$$

$$\triangleq e^{x^2}\frac{2^{n+1}}{\sqrt{\pi}}\int_0^{\infty} e^{-t^2} t^n (-1)^{(n-1)/2}\sin 2xt\,dt \qquad (9\text{-}45)$$

To obtain the second integral in (9-45), use $\cos(2xt - n\pi/2) = \sin 2xt(-1)(n-1)/2$ for n odd. Thus, the expressions for h_k in (9-44) can be rewritten by substituting $n = k - 2$, $x = c/\sqrt{2}\sigma$ as

$$h_k = 2\left(\frac{\sqrt{2}}{\pi\sigma}\right)\left(\frac{\sqrt{2}}{\sigma}\right)^{k-2}\left(\frac{e^{-c^2/2\sigma^2}}{2^{k-1}}\sqrt{\pi}\right)\mathcal{H}_{k-2}\left(\frac{c}{\sqrt{2}\sigma}\right)$$

$$= \frac{2}{\sqrt{\pi}}\frac{e^{-c^2/2\sigma^2}}{2^{(k-1)}/2}\frac{\mathcal{H}_{k-2}(c/\sqrt{2}\sigma)}{\sigma^{k-1}} \qquad (9\text{-}46)$$

The Hermite polynomials can be expressed as the derivative

$$\mathcal{H}_n(x) = (-1)^n 2^{n/2} e^{x^2}\frac{d^n}{dx^n}(e^{-x^2}) \qquad (9\text{-}47)$$

$$\mathcal{H}_1(x) = \frac{\sqrt{\pi}}{2}e^{x^2}\operatorname{erf} x \qquad (9\text{-}48)$$

where $\operatorname{erf} x = \dfrac{2}{\sqrt{\pi}}\displaystyle\int_0^x e^{-t^2 dt}$

For small x, the Hermite polynomials are approximately equal to

$$\mathcal{H}_1(x) \cong 2x \qquad \mathcal{H}_3(x) \cong -12x \qquad \mathcal{H}_5(x) \cong 120x \tag{9-49}$$

$$\mathcal{H}_m(x) \cong (2x)(-1)^{(m-1)/2} \frac{m!}{\left(\frac{m-1}{2}\right)!} \tag{9-50}$$

Thus, the output autocorrelation function of the piecewise-linear limiter from (9-42), (9-46) and (9-48) is the summation

$$R_z(\tau) = e^{-c^2/\sigma^2}\left[R(\tau)e^{c^2/\sigma^2} \operatorname{erf}^2\left(\frac{c}{\sqrt{2}\,\sigma}\right)\right]$$

$$+\frac{4}{\pi}\sum_{k=3,5,\ldots}^{\infty} \frac{R^k(\tau)}{\sigma^{2(k-1)}k!} \frac{\mathcal{H}_{k-2}^2\left(\frac{c}{\sqrt{2}\,\sigma}\right)}{2^{k-1}} \qquad k \text{ odd}$$

$$R_z(\tau) = R(\tau) \operatorname{erf}^2\left(\frac{c}{\sqrt{2}\,\sigma}\right) + \frac{4}{\pi} e^{-c^2/\sigma^2} \sum_{\ell=1}^{\infty} \frac{R^{2\ell+1}(\tau)}{\sigma^{4\ell}(2\ell+1)} \frac{\mathcal{H}_{2\ell-1}^2\left(\frac{c}{\sqrt{2}\,\sigma}\right)}{2^{2\ell}} \tag{9-51}$$

where we have substituted $k = 2\ell - 1$ for k odd.

We define the bandpass input autocorrelation of a narrow input process as $R(\tau) = \sigma^2\rho(\tau)\cos\omega_0\tau$ for $|\tau| < 2\pi/\omega_0$. Then the terms $R^k(\tau) = \sigma^{2k}\rho^k(\tau)$ $\cos^k \omega_0\tau$ represent a fundamental frequency output power

$$P_{Fk} = \frac{2\omega_0}{2\pi}\int_0^{2\pi/\omega_0} \rho^k(\tau)\cos^k\omega_0\tau\cos\omega_0\tau\, d\tau$$

$$= \rho^k(0)\frac{1}{\pi}\int_0^{2\pi} \cos^{k+1}\theta\, d\theta \triangleq \rho(0)C(k) \tag{9-52}$$

where we define the constant

$$C(k) = \frac{k!}{\left(\frac{k-1}{2}\right)!\left(\frac{k+1}{2}\right)!\,2^{k-1}}, \quad C(1) = 1,\ C(3) = \tfrac{3}{4},\ C(5) = \tfrac{5}{8} \tag{9-53}$$

Hence, we can represent the kth power of the input autocorrelation in terms of its component in the fundamental zone:

$$\frac{R^k(\tau)}{\sigma^{2k}} = \rho^k(\tau)C(k)\cos\omega_0\tau + \text{harmonic terms} \qquad k \text{ odd} \tag{9-54}$$

Baum [1957] has shown the results for the autocorrelation function of smoothly-limited and hard-limited Gaussian inputs. The resultant input/output transfer function of this limiter (not bandpass) is

$$v(u) = \operatorname{erf}\frac{u}{\sigma_0} \triangleq \frac{1}{\sqrt{2\pi}\sigma_0}\int_0^u \exp\left(\frac{-z^2}{2\sigma_0^2}\right) dz$$

Let the rms value of the input be σ and define $\alpha = \sigma_0/\sigma$ and the input autocorrelation is ρ_τ. Then the output autocorrelation is

$$\rho_v(\tau) = \frac{\sin^{-1}[\rho_\tau/(1+\alpha)]}{\sin^{-1}[1/(1+\alpha)]} \tag{9-55a}$$

If the system is hard limiting, $\sigma \gg \sigma_0$ and $\alpha \cong 0$, then the output autocorrelation for $z(t) = \operatorname{sgn} x(t)$ is [Sunde 1969]

$$\rho_v(\tau) = \frac{2}{\pi} \sin^{-1} \rho_\tau \tag{9-55b}$$

Fundamental Zone Output Autocorrelations

Using the definition of $C(k)$ in (9-53), the autocorrelation terms in the fundamental zone in (9-51) are $R_F(\tau) \triangleq \sigma^2 \rho_z(\tau) \cos \omega_0 \tau$, where σ^2 is the input power and can be written in terms of the useful signal and intermodulation components

$$\rho_z(\tau) = \rho(\tau) \operatorname{erf}^2\left(\frac{c}{\sqrt{2}\sigma}\right) + \frac{4}{\pi} e^{-c^2/\sigma^2} \sum_{\ell=1}^{\infty} \frac{\rho^{2\ell+1}(\tau)\mathcal{H}^2_{2\ell-1}\left(\frac{c}{\sqrt{2}\sigma}\right)}{(2\ell+1)!\,2^{2\ell}} \underbrace{\left(\frac{(2\ell+1)!}{\ell!(\ell+1)!\,2^{2\ell}}\right)}_{C(2\ell+1)}$$

$$\rho_z(\tau) = \underbrace{\rho(\tau) \operatorname{erf}\left(\frac{c}{\sqrt{2}\sigma}\right)}_{\text{useful signal}} + \underbrace{\frac{4}{\pi} e^{-c^2/\sigma^2} \sum_{\ell=1}^{\infty} \frac{\rho^{2\ell+1}(\tau)\mathcal{H}^2_{2\ell-1}\left(\frac{c}{\sqrt{2}\sigma}\right)}{\ell!(\ell+1)!\,2^{4\ell}}}_{\text{intermodulation distortion}} \tag{9-56}$$

The ratio of the power in the first term (set $\tau = 0$ in (9-56)) to the power in the second term is a signal-to-total-distortion ratio. This ratio does not account for the spectral spreading of the distortion as described for Gaussian signals with a hard limiter. Nevertheless, the total power-distortion ratio is still useful and can be written from (9-56) as

$$\frac{P_{so}}{P_{tl}} = \frac{\pi}{4} \frac{e^{c^2/\sigma^2} \operatorname{erf}^2 (c/\sqrt{2}\,\sigma)}{\sum_{l=1}^{\infty} \frac{\mathcal{H}_{2l-1}^2\left(\frac{c}{\sqrt{2}\,\sigma}\right)}{l!(l+1)!\,2^{4l}}} \triangleq \frac{P_{so}}{\sum_{l=1}^{\infty} P_{tl}} \tag{9-57}$$

where P_{tl} is the total intermodulation distortion power, and P_{so} is the output signal power.

Spectral-Density Ratio

The ratio of signal-spectral-density to intermodulation-density represents the true signal-to-distortion power ratio of individual contiguous narrow band carriers and must be computed by taking into account the bandpass nature of the input waveforms. Assume an input signal having a Gaussian power spectral density. The autocorrelation function of this input is $\rho(\tau) = e^{-\alpha\tau^2}$. The ratio of signal to total intermodulation spectral density at midband is then obtained using the transform (zero offset frequency) of $\rho^{2l+1}(\tau)$ from the second term of (9-56)

$$\int_0^\infty \rho^{2l+1}(\tau)\, d\tau = \frac{1}{\sqrt{2l+1}} \int_0^\infty \rho(\tau)\, d\tau \qquad \text{for } \rho(\tau) = \exp(-\alpha\tau^2) \tag{9-58}$$

Hence, the ratio of the $2l + 1$ intermodulation spectral density term to the fundamental is $1/\sqrt{2l+1}$ at band center. The output inverse NPR or output signal-to-intermodulation-distortion ratio at the center frequency f_α is obtained by summing these IM spectral density terms.

$$(\text{SNR})_{f_0} = \frac{P_{so}}{\sum_{l=1}^{\infty} \frac{P_{tl}}{\sqrt{2l+1}}} = \frac{\pi}{4} \frac{e^{c^2/\sigma^2} \operatorname{erf}^2 (c/\sqrt{2}\,\sigma)}{\sum_{l=1}^{\infty} \frac{\mathcal{H}_{2l-1}^2(c/\sqrt{2}\,\sigma)}{l!(l+1)!\,2^{4l}\sqrt{2l+1}}}$$

$$\cong \frac{\pi}{4} \frac{e^{c^2/\sigma^2} \operatorname{erf}^2 (c/\sqrt{2}\,\sigma)}{\frac{\mathcal{H}_1^2(c/\sqrt{2}\,\sigma)}{2^5\sqrt{3}} + \frac{\mathcal{H}_3^2(c/\sqrt{2}\,\sigma)}{3(2^{10})\sqrt{5}}} \tag{9-59}$$

This ratio has been plotted in Fig. 9-18 versus the ratio c/σ. Note that for hard limiting $c/\sigma \longrightarrow 0$, the output SNR decreases to a minimum of approxi-

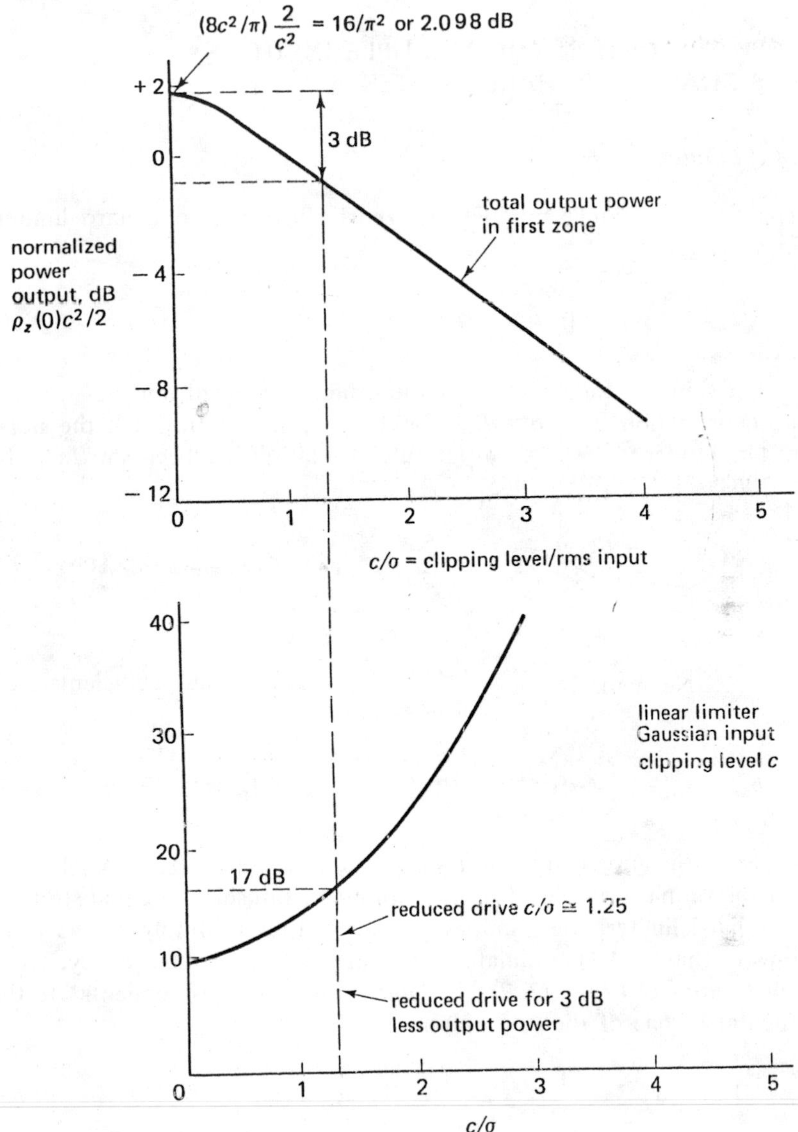

Fig. 9-18 Normalized output power and signal-to-intermodulation spectral-density ratio for a linear limiter with Gaussian input

mately 9 dB. The predominant effect is caused by the third- and fifth-order intermodulation distortion—the first two terms in the denominator. However, for 3-dB output power backoff, the SNR increases to 17 dB.

9-7 IM DISTORTION FOR MULTIPLE INPUT SIGNALS TO A HARD LIMITER

Hard Limiter

Apply a set of angle modulated signals plus noise to a hard limiter. This input is

$$x(t) \doteq \sum_{i=1}^{n} A_i \cos [\omega_i t + \phi_i(t)] + n(t) \tag{9-60}$$

where each signal is independent of the other, the signal and noise terms have autocorrelation functions $R_{si}(\tau)$ and $R_n(\tau)$, and $R_n(0) = \sigma^2$, the noise power. The limiter $y = \text{sgn}\, x$ output autocorrelation has been shown to be [Jones, 1963, 34–42; Shaft, 1965]

$$R_y(\tau) = \sum_{k=0}^{\infty} \frac{R_n^k(\tau)}{k!} \sum_{m_1=0}^{\infty} \cdots \sum_{m_n=0}^{\infty} \epsilon_{m_1}\epsilon_{m_2} \cdots \epsilon_{m_n} h^2_{km_1 \cdots m_n} R_{m_1 s_1}(\tau) \cdots R_{m_n s_n}(\tau) \tag{9-61}$$

where $\epsilon_m \triangleq$ Neumann factor $\begin{Bmatrix} 1 \text{ if } m = 0 \\ 2 \text{ if } m \neq 0 \end{Bmatrix}$, and the series coefficients are

$$h_{km_1,\ldots,m_n} = \frac{1}{\pi}(-1)^{k+m_1+\cdots+m_n} \int_{-\infty}^{\infty} x^{k-1} J_{m_1}(A_1 x) e^{-\sigma^2 x^2/2}\, dx \tag{9-62}$$

and $R_{m_i s_i}(\tau)$ is the autocorrelation function of $\sqrt{2} \cos m_i[\omega_i t + \phi_i(t)]$.

For the simple example where two noise-free sinusoids of equal strength enter the hard limiter, the bandpass limiter output is simply a sine wave square-wave that is (± 1) modulated (PSK) at the difference frequency. Thus, the power ratios of (2, 1), (3, 2), . . . , intermodulation components* to the fundamental of one of the two output signals are

$$\frac{s_{21}}{s_2} = \frac{1}{9}, \frac{s_{32}}{s_2} = \frac{1}{25}, \frac{s_{43}}{s_2} = \frac{1}{49}, \ldots \tag{9-63}$$

and the third-order IM product s_{21} is down by 9.54 dB below the fundamental. More generally, the output contains IM products of the form

$$\cos [m_1\theta_1(t) \pm m_2\theta_2(t) \pm \cdots \pm m_n\theta_n(t)] \tag{9-64}$$

---

*The notation (2,1) refers to an intermodulation product corresponding to the frequency $2\omega_j - \omega_k$. Note that the fundamental output power is $8/\pi^2$.

where $\theta_i(t) \triangleq \omega_1 t + \phi_i(t)$ at all multiples of the sum and difference frequencies, etc.

The same result in (9-61), (9-62) applies to a form of soft limiter—the error function limiter $f(x) = \text{erf}(x/c)$, $x > 0$—if one substitutes $-(\sigma^2 + c^2)x^2/2$ in the exponent in the h function integral (9-62) in place of $-\sigma^2 x^2/2$. Thus the results for the hard limiter can be easily extended to this form of soft limiting.

Examine the third-order IM products—that is, where $\sum m_i = 3$. Note that if the inputs are antipodal, biphase-modulated signals (BPSK), the (2, 1) IM product is also BPSK and is identical to one of the signals but is offset in frequency:

$$\cos[2\theta_i(t) + \theta_j(t)] = \cos[2(\omega_i - \omega_j/2)t + \phi_j(t)] \tag{9-65}$$

since $2\phi_i = 2\pi$. Thus, the effect of the ϕ_j modulation has been removed. The net effect is to make this IM product more disturbing when it falls within a signal passband since it is not spread in bandwidth as would be an FM signal. Similar effects for BPSK signals occur where any of the phases in the IM product are multiplied by an even number.

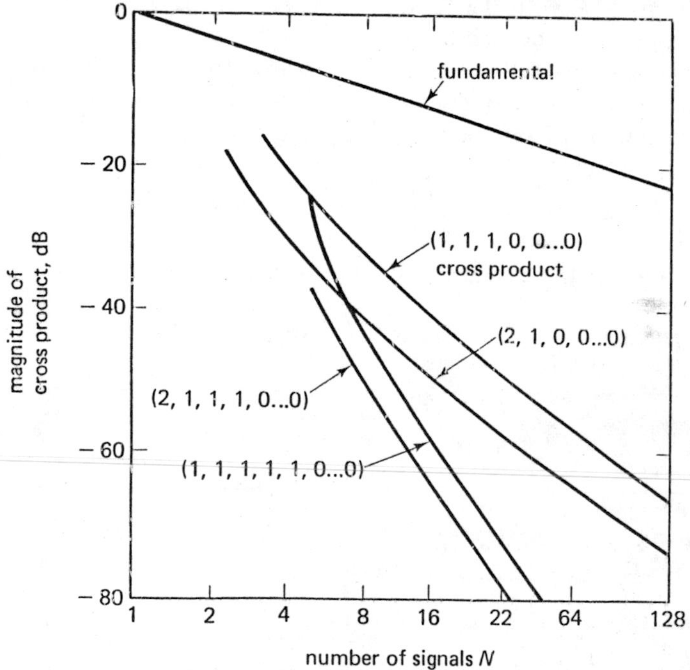

Fig. 9-19 IM product levels versus number of signals for a hard limiter. Each sinusoidal/input signal is of equal amplitude to the others. [From Shaft, 1965]

The magnitude of the third- and fifth-order cross products for equal amplitude inputs is given in Fig. 9-19; no noise is assumed here.

Approximate Analysis for a Large Number of Input Sinusoids

The previous discussion of intermodulation products for multiple input sinusoids is generally difficult to utilize without the use of a computer. Approximate analyses are developed in this section for both hard and soft limiters which are valid for large numbers ($N > 7$) of input sinusoids.

Define the intermodulation products of type $k_1 f_1 + k_2 f_2 + k_3 f_3 + \cdots + k_L f_L$ where the k_i are integers and $f_1, f_2, f_3, \ldots f_N$, represent the input frequencies. This IM product is denoted as $(k_1, k_2, k_3, \ldots k_n)$ where some of these k_i may be zero. The number of coefficients k_i with absolute value v for $v \neq 0$ is defined as r_v, and the sum $\sum r_v = L$. Thus L is the total number of input sinusoids involved in this IM product. For example, $100 \ldots 0$ indicates that the term is the signal at f_1. Similarly $2100 \ldots 0$ is the intermodulation product (IM) consisting of two times frequency f_1 minus the fundamental of signal 2 and no other terms contribute. For outputs in the fundamental zone, the sum, $\sum k_i = 1$. There are L nonzero values of k_i, and the order of the IM product is defined as $\sum |k_i| = m \geq L$.

One can use this definition of r_v and (9-61) to express the IM product amplitude. The output components for N equal amplitude inputs to a hard limiter with output power P are of amplitude

$$a_{k_1, k_2, \ldots, k_L} = (2P)^{1/2} \int_0^\infty \frac{dx}{x} J_1^{r_1}(x) J_2^{r_2}(x) \cdots J_L^{r_L}(x) [J_0(x)]^{N-L} \tag{9-66}$$

Table 9-4 lists the number of IM products of various orders. Note that

Table 9-4 NUMBER OF IM PRODUCTS OF VARIOUS ORDERS N INPUT SINUSOIDS*

IM Form	*Order* *m*	*Number of Frequencies L in this IM*	*Total Number of IM Products of this Form*
$2A - B$	3	2	$N(N-1)$
$\rightarrow A + B - C$	3	3	$T_3 \triangleq N(N-1)(N-2)/2$
$3A - B - C$	5	3	$N(N-1)(N-2)/2$
$A + 2B - 2C$	5	3	$N(N-1)(N-2)$
$\rightarrow A + B + C - D - E$	5	5	$T_5 \triangleq N(N-1)(N-2) \cdot (N-3)(N-4)/12$

*The notation $A, B, C, D, \ldots$, represents any of the input frequencies—for example, $A = f_i$, but $A \neq B$. Arrows indicate the dominant form of the third- and fifth-order cross products.

the error products of the form $A + B + C - D - E$ or in the previous notation 11111000 or 01110110 are of order $m = 5$ and occur most frequently. Thus these IM products are the "dominat form" of error product. The same is true for $m = 3$. The number of times the dominant term of order m occurs is denoted as T_m in Table 9-4. The value of T_m is given, in general, by Bond and Meyer, [1970] for N input sinusoids is

$$T_m = \frac{N^m}{[(m+1)/2]!\,[(m-1)/2]!} \qquad N \gg m \tag{9-67}$$

For other forms of IM products with at least one $k_i \geq 2$, the total number of products is proportional to N^L, $L < m$. Hence, for large N, the only significant contribution to the number of products is the dominant form.

The dominant IM products of order m have amplitude a_m from (9-66) [Bond and Meyer, 1970]

$$\begin{aligned} a_m &= (2P)^{1/2} \int_0^\infty \frac{dx}{x} J_1^m(x) J_0^{N-m}(x) \\ &\cong \frac{(2P)^{1/2}}{2^m} \int_0^\infty x^{m-1} \exp\left(-\frac{Nx^2}{4}\right) dx \\ &\cong \begin{cases} \left(\dfrac{P\pi}{2}\right)^{1/2} N^{-1/2} & \text{for } m = 1 \\ \left(\dfrac{P\pi}{2}\right)^{1/2} \left[\dfrac{1}{2} \cdot \dfrac{3}{2} \cdot \dfrac{5}{2} \cdots \left(\dfrac{m}{2} - 1\right)\right] N^{-m/2} & \text{for } m > 1 \end{cases} \end{aligned} \tag{9-68}$$

where we have used the approximations for small x

$$[J_0(x)]^N \cong \exp\left(-\frac{Nx^2}{4}\right) \text{ and } J_\nu(x) \cong \left(\frac{1}{\nu!}\right)\left(\frac{x}{2}\right)^\nu \tag{9-69}$$

Similarly, one can compute the IM product amplitude where one of the coefficients $k_i = 2$ and the remainder are ± 1 or 0. This value is

$$\begin{aligned} a_{m2} &= (2P)^{1/2} \int_0^\infty \frac{dx}{x} J_1^{m-2}(x) J_2(x) J_0^{m-1}(x) \\ &= (2P)^{1/2} 2^{-(m+1)} \int_0^\infty x^{m-1} \exp\left(-\frac{Nx^2}{4}\right) dx = \frac{a_m}{2} \end{aligned} \tag{9-70}$$

Note that this IM product has only one-half the amplitude and one-fourth the power of the dominant IM product. Hence, the dominant IM product not only occurs the most frequently but also has four times the power of the 2100 . . . 000 type IM product.

Using the same technique, the amplitude for any other form of IM product can be shown to be less than that of the dominant form. Since the dominant forms have the highest amplitudes as well as the only significant contribution to the total number of products for large N, they are the only forms that need be considered in estimating the total IM power.

Total Power for Each Order IM Product

The total power in the fundamental zone for each order IM product (T_m dominant term occurrences) is given by

$$P_m \cong T_m(a_m^2/2) \tag{9-71}$$

This power may or may not fall in the original input band, depending on the shape of the input spectrum and the type of modulation. From the previously computed value of T_m and a_m, this total IM power can be computed as

$$P_m = \frac{N^m}{[(m+1)/2]!\,[(m-1)/2]!}\left(\frac{1}{2}\right)\left[\frac{1}{2}\cdot\frac{3}{2}\cdot\frac{5}{2}\cdots\left(\frac{m}{2}-1\right)\right]^2 \cdot (2P)\frac{\pi}{4}N^{-m} \tag{9-72}$$

or in normalized form (9-72) becomes

$$\frac{P_m}{P} = \frac{\pi/4}{[(m+1)/2]!\,[(m-1)/2]!}\left[\frac{1}{2}\cdot\frac{3}{2}\cdot\frac{5}{2}\cdots\left(\frac{m}{2}-1\right)\right]^2 \tag{9-73}$$

For $m \geq 3$, Sterling's approximation for the factorial gives

$$\frac{P_m}{P} \simeq \frac{1}{m^2} \tag{9-74}$$

This intermodulation power is a function only of the limiter characteristic. The intermodulation power computed above for the hard limiter can also be computed for a soft limiter operating at various power back-off levels. Table 9-5 shows the dominant IM power P_m in the first few IM products, as well as the approximations, for various orders. These results should be correct to within 1 dB if the number of input signals $N \geq 7$. The IM products up to order 15 account for more than 85 percent of the total IM power.

Table 9-5 RELATIVE POWER IN THE DOMINANT INTERMODULATION PRODUCTS FOR HARD LIMITING AMPLIFIERS AND ERROR FUNCTION AMPLIFIERS BACKED OFF 3 dB IN OUTPUT POWER. THE APPROXIMATION $P_m/P \simeq 1/m^2$ IS SHOWN FOR COMPARISON FOR THE HARD LIMITER. THE NUMBER OF INPUT SIGNALS $N \gg 1$.

	Hard Limiter		*3-dB Backoff Error-Function Limiter*
IM Order	P_m/P	*Approximation* $P_m/P \simeq 1/m^2$	$P'_m/P = P_m 2^m/P$
3	0.0980	0.1111	0.0123
5	0.0367	0.250	0.00115
7	0.0191	0.0204	0.000149
9	0.0117	0.0123	0.000023
11	0.0079	0.00826	—
13	0.00568	0.00592	—
15	0.00429	0.00444	—
Total	0.215	$(\pi^2/8 - 1$ $= 0.2336$	0.0136

Error-Function Limiter

An error-function characteristic has been mentioned previously as a useful approximation to a "soft" limiter. The input/output characteristic is then

$$v_0 = (2P)^{1/2}\,\text{erf}'\,|v_{in}|\,(\text{sgn}\,v_{in}) \tag{9-75}$$

where the alternate error function is defined as

$$\text{erf}'\,\frac{x}{\sigma} = \left(\frac{1}{2\pi}\right)^{1/2}\int_0^x \exp\left(-\frac{t^2}{2\sigma^2}\right)dt \tag{9-76}$$

The same approach used for the hard limiter gives amplitude coefficients a'_m for the dominant IM products for the error function limiter

$$a'_m = \frac{a_m}{(1+y^2)^{m/2}} \tag{9-77}$$

where a_m is the amplitude value for the hard limiter from (9-68) and

$$y^2 \triangleq \frac{8\pi^2\sigma^2}{N} \tag{9-78}$$

is a measure of limiter hardness. At a value of $y = 0$, the results are the same as for a hard limiter since $\sigma^2 = 0$ and $a'_m = a_m$. For $y = 1$, there is a 3-dB backoff in the output power and from (9-77) $a'_m = a_m/2^{m/2}$.

The values of the dominant IM power coefficients, a'_m, are normalized to the peak output power P and are also given in Table 9-5. Notice that in this 3-dB backed-off mode of Table 9-5, the IM products beyond the fifth order are relatively negligible. Notice that the power backoff has reduced the total IM power by a factor of 0.215/0.0136 = 15.8, or 12.0 dB.

The relationship between intermodulation power and signal backoff can be expressed in somewhat different terms by noting that the output power is $P/(1 + y^2)$ and the mth-order intermodulation power is, from (9-74), (9-77)

$$P_m \cong \frac{P}{m^2(1 + y^2)^m} \tag{9-79}$$

Thus, the ratio of total output signal power $P_S = P_1$ to total third-order intermodulation power is

$$\frac{P_S}{P_3} \cong m^2(1 + y^2)^{m-1}\Big|_{m=3} = 9(1 + y^2)^2 \qquad m = 3 \tag{9-80}$$

or denoting $(1 + y^2) = B_0$ as the output power backoff factor, this expression becomes

$$\frac{P_S}{P_3} \cong 9.54\text{ dB} + 2B_0\text{ dB} \tag{9-81}$$

when power backoff B_0 is expressed in decibels. For $B_0 = 3$ dB, $P_S/P_3 = 15.5$ dB.

9-8 FREQUENCY SELECTION TO REDUCE INTERMODULATION DISTORTION EFECTS

We have dealt primarily with the power levels in each order IM product and the total IM power in the entire fundamental zone. However, the input spectral density and input frequency selection have a substantial impact on the IM power levels falling on each signal. Frequency selection can reduce IM effects substantially.

Effects of Input Spectral Characteristics

For a large number of equally spaced carriers, the hard-limiter intermodulation power is essentially the same as that shown in Table 9-3 for Gaussian inputs. That is, for large N, the values of the signal output line components are $0.785P/N$, and the intermodulation power at the center channel is $0.128P/N$ and at the edge channel it is $0.0912P/N$. Hence, the signal-power-to-intermodulation-power ratios P_S/P_{IM} are 7.8 and 9.35 dB for the center and edge channels, respectively, when the input signals are so closely spaced that they are essentially continuous. These results are valid to within 1 dB for values of $N \geqq 7$.

A more generalized set of inputs is shown in Fig. 9-20 where there are

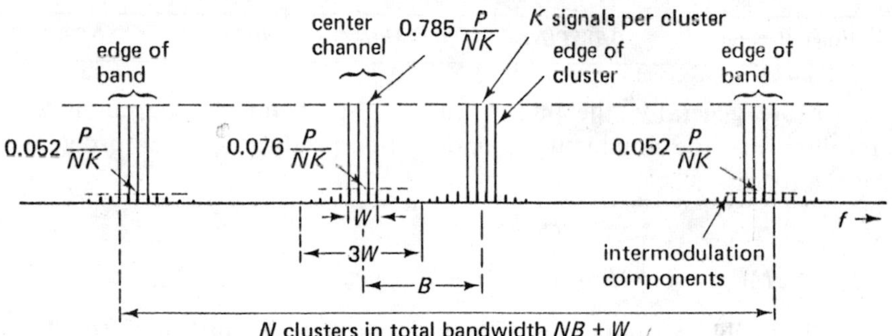

Fig. 9-20 Intermodulation spectrum for N clusters of K carriers each. The spacing between clusters is B Hz, each cluster has bandwidth W, and a cluster spacing of $B \geqq 3W$ is assumed. The spacing between carriers in a cluster is then $W/(K-1)$.

gaps in the signal spectrum. The N clusters of signals are spaced in frequency by B Hz from center to center, and there are K discrete signals in each cluster with spacing between signals $W/(K-1)$. Thus, the bandwidth of each cluster of signals is W. Clearly, if there is no gap between clusters and $W = B$, then the NK signals are all equally spaced, and the results already described apply.

On the other hand, if the signal clusters are spaced by a larger amount $B \geqq 3W$ and there is a gap between clusters, the intermodulation spectral line components have exactly the same shaped envelope for a single cluster as the envelope for the equally spaced carriers. However, the total power in each cluster of intermodulation products is weighted in exactly the same proportion as if each cluster were itself treated as a line component. We can use Table 9-3 to show that the IM power density at the center of the band or, more specifically, the center of the cluster is $(P/NK)[(0.128)/(0.215)](0.128) = 0.076P/NK$. This density has been reduced below the $0.128P/NK$ level because some of the IM product has been convolved outside the entire signal band NB. The total spectrum appears as shown in Fig. 9-20; power densities are listed in Table 9-6.

Table 9-6 INTERMODULATION POWER AND POWER DENSITY FOR SIGNAL CLUSTERS SPACED BY MORE THAN $3W$ Hz WHERE W IS THE SIGNAL BANDWIDTH OF EACH CLUSTER AS SHOWN IN FIG. 9-17.

	Signal Power Each Carrier	*Total IM Power in Cluster*	*IM Power Density*	
			Edge of Cluster	*Center of Cluster*
Center of band	$0.785P/NK$	$0.128P/N$	$0.076P/NK$	$0.052P/NK$
Edge of band	$0.785P/NK$	$0.128P/N$	$0.052P/NK$	$0.038P/NK$

More generally, the intermodulation spectrum can be computed by recalling that the dominant intermodulation products of each order are all of the form

$$IM_3 = A + B - C$$

$$\text{or} \quad IM_5 = A + B + C - D - E, \quad \text{etc.} \tag{9-82}$$

for all possible permutations of input frequencies if substituted for A, B, C, D, E. If input frequencies are translated from IF to baseband, the same relationships apply and the frequencies now occupy the positive and negative frequency range $-(NB + W)/2$ to $+(NB + W)/2$. Dominant intermodulation products can be expressed now as $A + B + C + D + E$, where both positive and negative frequencies $f_1, f_2, f_3, \ldots, f_{NK}, |f_i| \leqq (NB + W)/2$ are selected without replacement (since $A \neq B \neq C \ldots$), and substituted for A, B, C, D, E.

Thus, determination of the spectral distribution of intermodulation products of each order M for equal-amplitude inputs is reduced to exactly the same problem as calculating the amplitude probability distribution of the sum of M random inputs taken from NK locations without replacement.

For N uniformly spaced equal-amplitude signals within a cluster, the discrete intermodulation spectrum of order M at frequency $f_n = n$, for $|n| \leqq NM/2$ and unity frequency spacing, is

$$Q_M(n) = \sum_{i=0}^{I(y)} (-1)^i \frac{M!}{(M-i)!\,(i!)} \frac{(y-i)^{M-1}}{N(M-1)} \tag{9-83}$$

where $y \triangleq (M/2 - n/N)$ and the upper limit of the summation $I(y)$ is the largest integer $< y$.

Define $Q_c(x)$ as the input **cluster** distribution for N clusters of discrete line components. If each cluster is spaced by B Hz, then the total input spectral distribution for N odd at frequency n is

$$Q_T(n) = \frac{1}{N} \sum_{k=-(N-1)/2}^{(N-1)/2} Q_c(n - kB) \tag{9-84}$$

The resulting discrete line spectrum of intermodulation components of order M for this set of signal clusters is obtained using the characteristic function method,

$$Q_M(n) = \frac{1}{N^M} \sum_{k=-M(N-1)/2}^{M(N-1)/2} A_{Mk} Q_{cM}(n - kB) \tag{9-85}$$

where Q_{cM} is the Mth-order intermodulation component distribution of an individual cluster. The coefficients A_{Mk} are derived from the generating functions

$$\left[\sum_{l=-(N-1)/2}^{(N-1)/2} x^l \right]^M = \sum_{k=-M(N-1)/2}^{M(N-1)/2} A_{Mk} x^k \tag{9-86}$$

and

$$\sum_{k=-M(N-1)/2}^{M(N-1)/2} A_{Mk} = N^M \tag{9-87}$$

The distribution of intermodulation products thus obtained from (9-85) is the weighted sum of the cluster intermodulation products each offset by B Hz from the others.

Figure 9-21 shows the variation of total intermodulation power at the center frequency channel in the central cluster as a function of cluster frequency spacing B. The cluster frequency spacing is uniform. This central channel is the worst channel in the band. The power in units of P/NK varies from 0.128 for equal spacing $B = W$ and no guard space between channels, to 0.077 for large spacing $B < 3W$ between channels. Spacing of $B = 2W$ yields nearly full improvement with a 2-to-1 increase in bandwidth. Thus, as channel spacing increases above $B = W$, the inband signal-to-total intermodulation-power ratio increases from 8.9 dB to a maximum of 11.19 dB at the center channel. Almost all of the 2.29-dB improvement comes by increasing B to $2W$; very little further improvement is gained by larger increases in the frequency spacing B.

Consider now a set of contiguous channels of bandwidth B. If the clusters are of bandwidth W where $W < B$ but the clusters are randomly, rather than uniformly, selected in center frequency from one edge of the B-Hz channel to the other, the intermodulation spectrum remains exactly the same as above for a sufficiently large number of clusters N. However, only a fraction W/B of the intermodulation power is passed through the receive channel bandpass filter (bandwidth W). Hence, by increasing the total bandwidth of each channel B above W Hz and randomly selecting the position within the

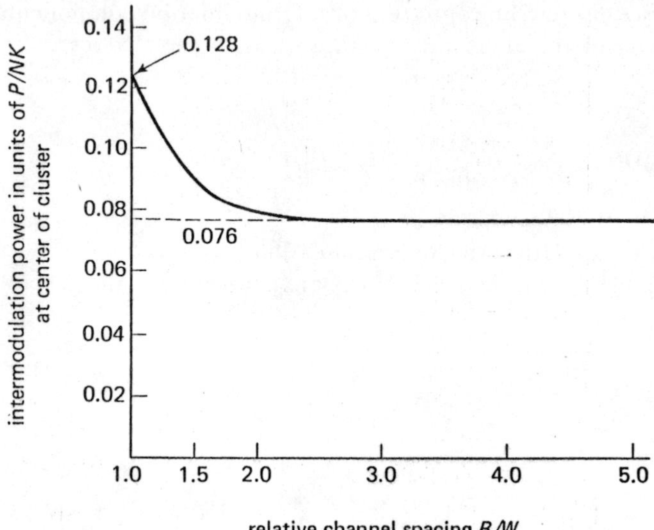

Fig. 9-21 Intermodulation power at the central channel in the center cluster (worst case) for a hard-limiter channel. Channel spacing is B, the cluster bandwidth is W, and there are a total of NK equal-amplitude input sinusoids.

channel, the center-channel-signal-to-IM-power ratio increases by the factor B/W to

$$\frac{P_s}{P_{\mathrm{IM}}} = 7.8 \text{ dB} + 10 \log\left(\frac{B}{W}\right) \text{dB} \tag{9-88}$$

for a total transmission bandwidth of NB Hz and a total input information bandwidth of NW Hz. As an example, if $B/W = 4$, then $P_s/P_{\mathrm{IM}} = 13.8$ dB. Thus this approach gives a valuable improvement in performance.

Deterministic Frequency Assignments

Suppose that N fixed-frequency sinusoids are to be packed in a channel of bandwidth MB where B Hz is assigned to each signal channel. By neglecting IM-product spreading and testing all input frequency choices, one can compute the minimum bandwidth required to avoid completely third- or third- and fifth-order cross products. These results for the minimum bandwidth required were obtained by Babcock [1953, 63–73], who also determined the frequency spacing required. The results are plotted in Fig. 9-22. Note that for ten channels to be free of third-order cross products one requires an RF

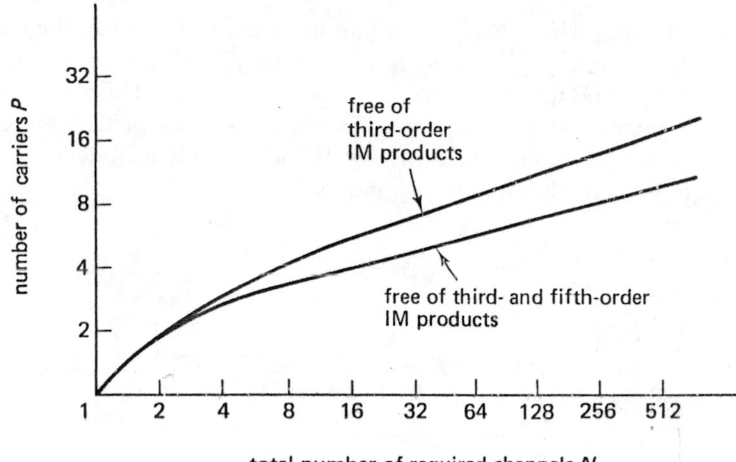

Fig. 9-22 Required number of frequency channels versus number of carriers for a nonlinear transponder

bandwidth $60B$, even when IM-product spreading is neglected. Thus for a transponder channel occupied by a large number of carriers, complete avoidance of third-order IM products is often impractical.

Specific Frequency Assignments

Table 9-7 lists specific channel frequency assignments. These two sets of assignments avoid third-order IM products for (1) no IM-product spreading, or (2) IM-product spreading equal to $3B$ [Sevy, 1966, 568–578].

Table 9-7 FREQUENCY PLANS TO AVOID THIRD-ORDER IM PRODUCTS WITH AND WITHOUT SPREADING

	Signals P	*Channels N*	*Frequencies f_i*
IM product spreading	3	4	1, 2, 4
	4	7	1, 2, 5, 7
	5	12	1, 2, 5, 10, 12
	6	18	1, 2, 5, 11, 13, 18
	7	26	1, 2, 5, 11, 19, 24, 26
	8	35	1, 2, 5, 10, 16, 23, 33, 35
	9	46	1, 2, 5, 14, 25, 31, 39, 41, 46,
	10	62	1, 2, 8, 12, 27, 46, 48, 57, 60, 62
IM product spreading	3	7	1, 3, 7
	4	15	1, 3, 7, 15

Note that this frequency channel is a bandpass channel; hence, the channel assignments are at $f_0 + f_i = f_0 + iB$ where $f_0 \gg f_i$. The bandpass nature of the channel must be taken into account when computing IM products. The channel assignment for the four-signal example with IM-product spreading taken into account is illustrated in Fig. 9-23. It yields a SNR > 21 dB in all channels for hard-limiting transponders.

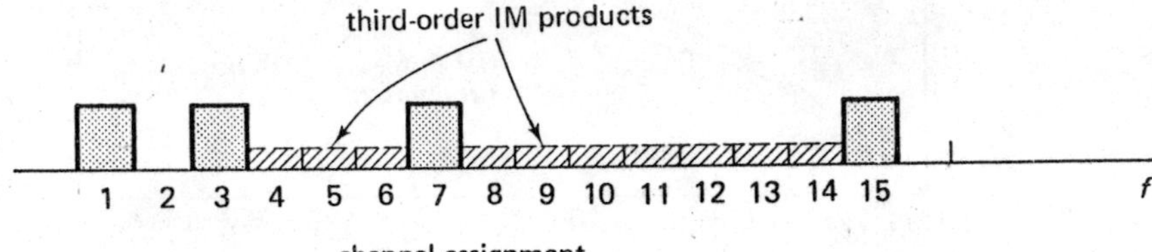

Fig. 9-23 Channel frequency assignment for four signals avoiding third-order IM products with IM-product spreading

Random Spacing

The third-order IM product has 57.4 percent, and the third- and fifth-order IM products together 74.6 percent, of the total intermodulation power at midchannel. (Refer to Table 9-5.) Hence, frequency selection to eliminate third- or both third- and fifth-order IM products increases the signal-to-IM power ratio (worst case) for a large number of signals by approximately 3.71 dB and 5.95 dB, respectively. This improvement is obtained, however, at the expense of a very large increase in total bandwidth (Fig. 9-20) if the number of signals N is large.

For N large the random spacing technique of (9-88) is more feasible. The signals of bandwidth W Hz are spaced on the average by H Hz for $H \gg W$ and are randomly centered within this bandwidth as described earlier. The input frequency plan over a small portion of the input spectrum then appears as in Fig. 9-24.

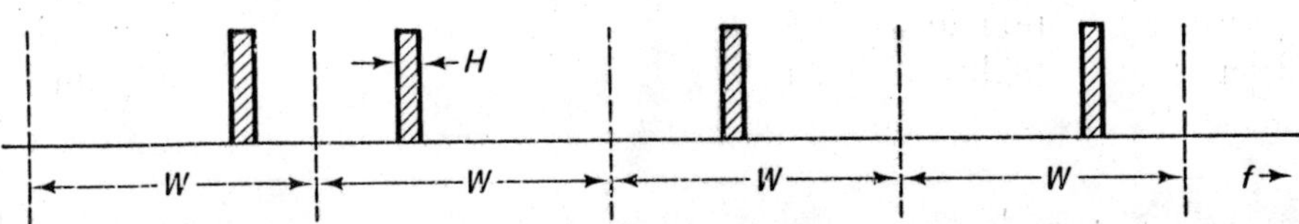

Fig. 9-24 Frequency plan for a large number N of carriers of bandwidth W; each signal center frequency is randomly selected within bandwidth H.

Performance improvement for this frequency plan from (9-88) is

$$10 \log \frac{B}{W} \quad \text{dB} \tag{9-89}$$

Hence, for a H/W of 10, a 10-dB improvement is obtained. Neither the third- nor the fifth-order IM products are eliminated completely. Instead, the IM power of all orders is uniformly distributed (on the average) across the frequency band because of random or pseudorandom frequency selection, and only a small fraction W/H of the IM power is accepted by individual receive filters of bandwidth W. For any specific pseudorandom frequency plan selected with only a moderate number of carriers, it is appropriate to check the actual IM-product spectrum since the above concept is based on average power-density results for a large number of carriers.

9-9 AM/PM CONVERSION EFFECTS

In addition to the instantaneous amplitude nonlinearity described in the preceding paragraphs, most amplifying devices also exhibit AM/PM conversion [Sunde, 1969, Chap. 8]. That is, a change in the envelope of a multicarrier input causes a change in the output phase of each signal component. In this section, we determine the intermodulation distortion caused by AM/PM conversion effects for multiple sine-wave inputs, and compare these effects with the intermodulation products of an instantaneous nonlinearity. It is shown that AM/PM effects often dominate instantaneous amplitude nonlinearities when the drive level to the nonlinear element is reduced well below saturation.

AM/PM Model

Figure 9-25 is a model of the AM/PM effect, which is a reasonable model for TWT and other amplifiers. A typical AM/PM characteristic $\theta(A)$ for a TWT amplifier [Berman and Mahle, 1970, 37–48] is shown in Fig. 9-26, where $A(t)$ is the input envelope. Note that for small input drive levels, the phase modulation induced by the envelope fluctuation is approximately proportional to the envelope squared—that is, proportional to the input power level

$$\theta(A) \cong \frac{K}{2} A^2(t) \tag{9-90}$$

for A sufficiently small. This behavior is typical of a large class of ampli-

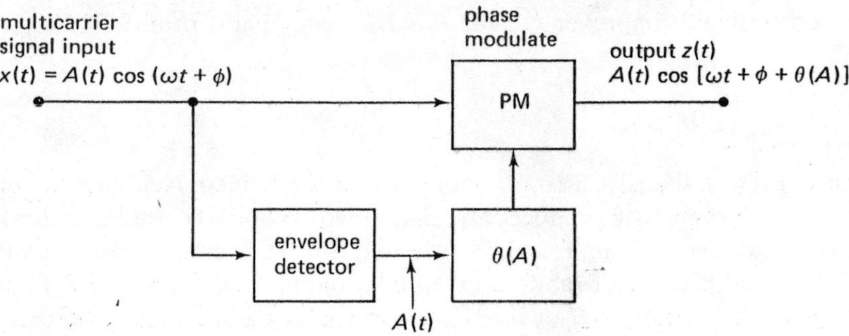

Fig. 9-25 Model of the AM/PM nonlinearity $\theta(A)$

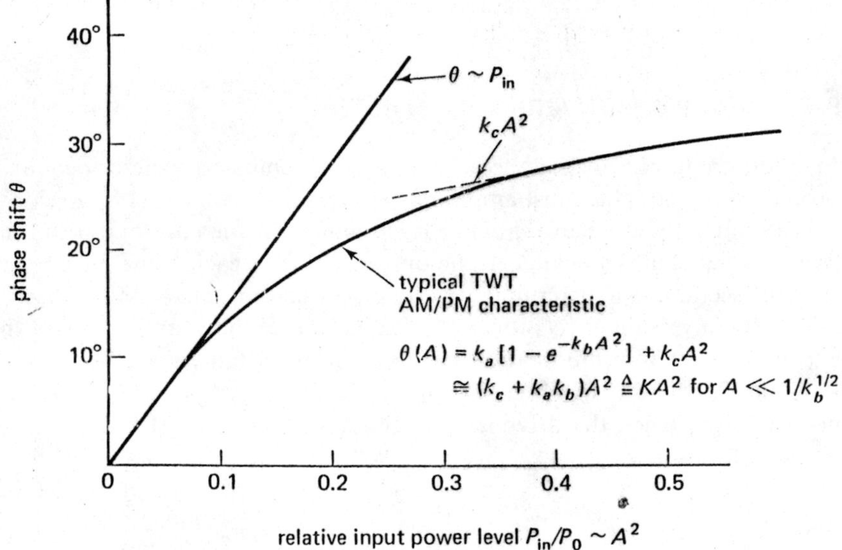

Fig. 9-26 Typical TWT AM-PM conversion phase shift θ plotted versus drive level P_{in}

fiers, although the value of the constant K changes from one amplifier to the next. In practice this AM/PM conversion factor K is often expressed in degrees/decibel because of measurement convenience.

In a typical measurement one can apply a sinusoidal input with a small amount of AM:

$$x(t) = A(1 + m \cos \omega_m t) \cos \omega_0 t \tag{9-91}$$

and $A(t) = A(1 + m \cos \omega_m t)$ is the input envelope. This input results in an output phase modulation, which for small A is approximately square law:

$$\begin{aligned}\theta(t) &= KA^2(1 + 2m\cos\omega_m t + m^2\cos^2\omega_m t) \qquad \text{radians}\\ &\cong KA^2(1 + 2m\cos\omega_m t) \qquad m \ll 1\end{aligned} \tag{9-92}$$

and the peak deviation from the mean phase is $\theta_p \cong KA^2 2m$. This peak phase error can be expressed in degrees/decibel of AM as

$$\begin{aligned}K_p &= \frac{\theta_p(180°/\pi)}{20\log_{10}(1+m)\,(\text{dB})} \simeq \frac{KA^2 2m(180°/\pi)}{8.69m\,(\text{dB})} = \frac{K2A^2 180°}{8.69\pi\,(\text{dB})}\\ &\cong 26.4KP_s(°/\text{dB}) \qquad \text{for } m \ll 1\end{aligned} \tag{9-93}$$

where $P_s \triangleq A^2/2$. Thus K_p is approximately linearly proportional to the input power level, P_s for a square-law AM/PM characteristic.

Multiple-Input Sinusoids

To compute the intermodulation products resulting from the AM/PM nonlinearity, we let the input be a summation of sinusoids (Fig. 9-27), which can be represented as

$$x(t) = \sum_{i=1}^{n} A_i \cos[\omega_0 t + \phi_i(t)] \triangleq A(t)\cos[\omega_0 t + \phi(t)] \tag{9-94}$$

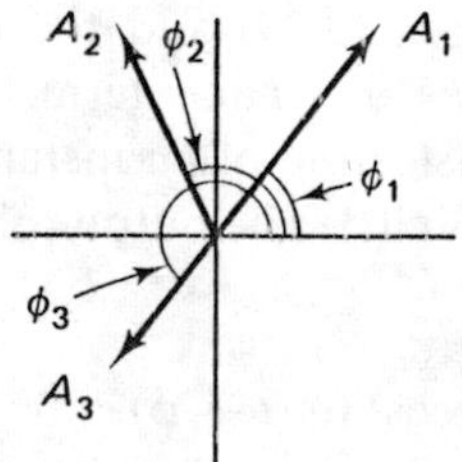

Fig. 9-27 Input amplitudes and phases of multiple sinusoids

Then the squared envelope is expressed as

$$\begin{aligned}A^2(t) &= (\textstyle\sum A_i\cos\phi_i)^2 + (\sum A_i\sin\phi_i)^2\\ &= \sum_i\sum_j A_iA_j\cos(\phi_i - \phi_j)\\ &= \sum_i A_i^2 + \sum\sum_{i\neq j} A_iA_j\cos(\phi_i - \phi_j)\end{aligned} \tag{9-95}$$

and the phase function of the resultant in (9-94) is

$$\phi(t) = \tan^{-1}\left(-\frac{\sum A_i \sin \phi_i}{\sum A_i \cos \phi_i}\right) \tag{9-96}$$

The resulting AM/PM output is a summation of sinusoids, each phase modulated by $\theta(A)$:

$$\begin{aligned} z(t) &= \sum A_i \cos \{\omega_0 t + \phi_i(t) + \theta[A(t)]\} \\ &= \sum A_i \cos [\omega_0 t + \phi_i(t)] \cos \theta(A) \\ &\quad - \sum A_i \sin [\omega_0 t + \phi_i(t)] \sin \theta(A) \end{aligned} \tag{9-97}$$

For small values of AM/PM distortion, the phase modulation is $\theta \ll 1$, and the output from (9-95), (9-96), (9-97) becomes

$$\begin{aligned} z(t) &\cong \sum A_i \cos (\omega_0 t + \phi_i) - \theta(A) \sum A_i \sin (\omega_0 t + \phi_i) \\ &= A(t) \cos [\omega_0 t + \phi(t)] + d(t) \end{aligned} \tag{9-98}$$

Since $\theta(A) \cong KA^2(t)$ the second term, which represents the distortion effect $d(t)$, can also be represented as

$$\begin{aligned} d(t) &\cong -KA^2(t)\{A(t) \sin [\omega_0 t + \phi(t)]\} \\ &\cong -KA^3(t) \sin [\omega_0 t + \phi(t)] \end{aligned} \tag{9-99}$$

There is a similarity between (9-99) and the corresponding result for an amplitude nonlinearity. This distortion term is 4/3 the amplitude and 90° shifted in phase from the distortion of an instantaneous cube-law amplitude nonlinearity $z(t) = x(t) + Kx^3(t)$. The output of the cube-law nonlinearity is then

$$\begin{aligned} z(t) &= A(t) + KA^3 \cos^3 (\omega_0 t + \phi) \\ &= A(t) + K(3/4)A^3 \cos [\omega_0 t + \phi(t)] + \text{term at } 3(\omega_0 t + \phi) \end{aligned} \tag{9-100}$$

The term A^3 is not all distortion; it contains a signal component since $\bar{A}^2 \neq 0$. Thus, the distortion IM products occur at precisely the same frequencies as the third-order IM products for an instantaneous amplitude nonlinearity but have a different amplitude and are 90° shifted in phase.

EXAMPLE Assume that there are three sine waves at the input. We can neglect the constant $\sum A_i^2$ of (9-95) because the constant phase shift, θ_0, it contributes is of no consequence. Then the AM/PM of (9-95) gives

$$\theta(A) - \theta_0 = K(A^2 - \sum A_i^2) = 2K[A_1 A_2 \cos(\phi_1 - \phi_2) + A_2 A_3 \cos(\phi_2 - \phi_3) + A_1 A_3 \cos(\phi_1 - \phi_3)] \quad (9\text{-}101)$$

Thus, the distortion from (9-98) and (9-99) contains nine components:

$$\begin{aligned}[\theta(A) - \theta_0] \sum A_i \sin(\omega_0 t + \phi_i) &= K[A_1^2 A_2 \sin(\omega_0 t + 2\phi_1 - \phi_2) \\ &+ A_2^2 A_1 \sin(\omega_0 t + 2\phi_2 - \phi_1) + A_1^2 A_3 \sin(\omega_0 t + 2\phi_1 - \phi_3) \\ &+ A_3^2 A_1 \sin(\omega_0 t + 2\phi_3 - \phi_1) + A_2^2 A_3 \sin(\omega_0 t + 2\phi_2 - \phi_3) \\ &+ A_3^2 A_2 \sin(\omega_0 t + 2\phi_3 - \phi_2)] \\ &+ 2KA_1 A_2 A_3 [\sin(\omega_0 t + \phi_1 + \phi_2 - \phi_3) \\ &+ \sin(\omega_0 t + \phi_1 - \phi_2 + \phi_3) + \sin(\omega_0 t + \phi_2 - \phi_1 + \phi_3)]\end{aligned} \quad (9\text{-}102)$$

Just as with a cubic instantaneous nonlinearity, the (1, 1, 1) cross product $\phi_i + \phi_j - \phi_\ell$, $i \neq j \neq \ell$ terms are twice the amplitude of the (2, 1) cross products because two terms in the product $\theta(A) \sum A_i \cos(\omega_0 t + \phi_i)$ give the same output frequency.

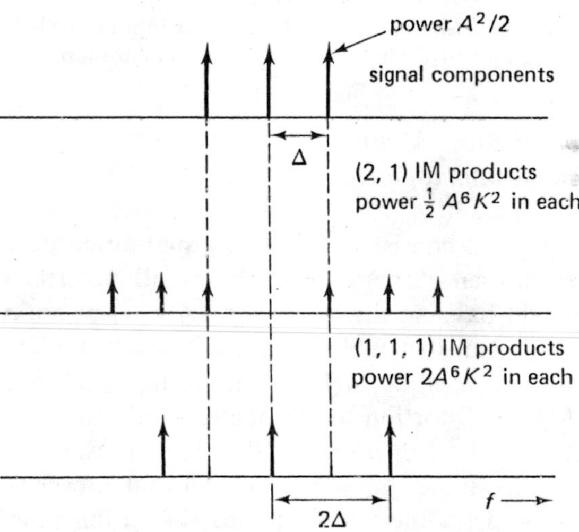

Fig. 9-28 Frequency spectrum of quadratic AM/PM IM products for an input of three sinusoids

The output spectrum for three equally spaced sinusoids of equal amplitude, $A_1 = A_2 = A_3 = A$ is shown in Fig. 9-28. The center channel clearly has the lowest signal-to-distortion ratio. For small distortion, it is approximately equal to

$$\frac{P_s}{P_{(111)}} = \frac{A^2/2}{2A^4(A^2K)} = \frac{1}{4}\frac{13.3}{K_p^2A^2} \tag{9-103}$$

where from (9-93) $K_p = 13.3\ KA^{2\circ}/\text{dB}$ and (9-103) is as shown in Fig. 9-29.

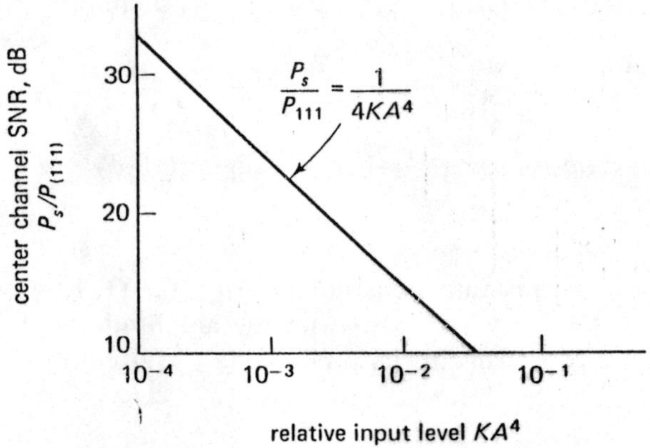

Fig. 9-29 Center channel SNR versus input drive level for an amplifier with quadratic AM/PM distortion

Comparison of AM/PM and Instantaneous Amplitude Nonlinearities

Consider an input consisting of three equal-amplitude sinusoids. The AM/PM distortion then decreases $\sim A^3$ for small A. Clearly, the piecewise linear limiter would have no distortion for $3A < c$, where c is the limiting level. Even for more realistic amplifiers, the amplitude nonlinearity distortion decreases rapidly at low input levels. For small input drive levels, therefore, one expects AM/PM distortion to dominate. This low drive level situation would be characteristic of operation with a high power TWT where bandwidth limitations are more important than output power. If the reverse is true and one must drive the amplifier into saturation (hard limiting), the AM/PM distortion effect increases with input power at a relatively slow rate, whereas nonlinear amplitude distortion increases rapidly in the limiting region and often dominates the AM/PM effect completely.

9-10 COMBINED EFFECT OF AM/PM AND AMPLITUDE NONLINEARITIES

An approximate model of a TWT or other nonlinear amplifier is shown in Fig. 9-30. The AM/PM effect appears first in the block diagram; it seems to be essentially evenly distributed along the beam, whereas the amplitude nonlinearity originates in the last part of the beam/slow-wave-structure interaction region. The order in which the nonlinearities appear does have some

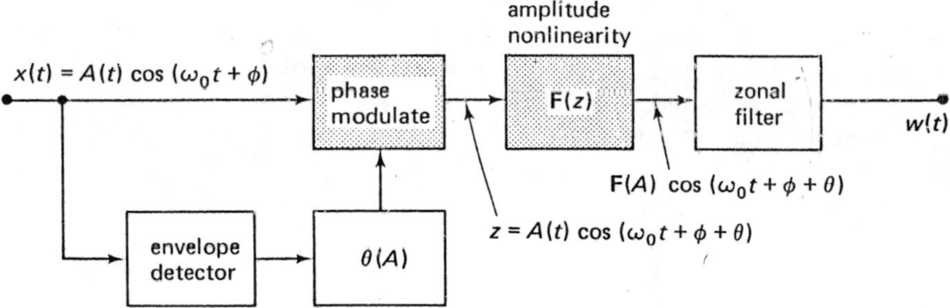

Fig. 9-30 Model of AM/PM nonlinearities in a TWT amplifier

effect on the output. Let us assume that $\theta(A) = KA^2$ and that the instantaneous amplitude nonlinearity is a simple cube law $\mathrm{F}(z) = 1 + K_3 z^3$. The $\mathrm{F}(z)$ output for an input $z(t) = A\cos(\omega_0 t + \phi)$ is then

$$\begin{aligned}\mathrm{F}(z) &= A\cos(\omega_0 t + \phi + \theta) + K_3 A^3 \cos^3(\omega_0 t + \phi + \theta)\\ &= A\cos(\omega_0 t + \phi + \theta) + K_3 A^3 \tfrac{3}{4}\cos(\omega_0 t + \phi + \theta)\\ &\quad + \text{term at } 3\omega_0 t \end{aligned} \tag{9-104}$$

We neglect the term at $3\omega_0 t$. Hence, the output of the zonal filter is

$$\begin{aligned} w(t) &= (1 + \tfrac{3}{4}K_3 A^2)A\cos[\omega_0 t + \phi + \theta(A)]\\ &\cong \left(1 + \frac{3}{4}K_3 A^2\right)A\left[\cos(\omega_0 t + \phi)\left(1 - \frac{\theta^2}{2}\right) - \theta\sin(\omega_0 t + \phi)\right]\\ &\qquad \theta \ll \frac{\pi}{2} \end{aligned} \tag{9-105}$$

If the AM/PM effect $\theta(A) = KA^2$, then the output becomes

$$
\begin{aligned}
w(t) \cong A(t) \cos(\omega_0 t + \phi) &+ \overbrace{\tfrac{3}{4} K_3 A^2 [A \cos(\omega_0 t + \phi)]}^{\text{amplitude nonlinearity (AN)}} \\
&- \overbrace{K A^2 [A \sin(\omega_0 t + \phi)]}^{\text{AM/PM}} + \underbrace{\tfrac{3}{4} K_3 K A^4 [A \sin \omega_0 t + \phi)]}_{\text{AN interacting with AM/PM}} \\
&- \underbrace{\tfrac{1}{2} K^2 A^4 [A \cos(\omega_0 t + \phi)]}_{\text{AM/PM}}
\end{aligned} \tag{9-106}
$$

The second term is the resulting distortion for an amplitude nonlinearity (AN) by itself; the third and fifth are the effects of AM/PM, and the fourth term represents the interaction between AM/PM and AN effects. First-order (A^2 dependent) effects are illustrated in the phase plot of Fig. 9-31 where the K^2A^4 term is neglected. Notice that the dominant AM/PM effect is a 90° phase offset from the AN effect.

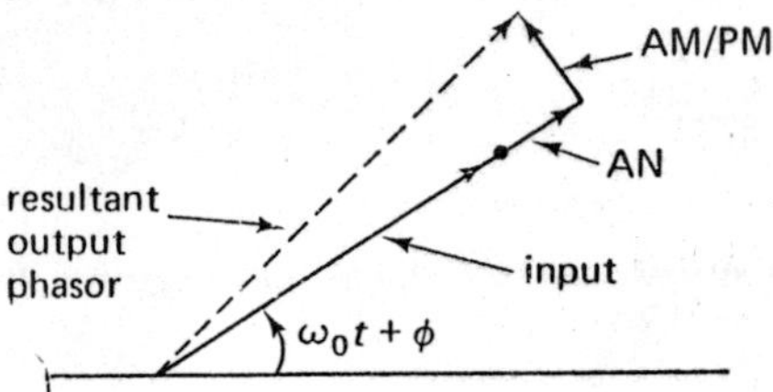

Fig. 9-31 First-order effects of combined AM/PM and AN nonlinearities

EXAMPLE Consider an input with two sinusoids $x = A_1 \cos(\omega_0 t + \phi_1) + A_2 \cos(\omega_0 t + \phi_2)$ where A_1, A_2 are constant. The squared envelope is then

$$
A^2(t) = (A_1^2 + A_2^2) + 2A_1A_2 \cos(\phi_1 - \phi_2) \tag{9-107}
$$

The first-order distortion-producing terms from (9-106), (9-107) are, therefore,

$$
\begin{aligned}
2A_1A_2 \cos(\phi_1 - \phi_2)\{\tfrac{3}{4} K_3 [A_1 \cos(\omega_0 t + \phi_1) + A_2 \cos(\omega_0 t + \phi_2)] \\
- K[A_1 \sin(\omega_0 t + \phi_1) + A_2 \sin(\omega_0 t + \phi_2)]\}
\end{aligned} \tag{9-108}
$$

Thus, we obtain terms at phases (frequencies) $2\phi_1 - \phi_2$, $2\phi_2 - \phi_1$:

$$
\begin{aligned}
A_1^2 A_2 [\tfrac{3}{4} K_3 \cos(\omega_0 t + 2\phi_1 - \phi_2) - K \sin(\omega_0 t + 2\phi_1 - \phi_2)] \\
+ A_2^2 A_1 [\tfrac{3}{4} K_3 \cos(\omega_0 t + 2\phi_2 - \phi_1) - K \sin(\omega_0 t + 2\phi_2 - \phi_1)]
\end{aligned} \tag{9-109}
$$

where $\phi_1 = \omega_1 t$, $\phi_2 = \omega_2 t$

and a IM-product spectrum is generated as shown in Fig. 9-32 where each IM product has a fixed phase offset from $2\phi_2 - \phi_1$, $2\phi_1 - \phi_2$, namely, $-\tan^{-1} 4K/3K_3$. Again note that the AM/PM effect with the coefficient K is 90° phase offset from the AN effect with coefficient K_3.

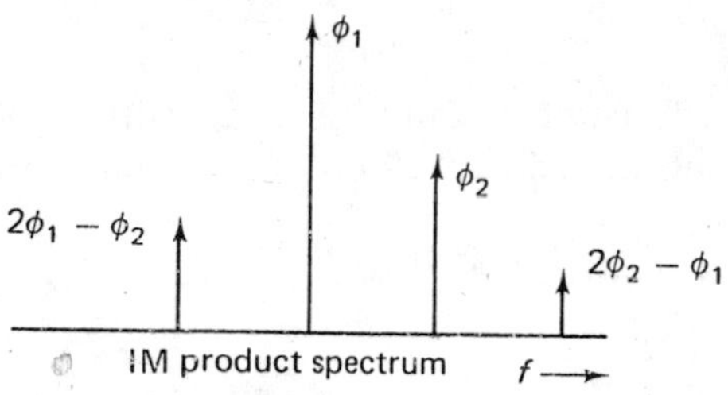

Fig. 9-32 IM-product spectrum for combined AM/PM and AN nonlinearities for a two-signal input

Intelligible Cross-Talk

The intermodulation distortion discussed above causes an unintelligible or noise-like distortion when an FM signal is transmitted. However, if a filter precedes the AM/PM distorting element, it is possible to generate intelligible cross talk where the frequency modulation $\dot{\phi}_1$ on one channel is added to another. Thus we have in effect a two step process—FM conversion to AM by the filter followed by AM conversion to PM by the amplifier nonlinearity. This effect can occur if the filter has a sloping amplitude-transfer function, thereby causing an envelope modulation at the input to the TWT, which is proportional to $\dot{\phi}_i$:

$$A_1(1 + \alpha\dot{\phi}_1) \cos (\omega_0 t + \phi_1) + A_2(1 + \alpha\dot{\phi}_2) \cos (\omega_0 t + \phi_2) \quad (9\text{-}110)$$

For two-carrier inputs, the PM induced by the AM/PM conversion is then

$$\begin{aligned} \theta(A) = KA^2 = K[&A_1^2(1 + \alpha\dot{\phi}_1)^2 + A_2^2(1 + \alpha\dot{\phi}_2)^2 \\ &+ 2A_1A_2(1 + \alpha\dot{\phi}_1)(1 + \alpha\dot{\phi}_2) \cos (\phi_1 - \phi_2)] \end{aligned} \quad (9\text{-}111)$$

If $\alpha \ll 1$, each carrier is modulated with a phase modulation

$$\begin{aligned} \theta(A) \cong K[&A_1^2(1 + 2\alpha\dot{\phi}_1) + A_2^2(1 + 2\alpha\dot{\phi}_2) \\ &+ 2A_1A_2(1 + \alpha\dot{\phi}_1 + \alpha\dot{\phi}_2) \cos (\phi_1 - \phi_2)] \end{aligned} \quad (9\text{-}112)$$

Clearly, then, each carrier will be PM modulated with $KA_i^2\, 2\alpha\dot{\phi}_i$, a term proportional to the modulation on the adjacent channel, and the carrier produces intelligible cross-talk.

In-Phase/Quadrature Model of Amplifier Nonlinearities

The combined effects of instantaneous amplitude nonlinearity and AM/PM conversion shown in Fig. 9-29 produce an output waveform

$$w(t) = g[A(t)] \cos \{\omega_0 t + \theta[A(t)] \phi\} \tag{9-113}$$

for an input waveform

$$x(t) = A(t) \cos (\omega_0 t + \phi) \tag{9-114}$$

This output can be split into in-phase and quadrature terms [Kay et al., 1972]:

$$w_p(t) = g_p[A(t)] \cos (\omega_0 t + \phi) \tag{9-115}$$

and

$$w_q(t) = g_q[A(t)] \sin (\omega_0 t + \phi) \tag{9-116}$$

Thus, the quadrature envelope nonlinearities of (9-113) are related to the AM/PM and AN nonlinearities, $\theta(A)$ and $g(A)$ by

$$g_p[A] = g(A) \cos \theta(A) \tag{9-117}$$

and

$$g_q[A] = g(A) \sin \theta(A) \tag{9-118}$$

and can be represented as shown in Fig. 9-33. This in-phase/quadrature model

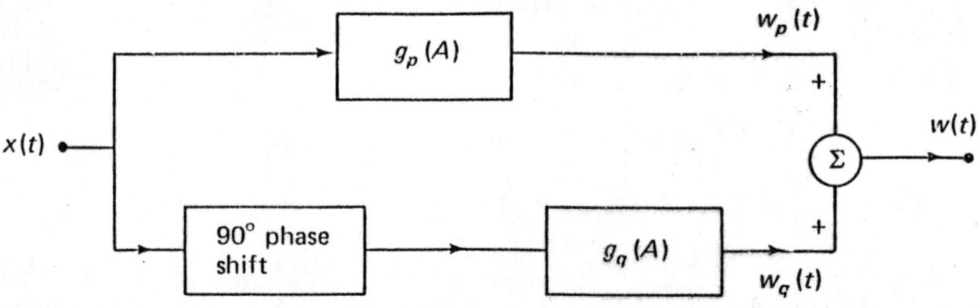

Fig. 9-33 In-phase/quadrature representation of a bandpass nonlinearity

is useful in developing the optimal predistortion network characteristics. The model is an attractive means for implementing predistortion compensation. It also permits the previously developed results on intermodulation products for amplitude nonlinearities to be extended to include the effects of AM/PM conversion.

The in-phase/quadrature model has several advantages over the model in Fig. 9-25. First, the outputs of the two quadrature channels are linearly independent; thus, the spectrum of the total nonlinearity is the sum of the spectra of individual channel outputs. Second, since each of the parallel channel nonlinearities exhibits no phase distortion, the instantaneous voltage transfer functions and the instantaneous envelope transfer functions are related by the Chebyshev transform (Sec. 9-5)

$$g_p(\sigma) = \frac{1}{\pi} \int_0^{2\pi} G_p(\sigma \cos \theta) \cos \theta \, d\theta \qquad (9\text{-}119)$$

where $G_p(\ )$ is the instantaneous amplitude (in-phase) nonlinearity.

Finally, for small AM/PM distortion, the AM/PM effect is totally exhibited in the quadrature channel, using the approximation for (9-118)

$$g_q(\sigma) \cong g(\sigma)\theta(\sigma) \qquad \text{for small } \theta(\sigma) \qquad (9\text{-}120)$$

Typical TWT in-phase and quadrature envelope nonlinearities are shown in Fig. 9-34.

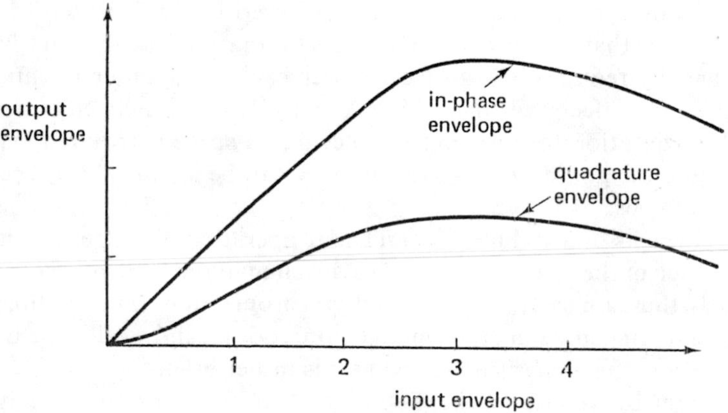

Fig. 9-34 Typical TWT in-phase and quadrature envelopes

CHAPTER 10

TIME-DIVISION MULTIPLE ACCESS

10-1 INTRODUCTION

Time-division multiple access (TDMA) is the primary alternate to frequency-division multiple access (FDMA), just as time-division multiplex (TDM) is the primary alternate to frequency-division multiplex (FDM) in multiplexing techniques. Time-division multiple access can achieve efficiencies in satellite-power utilization of 90 percent or more compared to the 3- to 6-dB loss in power efficiency that is typical of FDMA operation. As described in Chap. 9, FDMA usually requires a 3- to 6 dB power backoff in order to minimize intermodulation effects. Further, TDMA can achieve similar efficiencies in bandwidth utilization because no frequency guard space is required between channels, and the guard time loss in efficiency can be kept small by accurate timing techniques.

The concepts and techniques of TDMA operation are described in this chapter. Much of the discussion relies heavily on material from other chapters, particularly those on carrier recovery, bit synchronization, time multiplexing, baseband data transmission, PSK modulation, coding, and worldwide timing. For this reason, the analysis presented here is rather brief.

We begin by describing the TDMA system concept and some typical TDMA configurations. Key system elements described include: TDMA buffer storage and timing, coder/decoder configurations, TDMA frame format and efficiency, and TDMA carrier recovery. The chapter concludes with a brief discussion of satellite-switched TDMA, which combines the efficiencies of TDMA with those of satellite onboard processing and switched narrow-beam satellite antennas.

10-2 THE SYSTEM CONCEPT AND CONFIGURATION

Time-division multiple access (TDMA) is a multiple-access technique that permits individual earth terminal transmissions to be received by the satellite in separate nonoverlapping time slots, thereby avoiding the generation of intermodulation products in a nonlinear transponder. Each ground terminal must determine satellite system time and range so that the transmitted signals are timed to arrive at the satellite in the proper time slots. Signal timing and details of signal formats are discussed later.

Figure 10-1 illustrates the typical configuration of a TDMA network in which each high velocity burst of RF energy, typically quadriphase modulated, arrives at the satellite in its assigned time slot. There is little or no

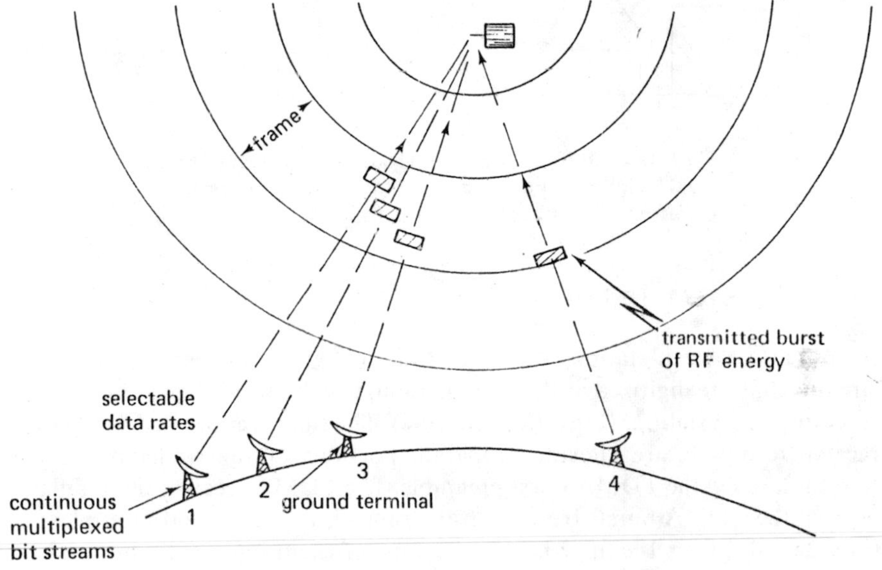

Fig. 10-1 Typical TDMA configuration

intermodulation caused by instantaneous nonlinearity, because only one signal enters the satellite transponder at a time. Note that the bit rates of the transmitted bursts are generally many times higher than that of the continuous input bit streams to the ground terminal.

Time-division multiple access permits the output amplifier to be operated in full saturation, often resulting in a significant increase in useful power output. IM-product degradation is largely avoided by transmitting each signal with sufficient guard time between time slots to accommodate any timing inaccuraçies, while preventing the "tails" of the pulsed previous and next signal from causing significant interference in the present time slot. The amplitude of these tails depends, of course, on the transient response and, in turn, on the amplitude and phase response of the filters, both in the satellite-transponder receive section and in the earth-terminal transmit filters.

If the transponder is operated in the "hard-limiting" mode and limits on noise input alone, the output envelope is essentially constant, even during the guard-time interval (Fig. 10-2). Typically, guard times can be made sufficiently small that the total guard-time frame consumes less than 10 percent of the usable signal power and the transponder is utilized to greater than 90% efficiency.

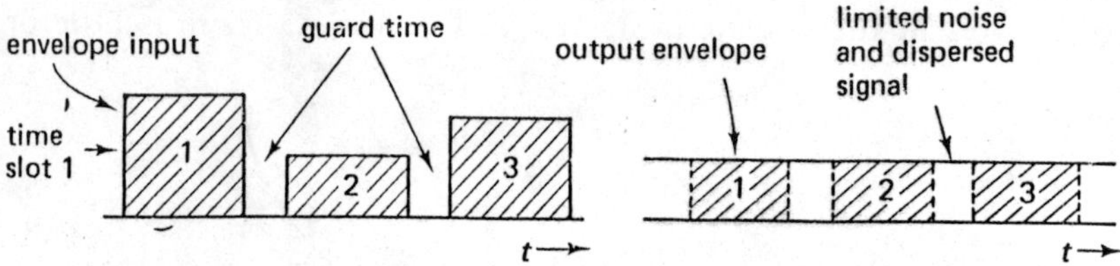

Fig. 10-2 Typical input and output envelopes for TDMA signals. The constant output envelope shown corresponds to a hard-limiting transponder.

10-3 SYSTEM TIMING

Each TDMA earth station has parallel input digital bit streams, or analog streams that are digitized at the earth station, which are addressed to separate receiving earth stations (Figs. 10-3 and 10-4). The signals addressed to separate receive terminals are allocated separate portions of the transmit TDMA burst following the TDMA burst preamble. The TDMA receiver demodulates each of the TDMA bursts from separate transmit terminals and multiplexers, then demultiplexes the appropriate portions of them into separate serial bit streams.

TDMA system timing is such that if all earth stations transmitted at the beginning (epoch) of their respective frames, all signals would arrive simultaneously at the satellite (Chap. 19). If the frame rate is $f_f = 1/T_F$, all input data rates f_{di} must be exact integral multiples of f_f—that is, $f_{di} = n_i f_f$. Otherwise, an integral number of bits could not be transmitted during each frame (or superframe). The burst rate f_{bi} is usually integrally related to the frame rate because $f_{di}\, T_f$ data bits are in each burst, and each burst duration is a natural fraction of the frame duration. Ordinarily, the burst rate should be

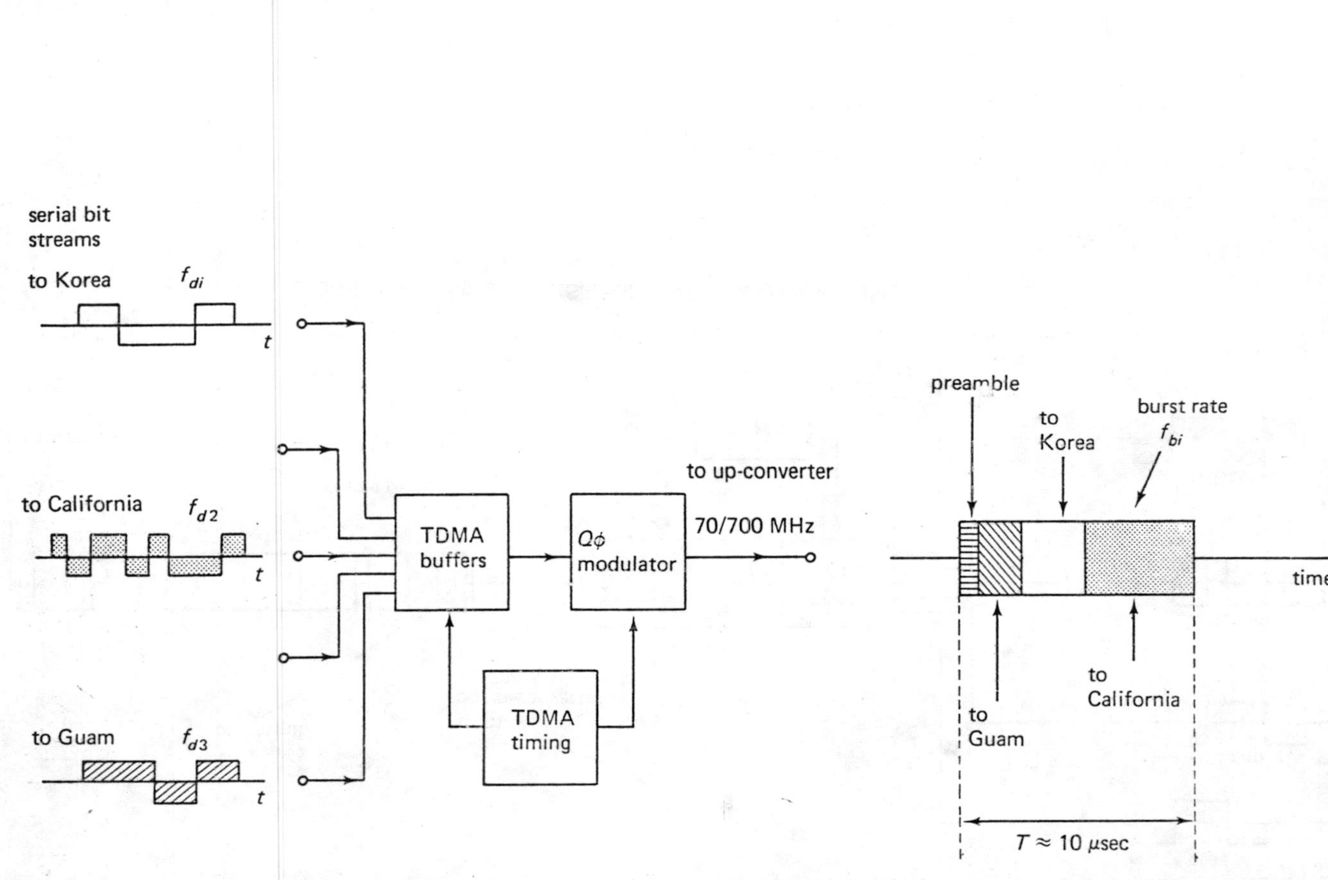

Fig. 10-3 Simplified TDMA transmit data formats. The TDMA buffers convert the serial bit stream to bursts of data bits at the burst rate which, in turn, are converted to IF signals by the quadriphase modulator.

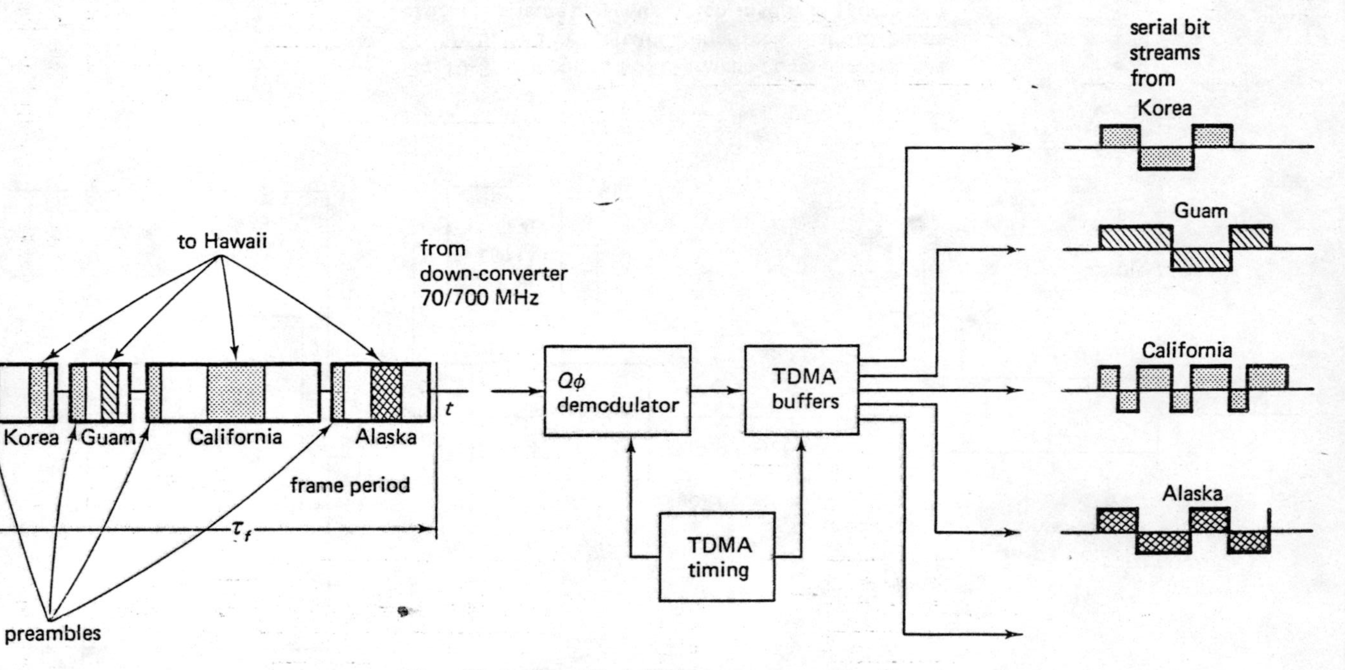

Fig. 10-4 Simplified TDMA receive data formats received at Hawaii

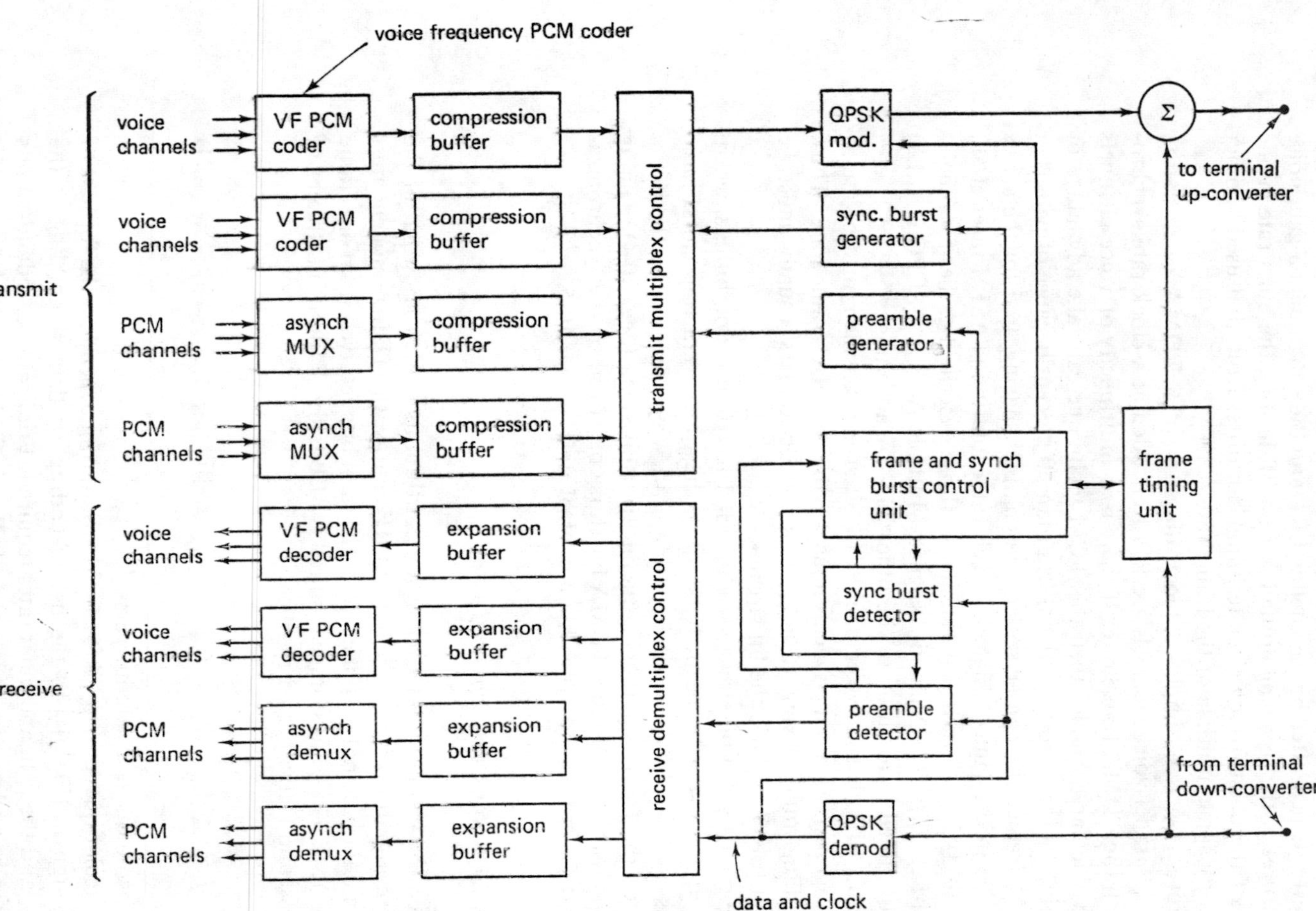

Fig. 10-5 A simplified TDMA earth-station terminal with analog voice channel inputs and nonsynchronous parallel input bit streams

the largest rate permitted by the satellite ERP and the $G/\mathfrak{T}$ (antenna gain/noise temperature) of the receiving ground station. If desired, the burst rate in one portion of the burst from a given terminal can differ from that used in other portions by some multiple of the frame rate.

Figure 10-5 is a simplified block diagram of a TDMA earth-station terminal. Parallel voice channels are PCM encoded at a clock rate synchronous with the TDMA frame rate. If there is a multiplicity of voice channels at the particular terminal, then this PCM technique can take advantage of the loading factor of voice, which is less than 50 percent because of pauses in speech. Voice activity can be sensed and channel sharing used for the active channels as in the Time-Assigned Speech Interpolation [TASI; Fraser et al., 1962, 1439–1454]. Thus, benefits similar to those described earlier for Single-Channel-Per-Carrier Demand-Assigned-Multiple-Access (SCPC-DAMA) can be obtained in addition to the other TDMA advantages.

Figure 10-5 also shows parallel PCM bit streams, which must be multiplexed together at a rate synchronous to the TDMA frame rate. Thus, pulse-stuffing multiplexing (Chap. 5) or elastic buffers usually must be employed to synchronize the bit streams.

The coded serial bit streams are then fed to compression buffers, where bits stored during one frame are burst out in the appropriate time slot. Frame timing is controlled by a separate timing unit, which may utilize the initial portion of the frame for ranging/timing transmissions (Sec. 10-4). Timing within the frame, or within a TDMA burst, is controlled by the synchronization burst generator and synch-burst control unit in Fig. 10-5.

TDMA Buffers and Timing Control

Since the incoming bit streams are continuous while the output of the TDMA modulator is a periodic burst of RF energy, the TDMA modem must contain a data buffer. This buffer stores the data bits received from one frame until the next. The total storage required is M bits for N input bit streams of bit rate f_{di} and frame period τ_f where

$$M = \sum_{i=1}^{N} f_{d_i}\tau_f \qquad (10\text{-}1)$$

and the products $f_{d_i}\tau_f$ are integers

The techniques employed for TDMA buffering depend on the formats of received signals (Fig. 10-4), the use of coding or data scrambling at the terminal, and the data and burst rates required. For example, redundant bits introduced by coding can be introduced after buffering, thus reducing buffer storage requirements.

TDMA timing at an earth station can be slaved either to an actual clock onboard the satellite or to an earth terminal clock at a terminal designated as the master. The master earth station generates a clock signal which is relayed by the satellite appears as if generated by the satellite.

Figure 10-6 illustrates the timing at the satellite and at the earth station

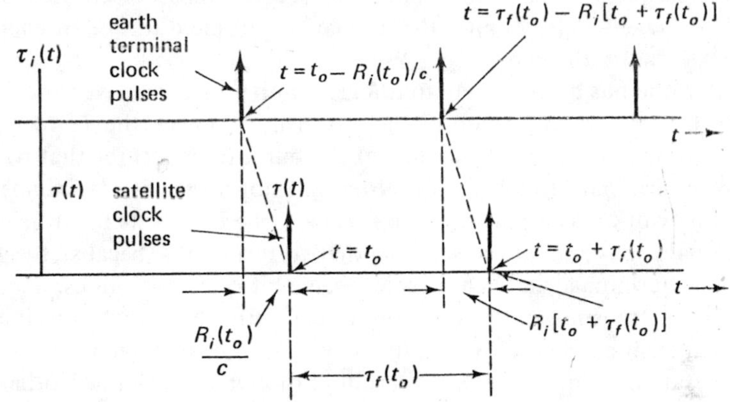

Fig. 10-6 Timing at satellite and at earth station slaved to the satellite. The range from earth terminal-to-satellite, measured at the arrival time at the satellite, is $R_i(t)$. The frame time at the satellite is $\tau_f(t)$. We define $\Delta R_i(t_0) = R_i(t_0) - R_i[t_0 + \tau_f(t_0)]$

slaved to the satellite clock where the earth terminal is assumed to be precisely slaved to the satellite clock. Table 10-1 defines the various types of time used in this discussion. The satellite clock is $\tau(t)$, the earth terminal clock at station i is $\tau_i(t)$, and "true" or universal time is t. The first satellite clock pulse shown in Fig. 10-6 occurs at $t = t_0$. The earth terminal slaves its clock and transmissions so that they arrive at the satellite in synchronism with the satellite frame clock pulses spaced by $\tau_f(t)$ sec. Thus, the earth

Table 10-1 DEFINITION OF CLOCK TIMING TERMS

Term	*Definition*
t	Universal or System Time
$\tau(t)$	Satellite Clock Pulse Sequence
$\tau_i(t)$	Clock Pulse Sequence at Earth Terminal i
$\tau_f(t)$	Frame Time at Satellite Clock Measured in Universal Time relative to frame start at t, $\tau_f(t) \cong \tau_f$
$R_i(t)$	Propagation Distance from Satellite to Terminal i Measured at the Arrival Time t
c	Velocity of Light

terminal clock pulses must really be transmitted earlier by $R_i(t)/c$, where $R_i(t)$ is the propagation distance between earth terminal and satellite, and c is the velocity of light. Thus the first pulse in Fig. 10-6 is emitted from the earth terminal at true time $t = t_0 - R_i(t_0)/c$ since the pulse arrival time at the satellite is $t = t_0$. The earth terminal clock is slaved to the satellite and since the range $R_i(t)$ is, in general, varying with time, $\tau_i(t)$ cannot be a fixed constant frequency pulse rate. Thus the second pulse is emitted at time $t = t_0 + \tau_f(t_0) - R_i[t_0 + \tau_f(t_0)]/c$. Hence the frame duration at each earth terminal varies with time.

After one has assured that an integer number of bits arrive in each frame one must also store these data bits and transmit them to the TDMA modulator for transmission to the satellite at the burst rate $f_{ij}(t)$ for that transmission from terminal i to terminal j. Note that not all receive terminals in the TDMA network may be the same size. Hence, the burst rate for transmission to terminal j may not be the same as that to terminal k because the $G/\mathfrak{J}$ of the receive terminals, and hence their receive bit rate capabilities, may differ.

TDMA transmit and receive timing may differ in phase as well as rate. The earth terminal TDMA transmit clock is slaved to arrive in synchronism with the satellite clock. The earth terminal receive clock, on the other hand, is slaved to the arrival time of clock pulses generated at the satellite. Thus these two earth terminal clocks differ by the round trip propagation time. Since the propagation path may be varying with time, the earth terminal transmit and receive clocks can differ in both phase and frequency. The more precise TDMA receive bit-timing clocks are generated from the TDMA demodulator bit synchronizer.

During this earth terminal frame time interval, precisely n_i bits of data must be transmitted, where n_i represents the total number of data bits to be transmitted in a frame. The actual average transmit bit rate $f_i^* \triangleq n_i/[\tau_{fi}(t) + \Delta R_i(t)/c]$ over a frame varies with time by a small fractional amount ($<10^{-3}$), and there must be some means for accommodating the difference between this rate and the actual input rate $f_i(t)$ received at the terminal by the user. Possible techniques for this purpose include:

1. Place a nonsynchronous multiplexer or pulse-stuffing buffer at the earth terminal to increase the actual input rate $f_i(t)$ to the TDMA transmit rate, $f_i^*(t)$; then $f_i^*(t) - f_i(t)$ is the pulse stuffing rate of the buffer.
2. Transmit the TDMA clock $f_i^*(t)$ or a clock-correction signal back to the users, or to the user multiplexers, so that received user bit streams arrive at the earth terminal TDMA transmitter in synchronism with the TDMA transmit clock.
3. Sample input analog waveforms at the earth terminal in synchronism with the TDMA transmit clock. This approach,

of course, assumes that the analog signals are in fact available at the earth terminal and that the sampling and quantizing are part of the TDMA equipment operation. This approach places only minimal requirements on the TDMA buffers if the TDMA frame rate is a multiple of or equal to the desired sampling rate. In this situation one can sample and quantize each of the analog input waveforms at the time that the TDMA transmit burst is to contain the bits from that input channel. The slight nonuniformity in sampling rate has negligible impact on the analog quantizing and reconstruction operation.

The three TDMA user interface techniques described above are illustrated in the simplified block diagrams of Fig. 10-7. In Fig. 10-7*a*, the analog waveform itself is used essentially to store information (for example, analog voice) until the time arrives in the frame for those data bits to be transmitted. If necessary, a sample-hold circuit can be employed to store the analog waveform. Variations in the sampling rate ≈ 1 in 10^3 do not usually affect quality of the analog transmission when the transmit clock rate slowly changes because of satellite motion and clock drifts.

In Fig. 10-7*b* the transmit data streams arrive at the earth terminal in digital form, perhaps in a bipolar form. In many cases, this data stream can be phase locked to a multiple of the TDMA transmit frame rate by comparing the phase of the received bit stream with that of the TDMA transmit clock at the earth terminal. Any timing error can be fed back to the data clock at the source as a frequency or phase correction. Thus we have a remote phase-locked oscillator. If data streams to and from the users are in bipolar format (Chap. 16), the clock or clock correction can be transmitted as a modulated carrier in the spectral null at the clock frequency of a bipolar waveform.

In Fig. 10-7*c* the received data streams are at some nominal value less than the TDMA data rate for that channel. The data streams are multiplexed together using pulse stuffing or word stuffing (Chap. 5).

Forward error-correction coding at the TDMA earth terminal can be used with TDMA buffers in at least three different forms:

1. Each data channel can be independently coded and decoded at the data channel level. (See Fig. 10-7(b).) If rate 1/2 coding is used, the data buffers must then double in size compared to those with no coding. Also, a separate coder/decoder is required for each data channel. For example, if there are $N = 10$ channels of 64 kbps each, then 10 coder/decoders are required, and $(10 \times 64 \times 10^3) \times 2 = 1.28 \times 10^5$ bits must be stored in high-speed burst buffers. Clearly a rate 3/4 code

analog inputs
only one word storage required
A/D convert
TDMA buffer
QPSK modulator
to IF
sample control
gate envelope
cables
terminal
time control
transmit frame time and clock from TDMA timing ranging unit

(a)

bit sync and clock contr.
possible error correction coders at data rate
data
user A/D or MUX
TDMA buffer
QPSK modul.
user A/D or MUX
TDMA buffer
microwave link
terminal
data
gate envelope
time control
transmit frame and clock

(b)

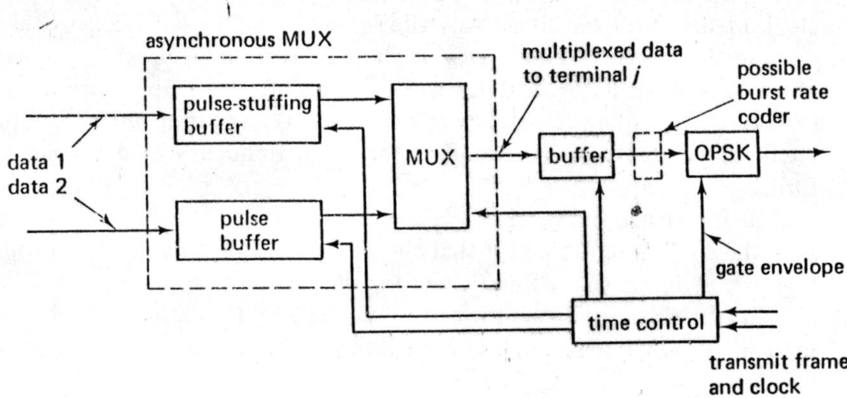

(c)

Fig. 10-7 User/TDMA interface configurations. In (b) analog inputs addressed to terminal *j* are A/D converted at the terminal itself to minimize buffer storage requirements.

would have an advantage over a rate 1/2 code in that it does not require as large a buffer (although its error rate performance is not as good).

2. The coder/decoder operates at burst rate and is placed between the burst buffer and the modulator/demodulator. It must operate at a high speed (perhaps 40 Mbps to 100 Mbps), and the TDMA burst may have to transmit the preamble and tail of the code (or the complete block) in addition to coded data bits. Only a single coder/decoder is required here, and the TDMA buffer storage is no greater than that for no coding. Only the burst rate and burst timing must change when coding is introduced. For some types of decoders, however, the required burst rate speed may be beyond the state of the art or too costly.
3. The coder can also be operated at the average or aggregate rate of n bits/frame. Low-rate buffering is used to insert each separate data channel at a separate nonoverlapping time interval in each TDMA frame into a coder/decoder operating at the average or aggregate rate (Fig. 10-8). The aggregate coded output is then increased to the TDMA burst rate by a burst rate buffer. This aggregate-rate approach was first discussed by Jacobs [1971].

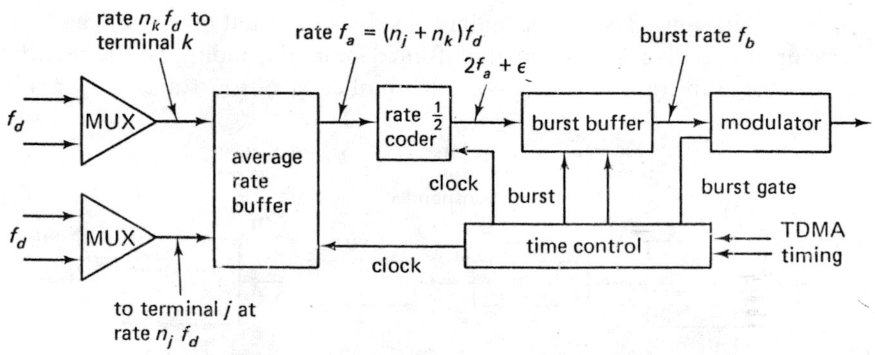

Fig. 10-8 Buffers and coder. In an aggregate-rate configuration. The coder output rate $2f_a + \epsilon$ has been increased to account for the tails in the code.

The aggregate rate coder/decoder operates at a duty factor much higher than the duty factor of the TDMA burst transmissions from this terminal. In fact, the coder/decoder duty factor can approach 100 percent, and the coder operating speed therefore can be substantially less than the TDMA burst rate and approaches a rate as low as the sum total input data rate. If there are 10 channels at 64 kbps data rate each, the coder operates at an output rate just above $10 \times 64 \times 10^3 \times 2 = 128$ kbps for rate 1/2 coding. Note that a second TDMA buffer is needed at the coder output to increase the coder output rate from the aggregate rate to the burst rate.

Elastic Buffers

If stable clocks are used for the data sources and sinks at each terminal in a TDMA network and the frame rate at the synchronous satellite transponder is held constant, it is possible to avoid the use of pulse-stuffing buffers altogether. An elastic buffer of varying fullness level can be used to accommodate the finite changes in the path delay caused by satellite motion. The path delay is in essence a variable length bit storage delay line. An elastic buffer simply stores data bits as they enter until they are called for by the transmit clock; hence, there are no pulse-stuffing inefficiencies. However, buffer size must be sufficient to prevent overflow or underflow. Typically, an elastic buffer must be reset periodically to accommodate clock drifts in imperfect earth terminal clocks.

A synchronous satellite is not perfectly stable relative to earth terminals, since there is a nonzero eccentricity* and nonzero orbital inclination in any realistic orbit (see Chap. 6). Thus, the number of bits in transit varies diurnally with time. The elastic buffer is fed by a constant bit stream, and fills and empties relative to a nominal half-filled state, depending on the number of bits stored in transit (Fig. 10-9). The number of bits in transit appears in

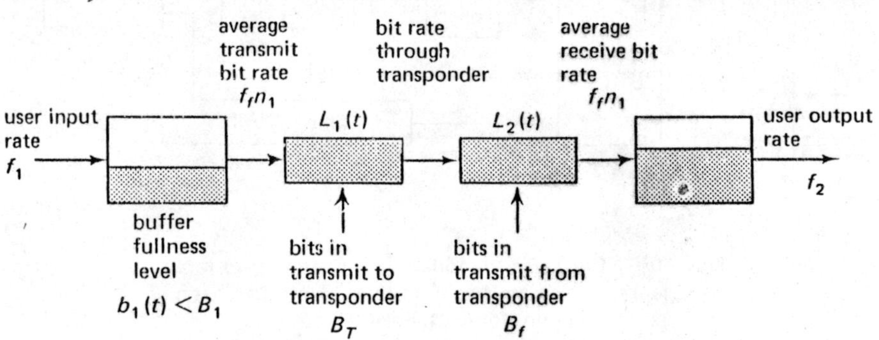

Fig. 10-9 Model of the elastic buffer for TDMA. The quantities L_1, L_2 denote distances from terminals 1 and 2, respectively, to the transponder.

effect as a "piston" storage buffer, which always fills and empties at a constant bit rate that corresponds to the frame rate at the satellite transponder. Figure 10-10 shows alternate frame timing techniques: the frame rate is

*An equatorial station observes a $\pm e(H + r_e)$ change in path delay for an orbit of eccentricity e and altitude H. For $e = 0.01$ and $H = 19{,}300$ nm, the variation is ± 228 nm.

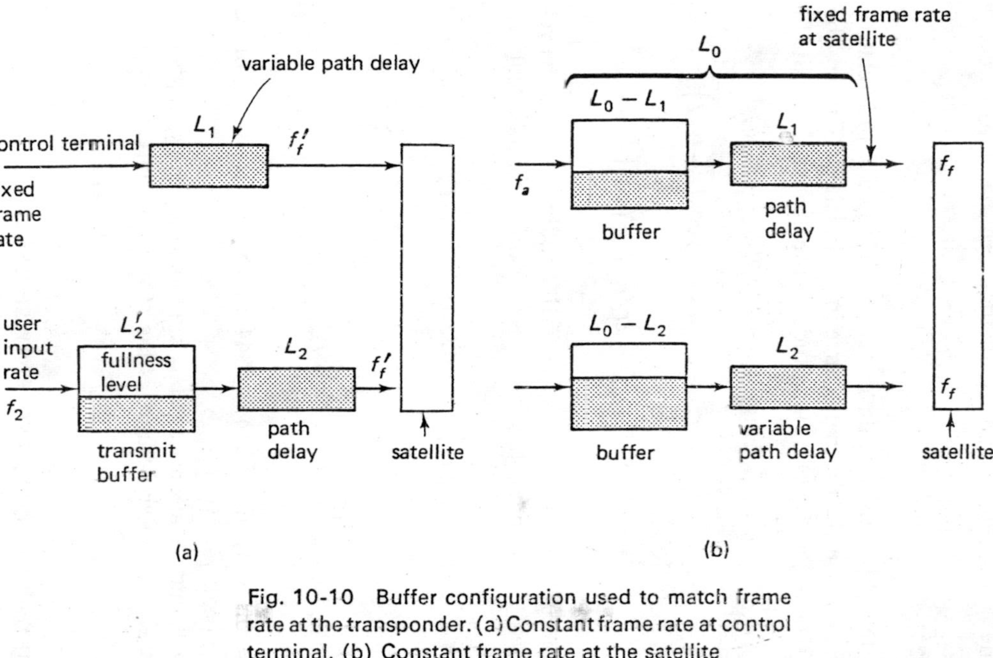

Fig. 10-10 Buffer configuration used to match frame rate at the transponder. (a) Constant frame rate at control terminal. (b) Constant frame rate at the satellite

constant at the control terminal (Fig. 10-10*a*), and all other terminals are slaved to that rate; or the frame rate is held constant at the satellite itself (Fig. 10-10*b*), and all other terminals are slave to that rate.

Define the TDMA signal frame rate as f_f frames/sec. Each earth terminal transmits (or receives) n_i bits/frame. If the path delay then changes by $\pm\Delta T_i$ sec, the elastic buffer capacity for that terminal must exceed

$$2f_f n_i \Delta T_i \text{ bits} \qquad \text{when } f_f = 1/\tau_f \tag{10-2}$$

If the satellite orbital inclination drifts to an angle I rad relative to the equatorial plane, then the diurnal change in path delay ΔT to an observer at the pole (worst case) at the same longitude as the mean subsatellite longitude is given by (see Fig. 6-13)

$$\begin{aligned}(\Delta T) = \frac{1}{c}\left[(h + r_e)^2 + r_e^2 - 2r_e(h + r_e)\cos\left(\frac{\pi}{2} + I\right)\right] \\ - \frac{1}{c}\left[(h + r_e)^2 + r_e^2 + 2r_e(h + r_e)\cos\left(\frac{\pi}{2} - I\right)\right]\end{aligned} \tag{10-3}$$

where h is the satellite altitude, r_e is the earth's radius, and c is the velocity of light. The satellite is assumed to have a circular orbit in this calculation. Then

$$(\Delta T) = \frac{2r_e}{c}(h + r_e)[2 \sin I] = \frac{4r_e}{c}(h + r_e) \sin I \tag{10-4}$$

and if $I \ll 1$

$$(\Delta T) \cong \frac{4I(h + r_e)r_e/c}{\sqrt{(h + r_e)^2 + r_e^2}} \cong 4Ir_e/c \tag{10-5}$$

Thus if $4I = 0.1$ rad, then $(\Delta T)c \cong 344$ nm $\cong 2.09 \times 10^6$ ft, or $\Delta T \cong 2.1$ msec peak-to-peak variation.

For any frame duration less than 2 msec, the storage-capacity requirement of this elastic buffer dominates the storage requirements of the burst buffer.

10-4 TDMA FRAME RATES AND FORMATS

Frame Format

The format of a TDMA transmission can have many variations within the basic structure shown in Fig. 10-11. A superframe of N (perhaps 2^8) frames can be used to allow for some very low data-rate users desiring to transmit at a rate below the frame rate. The frame rate, for example, might be

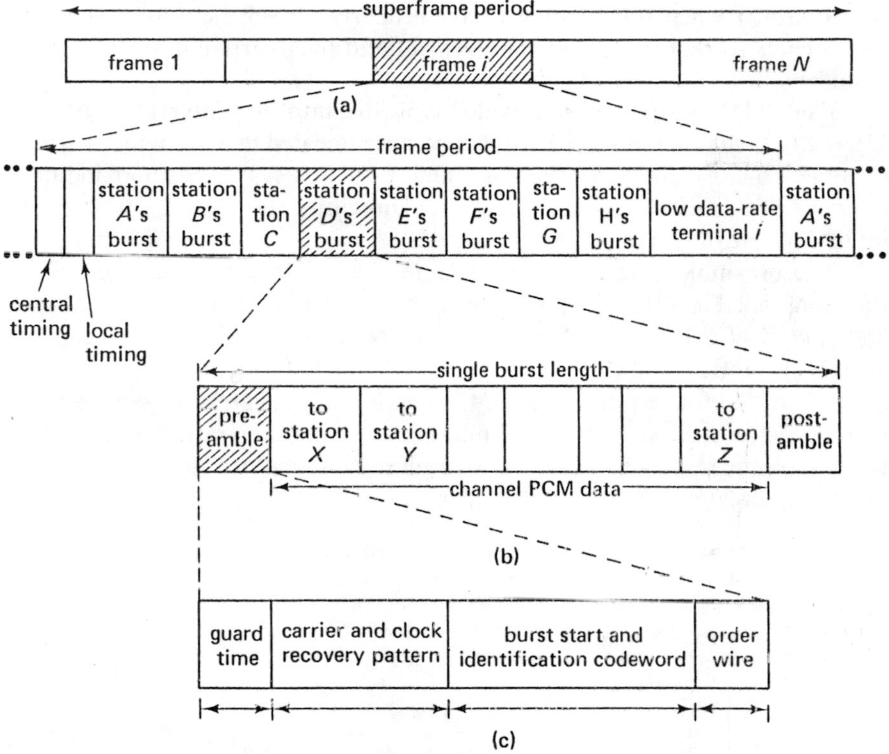

Fig. 10-11 Format structure of a typical TDMA system; (a) frame format, (b) burst format, and (c) preamble format

1200 frames/sec, and a user at terminal *i* who desires to transmit at 150 bits/sec would transmit on the average of one bit per 8 frames, or 8 bits per superframe of 64 frames. However, most users transmit one data burst per frame plus perhaps a timing burst used for synchronization.

Each TDMA burst is subdivided into a preamble for receiver synchronization followed by data bits addressed to various receive earth stations. The TDMA burst concludes with a postamble that identifies the end of the burst, and can be used to resolve carrier phase and frequency ambiguities.

The time slots shown for central and local timing represent one method of obtaining TDMA timing. In this method, one station is assigned as a central timing reference for each satellite. The "local" stations transmit their own timing signal in the adjacent time slot and receive both central timing and their own timing signal as relayed from the satellite. Each timing signal can be a time-gated PN-coded signal (Chap. 18). Timing is successfully accomplished when the terminals have properly placed the timing of their own signals so that the two signals (central and local) arrive in relative synchronism.

Each TDMA frame is subdivided typically into unit intervals, perhaps $2^{13} = 8192$ unit intervals, so that the time slot allocated to a given user can be identified by two integers, the start and stop times as measured in unit intervals. Table 10-2 lists the number of unit intervals needed for various input data bit rates and output burst rates.

The preamble interval comprises a number of elements that accomplish functions listed in Table 10-3. The postamble identifies the end of the message and is used also to assure that the carrier recovered is the correct carrier frequency, not a sideband of the carrier that has been offset by an odd multiple of the frame rate. Periodic time gating of the carrier of course generates spectral lines spaced at odd multiples of the frame rate. Transponder filter transient response affects TDMA guard-time requirements in that the tail of the previous time slot postamble can overlap the preamble and even

Table 10-2 NUMBER OF UNIT INTERVALS REQUIRED FOR A GIVEN DATA RATE AND BURST RATE. OVERHEAD NEEDED FOR PREAMBLE AND POSTAMBLE IS NEGLECTED. NOTE THAT TWO OF THE COMBINATIONS SHOWN PRODUCE 100% DUTY FACTORS AND ARE LISTED AS CONTINUOUS WAVE (CW).

Burst Rate (*Mbps*) 75×2^m	*Data Rate* (*Kbps*) $= 75 \times 2^n/1000$							
	38.4	76.8	153.6	307.2	614.4	1228.8	2457.6	4915.2
	$n = 9$	10	11	12	13	14	15	16
2.4576 $m = 15$	64	128	256	512	1024	2048	CW 4096	—
4.9152 $m = 16$	32	64	128	256	512	1024	2048	CW 4096
9.8304 $m = 17$	16	32	64	128	256	512	1024	2048
19.6608 $m = 18$	8	16	32	64	128	256	512	1024
39.3216 $m = 19$	4	8	16	32	64	128	256	512
78.6432 $m = 20$	2	4	8	16	32	64	128	256

N = Length of time slot required (in unit intervals) for $4096 = 2^{12}$ subintervals, or specifically N = Number of unit intervals required $= \left(\frac{\text{Data rate}}{\text{Burst rate}}\right) 4096 = \frac{75 \times 2^n}{75 \times 2^m} 2^{12}$.

Table 10-3 FUNCTIONS OF THE TDMA PREAMBLE

Preamble Element	*Function*
Guard time T_g	Prevent overlap of adjacent bursts from different ground terminals. This guard time* must be sufficient to account for system timing inaccuracies and tails from adjacent bursts caused by finite filter-response times.
Carrier recovery	Provide a preamble sequence of consecutive "1s" or an alternate pattern to permit coherent carrier recovery for a synchronous demodulator. Carrier recovery can be accomplished either by rapidly acquiring carrier on each new carrier burst, or by using the frame-frame coherence in consecutive bursts from the same transmit terminal in a narrow bandwidth carrier recovery loop.
Bit timing	This bit synchronization function can be accomplished with one or more transitions in the preamble bit stream. This function is accomplished after and during carrier synchronization.
Burst start and identification	The burst start symbol identifies the last bit in the preamble or the first bit of actual data. The identification symbol identifies the address of the data and the transmitting terminal. In some applications, this address information is redundant because the users know by way of parallel order wires which time slots are allocated to which users—that is, the time slots are, in effect, the addresses.
Order wire	This order wire is for terminal-terminal communications—for example, for setting up the TDMA time slots power levels.

*TDMA guard times typically range from 30 nsec to 300 nsec for high-bit-rate systems employing worldwide timing.

the data bits of the present time slot. This effect is accentuated if the previous burst is significantly larger in amplitude than the present burst and can cause data bit errors in the initial portion of the time slot.

Frame Rate Selection

The TDMA frame rate affects several important system parameters and should be selected with care. Some specific factors bearing on frame-rate selection are:

1. Primary data rates are limited to integral multiples of the frame rate (unless superframe multiplexing is used). Thus, the lower the frame rate, the greater the flexibility in data rates.
2. Long frame periods lead to greater efficiency relative to a given guard time T_g and a fixed preamble duration.
3. An increase in the frame period decreases the frame-frame coherence of the carrier. This decrease in coherence introduces added phase noise in the demodulation process when frame-frame coherent carrier recovery is used. This efficient method of carrier recovery is degraded by too large a frame period or too much carrier phase noise.
4. Increasing the frame period T_f causes the buffer storage requirement (Sec. 10-3) to increase since buffer storage memory is directly proportional to frame period (with the exception of overhead functions). Hence, increasing the frame period adds to the memory cost.
5. Increasing the frame period beyond 125 μsec (8-kHz sampling rate) makes it more difficult to use analog storage—that is, let the waveform remain analog until needed, then sample the analog waveform at the burst transmission time and transmit the quantized samples as described in Sec. 10-3.
6. An increase in the frame period byond 0.1 sec introduces a significant added delay to the transmission in addition to the satellite round trip delay of 0.25 sec. Nevertheless, the use of long frame times may have advantages for some applications where data are being transmitted at low rate rather than voice, and the simplifications in system timing associated with longer frame times outweigh the increased buffer storage costs.

10-5 TDMA SYSTEM EFFICIENCY

The power efficiency of the satellite transponder with TDMA inputs and hard limiting depends on the guard times between the transmissions T_{gi} of each terminal, the preamble and postamble time, used for example to provide addressing and carrier recovery, the addressing time required for each transmit/receive terminal pair T_{aij}, the time utilized for the timing-ranging function T_R, and the frame duration T_f. The maximum efficiency for all terminals fully occupying the frame is

$$\eta_{\max} = \left(\frac{T_f - [T_R + \sum_i (T_{gi} + \sum_j T_{aij})]}{T_f}\right) \tag{10-6}$$

where i is summed over all N terminals in the network and j is summed over all $N - 1$ which can be addressed by terminal. The preamble and postamble are included in T_{aij}. If all guard times and address times are identical, then the efficiency is

$$\eta_{max} = \frac{T_f - [T_R + NT_g + N(N-1)T_a}{T_f} \tag{10-7}$$

where it is assumed that all terminals are communicating with all other terminals and the frame is fully utilized. Since T_R and T_g tend to be approximately constant for a given channel bandwidth, efficiency increases as T_f increases.

Increasing T_f improves efficiency, but beyond a certain point it causes other difficulties as described in the preceding section. An example efficiency for a 125-μsec frame and $N = 5$ users is

$$\eta_{max} = \frac{125 - [2 + 5(0.1) + 20(0.025)]}{125} = \frac{125 - 3}{125}$$
$$= 97.6 \text{ percent} \tag{10-8}$$

for $T_R = 2$ μsec, $T_g = 0.1$ μsec, $T_a = 0.01$ μsec. Clearly, the actual efficiency is less than this because the data to be transmitted rarely just fill the total frame space available. A data channel to be transmitted may require more space than available, and it may be impossible to transmit only a fraction of that data channel. Hence a portion of the frame may go unused. In addition there are the possible inefficiencies of time division multiplexing the data channels together prior to entering the TDMA modulator (Chap. 5).

10-6 TDMA CARRIER RECOVERY USING FRAME-FRAME COHERENCE

The TDMA carrier bursts destined for a given receive terminal during one frame are received from many transmit terminals (Fig. 10-12). Each of these carrier bursts received from separate terminals has an independent carrier phase since the carrier phase at the satellite is not synchronized from one transmitting terminal to another.

Carrier recovery of these TDMA bursts of PSK signal can proceed in at least two different ways:

1. Use a rapid-acquisition carrier recovery phase-locked loop or narrow-band filter which is able to respond accurately to burst-to-burst phase transients. One carrier recovery loop can

then handle bursts from all earth terminals. However, the carrier recovery noise bandwidth must be relatively wide to provide a rapid transient response within the preamble time.

2. Use multiple time-gated carrier recovery loops or a single time-multiplexed phase-locked loop with phase memory from frame to frame. This approach permits the use of relatively narrow bandwidth carrier recovery loops which operate on the main spectral line component of a time-gated carrier.

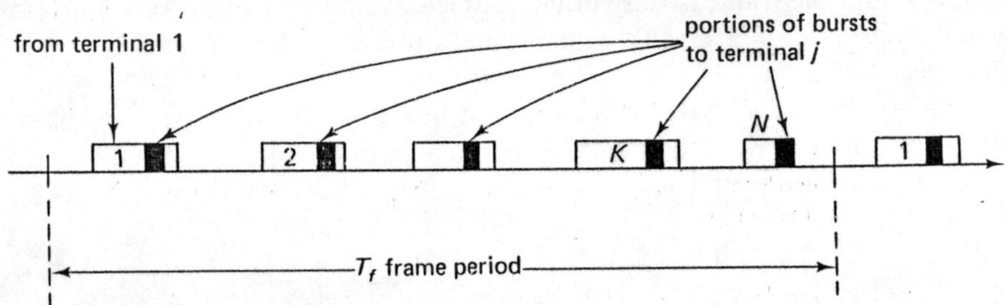

Fig. 10-12 Typical TDMA carrier bursts destined for terminal j from transmitting terminals $1, 2, \ldots, k \ldots N, k \neq j$). Shaded segments contain data to terminal j.

In method (2) one must avoid the potential false lockon to a sideband line component (Fig. 10-14) in the periodically time-gated carrier preamble spectrum. In addition, the carrier phase stability must be sufficiently good to provide phase coherence from frame to frame. Otherwise the phase stored in memory during the previous frame may differ substantially from that in the next frame even though it has been linearly extrapolated using the frequency estimate as a basis for extrapolation. The frame-frame coherent approach operates as shown in the simplified carrier recovery loop of Fig. 10-13. One carrier recovery loop can be used for each received carrier burst, or a single loop with n loop filter/VCOs can be time multiplexed for each carrier burst as shown. (There is a simple digital filter/NCO equivalent of Fig. 10-13.)

In order for the frame-frame coherent carrier recovery loop to operate satisfactorily, the carrier phase difference for a single transmit terminal caused by oscillator phase noise from one frame to the next must be small; that is,

$$|\phi(t) - \phi(t - T_f)| \triangleq |\Delta\phi(t)| \ll \pi/N \tag{10-9}$$

for N phase PSK and where the constant frequency part of the phase has been removed from $\phi(t)$. Satellite accelerations are usually negligible over these short frame durations and satellite doppler is tracked. The relationship of $\Delta\phi$ to oscillator phase-noise statistics is discussed in Chap. 12.

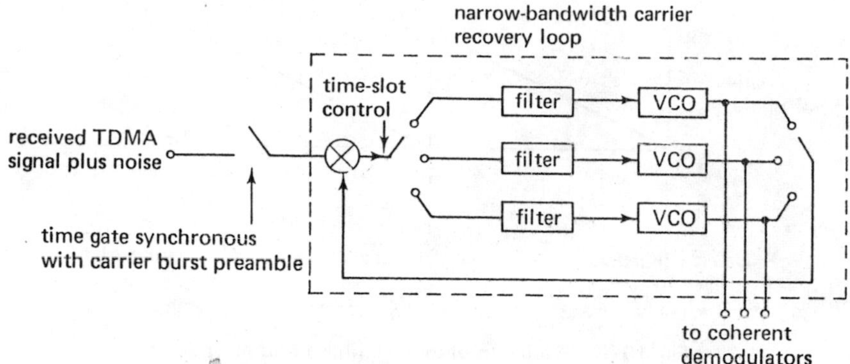

Fig. 10-13 Simplified time-gated frame-frame coherent carrier recovery loop for TDMA. The time slot control repeats its cyclic switching every frame.

The sidebands of a pulsed preamble of pure carrier used for carrier tracking are shown in Fig. 10-14. If a carrier tracking loop attempts to track the fundamental frequency of the carrier but instead tracks the carrier sideband at $1/\tau_f$ frequency offset, the carrier will be in phase at the middle of the carrier preamble $T_c/2$ sec after the beginning of the carrier (Fig. 10-15). A kth order offset carrier causes the cross correlation between reference and actual carrier to decrease at the edge of the burst to

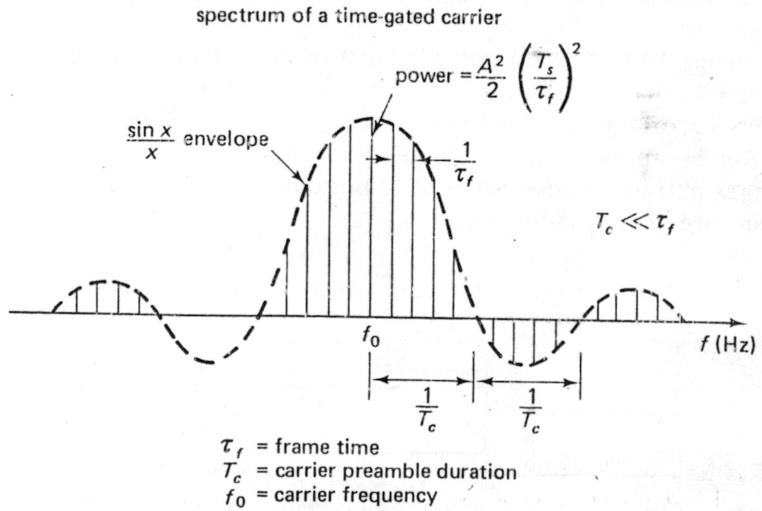

Fig. 10-14 Spectrum of the periodic pulsed, pure-carrier preamble used for carrier tracking

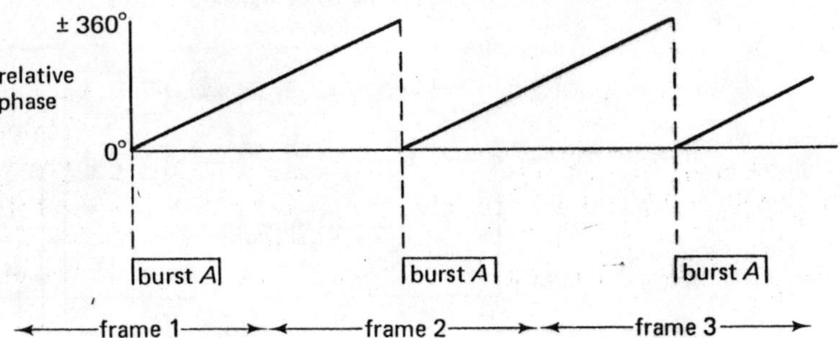

Fig. 10-15 Variation of relative phase and error signal between true carrier and reference when sideband lock occurs at ±1200 Hz

$$\cos\left[\left(\frac{2\pi k}{\tau_f}\right)\left(T_s - \frac{T_c}{2}\right)\right] \cong 1 - \frac{1}{2}\left(\frac{2\pi k}{\tau_f}T_s\right)^2$$

$$\text{for} \quad T_c \ll T_s,\ kT_s \ll \tau_f \tag{10-10}$$

where T_s is the total width of the time slot of RF energy and T_c is the duration of the TDMA preamble. The frequency offset is k/τ_f, and the time offset at the end of the signal burst is $T_s - T_c/2$. Clearly, if the order k of the sideband is sufficiently large for a given T_s, sideband lock can cause a substantial degradation in performance, which could be reduced or eliminated by sensing correlation using a burst postamble and retuning the carrier recovery loop if necessary.

Figure 10-16 shows a possible burst configuration for the inphase and quadrature channels of a quadriphase burst. The "**0**" represents a sequence of 16 serial zeros followed by a **1** representing 16 "ones." Multiplication in the receiver by an alternate **0**,**1** sequence of 16 bits each can then convert the preamble into a pure carrier of 32 bits duration. The postamble contains a sequence of consecutive zeros.

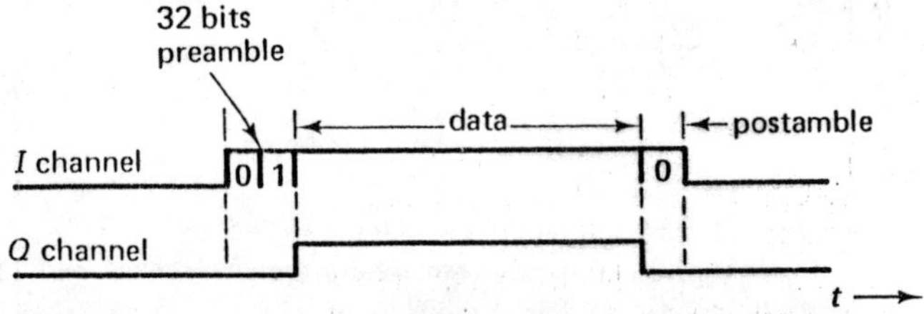

Fig. 10-16 TDMA burst modulation format for QPSK TDMA. Both a preamble and postamble are employed.

10-7 DELAYED REFERENCE CARRIER RECOVERY FOR TDMA

It is possible to shorten the carrier recovery portion of the TDMA burst preamble while still performing the carrier recovery on a burst-to-burst basis [Nosaki, 1970, 425–434]. Suppose that carrier recovery requires 0.2 μsec and that ordinarily a 0.2-μsec preamble is required just for carrier acquisition in addition to the other portions of the preamble used for station addressing and bit synchronization. If the first 0.2-μsec. portion of the preamble is deleted, carrier recovery can still be obtained by operating on the remaining portion of the preamble and the data modulation itself. Because pure carrier is not available during the data transmittal time, the carrier recovery circuit must operate on the modulated signal and may take somewhat longer, say, 0.3 μsec.

The entire preamble and data modulation can still be recovered, however, if the modulated carrier is delayed by $T_\Delta = 0.3$ μsec in parallel with the carrier recovery operation as shown in Fig. 10-17. The cable delay T_Δ corresponds to the carrier recovery acquisition time. When a TDMA burst begins, one of the two alternating carrier recovery loops is switched to that burst and the acquisition operation begins. The TDMA burst at IF frequency is being stored in a delay cable during that acquisition interval. At the end of T_Δ sec, the TDMA carrier recovery circuit for that burst switches to the

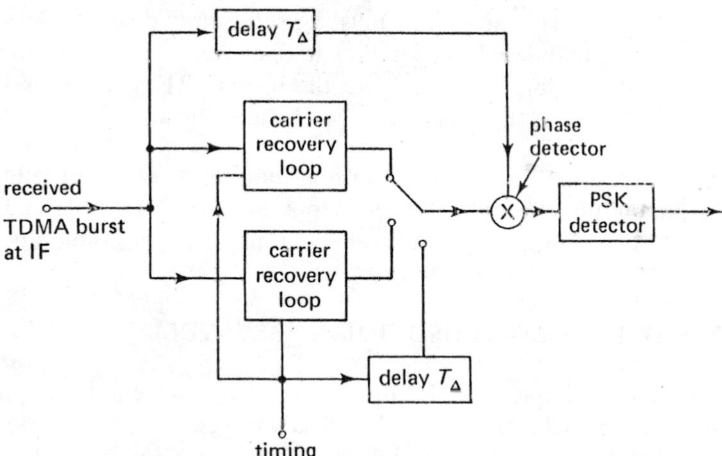

Fig. 10-17 Delayed reference carrier recovery for TDMA. [Two carrier recovery loops are shown so that while one is acquiring carrier for a new TDMA burst, the other can be holding the phase of the previous burst. The cable delay T_Δ corresponds to the carrier recovery acquisition time.]

coherent detector in coincidence with the TDMA burst exiting the delay cable. The carrier recovery circuit holds the phase of the recovered carrier until T_Δ sec after the end of the received TDMA burst. Time multiplexed carrier recovery loops are shown so that while one is acquiring carrier for a new TDMA burst, the other can be holding the phase of the previous burst. The alternate carrier recovery loop then is switched to the coherent phase detector, and the demodulation of the next TDMA burst begins.

In principle, therefore, no preamble is required for TDMA acquisition. In practice, however, one must also account for several other effects, including:

1. The delay T_Δ of the delay cable must be an integer number of IF cycles and must remain constant over the expected temperature fluctuations of the environment. Phase-stabilized cable may be required to hold the carrier phase shift to less than $\pm\epsilon°$ over the expected temperature range.
2. Doppler frequency changes or long-term frequency drifts must be sufficiently small that the phase change is also small from that cause. If the frequency drift is $\Delta f \pm 10^4$ Hz and the delay is $T_\Delta = 0.3\ \mu$sec, then the phase shift is $\Delta f T_\Delta = \pm 10^4 \times 3 \times 10^{-7} = \pm 3 \times 10^{-3}$ cycles, or $\Delta\phi = \Delta f T_\Delta \times 360° = \pm 1.08°$, an effect which would usually be negligible.
3. The delay cable itself must be sufficiently distortionless to the IF modulated carrier that it does not degrade performance. If T is large compared to the inverse RF bandwidth W, and $TW \gg 1$, then this effect can be a problem.

This reduction in preamble duration requires the use of additional circuitry, an additional carrier recovery loop, and the delay cable and driver amplifiers. For some applications, however, it may be a desirable approach.

10-8 SATELLITE-SWITCHED TDMA (SSW-TDMA)

A satellite with multiple spot beams or an MBA can employ high-speed signal switching to achieve the combined advantages of TDMA (no power backoff in the power amplifier) and the spot beams (high ERP and isolation of ground stations from one another). These combined features can yield both a power and bandwidth advantage over conventional TDMA.

Figure 10-18 shows a multibeam satellite/ground station configuration. The satellite has independent transmit and receive antennas and high speed onboard switching circuits which can interconnect the transmit and receive antennas. We assume that each link occupies the entire uplink and downlink

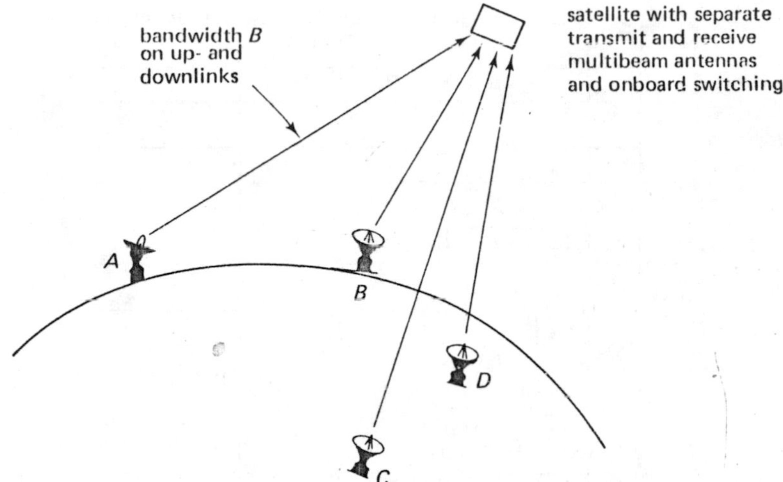

Fig. 10-18 Possible configuration of satellite communications systems with onboard satellite switching. Each ground station continuously transmits signal power of bandwidth B.

bandwidths $B_{\text{up}} = B_{\text{down}} = B$ of a given satellite transponder. Let us also assume that station C acts as the central timing source and transmits a central timing signal S_c to be received by all ground stations (Fig. 10-19).

The timing matrix for satellite switching can be controlled by ground command. Each format represents the envelope of the RF transmission. A signal preamble (just as in TDMA) precedes each signal burst for carrier recovery, bit synchronization, guard times, etc. The notation D_{ba} represents a signal from terminal b addressed to terminal a, and L_b is the local ranging signal from terminal b.

The satellite RF switching must be accomplished at high speed so that the guard-time requirements for the satellite switching transients added to those for earth station timing/ranging uncertainties are not large. Figure 10-20 shows one possible satellite transponder configuration in simplified form where the satellite contains a free-running but highly stable clock. This stable clock is transmitted to all ground terminals and is used by the terminal assigned to serve as the central timing station. The central timing station slaves its clock to the satellite.

Each separate TWT amplifier for each antenna beam can utilize the full downlink bandwidth available at full power output and therefore sees the same type of signal format as a single TWT does in conventional TDMA operation with an earth-coverage antenna. Just as with conventional TDMA, there are carrier phase and frequency discontinuities on the downlink when

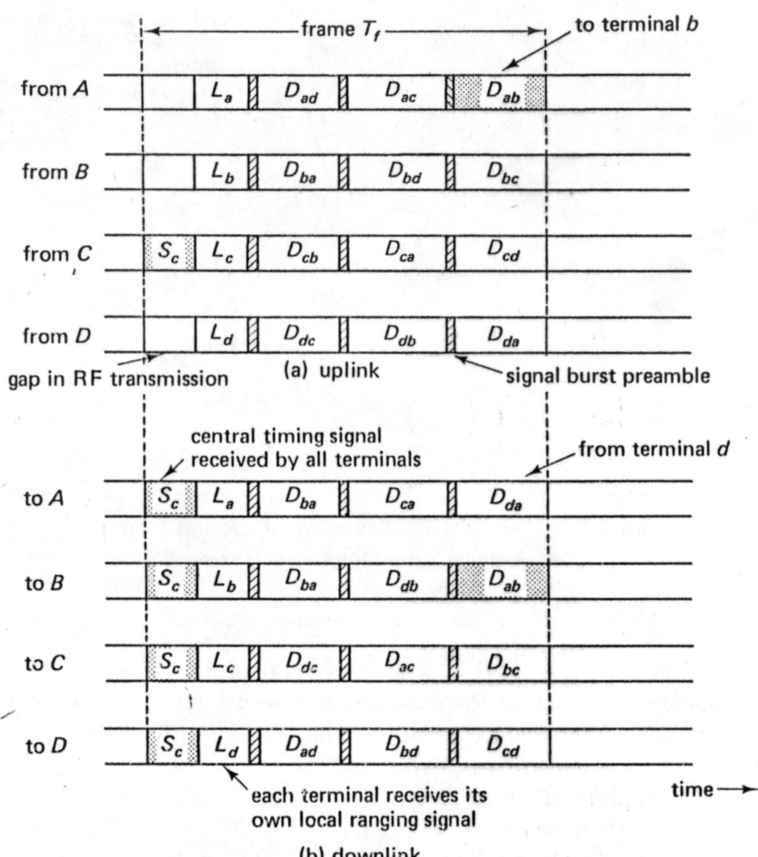

Fig. 10-19 Frame formats for satellite-switched TDMA; (a) uplink, (b) downlink. [The timing matrix for satellite switching can be controlled by ground command. Each format represents the envelope of the RF transmission. A signal preamble (just as in TDMA) precedes each signal burst for carrier recovery, bit synchronization, guard times, etc. The notation D_{ba} represents a signal from terminal b addressed to terminal a. L_b is the ranging signal from terminal b.]

transmission from one earth terminal ends and that from another begins. Note that if there are N properly isolated spot beams, each of bandwidth B, the total data rate passed by the satellite corresponds to NB, or N times as much as permitted by conventional TDMA with an earth-coverage antenna.

The lack of required TWT backoff (approximately 3 dB) and the absence of frequency guard bands and intermodulation products are the primary advantages of SSW-TDMA over FDMA systems that utilize uplink frequency

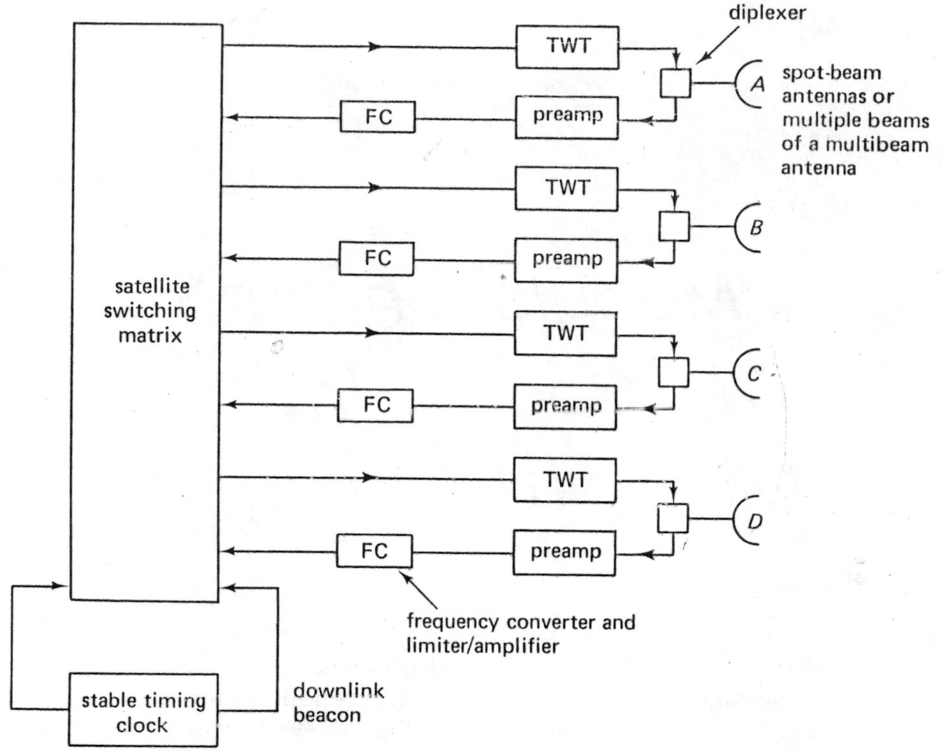

Fig. 10-20 Possible transponder for satellite-switched (SSW) TDMA operation. The downlink beacon carries satellite timing information to the central timing earth station.

to control which downlink beam is going to be used. FDMA would generally require a 3- to 6-dB power backoff in the TWT amplifier as described in Chap. 9.

Essentially the same up- and downlink formats as used for the SSW-TDMA frame formats can be used for FDMA switching on board the satellite if the time axis for one frame is relabeled "frequency." For this satellite-switched FDMA (SSW-FDMA) operation each TWT assigned to a multibeam antenna must still be operated in a power back-off mode. Thus, a higher overall data rate can be put through the same set ot TWT amplifiers and spot-beam antennas with SSW-TDMA than with multibeam FDMA. A higher order PSK modulation may be required, however, to take advantage of the increased power availability with a given RF bandwidth limitation.

PART 3

Modulation and Coding in Distorted Channels

The effects of stationary Gaussian noise on biphase shift-keyed (BPSK) and quadriphase (QPSK) signals has been widely discussed in many introductory texts on coherent modulation techniques. Theoretical error-rate results for BPSK and QPSK in this idealized channel are shown in Part 2. However, additional perturbing effects are present to some extent in every transmission channel and in many instances are dominating influences on system performance. In this part, quantitative results are given for the effect of the distortions generated by the satellite transponder or earth station, filters, amplifiers, and frequency converters.

The specific perturbations considered are: cochannel interference, multiplicative oscillator phase noise and frequency offsets, coherent carrier-recovery phase noise, bit-synchronization timing errors, and filter distortion (signal attenuation and intersymbol interference). Multiplicative phase noise is in turn related to the statistics of oscillator short-term stability. Transient errors and steady-state noise effects are reviewed for phase-locked oscillators employed in coherent carrier recovery.

Chapter 11 summarizes results for BPSK, QPSK, *M*ary PSK, minimal (frequency) shift keying (MSK), or other continuous-phase frequency-shift keying (CPFSK), and the phase coding and differential encoding related to each. Carrier phase tracking, doppler offsets, and oscillator noise models are discussed in Chap. 12, followed by an analysis of the filter distortion effects on PSK signals and bit synchronization techniques (Chaps. 13 and 14). A brief discussion of maximum-likelihood or Viterbi decoding of convolutional codes used for reduction in the required signal power is given in Chap. 15.

Part 3 concludes with a discussion of some digital baseband-modulation methods, bipolar, pair-selected ternary, and various partial-response signaling techniques. Digital-data scrambling, which randomizes the bit streams, is also discussed. These digital baseband-modulation techniques are applicable to FDM-FM transmission via satellite or microwave, or to transmission on coaxial cable.

CHAPTER 11

COHERENT AND DIFFERENTIALLY COHERENT TRANSMISSION TECHNIQUES

11-1 INTRODUCTION

A wide variety of digital modulation techniques have been discussed in the literature [for example, Bennett and Davey, 1965; and Schwartz, 1970]. These techniques include amplitude-shift keying (ASK), multiple-amplitude-shift keying (MASK), frequency-shift keying (FSK), multiple-frequency-shift keying (MFSK), phase-shift keying (PSK), multiple-phase-shift keying (MPSK), and composite modulation techniques involving both amplitude and phase-shift keying, as well as pulse-width modulation.

Of these techniques, however, only a limited subset is desirable for satellite communications. With TDMA the transponder nonlinearities and power-efficiency effects usually constrain the modulation format to have a constant envelope, thus excluding ASK and hybrid techniques involving ASK. With FDMA and quasi-linear transponder operation the choice of modulation type is somewhat more flexible. Even here, however, the presence of FDMA intermodulation products can restrict the use of MASK because of its greater sensitivity to cochannel interference. Neverthless, as bandwidth limitations become more severe, the use of hybrid MASK, such as employed in quadrature amplitude shift keying (QASK), can prove attractive.

Hence the primary interest is in PSK, MPSK, and a phase-continuous version of FSK known as minimal shift keying. All these forms of modulation are constant envelope and utilize relatively small amounts of bandwidth. MPSK for larger values of $M > 8$ has greater bandwidth efficiency than biphase or quadriphase shift keying but is less efficient in the use of power. MPSK for $M = 8$ is in active use because of its bandwidth saving, however. The error-rate performance for each of these modulation techniques is summarized and plotted versus the energy/bit-to-noise-density ratio for both coherent and differentially coherent detection.

Coherent detection requires the use of a phase-coherent carrier tracking loop for each of these modulation techniques, and several alternate approaches to the carrier recovery problem are discussed. The detailed effects of imperfect carrier recovery—that is, quasi-coherent or partially coherent detection—are discussed in quantitative terms in Chap. 9.

Staggered QPSK and MSK are two modulation techniques that avoid the 180° phase transients of BPSK and conventional QPSK. These phase reversals of BPSK cause envelope modulation nulls when the signal passes through a bandpass filter. This envelope fluctuation is undesirable because additional amplification by nonlinear devices can enhance the sideband energy, increase adjacent channel interference, and cause AM/PM distortion effects.

Cochannel interference effects are discussed for PSK and MPSK signals. This interference can be of particular importance in satellite communication because of potential interference from the sidelobes of an interfering earth terminal, the incomplete suppression of a cross-polarized signal, or intermodulation products produced by satellite transponder nonlinearities.

Finally, differentially coherent detection performance is summarized for both PSK and MPSK signals. Differential coherent detection of BPSK is nearly as efficient for an additive Gaussian noise channel and has a performance advantage over coherent detection in applications where the carrier phase-noise effects are severe.

11-2 BPSK DIGITAL TRANSMISSION

Biphase modulation, the simplest form of phase-shift keying, transmits binary data d_i by means of a carrier phase modulated by $\theta(t) = \sum d_i \theta_T(t + iT)$

$$\begin{aligned} s(t) &= A \sin [\omega_0 t + \theta(t)] \\ &= A \cos \theta(t) \sin \omega_0 t + A \sin \theta(t) \cos \omega_0 t \end{aligned} \tag{11-1}$$

where $\theta(t) = \pm\theta$ rad, and the phase shift changes with each new data bit of duration T. The pure carrier power in these $\sin \omega_0 t$ and $\cos \omega_0 t$ terms are $\frac{1}{2}A^2\{\mathrm{E}[\cos \theta(t)]\}^2$ and $\frac{1}{2}A^2\{\mathrm{E}[\sin \theta(t)]\}^2$ respectively. The remaining power is modulation power. The amount of signal energy useful in making a bit decision in a T sec interval depends on the value of θ and is maximized at $\theta = \pi/2$ rad.

Coherent Detection of BPSK

Coherent detection is the most efficient method of recovering the binary data contained in the two possible relative phases, $\pm\theta$, of the received signal. To coherently detect the two phases, the receiver must have a phase reference available. When less than $\pm\pi/2$ rad phase deviation is used, the received signal contains its own phase reference in the form of a residual carrier component. The strength of the residual carrier is $\mathrm{E}[A \cos \theta(t)] = A \cos \theta$. A modulation-restrictive phase-locked loop can then be used to track the carrier phase and to provide a coherent reference for product demodulation and matched-filter detection of the PSK data. This loop is sufficiently narrow in bandwidth that it cannot track the carrier phase modulation caused by the data.

For the antipodal signal ($\theta = \pm\pi/2$), on the other hand, the carrier component is zero; hence, the phase reference must be established by other means, two of which are (1) the frequency-doubling phase-locked loop, and (2) the Costas loop, also known as the inphase-quadrature loop or *IQ* loop. Both loops provide the same theoretical noise performance when their noise bandwidths are equal. When the loop is above threshold, this performance is approximately equal to the performance of a phase-locked loop operating on an unmodulated carrier having the same power as the PSK signal.

The probability of error for coherent detection of BPSK is given by [Nuttall, 1962, 305–314]

$$\mathcal{P}_b = \frac{1}{\sqrt{2\pi}} \int_{\sqrt{(1-\rho)(E_b/N_0)}}^{\infty} e^{-y^2/2}\, dy = \frac{1}{2} \operatorname{erfc} \sqrt{(1-\rho)(E_b/N_0)} \tag{11-2}$$

where ρ is the symbol correlation coefficient, and E_b/N_0 is the energy/bit-to-noise-density ratio. For PSK with $\pm\theta$ rad deviation, the symbol correlation coefficient is

$$\rho = \cos 2\theta \qquad 0 < \theta \leq \frac{\pi}{2} \tag{11-3}$$

Substituting for ρ, we obtain the bit error probability

$$\mathcal{P}_b = \frac{1}{\sqrt{2\pi}} \int_{\sin\theta\sqrt{2E_b/N_0}}^{\infty} e^{-y^2/2}\, dy = \frac{1}{2} \operatorname{erfc} [\sin\theta \sqrt{E_b/N_0}] \tag{11-4}$$

A peak phase deviation $\theta = 1$ rad is sometimes used. (The value for any particular application depends on the design requirements for the phase-locked loop.) For $\theta = 1$ rad, $[\sin 1]^{-2}$ or 1.5 dB more power is required to achieve the same error probability possible with $\theta = \pi/2$ because less power is devoted to the modulation for $\theta = 1$ rad. See (11-1).

The minimum binary error probability is achieved by coherent detection of PSK with $\theta = \pi/2$. This error probability is expressed by

$$\mathcal{P}_b = \frac{1}{\sqrt{2\pi}} \int_{\sqrt{2E_b/N_0}}^{\infty} e^{-y^2/2}\, dy \equiv \text{erfc}' \sqrt{\frac{2E_b}{N_0}} \triangleq \frac{1}{2} \text{erfc} \sqrt{\frac{E_b}{N_0}} \qquad (11\text{-}5)$$

where $E_b = (A^2/2)T$, as discussed in Chap. 3. See Figs. 3-4 and 11-25.

BPSK Spectrum

Glance [1971, 2857–2879] has calculated the continuous power spectrum for antipodal PSK modulations by a trapezoidal waveform (Fig. 11-1) with period T, rise and decay times s, and the top of the pulse having width $\tau = T - 2s$. The power spectral density of this signal is then

$$G(f) = \frac{T}{4}\left(\frac{\pi}{4}\right)^2 \left\{ \frac{\dfrac{\tau}{T} \dfrac{\sin (\omega_0 - \omega)\tau/2}{(\omega_0 - \omega)\tau/2} + \dfrac{s}{T} \cos (\omega_0 - \omega)\dfrac{T}{2}}{\left[(\omega_0 - \omega)\dfrac{s}{2}\right]^2 - \left(\dfrac{\pi}{4}\right)^2} \right\}^2 .$$

$$f > 0 \qquad (11\text{-}6)$$

where ω_0 is the carrier frequency $\omega = 2\pi f$. The spectrum has small line components because of the unequal top and bottom parts of the trapezoids.

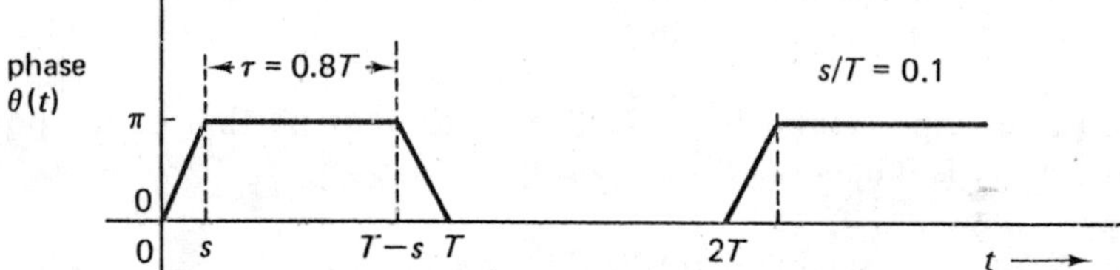

Fig. 11-1 Trapezoidal pulses of BPSK phase [Glance, 1971 Reprinted by permission from The Bell Telephone Laboratories, Inc. Copyright 1971. The American Telephone and Telegraph Company]

This spectrum decreases as $|f - f_0|^{-4}$ as the frequency departs significantly from the carrier center frequency and $|f - f_0| \longrightarrow \infty$ as shown in Fig. 11-2. The finite rise time degrades performance slightly but reduces the spectral sideband content. As the rise time decreases, $s \longrightarrow 0$, the spectrum approaches the commonly described $\text{sinc}^2\, x \triangleq (\sin x/x)^2$ spectrum where $x = (\omega_0 - \omega)T/2$.

Consider a raised-cosine phase-shift modulation where $\theta(t) = \pi/2(1 - \cos 2\pi t/T)$ in (11-1). The peak-peak phase deviation is $2\theta = \pi$

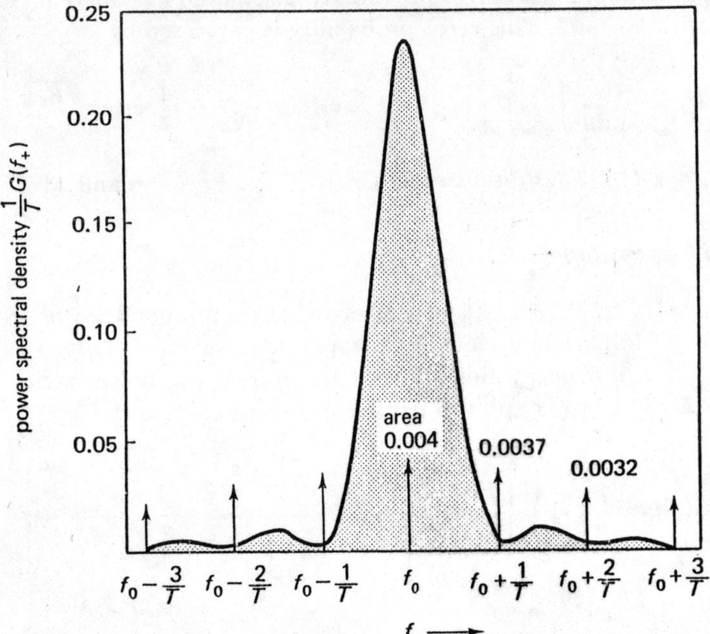

Fig. 11-2 Power spectral densities for BPSK with trapezoidal pulses [Glance, 1971 Reprinted by permission of the Bell Telephone Laboratories Inc. Copyright 1971. The American Telephone and Telegraph Company]

and the pulse slope $g(t) = \sin^2\Omega t/2$. The continuous portion of the power spectral density is [Glance, 1971 pp. 2873]

$$G(f) \cong \frac{T}{8}\left[\frac{\sin(\omega_0 - \omega)T/2}{(\omega_0 - \omega)T/2}\right]^2 \left\{J_0\left(\frac{\pi}{4}\right) - 2(\omega_0 - \omega)^2 \times \left[\frac{J_1(\pi/4)}{(\omega_0 - \omega)^2 - \Omega^2} + \frac{J_2(\pi/4)}{(\omega_0 - \omega)^2 - 4\Omega^2}\right]\right\}^2 \tag{11-7}$$

where $\Omega \triangleq 2\pi/T$, and T is the bit duration. Note $\frac{1}{2}J_0(\pi/4) \cong J_1(\pi/4) + J_2(\pi/4)$.

As $|f - f_0| \to \infty$, the power spectral density $G(f) \sim 1/|f - f_0|^6$. The discrete portion of the power spectral density caused by the nonzero average $E[\sin\theta(t)]$ in (11-1) is then

$$G_d(f) \cong \frac{1}{8}\sum_{m=-2}^{+2} J_m^2\left(\frac{\pi}{4}\right)\delta\left(f_0 - f + \frac{m}{T}\right) \tag{11-8}$$

and the main line component at $\omega = \omega_0$ has power 0.0907.

Differential Encoding of BPSK

Antipodal BPSK signaling is ideally of the form

$$A(t) \sin \omega_0 t \triangleq A \sin [\omega_0 t + \theta(t)], \; A_i = a_i + A_{i-1} \tag{11-9}$$

where $A(t) = \pm A$ and $\theta = (0, \pi)$. The sign of A for transmission is determined by the data. The data can be differentially encoded to resolve carrier phase ambiguities. In addition the data can be coded by a forward error correction (FEC) coder. Figure 11-3 shows a rate 1/2 coder that emits a coded bit stream at twice the input bit rate. Differential decoding then decodes the differences between bits; that is, $A_i \oplus A_{i+2}$ determines the sign of $\pm a_i$:

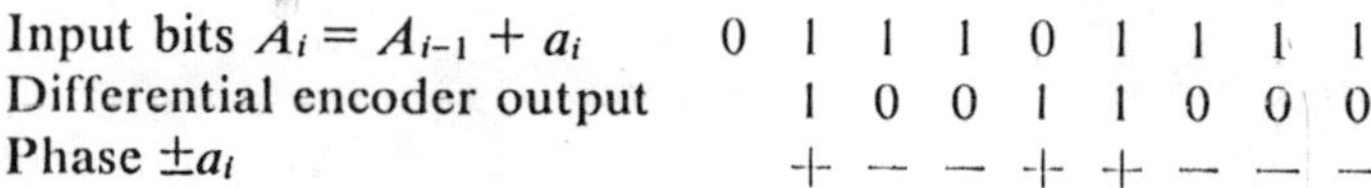

Input bits $A_i = A_{i-1} + a_i$	0	1	1	1	0	1	1	1	1
Differential encoder output		1	0	0	1	1	0	0	0
Phase $\pm a_i$		+	−	−	+	+	−	−	−

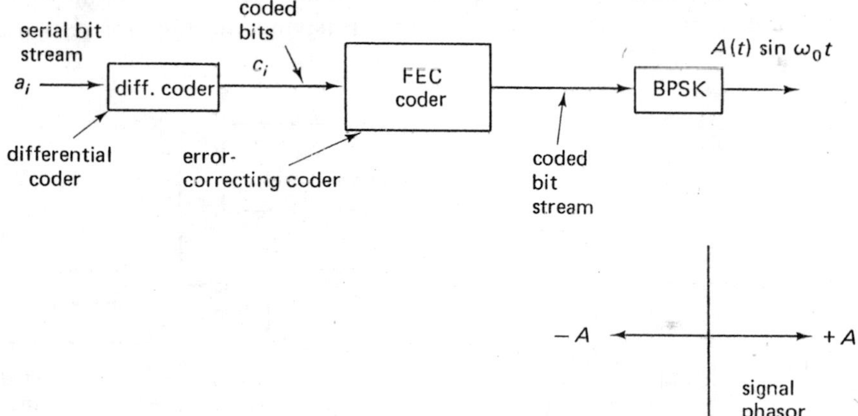

Fig. 11-3 Typical BPSK modulator for a coded bit stream $A(t) = \pm A$

Performance is usually best if the differential coder precedes the coder and follows the decoder. Decoders can be built to operate with either normal or inverted bit streams and produce normal or inverted outputs. These decoders are then *transparent* to ambiguities in the sign of the input bit stream. The decoded output bit stream has the same sign ambiguity as the input. However, the decoder output error rate is greatly reduced now (errors occur in bursts), and ambiguity removal by differential decoding can be performed at only a small increase in output error rate. (Error bursts are extended by 1 bit.)

While differential encoding with coherent detection essentially doubles the error rate, in the absence of FEC coding it can be used to eliminate problems of carrier-phase ambiguity in coherent detection. Ambiguity in the sign of the reference carrier, which always exists for the carrier recovered in antipodal modulation, produces no data ambiguity for this coding because there is no ambiguity as to whether or not a phase *change* occurred. Ambiguities may also be removed by transmitting a known pattern or sync word for synchronization (which is not ambiguous to a time shift). This sync pattern may be required anyway for frame synchronization.

Figure 11-4 is a block diagram of a coherent matched-filter detector for BPSK [Viterbi, 1966, Chaps. 7, 10; Sunde, 1969, Chap. 5; Stein and Jones, 1967, Chap. 12]. The noisy BPSK signal is first bandpass filtered to aid in recovery of a coherent (partially coherent) reference. The direct input is then multiplied by the partially coherent reference generated in the carrier recovery loop to recover

$$A(t) \cos \phi_n + n'(t) \tag{11-10}$$

where $\phi_n(t)$ is the phase noise in the carrier recovery operation. For a white noise input and an undistorted signal, the optimum detector is an integrate-and-dump filter or finite-memory integrator having the impulse response shown in Fig. 11-5.

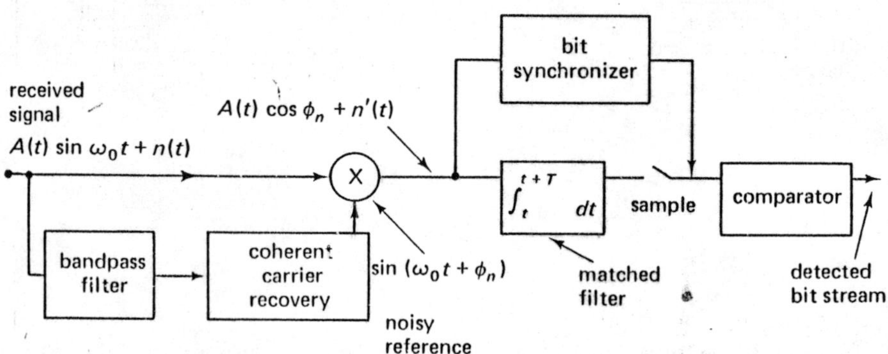

Fig. 11-4 Coherent matched-filter receiver for BPSK

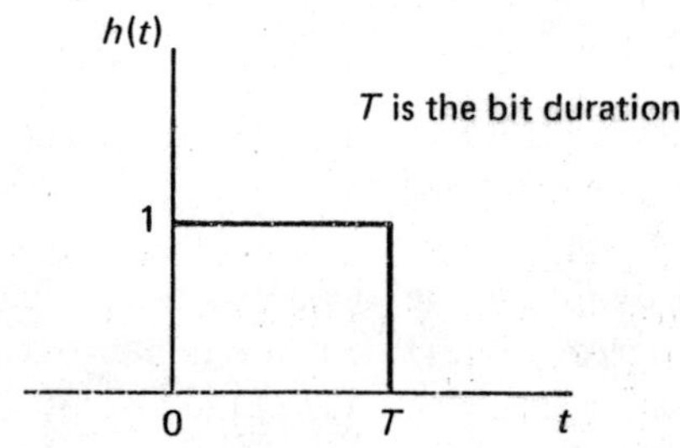

Fig. 11-5 Impulse response of the finite-memory integrator

BPSK Carrier Recovery

Carrier recovery for this suppressed carrier signal ($\theta_m = (0, \pi)$) [Lindsay, 1966, 393–401] can be obtained by a number of methods; the squaring or frequency-doubling loop shown in Fig. 11-6 is perhaps the most straightforward. Other carrier-recovery techniques are inphase/quadrature multiplication as in the Costas loop, and decision-directed carrier recovery using the detected bit stream to remove the modulation from the carrier.

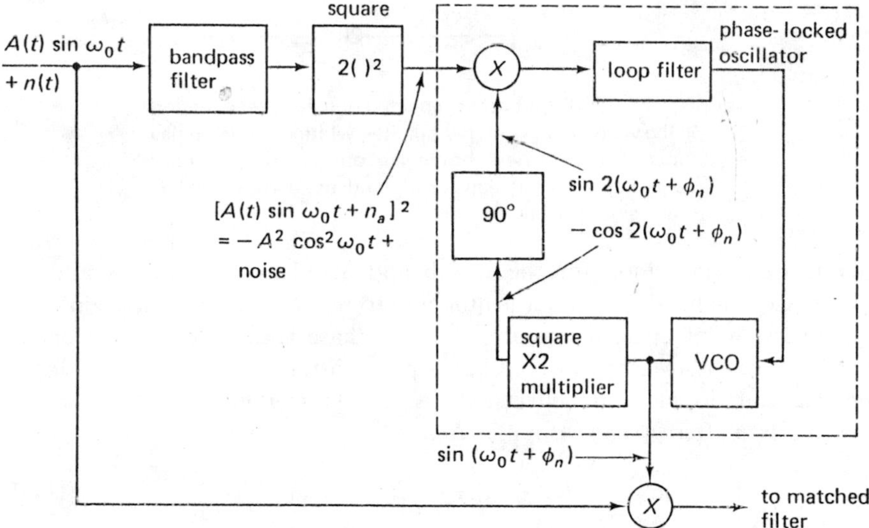

Fig. 11-6 Block diagram of the square-law carrier-recovery technique

The squaring operation effectively removes the modulation $\pm A$ and creates a line component in the spectrum at double the carrier frequency $2\omega_0$. The squared output for an input $r(t)$ can be represented as

$$\begin{aligned} r^2(t) &= 2\{A(t) \sin \omega_0 t + N_a(t) \sin [\omega_0 t + \theta_n(t)]\}^2 \\ &= -A^2 \cos 2\omega_0 t - 2A(t)N_a(t) \cos (2\omega_0 t + \theta_n) \\ &\quad - N_a^2(t) \cos (2\omega_0 t + 2\theta_n) + \text{dc baseband terms} \end{aligned} \tag{11-11}$$

If the bandpass filter has a rectangular passband of bandwidth W Hz and the input noise is white Gaussian noise, then the output spectrum in the vicinity of $f = 2f_0$ is as shown in Fig. 11-7. The phase-locked oscillator (Chap. 12) acts as a narrowband filter about the carrier component. Ideally, of course, the PLO would have as small a bandwidth as the search and acquisition time (lock-on time) requirement of the user can permit. One often employs a mod-

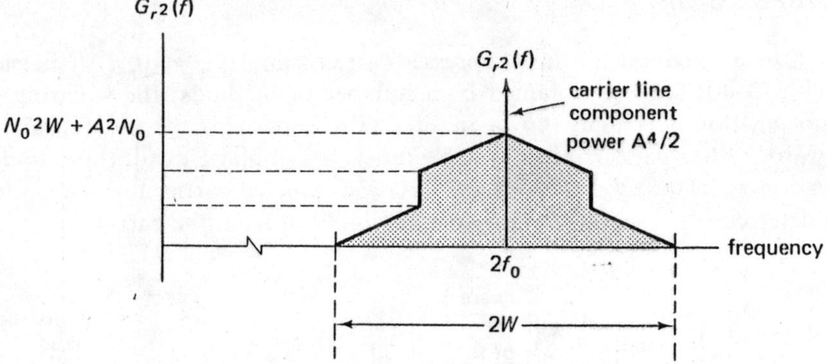

Fig. 11-7 Output power spectral density after squaring in the vicinity of $2f_0$ for bandlimited white Gaussian noise input. The densities shown are one-sided. The input noise has an input density N_0 and bandwidth W. The input sinusoid has power $A^2/2$.

erate bandwidth loop for the search and acquisition operation and then decreases the bandwidth after acquisition to reduce the phase noise in carrier recovery. In practice, the received carrier phase is not a constant slope $\omega_0 t$ as indicated above, but instead has a phase jitter and has possible doppler offset and doppler rate effects. Thus, the carrier component is received with phase jitter $\psi(t)$ and doppler $\Delta\omega t$:

$$A(t)\sin[\omega_0 t + \theta(t)] + n(t) \qquad \theta(t) = \Delta\omega t + \psi(t) \tag{11-12}$$

The PLO also must be sufficiently wide to track accurately the squared component phase jitter $2\psi(t)$, but not so wide as to track the BPSK modulation.

If perfect phase tracking and bit timing are performed, the system performs with an error rate $\mathcal{P}_b$ (without differential decoding):

$$\mathcal{P}_b = \frac{1}{2}\operatorname{erfc}\sqrt{\frac{E_b}{N_0}} \tag{11-13}$$

$$\text{where} \quad \operatorname{erfc} x \triangleq \frac{2}{\sqrt{\pi}}\int_x^\infty e^{-y^2}\,dy$$

$E_b = P_s T = (A^2/2)T$, and N_0 is the one-sided noise density. Imperfect carrier-phase tracking degrades this performance (Chap. 12).

Another carrier-recovery technique uses the Costas loop (Fig. 11-8), which generates a coherent phase reference independent of the binary modulation by use of both inphase and quadrature channels [Costas, 1956, 1713–1718]. The inphase channel removes the modulation effects from the quadrature error channel and does not involve much additional circuit complexity since it is needed in any event for data demodulation.

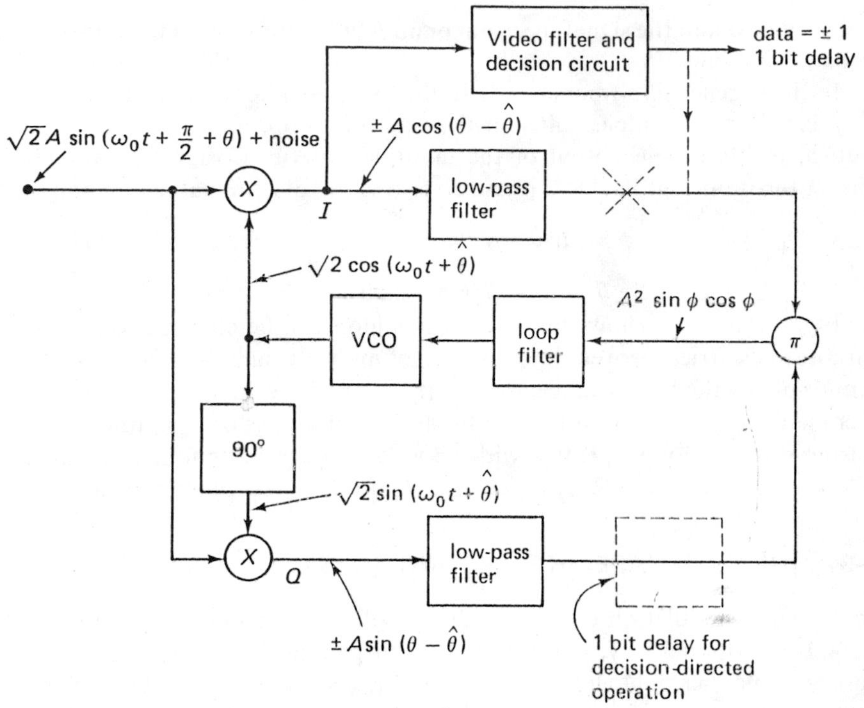

Fig. 11-8 Costas-loop PSK carrier-recovery circuit and bit detector; the phase error is defined as $\phi \triangleq \theta - \hat{\theta}$. The decision-directed configuration is shown in dashed lines.

This performance can be improved by use of a decision-directed or data-aided carrier tracking loop, where hard $D = \pm 1$ decisions are made with an integrate-and-dump circuit in the inphase channel. See Fig. 11-8. The quadrature error channel then must be delayed by one bit period before forming the product $D(t)A(t) \sin \phi$, which has an average value $(1 - 2\mathcal{P}_b) \sin \phi$ where $\mathcal{P}_b$ is the bit error probability [Lindsay and Simon, 1971, 152–169].

The Costas loop is often preferred over the squaring loop because its circuits are less sensitive to center-frequency shifts and are generally capable of wider bandwidth operation. A wide frequency drift can be tolerated without widening the low-pass filters in each channel because the closed-loop operation of the VCO can track and remove the slow frequency drift before the signal enters the low-pass filter. One must be careful to match the group delay in the two channels, however. These I and Q baseband channel filters perform the equivalent operation to a bandpass filter which tracks the signal in center frequency. The open-loop squaring technique, on the other hand, must be preceded by a bandpass filter wide enough to accommodate both the frequency drift and the signal spectrum or an AFC loop must be used around the bandpass filter.

In the absence of a frequency drift, the low-pass filters are the baseband equivalent to the bandpass filter in the squaring loop, and performance is identical in both. The output of the multiplier in the Costas loop for no noise at the input but having a phase error ϕ is the discriminator function

$$A^2(t) \sin \phi \cos \phi = \sin \phi \cos \phi \tag{11-14}$$

since $A^2 = 1$ and where $\theta - \hat{\theta} \triangleq \phi$ is the phase error.

Figure 11-8 also shows how the Costas loop can be changed to a decision-directed carrier recovery loop. Instead of multiplying by $\pm A \cos (\theta - \hat{\theta})$, we multiply by the binary bit decisions $\hat{A}(t)$. A one-bit delay is inserted in the other channel to account for the one-bit delay in the decision operation. This decision-directed loop operates with very small degradation. Even with a 5% error rate, only $(1 - 2\mathcal{P}_b) = 0.9$ or 0.46 of the carrier power is lost.

11-3 QUADRIPHASE-SHIFT KEYING (QPSK)

Quadriphase modulation encodes each pair of bits into one of four phases (Fig. 11-9). Alternatively, one can code each pair of bits into a change in phase in a manner analogous to differential encoding of BPSK. One of the principal advantages of QPSK over BPSK is that, under certain transmission conditions (Sec. 11-5), QPSK achieves the same power efficiency as BPSK using only half the bandwidth.

The ideal QPSK signal waveform can be represented in two equivalent forms

$$\begin{aligned} A \sin [\omega_0 t + \theta'_m(t)] = {} & \frac{A}{\sqrt{2}} u_s(t) \sin \left(\omega_0 t + \frac{\pi}{4}\right) \\ & + \frac{A}{\sqrt{2}} u_c(t) \cos \left(\omega_0 t + \frac{\pi}{4}\right) \end{aligned} \tag{11-15}$$

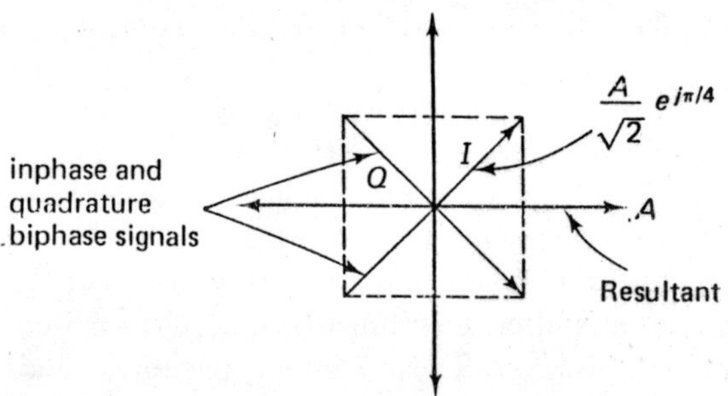

Fig. 11-9 Representation of QPSK modulation in phasor form when the *I* and *Q* channels are of equal amplitude and in exact phase quadrature

where $\theta_m = (0, \pi/2, \pi, 3\pi/2)$, and $u_s, u_c = \pm 1$ represents the data modulation. Each symbol lasts T sec.

An alternate form, *staggered quadriphase modulation* (SQPSK), delays the inphase bit stream by $T/2$ sec relative to the quadrature Q bit stream, giving a 0°, 90° phase transition each T sec. This modulation has the advantage that no envelope nulls occur as do with 180° phase transitions, because only one channel—the I or the Q—changes phase at each bit transition. Hence, performance with bandpass filtering can be improved over that of conventional QPSK because the phase changes occur in smaller step sizes.

The quadriphase signal thus is the sum of two BPSK signals that are inphase quadrature with one another. With symmetrical bandpass Gaussian noise, the inphase and quadrature noise terms are independent, and the performance of QPSK is identical to that of BPSK; that is, the error rate is

$$\mathcal{P}_b = \text{erfc}' \sqrt{\frac{2E_b}{N_0}} = \frac{1}{2} \text{erfc} \sqrt{\frac{E_b}{N_0}} \tag{11-16}$$

where the energy per bit is now $E_b = (1/2)[(A^2/2)T]$ since two bits are transmitted in each T sec interval.

Assuming the inphase and quadrature bit streams are independent, the power spectral density of QPSK is exactly half the width of that for BPSK for the same bit rate (Fig. 11-10).* Thus, the phase-chip duration T is related to the bit rate f_b by

$$T = \frac{2}{f_b} \tag{11-17}$$

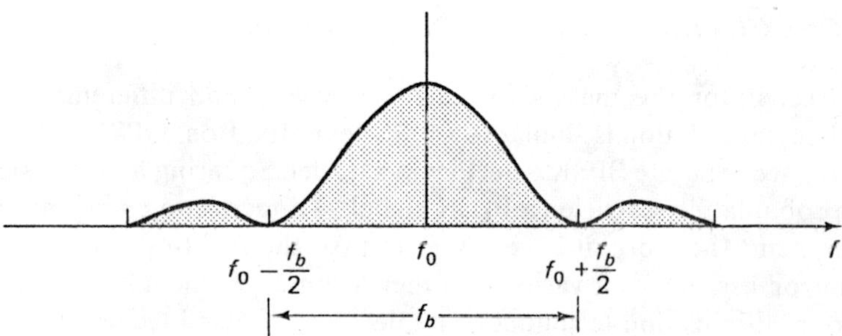

Fig. 11-10 Power spectrum of QPSK for independent random bit streams on the inphase and quadrature channels

---

*Correlation between bit streams or correlation with a delay offset can cause changes in spectrum shape.

Differential Encoding of QPSK

As with BPSK, it is possible to transmit QPSK without differential encoding. Ambiguities can be resolved by use of structure in error-correcting coding or by using particular unique synchronization words on a periodic basis. In the absence of these techniques, however, it is essential to differentially encode the data to resolve the fourfold phase ambiguity. Differential encoding of the bit pairs selects the phase change rather than the absolute phase. Figure 11-11 shows an example Gray coding technique for changes in phase. In this differential phase coding technique, a 90° phase error in the demodulator output carries only a one-bit output error at that time instant, whereas other coding schemes can cause both output bits to be in error.

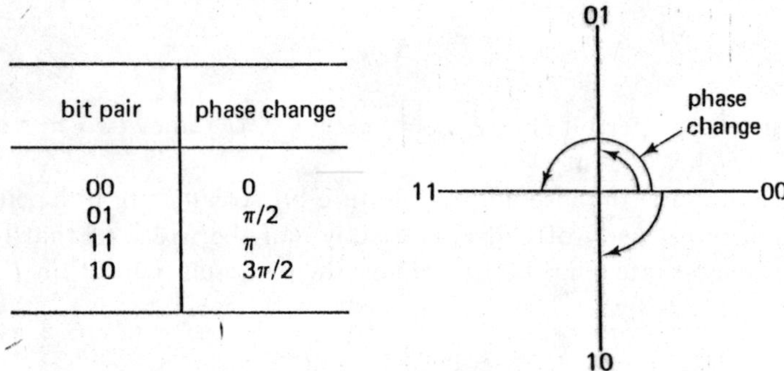

bit pair	phase change
00	0
01	$\pi/2$
11	π
10	$3\pi/2$

Fig. 11-11 Phase coding relationships for differential encoding of QPSK

QPSK Carrier Recovery and Data Detection

Except for the method of carrier recovery and differential decoding, QPSK demodulation is similar to BPSK demodulation. QPSK detection consists of two separate BPSK detectors in parallel. Squaring a QPSK signal with equiprobable phases yields a BPSK signal. To recover a carrier spectral line component, therefore, it is necessary to take the fourth power of the QPSK carrier or use one of a variety of other techniques, including inphase/quadrature multiplication techniques, and decision-directed techniques analogous to those of BPSK.

Figure 11-12 is a block diagram of a QPSK detector. The received signal plus noise is filtered and taken to the fourth power. Carrier recovery is accomplished by locking a phase-locked loop to the fourth harmonic of the received carrier obtained from

$$8[A^4 \cos^4 (\omega_0 t + \theta_m)] = A^4 \cos (4\omega_0 t + 4\theta_m) = A^4 \cos 4\omega_0 t \qquad (11\text{-}18)$$

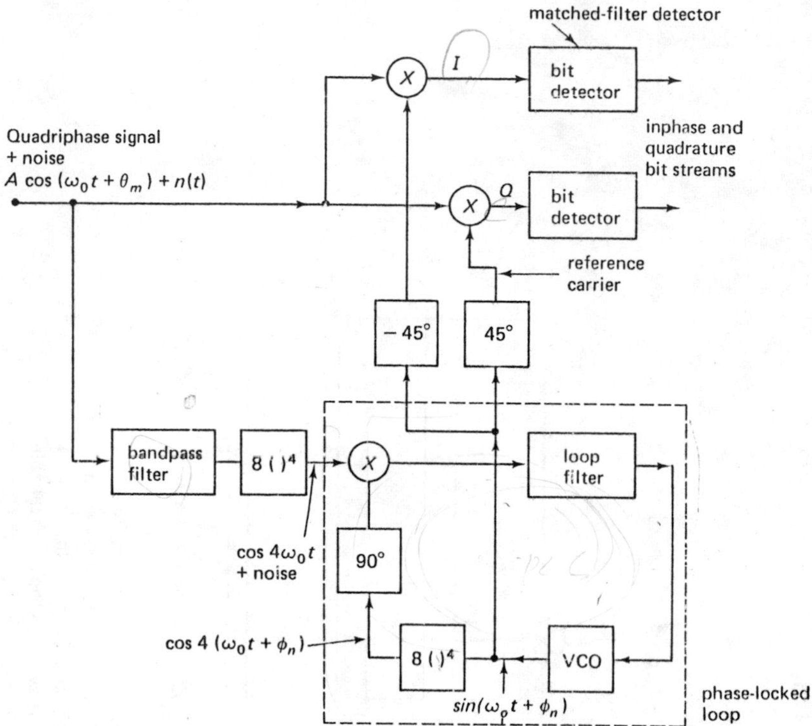

Fig. 11-12 Block diagram of a QPSK carrier-recovery technique

where we include only terms at $4\omega_0$ in the right-hand side of (11-18) and $\theta_m = m\pi/2$, $4\theta_m = m2\pi$ are assumed. Thus, a spectral line component is present at $4\omega_0$, and a phase-locked oscillator can be used to recover it. Noise effects on the phase-locked loop are related to QPSK error rate in Sec. 12-8.

Notice that if there is amplitude imbalance between the I and Q channels the signal is still quadriphase but $\theta \neq m\ 90°$.

$$A_s u_s(t) \sin \omega_0 t + A_c u_c(t) \cos \omega_0 t = A \cos [\omega_0 t + \theta(t)]$$

where $A = \sqrt{A_s^2 + A_c^2}$ and $\theta(t) = \pm \tan^{-1} A_s/A_c,\ \pi \pm \tan^{-1} A_s/A_c$.

A $\times 4$ carrier multiplication then produces a carrier $\cos 4[\omega_0 t + \theta(t)]$ which has phase modulation unless $\theta = n\pi/4$ rad or $n45°$. Thus the pure carrier component at $4\omega_0$ is attenuated by $\cos [\tan^{-1} A_s/A_c - \pi/4]$ if there is amplitude imbalance.

The Costas loop of Fig. 11-8 can be generalized to provide carrier recovery for QPSK. Figure 11-13 shows one such generalization. As shown, the outputs of the limiter circuits remove the biphase modulation from the

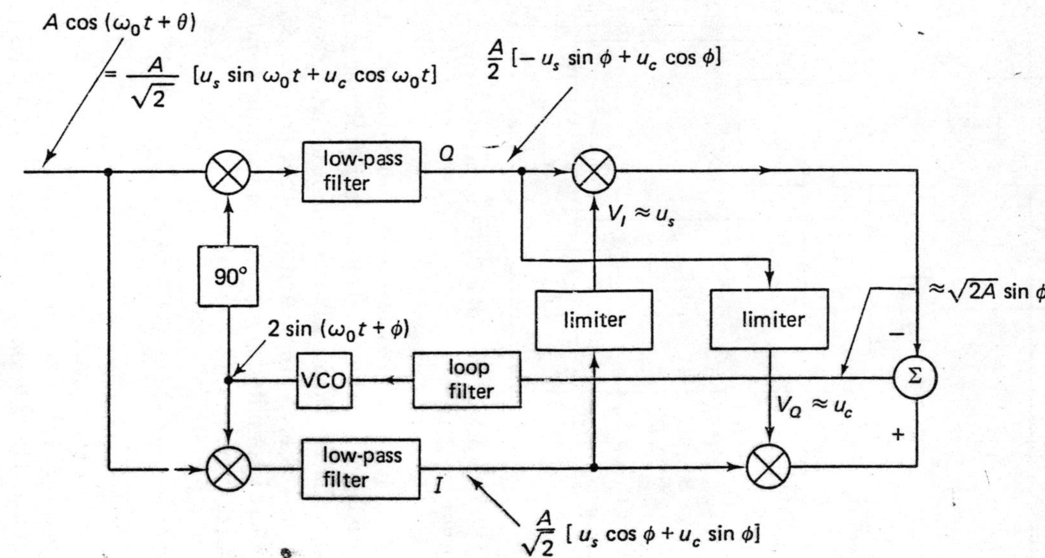

Fig. 11-13 Generalization of the Costas type carrier recovery loop to QPSK. Note that for small ϕ and high SNR the limiter outputs provide u_s, u_c for the I and Q channels, respectively. If symbol clock is available, as it might be in a TDMA receiver, then the low-pass filters can be replaced by integrate-and-dump filters.

sin ϕ correction terms. The subtraction operation removes the dc bias effects. The resulting phase detector characteristic is

$$D(\phi) = \frac{A}{\sqrt{2}}[(v_Q u_s - v_I u_c)\cos\phi + (v_Q u_c + v_I u_s)\sin\phi] \tag{11-19}$$

For small ϕ and moderate or high SNR, the limiter outputs are approximately $v_I = u_s$, $v_Q = u_c$, and the phase detector characteristic simplifies to

$$D(\phi) = \sqrt{2}\,A\sin\phi \tag{11-20}$$

where ϕ is the phase error.

The group delay and impulse response of each of these low-pass channels must be equal to prevent possible carrier-recovery problems. A disparity in impulse response can cause a bit-sequence-dependent bias error since v_I could be delayed by one symbol. A delay between channels creates a term similar to the product $u_s(t)u_c(t-\Delta)$, which has a line component at the bit rate (Chap. 14).

QPSK carrier-recovery technique using a reverse modulation method is shown in Fig. 11-14 [Yamamoto et al., 1972, 803–807]. This method avoids the use of baseband multipliers required in the Costas-type carrier-recovery loops. Filter delays must be compensated for by the matching delays as shown. Each of the two channels is reverse modulated and delayed by $T_1 = T_A = T_B = T$ to obtain the sum of products for no noise:

$$\begin{aligned}\frac{S}{\sqrt{2}\,A} &= [u_s(t-T)\sin\omega_0(t-T) + u_c(t-T)\cos\omega_0(t-T)]\\ &\quad\times[u_s(t-T)\cos\phi + u_c(t-T)\sin\phi]\\ &\quad+[u_s(t-T)\sin\omega_0(t-T) + u_c(t-T)\cos\omega_0(t-T)]\\ &\quad\times[-u_s(t-T)\sin\phi + u_c(t-T)\cos\phi]\end{aligned} \tag{11-21}$$

$$\begin{aligned}\frac{S}{\sqrt{2}\,A} &= \cos\phi\sin\omega_0(t-T) + \sin\phi\cos\omega_0(t-T)\\ &\quad+C(t)\,[\cos\phi\cos\omega_0(t-T) + \sin\phi\sin\omega_0(t-T)]\\ &\quad+\cos\phi\cos\omega_0(t-T) - \sin\phi\sin\omega_0(t-T)\\ &\quad+C(t)\,[\cos\phi\sin\omega_0(t-T) - \sin\phi\cos\omega_0(t-T)]\end{aligned} \tag{11-22}$$

where the product $u_s(t-T)u_c(t-T) \triangleq C(t) = \pm 1$ is a binary wave train. Defining $t' = t - T$, we obtain

$$S' = 2A\left[\cos\left(\phi + \frac{\pi}{4}\right)\sin\omega_0 t' + \cos\left(\phi - \frac{\pi}{4}\right)\cos\omega_0 t'\right] \tag{11-23}$$

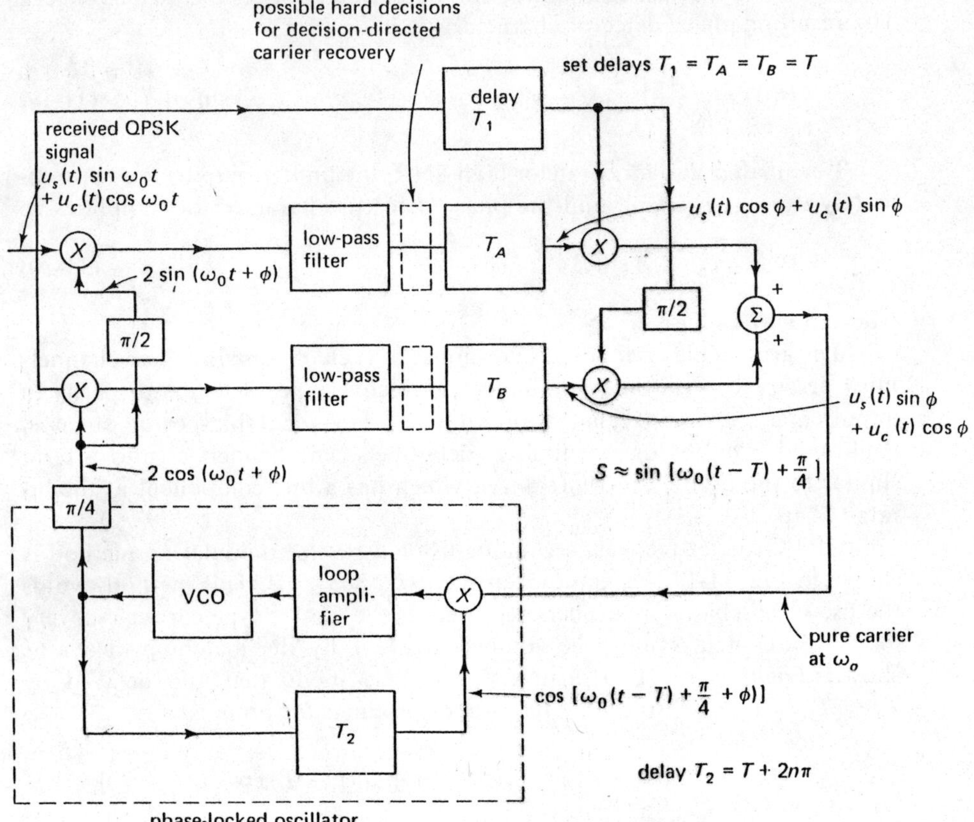

Fig. 11-14 QPSK carrier recovery using reverse modulation and a phase-locked oscillator [After Yamamoto et al., 1972, 803–807]. Decision-directed carrier recovery can be performed by making hard decisions as shown in the dashed boxes in the diagram.

Thus we have a pure carrier component at ω_0 rather than at $4\omega_0$ as we found in the $\times 4$ multiplier technique. The power P_c and phase θ in the pure carrier (11-23) are

$$P_c = 4A^2\left[\cos^2\left(\phi + \frac{\pi}{4}\right) + \cos^2\left(\phi - \frac{\pi}{4}\right)\right] = 4A^2,$$

$$\theta = \tan^{-1} \frac{\cos(\phi + \pi/4)}{\cos(\phi - \pi/4)}$$

This carrier component is then acquired by the phase-locked loop and used to generate the reference carrier $\sin[\omega_0 t' + \phi]$. If the delay $T_2 = T_1$

$+ 2n\pi$, then the output of the $\pi/4$ phase-shift network is $\cos(\omega_0 t + \phi)$ as assumed initially.

Both the Nth power loop and the Costas loop can be generalized to operate with MPSK signals [Lindsey and Simon, 1972*a*, 441–454]. The performance of the N-phase Costas loop is mathematically equivalent to the Nth power loop, except for possible doppler effects as noted above for BPSK.

Error Probability with Differential Decoding of QPSK

Carrier recovery produces four-phase ambiguities at $\pm n\pi/2$ increments. Ambiguities are avoided by differential encoding and decoding of the phase changes. Phase changes are decoded into one of four levels in differentially encoded quadriphase as shown in Fig. 11-15. Hence, the output bit-error probability is determined by examining two consecutive two-bit/symbol error patterns. For simplicity, one can assume that the true data are all zeros.

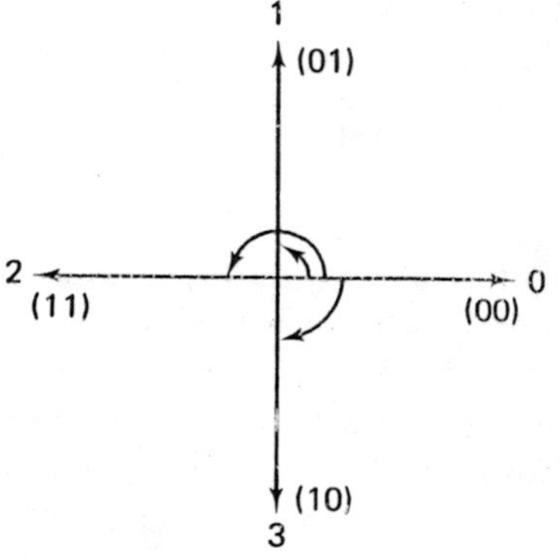

Fig. 11-15 Phase diagram for differential encoding of QPSK

The output bit error probability is then one-half of the average of $n(i, j)$

$$\mathcal{P}_{\text{out}} = \frac{1}{2} \sum_{i=1}^{4} \sum_{j=1}^{4} n(i, j)\mathcal{P}(i, j) \tag{11-24}$$

where $n(i, j)$ is the number of errors in the output generated by error pattern (i, j), and $\mathcal{P}(i, j)$ is the probability of error pattern (i, j), where i is the first symbol error and j is the second $(i, j = 0, 1, 2, 3)$. The factor of 1/2 in (11-24) appears because there are 2 bits/symbol.

The probability of error pattern (i, j) is $p(i, j) = p^k(1 - p)^{4-k}$ where k is the total number of decoder input bit errors; that is, $p(1, 3) = p(01, 10) = p^2(1 - p)^2$ since there are $k = 2$ bit errors. That is, the phase change from level 1 to 3 is interpreted as a 11 rather than the actual value 00; hence there are 2 bit errors caused by these 2 symbol errors. The input bit-error probability is $p = 1 - q$. Thus, the output-error probability is one-half the sum of all possible input-error-pattern probabilities times the number of resulting output bit errors for each pattern and (11-24) becomes:

$$\begin{aligned}\mathcal{P}_{\text{out}} &= \tfrac{1}{2}(1)[p(00, 01) + p(00, 10) + p(01, 00) + p(01, 11) \\ &\quad + p(11, 01) + p(11, 10) + p(10, 00) + p(10, 11)] \\ &\quad + \tfrac{1}{2}(2)[p(00, 11) + p(01, 10) + p(11, 00) + p(10, 01)] \\ &= \tfrac{1}{2}[4pq^3 + 4p^3q] + 4p^2q^2 \\ &= [2pq^3 + 4p^2q^2 + 2p^3q] = 2pq[q^2 + 2pq + p^2] \qquad (11\text{-}25)\end{aligned}$$

This result simplifies to

$$\mathcal{P}_{\text{out}} = 2pq = 2p(1 - p) \cong 2p \qquad p \ll 1 \qquad (11\text{-}26)$$

Thus the output error probability is essentially doubled at low error rates.

11-4 STAGGERED QPSK AND MINIMAL SHIFT KEYING (MSK)

Two other related modulation techniques have improved bandwidth efficiency and other advantages over QPSK when hard limiting occurs in the system. Hard or soft limiting in a TWT amplifier can build back up the MPSK spectral sidebands previously attenuated by bandpass filtering. Robinson et al. [1973, 227–256] have shown that the sideband buildup of a TWT with 0-dB power backoff can exceed 15 dB and is largely caused by amplitude nonlinearities rather than AM/PM effects.

Staggered QPSK, or SQPSK (also termed offset QPSK), has phase relationships identical to those of QPSK, except that the inphase and quadrature bit streams are offset in time by half a symbol period as shown in Fig. 11-16. Each of these bit streams can be separately differentially encoded as

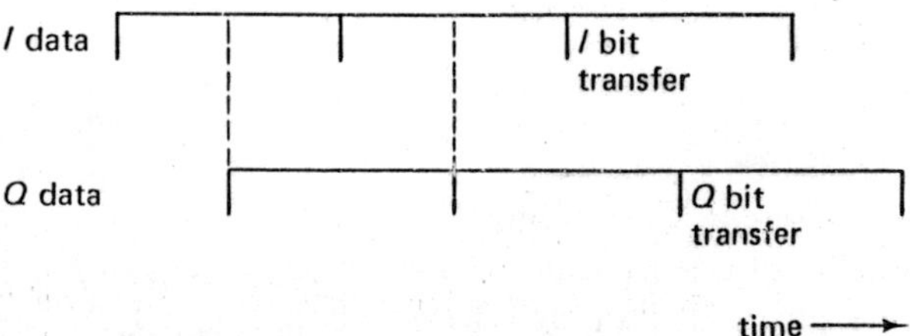

Fig. 11-16 Clock and data relationships in SQPSK

bit sequence				phasor
I_{t_0}	Q_{t_1}	I_{t_2}	phase change at t_2	
0	0	0	0°	t_1, t_2
0	0	1	+ 90°	t_1 → t_2
0	1	1	− 90°	t_1 → t_2
0	1	0	0°	t_1, t_2
1	0	1	0°	t_1, t_2
1	0	0	− 90°	t_1 → t_2
1	1	0	+ 90°	t_1 → t_2
1	1	1	0°	t_1, t_2

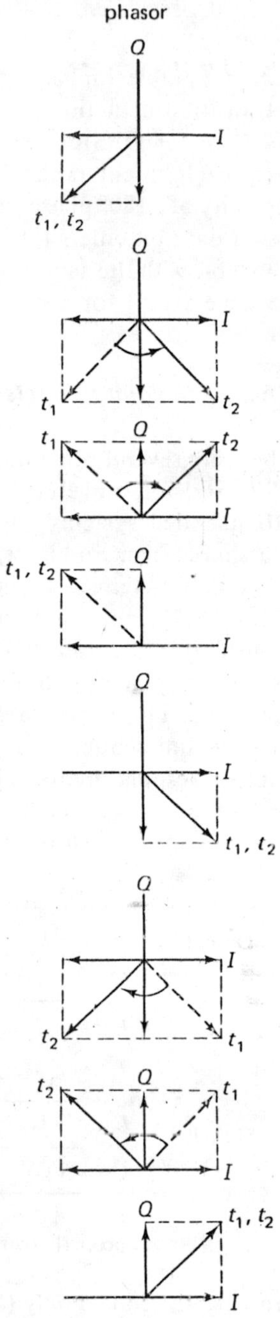

Fig. 11-17 Phase transitions for staggered QPSK. The data bit stream alternates between inphase I_t and quadrature Q_t channels.

with two separable biphase PSK channels. Since only one channel has a phase transition at a time, the differential decoding is slightly simpler for SQPSK than for QPSK. As a result, SQPSK has a 0, +90°, −90° phase change as each bit enters the I or Q channel of the modulator. Thus there is no possibility of a 180° phase transition. Figure 11-17 shows the phase changes as a new I data bit enters the modulator for various initial conditions; for a Q channel bit with the same initial conditions, the phase-change relationships are the same except for a sign reversal.

Phase Ambiguity Effects

There are several other important points regarding the phase ambiguities of QPSK, SQPSK, and BPSK. These points are discussed here with reference both to uncoded systems and to systems employing transparent decoders where a simple sign ambiguity does not affect decoding. Furthermore with rate 1/2 coded systems, the automatic pairing of data and parity bits possible with QPSK is an important consideration.

Consider QPSK and SQPSK where the reference carrier is in one of 4 phases, ϕ_A, in phase with the original carrier reference, $\phi_B = \phi_A + 90°$, $-\phi_A$ or $-\phi_B$. For these 4 reference phases the I and Q channel outputs appear as a time sequence of paired bits as shown below for QPSK. The bit sequence corresponding to ϕ_A is, of course, the correct one.

QPSK Output Bits in Time Sequence

Output Channel	ϕ_A			ϕ_B		
I	I_1	I_2	I_3	Q_1	Q_2	Q_3
Q	Q_1	Q_2	Q_3	$-I_1$	$-I_2$	$-I_3$
$t \longrightarrow$						$t \longrightarrow$

Output Channel	$-\phi_A$			$-\phi_B$		
I	$-I_1$	$-I_2$	$-I_3$	$-Q_1$	$-Q_2$	$-Q_3$
Q	$-Q_1$	$-Q_2$	$-Q_3$	I_1	I_2	I_3
$t \longrightarrow$						$t \longrightarrow$

Bits are properly paired for ϕ_A and $-\phi_A$

For transparent codes, only $(\phi_A, -\phi_A)$ and $(\phi_B, -\phi_B)$ must be separated for proper decoding, since ϕ_A and $-\phi_A$ both produce properly paired bits. Likewise ϕ_B and $-\phi_B$ reverse the roles of I and Q. The only difference between ϕ_A and $-\phi_A$ is the sign of the output bits.

The output bits for SQPSK are in the correct output sequence for uncoded operation. In this uncoded operation there are only the sign ambiguities with which to contend. Note, however, that the sign ambiguity of the I and Q channels can be different as shown below.

STAGGERED QPSK Output bits

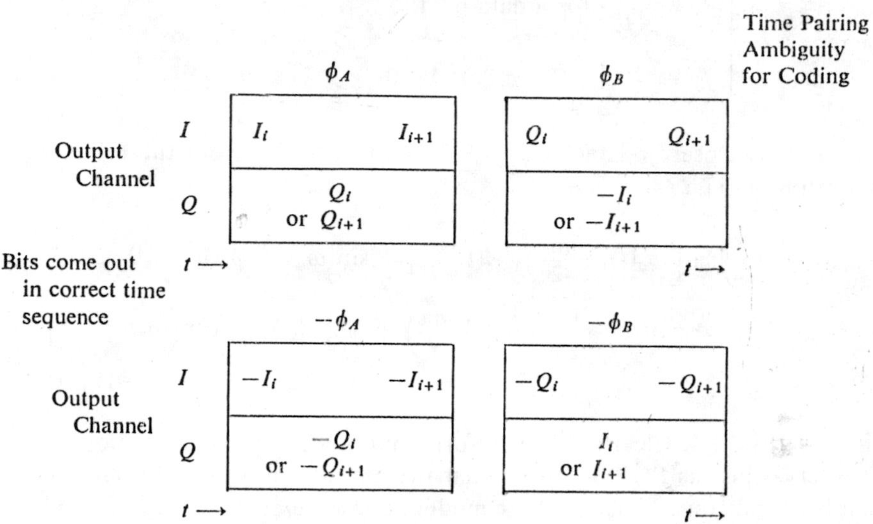

Hence there is still a 4-fold ambiguity. If rate 1/2 coding is employed, the I and Q bits must be properly paired for decoding. Modulator and demodulator pairing must match. The I channel bit can pair with the Q channel bit ahead or behind it. Coded BPSK (with rate 1/2 coding) also has a four-fold ambiguity as shown below.

BPSK Output bits

ϕ_A

Output Bit Stream $\quad I_1 \quad I_2 \quad I_3$

$t \longrightarrow$ Time Pairing and Sign Ambiguity

$-\phi_A$

Output Bit Stream $\quad -I_1 \quad -I_2 \quad -I_3$

BPSK has a time pairing as well as a sign ambiguity.

Minimum-shift keying (MSK) is a form of frequency-shift keying in which the frequency is shifted or kept the same at every new data bit period T [Sullivan, 1972, 23-4–23-8; de Buda, 1972, 41-25–41-27]. The frequency shift precisely increases or decreases the phase by 90° each T sec. Thus the signal waveform is

$$s(t) = \sin\left(\omega_0 t + 2\pi \int f_i \, dt + \frac{n\pi}{2}\right) \qquad 0 < t < T \qquad \text{(11-27)}$$

$$\text{where} \quad f_i = \begin{cases} f_1 = \dfrac{1}{4T} & \text{for a data bit 1} \\ f_2 = -\dfrac{1}{4T} & \text{for a data bit 0} \end{cases}$$

MSK is closely related to SQPSK as can be seen from the following expansion of (11-27)

$$s(t) = \sin\left[\omega_0(t) + \frac{n\pi}{2} + A(t)\frac{\pi t}{2T}\right] = \sin[\omega_0 t + \phi)t)] \qquad 0 \leq t \leq T$$

$$= \cos\left[A(t)\frac{\pi t}{2T}\right]\sin\left(\omega_0 t + \frac{n\pi}{2}\right) + \sin\left[A(t)\frac{\pi t}{2T}\right]\cos\left(\omega_0 t + \frac{n\pi}{2}\right) \qquad \text{(11-27a)}$$

where $A(t) = \pm 1$. Clearly this waveform makes a 90° phase transition each bit interval just as SQPSK except that the phase transition is linear rather than instantaneous. Thus the phase modulation waveforms of the I, Q modulators of SQPSK are modulated by $\sin x$ and $\cos x$ waveforms as in (11-27a), the output will be identical to MSK. Note that it is necessary to modulate both the I and Q channels during each bit interval to retain the constant envelope of $s(t)$.

Since the phase is continuous from bit to bit, the spectral sidebands fall off more rapidly than PSK or QPSK, even though QPSK has a significantly smaller 3-dB bandwidth. The integrated power spectra are compared for PSK, QPSK, and MSK in Fig. 11-18.

The MSK phase over a 2-bit interval of $\Delta t = 2T$ duration is shown in Fig. 11-19 for all possible 2-bit patterns. In this example, we begin at $t = 0$ with $\phi = 0$ and end at $\phi(2T) = \pm\pi, 0$. If one correlates the received signal with the reference carrier $\sin(\omega_0 t + \pi/2)$, then the low-pass portion of the cross-correlator multiplier output is

$$s(t)\sin(\omega_0 t + \pi/2) = \sin[\omega_0 t + \phi(t)]\sin(\omega_0 t + \pi/2)$$
$$= \cos[\phi(t) - \pi/2] + \text{terms at } 2\omega_0$$

Thus the cross-correlator multiplier output over the $2T$ sec interval is one of the 2 antipodal waveforms shown in Fig. 11-19.

$$v(t) = \pm \cos \pi t/T$$

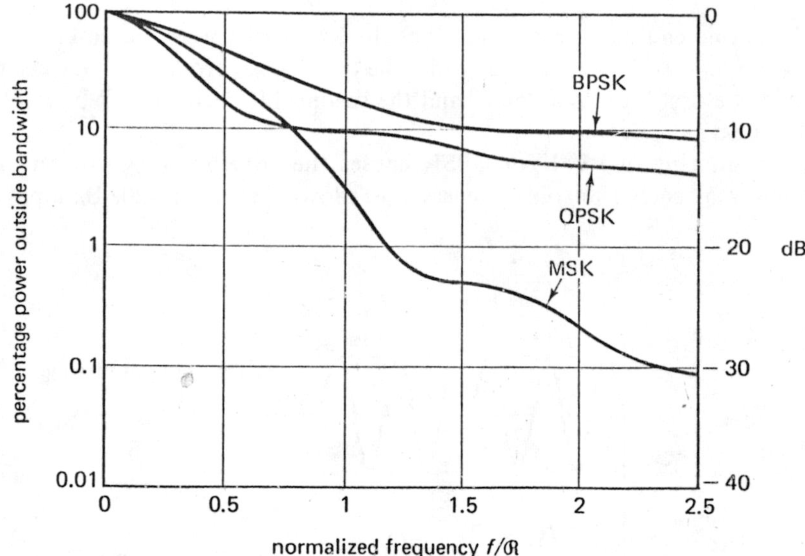

Fig. 11-18 Integrated power spectra for PSK, QPSK, and MSK; frequency is normalized to the bit rate $\mathcal{R}$.

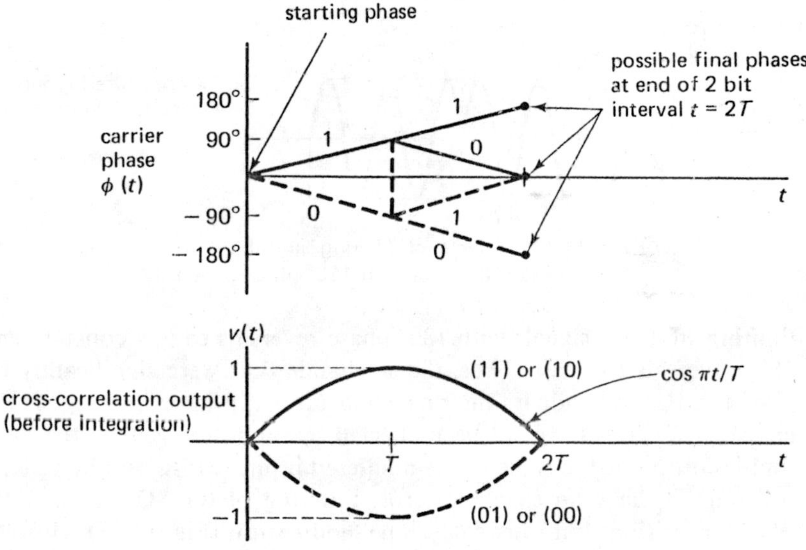

Fig. 11-19 Phase of MSK modulated carrier and cross-correlator output for two-bit intervals showing antipodal characteristics of the modulations

Thus one can make a decision that the last 2 bits were either 11 or 10, for example, if the $+\cos \pi t/T$ provides the best cross correlation. One can decide that the first bit was a "one," and the second bit decision will be resolved at the next bit interval.

Filtering of BPSK or QPSK causes the envelope to go to zero immediately at each 180° phase reversal as shown in Fig. 11-20. Bandpass hard

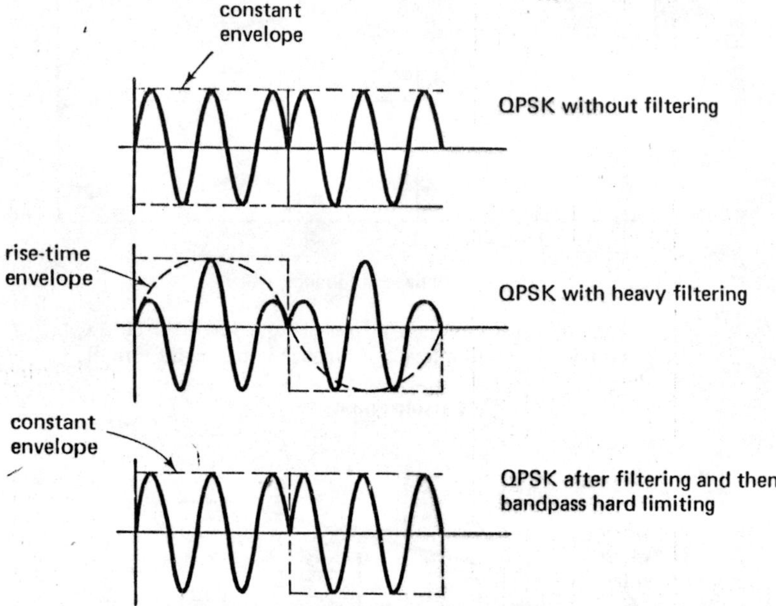

Fig. 11-20 Effects of filtering and hard limiting on BPSK or QPSK signal with 180° phase reversals

limiting of these signals with 180° phase reversals causes constant envelope 180° reversals to recur. Thus, the sidebands that were significantly reduced after the heavy bandpass filtering build up again after limiting. If the zero crossing intervals have not been affected by the filtering, then the bandpass-limited output is the same as the unfiltered input. MSK and SQPSK, on the other hand, have no envelope zeros. Filtering of the SQPSK smoothes the phase transitions but causes envelope modulation (Fig. 11-21). However, the minimum value of the envelope for reasonable filtering is no less than 0.707 of its peak value. Hard-limiting SQPSK thus restores the constant envelope, but it also changes the smoothed phase response shown by the dotted lines in Fig. 11-21.

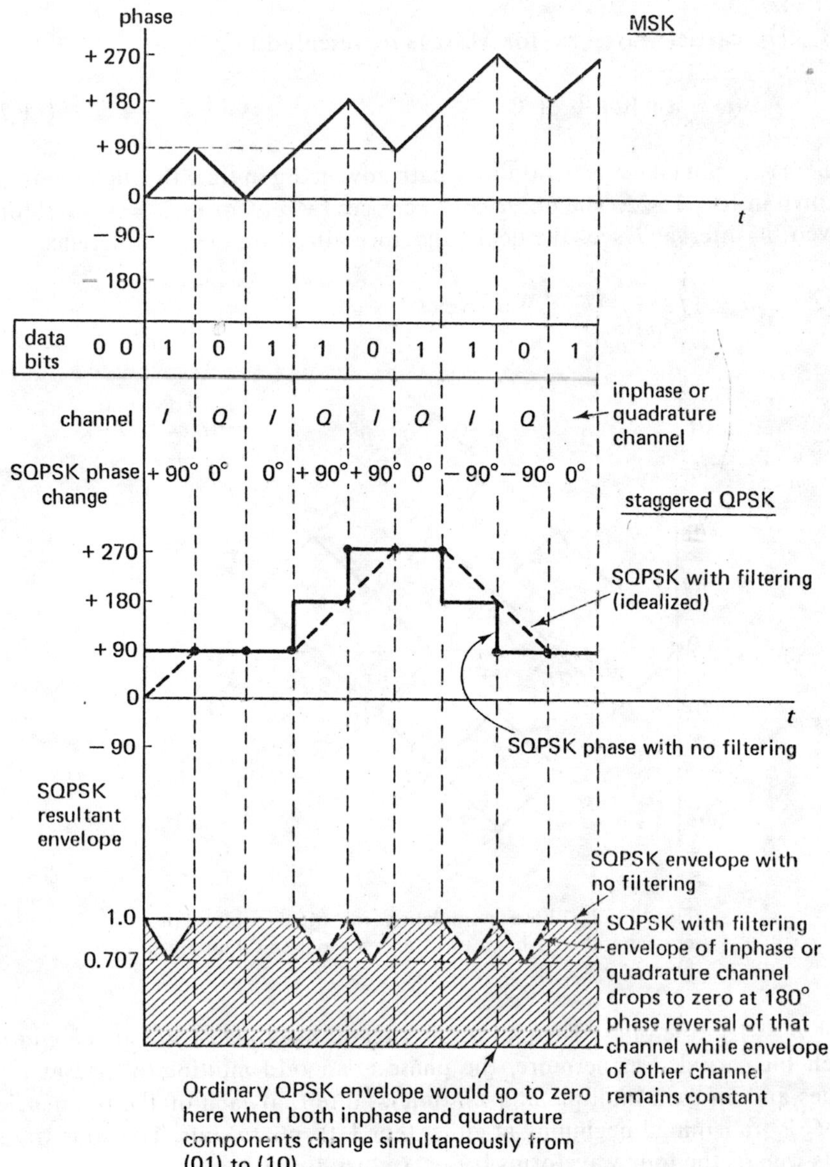

Fig. 11-21 Comparison of phase transitions for minimum shift keying (MSK) and staggered QPSK (SQPSK). Differential encoding is not employed in this example. The SQPSK phase transitions are dependent on the previous phase.

MSK Carrier Recovery

The carrier waveform for MSK is represented by

$$s(t) = \sin\,[\omega t + \phi(t)] \tag{11-28}$$

where $\phi(t)$ can take any continuous path advancing in time through the trellis shown in Fig. 11-22. The phase $\phi(t)$ increases by $\omega_1 T$ or $\omega_2 T = -\omega_1 T$ for a given bit interval T sec. We define the normalized frequency difference

$$h \triangleq (f_1 - f_2)T = 2f_1 T \text{ cycles}$$

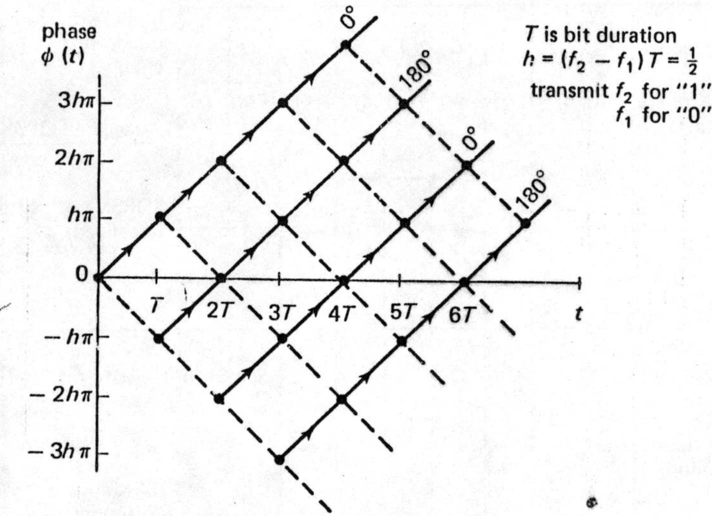

Fig. 11-22 Phase modulation for MSK; solid lines correspond to a binary "1" and the dashed lines for a binary "0".

For $h = 1/2$, the phase is equal to an integral multiple of $\pi/2$ at the end of each bit period. Furthermore, the phase is an odd multiple of $\pi/2$ at odd times and an even multiple at even times. Note that each of the frequencies f_1, f_2 is transmitted beginning at either 0 or 180° phase shift. Thus one transmits one of the four waveforms $[nT < t < (n + 1)T]$

$$\text{Data "1"} \longrightarrow \pm\sin\,(\omega_0 t + 2\pi f_1 t) = \pm\sin\left(\omega_0 t + \frac{\pi t}{2T}\right) \tag{11-29}$$

$$\text{Data "0"} \longrightarrow \pm\sin\,(\omega_0 t + 2\pi f_2 t) = \pm\sin\left(\omega_0 t - \frac{\pi t}{2T}\right) \tag{11-30}$$

Thus one has either one of two biphase modulated carriers with a difference frequency π/T rad/sec equal to half the bit frequency. This difference frequency can be recovered from the transmitted signal and is then used as the bit-synchronization clock.

Carrier recovery and coherent demodulation can be accomplished by the squaring and phase-locked loop operations shown in Fig. 11-23. There is a

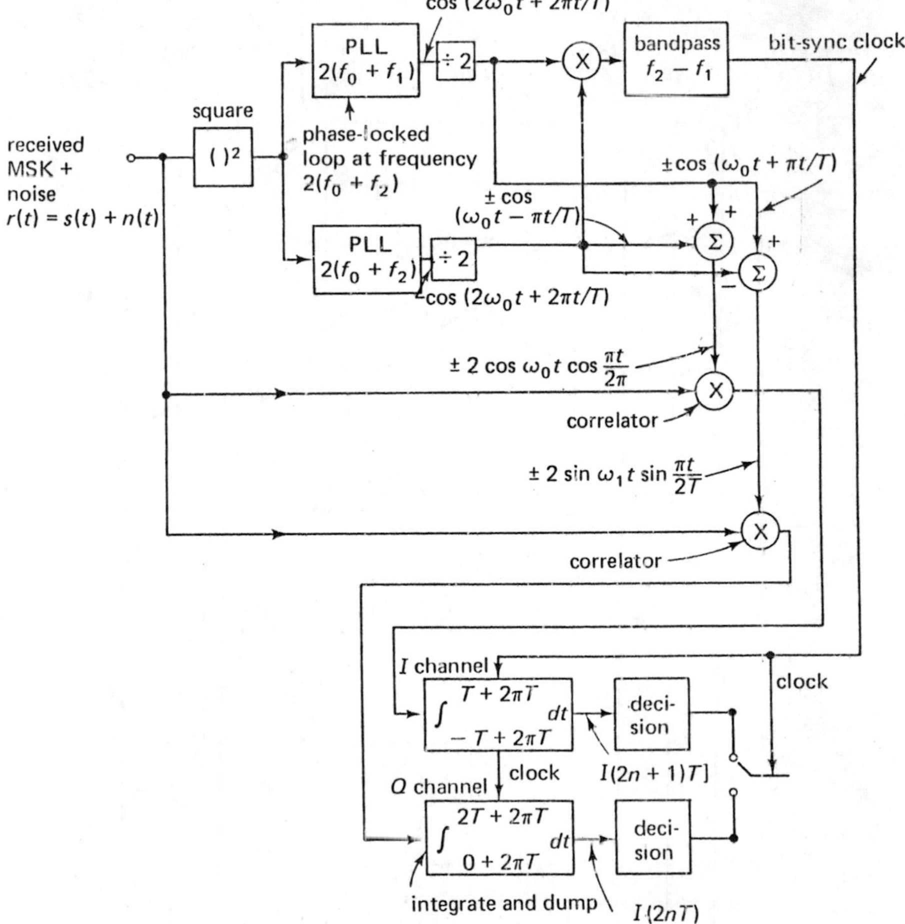

Fig. 11-23 Block diagram of an MSK receiver showing phase-locked recovery of the two BPSK carriers, clock recovery, and bit decisions made by integrating over two consecutive bits using alternate integrate-and-dump circuits. The I and Q channels can be as shown or reversed, depending on the sign ambiguity in the carrier recovery. The frequency difference $f_1 - f_2 = \frac{1}{2}T$ for $h = \frac{1}{2}$.

pure carrier at the output of the squaring device both at frequencies $2(f_0 + f_1)$, and $2(f_0 + f_2)$. Both carriers, each with 50% duty factor, are recovered by separate phase-locked loops. Individual bit decisions are then made on the biorthogonal waveforms by integration over two-bit intervals (rather than over a single bit as in PSK) by the correlation operations:

$$I(2nT) = \int_{-T+2nT}^{T+2nT} \cos \omega_0 t \cos \frac{\pi t}{2T}\, r(t)\, dt \tag{11-31}$$

$$I[(2n + 1)T] = \int_{0+2nT}^{2T+2nT} \sin \omega_0 t \sin \frac{\pi t}{2T}\, r(t)\, dt \tag{11-32}$$

Integrate-and-dump outputs (11-31), (11-32)

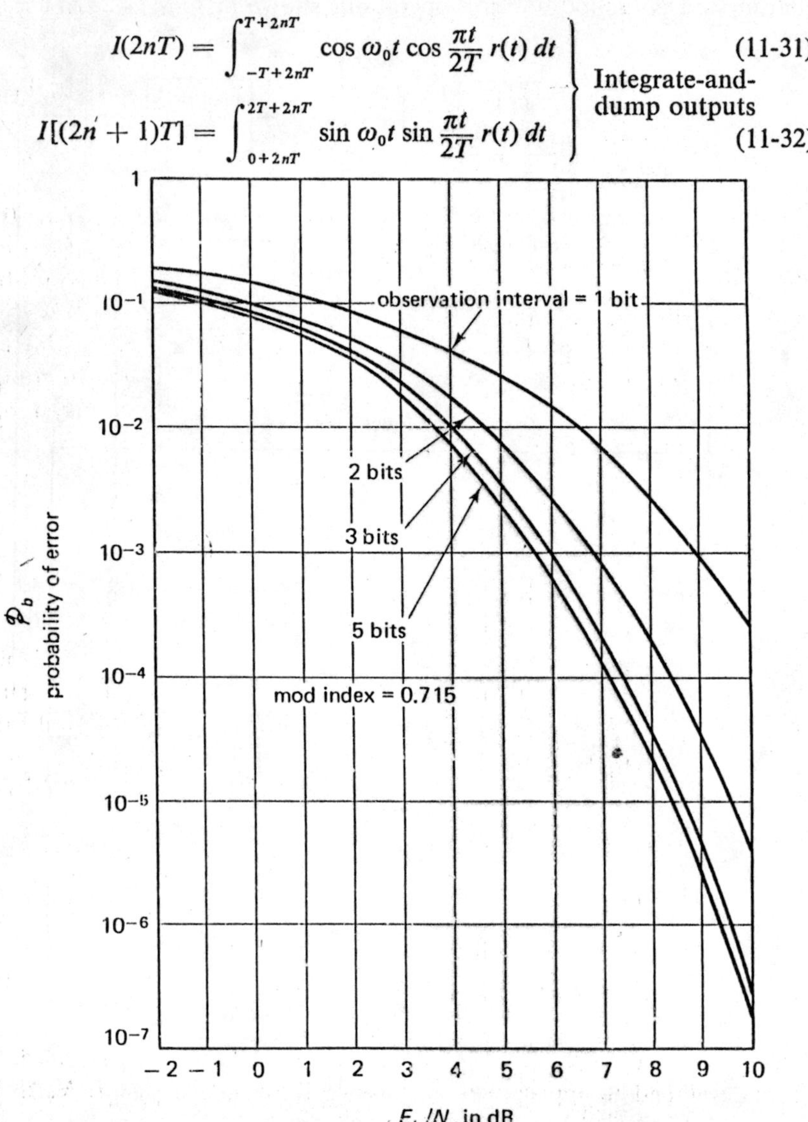

Fig. 11-24 Upper bound on error rate for optimum detection of Continuous Phase Frequency Shift Keying (CPFSK) where single bit decisions are made over an n bit observation interval. The modulation index is 0.715. [From Osborne and Lutz, 1974]

and depend on $I(nT) \gtrless 0$. Because the carriers are recovered with a sign ambiguity, the output bit decisions have the same ambiguities. Differential coding and decoding can be used to removed the sign ambiguity as discussed for QPSK (Sec. 11-3).

MSK or continuous phase frequency shift keying (CPFSK) may be generalized to include other values of h, the normalized frequency separation, and a longer n bit memory before the decision has to be made. If a longer memory is allowed, the optimum detection at high SNR has been shown by Osborne and Luntz [1974], to be a correlation bank where all possible 2^n bit patterns are correlated with the received waveform. The bit pattern that provides the largest correlation then provides the bit decision for the bit received $n - 1$ bits ago.

The optimum value of $h = 0.715$, slightly longer than the $h = 0.5$ just discussed. The upper bound on error rate for various values of the observation interval nT sec is shown in Fig. 11-24. As can be seen, a two-bit observation interval provides equivalent performance to coherent PSK. A three-bit observation interval provides a 0.9 dB improvement over PSK at high SNR. However, the improvement in performance is at some expense in equipment complexity, and the use of forward error correction coding can provide a potentially larger gain in performance as shown in Chap. 15.

11-5 MULTIPLE-PHASE-SHIFT KEYING (MPSK)

Multiple-phase-shift keying has a signaling alphabet consisting of M equally spaced carrier phase shifts in increments of $2\pi/M$ rad BPSK and QPSK, for which the values of $M = 2, 4$, respectively, have been discussed in Secs. 11-2 and 11-3. The next higher level of $M = 2^n = 8$ has phase shifts in increments of 45° or $\pi/4$ rad. As M increases above 2, the symbol rate and hence the signal bandwidth decrease for a given bit rate requirement.

The symbol-error probability for MPSK has been calculated by Cahn [1959, 3–6] to be

$$\mathcal{P}_s = 1 - \frac{1}{2\pi} \int_{-\pi/M}^{+\pi/M} d\theta \left\{ \exp\left(-\frac{\gamma}{2}\right) + \exp\left(-\frac{\gamma}{2}\sin^2\theta\right) \right. \\ \left. \times \sqrt{\gamma}\cos\theta \int_{-\sqrt{\gamma}\cos\theta}^{\infty} \exp\left(-\frac{t^2}{2}\right) dt \right\} \qquad (11\text{-}33)$$

where $\gamma = E/N_0$; E is the symbol energy. Arthurs and Dym [1962, 336–372] have developed an approximate expression for $\mathcal{P}_s$ where $\gamma = E/N_0 \gg 1$:

$$\mathcal{P}_s \simeq \frac{2}{\sqrt{2\pi}} \int_{\sqrt{2\gamma}\sin\pi/M}^{\infty} \exp\left(-\frac{y^2}{2}\right) dy \qquad (11\text{-}34)$$

Bit error probability $\mathcal{P}_b$ is related to symbol-error probability $\mathcal{P}_s$ by $\mathcal{P}_b \cong \mathcal{P}_s/\log_2 M$ for $\mathcal{P}_s \ll 1$. Figure 11-25 shows results for $\mathcal{P}_s$ as a function of $E_b/N_0 = E/nN_0$. The error probability increases versus E_b/N_0 as M exceeds 4. Clearly, the minimum cross-correlation between signals is increasing. As bandwidth limitations become more severe and more signal power is available, however, MPSK offers some real advantages because of its improved bandwidth efficiency. If bandwidth becomes a more severe constraint than

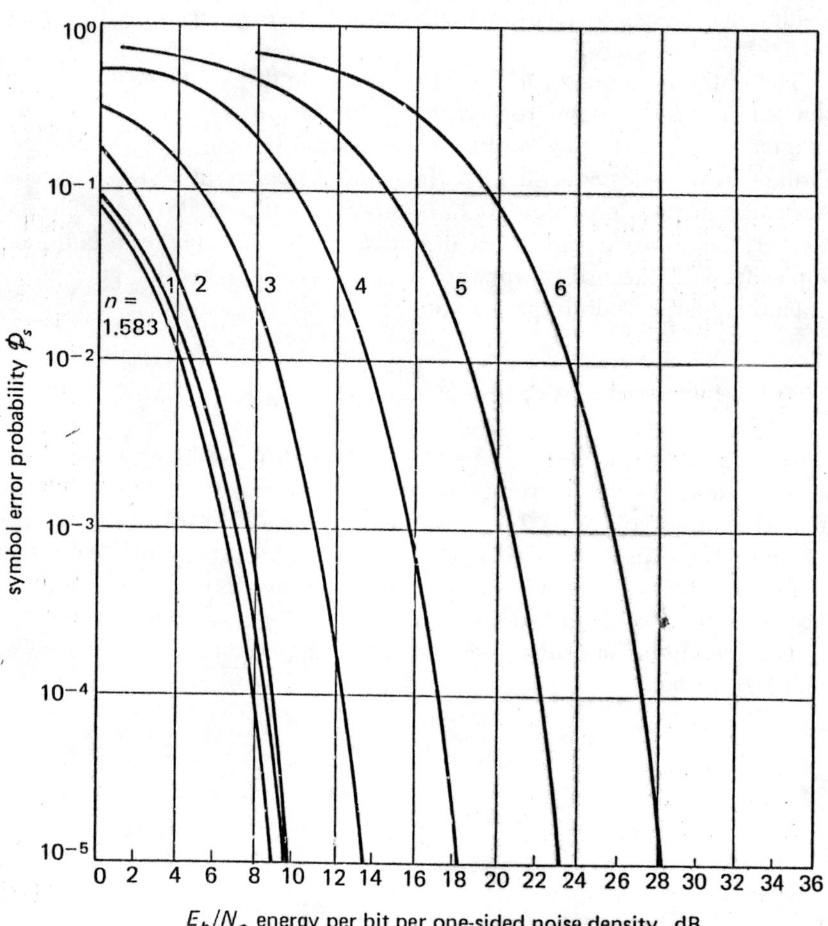

Fig. 11-25 Symbol-error probability $\mathcal{P}_s$ for matched filter detection of MPSK where $2^n = M$

signal power, higher values of $M = 8$, become attractive since the symbol rate and hence the bandwidth decrease as $1/\log_2 M = 1/n$. Furthermore, the bandwidth requirements for a given value of M can be reduced by appropriate signal filtering and equalization as discussed in Chap. 13.

11-6 EFFECTS OF COCHANNEL INTERFERENCE ON PSK SIGNALS

MPSK signals that are transponded by a satellite must contend with some amount of cochannel interference, which is independent of the signal that is desired. The source of this interference could be a cross polarized signal emanated from the same satellite, sidelobe interference from a ground station transmitting to a nearby satellite, an intermodulation product, or a signal from a ground microwave link operating in the same band as the receive earth terminal (Chap. 6).

Here we calculate the degradation in bit error rate caused by small or moderate cochannel interference signals on coherently detected MPSK signals. Results are expressed in terms of the loss in effective signal power or equivalently the increase in input E_b/N_0 required for the same symbol-error rate. Performance curves are based on the work of Prabhu [1969, 743–767]; see also Rosenbaum [1969, 413–442]. Results are given for 2-, 4-, 8-, and 16-phase signals.

The interference considered here is restricted to a single sinusoid. It has been shown that for small error rates, degradation increases monotonically as the number K of interfering signals increases; as $K \longrightarrow \infty$, the effect of interfering noise power approaches that of an equivalent amount of Gaussian noise power, a result already known and discussed in the previous section.

Assume that the MPSK signal has a symbol duration T sec, is received at a power level P_s, and can be represented by

$$S_N(t) = \sqrt{2P_s} \cos(\omega_0 t + \theta), \qquad NT \leq t \leq (N+1)T \tag{11-35}$$

where the phase modulation $\theta = 2\pi k/M$, $k = 1, 2, 3, \ldots, M$; $M = 2^n$, and n is the number of bits/symbol. The interference sinusoid of power P_I has the same frequency as the signal and is represented by

$$I_n(t) = \sqrt{2P_I} \cos(\omega_0 t + \theta + \eta), \qquad NT \leq t \leq (N+1)T \tag{11-36}$$

where η is the random reference phase variable which is uniformly distributed over the interval $(0, 2\pi)$. The total received signal during the Nth time interval is then

$$r_N(t) = S_N(t) + I_n(t) + n(t) \tag{11-37}$$

where $n(t)$ is white Gaussian noise of density N_0 (one sided):

$$n(t) = N_c(t) \cos(\omega_0 t + \theta) + N_s(t) \sin(\omega_0 t + \theta) \tag{11-38}$$

and the inphase and quadrature noise components $N_c(t)$, $N_s(t)$ are also Gaussian with density N_0.

The received signal is passed through inphase and quadrature integrate-and-dump filters. The sampled output gives a phase measurement, which is then compared with the phase decision boundaries for the value of M being used. There is no loss in generality if the carrier phase offset θ is set at 0 for the remaining analysis. The integrator in the detector computes the finite memory integral $(1/T) \int_{NT}^{(N+1)T} dt$. The inphase and quadrature outputs of the integrate-and-dump filters are then

$$X_I(t) = \sqrt{2P_s} + \sqrt{2P_I} \cos \eta + n_c(t) \tag{11-39}$$

$$X_Q(t) = \sqrt{2P_I} \sin \eta + n_s(t) \tag{11-40}$$

where n_c and n_s each are Gaussian of variance $\sigma^2 = N_0/T$.

We define the symbol energy-to-noise-density ratio (signal-to-noise ratio, SNR) as

$$\rho^2 \triangleq \frac{P_s}{\sigma^2} = \frac{P_s T}{N_0} = \frac{E_s}{N_0} \tag{11-41}$$

and the ratio of interference to signal power is $R^2 \triangleq P_I/P_s$.

Figure 11-26 shows the phasor representation of X_I, X_Q and the decision region for an MPSK signal. The minimum distances to the error-decision thresholds A, B are D_A and D_B, respectively:

$$\begin{aligned} D_A &= \sqrt{2P_s} \sin \frac{\pi}{M} + \sqrt{2P_I} \sin\left(\frac{\pi}{M} - \eta\right) \\ &= \sqrt{2}\sigma\rho\left[\sin \frac{\pi}{M} + R \sin\left(\frac{\pi}{M} - \eta\right)\right] \end{aligned} \tag{11-42}$$

$$\begin{aligned} D_B &= \sqrt{2P_s} \sin \frac{\pi}{M} + \sqrt{2P_I} \sin\left(\frac{\pi}{M} + \eta\right) \\ &= \sqrt{2}\sigma\rho\left[\sin \frac{\pi}{M} + R \sin\left(\frac{\pi}{M} + \eta\right)\right] \end{aligned} \tag{11-43}$$

For any given value of the interference phase η (Fig. 11-26), the probability of an error is the sum of the probability of being in decision-error regions A, B minus the probability of being in region C (since regions A, B both

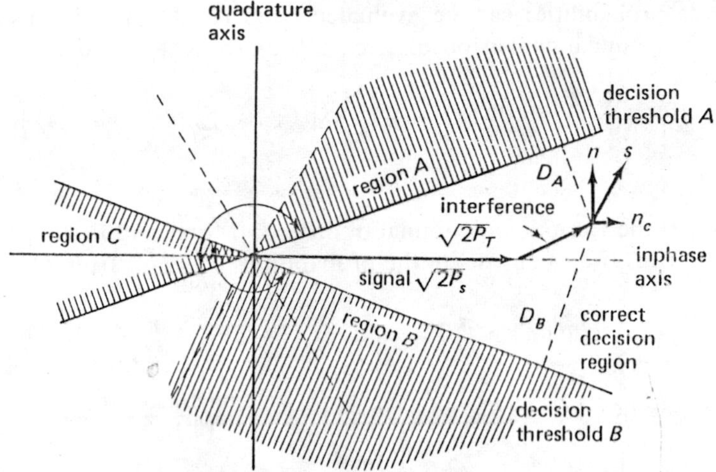

Fig. 11-26 Phasor representation of an MPSK signal with cochannel interference and additive noise

include region C). The probabilities of a noise component causing region A, B errors are π_A, π_B, respectively, for a given value of η where

$$\pi_A(\eta) = \text{Prob}\,[D_A < n_s < \infty] = \frac{1}{2}\,\text{erfc}\,\frac{D_A}{\sigma}$$

$$= \frac{1}{2}\,\text{erfc}\,\rho\left[\sin\frac{\pi}{M} + R\sin\left(\frac{\pi}{M} - \eta\right)\right] \tag{11-44}$$

$$\pi_B(\eta) = \frac{1}{2}\,\text{erfc}\,\rho\left[\sin\frac{\pi}{M} + R\sin\left(\frac{\pi}{M} + \eta\right)\right] \tag{11-45}$$

and where

$$\text{erfc}\,x \triangleq \frac{2}{\sqrt{\pi}}\int_x^\infty e^{-u^2}\,du$$

The probabilities of each of these two decision errors are $\mathcal{P}_A$, $\mathcal{P}_B$, taken as the expectation of π_A, π_B, averaged over all values of the cochannel interference phase η; that is,

$$\mathcal{P}_A = \mathbf{E}[\pi_A] \qquad \mathcal{P}_B = \mathbf{E}[\pi_B] \tag{11-46}$$

The probability of symbol error is then

$$\mathcal{P}_s = \mathcal{P}_A + \mathcal{P}_B - \mathcal{P}_c = 2\mathcal{P}_A - \mathcal{P}_C \tag{11-47}$$

since $\mathcal{P}_A = \mathcal{P}_B$.

These probabilities can be evaluated from (11-44), (11-45) using the Hermite polynomial expansion of erfc $(x + y)$, for $x + y > 0$.

$$\operatorname{erfc}(x + y) = \operatorname{erfc}(x) + \frac{2}{\sqrt{\pi}} \exp(-x^2) \sum_{i=1}^{\infty} (-1)^i \mathcal{H}_{i-1}(x) \frac{y^i}{i!} \tag{11-48}$$

where $\mathcal{H}_k$ is the Hermite polynomial of order k [Morse and Feshbeck, 1953, 786–787]. Thus, we can evaluate the probabilities $\mathcal{P}_A = \mathcal{P}_B$ from (11-46) as

$$\begin{aligned}
\mathcal{P}_A &= \frac{1}{2} \operatorname{erfc}\left(\rho \sin \frac{\pi}{M}\right) + \frac{1}{\sqrt{\pi}} \exp\left(-\rho^2 \sin^2 \frac{\pi}{M}\right) \\
&\quad \times \sum_{i=1}^{\infty} (-1)^i \mathcal{H}_{2i-1}\left(\rho \sin \frac{\pi}{M}\right) \frac{\mathbf{E}[\rho R \sin(\pi/M - \eta)]^{1/2}}{(2i)!} \\
&= \frac{1}{2} \operatorname{erfc}\left(\rho \sin \frac{\pi}{M}\right) + \frac{1}{\sqrt{\pi}} \exp\left(-\rho^2 \sin^2 \frac{2\pi}{M}\right) \\
&\quad \times \sum_{i=1}^{\infty} (-1)^i \mathcal{H}_{2i-1}\left(\rho \sin \frac{\pi}{M}\right) \frac{(\rho R)^{2i}}{2^{2i}(i!)^2}
\end{aligned} \tag{11-49}$$

since $\quad \mathbf{E}\left[\sin^{2i}\left(\frac{\pi}{M} + \eta\right)\right] = \frac{(2i)!}{2^{2i}(i!)^2} \triangleq \mu_{2i}, \ \mu_{2i-1} = 0$

The newly defined quantity μ_{2i} is used below. Hence, the symbol-error probability is

$$\mathcal{P}_s = 2\mathcal{P}_A - \mathcal{P}_C < 2\mathcal{P}_A \tag{11-50}$$

The value of $\mathcal{P}_C$ cannot easily be evaluated for $M > 4$. Only the bound $\mathcal{P}_s < 2\mathcal{P}_A$ is easily obtainable for $M > 4$, but it is quite accurate ($\simeq 5$ percent) for relatively high or moderate signal-to-noise and signal-to-interference ratios. For $M = 2$, $\mathcal{P}_s = \mathcal{P}_A = \mathcal{P}$, and

$$\mathcal{P}|_{M=2} = \frac{1}{2} \operatorname{erfc}(\rho) + \frac{1}{\sqrt{\pi}} \exp(-\rho^2) \sum_{i=1}^{\infty} \mathcal{H}_{2i-1}(\rho) \left(\frac{\rho R}{2}\right)^{2i} \frac{1}{(i!)^2} \tag{11-51}$$

since there is only one decision boundary. For $M = 4$,

$$\begin{aligned}
\mathcal{P}_s &= \operatorname{erfc} \rho_0 - \frac{1}{4} \operatorname{erfc}^2 \rho_0 \\
&\quad + \frac{1}{\sqrt{\pi}} \exp(-\rho_0^2)[2 - \operatorname{erfc} \rho_0] \sum_{i=1}^{\infty} \mathcal{H}_{2i-1}(\rho_0) \frac{\rho^{2i} \mu_{2i}}{(2i)!} \\
&\quad - \frac{1}{\pi} \exp(-2\rho_0^2) \sum_{i=1}^{\infty} \sum_{j=1}^{\infty} \frac{\mathcal{H}_{2i-1}(\rho_0)\mathcal{H}_{2j-1}(\rho_0)}{(2i)!\,(2j)!} \rho^{2(i+j)} \mu_{2i,2j}
\end{aligned} \tag{11-52}$$

where $\rho_0 \triangleq \rho \sin\frac{\pi}{4}, \; \mu_{2i,2j} \triangleq R^{2(i+j)} \frac{(2i)!\,(2j)!}{2^{2(i+j)}(i!)(j!)(i+j)!}$ (11-53)

and $R^2 \triangleq P_I/P_s$ is the interference-to-signal power ratio.

Figure 11-27 shows the degradation in performance caused by cochannel interference for a symbol-error probability $\mathcal{P}_s = 10^{-6}$. The quantity plotted is the required increase in signal power for a noise environment as a function of signal-to-interference ratios. Results are given for $M = 2, 4, 8, 16$. The results are somewhat insensitive to the exact value of $\mathcal{P}_s$ selected. Note that for a P_s/P_I of 20 dB, the degradation ranges from only 0.5 dB, 0.8 dB for $M = 2, 4$, respectively, up to 5.3 dB for $M = 16$. For a degradation of only 0.8 dB, the interference level must decrease to 30 dB below the signal level for $M = 16$ MPSK. Thus the use of $M \geq 16$ introduces some important constraints that limit the usefulness of such a system.

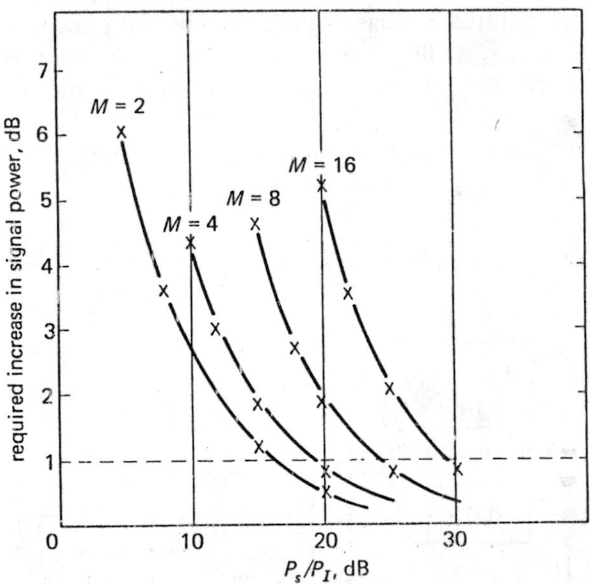

Fig. 11-27 Degradation in MPSK signals caused by cochannel interference (single sinewave interference) for $\mathcal{P}_s = 10^{-6}$; the ratio of signal power to interference power is P_S/P_I.

Low-level interference in the form of a single sinusoid is not nearly as damaging as interference in the form of an equal amount of power in Gaussian noise, or equivalently the interference power equally distributed among an arbitrarily large number of carriers. For example, for QPSK, $M = 4$ and $P_s/P_I = 10$ dB, there is a 4.5 dB loss in signal performance at $\mathcal{P}_s = 10^{-6}$.

For Gaussian noise interference, however, at an SNR of 10 dB, the error rate would be in excess of $\mathcal{P}_e = 10^{-3}$ with no additional additive noise.

11-7 DIFFERENTIAL COHERENT DETECTION OF PSK (DCPSK)

The problems of recovering a reference carrier for coherent detection can be avoided by using differentially coherent detection of PSK or MPSK at the expense of a degradation in error rate. Furthermore, differentially coherent detection avoids the use of coherent carrier recovery loops and can have a smaller acquisition or reacquisition time. There is also a smaller effect caused by carrier phase noise when differentially coherent detection is employed, since one is then mainly concerned with the change in phase over a single bit interval. The transmitted M-phase signal is differentially encoded as described earlier to encode $\log_2 M$ bits in each change in phase. Thus, $N + 1$ MPSK symbols are transmitted to carry N Mary symbols or $N \log_2 M$ bits of information.

Figure 11-28 is a block diagram of the differentially coherent receiver.

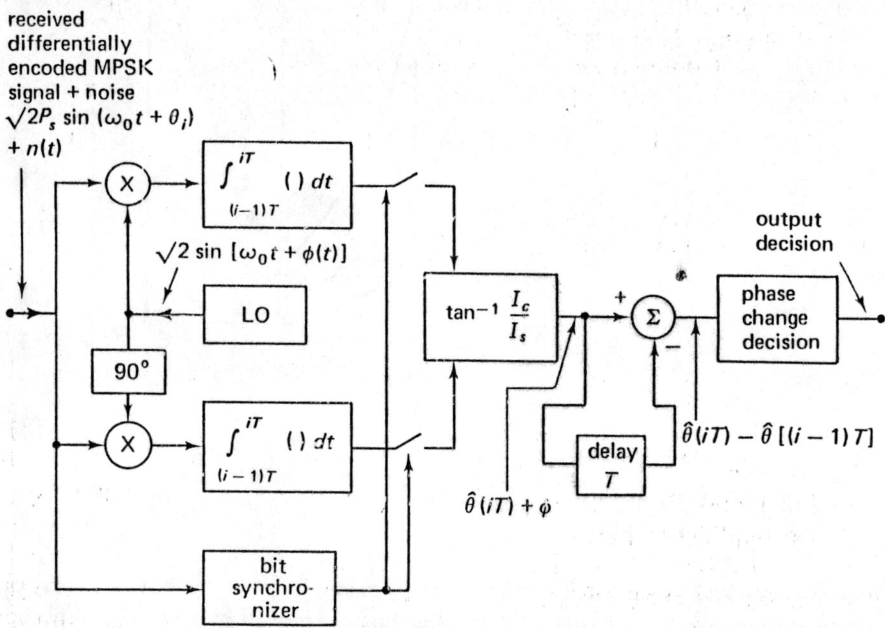

Fig. 11-28 Block diagram of a differentially coherent detector for *M*ary differentially encoded MPSK (MDPSK)

The reference carrier $V_{\mathrm{LO}I}$, $V_{\mathrm{LO}Q}$ is not coherent with the received signal at frequency $\omega = \omega_0$ and is supplied by the local oscillator (LO).

$$V_{\mathrm{LO}I} = \sqrt{2}\sin[\omega_0 t + \phi(t)] \tag{11-54}$$

$$V_{\mathrm{LO}Q} = \sqrt{2}\cos[\omega_0 t + \phi(t)] \tag{11-55}$$

is at approximately the same frequency as the received signal, and $\phi(t)$ represents the phase difference between the LO and the received carrier. Note that the LO is not assumed to be coherent with the input carrier; however, the phase difference of the LO from the received carrier $\phi(t)$ is assumed to be essentially constant over a bit interval. The oscillator phase change over a two-symbol transmission period of $2T$ is assumed to be small, namely, $|\phi(t) - \phi(t - 2T)| \ll 2\pi/M$ rad. Thus the LO is assumed to be at approximately the same frequency as the received carrier. In practice, this requirement is often satisfied by the use of automatic frequency control (AFC) around the LO.

In particular, the outputs of the two phase detectors are integrated and sampled to produce $I_c(iT)$ and $I_s(iT)$. Sampling is controlled by means of the bit (symbol) synchronizer.

A $\tan^{-1} x$ computation provides the estimate $\hat{\theta}(iT) + \phi(iT)$. The phase estimate difference then provides $\hat{\theta}[iT] - \hat{\theta}[(i-1)T]$, which is compared with $k2\pi/M$ modulo 2π to provide the symbol decision output. The output-symbol-error probability has been computed to be [Lindsey and Simon, 1972]

$$\mathcal{P}_s(M) = 1 - \int_{-\pi/M}^{\pi/M} P_\psi(\psi)\,d\psi = 2\int_{\pi/M}^{\pi} P_\psi(\psi)\,d\psi \tag{11-56}$$

The phase-error difference $\psi(iT) = \hat{\theta}(iT) - \hat{\theta}[(i-1)T] - \{\theta(iT) - \theta[(i-1)T]\}$ has a probability density

$$P(\psi) = \frac{1}{2\pi}\int_0^{2\pi} \sin\alpha[1 + R_d(1 + \cos\psi\sin\alpha)]\exp[-R_d(1 - \cos\psi\sin\alpha)]\,d\alpha \tag{11-57}$$

where $R_d = P_s T/N_0$, and the energy/bit/noise-density is $E_b/N_0 = P_s T/(N_0 \log_2 M)$ and N_0 is the one-sided noise density.

For binary BPSK, this error probability is the bit-error probability and (11-56) simplifies to [Arthurs and Dym, 1962, 336–372]:

$$\mathcal{P}_b = \tfrac{1}{2}\exp(-E_b/N_0) = \mathcal{P}_s \qquad \text{for } M = 2 \tag{11-58}$$

Plots of the symbol-error probability for differential coherent detection and coherent detection of MPSK are shown in Fig. 11-29. The performance of

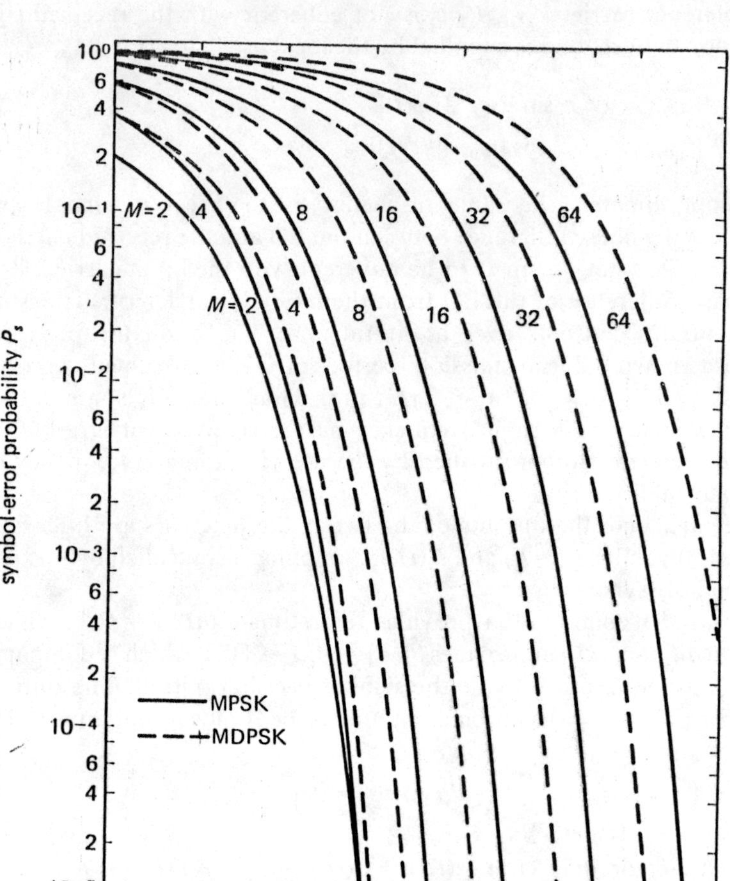

Fig. 11-29 Comparison of coherent detection of MPSK (solid curve) with differentially coherent detection of MDPSK (dashed curve) for various values of M; signal phase shifts are in increments of $2\pi/M$ rad.

the differentially coherent BPSK detector is within $\simeq 0.5$ dB of the coherent detector for $E_b/N_0 = 10$ dB and approaches a 0-dB degradation at large values of E_b/N_0. The degradation increases substantially as $M \geq 4$, however. For $M = 4$, the degradation at $E_b/N_0 = 10$ dB is 2.5 dB, and it is 3 dB for $M > 4$. Thus, this circuit simplification is obtained at the expense of degraded performance. However, the advantages of differentially coherent detection for BPSK can be substantial, particularly where oscillator phase noise is large or where there can be a momentary loss of signal.

Differentially coherent detection has a more serious degradation if error correction is employed. The degradation is more serious at low values of $E_b/N_0 \cong 5$ dB, and error pairs or bursts are more likely because of the differential detection, unless appropriate interleaving of bits is provided to reduce this correlated error effect.

Higher order MPSK with differentially coherent detection also is more sensitive to cochannel interference. For example, Rosenbaum [1969, 413–442]

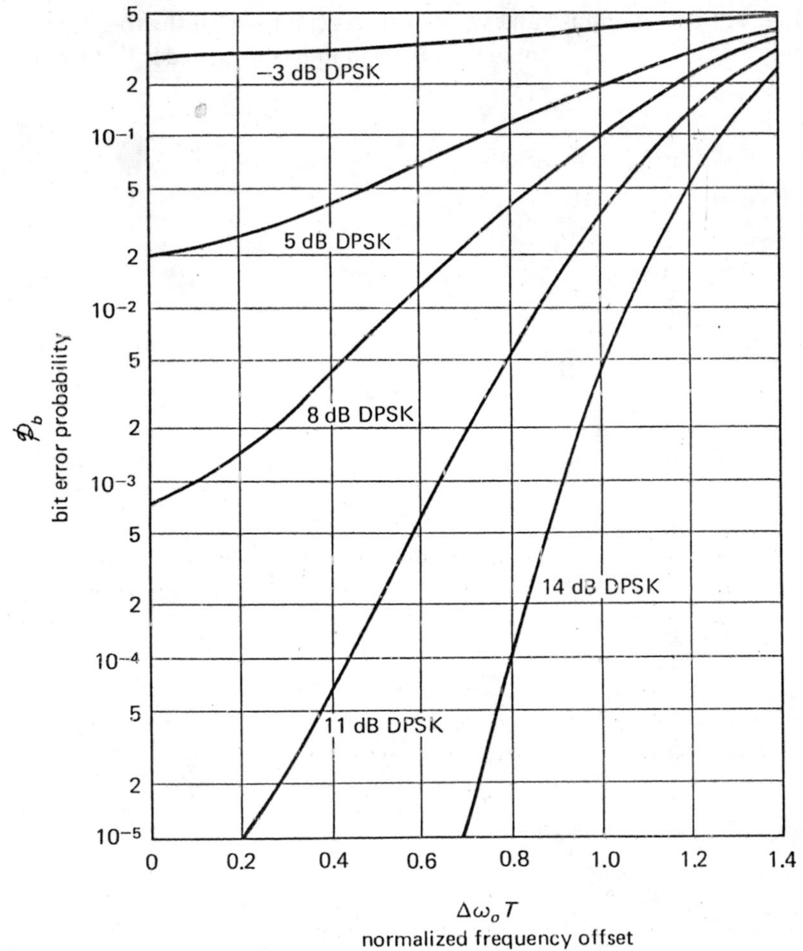

Fig. 11-30 Bit error probability for differentially coherent detection of PSK(DCPSK) for various values of frequency offset $\Delta\omega_0 T$ normalized to bit duration T sec [From Henry, 1970]

has shown that at $E_b/N_0 = 12$ dB, a carrier-to-interference (C/I) ratio of 10 dB produces approximately 2.6- to 3-dB degradation in BPSK, depending on the relative phase of the interference, whereas it can be shown that the same C/I and E_b/N_0 produce a 5.4-dB degradation in QPSK. If the C/I increases to 13 dB for QPSK, consistent with the two bits/symbol, the degradation decreases to 3.5 dB, but is still larger than that for BPSK.

The effects of doppler frequency offset effects on DCPSK have been studied by Henry [1970]. For a bit duration T sec and a frequency offset $\Delta\omega T$ radians/sec, the curves of bit error probability $\mathcal{P}_b$ versus $\Delta\omega T$ are given in Fig. 11-30 for various values of E_b/N_0. As can be seen, the frequency offset should be $\Delta\omega T < 0.2$, if small degradation in performance is to be obtained.

CHAPTER 12

CARRIER-PHASE TRACKING AND OSCILLATOR-PHASE NOISE

12-1 INTRODUCTION

Coherent demodulation techniques and the associated carrier-tracking loops were described in block diagram form in Chap. 11. Carrier-tracking accuracy, the subject of this chapter, depends on several system parameters, including the phase noise in the carrier induced by various oscillator short-term stabilities, carrier-frequency drifts, carrier-tracking-loop dynamics, transient response and acquisition-performance requirements, and signal-to-noise ratios in the carrier-tracking loop.

The unmodulated carrier as received in the earth terminal has phase noise and frequency drifts induced by each of the oscillators and frequency converters in the communications link. Here we develop a model of the received carrier-phase noise and frequency drifts. This model can be described both in terms of the phase noise power spectral density in the frequency domain and a measure of frequency variance versus averaging time in the time domain (long- and short-term stability). These stability measures can then be related to the required bandwidth of carrier-tracking loops. Section 17-7 on clock timing errors supplements this discussion by concentrating on long-term clock stability measures and reviews the performance of several types of atomic frequency standards.

Each carrier-tracking loop requires the use of some form of phase-locked-oscillator tracking loop. In addition to tracking the oscillator phase noise, the loop must also acquire the carrier in a reasonable acquisition time and operate over the required range of oscillator-frequency drifts.

Once the loop bandwidth has been constrained by the above signal statistics, the effects of received noise are then computed. The phase-locked-loop analysis begins with the linear and quasi-linear noise analyses and concludes with a discussion of the nonlinear analysis of the loop and the associated cycle slipping phenomena.

Finally, the implications of these results for the performance of the PSK or QPSK detectors themselves are considered in the practical situation where the detection is only partially coherent because the various loop noise effects cause imperfect carrier tracking. These carrier reference phase noise statistics are then related to the loss in effective signal power. Section 15-5 discusses additional constraints placed by coded data links on phase noise in carrier tracking loops if full performance is to be achieved.

12-2 FREQUENCY STABILITY AND OSCILLATOR NOISE

To determine the required PLL bandwidth for carrier recovery, one must first examine the input carrier-phase statistics. The oscillator phase statistics are modeled and related to commonly measured frequency stability measures. We assume that the recovered carrier has the form

$$s(t) = \sqrt{2}A[1 + a(t)]\sin\left[\omega_0 t + \phi(t) + \frac{dt^2}{2}\right] \tag{12-1}$$

where d represents the aging effect (long-term drift) of an oscillator, and $\phi(t)$ represents the phase jitter [Barnes, 1966, 207–220]. The amplitude noise $a(t)$ is assumed to be negligible in this discussion. Several oscillators in the satellite link may have contributed to this phase noise. As discussed in Part 2, the local oscillators or frequency synthesizers in the earth terminal upconverter, the satellite frequency translator, and the receive earth terminal downconverter all can contribute to the phase noise.

Frequency Stability Models

The phase noise $\phi(t)$ is a superposition of causally generated signals and nondeterministic random noise. The causally generated effects are created by changes in the oscillator temperature, supply voltage or power line fluctuations, magnetic field, humidity, physical vibration, or output load impedance. Our interest focuses on the random noise effects in $\phi(t)$, however, and we largely ignore the causal effects except for drifts.

We wish to obtain a description of the oscillator phase noise in both the frequency domain and the time domain. A frequency domain description in terms of power spectral density is most useful when determining the required carrier-tracking loop bandwidth. A time domain description is needed when determining the accuracy of clocks from a clock timing error standpoint or when using the measurement of fractional frequency stability. We use both frequency and time domain representations in this chapter and also show the transformation between the frequency and time domain descriptions. Chapter 17 on Satellite Timing Concepts concentrates more fully on the time domain representation.

The power spectral density representation of phase noise relies on an assumption of a wide sense stationary model for the process that produced the collected oscillator phase noise statistics. However, one need only assume that the measured data are consistent with a stationary model over the time interval of interest and not that the oscillator physical noise sources are all completely stationary. Very few sources in nature are perfectly stationary.

A spectral model that has been found very useful [Barnes et al., 1970] represents the phase-noise power spectral density (one-sided) as:

$$G_\phi(f) = \underbrace{k_{-4}\frac{f_0^2}{f^4}}_{\substack{\text{random}\\ \text{frequency}\\ \text{walk}}} + \underbrace{k_{-3}\frac{f_0^2}{f^3}}_{\substack{\text{flicker}\\ \text{frequency}\\ \text{noise}}} + \underbrace{k_{-2}\frac{f_0^2}{f^2}}_{\substack{\text{random}\\ \text{phase}\\ \text{walk or}\\ \text{white}\\ \text{frequency}\\ \text{noise}}} + \underbrace{k_{-1}\frac{f_0^2}{f}}_{\substack{\text{flicker}\\ \text{phase}\\ \text{noise}}} + \underbrace{k_0 f_0^2}_{\substack{\text{white}\\ \text{phase}\\ \text{noise}}}, \quad f_\ell \le f \le f_h \tag{12-2}$$

The phase-noise spectral density is zero outside of this frequency range. The lower frequency cutoff f_ℓ may be selected as a frequency below the fundamental frequency of the longest observation period. Notice that the low-frequency cutoff is necessary if $G_\phi(t)$ is to have finite power. However, much of the following discussion is meaningful in the limiting case as $f_\ell \to 0$, and indeed some noise spectral effects have been found at as low a frequency as 1 cycle/year. Figure 12-1 illustrates a typical measured power spectral density for phase noise that exhibits the general characteristics of (12-2).

Figure 12-2 illustrates a typical phase noise characteristic in the time domain. A measure of the fractional frequency stability can be defined as

$$\delta_j \triangleq \frac{1}{T}\frac{\phi[(j+1)T] - \phi(jT)}{\omega_0} \tag{12-3}$$

where j is an integer and ω_0 is the nominal center frequency. We note that conventional frequency counters measure $\{\phi[(j+1)T] - \phi(jT)\}/2\pi$ in a

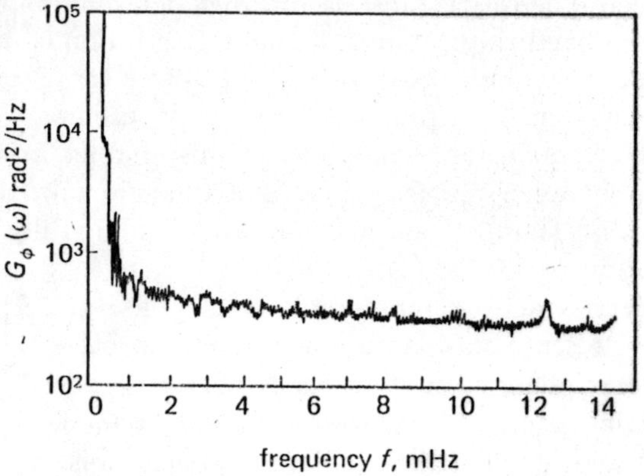

Fig. 12-1 Typical measured oscillator phase-noise power spectral density

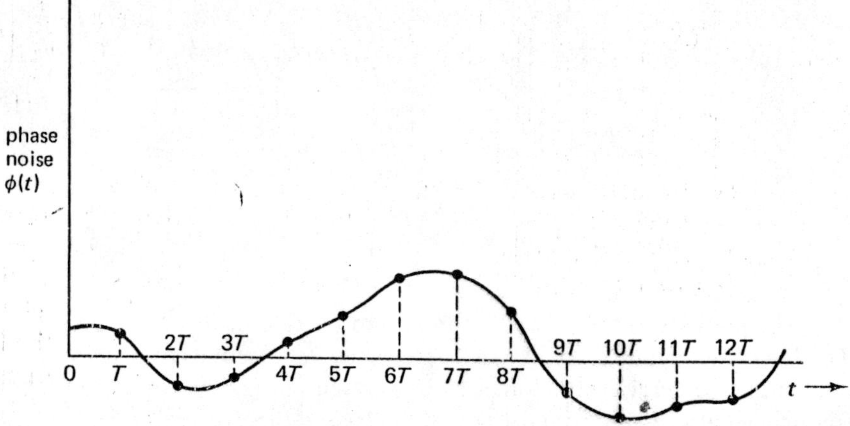

Fig. 12-2 Typical phase noise characteristic in the time domain with samples taken every T seconds

period T sec. Note that δ_j is dimensionless. The Allan variance can be defined as an infinite average of NT sec segment variances [Allan, 1966]

$$\begin{aligned}\langle\sigma^2(N, T)\rangle &= \left\langle\frac{1}{N-1}\sum_{n=1}^{N}(\delta_n - \delta_{\text{mean}})^2\right\rangle \\ &= \left\langle\frac{1}{N-1}\left\{\sum_{n=1}^{N}\delta_n^2 - N\delta_{\text{mean}}^2\right\}\right\rangle \qquad (12\text{-}4)\end{aligned}$$

where $\delta_{\text{mean}} = (1/N)\sum_{j=1}^{N}\delta_j$ is the mean fractional frequency stability measure, and $\langle\sigma^2\rangle$ represents the infinite time average of σ^2.

For $N = 2$, the Allan variance simplifies to:

$$\sigma^2(T) \triangleq \langle\sigma^2(2, T)\rangle = \left\langle\frac{[\phi(t+2T) - 2\phi(t+T) + \phi(t)]^2}{2T^2\omega_0^2}\right\rangle$$
$$= \lim_{m\to\infty} \frac{1}{2m} \sum_{j=1}^{m} \frac{\{\phi[(2j+2)T] - 2\phi[(2j+1)T] + \phi(2jT)\}^2}{T^2\omega_0^2} \tag{12-5}$$

where the latter equation represents the mean over m time segments.

It is easily seen that this Allan variance is zero if $\phi(t)$ is varying linearly with time since the sample mean δ_{mean} and the values of fractional frequency δ_j are all the same in (12-4).

The Allan variance measured in the time domain by frequency counter type measurements can be related to the power spectral density of the phase noise by the relationship [Barnes et al., 1970]

$$\langle\sigma^2(N, T)\rangle = \frac{N}{N-1}\int_0^\infty \omega^2 G_\phi(f)\,\text{sinc}^2(\pi f T)\left[1 - \frac{\sin^2 \pi f T N}{N^2 \sin^2 \pi f T}\right] df$$

where $\quad \text{sinc}\, x \triangleq \dfrac{\sin x}{x}$ (12-6)

Table 12-1 gives the relationship between each of the phase noise contributors and the Allan variance. Notice that we have let $f_\ell \to 0$ in Table 12-1, and yet the Allan variance still exists for finite T and N. Thus, given the power spectral density, there is a unique transformation to the time domain stability measure. The reverse is not true, however. Given a plot of the Allan variance vs T, there is no general simple translation to the frequency domain. Thus the use of $G_\phi(f)$ might be preferred for some purposes.

Measurement of Phase-Noise Spectral Density

The phase-noise spectral density can be measured by using a technique similar to that of Fig. 12-3. The oscillator to be tested is connected to a phase detector where a reference oscillator is used to track low-frequency phase-noise fluctuations. The phase-locked loop effectively removes all frequency components less than some very low frequency f_ℓ. The remaining phase-noise term $\sin \phi(t)$ then enters a low-pass filter of bandwidth f_h where f_h is above the frequency range of the modulation and half-bandwidth of the IF.

If residual phase noise is small, that is, $|\phi| \ll 1$, then the quantity fed to the low-pass filter and spectrum analyzer is approximately equal to $\phi(t)$.

Table 12-1 ALLAN VARIANCE ($N = 2$) FOR VARIOUS TYPES OF PHASE-NOISE POWER SPECTRAL DENSITIES

Noise Type	Phase Spectral Density $G_\phi(f)$ $0 \le f \le f_h$	Allan Variance $\sigma^2(T) = \left\langle \frac{[\phi(t+2T) - 2\phi(t+T) + \phi(t)]^2}{2T^2\omega_0^2} \right\rangle$	$\sigma_y^2(N,T) = \left\langle \frac{1}{N-1} \sum_{n=1}^{N} \left\{ \frac{\phi[(n+1)T] - \phi(nT)}{\omega_0 T} - \frac{1}{N} \sum_{k=1}^{N} \frac{\phi[(k+1)T] - \phi(kT)}{\omega_0 T} \right\}^2 \right\rangle$
Random Frequency Walk	$\frac{k_{-4}}{(2\pi)^2 f^4}$	$\frac{k_{-4}}{\omega_0^2} \frac{(2\pi)^2 T}{6}$	$\frac{k_{-4}}{\omega_0^2} \frac{(2\pi)^2 TN}{12}$
Flicker Frequency Noise	$\frac{k_{-3}}{(2\pi)^2 f^3}$	$k_{-3}\left(\frac{2 \log 2}{\omega_0^2}\right)$	$\frac{k_{-3}}{\omega_0^2}\left(\frac{N \log N}{N-1}\right)$
White Frequency Noise (Random Walk Phase Noise)	$\frac{k_{-2}}{(2\pi)^2 f^2}$	$\frac{k_{-2}}{2\omega_0^2 T}$	$\frac{k_{-2}}{2\omega_0^2 T}$
Flicker Phase Noise	$\frac{k_{-1}}{(2\pi)^2 f}$	$\frac{k_{-1}}{T^2(2\pi)^2\omega_0^2}\{3[2 + \log(2\pi f_k T)] - \log 2\}$	$\frac{k_{-1}}{\omega_0^2} \frac{2(N+1)}{NT^2(2\pi)^2}\left[2 + \log(2\pi f_h T) - \frac{\log N}{N^2 - 1}\right]$
White Phase Noise	$\frac{k_0}{(2\pi)^2}$	$\frac{k_0}{\omega_0^2} \frac{3 f_h}{(2\pi)^2 T^2}$	$\frac{k_0}{\omega_0^2} \frac{N+1}{N(2\pi)^2} \frac{2 f_h}{T^2}$

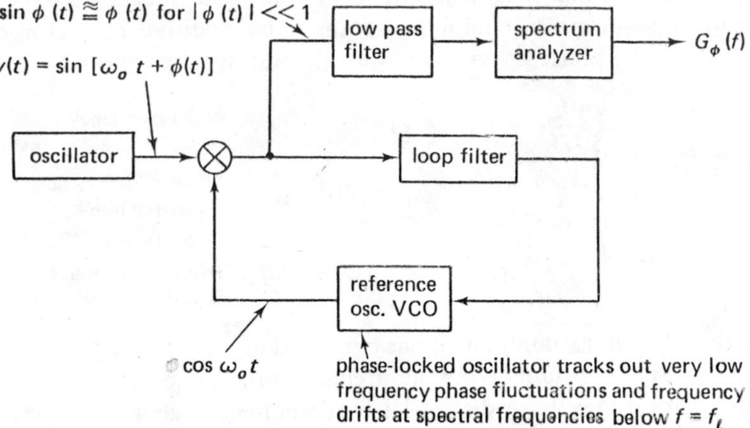

Fig. 12-3 Measurement of phase-noise power spectral density

Counted Frequency Measurements

The spectral density coefficients must be related to measurable parameters of the oscillator or commonly specified characteristics, such as short-term fractional frequency stability [Cutler and Searle, 1966, 136–154; Gray and Tausworth, 1971, 21–30]. A frequency counter measures the average frequency $\hat{\omega}$ in T sec by counting the phase cycle increases. If roundoff errors and the drift term $dt^2/2$ are neglected, the $\hat{f}$ becomes

$$\hat{f} = \frac{\hat{\omega}}{2\pi} = \frac{\omega_0 T + \phi(T) - \phi(0)}{2\pi T} \text{ Hz} \tag{12-7}$$

The variance of this frequency measurement depends on the autocorrelation $R(\tau)$ of the phase jitter $\phi(t)$:

$$\begin{aligned} \sigma_T^2 &= \frac{1}{T^2} \mathrm{E} \frac{[\phi(T) - \phi(0)]^2}{(2\pi)^2} = \frac{2}{T^2} \frac{[R(0) - R(T)]}{(2\pi)^2} \qquad (\text{rad/sec})^2 \\ &= \frac{2}{T^2} \int_0^\infty G_\phi(\omega) \frac{(1 - \cos\omega T)}{(2\pi)^2}\, df = \frac{2}{2\pi} \int_0^\infty d\omega G_\phi(\omega) \frac{(1 - \cos\omega T)}{(2\pi)^2 T^2} \end{aligned} \tag{12-8}$$

where $G_\phi(\omega)$ is the power spectral density of $\phi(t)$. This measurement of average frequency can have infinite variance unless the phase-noise spectrum increases less rapidly than $1/\omega^3$ as $\omega \longrightarrow 0$.

Simplified Phase-Noise Models

In the next parts of this section we utilize simplified noise models.

It is assumed that the phase jitter is the sum of a "flicker-noise" component and a white noise component (caused by additive oscillator output noise). Thus, the phase noise has power spectral density (one-sided)

$$G_\phi(f) = \begin{cases} \dfrac{k_a}{f_\ell f^2} + N_0 & \omega < \omega_\ell \quad \text{white frequency noise + white phase noise} \\ \dfrac{k_a}{f^3} + N_0 & \omega_\ell < |\omega| < \eta \quad \text{“flicker” frequency noise + white phase noise} \\ N_0 & \eta < |\omega| < 2\pi f_h \quad \text{white phase noise (band limited)} \end{cases} \tag{12-9}$$

where the white noise component has one-sided density N_0, two-sided density $N_0/2$, and k_a is a constant giving the strength of the flicker noise. In addition, the phase noise may have discrete line components (incidental FM) at the power line frequency and its harmonics.

Frequency Flicker-Noise Contribution

The contributions of the white-noise and flicker-noise terms are additive and can be evaluated separately. The flicker-noise spectrum $G_{\phi F}(f) = k_a/f^3$ for $f_\ell < |f| < \eta$, and $G_{\phi F}(f) = k_a/f_\ell f^2$ for $|f| < f_\ell$ in (12-9) produce a frequency-noise variance of the RF carrier from (12-8)

$$\sigma_{TF}^2 \triangleq 2\left[\int_0^{\omega_\ell} d\omega \frac{k_a}{\omega_\ell \omega^2} \frac{(1-\cos\omega T)}{T^2} + \int_{\omega_\ell}^{\eta} d\omega \frac{k_a}{\omega^3} \frac{(1-\cos\omega T)}{T^2}\right] \tag{12-10}$$

where $\omega_\ell \triangleq 2\pi f_\ell$. By setting $x \triangleq \omega T$, we obtain a frequency-noise variance

$$\sigma_{TF}^2 = 2k_a\left[\int_0^{\omega_\ell T} \frac{(1-\cos x)}{\omega_\ell T x^2}\,dx + \int_{\omega_\ell T}^{\eta T} \frac{(1-\cos x)}{x^3}\,dx\right] \tag{12-11}$$

By integrating by parts and defining $x_\ell \triangleq \omega_\ell T$ and $\eta T = x_\eta$, we find

$$\begin{aligned} \sigma_{TF}^2 &\cong 2k_a + 2k_a\left[\frac{-(1-\cos x)}{x^2}\bigg|_{x_\ell}^{x_\eta} + \int_{x_\ell}^{x_\eta} \frac{\sin x}{x^2}\,dx\right] \qquad x_\ell \ll 1 \\ &\cong 2k_a + 2k_a\Bigg[\underbrace{\frac{1-\cos x_\ell}{(x_\ell)^2}}_{1/2} - \underbrace{\frac{(1-\cos x_\eta)}{(x_\eta)^2}}_{0} - \underbrace{\frac{\sin x}{x}\bigg|_{x_\ell}^{x_\eta}}_{-1} \\ &\quad + \int_{x_\ell}^{x_\eta} \frac{\cos x}{x}\,dx\Bigg] \qquad x_\eta \gg 1 \end{aligned} \tag{12-12}$$

$$\cong 2k_a\left[\frac{5}{2} + \int_{x_\ell}^{\infty} \frac{\cos x}{x}\,dx\right] = k_a\left[\frac{5}{2} - Ci(x_\ell)\right] \tag{12-13}$$

where $Ci(x)$ is the cosine integral

$$Ci(x) \triangleq \int_{-\infty}^{x} \frac{\cos t}{t}\, dt = \gamma + \log x + \sum_{n=1}^{\infty} \frac{(-1)^n x^{2n}}{2n(2n)!} \tag{12-14}$$

where $\gamma = 0.577216$ and $\log x$ is the natural logarithm of x. Hence, this frequency variance is

$$\sigma_{TF}^2 \cong 2k_a\left[\frac{5}{2} - \gamma + \log\frac{1}{x_\ell}\right] = k_a\left(1.92 + \log\frac{1}{x_\ell}\right) \tag{12-15}$$

Note that σ_{TF}^2 increases slowly as the value of the cutoff frequency ω_ℓ decreases. As an example, suppose $\omega_\ell = 10^{-4}$ r/s, $T = 10^{-1}$ sec, and $x_\ell = 10^{-5}$; then

$$\sigma_{TF}^2 \cong 2k_a[1.92 + \underbrace{5 \log 10}_{11.5}] = 26.86k_a \tag{12-16}$$

Table 12-2 summarizes the effects of phase noise on the frequency variance of (12-8). Notice that the $\sigma_{TF}^2 \longrightarrow \infty$ in this table as $\omega_\ell \longrightarrow 0$, while the Allan variance of Table 12-1 is well behaved as $\omega_\ell \longrightarrow 0$.

Table 12-2 VARIANCE OF FREQUENCY COUNTER MEASUREMENTS

The quantity $\gamma = 0.577216$ and sinc $x \triangleq (\sin x)/x$

Type of Phase Noise	*Phase-Noise Spectral Density* (one-sided)	*Frequency Variance* $\sigma_T^2 \triangleq \frac{1}{T} \mathrm{E} \frac{[\phi(T) - \phi(0)]^2}{(2\pi)^2}$ Hz^2 $= \frac{1}{\pi}\int_0^\infty d\omega\, G_\phi(\omega) \frac{1 - \cos \omega T}{(2\pi)^2 T^2}$ Hz^2
Flicker frequency noise	$G_\phi(f) = \frac{k_a}{f^3}, f_\ell < f$	$2k_a\left[\frac{5}{2} - \gamma + \log \frac{1}{\omega_\ell T}\right]$
White frequency noise	$G_\phi(f) = \frac{k_b}{f^2}$	$\frac{k_b}{2T}$
White phase noise	$G_\phi(f) = N_0, f_\ell < f_h$	$\frac{2f_h N_0}{(2\pi T)^2}(1 - \operatorname{sinc} 2\pi f_h T)$

If $T = 1$ sec, this expression changes only slightly to

$$\sigma_{TF}^2 = 31.48k_a \tag{12-17}$$

The short-term frequency stability of an oscillator measured in $T = 0.1$ sec might be measured as

$$\frac{\Delta f}{f_0} \triangleq \delta = 10^{-11} \tag{12-18}$$

and at $f_0 = 10^{10}$ Hz, the differential frequency is $\Delta f = f_0\delta = 10^{-1}$ Hz. Assume as a special example that frequency flicker noise for this oscillator is the dominant effect. By setting $\Delta f = \sigma_{TF}$, then one can compute the stability coefficient k_a for this center frequency f_0 based on the measured oscillator-frequency stability; thus for this example,

$$k_a = 28.86(\delta f_0)^2 = 3.72 \times 10^{-3} \qquad \delta f_0 = 10^{-1}\text{ Hz},\ x_\ell = 10^{-5} \tag{12-19}$$

Effect of White Frequency Noise on Frequency Noise Variance

Consider an oscillator-phase noise spectral density that is created by white frequency noise. The phase-noise spectral density is then represented by the one-sided density

$$G_{\phi b}(f) = \frac{k_b}{f^2} \qquad f > 0 \tag{12-20}$$

Using (12-8) the frequency-noise variance for an averaging time T sec is then

$$\begin{aligned}\sigma_{Tb}^2(T) &= \frac{2}{T^2}\int_0^\infty G_{\phi b}(f)\frac{(1-\cos\omega T)}{(2\pi)^2}\,df \\ &= \frac{2}{T^2}\int_0^\infty \frac{G_{\phi b}(f)}{(2\pi)^2}2\sin^2\frac{\omega T}{2}\,df\end{aligned} \tag{12-21}$$

since $1 - \cos x \equiv 2\sin^2(x/2)$. Insert the function $G_{\phi b}(f)$ from Eq. (12-20) to obtain

$$\begin{aligned}\sigma_{Tb}^2(T) &= 4\,k_b\int_0^\infty \frac{1}{2}\,\frac{2\sin^2(\omega T/2)}{(\omega T/2)^2}\,\frac{d\omega}{2\pi} \\ &= \frac{k_b}{2\pi}\int_0^\infty \frac{\sin^2(\omega T/2)}{(\omega T/2)^2}\,\frac{d\omega T}{2}\,\frac{2}{T}\end{aligned} \tag{12-22}$$

Set $x \triangleq \omega T/2$ to obtain

$$\sigma_{Tb}^2(T) = \frac{k_b}{2\pi}\left(\frac{2}{T}\right)\underbrace{\int_0^\infty \frac{\sin^2 x}{x^2}\,dx}_{\pi/2} = \frac{k_b}{2T} \tag{12-23}$$

As an example, if $k_b = 7 \times 10^{-4}$ and $T = 0.1$ sec, then

$$\sigma_{Tb}^2(0.1) = \frac{(7 \times 10^{-4})}{2 \times 0.1} = 3.5 \times 10^{-3} \tag{12-24}$$

and $$\sigma_{Tb}(0.1) = 5.9 \times 10^{-2}\text{ Hz} \tag{12-25}$$

If the signal center frequency $f_0 = 10^9$ Hz and then the fractional frequency instability resulting from white frequency noise is

$$\frac{\Delta f}{f_0} = \frac{\sigma_{Tb}^2(T)}{\omega_0} = \frac{1}{f_0}\sqrt{\frac{k_b}{2T}}$$
$$= 5.9 \times 10^{-11} \tag{12-26}$$

for $T = 0.1$ sec. Thus, the fractional stability resulting from white frequency noise varies in general as the inverse square root of the integration time

$$\frac{\Delta f}{f_0} = \frac{1}{f_0}\sqrt{\frac{k_b}{2T}} \tag{12-27}$$

and is proportional to the square root of the white frequency noise power spectral density coefficient k_b.

White Phase-Noise Contribution

The white phase-noise contribution for $G_{\phi w}(f) = N_0$ (one-sided) for $\omega < \omega_h$ yields a frequency component of variance

$$\sigma_{TW}^2 = \int_0^\infty d\omega G_{\phi w}(\omega)\frac{(1 - \cos \omega T)}{(2\pi)^2 T^2} = \int_0^{\omega_h} \frac{N_0}{T^2}\frac{(1 - \cos \omega T)}{(2\pi)^2}\, d\omega$$
$$= \frac{2f_h N_0}{(2\pi T)^2}\left(1 - \frac{\sin \omega_h}{\omega_h T}\right) = \frac{2P_n}{(2\pi T)^2}(1 - \operatorname{sinc} \omega_h T) \tag{12-28}$$

where $P_n = f_h N_0$ is the phase-noise power. If $f_h = 10^3$ Hz and $T = 0.1$ sec, then $\sigma_{TW}^2 \cong 5.0\, P_n$. Again, this variance can be related to frequency stability using $\Delta f = \sigma_{TW}$, or if both flicker noise and white phase noise are present, use $(\Delta f)^2 = (\sigma_{TW}^2 + \sigma_{TF}^2)$.

Oscillator Frame-To-Frame Coherence for TDMA

TDMA carrier recovery using frame-to-frame phase coherence relies on an essentially constant oscillator-phase noise for a given transmitting terminal from one frame to the next. That is, the phase-noise change from frame-to-frame (excluding doppler effects) must have a small variance:

$$\sigma_\epsilon^2 = (2\pi)^2 T^2 \sigma_T^2 = \mathbf{E}[\phi(t) - \phi(t - T)]^2 = [2R(0) - 2R(T)]$$
$$(\text{rad/sec})^2 \ll 1 \tag{12-29}$$

Typical values of the TDMA frame period T are $10^{-4} < T < 10^{-2}$ sec, and the oscillator stability of interest is the short-term stability measured over this frame period.

The phase error variance of (12-29) is related to the variance of frequency error (12-8) by

$$\sigma_\epsilon^2 = (2\pi)^2 T^2 \sigma_T^2 \text{ rad}^2 \tag{12-30}$$

If the short-term stability δ is measured using the frequency variance over this same time interval T, then we have $\sigma_T^2 = (\delta f_0)^2$ and (12-30) becomes

$$\sigma_\epsilon^2 = (2\pi)^2 T^2 (\delta f_0)^2 \text{ rad}^2 \qquad \delta \triangleq \frac{\Delta f}{f_0} \tag{12-31}$$

As an example, let $\delta = 10^{-11}, f_0 = 10^{10}, T = 0.1$ sec, then

$$\sigma_\epsilon^2 = (2\pi)^2(10^{-2})(10^{-2}) = 3.94 \times 10^{-3} \text{ rad}^2$$

or $$\sigma_\epsilon = 6.28 \times 10^{-2} \text{ rad or } 3.60° \tag{12-32}$$

This value of rms phase error from one frame to the next would result in a small degradation. The degradation for small phase error is on the order of $\cos \sigma_\epsilon$ and is beginning to be significant at $\sigma_\epsilon = 0.1$ rad.

12-3 THE PHASE-LOCKED LOOP

As discussed in Chap. 11, the phase-locked loop operates as part of the carrier recovery circuit. Since there is no pure carrier component in some of the digital signals, a nonlinearity must be introduced to generate this pure carrier. The phase-locked loop attempts to track the input phase ϕ representing the pure carrier component at the output of the second- or fourth-power multiplier in the carrier recovery loop. Figure 12-4 is a block diagram of a phase-locked oscillator [Viterbi, 1966, Chap. 2].

Assume at first that the phase-locked loop is tracking the carrier with small phase error $|\epsilon| \ll 1$ where $\epsilon \triangleq \phi - \hat{\phi}$ is the phase error in the loop. A sudden small increase in ϕ then produces an increase in $\sin \epsilon$, which in turn increases the frequency of the VCO and the phase estimate $\hat{\phi}$. Thus the phase estimate $\hat{\phi}$ tends to track ϕ.

Given the phase jitter statistics of the oscillator discussed in the previous section, the required minimum phase-locked-loop bandwidth can then be determined. The loop bandwidth must be sufficient to track the oscillator phase jitter. These requirements are determined here in terms of the residual phase error in the carrier tracking loop. Noise effects that must be considered in determining the optimum loop bandwidth are discussed in Sec. 12-7.

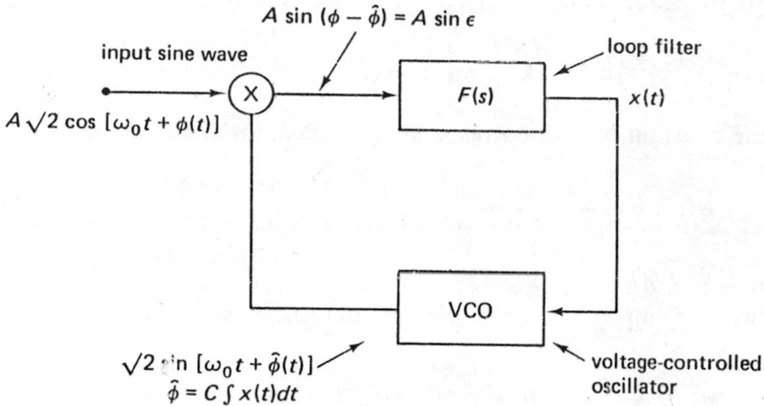

Fig. 12-4 Simplified block diagram of a phase-locked oscillator

Linearized Models of the Phase-Locked Loop

If the phase error is small, then we have $\sin \epsilon \cong \epsilon$ and the loop of Fig. 12-4 reduces to the commonly used linearized model shown in Fig. 12-5. Use operator notation to represent $s \triangleq d/dt$. The linearized closed-loop transfer function is then expressed in terms of the loop filter $F(s)$:

$$H(s) \triangleq \frac{\hat{\phi}(s)}{\phi(s)} = \frac{AKF(s)}{AKF(s) + s} \tag{12-33}$$

If $F(s) = (s + a)/(s + b)$, then the transfer function of this second-order

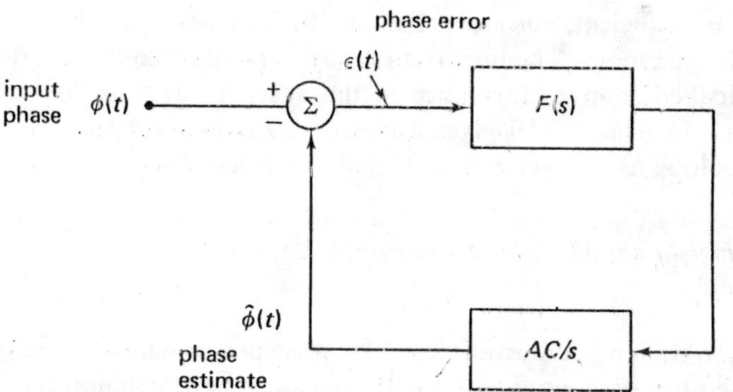

Fig. 12-5 Linearized equivalent model of the phase-locked oscillator

loop is

$$H(s) = \frac{AK(s + a)}{s^2 + (AK + b)s + AKa} \tag{12-34}$$

Let the constant $b \longrightarrow 0$, corresponding to ideal integration; then

$$H(s) = \frac{AK(s + a)}{s^2 + AKs + AKa} = \frac{\omega_n^2 + 2\zeta\omega_n s}{\omega_n^2 + 2\zeta\omega_n s + s^2} \tag{12-35}$$

where $\omega_n^2 = AKa$, $\zeta = \frac{1}{2}\sqrt{AK/a}$. When we apply an input phase modulation with an input spectrum $G_\phi(\omega)$, the output phase-error variance is

$$\sigma_\epsilon^2 = 2\int_0^\infty G_\phi(\omega)\,|1 - H(j\omega)|^2\, df \tag{12-36}$$

$$= 2\int_0^\infty G_\phi(\omega)\frac{\omega^4}{(AKa - \omega^2)^2 + (AK\omega)^2}\, df$$

and $$\sigma_\epsilon^2 = 2\int_0^\infty \frac{\omega^4/\omega_n^4}{[1 - (\omega/\omega_n)^2]^2 + [2\zeta(\omega/\omega_n)]^2} G_\phi(\omega)\, df \tag{12-37}$$

since $2\zeta/\omega_n = AK$. The closed-loop noise bandwidth (one-sided) for this loop is

$$B_n = \int_0^\infty |H(j\omega)|^2\, df = \int_0^\infty \frac{1 + (2\zeta\omega/\omega_n)^2}{[1 - (\omega/\omega_n)^2]^2 + [2\zeta(\omega/\omega_n)]^2}\, df \tag{12-38}$$

$$B_n = \frac{\omega_n}{8\zeta}(1 + 4\zeta^2) = \frac{3\omega_n}{4\sqrt{2}} = 0.530\omega_n = 3.33 f_n \qquad \text{if } \zeta = \frac{1}{\sqrt{2}} \tag{12-39}$$

Table 12-3 gives the loop filter, closed-loop response, natural frequency, damping coefficient, and noise bandwidth for first-, second-, and third-order phase-locked loops. The linearized model is most often used in describing the phase-locked loop performance in the absence of noise at high signal-to-noise levels. This model gives a useful approximation to performance with noise as long as the phase error magnitude is less than 0.1 rad.

Quasi-linear Models of the Phase-Locked Loop

To obtain a good estimate of the noise performance of the phase-locked loop we must use a nonlinear model and analysis as given in Sec. 12-5. However, there is also a quasi-linear approximation that is useful to show the decrease in effective loop gain as the noise level increases. Boonton [1953,

Table 12-3 Phase-Locked Loop Filters, Closed-Loop Transfer Functions, Loop Error Functions, and Closed-Loop Noise Bandwidths

Loop Filter $KF(s)$	Natural Frequency ω_n rad/sec	Damping Constant ζ	Closed-Loop Transfer Function $H(s)$	Loop Error Function $E(j\omega) = [1 - H(j\omega)]$	Noise Bandwidth $B_n = \int_0^\infty \lvert H(j\omega)\rvert^2\, df$
K	K	—	$\dfrac{K}{K+s}$	$\dfrac{-j\omega}{K+j\omega}$	$\dfrac{K}{4}$
$\dfrac{K}{1+bs}$	$\sqrt{K/b}$	$1/2\sqrt{Kb}$	$\dfrac{K}{K+s+s^2}$	$\dfrac{-(\omega/\omega_n)^2}{1+j2\zeta(\omega/\omega_n)-(\omega/\omega_n)^2}$	$\dfrac{\omega_n}{8\zeta}(1+4\zeta^2)$
$\dfrac{C(1+as)}{s}$	$\sqrt{C}$	$a\sqrt{C/2}$	$\dfrac{\omega_n^2+2\zeta\omega_n(1-\omega_n/2\zeta K)s}{\omega_n^2+2\zeta\omega_n s+s^2}$	$\dfrac{-(\omega/\omega_n)^2+j2\zeta(\omega/\omega_n)(\omega_n/2\zeta K)}{1+j2\zeta(\omega/\omega_n)-(\omega/\omega_n)^2}$	$\dfrac{\omega_n}{8\zeta}\left[1+4\zeta^2+4\zeta\left(\dfrac{\omega_n}{K}\right)+\left(\dfrac{\omega_n}{K}\right)^2\right]$
$\dfrac{K(1+as)}{1+bs}$	$\sqrt{K/b}$	$(1/\sqrt{bK})(1+aK/2)$			
$\dfrac{B_2^2+\sqrt{2}B_2 s}{s}$	B_2	$1/\sqrt{2}$	$\dfrac{\omega_n^2+\sqrt{2}\,\omega_n s}{\omega_n^2+\sqrt{2}\,\omega_n s+s^2}$	$\dfrac{-(\omega/\omega_n)^2}{1+j\sqrt{2}(\omega/\omega_n)-(\omega/\omega_n)^2}$	$\dfrac{3\omega_n}{4\sqrt{2}} = 0.53\omega_n = 3.33 f_n$
$\dfrac{B_2^2+B_2 s}{s}$	B_2	$1/2$	$\dfrac{\omega_n^2+\omega_n s}{\omega_n^2+\omega_n s+s^2}$	$\dfrac{-(\omega/\omega_n)^2}{1+j(\omega/\omega_n)-(\omega/\omega_n)^2}$	$\dfrac{\omega_n}{2} = 0.50\omega_n = 3.14 f_n$
$\dfrac{2B_3 s^2+2\sqrt{2}B_3^2 s+B_3^3}{s^2}$	B_3	—	$\dfrac{\omega_n^3+2\sqrt{2}\,\omega_n^2 s+2\omega_n s^2}{\omega_n^3+2\sqrt{2}\,\omega_n^2 s+2\omega_n s^2+s^3}$	$\dfrac{-(\omega/\omega_n)^3}{1+j2\sqrt{2}(\omega/\omega_n)-2(\omega/\omega_n)^2-j(\omega/\omega_n)^3}$	$0.949\omega_n = 5.91 f_n$

369–391] introduced the concept of equivalent gain as a "best" linear approximation, in the mean-squared sense, to the actual nonlinearity. More specifically, an equivalent gain K_{eq} is set at a value that minimizes $E(K \sin \epsilon - K_{eq}\epsilon)^2$. If the phase error is assumed Gaussian, then the best $K_{eq} = Ke^{-\sigma^2} \cos \epsilon_0$ where $E\epsilon = \cos \epsilon_0$ and var $\epsilon = \sigma^2$ are the bias and variance in phase error, respectively.

Figure 12-6 shows a quasi-linear model of the loop using this equivalent gain factor, which increases the effective input noise spectral density from N_0 to N_0/K_{eq}. Thus, the mean-squared phase error (tracking and noise errors) is

$$\sigma_\epsilon^2 = \int_0^\infty G_\phi(\omega)\,|1 - H(j\omega)|^2\,df + \int_0^\infty N_0 \left[\frac{K}{K_{eq}}\right]^2 |H(j\omega)|^2\,df \qquad \text{(12-40)}$$

where the closed-loop transfer function $H(j\omega)$ deqends on K_{eq}, and $G_\phi(\omega)$ is the phase-modulation spectrum. Develet [1953, 349–356] has shown that there is maximum value of N_0 for which this equation can be satisfied. This maximum noise level is the threshold of the model.

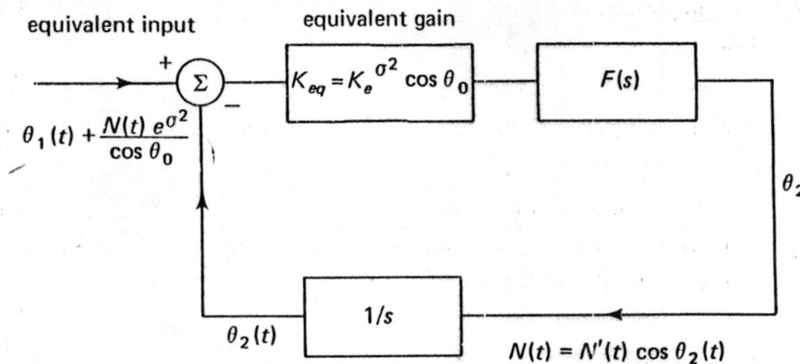

Fig. 12-6 Quasi-linear model of the phase-locked loop using the concept of equivalent gain, which is dependent on the phase error mean ϵ_0 and variance σ^2

12-4 PHASE-LOCKED TRACKING OF OSCILLATOR-PHASE NOISE

The phase-locked loop used in carrier tracking must have sufficient noise bandwidth to track the phase noise of the carrier with sufficient accuracy (rms phase error $\sigma_\epsilon \leq 0.1$ rad) to produce high correlation in coherent demodulation. Thus, for a given carrier phase-noise spectrum, one can compute the phase-locked loop noise bandwidth B_n required to track the carrier. Clearly, too large a noise bandwidth B_n can permit the thermal noise effects of the phase-locked loop to cause a severe degradation. Hence, it is desirable to minimize B_n from a thermal noise standpoint.

Alternatively, if a given noise bandwidth B_n is desired, one can compute various constraints on the phase noise power spectral density. These constraints are on the phase-noise power spectral density, which includes flicker noise, white frequency noise, and white phase-noise components. The phase noise discussed here is the carrier phase noise at the input to the phase-locked loop. If a $\times 2$ multiplier is required to generate the pure carrier, the phase noise in the modulated carrier is doubled by the multiplier. Similarly, the phase noise for a $\times 4$ multiplier for QPSK is quadrupled by this multiplier.

In the discussion below we assume that the phase noise consists of a sum of frequency flicker noise, with one-sided spectrum $G_{\phi a}(f)$, white frequency noise $G_{\phi b}(f)$, and white phase noise $G_{\phi c}(f)$, namely

$$\begin{aligned} G_\phi(f) &= G_{\phi a}(f) + G_{\phi b}(f) + G_{\phi c}(f) \quad f > 0 \\ &= \frac{k_a}{f^3} + \frac{k_b}{f^2} + k_c \qquad f > 0 \end{aligned} \tag{12-41}$$

The phase-locked-loop tracking errors are computed for each of these terms in succession. The tracking error for the phase-locked loop is

$$\sigma_\epsilon^2 = \int_0^\infty G_\phi(f)|1 - H(j\omega)|^2\, df = \sigma_{\epsilon a}^2 + \sigma_{\epsilon b}^2 + \sigma_{\epsilon c}^2 \tag{12-42}$$

where $H(j\omega)$ is the closed-loop transfer function of the phase-locked loop. Then the output mean-square phase error for the second-order loop described above with damping $\zeta = 1/\sqrt{2}$ is

$$\begin{aligned} \sigma_\epsilon^2 &= \int_0^\infty \frac{\omega^4/\omega_n^4}{[1 - (\omega/\omega_n)^2]^2 + [2\zeta(\omega/\omega_n)]^2} \frac{K_a}{\omega^3}\, df \\ &= \frac{1}{2\pi} \int_0^\infty \frac{K_a\omega/\omega_n^4\, d\omega}{[1 - (\omega/\omega_n)^2]^2 + 2(\omega/\omega_n)^2} = \frac{K_a}{2\pi\omega_n^2} \int_0^\infty \frac{x\, dx}{1 + x^4} \end{aligned} \tag{12-43}$$

where $x \triangleq \omega/\omega_n$. Each of these phase-noise contributors is evaluated separately in the next paragraphs.

Flicker-Noise Effects

Consider first a phase-noise component with a one-sided flicker spectral density

$$G_{\phi a}(f) = \frac{k_a}{f^3} \qquad f > 0 \tag{12-44}$$

Use a second-order phase-locked loop with damping $\zeta = 1/\sqrt{2}$ and tracking error

$$|1 - H(j\omega)|^2 = \frac{(\omega/\omega_n)^4}{1 + (\omega/\omega_n)^4} \tag{12-45}$$

where $H(j\omega)$ is the closed-loop transfer function of the phase-locked loop and ω_n is the natural frequency. The residual phase error in the phase-locked loop has variance

$$\sigma_{ea}^2 = \int_0^\infty \frac{k_a}{f^3}\left[\frac{(\omega/\omega_n)^4}{1 + (\omega/\omega_n)^4}\right] df = \int_0^\infty \frac{k_a(2\pi)^3}{x^3\omega_n^3}\,\frac{x^4}{1 + x^4}\,\frac{dx}{2\pi}\,\omega_n \tag{12-46}$$

where $x \triangleq \omega/\omega_n$. Use the definite integral result

$$\int_0^\infty \frac{x\,dx}{1 + x^4} = \frac{\pi}{4} \tag{12-47}$$

Thus, from (12-46), (12-47), we have

$$\sigma_{ea}^2 = \frac{k_a(2\pi)^2}{\omega_n^2} \underbrace{\int_0^\infty \frac{x}{1 + x^4}\,dx}_{\pi/4} = \frac{k_a(2\pi)^2}{\omega_n^2}\,\frac{\pi}{4} = \frac{k_a\pi^3}{\omega_n^2} \tag{12-48}$$

$$\sigma_{ea}^2 = \frac{k_a\pi^3}{3.56B_n^2} = \frac{8.71\,k_a}{B_n^2} \qquad \text{for } \zeta = 1/\sqrt{2} \tag{12-49}$$

since $\omega_n = B_n/(0.53)$ rad/sec. Then the rms phase error is inversely proportional to loop noise bandwidth. The result is shown in Table 12-4. Set the

Table 12-4 PHASE-LOCKED TRACKING OF PHASE NOISE FOR A PHASE-LOCKED LOOP WITH DAMPING $\zeta = 0.707$, AND NOISE BANDWIDTH $B_n = 0.53\omega_n$

Type of Phase Noise	*Phase-Noise Spectral Density*	*Phase Error — Second-Order Phase-Locked Loop,* $\zeta = 0.707$ $\sigma_\epsilon^2 = \int_0^\infty \frac{\omega^4/\omega_n^4}{1 + (\omega/\omega_n)^4} G_\phi(f)\,df$
Frequency flicker noise	$\frac{k_a}{f^3}$	$\frac{k_a\pi^3}{\omega_n^2} = \frac{k_a\pi^3}{(1/0.53)^2B_n^2} = \frac{8.71k_a}{B_n^2}$
White frequency noise	$\frac{k_b}{f^2}$	$\frac{3.70k_b}{B_n}$
White phase noise	$k_c, f < f_h$	$k_c f_h$

rms phase error from the flicker-noise contribution equal to

$$\sigma_{\epsilon a} = 0.05 \text{ rad} \tag{12-50}$$

Then the maximum allowed value of k_a is related to B_n by

$$k_a = \frac{(25 \times 10^{-4})}{\pi^3}(3.95)B_n^2 = 3.19 \times 10^{-4} B_n^2 \tag{12-51}$$

Thus the maximum value of k_a is strongly dependent upon B_n.

For a noise bandwidth of $B_n = 1$ Hz (one-sided), this maximum value of k_a is

$$k_{a\,\max} = 3.19 \times 10^{-4} \tag{12-52}$$

White Frequency-Noise Contribution

Consider a phase-noise component with a white frequency phase-noise spectrum (one-sided)

$$G_{\phi b}(f) = \frac{k_b}{f^2} \qquad \text{for } f > 0 \tag{12-53}$$

The residual phase noise in a second-order $\zeta = 1/\sqrt{2}$ phase-locked loop is then

$$\sigma_{\epsilon b}^2 = \int_0^\infty \frac{k_b}{f^2}\left[\frac{(\omega/\omega_n)^4}{1 + (\omega/\omega_n)^4}\right]df = \int_0^\infty \frac{k_b(2\pi)^2}{x^2\omega_n^2}\left[\frac{x^4}{1 + x^4}\right]\frac{dx}{2\pi}\omega_n \tag{12-54}$$

$$\sigma_{\epsilon b}^2 = \frac{k_b 2\pi}{\omega_n}\underbrace{\int_0^\infty \frac{x^2}{1 + x^4}dx}_{\pi/(2\sqrt{2})} = \frac{k_b 2\pi}{\omega_n}\frac{\pi}{2\sqrt{2}} = \frac{k_b}{\sqrt{2}}\frac{\pi^2}{\omega_n} \tag{12-55}$$

$$\sigma_{\epsilon b}^2 = \frac{3.70\, k_b}{B_n} \tag{12-56}$$

Set $\sigma_{\epsilon b} = 0.05$ rad. Then the maximum allowed value of k_b is

$$k_b = \frac{1}{3.70}(25 \times 10^{-4})B_n = 6.76 \times 10^{-4} B_n \tag{12-57}$$

For a noise bandwidth $B_n = 1$ Hz, the value of $k_b = 6.76 \times 10^{-4}$.

White Phase-Noise Contribution

Finally, consider the phase-noise component with a white phase-noise spectral density (one-sided)

$$G_{\phi c}(f) = k_c \qquad \text{for } 0 < f < f_{\max} \tag{12-58}$$

where $f_{\max}$ may be limited by IF filters in the modulator. Assume also that $f_{\max} \gg B_n$, the closed-loop noise bandwidth of the phase-locked loop. The residual phase error is then of variance

$$\sigma_{\epsilon c}^2 = \int_0^{f_{\max}} k_c \, df = k_c f_{\max} \tag{12-59}$$

Assume that $f_{\max}$ is governed by the IF bandwidth and that $f_{\max} = 10^3$ Hz. The maximum allowed value of k_c for $\sigma_{\epsilon c} = 0.05$ rad is then

$$k_c = \frac{25 \times 10^{-4}}{f_{\max}} = \frac{25 \times 10^{-4}}{10^3} = 2.5 \times 10^{-6} \tag{12-60}$$

Phase-Noise Spectrum Constraints

The total phase error results from the composite oscillator-phase noise spectrum of

$$\begin{aligned} G_\phi(f) &= G_{\phi f}(f) + G_{\phi b}(f) + G_{\phi c}(f) \\ &= \frac{k_a}{f^3} + \frac{k_b}{f^2} + k_c \end{aligned} \tag{12-61}$$

If each of the k_i components is set to produce an rms phase error $\sigma_{\epsilon i} = 0.05$ rad as discussed above, the composite phase-noise spectrum for a $B_n = 1$ Hz from (12-51), (12-57), (12-60), is then

$$G_\phi(f) = \frac{3.19 \times 10^{-4}}{f^3} + \frac{6.77 \times 10^{-4}}{f^2} + 2.5 \times 10^{-6} \text{ rad}^2/\text{Hz} \tag{12-62}$$

A satisfactory oscillator is one with a phase-noise spectral density $G_{\phi\,\text{actual}}(f) < G_\phi(f)$. The resulting rms phase error for the phase-locked loop is

$$\sigma_\epsilon = 0.05(\sqrt{3}) = 0.0866 \text{ rad}$$

This spectrum is plotted component by component on a logarithmic scale

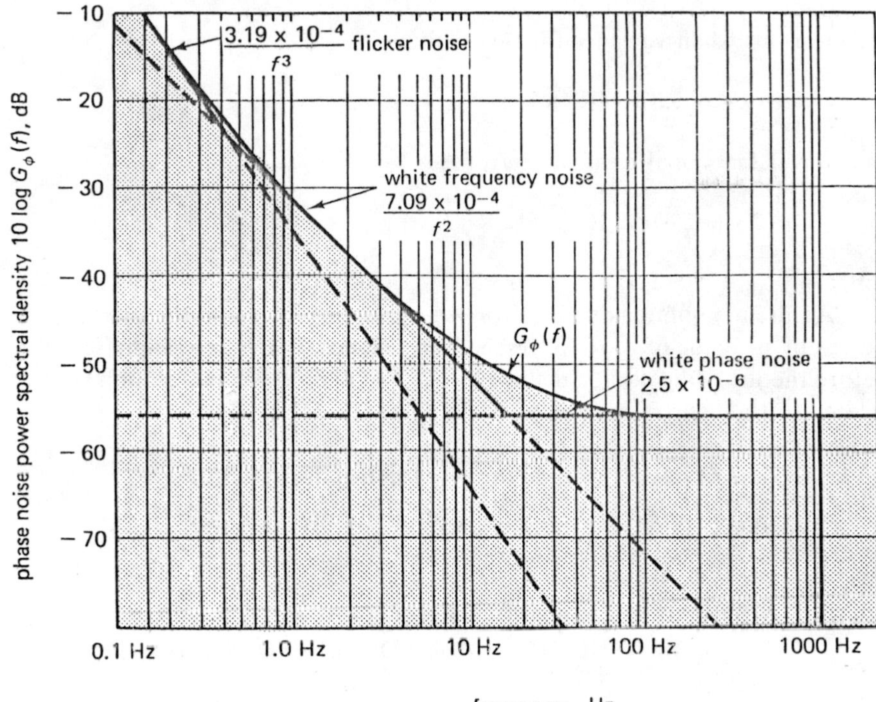

Fig. 12-7 Example constraints on oscillator-phase noise components for a 1-Hz phase-locked loop

10 log $G_\phi(f)$ in Fig. 12-7. Note that the flicker-noise effect is the dominant constraint for frequencies below 0.57 Hz. Between 0.57 and 16 Hz, white frequency noise dominates. Note also that the spectral density $G_\phi(f)$ is the summation of these components. Hence, the actual constraint is 3 dB higher at the point where these various components cross over.

If we take the oscillator center frequency as $f_0 = 10^9$ Hz and use the phase-noise spectrum of (12-62), then the Allan variance can be computed from Table 12-1 as

$$\begin{aligned}(2\pi)^2\sigma^2(T) &= \frac{(2\pi)^2}{\omega_0^2}k_{-3}2\log 2 + (2\pi)^2\frac{k_{-2}}{2\omega_0^2 T} + (2\pi)^2 k_0 \frac{3f_h}{\omega_0^2 T^2(2\pi)^2} \\ &= \frac{1}{f_0^2}\left[\frac{3.19\times 10^{-4}}{(2\pi)^2}(2\log 2) + \frac{7.09\times 10^{-4}}{(2\pi)^2 2T} \right. \\ &\quad \left. + \frac{2.5\times 10^{-6}}{(2\pi)^4}\left(\frac{3f_h}{T^2}\right)\right] \qquad (12\text{-}63)\end{aligned}$$

where $(2\pi)^2 k_{-3} = k_a$, $(2\pi)^2 k_{-2} = k_b$, $(2\pi)^2 k_0 = k_c$ in (12-61), (12-62). Thus we have the Allan variance for this example

$$\sigma^2(T) = 5.82 \times 10^{-4}$$

or the rms fractional frequency error is

$$\sigma(T) = 2.41 \times 10^{-2}$$

for $f_h = f_{max} = 10^3$ Hz, $T = 0.1$ sec.

As already pointed out, a $\times 2$ or $\times 4$ multiplier increases the phase-noise power by a factor of 4 or 16, respectively. Hence, if the carrier-phase noise before the multiplication is defined as $G_{\phi RF}(f)$, then the constraint on $G_{\phi RF}$ is related to the constraining value of $G_\phi(f)$ in (12-61) by

$$\begin{aligned} G_{\phi RF}(f) &< \tfrac{1}{16} G_\phi(f) \qquad \text{for a } \times 4 \text{ multiplier (QPSK)} \\ G_{\phi RF}(f) &< \tfrac{1}{4} G_\phi(f) \qquad \text{for a } \times 2 \text{ multiplier (BPSK)} \end{aligned} \tag{12-64}$$

12-5 ACQUISITION AND DOPPLER TRACKING PERFORMANCE OF PHASE-LOCKED LOOPS (PLL)

A phase-locked loop that is switched on to track a single sine-wave input may or may not lock on (reach a condition of small steady-state phase error) to that sine wave. If the loop does not lock-on or acquire, that is, produce a steady state condition where $|\epsilon(t)| \ll 1$, then the phase-locked oscillator is of little or no value in coherent demodulation. This carrier **acquisition** time is important in a PSK carrier-recovery circuit, such as a squaring loop, or the carrier recovery of a TDMA time-gated carrier preamble. Whether or not the PLL locks on depends on the offset between the quiescent VCO frequency and the input signal frequency, the frequency search rate of the VCO, the loop filter, and the loop initial conditions—for example, the initial phase-locked-loop phase error. Furthermore, for the loop to remain in lock, there must be constraints placed on the doppler frequency, and doppler rate and oscillator-frequency drifts relative to the loop filter.

Here we review the PLL results of Viterbi. The pull-in range, the maximum search rate, hold-in range, and lock-on time are discussed for the first-, second-, and third-order loops with a select set of transfer functions. An approximate method of analysis also is reviewed that appears to give good agreement with computer simulations.

Acquisition behavior is described in terms of the carrier-tracking application where the center frequency of the carrier is fixed or changing at a rate that is very slow compared to the closed-loop dynamics of the phase-locked loop. Short-term phase jitter is assumed absent in this discussion.

Terms used in discussing acquisition of phase-locked loops are:

Quiescent VCO frequency ω_0 The steady-state frequency of the voltage-controlled oscillator (VCO) in the phase-locked loop when the control voltage is zero.

Hold-in range Maximum amount the input frequency can deviate from the quiescent VCO frequency and have the loop remain in lock. This term assumes the phased-locked loop is initially in a locked-on state; that is, initial steady-state phase error $\ll \pi/2$ rad.

Pull-in range Maximum amount the input frequency can deviate from the quiescent VCO frequency while permitting the phase-locked loop to converge to a locked-on condition. Pull-in occurs with the loop initially unlocked; the initial VCO and signal frequencies are different.

Lock-on time The time interval after an input carrier is applied which is required for the phase-locked loop to achieve "frequency lock"—that is, stop skipping cycles. Once the loop has achieved this "frequency lock," the phase error gradually decays towards its steady-state value. (The decay time is not included in the lock-on time definition.)

Search rate The frequency ramp rate of the VCO in Hz/sec during the acquisition mode.

Viterbi [1960, 583–619] has published the most complete set of acquisition curves for the phase-locked loop. Some conclusions based on his analog computer runs are described below, with emphasis given to the second-order loop. Noise effects are not considered.

The hold-in range of phase-locked loops with a sinusoidal phase detector phase characteristic is simply determined by the dc loop gain, $KF(0)$. Since the phase-detector output has a maximum output $\sin \pi/2 = 1$, the hold-in range of the PLL is equal to the dc loop gain K rad/sec.

First-Order Loop

Here the loop filter is simply a gain constant, $KF(s) = K$ and the closed-loop transfer function is given by

$$H(p) = \frac{KF(s)}{s + KF(s)} = \frac{K}{K + s} \tag{12-65}$$

and the noise bandwidth (one-sided) is simply $B_n = K/4$ where

$$2B_n \triangleq \int_{-\infty}^{\infty} |H(j\omega)|^2 \, df \tag{12-66}$$

The pull-in range of the first-order loop is $\Delta\omega = K$ rad/sec. For carrier offsets less than $\Delta\omega = K$ rad/sec, the loop will lock on within one cycle of carrier phase change. This statement assumes a first-order loop in the strict sense. In practice parasitic effects cause the order of the loop to be greater than one. The exact lock-on time is determined by the initial phase error when the carrier is first applied.

Second-Order Loop

Three types of second-order loop acquisition behavior are discussed:

1. Loop with a perfect integrator and a constant input carrier frequency.
2. Loop with a perfect integrator and a linearly varying carrier frequency input.
3. Loop with an imperfect integrator and a constant input carrier frequency.

Consider first the loop with a perfect integrator where the loop filter is $F(s) = K(a + s)/s$. The closed-loop natural frequency is $\omega_n = \sqrt{aK}$ and the damping is $\zeta = K/2\omega_n$. Because of the infinite dc loop gain, the hold-in range is theoretically infinite for a fixed input frequency. The pull-in range is also infinite given enough skipped cycles and sufficient lock-on time. Viterbi [1960, 583–619] has shown that the region of "frequency lock," within which the loop stops skipping cycles after pull-in, is approximately given by

$$\Delta\omega_m \cong 2\omega_n(\zeta + 0.6) \qquad \zeta > 0.3 \tag{12-67}$$

That is, when the VCO frequency pulls to within $\Delta\omega$ rad/sec of the input carrier, the loop begins to pull into a diminishing phase error. Note that the pull-in range is infinite and not equal to $\Delta\omega_m$. For $\zeta = 0.707$, the region of frequency lock is $\Delta\omega_m = 2.614\omega_n$ or $\Delta f_m = 0.785 B_n$.

Frequency Search and Acquisition

In practice, acquisition usually occurs by sweeping the VCO frequency as shown in Fig. 12-8. The frequency search shown in this figure is a linearly varying sawtooth. The transient response results for this ramp frequency input are also reviewed. The net effect is just the same as if the input frequency were varying as a frequency ramp. Assume that the input frequency is linearly varying according to

$$\omega = \omega_s + Dt = \omega_0 - \Delta\omega + Dt$$

where D is the frequency sweep rate and $\Delta\omega = \omega_s - \omega_0$ is the initial frequency offset when the carrier is applied at $t = 0$. It is difficult to make general

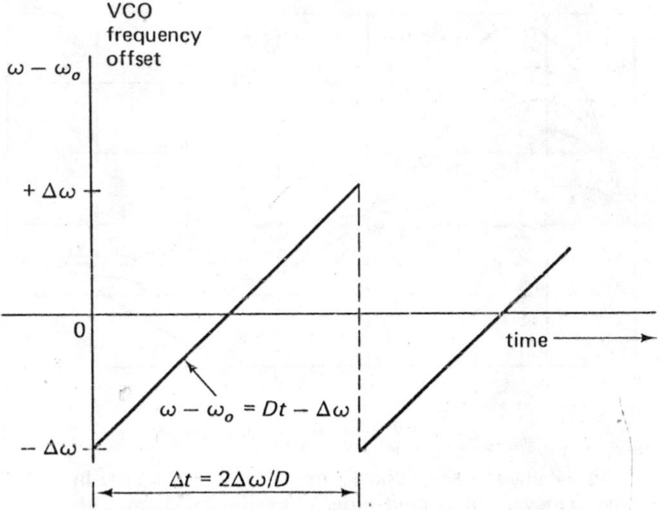

Fig. 12-8 VCO frequency sweep during acquisition mode

statements regarding the acquisition performance for this input, even for the second-order loop. However, if $D \geq \omega_n^2$ rad/sec², it can be stated that if the loop is initially unlocked, it can never achieve "lock," and if it is initially locked on, the loop will fall out of lock. Acquisition occurs or is missed, depending on the relative phases of the oscillators, as the VCO frequency sweeps toward the carrier frequency. The loop can acquire with certainty only if $D \leq \omega_n^2/2$ rad²/sec². The frequency sweep is sometimes implemented as a very fine-grained digitally controlled staircase. Once a correlation detector senses that the loop is in-lock, frequency sweep is then easily stopped by stopping the digital clock that drives the staircase generator. Note that for damping $\zeta = 1/2$, $\omega_n = 2B_n$, where B_n is the noise bandwidth of the loop then $D \leq 2B_n^2$ or

$$\frac{D}{2\pi} = \frac{B_n^2}{\pi} \text{ Hz/sec} \tag{12-68}$$

is the maximum search rate of the loop.

If the loop filter has an imperfect integrator in the loop filter—that is, $F(s) = K(1 + as)/(1 + bs)$, the loop natural frequency and damping constant are known to be

$$\omega_n = \sqrt{\frac{K}{b}} \qquad \zeta = \frac{a}{b\omega_n} + \frac{Ka}{2b\omega_n} \tag{12-69}$$

In Fig. 12-9, we show the transient performance of a second-order loop

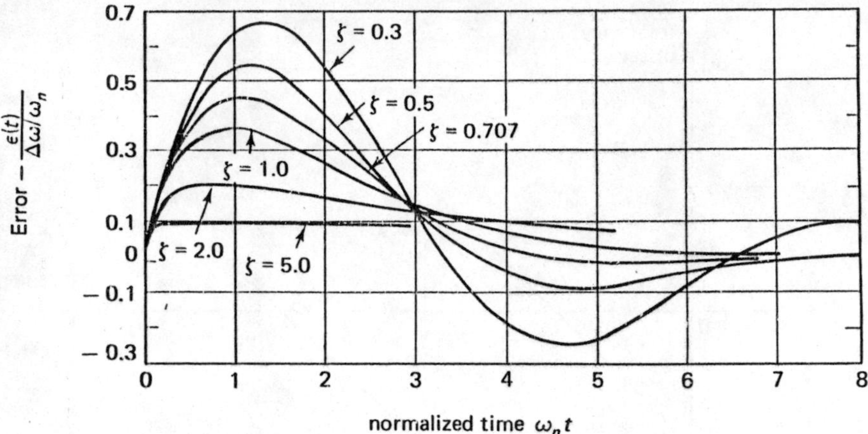

Fig. 12-9 Transient phase error $\epsilon(t)$ due to a step in frequency for a second-order phase-locked loop with various damping factors ζ (infinite dc gain) [From Gardner, 1966]

(perfect integrator) to a step in frequency when the loop is already locked-on [Gardner, 1966]. The quantity plotted is phase error versus time. Because the loop has infinite dc gain, there is no steady-state error to a step in frequency. The approximate pull-in range for this second-order loop for a constant frequency input, $D = 0$, is bounded by

$$\Delta\omega < 2\omega_n\sqrt{\zeta\omega_n b + 1} \tag{12-70}$$

given sufficiently long acquisition time. Note that $\Delta\omega \longrightarrow \infty$ as $b \longrightarrow \infty$ with ω_n, ζ held constant. Thus, the pull-in range becomes infinite as the loop filter approaches a perfect integrator. For $\Delta\omega/2\zeta\omega_n = 0.1$, the above result (12-70) is within 10 percent of the exact result.

For noisy inputs with an SNR of 14 dB in the noise bandwidth, Frazier and Page [1962, 210–227] have described experimental results for $\zeta = 1/2$ where the maximum sweep rate dropped to 95 percent of its value with no noise (for a 90 percent probability of acquisition).

More generally, it can be estimated that with a noisy input and a damping constant $\zeta \geq 1/2$, the empirically derived maximum frequency search rate for high confidence $\geq$ 90 percent of achieving lock is [Lindsey, 1972, p. 477; Frazier and Page, 1962]*

$$D = \frac{\omega_n^2}{2}\left(1 - \frac{1}{\sqrt{\alpha}}\right) \text{rad/sec}^2 \tag{12-71}$$

*Frazier and Page actually presented a result $D = \omega_n^2(1 - 1/\sqrt{\alpha})$.

or $$\frac{D}{2\pi} = \frac{\omega_n f_n}{2}\left(1 - \frac{1}{\sqrt{\alpha}}\right) \text{ Hz/sec} \tag{12-72}$$

where $\alpha = P_s/N_0 B_n$ is the signal-to-noise ratio in the closed-loop noise bandwidth, for $\zeta = 1/2$, $\omega_n = 2B_n$ and

$$\frac{D}{2\pi} = \frac{1}{2\pi}\frac{(4B_n^2)}{2}\left(1 - \frac{1}{\sqrt{\alpha}}\right) = \frac{B_n^2}{\pi}\left(1 - \frac{1}{\sqrt{\alpha}}\right) \text{ Hz/sec} \tag{12-73}$$

Note that (12-73) is equivalent to (12-68) if $\alpha = \infty$.

Third-Order Loop

The acquisition behavior of third-order loops has been studied by Viterbi [1960] for a loop-filter transfer function

$$F(s) = K\left[1 + \frac{\alpha}{s} + \frac{\beta}{s^2}\right] \tag{12-74}$$

By limiting the calculation to a limited set of initial conditions—namely, the locked-on condition—a meaningful projection to the phase plane can be made. The initial condition for a frequency ramp input signal with instantaneous frequency $\omega = \omega_0 + Dt$, $\phi(t) \triangleq \int_0^t \omega \, dt + \phi_0$ is $\phi_0 = 0$, $\ddot{\phi} = D$, where ϕ_0 is the initial phase of the VCO.

For linear frequency shifts and $D/\omega_n^2 < 1$, the third- and second-order loops perform in a similar manner where $\zeta = 1/\sqrt{2}$. The third-order loop can extend the tracking range to approximately twice the value of D, but at the expense of loop stability. As D increases toward $D = 2\omega_n^2$, the required β/ω_n^2 nears unity. By examining the characteristic polynomial, and applying the Routh-Hurwitz test, one finds that for

$$\frac{\beta}{\omega_n^2}\left(\frac{\sin \epsilon}{\epsilon}\right) < 1 \qquad \text{where } \epsilon \text{ is the phase error} \tag{12-75}$$

the loop is unstable. Thus, when the loop is not initially in lock, its pull-in behavior does not seem to be as stable as for the second-order loop. Hence, one sometimes designs the loop to acquire as a second-order loop. Once acquisition is sensed, the loop filter can be switched to a third-order loop mode, which might employ a smaller noise bandwidth.

Although the third-order loop has zero steady-state error in response to a constant value of $\dddot{\phi} = \dot{\omega}$, there is a nonzero transient error to a sudden step of doppler rate. The transient response of a third-order loop to a step of constant doppler rate (caused by a constant acceleration) is shown in Fig.

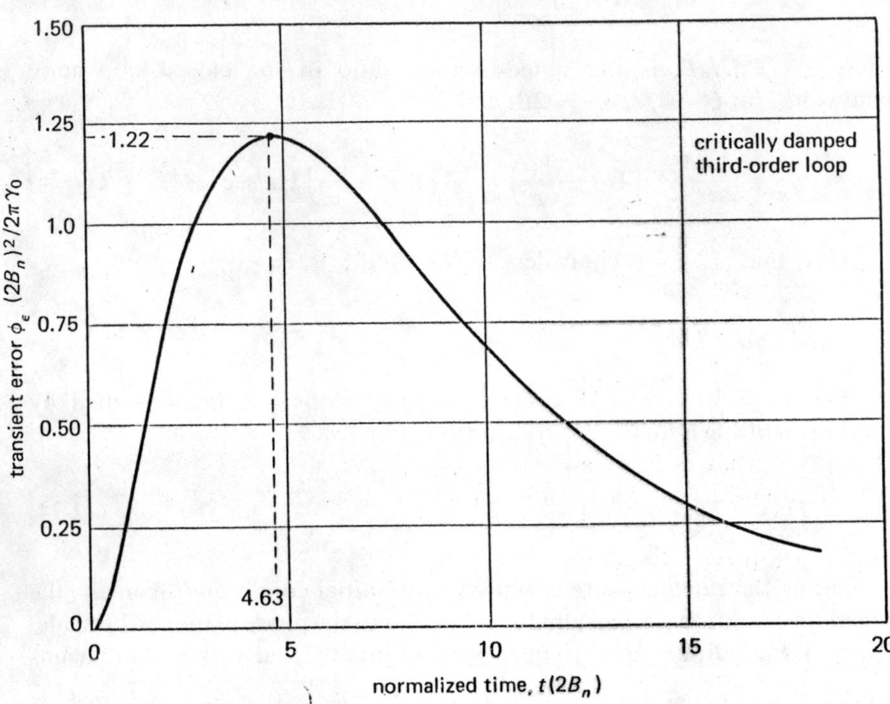

Fig. 12-10 In-lock transient response of third-order loop to a sudden frequency ramp excitation, γ_0 Hz/sec, [Tausworthe, Crow, 1971. Furnished through the courtesy of the Caltech/JPL.]

12-10 [Tausworthe, Crow, 1971]. The maximum transient error is shown to be

$$\epsilon_T(\max) = \frac{1.22(2\pi)}{4}\frac{\gamma_0}{B_n^2} \tag{12-76}$$

where γ_0 = doppler rate in Hz/sec
B_n = noise bandwidth of the third-order loop (one-sided)

This result is for a critically damped loop employing perfect integrators.

Thus the B_n required for a given ϵ_T (max) is

$$B_n = \sqrt{1.92\frac{\gamma_0}{\epsilon_T(\max)}} \tag{12-77}$$

As an example, if we allow a maximum phase error of $\epsilon_T = 0.5$ rad, and assume $\gamma_0 = 0.5$ Hz/sec, (12-77) gives $B_n \approx 1.4$ Hz. A signal power-to-noise density ratio of $C/N_0 \approx 11.4$ dB-Hz is required for a 10-dB loop SNR, $\text{SNR} = C/N_0B_n$.

Pull-In Time for a Second-Order Loop

Gruen [1953, 1043–1048] has made approximate calculations for pull-in or unaided acquisition by applying quasi-steady-state analysis as follows. When the input sinusoid of frequency ω_s is applied and the VCO frequency **average** frequency is ω_1 (which may differ from the quiescent frequency ω_0), a dc voltage component can appear at the output of the phase detector. If bias is artificially applied to the VCO so that the average input frequency is ω_1, and the instantaneous VCO frequency is allowed to vary, one can then calculate the dc phase-detector output to see whether the VCO input voltage builds up to drive the VCO average frequency closer to ω_s. The pull-in range is limited by the maximum offset in ω_1 from ω_0 that permits the input voltage buildup.

This lock-on operation can be visualized in an approximate sense if the VCO average frequency builds up in small increments towards the input signal frequency ω_s. If, for example, the first increment towards ω_s is to an average frequency $\omega_0 + \Delta$, next to $\omega_0 + 2\Delta$, then at each voltage increment, one examines the dc component at the VCO input to see if the VCO builds up to that average frequency.

For the second-order loop, Gruen has calculated the pull-in range as

$$\Delta\omega \cong 2\omega_n\sqrt{\zeta\omega_n b - \tfrac{1}{2}} \tag{12-78}$$

for a loop filter.

$$F(s) = K\left(\frac{1 + as}{1 + bs}\right) \tag{12-79}$$

Note that this pull-in range satisfies the inequality

$$\Delta\omega_n \cong 2\omega_n\sqrt{\zeta\omega_n b - \tfrac{1}{2}} < 2\omega_n\sqrt{\zeta\omega_n b + 1},$$

which is the bound obtained by Viterbi (12-70).

Acquisition occurs in an approximate lock-on time [Gruen, 1953]

$$T_{\text{acq}} \cong \frac{(\Delta\omega)^2}{4\zeta\omega_n^3 - [(\Delta\omega)^2\omega_n^2/K] - (2\omega_n^4/K)} \cong \frac{(\Delta\omega)^2}{2\zeta\omega_n^3} \qquad \omega_n/K \to 0 \tag{12-80}$$

For a damping constant $\zeta = 1/\sqrt{2}$ and $\omega_n \cong 2B_n$, this frequency-acquisition time becomes approximately [Tausworthe, 1966]

$$T_{\text{acq}} \cong \frac{3.5(\Delta f)^2}{B_n^3} \tag{12-81}$$

where Δf is the frequency offset. The additional time T_s for the loop to settle from $\pi/2$ rad phase error to a phase error less than $\epsilon_{T\delta}$ for zero steady-state error (no frequency offset) has been computed to be

$$T_s \cong \frac{1}{2B_n} \log \frac{2}{\epsilon_{T\delta}}$$
$$\cong \frac{1}{2B_n} \log 20 \cong \frac{1.5}{B_n} \qquad \epsilon_{T\delta} = 0.1 \text{ rad} \tag{12-82}$$

where B_n is the one-sided noise bandwidth, and log is the natural logarithm.

Magill [1968] has given an example for the acquisition transient for a second-order loop with the following parameters

$$B_n = 5 \text{ Hz} \qquad \zeta = \frac{1}{\sqrt{2}} \qquad (\Delta f) = 10 \text{ Hz} \tag{12-83}$$

and an initial phase error at $t = 0$ of $-\pi/2$ rad. The transient response is plotted versus time in Fig. 12-11. The measured and computed acquisition and

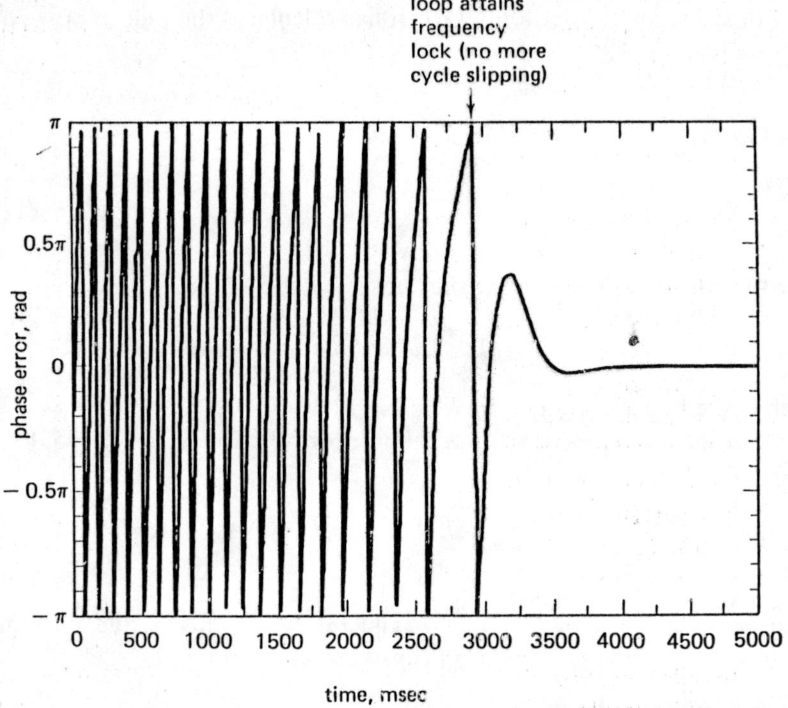

Fig. 12-11 Transient response of a second-order phase-locked loop with an initial frequency offset of $\Delta f = 10$ Hz and a noise bandwidth of $B_n = 5$ Hz [From Magill, 1968]

Table 12-5 FREQUENCY AND PHASE ACQUISITION TIMES, T_{acq} AND T_s FOR A SECOND-ORDER PHASE-LOCKED LOOP

	Computed	Measured
T_{acq}	$3.5(\Delta f)^2/B_n^3 = 2.8$	2.9
T_s	$\frac{1}{2B_n}\log\frac{2}{\delta\phi} = 1.5/B_n = 0.3$ for $\delta\phi = 0.1$ rad	0.5

settling times, in seconds, from (12-81) and (12-82) compare as shown in Table 12-5 for $B_n = 5$ Hz, $\Delta f = 10$ Hz.

TDMA Burst-Burst Carrier Acquisition

Carrier acquisition for a time division multiple access can be performed either on a burst-to-burst basis or on a frame-to-frame basis. In burst-to-burst operation a single carrier reconstruction loop receives consecutive TDMA carrier bursts from different ground terminals that have a random phase relationship to one another. In particular, the phase differential between the carrier in one burst to another is ϕ radians, which is uniformly distributed $(0,2\pi)$. The carrier frequency difference $\dot{\phi}$ between consecutive bursts is often small compared to the carrier reconstruction loop bandwidth.

In frame-to-frame operation, on the other hand, multiple carrier recovery loops operate on each separate ground terminal carrier. The consecutive TDMA carrier bursts from one frame to the next from a single earth station are presented to the carrier recovery loops, which can track slow variations in phase and frequency from frame to frame. The phase change from one frame to the next is dependent on the phase noise in the carrier and the frame duration. If the oscillator is well designed and the frame duration is not too long, there is only a small phase discontinuity normally.

If a phase-locked loop is employed in a burst-to-burst mode, and the phase differential is exactly $\epsilon_T = \pi$ radian, then the phase-locked loop initial operating point is a conditionally stable point of equilibrium, and acquisition time can be arbitrarily long unless noise is present to perturb the phase-locked loop from this equilibrium phase error. For this reason it may be desired to avoid the use of phase-locked loops in a burst-to-burst carrier recovery mode. On the other hand, a small sinusoidal phase jitter or dither can be added purposely to the VCO drive during the initial acquisition portion of the TDMA burst to alleviate this difficulty. Alternatively, a narrow bandpass filter-limiter can be employed.

The acquisition performance of an ideal second-order loop (infinite integration time) to a noisy carrier with zero frequency offset but arbitrary phase offsets has been studied using computer simulation [Goldman, 1973, 297–300]. The phase-locked loop damping constant is $\zeta = 0.707$. Hence the noise bandwidth (one-sided) is $B_n = 3.33 f_n$, where f_n is the loop natural frequency.

Figure 12-12 shows the computer simulation results for an initial phase error $\epsilon_T = \pi$ radians, the worst case at various values of signal-to-noise ratios of 4 to 50 dB in the closed-loop noise bandwidth. The results are plotted in terms of the percent of the simulations which had acquisition times less than T_{acq} sec. Acquisition time is normalized to the loop natural frequency f_n. Phase acquisition is said to occur when phase error drops below 37.3°.

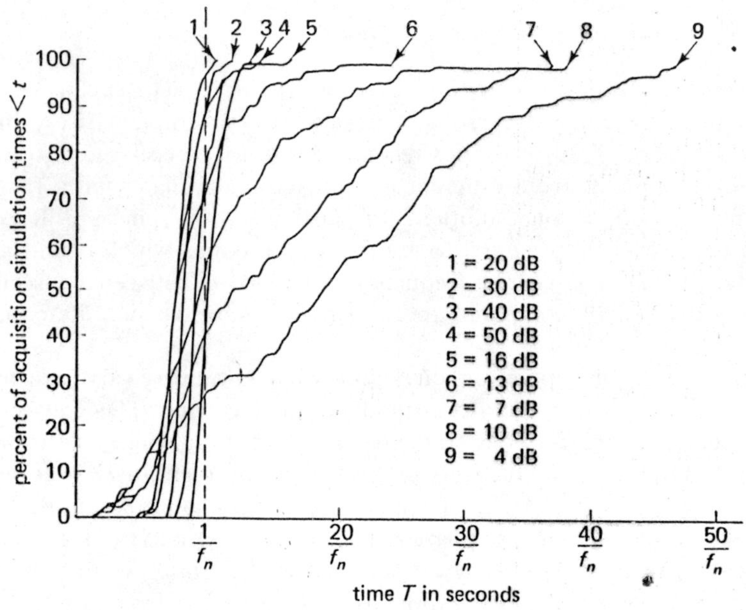

Fig. 12-12 Computer simulations of second-order phase-locked loop acquisition time for zero frequency offset and an initial phase error of $\epsilon_T = \pi$ rad [From Goldman, 1973]

Note that for this initial phase error the acquisition time for 90 percent acquisition decreases as signal-to-noise ratio decreases until the SNR drops to about 20 dB. Minimum value of acquisition time is approximately $T_{acq} = 10/f_n = 2.99/B_n$. If the phase-locked loop SNR decreases further to 10 dB, the T_{acq} approximately doubles.

Thus, a small amount of phase noise actually aids acquisition. As already noted above, a small 0.1 rad phase dither can similarly aid acquisition for the worst case initial phase error.

12-6 NONLINEAR ANALYSIS OF PHASE-LOCKED LOOPS

In this section, we determine the nonlinear effects of noise on the partially coherent detection of PSK signals using phase-locked-loop carrier recovery. This analysis proceeds in three steps: (1) Fokker-Planck equations are derived for a Markov process representing phase tracking noise in a phase-locked loop; (2) the phase-error probability density for a phase-locked loop is determined; and (3) the BPSK and QPSK error probabilities are computed versus E_b/N_0 when noisy inputs are fed to the carrier reconstruction loop.

*Smoluchowski and Fokker-Planck Equations**

The statistics of a first-order phase-locked loop output phase with a white noise input are shown below [Viterbi, 1966] to be a first-order Markov process. Thus the statistics of the loop analysis can draw upon the previous results for Brownian motion [Chandrasekar, 1943; Wang and Uhlenbeck, 1945.] Consider a Markov process $y(t)$ having an initial condition $y(0) = y_0$ and varying with time so as to reach y at t_1 (Fig. 12-13).† This process $y(t)$ is used later to represent the phase error in a phase-locked loop for BPSK and QPSK receivers.

We define the conditional probability density of y gives y_0, t as

$$p(y \mid y_0, t)\, dy \triangleq p[y(t) \mid y(0), t]\, dy \tag{12-84}$$

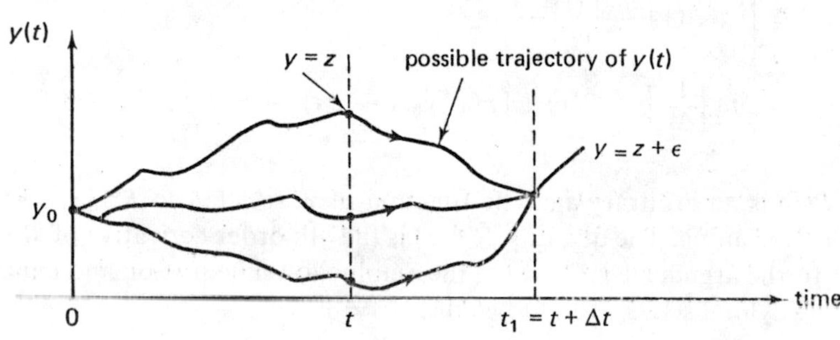

Fig. 12-13 Possible transitions of a Markov process from y_0 to y in a Δt-sec time interval

---

*Those readers who are only interested in the quantitative results can skip the first paragraphs of this section and proceed to Equation (12-134), where the phase noise probability density is given for a first-order phase-locked loop.

†An introduction to Markov processes is given in Bharucha-Reid [1960], and in Kemeny and Snell [1963].

as the probability that at time t, the value of $y(t)\epsilon(y, y + dy)$ given that t sec ago it had value $y(0) = y_0$.

Because $y(t)$ is the output of a Markov process (one-dimensional), $y(t) = y$ depends only on its previous value $y(t_1) = z$, $t = t_1 + \Delta t$, and we have

$$p(y \mid z, \Delta t; y_0, t_1) = p(y \mid z, \Delta t) \tag{12-85}$$

That is, the conditional density conditioned on all past samples is equal to the density conditioned on only the most recent sample z. Thus, we can write the joint probability density conditioned on y_0, omitting the time variable for simplicity in notation, as

$$p(y, z \mid y_0) = p(y \mid z, y_0)p(z \mid y_0) = p(y \mid z)p(z \mid y_0) \tag{12-86}$$

We integrate (12-86) over all possible values of the intermediate value z defined in Fig. 12-13 to obtain

$$p(y \mid y_0) = \int_{-\infty}^{\infty} p(y \mid z)p(z \mid y_0)\, dz \equiv \int_{-\infty}^{\infty} p(y \mid z, \Delta t)p(z \mid y_0, t_1)\, dz \tag{12-87}$$

which is the Smoluchowski (or Chapman-Kolmorgov) equation.

Now define the integral I

$$\begin{aligned} I &\triangleq \int_{-\infty}^{\infty} R(y)\frac{\partial p(y \mid y_0, t)}{\partial t}\, dy \\ &= \lim_{\Delta t \to 0}\left\{\frac{1}{\Delta t}\int_{-\infty}^{\infty} R(y)\, dy[p(y \mid y_0, t + \Delta t) - p(y \mid y_0, t)]\right\} \end{aligned} \tag{12-88}$$

where $R(y)$ is an arbitrary analytic function of y, selected so $R^{(n)}(y) \rightarrow 0$ as $|y| \rightarrow \infty$ for any n. The quantity $R^{(n)}(y)$ is the nth order derivative of R with respect to the argument y. By using the Smoluchowski equation and expanding in a Taylor's series, (12-88) becomes

$$\begin{aligned} I &= \lim_{\Delta t \to 0}\frac{1}{\Delta t}\left[\int R(y)\, dy \int p(z \mid y_0, t)p(y \mid z, \Delta t)\, dz - \int R(z)p(z \mid y_0, t)\, dz\right] \\ &= \lim_{\Delta t \to 0}\frac{1}{\Delta t}\left[\int p(z \mid y_0, t)\sum_{n=1}^{\infty}\frac{R^{(n)}(z)}{n!}\int \varepsilon^n p(\varepsilon \mid z, \Delta t)\, d\epsilon\right] \end{aligned} \tag{12-89}$$

where $y - z \triangleq \varepsilon$ and $R(y) = \sum_{n=0}^{\infty}\frac{\varepsilon^n}{n!}R^{(n)}(z)$

Define the conditional moments of ε as

$$A_n(z) \triangleq \lim_{\Delta t \to 0} \frac{1}{\Delta t} \int_{-\infty}^{\infty} \varepsilon^n p(\varepsilon \mid z, \Delta t)\, d\epsilon \qquad n \geq 1 \tag{12-90}$$

The integral (12-89) then can be rewritten and integrated by parts:

$$I = \sum_{n=1}^{\infty} \frac{1}{n!} \int \underbrace{A_n(z)p(z \mid y_0, t)}_{u}\, \underbrace{R^{(n)}(z)\, dz}_{dv} = \sum \frac{1}{n!} uv \Big|_{-\infty}^{\infty} - \sum \frac{1}{n!} \int v\, du \tag{12-91}$$

By proper selection of the arbitrary function $R(z)$ and its derivatives as $|z| \longrightarrow \infty$, we have

$$R^{(n-j)}(z)A_n(z)p(z \mid y_0, t)\Big|_{-\infty}^{\infty} = uv\Big|_{-\infty}^{\infty} = 0 \tag{12-92}$$

We then subtract the expression (12-91) for I from the definition of I (12-88) after n successive integrations by parts to obtain

$$I - I = 0 = \int R(y)\, dy \left\{ \frac{\partial p(y \mid y_0, t)}{\partial t} - \sum_{n=1}^{\infty} \frac{(-1)^n}{n!} \frac{\partial^n}{\partial y^n} [A_n(y)\, p(y \mid y_0, t)] \right\} \tag{12-93}$$

Since $R(y)$ is arbitrary, the quantity in braces must be zero, and the partial derivative must be

$$\frac{\partial p(y \mid y_0, t)}{\partial t} = \sum_{n=1}^{\infty} \frac{(-1)^n}{n!} \frac{\partial^n}{\partial y^n} [A_n(y)p(y \mid y_0, t)] \tag{12-94}$$

where $p(y \mid y_0, 0) = \delta(y - y_0)$ in which $\delta(x)$ is the Dirac delta function. This partial differential equation is the Fokker-Planck (FP) equation. It can be shown that for a first-order differential equation process, $A_n(y)$ vanishes for $n > 2$, and there are only two terms in the above series.

$$\frac{\partial p(y, t)}{\partial t} = \frac{-\partial}{\partial y} [A_1(y)p(y, t)] + \frac{1}{2} \frac{\partial^2}{\partial y^2} [A_2(y)p(y, t)] \tag{12-95}$$

Noise Performance of the Phase-Locked Loop

In general, noise affects the phase-locked loop in a nonlinear manner since the differential equation for the PLL is nonlinear. This effect can be analyzed using the FP equations derived above. Consider the phase-locked

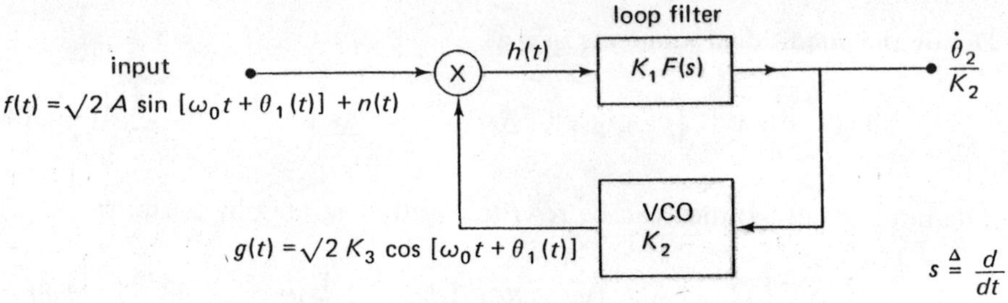

loop filter
input
$f(t) = \sqrt{2}A \sin[\omega_0 t + \theta_1(t)] + n(t)$
$h(t)$
$K_1 F(s)$
$\dot{\theta}_2 / K_2$
VCO
K_2
$g(t) = \sqrt{2} K_3 \cos[\omega_0 t + \theta_1(t)]$
$s \triangleq \frac{d}{dt}$

Fig. 12-14 Block diagram of phase-locked loop

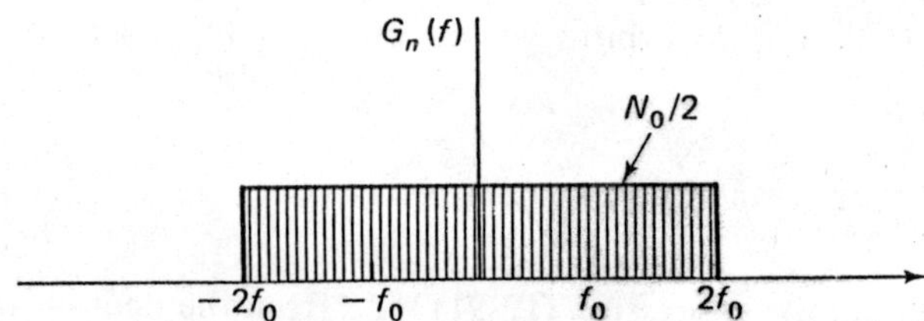

Fig. 12-15 Power spectral density of the bandlimited white input noise $n(t)$

loop of Fig. 12-14, and assume that the noise $n(t)$ is Gaussian, band-limited, white noise with power spectral density $G_n(f)$ (Fig. 12-15).

The noise $n(t)$ can be broken into its sine and cosine components, which are independent of one another

$$n(t) = \sqrt{2}\,n_s(t) \sin \omega_0 t + \sqrt{2}\,n_c(t) \cos \omega_0 t \tag{12-96}$$

where both $n_s(t)$ and $n_c(t)$ are Gaussian and have the same spectrum $G_n(f)$. We define $h(t)$ as the product of the input $f(t)$ and the VCO outputs $g(t)$. If we assume that the loop does not respond to the frequencies as high as ω_0 and above, then the multiplier output $h(t)$ can be written as

$$h(t) = AK_3 \sin \epsilon(t) + K_3 n_a(t) \tag{12-97}$$

where $g(t) = \sqrt{2}\,K_3 \cos[\omega_0 t + \theta(t)]$ and the phase error is

$$\epsilon(t) \triangleq \theta_1(t) - \theta_2(t) \tag{12-98}$$

and the noise component is defined as

$$n_a(t) = -n_s(t) \sin \theta_2(t) + n_c(t) \cos \theta_2(t) \equiv f_1(n_s, n_c, \theta_2) = n_{a\theta}(n_s, n_c) \tag{12-99}$$

Phase-Error Probability Density

PROBABILITY DISTRIBUTION OF n_a, n_b We also define the noise component

$$n_b = n_s \cos\theta_2 + n_c \sin\theta_2 \equiv f_2(n_s, n_c, \theta_2) = n_{b\theta}(n_s, n_c) \tag{12-100}$$

To find the joint density $p(n_a, n_b, \theta_2)$ in terms of $p(n_s, n_c, \theta_2)$, we first assume that θ_2 is independent of n_s and n_c, an assumption that is valid when the noise is white and there is a small time delay in the loop. Thus, this density can be expressed as $p(n_s, n_c, \theta_2) = p(n_s, n_c)p(\theta_2)$, and since n_s and n_c are Gaussian and independent, the joint density of the input noise is

$$p(n_s, n_c, \theta_2) = \frac{1}{(\sqrt{2\pi}\,\sigma)^2} e^{-(n_s^2+n_c^2)/2\sigma^2} p(\theta_2) \tag{12-101}$$

The n_a, n_b density, in terms of the n_s, n_c density, is written

$$\begin{aligned} p(n_a, n_b, \theta_2) &= p(n_a, n_b \mid \theta_2)p(\theta_2) \\ &= p[n_{s\theta_2}(n_a, n_b), n_{c\theta_2}(n_a, n_b)] \left|\frac{\partial(n_s, n_c)}{\partial(n_a, n_b)}\right| p(\theta_2) \end{aligned} \tag{12-102}$$

where the Jacobian

$$\left|\frac{\partial(n_s, n_c)}{\partial(n_a, n_b)}\right| = \begin{vmatrix} -\sin\theta_2 & \cos\theta_2 \\ \cos\theta_2 & \sin\theta_2 \end{vmatrix} = 1$$

We write simply

$$\begin{aligned} p(n_a, n_b, \theta_2) &= \frac{1}{2\pi\sigma^2} e^{-(n_s^2+n_c^2)/2\sigma^2} p(\theta_2) \\ &= \frac{1}{2\pi\sigma^2} e^{-(n_a^2+n_b^2)/2\sigma^2} p(\theta_2) = p(n_a, n_b)p(\theta_2) \end{aligned} \tag{12-103}$$

Hence, n_a, n_b are Gaussian and are independent of θ_2.

EQUATION FOR FREQUENCY ERROR $\dot{\epsilon}(t)$ We define $\dot{\theta}_2/K_2$ as the output of the loop filter $K_1F(s)$; θ_2 is an estimate of θ_1. Recalling that s is the operator d/dt, we obtain the differential equation

$$\frac{s\theta_2(t)}{K_2} = \frac{\dot{\theta}_2}{K_2} = K_1F(s)[AK_3 \sin\epsilon(t) + K_3n_a(t)] \tag{12-104}$$

where ϵ is the phase error.

Define an equivalent gain $K \triangleq K_1 K_2 K_3$; then the differential equation simplifies to

$$\dot{\theta}_2 = s\theta_2 = KF(s)[A \sin \epsilon + n_a] \tag{12-105}$$

(not writing t for convenience). Since the phase error $\epsilon \triangleq \theta_1 - \theta_2$, $\dot{\epsilon} = \dot{\theta}_1 - \dot{\theta}_2$, the phase-error differential equation is

$$\dot{\epsilon} = \dot{\theta}_1 - KF(s)[A \sin \epsilon + n_a] \tag{12-106}$$

This expression gives the derivative of the phase error in terms of the derivative of the original phase $\dot{\theta}_1$, the phase error ϵ, and the noise n_a. The nonlinear phase-locked-loop model that satisfies the differential equation (12-106) is shown in Fig. 12-16. As an example, we let $F(s) = 1$ and $\dot{\theta}_1 = 0$; then the derivative of phase error is given by the first-order differential equation

$$\dot{\epsilon} = -K[A \sin \epsilon + n_a] \tag{12-107}$$

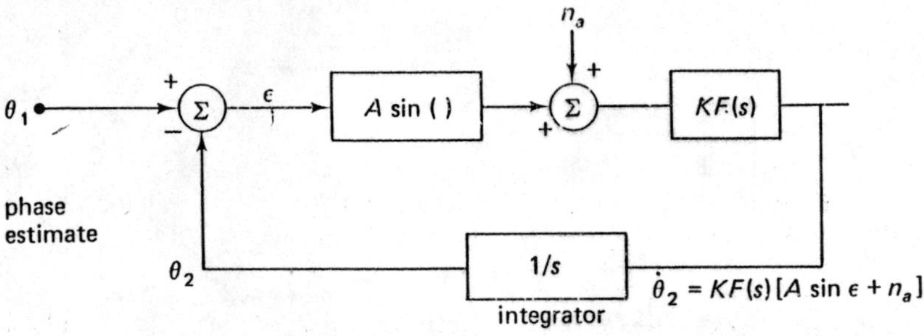

Fig. 12-16 Nonlinear model of the phase-locked oscillator in a noise environment

Evaluation of θ_2 for Small ϵ and Arbitrary θ_1 Assume a small phase error $|\epsilon| \ll \pi/2$. Then $\sin \epsilon \cong \epsilon$, and the differential equation (12-105) becomes linear:

$$\dot{\theta}_2 = s\theta_2 \cong KF(s)[A(\theta_1 - \theta_2) + n_a] \tag{12-108}$$

and the approximate operator equation for θ_2 is

$$\theta_2 = \frac{AKF(s)}{AKF(s) + s}\left[\theta_1 + \frac{n_a}{A}\right] \tag{12-109}$$

The closed-loop transform is defined as [See (12-33)]

$$H(s) = \frac{AKF(s)}{AKF(s) + s}$$

and yields a noise bandwidth consistent with Table 12-3:

$$B_n = \int_0^{\infty} |H(j\omega)|^2 \, df = \frac{KA}{4} \tag{12-110}$$

The operator equation (12-109) thus simplifies to

$$\theta_2 = H(s)\left[\theta_1 + \frac{n_a}{A}\right] \tag{12-111}$$

and is the linearized equivalent of the phase-locked loop for small phase error.

PROBABILITY DISTRIBUTION OF ϵ The operation of the phase-locked loop with a white Gaussian noise input is a Markov process, as is shown below. Since the time varying probability density for any Markov process can be described by the Fokker-Planck equation, equation (12-95) will be used here. Define the conditional moments

$$A_1(\epsilon) = \lim_{\Delta t \to 0} \frac{1}{\Delta t} \overline{\Delta \epsilon} \qquad \text{and} \qquad A_2(\epsilon) = \lim_{\Delta t \to 0} \overline{(\Delta \epsilon)^2} \tag{12-112}$$

Then an FP equation that describes the development in time of the probability density of phase error ϵ is

$$\frac{\partial p(\epsilon, t)}{\partial t} = \frac{\partial}{\partial \epsilon}[A_1(\epsilon)p(\epsilon, t)] + \frac{1}{2}\frac{\partial^2}{\partial \epsilon^2}[A_2(\epsilon)p(\epsilon, t)] \tag{12-113}$$

When the loop filter $F(s) = 1$ and the input frequency offset $\dot{\theta}_1 = 0$ (interchanging integration and expectation), the mean phase-error change in a small time offset Δt becomes, using (12-106),

$$\begin{aligned} \overline{\Delta \epsilon} &\triangleq \overline{\int_0^{\Delta t} \dot{\epsilon} \, dt} = \int_0^{\Delta t} \overline{[-KA \sin \epsilon + n_a]} \, dt \\ &= -KA(\sin \epsilon)\, \Delta t + \int_0^{\Delta t} \overline{Kn_a(t)} \, dt \end{aligned} \tag{12-114}$$

Since $n_a = N(0, \sigma^2)$. Normal or Gaussian probability density, zero mean, variance σ^2, and $\overline{n_a(t)} = 0$, the expected value of phase change in Δt is

$$\overline{\Delta \epsilon} = -KA \, \Delta t \sin \epsilon \tag{12-115}$$

The value of $\overline{\Delta\epsilon^2}$ for (12-112) is computed as

$$\overline{(\Delta\epsilon)^2} = \overline{\left[\int_0^{\Delta t} \epsilon\, dt\right]^2} = \overline{\left[\int_0^{\Delta t} -K(A \sin \epsilon + n_a)\, dt\right]^2} \tag{12-116}$$

$$\overline{(\Delta\epsilon)^2} = (KA)^2(\Delta t)^2 \sin^2 \epsilon + 2K^2 A\, \Delta t \sin \epsilon \int_0^{\Delta t} \overline{n_a(t)}\, dt + K^2 \int_0^{\Delta t}\int_0^{\Delta t} \overline{n_a(t) n_a(u)}\, dt\, du \tag{12-117}$$

Since $n_a(t)$ is essentially white, the autocorrelation of the noise is

$$\overline{n_a(t) n_a(u)} = R_n(t-u) = \frac{N_0}{2}\delta(t-u) \tag{12-118}$$

under the assumption that f_0 is large compared to the noise bandwidth, and the noise term in (12-117) is

$$\overline{(\Delta\epsilon)_n^2} \triangleq K^2 \int_0^{\Delta t}\int_0^{\Delta t} \overline{n_a(t) n_a(u)}\, dt\, du = K^2 \int_0^{\Delta t}\int_0^{\Delta t} \frac{N_0}{2}\delta(t-u)\, dt\, du$$
$$= K^2 \int_0^{\Delta t} \frac{N_0}{2}\, dt = \frac{K^2 N_0}{2}\Delta t \tag{12-119}$$

Thus, the mean-square phase change in (12-117) is

$$\overline{(\Delta\epsilon)^2} = (KA\, \Delta t)^2 \sin^2\epsilon + \frac{K^2 N_0}{2}\Delta t \tag{12-120}$$

From (12-112), (12-115), and (12-120), the conditional moments are then

$$A_1(\epsilon) = -KA \sin \epsilon \qquad \text{and} \qquad A_2(\epsilon) = \frac{K^2 N_0}{2} \tag{12-121}$$

can be calculated. The FP equation (12-113) thus becomes

$$\frac{\partial p(\epsilon, t)}{\partial t} = \frac{\partial}{\partial \epsilon}[p(\epsilon, t) KA \sin \epsilon] + \frac{1}{4} K^2 N_0 \frac{\partial^2}{\partial \epsilon^2} p(\epsilon, t) \tag{12-122}$$

where the $\sin \epsilon$ term indicates a periodicity in ϵ. Figure 12-17 shows the typical development of $p(\epsilon, t)$ with time, as expressed by the FP equation (12-122). The modes occur at integral multiples of 2π since (12-122) is periodic in ϵ and is indicative of cycle-slipping events.

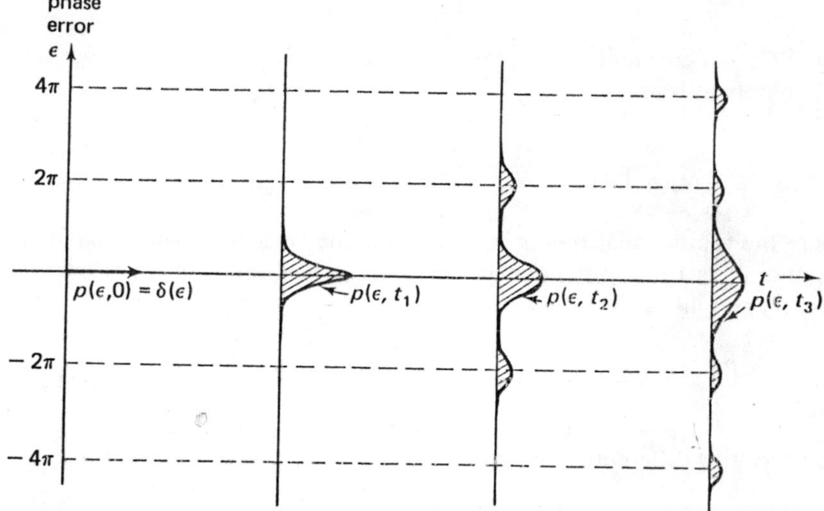

Fig. 12-17 Typical variation of $p(\epsilon, t)$ with time

COMPUTATION OF $p(\epsilon)$ IN THE STEADY STATE In the steady state, $p(\epsilon)$ is spread over all ϵ since cycle slipping events (lock-on at $2n\pi$) have occurred. To get a meaningful steady-state probability density, ambiguities of 2π must be neglected and a new density $P(\epsilon)$ must be considered:

$$P(\epsilon) \triangleq \sum_{n=-\infty}^{\infty} p(\epsilon + 2n\pi) \tag{12-123}$$

Since the FP equation is periodic in ϵ, $P(\epsilon)$ is a solution of (12-122) if $p(\epsilon)$ is a solution. Note that $P(\epsilon)$ has a period of 2π rad. Consider only the interval $-\pi < \epsilon < \pi$ and set the boundary conditions

$$P(\pi) = P(-\pi) \qquad \text{and} \qquad \int_{-\pi}^{\pi} P(\epsilon)\, d\epsilon = 1 \tag{12-124}$$

The steady state occurs when $t \longrightarrow \infty$. As $t \longrightarrow \infty$,

$$\partial p/\partial t \longrightarrow 0 \tag{12-125}$$

and (12-122) becomes

$$\frac{N_0 K^2}{4} \frac{\partial^2 P(\epsilon, t)}{\partial \epsilon^2} = \frac{-\partial[P(\epsilon, t)\, KA \sin \epsilon]}{\partial \epsilon} \qquad t \longrightarrow \infty \tag{12-126}$$

We integrate this equation, with respect to ϵ, to obtain

$$\frac{N_0 K^2}{4} \frac{\partial P(\epsilon, t)}{\partial \epsilon} = -P(\epsilon, t) KA \sin \epsilon + C_0 \tag{12-127}$$

where C_0 is a constant of integration. We define the SNR in the closed-loop PLL bandwidth as

$$\alpha \triangleq \frac{A^2}{N_0(KA/4)} = \frac{A^2}{N_0 B_n} = \frac{P_s}{P_n} \tag{12-128}$$

where the input signal power is P_s, $KA/4$ is the closed-loop noise bandwidth B_n, and P_n is the input noise power in B_n. Then the differential equation (12-127) simplifies to [Viterbi, 1963]:

$$\frac{\partial P(\epsilon, t)}{\partial \epsilon} = -P(\epsilon, t)\,\alpha \sin \epsilon + C_0 \tag{12-129}$$

We solve this differential equation and apply the boundary conditions

$$P(\pi) = P(-\pi) \tag{12-130}$$

to obtain $C_0 = 0$. Then, by solving (12-129), the probability density becomes

$$P(\epsilon) = Be^{\alpha \cos \epsilon} \tag{12-131}$$

Evaluate the constant B by using the constraint

$$\int_{-\pi}^{\pi} P(\epsilon)\, d\epsilon = 1 \tag{12-132}$$

which is evaluated by integration of (12-131) as

$$B \int_{-\pi}^{\pi} e^{\alpha \cos \epsilon}\, d\epsilon = 1 = B[2\pi I_0(\alpha)] \tag{12-133}$$

Thus, the constant is $B = 1/2\pi I_0(\alpha)$.

Phase-Error Probability Density for a PLL

The probability density for the phase error ϵ in the steady state for a first-order loop is then given by using (12-131) and (12-133) to obtain

$$P(\epsilon) = \frac{1}{2\pi I_0(\alpha)} e^{\alpha \cos \epsilon} \tag{12-134}$$

For large values of the equivalent input signal-to-noise ratio α, use the approximation

$$I_0(\alpha) \sim \frac{e^{\alpha}}{\sqrt{2\pi\alpha}}$$

to obtain the approximation to phase-error probability density,

$$P(\epsilon) \cong \frac{1}{\sqrt{2\pi/\alpha}} e^{-\alpha\epsilon^2/2} \qquad \alpha \gg 1 \tag{12-135}$$

which shows that the phase error is asymptotically Gaussian with variance $1/\alpha = \sigma_\epsilon^2$.

Phase Error in a Bandpass Limiter

An alternative phase estimate of the carrier in the presence of noise can be obtained simply by the use of a bandpass filter followed by a bandpass limiter. In this paragraph we compare the probability density of phase error for the phase-locked loop with that for a sine wave plus Gaussian noise in the output of a bandpass limiter. The phase error for the bandpass limiter gives [Stein and Jones, 1967, 136–141]

$$P(\epsilon) = \frac{1}{2\pi} e^{-\alpha/2} \left\{1 + \sqrt{\frac{\pi\alpha}{2}} \cos\epsilon \left[1 + \operatorname{erf}\left(\sqrt{\frac{\alpha}{2}} \cos\epsilon\right)\right] e^{\alpha/2 \cos^2 \epsilon}\right\} \tag{12-136}$$

where $\alpha/2$ is the input SNR in an IF bandwidth $2B_n$. One can simplify (12-136) by using the approximation for high SNR

$$\operatorname{erf} x = \frac{2}{\sqrt{\pi}} \int_0^x e^{-t^2}\, dt \cong 1 - \frac{e^{-x^2}}{x\sqrt{\pi}} \qquad x \gg 1 \tag{12-137}$$

The phase error of (12-138) then exhibits asymptotic behavior similar to that of the phase-locked loop—namely,

$$P(\epsilon) \cong \frac{1}{\sqrt{2\pi/\alpha}} \cos\epsilon (e^{-\alpha/2 \sin^2 \epsilon}) \cong \frac{1}{\sqrt{2\pi/\alpha}} e^{-\alpha\epsilon^2/2} \qquad |\epsilon| < \frac{\pi}{2} \tag{12-138}$$

Thus, the phase-noise statistics of the phase-locked loop and bandpass limiter are the same at high SNRs for equivalent noise bandwidths. One of the main advantages of the phase-locked loop is its ability to track over a wide range of carrier frequencies caused by doppler shifts and frequency drifts while keeping B_n small. The bandpass limiter, on the other hand, must have an IF bandwidth that is sufficiently broad to track the doppler effects. A second advantage of the phase-locked loop is the ability to obtain relatively small B_n with realistic component values. Conversely, very small bandwidths in a bandpass filter require very large filter Q values.

Phase-Locked-Loop Error Variance

The variance of phase error in a first-order phase-locked oscillator is computed using the expression (12-134) for $P(\epsilon)$ as

$$\sigma_\epsilon^2 = \frac{1}{2\pi I_0(\alpha)} \int_{-\pi}^{\pi} \epsilon^2 e^{\alpha \cos \epsilon} \, d\epsilon$$

Using the Jacobi-Anger formula, we obtain a series expansion for the variance

$$\sigma_\epsilon^2 = \frac{1}{2\pi I_0(\alpha)} = \int_{-\pi}^{\pi} \epsilon^2 \left[I_0(\alpha) + 2 \sum_{n=1}^{\infty} I_n(\alpha) \cos n\epsilon \right] d\epsilon$$

$$= \frac{\pi^2}{3} + 4 \sum_{n=1}^{\infty} \frac{(-1)^n I_n(\alpha)}{n^2 I_0(\alpha)} \qquad (12\text{-}139)$$

and for $\alpha = 0$, the variance reaches its maximum:

$$\sigma_\epsilon^2 = \frac{\pi^2}{3} = 3.29 \quad \text{for } P(\epsilon)$$

This relationship is plotted in Fig. 12-18 and compared with $\sigma_\epsilon^2 = 1/\alpha$ for the linear approximation.

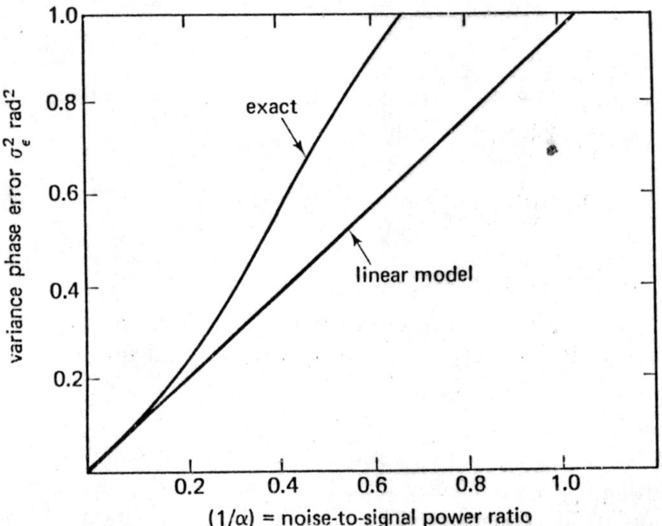

Fig. 12-18 Variance of phase error versus input noise-to-signal ratio for a phase-locked loop

Cycle Slipping in Phase-Locked Oscillators

Cycle slipping in a phase-locked oscillator occurs where the dynamic phase-error magnitude $|\epsilon| \geq 2\pi$. A typical event is shown in Fig. 12-19 where the phase error begins at $\epsilon(0) = 0$ and exceeds 2π at $t = t_1$. We define $q(\epsilon, t)$ as the density function for phase error for those phase-error trajectories where cycle clipping has *not* yet occurred. The density $q(\epsilon, t)$ is also shown in Fig. 12-19. Since cycle slipping always occurs eventually if noise is present, the integral of this density becomes

$$\int_{-2\pi}^{2\pi} q(\epsilon, t)\, d\epsilon \longrightarrow 0 \text{ as } t \longrightarrow \infty \tag{12-140}$$

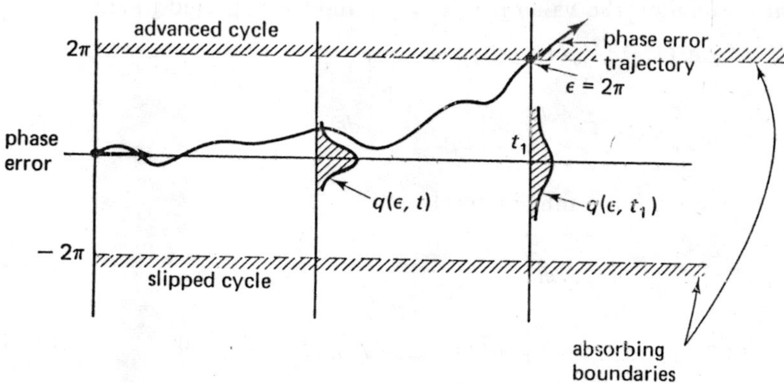

Fig. 12-19 Typical phase-error trajectory versus time

The density $q(\epsilon, t)$ must satisfy the FP equation of (12-122)

$$\frac{\partial q(\epsilon, t)}{\partial t} = \frac{\partial}{\partial \epsilon}[AK \sin \epsilon q(\epsilon, t)] + \frac{K^2 N_0}{4} \frac{\partial^2 q(\epsilon, t)}{\partial \epsilon^2} \tag{12-141}$$

where $q(\epsilon, 0) = \delta(\epsilon)$ is the initial condition.

To determine the expected value of t_1, we set boundaries at $\epsilon = 2\pi, -2\pi$ to absorb any trajectory reaching them. Thus, the boundary conditions become

$$q(2\pi, t) = q(-2\pi, t) = 0 \qquad \text{all } t \tag{12-142}$$

The probability $\psi(t)$ that a random trajectory is not absorbed is then

$$0 \leq \psi(t) \triangleq \int_{-2\pi}^{2\pi} q(\epsilon, t)\, d\epsilon \leq 1 \tag{12-143}$$

and the probability that a cycle has slipped is $\Phi(t) \triangleq 1 - \psi(t)$. The probability of cycle slipping for the first time in the interval $(t, t + d\epsilon)$ is then

$$\lim_{\Delta t \to 0} \frac{\psi(t) - \psi(t + \Delta t)}{\Delta t} = \frac{d\Phi}{dt} dt = \frac{-d\psi}{dt} dt \tag{12-144}$$

The expected time T to slip a cycle is found by averaging t, using (12-144)

$$T = \int_0^\infty -t\frac{d\psi}{dt} dt = -t\psi(t)\Big|_0^\infty + \int_0^\infty \psi(t)\, dt \tag{12-145}$$

where the second equation is obtained by integration by parts.

Since $\psi(0) = 1$ and $\psi(t) \lesssim t^{-(1+\epsilon)}$ asymptotically (or else the second integral is infinite), the value $t\psi(t)\Big|_0^\infty = 0$, and the expected time to a cycle-slip event from (12-143), (12-145) becomes

$$T = \int_0^\infty \psi(t)\, dt = \int_0^\infty dt \int_{-2\pi}^{2\pi} q(\epsilon, t)\, d\epsilon \tag{12-146}$$

If we define the infinite time integral

$$Q(\epsilon) \triangleq \int_0^\infty q(\epsilon, t)\, dt \tag{12-147}$$

then $\quad T = \int_{-2\pi}^{2\pi} Q(\epsilon)\, d\epsilon \qquad$ where $Q(2\pi) = Q(-2\pi) = 0 \qquad$ (12-148)

By integrating the FP equation (12-141) with respect to t, we can solve for Q:

$$\int_0^\infty \frac{\partial q}{\partial t} dt = \underbrace{q(\epsilon, \infty)}_{0} - \underbrace{q(\epsilon, 0)}_{\delta(\epsilon)} = \frac{d}{d\epsilon}[AK \sin \epsilon Q(\epsilon)] + \frac{N_0 K^2}{4} \frac{d^2 Q(\epsilon)}{d\epsilon^2} \tag{12-149}$$

Integrate both sides of the equation (12-149) with respect to ϵ to obtain

$$C - u(\epsilon) = AK \sin \epsilon Q(\epsilon) + \frac{N_0 K^2}{4} \frac{dQ(\epsilon)}{d\epsilon} \tag{12-150}$$

where u is the unit step function and C is a constant from the indefinite integral. Using an integrating factor [Hildebrand, 1949, 2–10], we obtain the solution to this first-order differential equation:

$$Q(\epsilon) = De^{\alpha \cos \epsilon} + \frac{e^{\alpha \cos \epsilon}}{\gamma} \int_{-2\pi}^{\epsilon} e^{-\alpha \cos x}[C - u(x)]\, dx \tag{12-151}$$

where $\alpha \triangleq \dfrac{4A^2}{N_0AK} = \dfrac{A^2}{N_0B_n}$ is the closed-loop SNR

and $\gamma \triangleq \dfrac{N_0K^2}{4} = \dfrac{4B_n}{\alpha}$ is proportional to noise density

Applying the boundary conditions of (12-148), we then evaluate the constants as $D = 0$ and $C = 1/2$. Thus, (12-151) becomes

$$Q(\epsilon) = \frac{e^{\alpha \cos \epsilon}}{\gamma} \int_{-2\pi}^{\epsilon} e^{-\alpha \cos x}\left[\frac{1}{2} - u(x)\right] dx \tag{12-152}$$

where $(1/2) - u(x)$ is defined in Fig. 12-20a and (12-148) becomes

$$\begin{aligned} T &= \int_{-2\pi}^{2\pi} Q(\epsilon)\, d\epsilon \\ &= \frac{1}{2\gamma}\int_{-2\pi}^{2\pi} d\epsilon \int_{-2\pi}^{\epsilon} \exp \alpha(\cos \epsilon - \cos x)\left[\frac{1}{2} - u(x)\right]dx \\ &= \frac{1}{2\gamma}\int_{-2\pi}^{2\pi} d\epsilon \int_{\epsilon}^{2\pi} \exp \alpha(\cos \epsilon - \cos x)\, dx \end{aligned} \tag{12-153}$$

using integration over the area shown in Fig. 12-20b.

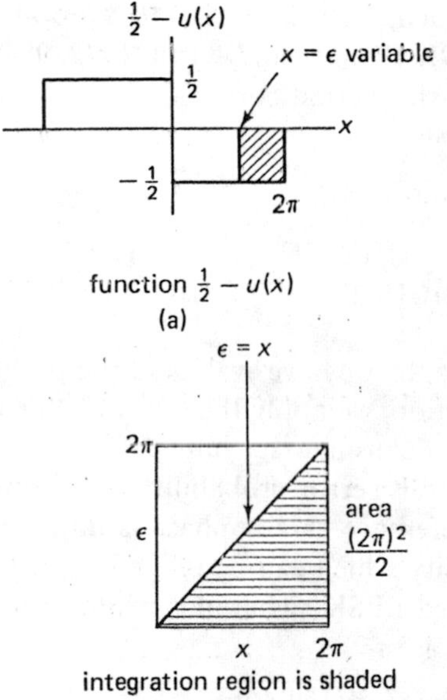

Fig. 12-20 Graphical description of quantities in the integral representation (12-152) of $Q(\epsilon)$

The product exponentials can be expanded using for each

$$e^{-\alpha \cos \epsilon} = I_0(\alpha) + 2 \sum_{n=1}^{\infty} (-1)^n I_n(\alpha) \cos n\epsilon \tag{12-154}$$

and is to be integrated over $(-2\pi, 2\pi)$. Only the first term $I_0(\alpha)$ is nonzero in the above integral (12-153). The expected time between skipping cycles is

$$\begin{aligned} T &= \frac{1}{\gamma} \int_0^{2\pi} d\epsilon \int_\epsilon^{2\pi} I_0^2(\alpha)\, dx = \frac{I_0^2(\alpha)}{\gamma} \frac{(2\pi)^2}{2} = \frac{\pi^2 \alpha I_0^2(\alpha)}{2B_n} \\ &\cong \frac{\pi^2 \alpha}{2B_n} \left(\frac{e^{2\alpha}}{2\pi\alpha} \right) = \frac{\pi e^{2\alpha}}{4B_n} \qquad \alpha \gg 1 \end{aligned} \tag{12-155}$$

or at high signal-to-noise ratio, $\alpha \gg 1$, we have

$$TB_n \cong \frac{\pi}{4} e^{2\alpha}$$

Thus, the expected time to a skipped cycle increases exponentially with 2α. This behavior is important for digital communications systems because many bit errors are possible when a cycle-skipping event occurs. Experimental results on phase-locked recovery of BPSK and QPSK carriers have shown that the mean duration of each cycle slip lasts 2 to 3 loop-time constants for loop SNR $\geq$ 16 dB for BPSK [Palmer and Klein, 1972, 984–991]. For small loop bandwidths, this cycle-slip period corresponds to many symbols and cause a substantial error burst.

12-7 PARTIALLY COHERENT RECEPTION OF BPSK SIGNALS-BIT ERROR EFFECTS

In the previous paragraphs we have evaluated the probability density of PLL phase error. In this section, we apply the effect of this phase error to carrier-recovery circuits and determine its impact on bit-error probability. The biphase-shift-keyed (BPSK) error probability is determined for matched-filter detection where the reference carrier phase is imperfectly known and has a given probability density which can be related to PLL probability density.

Consider a received BPSK signal plus white Gaussian noise of the form

$$s(t) + n(t) = \sqrt{2}\, A(t) \sin(\omega_0 t + \theta) + n(t) \tag{12-156}$$

where the data $A(t) = \pm A$, and $n(t)$ has zero mean and power spectral density (one-sided) N_0. Assume the signal is detected using a reference carrier

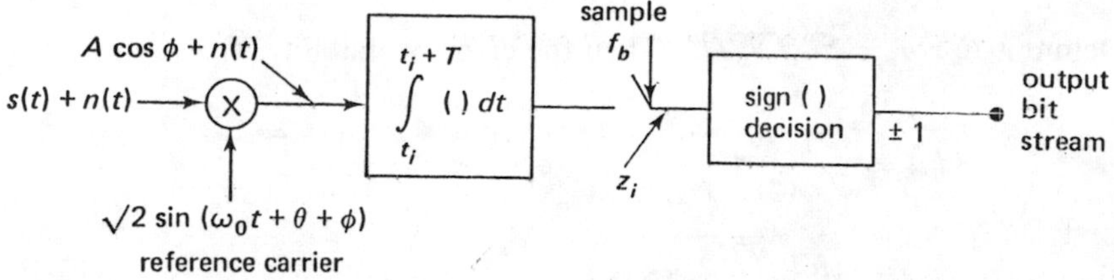

Fig. 12-21 PSK signal detection with a noisy phase reference with error ϕ_e; the bit rate $f_b = 1/T$

with phase error ϕ as shown in Fig. 12-21. The noise is represented as

$$n(t) = \sqrt{2}\, n_s(t) \sin(\omega_0 t + \theta + \phi_e) + \sqrt{2}\, n_c(t) \cos(\omega_0 t + \theta + \phi_e) \tag{12-157}$$

where $n_s(t)$ has power spectral density (one-sided) N_0. The signal and noise terms are integrated and dumped every T sec corresponding to one bit period to produce an output for $A(t_i) = +A$ with value

$$z_i = \int_{t_i}^{t_i+T} [A\cos\phi_e + n_s(t)]\, dt = AT\cos\phi_e + \int_{t_i}^{t_i+T} n_s(t)\, dt \tag{12-158}$$

where the reference carrier phase error $\phi_e(t)$ is essentially constant over $(t_i, t_i + T)$ and t_i is the beginning of a symbol period. The conditional mean value of z_i is $\bar{z}_i = \mathrm{E}(A\cos\phi_e) = A\mathrm{E}(z_i|\phi_e)$, and the variance of z_i is

$$\begin{aligned}\sigma_{z_i}^2 &= \int_{t_i}^{t_i+T} dt \int_{t_i}^{t_i+T} du\, \overline{n_s(t)n_s(u)} = \int_0^T dt \int_0^T du\, R_{n_s}(t-u) \\ &= \int_0^T dt \int_0^T du \frac{N_0}{2}\delta(t-u) = N_0 T\end{aligned} \tag{12-159}$$

Since z_i is Gaussian, $p(z_i)$ appears as shown in Fig. 12-22. The system makes an error when $p(z < 0)$. Hence, the bit error probability for a given value of ϕ is

$$\mathcal{P}_b(\phi_e) = \int_{-\infty}^{0} p(z)\, dz = \frac{1}{\sqrt{2\pi}\sigma_z} \int_{AT\cos\phi_e}^{\infty} e^{-z^2/2(N_0T/2)} dz \tag{12-160}$$

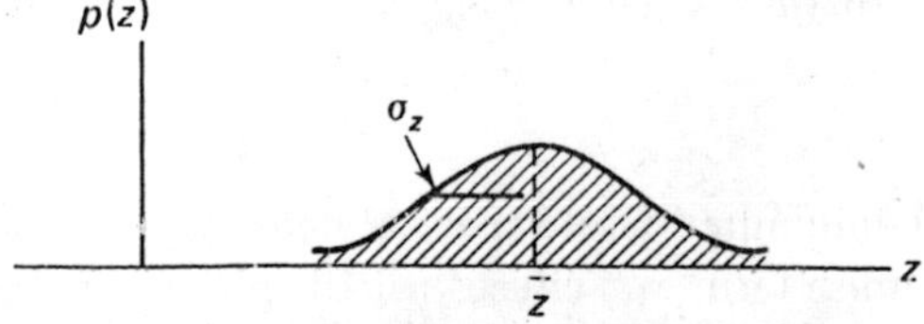

Fig. 12-22 Probability density of matched filter output z_i

Define $y \triangleq z/\sigma_z = z/\sqrt{N_0T/2}$. Then the error probability is written

$$\mathcal{P}_b(\phi_\epsilon) = \frac{1}{\sqrt{2\pi}\sigma_z}\int_{A\sqrt{2T/N_0}\cos\phi_\epsilon}^{\infty} e^{-y^2/2}\,dy = \frac{1}{\sqrt{2\pi}}\int_{\sqrt{2E_b/N_0}\cos\phi_\epsilon}^{\infty} e^{-y^2/2}\,dy$$
$$= \operatorname{erfc}\left(\sqrt{\frac{2E_b}{N_0}}\cos\phi_\epsilon\right) \tag{12-161}$$

where $E_b = A^2T$ is the energy per bit. Clearly, if $\phi = 0$, the error probability reduces to the standard result:

$$\mathcal{P}_b(0) = \operatorname{erfc}\sqrt{\frac{2E_b}{N_0}} \tag{12-162}$$

The average bit error probability for random phase error is then

$$\mathcal{P}_b = \int_{-\pi}^{\pi} \mathcal{P}_b(\phi_\epsilon)p(\phi_\epsilon)d(\phi_\epsilon)$$
$$= \int_{-\pi}^{\pi} \operatorname{erfc}\left(\sqrt{\frac{2E_b}{N_0}}\cos\phi_\epsilon\right)p(\phi_\epsilon)\,d\phi_\epsilon \tag{12-163}$$

where $p(\phi)$ is the probability density of the phase error in the carrier reconstruction loop. The probability density $p(\phi_\epsilon)$ is evaluated below and related to PLL phase error ϵ to calculate $\mathcal{P}_b$.

Square-Law Carrier-Tracking Loop

A BPSK signal contains no discrete carrier component, and the quasi-coherent reference carrier must be obtained by a square law or related nonlinear operation [Costas, 1956, 1713–1718]. Here we relate the phase noise induced by the carrier-tracking operation to the detected error probability. If the signal plus white noise is first filtered by a rectangular bandpass filter of bandwidth W, then the square-law device output in Fig. 12-23 is

$$[\sqrt{2}A(t)\sin(\omega_0 t + \theta) + n(t)]^2 = A^2 + A^2\cos 2(\omega_0 t + \theta) + v(t) \tag{12-164}$$

where $n(t)$ has a bandlimited white spectral density (one-sided) N_0 and power $P_n = N_0W$. The noise term $v(t)$ causes a PLL phase error ϵ and is

$$v(t) = 2\sqrt{2}A\sin(\omega_0 t + \theta)n(t) + n^2(t) \tag{12-165}$$

For stationary $A(t)$, $\sin(\omega_0 t + \theta)$, the autocorrelation function of $v(t)$ is

$$R_v(\sigma) = \mathbf{E}[2\sqrt{2}\,A(t)\sin(\omega_0 t + \theta)n(t) + n^2(t)]$$
$$\cdot\{\sqrt{2}\,2A(t+\sigma)\sin[\omega_0(t+\sigma)+\theta]n(t+\sigma) + n^2(t+\sigma)\} \tag{12-166}$$

The autocorrelation function of the input noise is

$$\mathbf{E}[n^2(t)n^2(t+\sigma)] = 2R_n^2(\sigma) + R_n^2(0) \tag{12-167}$$

and $\quad R_n(\sigma) = \rho_n(\sigma)\cos\omega_0\sigma \qquad (12\text{-}168)$

where $\rho_n(\sigma) \triangleq E[N(t)N(t+\sigma)]$ and $n(t) \triangleq \sqrt{2}\,N(t)\sin\omega_0 t$,

The noise autocorrelation (12-166) is then

$$\begin{aligned} R_v(\sigma) &= 4R_A(\sigma)R_n(\sigma)\cos\omega_0\sigma + 2R_n^2(\sigma) + R_n^2(0) \\ &= 2R_A(\sigma)\rho_n(\sigma)(1+\cos 2\omega_0\sigma) + \rho_n^2(\sigma)(1+\cos 2\omega_0\sigma) + \rho_n^2(0) \end{aligned} \tag{12-169}$$

The component of noise autocorrelation at frequency $2\omega_0$ is the only one passed by the bandpass filter

$$R_{v_{2\omega_0}}(\sigma) = [2R_A(\sigma)\rho_n(\sigma) + \rho_n^2(\sigma)]\cos 2\omega_0\sigma \tag{12-170}$$

If the bandpass filter causes envelope modulation at each phase reversal, then $|A(t)|$ has a pulse amplitude modulation component at the data clock rate. The squaring operation generates $A^2(t)\cos 2\omega_0 t$ and thereby produces a pair of line frequency components at frequency $2\omega_0 \pm 2\pi/T$ where $1/T$ is the bit rate. These bit frequency line components can cause a potential false lock problem if the frequency sweep must be relatively broad. The carrier recovery utilizes only terms at frequency $2\omega_0$ as shown in Fig. 12-23, since the square-law device is followed by a bandpass filter centered at $2\,\omega_0$. The power in the signal term at frequency $2\omega_0$ is $A^4/2 \triangleq P_s^2/2$ where $P_s = A^2$ is the input signal power. The one-sided noise spectral density at $2\omega_0$ for a filter bandwidth $W \gg B \triangleq 1/T$, where B is the bandwidth of $A(t)$. From (12-170) the noise-power spectral density at $2\omega_0$ is approximately equal to

$$\begin{aligned} N_0' &= \int_{-\infty}^{\infty} R_{v_{2\omega_0}}(\sigma)\cos 2\omega_0\sigma\,d\sigma = \int_{-\infty}^{\infty} [2R_A(\sigma)\rho_n(\sigma) + \rho_n^2(\sigma)]\,d\sigma \\ &= 2G_A(\omega)*G_N(\omega) + G_N(\omega)*G_N(\omega) \\ &= 2A^2N_0 + N_0^2W \end{aligned} \tag{12-171}$$

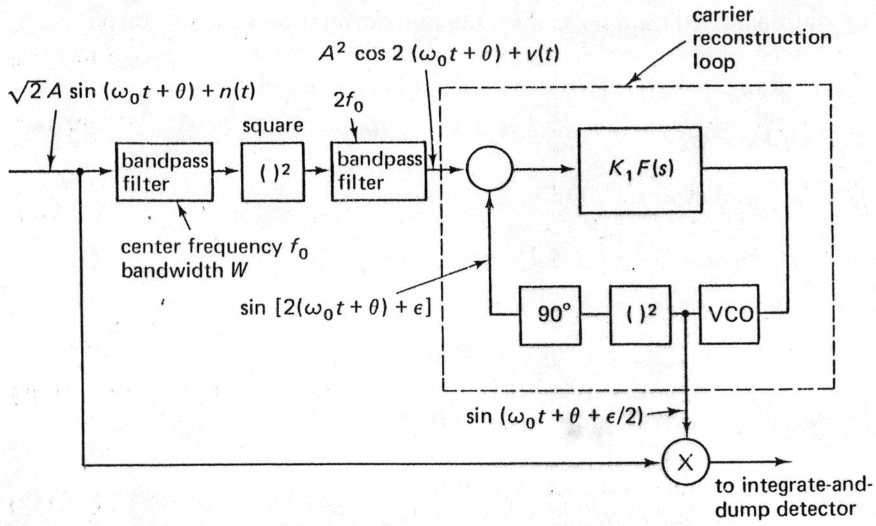

Fig. 12-23 Square-law carrier recovery for coherent phase detection of BPSK signals

where the notation $G(\omega) * G(\omega)$ represents the convolution of the two spectra. This result can be obtained by examining the spectral densities of $N(t)$, $n(t)$—that is, $G_N(\omega)$, $G_n(\omega)$, which are as shown in Fig. 12-24. Thus, the ratio of output signal to output noise power in a noise bandwidth B_n centered at $2\omega_0$ (for $B_n \ll W$) is

$$\alpha = \frac{A^4/2}{N_0' B_n} = \frac{P_s^2/2}{B_n[2A^2N_0 + N_0^2W]} = \frac{P_s}{4P_{n0}} \frac{1}{[1 + P_n/2P_s]}$$

$$= \frac{P_s}{4P_{n0}} \frac{1}{(1 + N_0/2E_b)} \tag{12-172}$$

where $P_{n0} \triangleq B_n N_0$ is the input noise power in a bandwidth B_n, $P_s = A^2$, and $P_n \triangleq N_0 W$. Note the 6-dB suppression (the factor of 4 in the denominator

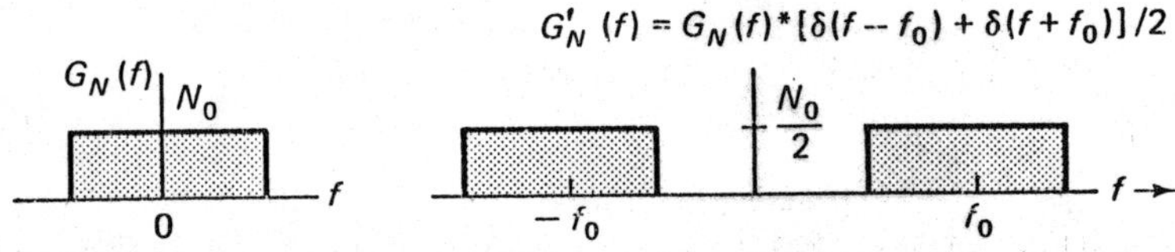

Fig. 12-24 Spectral density components of noise components; note that the equivalent one-sided noise density of $n(t)$ is N_0 watts/Hz.

of (12-172)) of the output-signal-to-noise ratio at high SNRs; that is, $\alpha \approx P_s/4P_{n0}$. However, since the VCO reference phase at $2\omega_0$ is divided by 2 for use in coherent detection, the variance of $\phi_\epsilon \triangleq \epsilon/2$ is $\sigma_\phi^2 = (1/4)\sigma_\epsilon^2 = 1/4\alpha$. Thus, the coherent reference phase error ϕ_ϵ does not have a 6-dB suppression for large α [Viterbi, 1966, 291]. Near and at threshold, however, this 6-dB suppression is important because it causes the PLL to act in a nonlinear manner. It is important that the carrier recovery threshold occurs at a SNR below the lowest E_b/N_0 which produces a usable error rate.

Output-Error Probability

We have determined that the steady-state phase-error probability density (12-134) for a phase-locked oscillator is

$$P(\epsilon) = \frac{e^{\alpha \cos \epsilon}}{2\pi I_0(\alpha)} \qquad |\epsilon| < \pi \tag{12-173}$$

where the signal-to-noise ratio α has been determined in (12-172). However, as mentioned above, this VCO phase error ϵ is divided by 2 to determine the reference phase error $\phi_\epsilon = \epsilon/2$. Thus, the output-error probability for BPSK is obtained by averaging the error probability for a given ϵ over all $|\epsilon| < \pi$:

$$\mathcal{P}_b = \int_{-\pi}^{\pi} \frac{e^{\alpha \cos \epsilon}}{2\pi I_0(\alpha)} \operatorname{erfc}\left(\sqrt{\frac{2E_b}{N_0}} \cos \frac{\epsilon}{2}\right) d\epsilon \tag{12-174}$$

where we have assumed no cycle slippage—that is, $P(\epsilon) = 0$ for $|\epsilon| \geq \pi$. This assumption implies that whenever a cycle slippage occurs, the system is immediately rephased so that succeeding bits are not inverted. Otherwise, the steady-state probability of error would be 1/2 because the system is certain to slip cycles eventually, and a cycle slip can reverse the sign of all data bits.

The error probability can be computed and plotted for various values of the normalized output-noise bandwidth factor δ:

$$\delta = \frac{1}{B_n T} \frac{1}{[1 + P_n/2P_s]} = \frac{4\alpha}{E_b/N_0} = \frac{1}{B_n T} \frac{1}{(1 + N_0 WT/E_b)}$$

or $$\alpha = \frac{E_b}{N_0} \frac{\delta}{4} \tag{12-175}$$

Note that δ is mostly dependent on the normalized noise bandwidth and that as $B_n \longrightarrow 0$, $\delta \longrightarrow \infty$. For large δ, the output-error probability approaches the value for no phase noise. Lindsey [1966, 399] gives an approximate expression for error probability, and the results shown in Fig. 12-25 are from his curves.

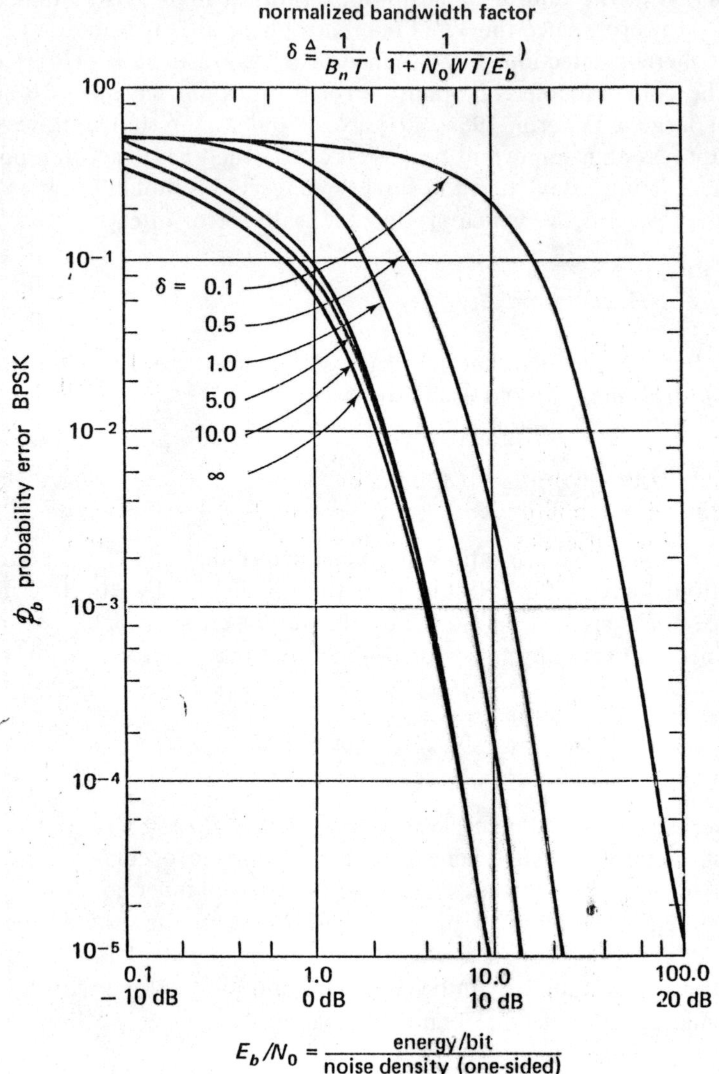

Fig. 12-25 BPSK error probability $\mathcal{P}_b$ versus E_b/N_0 for various values of δ, the normalized bandwidth factor [Lindsey, 1966]

Note that a value of $\delta \geq 5$ (an effective input SNR $\alpha_e \triangleq P_s/P_{n0} \geq 20$ or 13 dB) gives practically no degradation for $\mathcal{P}_b < 10^{-2}$, and at $E_b/N_0 > 10$ the required

$$\alpha = \frac{E_b}{N_0}\frac{\delta}{4} = 10\left(\frac{5}{4}\right) = 12.5 \qquad \text{or} \qquad 10.8 \text{ dB} \tag{12-176}$$

As an example, if the symbol rate (bit rate) is 10 Mbps, then the noise bandwidth must be $B_n \leq 1.33$ MHz in order to provide $\delta \geq 5$, for $WT = 2$ and $E_b/N_0 \geq 6$ dB.

12-8 PHASE NOISE IN QPSK CARRIER RECONSTRUCTION

The QPSK carrier-recovery circuit employing a fourth-power multiplier is shown in Fig. 12-26. We define the QPSK symbol period as T and the QPSK bit rate is $2/T$ bits/sec. The received signal, at f_0, plus noise is first filtered with a bandpass filter of bandwidth W Hz, where for low distortion it is assumed that $a \triangleq WT \geq 2$. Smaller values of a are feasible, but they cause larger envelope transients. The filtered signal is then passed through a $(\ )^4$ power device to generate a pure carrier component at $4f_0$ (Chap. 11), and a phase-locked loop of closed-loop bandwidth B_L is then locked to it. Just as with the $\times 2$ multiplier, envelope modulation in $|A(t)|$ caused by bandpass filtering causes clock components at $4\omega_0 \pm 2\pi n T$, where $1/T$ is now the symbol rate for QPSK. One must carefully avoid false carrier lock if the acquisition sweep is very wide (sweep $> 1/T$).

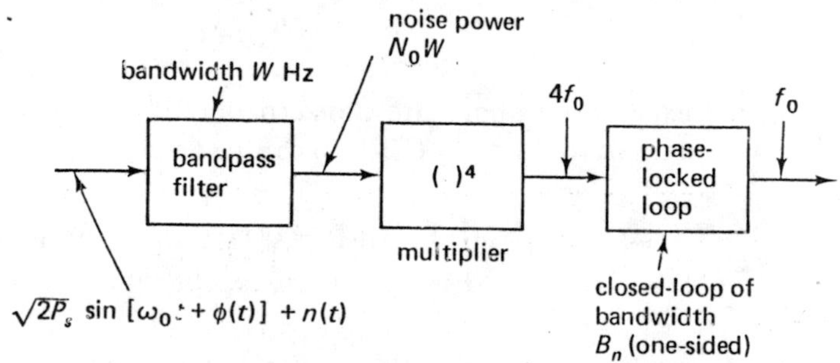

Fig. 12-26 Carrier recovery of QPSK using a fourth-power multiplier. The PLL has a built-in ×4 multiplier.

Here we analyze the output signal and noise statistics at the output of the $(\ )^4$ device, compute the output SNR for a phase-locked loop of bandwidth B_n, and then compute the detected bit-error probability obtained using this noisy phase reference.

Signal-to-Noise Ratio at the Output of a Fourth-Power Device

The input $y(t)$ to the fourth-power-law device consists of a QPSK or SQPSK signal $s(t)$, assumed undistorted by the bandpass filter operation, plus bandlimited white Gaussian noise $n(t)$ of one-sided density N_0 and IF bandwidth W; that is, the input to the 4th power device is

$$y(t) = s(t) + n(t) \tag{12-177}$$

where $s(t) = \sqrt{2}\,A \sin[\omega_0 t + \theta(t)]$ and $P_s = A^2$ is the input signal power. The output of the fourth-power device is then

$$z'(t) \triangleq y(t) = [s(t) + n(t)]^4 \tag{12-178}$$

Only the portion of $z'(t)$ in the immediate vicinity of frequency $4f_0$ is of significance because the phase-locked loop acts as a narrowband IF filter of bandwidth $2B_n$ and is tuned to $4f_0$. The first objective is to determine the signal power and noise spectral density in the vicinity of $4f_0$ and to compute the SNR in the filter of bandwidth $2B_n$.

The autocorrelation function of $z'(t)$ can be computed as

$$\begin{aligned} R_{z'}(\tau) &= \mathbf{E}[(s_1 + n_1)^4(s_2 + n_2)^4] \\ &= \mathbf{E}\left[\sum_{i,j,k,\ell=0}^{4} \alpha_{i,j,k,\ell}\, s_1^i s_2^j n_1^k n_2^\ell\right] \qquad i + j + k + \ell = 8 \end{aligned} \tag{12-179}$$

where $s_1 \triangleq s(t) \qquad s_2 \triangleq s(t+\tau) \qquad n_1 \triangleq n(t) \qquad n_2 \triangleq n(t+\tau)$

The terms in the expectation above are nonzero only if $i + j =$ even and $k + \ell =$ even. The nonzero terms in (12-179) can be shown to be

$$\begin{aligned} R_{z'}(\tau) = \mathbf{E}(&s_1^4 s_2^4 + 12 s_1^4 s_2^2 n_2^2 + 2 s_1^4 n_2^4 + 16 s_1^3 s_2^3 n_1 n_2 + 32 s_1^3 s_2 n_1 n_2^3 \\ &+ 36 s_1^2 s_2^2 n_1^2 n_2^2 + 12 s_1^2 n_1^2 n_2^4 + 16 s_1 s_2 n_1^3 n_2^3 + n_1^4 n_2^4) \end{aligned} \tag{12-180}$$

where we have used $\mathbf{E}(n_1^k) = \mathbf{E}(n_2^k)$, $\mathbf{E}(s_1^k) = \mathbf{E}(s_2^k) \ldots$, resulting from the stationarity of the input process. The moments of s and n are given in Table 12-6.

Table 12-6 MOMENTS OF s AND n FOR A FOURTH-POWER DEVICE WHERE P_s IS THE POWER IN THE SINUSOID, AND $N_0 W$ IS THE POWER IN THE INPUT CARRIER

$E(x_1^i, x_2^j)$	$x = s$ *Sinusoid*	$x = n$ *Gaussian*
$E(x_1 x_2)$	$P_s \psi(\tau) \cos \omega_0 \tau$	$W N_0 \rho(\tau)$
$E(x_1^2)$	P_s	$W N_0$
$E(x_1 x_2^3)$	$\frac{3}{2} P_s^2 \psi(\tau) \cos \omega_0 \tau$	$3 W^2 N_0^2 \rho(\tau)$
$E(x_1^2 x_2^2)$	$P_s^2[1 + \frac{1}{2}\psi(\tau) \cos 2\omega_0 \tau]$	$W^2 N_0^2 [1 + 2\rho^2(\tau)]$
$E(x_1^2 x_3^4)$	$P_s^3[\psi(\tau) \cos 2\omega_0 \tau + \frac{3}{2}]$	$3 W^2 N_0^3 [1 + 4\rho^2(\tau)]$
$E(x_1^3 x_2^3)$	$P_s^3/4[\psi(\tau) \cos 3\omega_0 \tau + 9\psi(\tau) \cos \omega_0 \tau]$	$3 W^3 N_0^3 \rho(\tau)[3 + 2\rho^2(\tau)]$
$E(x_1^4 x_2^4)$	$P_s^4[\frac{1}{8} \cos 4\omega_0 \tau + 2\psi(\tau) \cos 2\omega_0 \tau + \frac{9}{4}]$	$3 W^4 N_0^4[3 + 8\rho^4(\tau) + 24\rho^2(\tau)]$

$$\text{where}\quad \rho(\tau) \triangleq \frac{\sin \pi W\tau}{\pi W\tau}\cos \omega_0\tau \quad \text{and} \quad \psi(\tau) = 1 - \frac{|\tau|}{T} \qquad |\tau| < T$$
$$\triangleq \operatorname{sinc} \pi W\tau \cos \omega_0\tau \qquad\qquad \psi(\tau) = 0 \qquad |\tau| \geq T \tag{12-181}$$

The output of the fourth-power device is filtered at $4\omega_0$ to produce $z(t)$. Hence, only frequency components in the vicinity of $4\omega_0$ are of interest. The autocorrelation function of these components is then

$$R_z(\tau) = \frac{P_s^4}{8}\cos^4 \omega_0\tau + 4WN_0P_s^3\rho(\tau)\psi(\tau)\cos 3\omega_0\tau$$
$$+ 36W^2N_0^2P_s^2\rho^2(\tau)\psi(\tau)\cos 2\omega_0\tau$$
$$+ 32W^3N_0^3P_s\rho^3(\tau)\psi(\tau)\cos \omega_0\tau + 4W^4N_0^4\rho^4(\tau) \tag{12-182}$$

minus terms not at $4\omega_0$. By combining (12-181), (12-182) one can write

$$R_z(\tau) = 2P_s^4[\tfrac{1}{16} + R \operatorname{sinc} \pi W\tau\psi(\tau) + \tfrac{9}{2}R^2 \operatorname{sinc}^2 \pi W\tau\psi(\tau)$$
$$+ 6R^3 \operatorname{sinc}^3 \pi W\tau\psi(\tau) + \tfrac{3}{2}R^4 \operatorname{sinc}^4 \pi W\tau]\cos 4\omega_0\tau \tag{12-183}$$

where $R \triangleq N_0W/P_s$ is the input noise/signal power ratio. Clearly, the first term is the signal term, has power $P_s^4/16$, and suffers a factor of 16, or 12-dB attenuation relative to the first noise term, which is also the most significant at moderately high input SNRs $\alpha = 1/R \gg 1$.

Since the phase-locked-loop noise bandwidth B_n is small compared to W, it is adequate to evaluate the power spectral density of $z(t)$ at $4\omega_0$. The result for the noise-density terms is

$$\frac{G_z(4f_0)}{2P_s^4} = \frac{R}{W\pi}\mu(\pi a) + \frac{9R^2/W}{4a\pi^2}\eta(2\pi a) + \frac{7R^4}{8W}$$
$$+ 3\frac{R^3}{2\pi W}\frac{3\mu(a\pi)}{2} + \frac{5(\sin a\pi)}{\pi^2a^2} - \frac{a}{2}\mu(3a\pi) - \frac{2}{a\pi}$$
$$- \frac{\sin 3a\pi}{a^2\pi^2} + \frac{3}{a\pi}[\eta(3a\pi) - \eta(a\pi)] \tag{12-184}$$

where $a \triangleq TW$, T is the symbol duration, and $\mu(x)$, $\eta(x)$ are defined as

$$\mu(x) \triangleq \int_0^x \left[\frac{\sin u/2}{u/2}\right]^2 du \qquad \eta(x) \triangleq \int_0^x \mu(u)\, du \tag{12-185}$$

Several values of these two functions are given in Table 12-7.

Table 12-7 VALUES OF INTEGRALS $\mu(x)$ AND $\eta(x)$

x	$\mu(x)$	$\eta(x)$
$3\pi/2$		2.707π
2π	2.84	3.88π
3π	2.93	6.75π
4π	2.98	9.69π
5π	3.01	12.69π
6π	3.04	15.71π
9π	3.07	24.87π
12π	3.09	34.11π

Since a QPSK symbol carries 2 bits, the energy per bit $E_b = P_sT/2$. Thus, if the output noise power is B_n times the input noise density, the output SNR for this linearized model can be computed as

$$\alpha \triangleq SNR_{PLL} \triangleq \frac{E_b}{N_0}\beta\left(a, \frac{W}{B_n}, \frac{E_b}{N_0}\right) \tag{12-186}$$

For $a = TW = 2$, the quantity $\beta\ (2,\ W/B_n,\ E_b/N_0)$ is

$$\beta = \frac{(W/B_n)}{14.4 + 55.5(N_0/E_b) + 61.5(N_0/E_b)^2 + 14.02(N_0/E_b)^3} \tag{12-187}$$

and for $a = TW = 3$

$$\beta = \frac{W/B_n}{22.4 + 135(N_0/E_b) + 219.3(N_0/E_b)^2 + 70.8(N_0/E_b)^3} \tag{12-188}$$

For a given E_b/N_0 the value of β increases directly as W/B_n. Values of W/B_n for various values of E_b/N_0 are given in Table 12-8 for $\beta = 5$. The values of W/B_n for $\beta = 1$ are simply 1/5 of these values.

Effects of Carrier-Phase Noise on Bit-Error Rate

The probability density of phase noise ϵ in the phase-locked loop has been shown to be (12-132)

$$p(\epsilon) = \frac{\exp(\alpha \cos \epsilon)}{2\pi I_0(\alpha)} \qquad |\epsilon| \leq \pi \tag{12-189}$$

where α is the SNR in the loop bandwidth and $I_0(x)$ is the modified Bessel function. The value of $\alpha \triangleq \beta E_b/N_0$. Thus, one has recovered a carrier component at $4\omega_0 t$

$$\cos\,[4\omega_0 t + 2n\pi + \epsilon(t)]$$

After dividing the phase by 4 to convert to the original carrier frequency, the reference is

$$\cos\left[\omega_0 t + \frac{n\pi}{2} + \frac{\epsilon(t)}{4}\right] = \cos\left[\omega_0 t + \frac{n\pi}{2} + \phi_\epsilon(t)\right]$$

Thus, the phase noise power is reduced by a factor of 16, or 12 dB relative to the PLL error ϵ, exactly compensating the 12-dB degradation that occurs at high SNR in the fourth-power multiplier. This 12-dB improvement does not occur, of course, if the loop drops below threshold and responds in a nonlinear manner: for example, by slipping cycles.

The probability of a symbol error in the QPSK detector is the probability that the inphase and quadrature noise components cause the resultant vector to cross the boundaries shown in Fig. 12-27. As can be seen, a symbol error is caused when either

$$n_x > E_s \cos\left(\frac{\pi + \epsilon}{4}\right)$$

or

$$n_y > E_s \sin\left(\frac{\pi + \epsilon}{4}\right)$$

where the symbol energy $E_s = 2E_b$. The probability of a symbol error is then

$$\mathcal{P}_s = \frac{1}{2}\mathbf{E}\left\{\operatorname{erfc}\left[\frac{\sqrt{E_s}}{N_0}\cos\left(\frac{\pi + \epsilon}{4}\right)\right] + \operatorname{erfc}\left[\frac{\sqrt{E_s}}{N_0}\sin\left(\frac{\pi + \epsilon}{4}\right)\right]\right\} \tag{12-190}$$

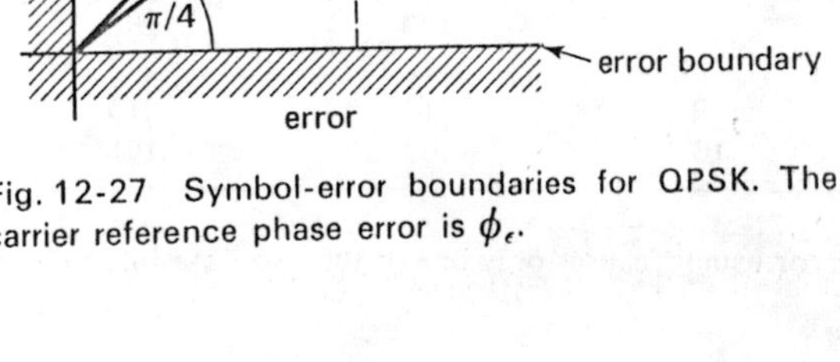

Fig. 12-27 Symbol-error boundaries for QPSK. The carrier reference phase error is ϕ_ϵ.

where **E** is the expectation operation averaged over all ϕ and

$$\operatorname{erfc} x \triangleq \frac{2}{\sqrt{\pi}} \int_x^\infty e^{-t^2}\, dt$$

The probability $\mathcal{P}_b$ of a bit error* for low symbol error rate $\mathcal{P}_s \ll 1$ is $\mathcal{P}_b \cong \mathcal{P}_s/2$ and from (12-189) and (12-190) is then

$$\mathcal{P}_b = \int_{-\pi}^{\pi} \frac{\exp(\alpha \cos \epsilon)}{8\pi I_0(\alpha)} \left[\operatorname{erfc} \frac{\sqrt{E_s}}{N_0} \cos\left(\frac{\pi + \epsilon}{4}\right) + \operatorname{erfc} \frac{\sqrt{E_s}}{N_0} \sin\left(\frac{\pi + \epsilon}{4}\right)\right] d\epsilon \qquad (12\text{-}191)$$

Computer results for $\mathcal{P}_b$ obtained by Sherman [1969, 45–52] are plotted in Fig. 12-28 versus E_b/N_0 for various values of β ($\alpha \triangleq \text{SNR}_{PLL} \triangleq \beta E_b/N_0$). For $\beta = 5$, the degradation is less than 0.2 dB. The maximum allowed values of phase-locked loop bandwidth B_n decrease as the values of E_b/N_0 decrease and the IF SNR decreases.

Table 12-8 shows the required values of W/B_n for various $a = WT$ and E_b/N_0 and $\beta = 5$. If a more rapid loop acquisition time is desired, a smaller value of β is required and the degradation increases. For the lowest E_b/N_0 expected, one may select W/B_n to minimize system degradation due to a noisy reference. As an example, if $E_b/N_0 = 2.1$ dB and $TW = 2$, then for $\beta = 5$

Table 12-8 W/B_n Required to Maintain a Normalized SNR Factor $\beta = 5$ Where $a = TW$. The Values of W/B_n for $\beta = 1$ Are 1/5 of This Value.

E_b/N_0	*Bandwidth Ratio* W/B_n	
(dB)	$a = 2$	$a = 3$
1	522	1517
2	387	1063
3	297	770
4	236	577
5	193	446
6	162	356
7	140	293
8	124	248
9	112	215
10	103	191

*A symbol error usually causes only one of the two data bits to be in error.

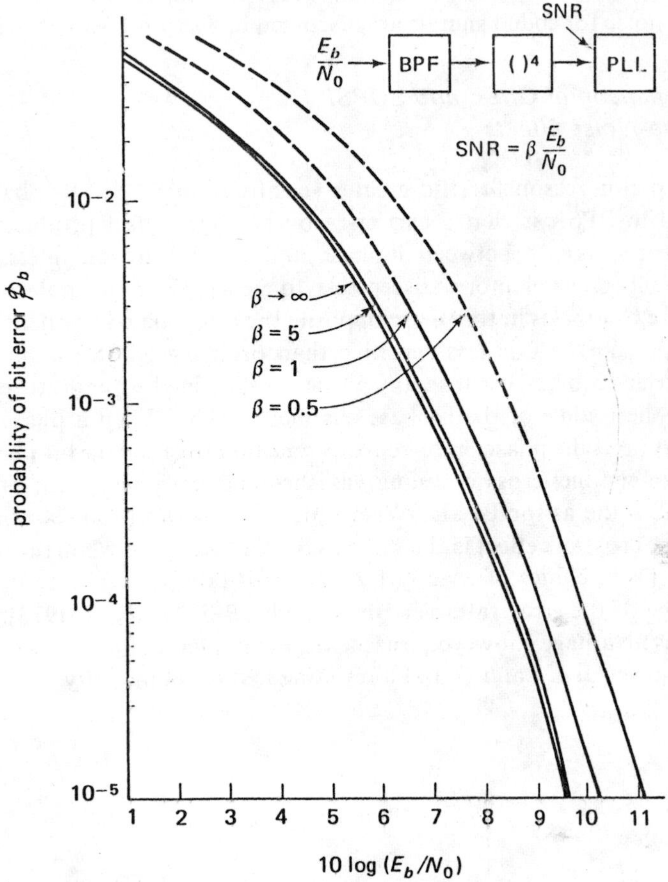

Fig. 12-28 Probability of bit error $\mathcal{P}_b$ versus E_b/N_0 for QPSK [Sherman, 1969]

we require $B_n = W/230$ or $B_n = 1/115T$. For coded operation one would typically design the carrier reconstruction loop for good operation at small $E_b/N_0 \approx 0$ dB, thus requiring a ratio of IF bandwidth to PLL bandwidth $W/B_n \approx 400$. For a phase-locked loop operating at moderately high $E_b/N_0 \approx 10$ dB, the SNR for this $W/B_n = 400$ is then approximately (for $a = 2$)

$$\alpha = SNR_{PLL} \triangleq \beta \frac{E_b}{N_0} \cong \frac{E_b}{N_0}\left(\frac{W/B_n}{20}\right) = \frac{10(400)}{20} = 200 \text{ or } 23 \text{ dB} \tag{12-192}$$

where we have used the value of β from (12-187) for $E_b/N_0 = 10$. Note that this loop SNR is significantly higher, by approximately 7 to 10 dB, than that required for BPSK [Chiu and Simpson, 1973, 945–948]. Additional constraints on phase noise for coded signals are discussed in Section 15-5.

Comparison of QPSK and SQPSK Phase-Noise Effects

The major reason for the greater sensitivity of QPSK to phase errors compared to BPSK is that a carrier-recovery phase offset produces a cross coupling or crosstalk between inphase and quadrature channels, thereby making each channel more susceptible to noise when the polarity of the quadrature channel is in the phase opposite the crosstalk component polarity.

Staggered QPSK is less sensitive than ordinary QPSK to phase noise in the carrier recovery because any phase reversal in the quadrature channel occurs in the middle of the inphase channel symbol. When a phase reversal does occur (and the phase error remains constant over a data bit period), the quadrature channel cross coupling vanishes and the phase-error effects are exactly the same as for BPSK. Where there is no transition, of course, the phase noise crosstalk effect is the same as for QPSK. Thus, when the data have transition probabilities of exactly 1/2, the error-rate performance is equal to the average of the error rates for BPSK and QPSK [Rhodes, 1973]. SQPSK has the disadvantage, however, that a transient phase slip of $n\pi/2$ radians can cause reversal of I and Q and a resulting loss of bit integrity.

CHAPTER 13

FILTER DISTORTION EFFECTS ON PSK SIGNALS

13-1 INTRODUCTION

There are numerous places in a satellite communications link where bandpass filters appear and can cause nonnegligible distortion. These include: the upconverter and IF amplifier in the earth terminal, earth terminal diplexer filter, satellite transponder, earth terminal downconverter and IF amplifier, and communications modems. As discussed in Chaps. 7 and 8, these filters are essential just to prevent overloads in the earth terminals, to avoid spurious, and to separate and to transpond properly the signals in the satellite transponder. Filtering also is necessary to avoid adjacent channel interference in FDMA systems and to make efficient use of the available frequency spectrum.

This chapter begins with a summary of the filter distortion effects in a typical satellite communications link. These effects include the signal distortion itself and intersymbol interference effects. Several means are available to characterize the filter distortion characteristic of the linear filters in the earth terminal-satellite-earth terminal link. Typical representations of the filter distortion include the amplitude ripple and various phase nonlinearities or group delay fluctuations. These phase nonlinearities are sometimes approximated by parabolic and cubic phase nonlinearities, or, equivalently, by linear and parabolic group delay variations with frequency. Alternatively there may be a random ripple in group delay. The parameters of group delay variation and phase nonlinearity can be easily related to one another for some simple forms of linear filter distortion.

The communication channel is considered linear in this chapter. Thus we assume that there is sufficient power back-off in the amplifiers in the link to avoid saturation or other nonlinear effects. In practice, however, there will often be some of both—linear filter and nonlinear amplifier—types of distortion.

The effects on error rate caused by some simple types of filter distortion are derived for BPSK, QPSK, 8-PSK, and 16-PSK. These effects depend not only on the filter distortion itself but on the position of the filter within the link—for example, before or after power amplification, and before or after additive thermal noise. The effects on correlation loss with ranging signals is discussed in Chapter 18.

Finally, the problem of filter and channel equalization is discussed. Most critical channels employ some form of phase and amplitude equalization either in the filters themselves or in the modems. The discussion here is limited to equalization techniques employing transversal equalizers.

13-2 FILTER DISTORTION MODEL IN A SATELLITE COMMUNICATION LINK

Figure 13-1 illustrates the typical sequence of filtering and amplification operations in a satellite communications link. The placement of these filters, for example, before or after power amplification or before or after the addition of noise in a low-noise receiver, can be as important in determining their effect as the type of filtering. Note that some filters are located ahead

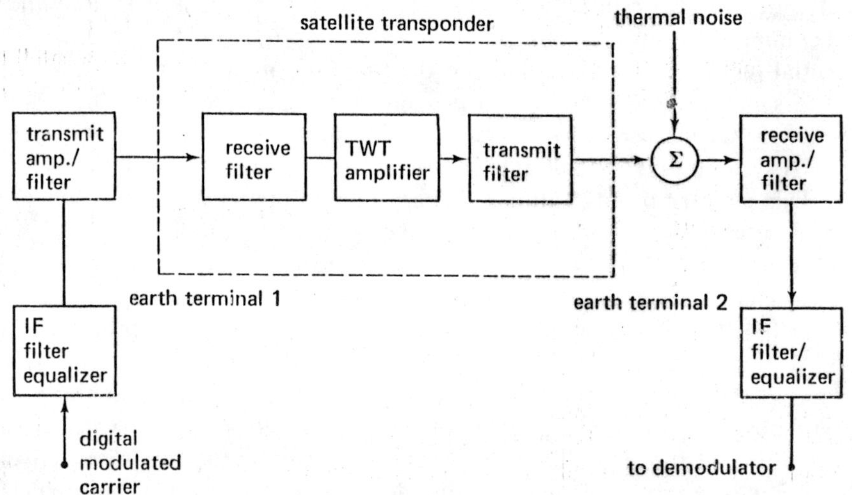

Fig. 13-1 Block diagram showing the sequence of filtering and amplifying operations that contribute to filter distortion in a satellite communication link

of the satellite TWT amplifier. Sidebands removed by filtering prior to this amplification, therefore, do not use TWT power (unless perhaps the TWT limits and regenerates some of these sidebands). Also note that some of the filtering is introduced after the noise is added and, hence, filtering removes some of the noise as well as the signal. Thus, the placement of the filter is important in determining its effect in degrading the error rate.

The output of the digital modulator at IF frequency, typically 70 MHz or 700 MHz, first enters the transmit IF filter/amplifier and equalizer. The equalizer is designed to linearize the phase and to flatten the amplitude response of the IF amplifier/upconverter assembly. Other earth terminal filters are provided to isolate the transmit and receive frequency channels as described in Chap. 8. Additional phase distortions can be introduced by waveguide reflections. Delay distortion is also introduced by optical fibers and coaxial cables which can be used as user-earth terminal links [Miller, 1974, 174–194; Marcuse, 1974, 195–216].

The earth-station transmission then travels to the satellite where the propagation medium generally introduces little additional phase distortion at microwave frequencies (> 1 GHz).* After the signal enters the satellite, it encounters receive channel filtering which is employed to separate the channels in a multichannel transponder. Additional filtering then takes place in the satellite as the signals are combined after frequency translation and amplified in the TWT or other power amplifier.

Finally, a second earth terminal receives the signal from the satellite, where it is filtered by a diplexer amplified in a low-noise amplifier, typically a paramp or tunnel diode amplifier, and downconverted to IF. Additional amplification filtering and equalization takes place at IF where the signals are amplified to a level suitable for use by the signal demodulators.

The linear filter transfer characteristics of an earth terminal or satellite can be specified by the amplitude flatness and by the variations of group delay with frequency or the deviation from linearity of the phase response. The phase linearity is a parameter more easily related to signaling performance while group delay flatness is more often measured. Thus, it is appropriate to relate the two quantities in some simple examples. Group delay τ_g is simply the derivative of phase shift, $\phi(j\omega)$, with respect to frequency:

$$\tau_g(j\omega) = \frac{d\phi(j\omega)}{d\omega} \tag{13-1}$$

Figure 13-2 shows examples of group delay and phase response for parabolic and linear group-delay functions of frequency. Appendix B shows the phase, amplitude, and group-delay responses of Butterworth, Chebyshev, Bessel,

*Some phase distortion is introduced by the frequency dependence of group delay in the ionosphere. These effects can become significant at signal bandwidths 10 percent of the RF center frequency (Chap. 17).

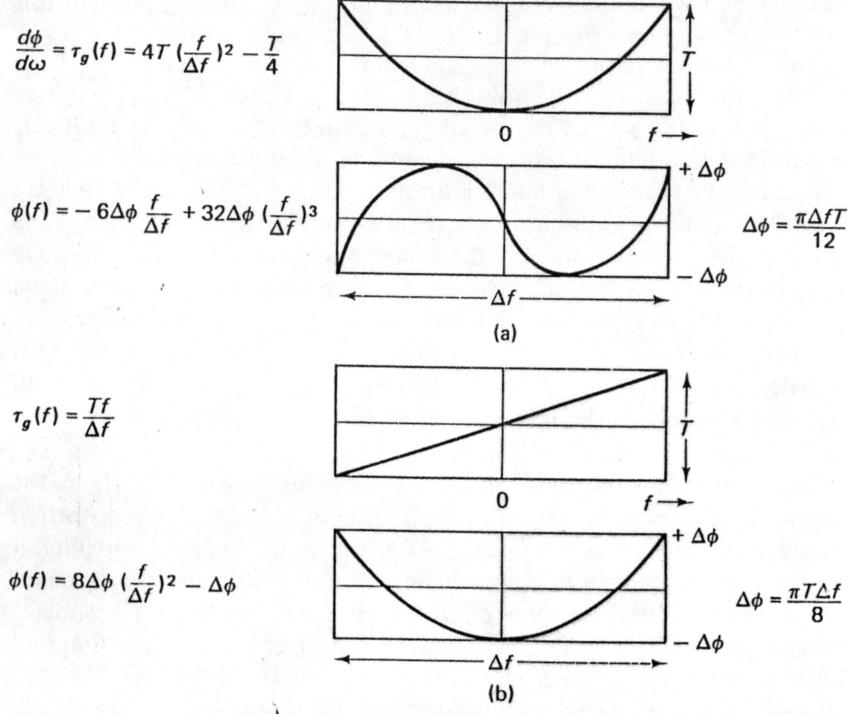

Fig. 13-2 Typical relationships between group-delay variations $\tau_g(j\omega)$ with frequency and phase $\Delta\phi(j\omega)$ directions from a best-fit linear phase. Shown here (a) parabolic group delay, and (b) linear group delay.

transitional Butterworth-Thompson, and Legendre filters for various numbers of poles.

13-3 FILTER DISTORTION EFFECTS ON PSK AND QPSK SIGNALS

Almost any transmission channel requires bandpass filtering in order to reduce received interference from an adjacent channel and to reduce transmission interference in an adjacent channel. This filtering causes a certain amount of signal distortion and intersymbol interference, thereby degrading the system noise performance.

We first compute the effect of linear transmission filtering on the error probability of a biphase or quadriphase signal. This effect is measured in terms of the E_b/N_0 performance degradation for a given error rate caused by the transmission filter. A typical demodulator for a filtered QPSK signal is shown in Fig. 13-3. (Note that integrate/dump detection is no longer

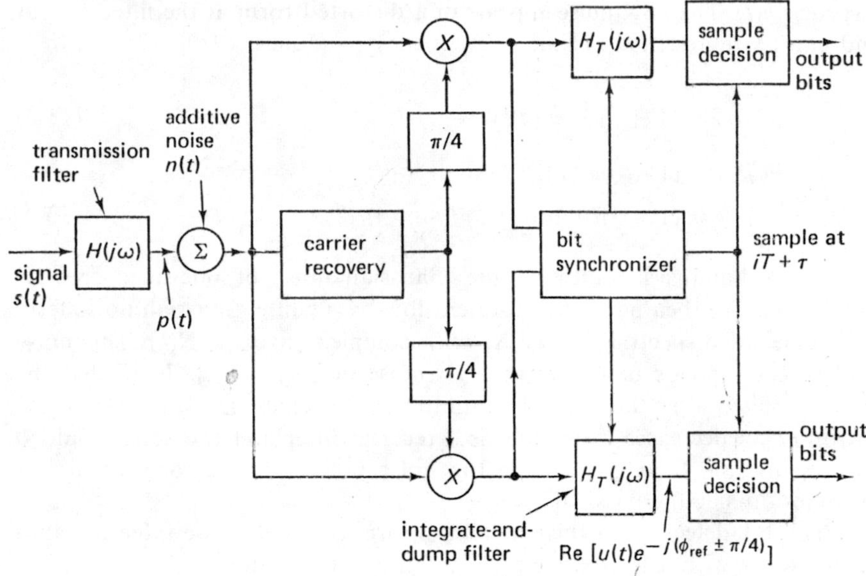

Fig. 13-3 Block diagram of a QPSK demodulator; the transmission filter represents filtering at the transmitter or in the satellite transponder.

exactly optimum.) We represent one of the four possible phases of a quadriphase signal pulse during the time interval $(0, T)$ by complex envelope notation

$$\begin{aligned} s(t) &= \text{Re}\,[u(t)e^{j\omega_0 t}] = \text{Re}\,[(1+j)u_r(t)e^{j\omega_0 t}] \qquad 0 < t < T \\ &= [\cos\omega_0 t + \sin\omega_0 t]u_r(t) \end{aligned} \tag{13-2}$$

which has unity power and where $u_r(t)$ is a unit amplitude rectangular pulse over $(0, T)$, $j \triangleq \sqrt{-1}$, and $u(t)$ is the complex envelope [Stein and Jones, 1967]. See Fig. 13-4. The other QPSK signal phases are $(1+j)(j)^{i_k}$, which has values $(-1+j)$, $(1-j)$, $(-1, -j)$. BPSK is a special form of this signal where we have only $(1+j)$, $(-1, -j)$ when $s(t)$ enters the filter.

The transmission filter is represented either by its low-pass equivalent frequency transfer function $H(j\omega)$ or by the low-pass equivalent impulse

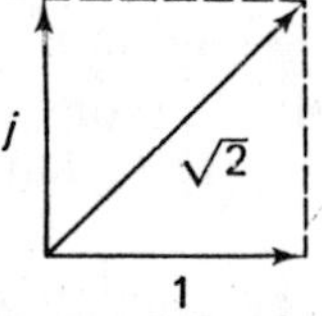

Fig. 13-4 Phasor representation of $(1+j)e^{j\omega_0 t}$ in (13-2)

response $h(t)$. The PSK pulse appears in a distorted form at the filter output and can be represented in time or frequency by [Jones, 1970]

$$\mathbf{p}(t) = (1 + j) \int_0^T u_r(\tau)h(t - \tau)\, d\tau \tag{13-3}$$

$$\begin{aligned} P(j\omega) &= (1 + j)u_r(f)H(j\omega) \\ &= (1 + j)H(j\omega)Te^{-j\pi fT} \operatorname{sinc} \pi fT \end{aligned} \tag{13-4}$$

Assume a bandpass additive white Gaussian noise of one-sided spectral density N_0. We then have an equivalent low-pass white Gaussian noise term of double-sided spectral density N_0 with complex envelope $\mathbf{N}(t)$. This noise density corresponds to a one-sided IF noise density of N_0. Note that the noise is added after the transmission filter in this example. The filter degradation effect decreases if the filter is a receiver filter and the noise is added before filtering, thereby reducing the noise power spectral density at frequencies outside the passband.

The bit detectors in this calculation are assumed to be integrate-and-dump filters with a low-pass equivalent transfer function

$$H_T(j\omega) = \frac{1 - e^{-j\omega T}}{j\omega} = Te^{-j\pi fT} \operatorname{sinc} (\pi fT) \tag{13-5}$$

It is assumed that the bit-synchronizer timing circuits compensate for the group delay of the filter at the signal center frequency. This delay compensates for the linear phase portion of the filter transfer characteristic.

Let us first examine one quadrature channel in detail. The composite signal output of the transmission filter (the present pulse plus "tails" from K previous and K future pulses) appears after phase detection in the form

$$\Big[\underset{\substack{\uparrow \\ \text{present pulse}}}{\mathbf{p}(t)} + \underbrace{\sum_{k=-K}^{K}{}' (-j)^{i_k}\mathbf{p}(t + kT)e^{j\omega_0 T}}_{\text{contribution from } 2K \text{ previous and future pulses}}\Big] e^{-j(\phi_{rel} \pm \pi/4)} \tag{13-6}$$

where the relative phases of the previous pulses are determined by the i_k in $(-j)^{i_k} = \pm 1, \pm j$ for $i_k = 0, 1, 2, 3$—the four phase possibilities. If we have only biphase signals (BPSK) then, $i_k = 0, 2$ and $(-j)^{i_k} = \pm 1$.

Assume for simplicity that $e^{j\omega_0 T} = 1$; that is, there is an integral number of carrier periods in a symbol period of T sec. We define the output of the integrate-and-dump filter (omitting the complex constant $e^{j(\phi_{rel} \pm \pi/4)}$ until later) at the sampling time $T + \tau$ using (13-5) and (13-6) as the complex quantity

$$\begin{aligned} \mathbf{v}(T + \tau | i_1, i_2, \ldots, i_K) &\triangleq \int_0^T \mathbf{p}(t + \tau)\, dt + \sum_{k=-K}^{K}{}' (-j)^{i_k} \int_{kT}^{(k+1)T} \mathbf{p}(t + \tau)\, dt \\ &= (1 + j)T \end{aligned} \tag{13-7}$$

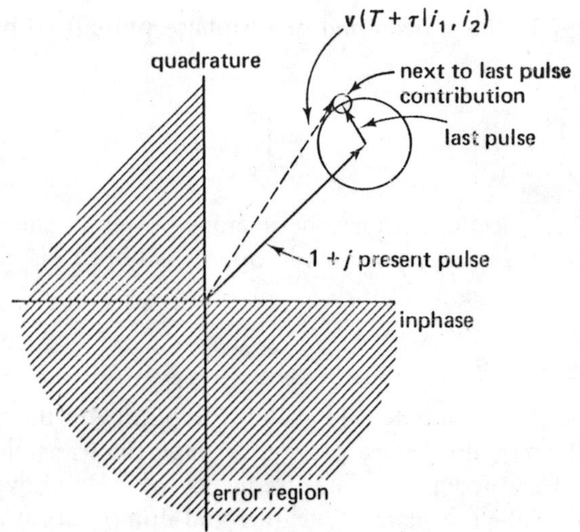

Fig. 13-5 Phasor representing integrate-and-dump filter output, showing intersymbol interference effects on a QPSK signal

if there is no distortion. The phasor representing v(   ) and the error boundaries are shown in Fig. 13-5. Use the identity for the integrated pulse

$$\int_0^{nT} \mathbf{p}(t+\tau)\,dt = \int_{-\infty}^{\infty} \underbrace{P(j\omega)}_{\text{distorted signal}} \underbrace{e^{j2\pi f\tau}}_{\text{signal}} \underbrace{\mathbf{H}_{nT}(j\omega)e^{j2\pi fnT}}_{\text{integrate and dump 0 to } nT}\,df \tag{13-8}$$

where $\mathbf{H}_{nT}(j\omega) \triangleq nTe^{-j\pi fnT}\,\text{sinc}\,(\pi fnT)$. The expression for $\mathbf{v}(T+\tau)$ can be rewritten by defining the series

$$\mathbf{z}_K(j\omega) \triangleq 1 + \sum_{k=-K}^{K}{}' (-j)^{i_k} e^{j\omega kT}, \tag{13-9}$$

which defines the sequence of the adjacent $2K$ symbols. Thus we obtain

$$|\, i_K, \ldots, i_{-1}, i_1, , , , i_K) = T\int_{-\infty}^{\infty} \underbrace{P(j\omega)}_{\text{distorted signal}} \underbrace{\mathbf{z}_K(j\omega)}_{\text{sum of } 2K \text{ pulse effects}} \underbrace{\text{sinc}\,(\pi fT)}_{\text{integrate and dump}} \underbrace{e^{j\pi f(T+2\tau)}}_{\text{delay in sampling}}\,df \tag{13-10}$$

where $\mathbf{z}_K$ takes into account the relative phases of the previous pulses.

The average noise power in the integrate-and-dump output is

$$\sigma_n^2 \triangleq N_0 \int_{-\infty}^{\infty} G_T^2(f)\,df = N_0T^2 \int_{-\infty}^{\infty} \text{sinc}^2\,(\pi fT)\,df = N_0T \tag{13-11}$$

and the total energy in the transmitted quadriphase pulse (two bits/symbol in T sec) is $2E_b$:

$$2E_b = \frac{1}{2}\int_0^T |u(t)|^2 = \frac{1}{2}|1+j|^2 \int_0^T [u_r(t)]^2\, dt = T \tag{13-12}$$

The factor of 1/2 occurs because the average power in the sine wave, $\sin \omega_0 t$, is 1/2. Thus, the energy-per-bit-to-noise-density ratio is

$$\frac{E_b}{N_0} = \frac{T}{2N_0} \tag{13-13}$$

where N_0 is the one-sided noise density at the IF input. Note that this quantity E_b is the signal energy/bit **before** passing through the transmission filter. If the signal is displaced from the center frequency by ΔHz above the filter center frequency, then the normalized integrate-and-dump output is obtained by using (13-10), (13-4) and by substituting the appropriate value of $P(j\omega)$:

$$\frac{\mathbf{v}}{\sqrt{2}\,\sigma_n} = \frac{1}{\sqrt{2N_0T}}T \int \underbrace{[(1+j)H[j2\pi(f+\Delta)]e^{-j\pi fT}T \operatorname{sinc} \pi fT]}_{P(j\omega)} \times \mathbf{z}_K(j\omega) \operatorname{sinc}(\pi fT)e^{+j\pi f(T+2\tau)}\, df \tag{13-14}$$

We note that $|\mathbf{v}|^2/2\sigma_n^2$ is the SNR on the inphase or quadrature data channel. Equation (13-14) reduces to

$$\begin{aligned}\frac{\mathbf{v}}{\sqrt{2}\,\sigma_n} &= \frac{\mathbf{v}(T+\tau\,|\,i_{-K},\ldots,i_{-1};i_1\ldots,i_K)}{\sqrt{2}\sigma_n} \\ &= \frac{Te^{j\pi/4}}{\sqrt{2}}\sqrt{\frac{2E_b}{N_0}}\int_{-\infty}^{\infty} H[j2\pi(f+\Delta)]\mathbf{z}_K(f)\operatorname{sinc}^2(\pi fT)e^{j\omega\tau}\, df\end{aligned} \tag{13-15}$$

where $(1+j) = \sqrt{2}\,e^{j\pi/4}$ and $\sigma_n^2 = N_0T$.

Probability of Error

The bit decision is based on a signal component at the integrate-and-dump output normalized to the rms noise. This ratio is the real part of

$$\begin{aligned}&\operatorname{Re}\left[\frac{\mathbf{v}(T+\tau\,|\,i_{-K},\ldots,i_{-1};i_1\ldots,i_K)}{\sqrt{2}\,\sigma_n}e^{-j(\phi_{ref}\pm\pi/4)}\right] \\ &\quad = \frac{|\mathbf{v}(T+\tau\,|\,i_{-K},\ldots,i_{-1};i_1\ldots,i_K)|}{\sqrt{2}\,\sigma_n}\cos\left(\theta_{i-K,\ldots,i_{-1};i_1,\ldots,i_K} - \phi_{ref} \pm \frac{\pi}{4}\right)\end{aligned} \tag{13-16}$$

where θ is the angle of the vector $\mathbf{v}$, and the factor $\sqrt{N}$ is the normalization. The probability of bit error for a given sequence of previous and future symbols $i_{-K}, \ldots, i_{-1}; i_1, \ldots, i_K$ is

$$\mathcal{P}_b = \frac{1}{2} \operatorname{erfc} \sqrt{\gamma_{i_{-K},\ldots,i_{-1};i_1,\ldots,i_K}} \geq \frac{1}{2} \operatorname{erfc} \sqrt{\frac{E_b}{N_0}} \tag{13-17}$$

where the SNR is

$$\begin{aligned} \gamma_{i_{-K},\ldots,i_{-1};i_1,\ldots,i_K} &\triangleq \frac{1}{2\sigma_n^2} |\mathbf{v}(T+\tau | i_{-K}, \ldots, i_{-1}; i_1, \ldots, i_K)|^2 \\ &\qquad \cos^2\left(\theta_{i_{-K},\ldots,i_{-1};i_1,\ldots,i_K} - \phi_{\text{ref}} \pm \frac{\pi}{4}\right) \\ &= \frac{|\mathbf{v}|^2 \cos^2 \theta}{\sigma_n^2} \end{aligned} \tag{13-18}$$

and

$$\operatorname{erfc} x \triangleq \left(\frac{2}{\sqrt{\pi}}\right) \int_x^{\infty} e^{-t^2}\, dt \tag{13-19}$$

The average bit-error probability for either data channel is obtained from (13-17) by averaging over all 4^K combinations of the sequences of i_k. Thus, the average bit-error probability is

$$\overline{\mathcal{P}}_b = \frac{1}{4^{2K}} \sum_{\text{all } i_k} \frac{1}{2} \operatorname{erfc} \sqrt{\gamma_{i_{-K},\ldots,i_{-1};i_1,\ldots,i_K}} \qquad i_k = 0, 1, 2, 3 \tag{13-20}$$

For purposes of this analysis, we take the reference phase equal to the square-law weighted average:

$$\phi_{\text{ref}} = \frac{\sum_{\text{all } i_k} |\mathbf{v}(T+\tau | i_{-K}, \ldots, i_{-1}; i_1, \ldots, i_K)|^2 \theta_{i_{-K},\ldots,i_{-1};i_1,\ldots,i_K}}{\sum_{\text{all } i_K} |\mathbf{v}(T+\tau | i_{-K}, \ldots, i_{-1}; i_1, \ldots, i_K)|^2} \tag{13-21}$$

Solutions for (13-20) are generally complex and require calculation by computer.

Filter distortion degradation is shown in Fig. 13-6 for filters with parabolic or cubic group delay and flat amplitude response. The quantity plotted is the increase in E_b/N_0 required for an error probability of 10^{-6}. The transmission symbol rate is $\mathcal{R}_s = \mathcal{R}/n$, where $\mathcal{R}$ is the information bit rate and n is the number of bits/symbol where no coding is employed. Notice that for parabolic phase, BPSK is degraded by 0.52 dB at $\Delta\phi = 10°$. For a symmetrical-filter-amplitude transfer characteristic about the signal center frequency and an asymmetrical phase function, the degradation for BPSK and QPSK is the same if the symbol rates are the same for both. In other words, for a given symmetrical filter function, exactly twice the data rate can be

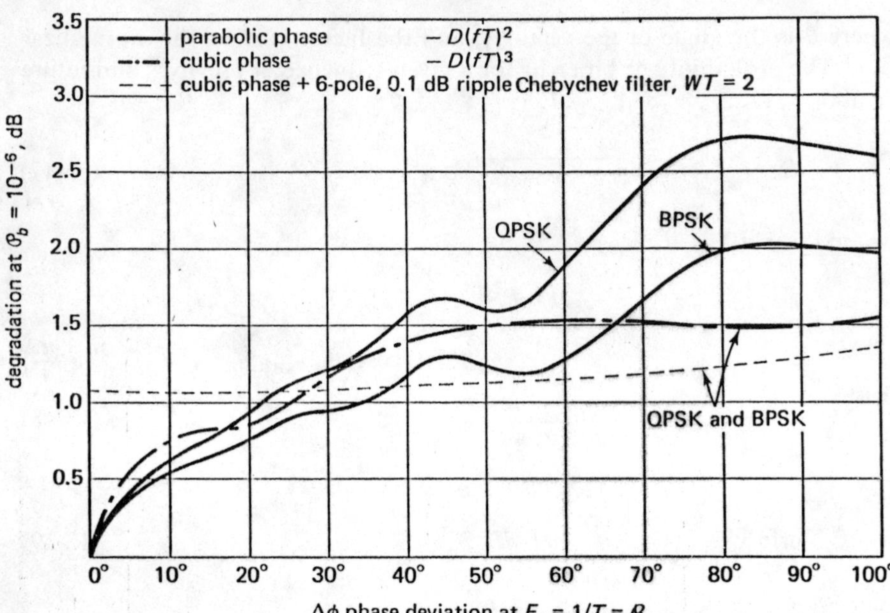

Fig. 13-6 Degradation in effective E_b/N_0 for parabolic and cubic phase nonlinearities and BPSK, QPSK modulation [From Jones, 1971, 132]

transmitted with QPSK as with BPSK for the same degradation in E_b/N_0. This result is easily shown because there is no crosstalk between inphase and quadrature channels for symmetrical filters. Hence, the inphase and quadrature channels of QPSK are affected just as if the other did not exist.

The absence of crosstalk between inphase and quadrature channels for a symmetrical filter can be shown by using a complex representation of bandpass signals and their Fourier transform

$$a(t) + jb(t) \longrightarrow A(j\omega) + jB(j\omega) \tag{13-22}$$

Hence, the effect of filtering $H(j\omega)$ is to produce an output with transform

$$H(j\omega)A(j\omega) \triangleq \hat{A}(j\omega) \triangleq |\hat{A}(j\omega)|^{j\alpha(j\omega)} \tag{13-23}$$

where $|A(j\omega)|$ is even in ω for a symmetrical filter.

If the filter output time waveform is real for real input $a(t)$, there is no crosstalk. The output is

$$\begin{aligned} a_0(t) &= \int |\hat{A}(j\omega)|\, e^{j\alpha(j\omega)} e^{-j\omega t} \\ &= \int |\hat{A}|[\cos(\alpha - \omega t) + j\sin(\alpha - \omega t)]\, d\omega \end{aligned} \tag{13-24}$$

If $|H(j\omega)|$ is even and $\angle H(j\omega)$ is odd in ω, where $\angle H$ represents the phase transfer functions of $H(j\omega)$, then

$$j \int |\hat{A}| \sin(\alpha - \omega t) \, d\omega = 0$$

Thus the output $a_0(t)$ is real in (13-24), and hence there is no crosstalk [Brayer, 1971].

Filter distortion losses have also been computed for the Chebyshev bandpass filters of Fig. 13-7. In this case, some of the signal power is removed by

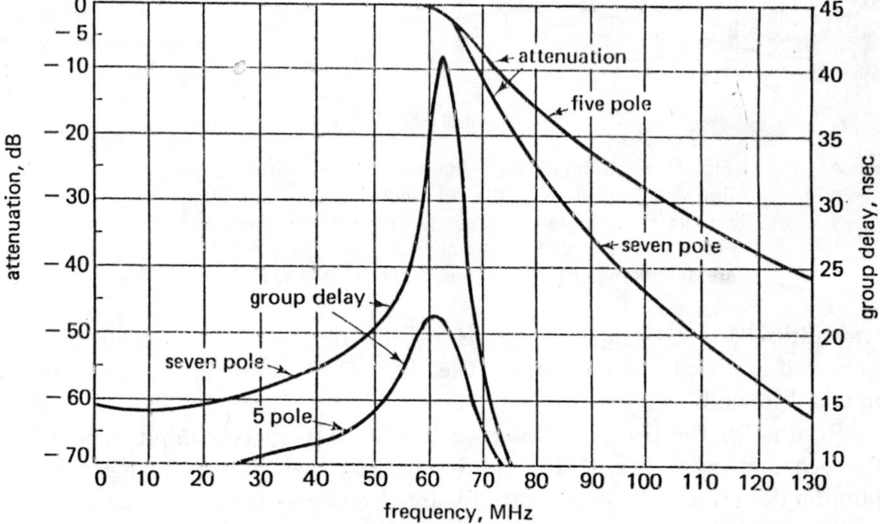

Fig. 13-7 Selectivity and phase response for Chebyshev filters

the filter. The filtering can occur either before or after power amplification; thus, the full power output may or may not be devoted to the filtered signal. If the signal is filtered before power amplification, the full power is in the filtered spectrum. It also is important to note whether the receiver noise is added before or after filtering. Figure 13-8 shows these results by Jones [1971, 120–132].

More general results for filter distortion are shown in Fig. 13-9 for Chebyshev bandpass filters placed after power amplification. (Noise is added after filtering.) The result is shown for two-pole Butterworth data detection filters with a 3 dB bandwidth $B_d = \mathcal{R}_s \triangleq 1/T$ in addition to the result for integrate-and-dump (I/D) detection. The symbol rate is $\mathcal{R}_s$ and the detection filter is sampled once per symbol. Note that the distortion loss for the two-pole Butterworth detection is less than that for I/D filters at narrow RF

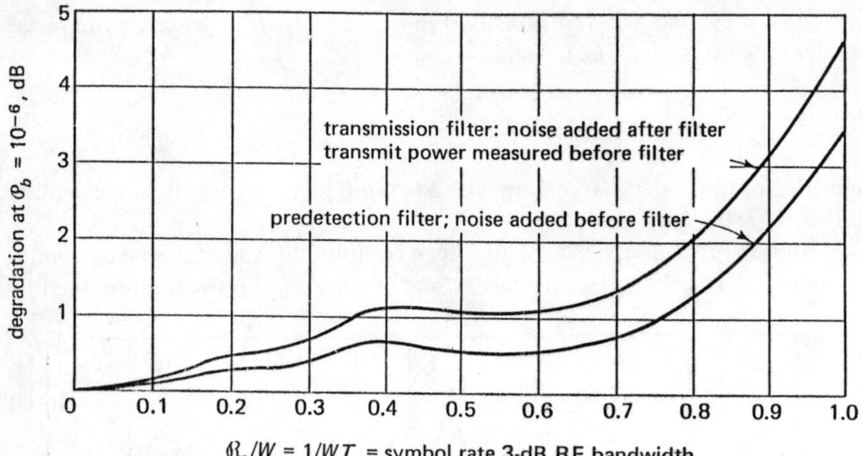

Fig. 13-8 Comparison of transmission and predetection filtering: 5-pole, 0.1-dB ripple Chebyshev filter integrate-and-dump detection. The results apply to either QPSK or BPSK symbol rates because the filters are symmetrical. [From Jones, 1971, 120–132]

bandwidths $W < 2/T$ Hz. This filter is a better match to the filter distorted signal and is simple to implement. Clearly, I/D detection is not optimum for this distorted channel.

Results for the two-pole Butterworth filter detection without filter distortion are given in Fig. 13-10, showing that $B_dT = 1$ is approximately the optimum detection filter bandwidth. Figure also shows the power truncation loss if the signal bandwidth is truncated to the bandwidth W. If this filtering (without phase distortion) is performed before power amplification and the power is increased to full level, then the additional distortion caused by the bandpass filter is the difference between the two curves. For example, if the $\mathcal{R}_s/W = 0.6$, the power truncation loss is 0.5 dB. If a five-pole bandpass filter ($\mathcal{R}_s/W = 0.6$) is added to the channel at the power amplifier output, the total is 1.1 dB. However, if the signal spectrum had been filtered before power amplification (and the sidebands do not build back up during amplification*), the net filter distortion loss is $1.1 - 0.5 = 0.6$ dB.

The effect of a Butterworth bandpass filter for I/D detection is shown in Fig. 13-9b. Note that if the transmitted signal is a bandlimited sequence of $\pm(\sin \pi Bt/\pi Bt) \sin \omega_0 t$ pulses, the optimum matched filter detector after coherent detection is an ideal rectangular filter. The filter output is sampled at the bit rate B. Clearly, I/D detection is nonoptimal for this bandlimited signal.

---

*Hard limiting of PSK during power amplification causes the sidebands to resume their original level, whereas limiting of staggered QPSK does not greatly increase the sidebands from their prefiltered value.

Fig. 13-9 Degradation caused by Chebyshev filtering on PSK and QPSK signals for integrate-and-dump and two-pole Butterworth filters; (a) filter degradation and loss in signal energy caused by bandwidth truncation, and (b) the degradation caused by Butterworth band-pass filters on BPSK and QPSK modulation with I/D detection. Noise is added after transmission filtering. [After Jones, 1975]. Adding receive filtering can improve performance by eliminating noise in the integrate-and-dump sidelobes (See Table B-3)

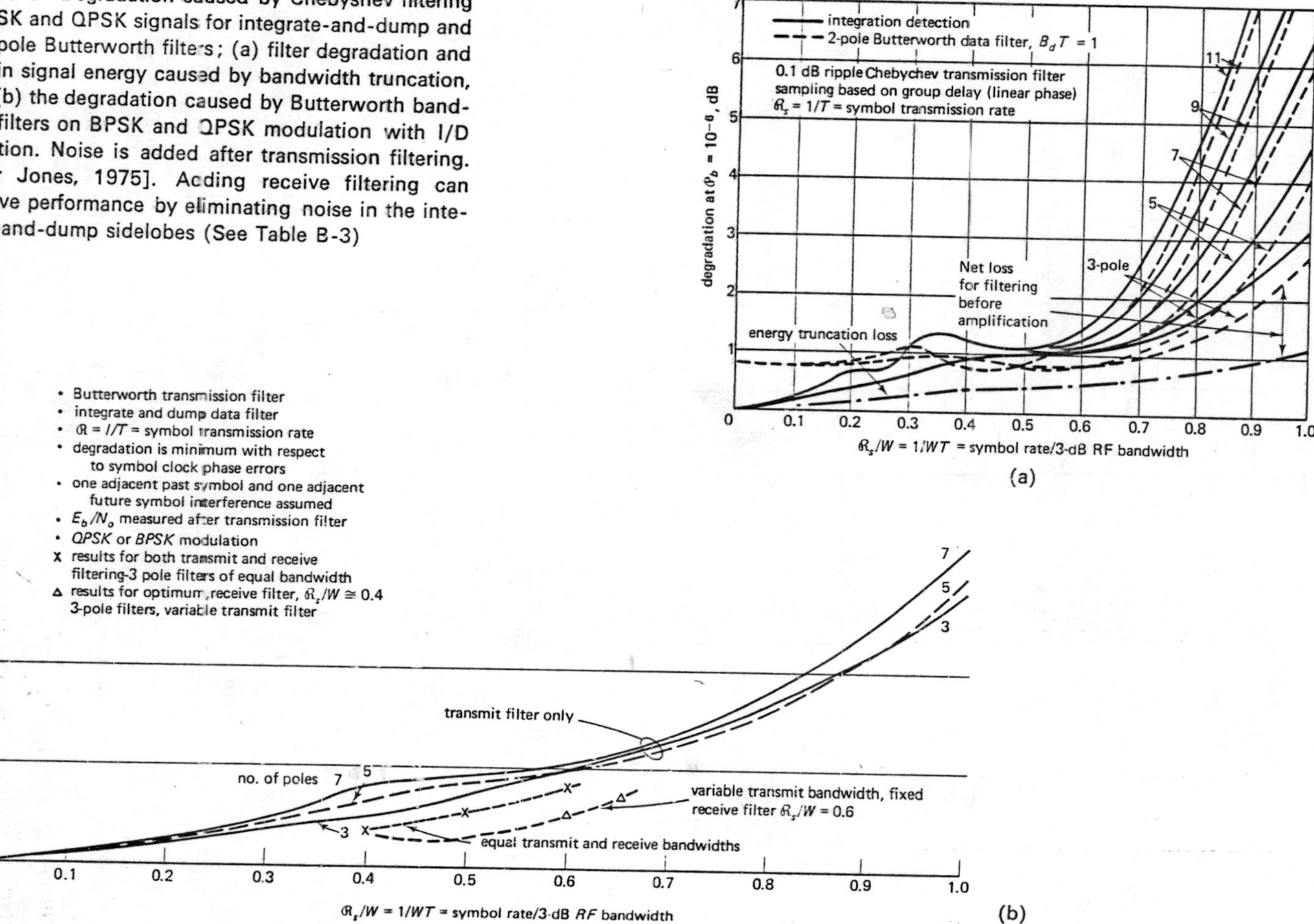

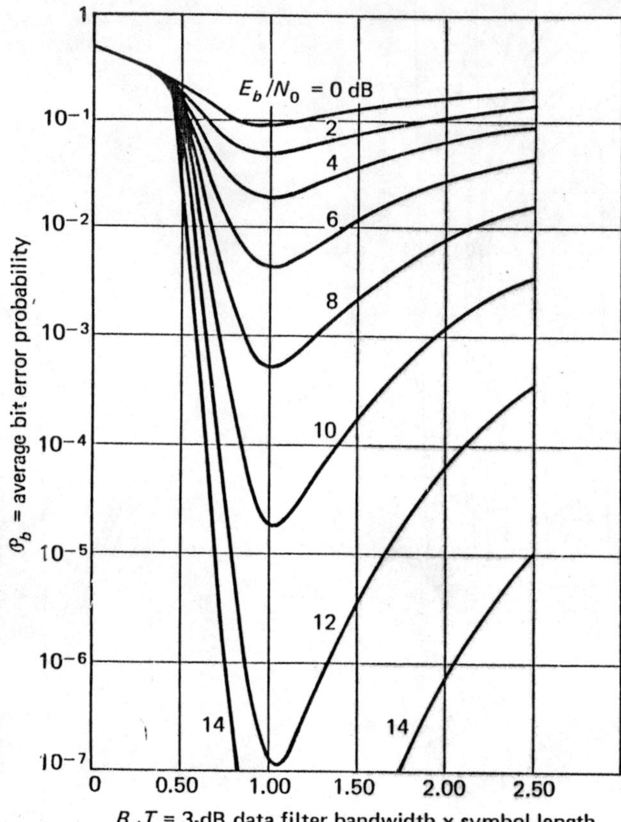

Fig. 13-10 Two-pole Butterworth data filter detection of coherent QPSK and BPSK [From Jones, 1971, 127]

13-4 FILTER DISTORTION EFFECTS ON QPSK AND 8-ARY PSK

Similar results for QPSK and 8-ary PSK have been obtained for three-pole maximally flat Butterworth filters of 3-dB bandwidth W. This receiver consists of a filter, coherent phase detector, sampler, and decision device. Both signal and noise levels are measured after filtering. Each symbol decision is based on a single phase detector output sample. Figure 13-11 shows the required increase in SNR for a symbol error probability $\mathcal{P}_s = 10^{-6}$. Note the rapid increase in loss for $M = 8$ PSK as the 3-dB bandwidth W approaches the symbol rate. The losses decrease rapidly for Butterworth filters as $\mathcal{R}_s/W$ decreases below $\mathcal{R}_s/W = 0.8$.

Note that for 8-ary PSK, and a ratio of symbol rate to filter bandwidth,

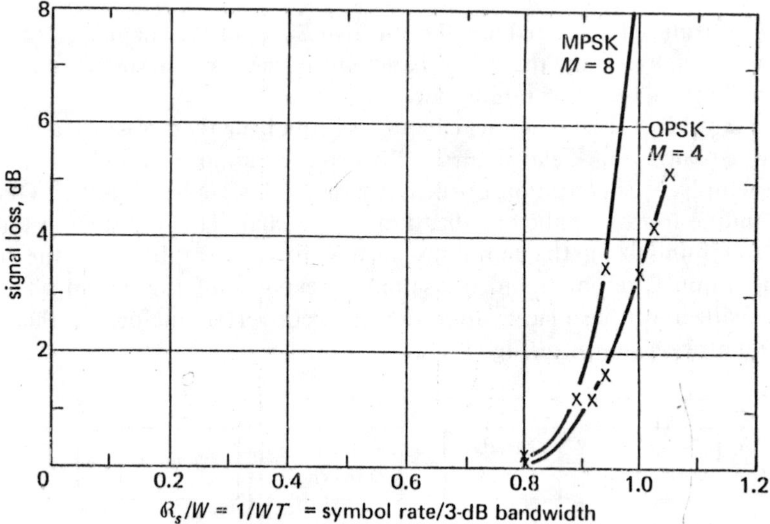

Fig. 13-11 Plot of signal degradation due to inter-symbol inteference caused by three-pole Butterworth filters for MPSK, $M = 4$, 8 and $\mathcal{P}_s = 10^{-6}$ at SNR = 13.6dB and 19.0 dB for QPSK and 8 PSK for $R_s/W = 0$ [After Prabhu, 1973]

$\mathcal{R}_s/W = 0.8$ corresponds to a ratio of information bit rate to bandwidth of $\mathcal{R}/W = 2.4$ because there are $\mathcal{R}_s/\mathcal{R}_s = 3$ bits/symbol. Thus there are 2.4 bits transmitted per Hz. However, this bandwidth efficiency is obtained with a decrease in SNR efficiency as noted in Chap. 11.

13-5 BANDPASS TRANSVERSAL EQUALIZERS

We have shown that the effects of distortion in the phase and amplitude transfer function of the channel can accumulate to the point where a serious degradation in performance results at high bit rates. At the same time, however, it often is too costly to reduce tolerance of the filter/amplifier performance beyond a given point, and it may be necessary to compensate for these distortions by a fixed or variable equalizer. In this manner, the channel once built and properly operating can be equalized to provide a flatter amplitude response and a more linear phase response with frequency. Because of circuit and system tolerances and variations with temperature, this equalization must be set after construction of the equipment and perhaps periodically adjusted with time.

Both linear filter equalizers and nonlinear decision-feedback equalizers can be designed. The linear equalizer can be designed to eliminate totally the intersymbol interference of the filter. In the process, however, additional noise is introduced which reduces the effective E_b/N_0 of the channel. Decision-feedback equalizers, on the other hand, introduce only a small additional noise as long as the error rate is low.

Here we first discuss a linear bandpass equalizer that uses a transversal filter to equalize a modulated carrier distorted by an arbitrary channel. Particular emphasis is placed on operation with PSK signals, especially QPSK. A technique for automatic equalization of the channel is considered that is based on minimizing the minimum mean-square error between the ideal channel impulse response and the sampled response of the actual channel. The equalizer can be placed after the downconverter but before the PSK demodulator as shown in Fig. 13-12.

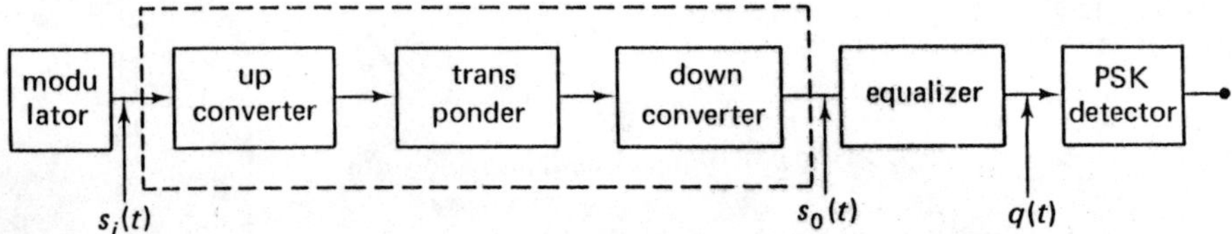

Fig. 13-12 Configuration of the bandpass equalizer in a satellite communication chain

Assume that the input MPSK signal is of the form

$$s_i(t) = \left[\sum_n a_n p(t - nT)\right] \cos \omega_0 t - \left[\sum_n b_n p(t - nT)\right] \sin \omega_0 t \tag{13-25}$$

where T represents the symbol duration and a_n, b_n are the data modulation components. For QPSK, a_n, $b_n = \pm a$, and $p(t)$ is the unit pulse function $p(t) = 1$, $0 \leq t \leq T$. For our purposes, $p(t)$ can represent any other shaped pulse response—for example, a sine-squared pulse or the sawtooth wave output of an integrate-and-dump filter.

The channel-response transfer function has an impulse response

$$h(t) = 2F_c(t) \cos \omega_0 t - 2F_s(t) \sin \omega_0 t \tag{13-26}$$

If the channel has an even amplitude response and an odd-phase symmetry about the carrier center frequency, then $F_s(t) = 0$ and there is no crosstalk between I and Q channels. The channel output for the input (13-25) is

$$\begin{aligned} s_0(t) &= \cos(\omega_0 t + \phi)\left[\sum_n a_n x(t - nT) - b_n y(t - nT)\right] \\ &\quad - \sin(\omega_0 t + \phi)\left[\sum_n b_n x(t - nT) + a_n y(t - nT)\right] + n(t) \\ &= s_I(t) \cos(\omega_0 t + \phi) + s_Q(t) \sin(\omega_0 t + \phi) \end{aligned} \tag{13-27}$$

where $x(t) \triangleq p(t)*F_c(t)$ and $y(t) \triangleq p(t)*F_s(t)$, and $*$ represents the convolution operation and $n(t)$ represents additive noise. The cross-modulation term $y(t)$ is zero for symmetric channels. However, even some channels with a nonzero $y(t)$ produce no distortion. For example, a channel that corresponds to a pure phase shift by μ rad causes no distortion and produces

$$\begin{bmatrix} x_\mu(t) \\ y_\mu(t) \end{bmatrix} = \begin{bmatrix} \cos\mu & \sin\mu \\ -\sin\mu & \cos\mu \end{bmatrix} \begin{bmatrix} x(t) \\ y(t) \end{bmatrix} \tag{13-28}$$

and results in the angle $\omega_0 t + \phi$ being changed to $\omega_0 t + \phi + \mu$. Since ϕ is arbitrary, this phase shift is irrelevent; the demodulator would recover it anyway.

Consider the equalizer structure [Gitlin et al., 1973, 219–238; Ryan and Ziemer, 1973, 18-4–18-9] shown in Fig. 13-13a and its alternate implementation which is a baseband equivalent as shown in Fig. 13-13b. The equalizer output $q(t)$ can be written

$$\begin{aligned} q(t) = \cos\omega_0 t[\textstyle\sum a_n g(t-nT) - b_n h(t-nT)] \\ - \sin\omega_0 t[\textstyle\sum b_n g(t-nT) + a_n h(t-nT)] \end{aligned} \tag{13-29}$$

where the a_n, b_n are the channel distortion coefficients and $g(t)$, $h(t)$ are the inphase and quadrature equalized signals. These equalized signals can in turn be described in terms of the equalizer coefficients c_n, d_n and the inphase, quadrature representation of $s_0(t)$ in (13-27) are

$$g(t) = \sum_{n=-N}^{N} c_n x(t-nT) - d_n y(t-nT) \tag{13-30}$$

$$h(t) = \sum_{n=-N}^{N} d_n x(t-nT) + c_n y(t-nT) \tag{13-31}$$

The value of phase shift $\omega_0 T = 2k\pi$ and the time reference has been taken as the center tap of the transversal equalizer.

If we change the tap weights to $\hat{c}_i$, $\hat{d}_i$, as in (13-28),

$$\begin{bmatrix} \hat{c}_i \\ \hat{d}_i \end{bmatrix} = \begin{bmatrix} \cos\theta & \sin\theta \\ -\sin\theta & \cos\theta \end{bmatrix} \begin{bmatrix} c_i \\ c_i \end{bmatrix} \qquad i = -N \text{ to } +N \tag{13-32}$$

the equalizer output is simply phase shifted by θ rad and the output is essentially unaffected because phase shift is not relevant at this point. Clearly, there must be both a c_0 and a d_0 tap to perform this rotation at the equalizer output, $i = 0$.

To adapt the equalizer coefficients, one must have a cost function to minimize. Ideally one would have the output at all time instants $g_i \triangleq$

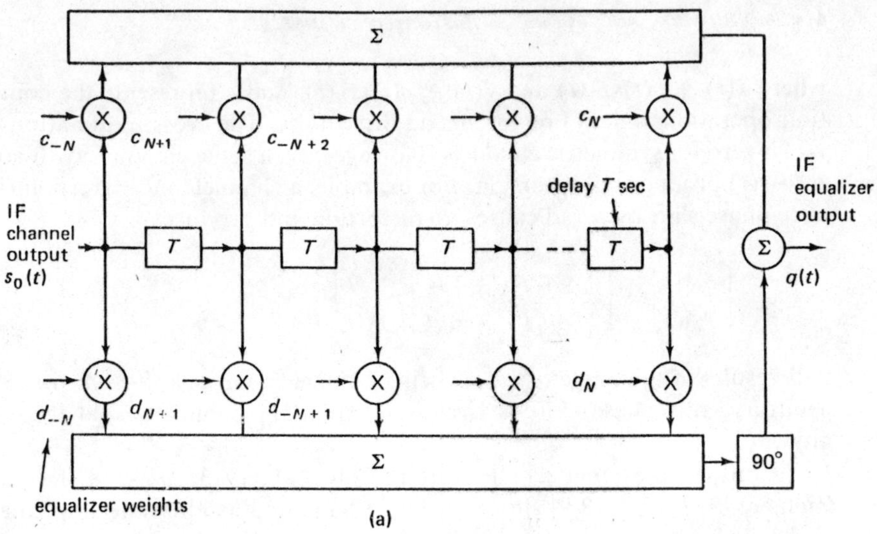

(a)

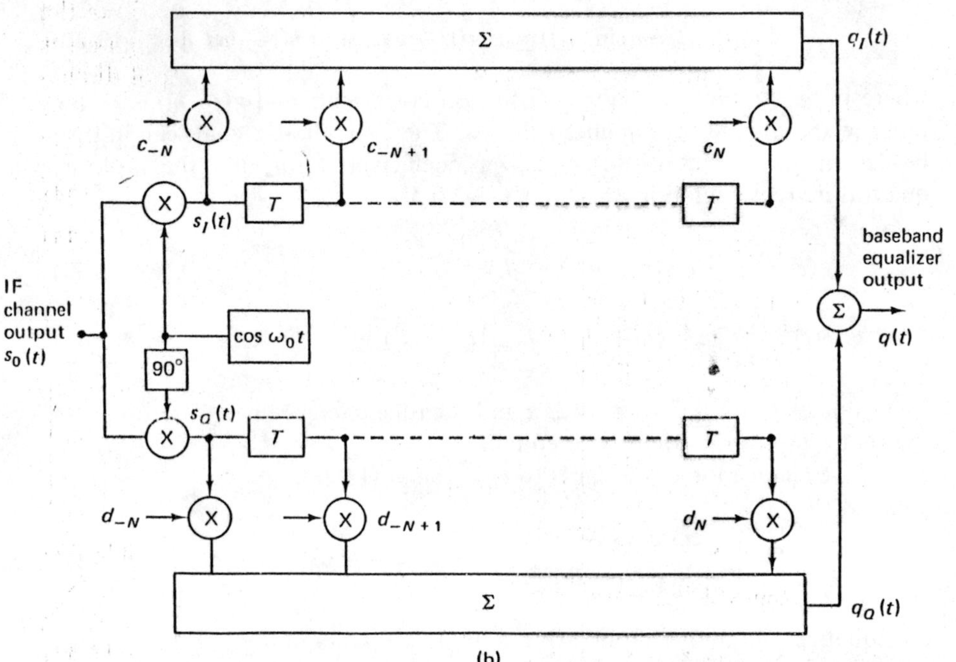

(b)

Fig. 13-13 Bandpass transversal equalizers; (a) transversal equalizer at IF, and (b) signals downconverted to baseband. If the cos $\omega_0 t$ is phaselocked to the carrier, the inphase and quadrature outputs $q_I(t)$ and $q_Q(t)$ can be fed to a bit detector.

$g(iT + t_0)$, $h_i \triangleq h(iT + t_0)$ equal to zero for $i \neq 0$ except at the desired time $g_0 = 1$, $h_0 = 0$. If there is channel distortion as in (13-29), this objective is generally not possible even if noise is absent. However, we can minimize the mean-square error between this desired outcome and the actual output—that is, the mean-square distortion or cost function

$$\begin{aligned} \epsilon &= \sum_{i \neq 0} (g_i^2 + h_i^2) + (1 - g_0)^2 + h_0^2 \\ &= \sum_{i=0} (g_i^2 + h_0^2) + 1 - 2g_0 \end{aligned} \tag{13-33}$$

Equalizer Tap Adjustment

To determine how well the adaptive equalizer operates, the optimum tap settings are calculated for the bandpass transversal equalizer. Iterative techniques are used to adjust the taps to their optimum values. These adaption techniques can be employed even with random input data by using the detected bits (decision-directed operation) rather than the analog output to form the equalizer error voltage, which is then connected with each of the equalizer tap outputs. This approach is feasible even at relatively high error rates $\mathcal{P}_b \leq 10^{-2}$. However, noise effects are neglected in the much of discussion that follows.

We denote the tap coefficients in vector form $(2N + 1)$ dimensions as

$$\mathbf{c}^T = (c_{-N}, c_{-N+1}, \ldots, c_0, \ldots, c_N) \tag{13-34}$$

$$\mathbf{d}^T = (d_{-N}, d_{-N+1}, \ldots, d_0, \ldots, d_N) \tag{13-35}$$

where the superscript T denotes the transpose. Introduce the channel correlation matrices $\mathbf{X}$, $\mathbf{Y}$, $\mathbf{K}$ with elements (taken at the sample time)

$$\left.\begin{aligned} x_{ij} &= \sum_n x_{n-i} x_{n-j} && \text{inphase channel autocorrelation} \\ y_{ij} &= \sum_n y_{n-i} y_{n-j} && \text{quadrature channel autocorrelation} \\ k_{ij} &= \sum_n y_{n-i} x_{n-j} - x_{n-i} y_{n-j} && \text{channel cross-correlation} \end{aligned}\right\} \tag{13-36}$$

and the transposed channel vectors are

$$\mathbf{x}^T = (x_{-N}, x_{-N+1}, \ldots, x_0, \ldots, x_N) \tag{13-37}$$

$$\mathbf{y}^T = (y_{-N}, y_{-N+1}, \ldots, y_0, \ldots, y_N) \tag{13-38}$$

The mean-square distortion (13-33) can then be written in matrix notation

$$\varepsilon = \mathbf{b}^T \mathbf{A} \mathbf{b} - 2\mathbf{b}^T \mathbf{z} + 1 \tag{13-39}$$

where the composite tap vector is $\mathbf{b}^T \triangleq (\mathbf{c}^T, \mathbf{d}^T)$, the augmented channel vector is $\mathbf{z} = (\mathbf{x}^T, -\mathbf{y}^T)$, and the bandpass correlation matrix is defined using (13-36) as

$$\mathbf{A} \triangleq \begin{bmatrix} \mathbf{X} + \mathbf{Y} & \mathbf{K} \\ -\mathbf{K} & \mathbf{X} + \mathbf{Y} \end{bmatrix} \tag{13-40}$$

The gradient of ε with respect to $\mathbf{b}$ is set to zero to minimize ε. This optimum value of $\mathbf{b}$ is given by the equation

$$\mathbf{A}\mathbf{b} - \mathbf{z} = 0 \qquad \text{or} \qquad \mathbf{b} = \mathbf{A}^{-1}\mathbf{z} \tag{13-41}$$

under the assumption that $\mathbf{A}$ is positive definite and that $\mathbf{A}^{-1}$ exists. The optimum setting of $\mathbf{b}$ gives $\varepsilon = 1 - \mathbf{z}^T\mathbf{A}^{-1}\mathbf{z}$.

Real-time adjustment of the taps can be accomplished by iterating the value of ε to zero (ε is a convex function with a single minimum) by a gradient—the steepest descent—algorithm. The value of the tap vector $\mathbf{b}^{(k)}$ at iteration $k + 1$ is related to the value at iteration k by the expression

$$\mathbf{b}^{(k+1)} = \mathbf{b}^{(k)} - \alpha^{(k)}\nabla\epsilon(\mathbf{b}^k) \tag{13-42}$$

where $\alpha^{(k)}$ is the gain or step size of the adaption algorithm, and $\nabla\epsilon$ is the gradient of ε. This algorithm converges as it approaches the minimum since the gradient $\nabla\varepsilon$ approaches zero as $\mathbf{b}^{(k)} \longrightarrow \mathbf{b}_{\text{opt}} = \mathbf{A}^{-1}\mathbf{z}$.

The components of the gradient of ϵ from (13-30), (13-31), (13-33) are

$$\frac{\partial\epsilon}{\partial c_i} = \sum_n (g_n x_{n-i} + h_n y_{n-i}) - x_{-i} \tag{13-43}$$

$$\frac{\partial\epsilon}{\partial d_i} = \sum_n (h_n x_{n-i} - g_n y_{n-i}) + y_{-i} \tag{13-44}$$

for $i = -N, \ldots, 0, \ldots, N$. The terms in parentheses in (13-43), (13-44) would be $g_0 x_{-i} = x_{-i}$ if there were no distortion effects remaining.

Notice that these components of the gradient include x_i, y_i, and g_i, h_i (which are dependent on x_i, y_i), and that x_i, y_i are not available at any point in the transversal filter of Fig. 13-12. They are always present as a term which is multiplied by the modulation terms a_n, b_n. Hence, one must generate a reference data waveform based on transmission of a known data preamble at the outset of transmission and periodically retransmitted thereafter to accommodate changes in the channel. Alternatively, if the channel can be assumed to operate at reasonably low error rates $\mathcal{P}_b < 10^{-2}$, the decisions $\hat{a}_i$, $\hat{b}_i$, which are then correct at least 99 percent of the time, can be used to adapt the transversal filter as shown in Fig. 13-14 for an MPSK signal.

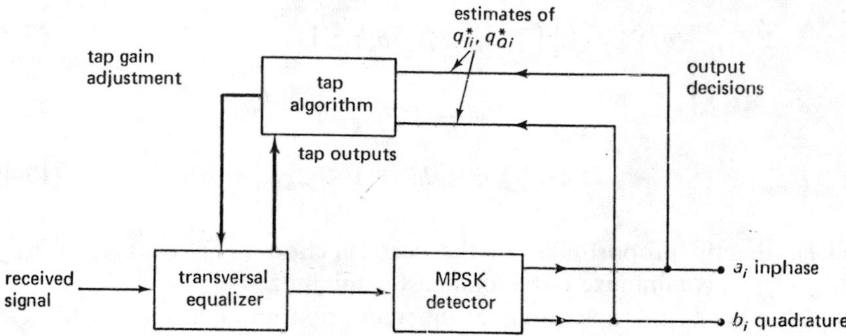

Fig. 13-14 Block diagram of an adaptive transversal equalizer using decision-directed adaption of the tap coefficients

To determine the mechanization for tap adjustment in greater detail, the cost ε must be rewritten as a function of terms available within the equalizer. We first rewrite the equalizer output at the sampling instants $jT + t_0$ at the end of each data symbol as

$$q(jT + t_0) \triangleq q_j = g_0 \cos(\omega_0 t_0 + \phi)\Big[\sum_n (a_n g_{j-n} - b_n h_{j-n})\Big] - g_0 \sin(\omega_0 t_0 + \phi)\Big[\sum_n (b_n g_{j-n} + a_n h_{j-n})\Big] \tag{13-45}$$

For perfect equalization (13-45) would become

$$q_j^* = g_0 a_j \cos(\omega_0 t_0 + \phi) - g_0 b_j \sin(\omega_0 t_0 + \phi) \qquad g_0 = 1 \tag{13-46}$$

Because of the tap rotation property of the equalizer, ϕ can be adjusted to any phase; for example, let us set $(\omega_0 t_0 + \phi) = \pi/4$ and then we have

$$q_j^* = (a_j - b_j) \qquad g_0 = \sqrt{2} \tag{13-47}$$

Define the mean-square value of a_i, b_i as

$$a^2 = \mathbf{E}(a_i^2) = \mathbf{E}(b_i^2) \tag{13-48}$$

For QPSK modulation, of course, $a_i^2 = b_i^2 = a^2$. Assume also that a_i, a_{i+j}, b_i, b_{i+k} are all independent. Thus, the mean-square error $e_i = q_i - q_i^*$ in the equalizer output can be written from (13-46), (13-47) as

$$\begin{aligned} \mathbf{E}[(q_j - q_j^*)^2] &= \mathbf{E}\Big[\Big(\sum_n a_n g_{j-n} - b_n h_{j-n}\Big) \\ &\quad - \Big(\sum_n b_n g_{j-n} + a_n h_{j-n}\Big) - \Big(a_j - b_j\Big)\Big]^2 \\ &= a^2\Big[\sum_n g_n^2 + h_n^2 + 1 - 2g_0\Big] = a^2\epsilon \end{aligned} \tag{13-49}$$

and is directly proportional to the cost function ϵ. Therefore, if $\sigma_\epsilon^2 \triangleq \mathbf{E}[(q_j - q_j^*)^2]$ is minimized, the total cost is minimized.

Interchange the operations of differentiation and expectation to determine the components of the gradient of $a^2\epsilon$ as

$$\frac{\partial}{\partial c_i}\mathbf{E}\Big[(q_j - q_j^*)^2\Big] = 2\mathbf{E}\Big[(q_j - q_j^*)\frac{\partial q_j}{\partial c_i}\Big] \tag{13-50}$$

$$\frac{\partial}{\partial d_i}\mathbf{E}\Big[(q_j - q_j^*)^2\Big] = 2\mathbf{E}\Big[(q_j - q_j^*)\frac{\partial q_j}{\partial d_i}\Big] \tag{13-51}$$

where $q_j - q_j^*$ is the error at the equalizer output at the jth sampling instant. The differentiated terms can be evaluated as [Gitlin, 1973, p. 234]

$$\begin{aligned} \frac{\partial q_j}{\partial c_i} &= \cos\alpha\Big[\sum_n a_n x_{j-i-n} - \sum_n b_n y_{j-i-n}\Big] \\ &\quad - \sin\alpha\Big[\sum_n b_n x_{j-i-n} + \sum_n a_n y_{j-i-n}\Big] \end{aligned} \tag{13-52}$$

$$\begin{aligned} \frac{\partial q_j}{\partial d_i} &= -\cos\alpha\Big[\sum_n a_n y_{j-i-n} + \sum_n b_n x_{j-i-n}\Big] \\ &\quad + \sin\alpha\Big[\sum_n b_n y_{j-i-n} - \sum_n a_n y_{j-i-n}\Big] \end{aligned} \tag{13-53}$$

where $\omega_0 t_0 + \phi \triangleq \alpha$. Note that $\partial q_j/\partial c_i$ is simply the jth output of the channel delayed by iT sec and thus is available at the output of the ith delay element. Furthermore, $\partial q_j/\partial d_i = \partial q_j/\partial c_i$ except for a 90° carrier phase shift. Thus, both of these elements are available at points within the equalizer. The tap adjustment algorithm of (13-50), (13-51) then is obtained by forming the error signal $e \triangleq q_j - q_j^*$ and correlating it in parallel with each of the tap outputs.

The equalizer adaption circuit then has the structure as shown in Fig. 13-15. The equalizer operates at baseband by coherent downconversion at the equalizer input, effectively incorporating the equalizer as part of the data-detection operation. A separate adaption unit is employed for each of the tap weights c_i, d_i, only one of which is shown in Fig. 13-15. In general, the data-detection unit requires a one-bit delay not indicated in the figure,

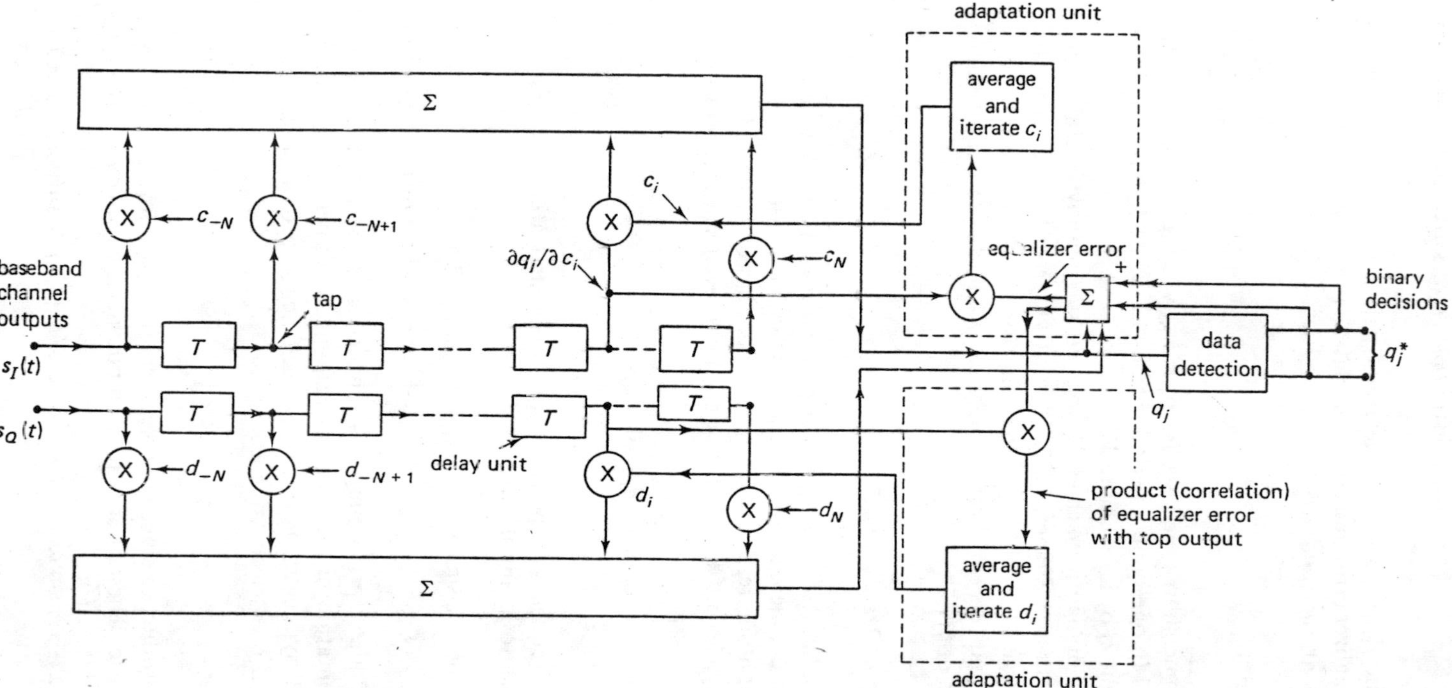

Fig. 13-15 Adaptive bandpass transversal equalizer block diagram. The adaption is based on sampling at the end of each data symbol—that is, at $t = jT$. The algorithm employs decision direction to recover the equalizer output error q_j.

and the terms q_j and $\partial q_j/\partial c_i$ should be delayed by the amount of the detecter delay before the equalizer error and cross-correlation functions are performed.

In the process *average and iterate*, the product

$$\frac{\partial q_j}{\partial c_i}(q_j - q_j^*) \tag{13-54}$$

is averaged over some short time τ sec to form Δ_{ji} where $\tau \gg T$ sec, that is, τ is large compared to one symbol period. From (13-42) each iteration increments the tap weight c_i by an amount defined as $\alpha^{(k)}\,\Delta_{ji}^{(k)}$.

This same type of algorithm can be employed successfully whether there is noise or not [Ryan and Ziemer, 1973, 18-4–18-9], provided the error rate does not get too large ($\mathcal{P}_b < 10^{-2}$) and the adaption is sufficiently slow. It is important to note that this equalizer automatically adjusts the output phase to match the phase of q_j^*.

Simple Forms of Transversal Equalizers

The same equalizer structure described above can be used in greatly simplified forms to compensate for parabolic or cubic phase nonlinearities or nonuniform amplitude response for either bandpass or baseband signals [Bennett and Davey, 1965, Chap. 15]. For example, even symmetry of three-tap coefficients

$$\left.\begin{array}{ll} c = c_{-1} = c_1 & \text{other } c_i, d_i = 0 \\ c_0 = 1 & c \ll 1 \end{array}\right\} \tag{13-55}$$

produces a cosine frequency response in amplitude and a constant phase. The output response to an input sinusoid $\sin \omega t$ is

$$\begin{aligned} q(t) &= \sin \omega t + c \sin \omega(t - T) + c \sin \omega(t + T) \\ &= (1 + 2c \cos \omega T) \sin \omega t \end{aligned} \tag{13-56}$$

The transversal equalizer performing this function is shown in Fig. 13-16. The phasor representation of the equalizer response is shown in Fig. 13-17.

Similarly, odd symmetry in the tap coefficients produces a phase equalization. Thus if one selects tap coefficients

$$c = c_{-1} = -c_1 \qquad c_0 = 1 \qquad c \ll 1 \tag{13-57}$$

then the equalizer produces a sinusoidal phase characteristic and little change in amplitude response. The response to an input sinusoid $\sin \omega t$ is $\omega T \ll 1$

$$q(t) = \sin \omega t + c \sin \omega(t - T) - c \sin \omega(t + T) \cong \sin(\omega t - 2\omega c T) \tag{13-58}$$

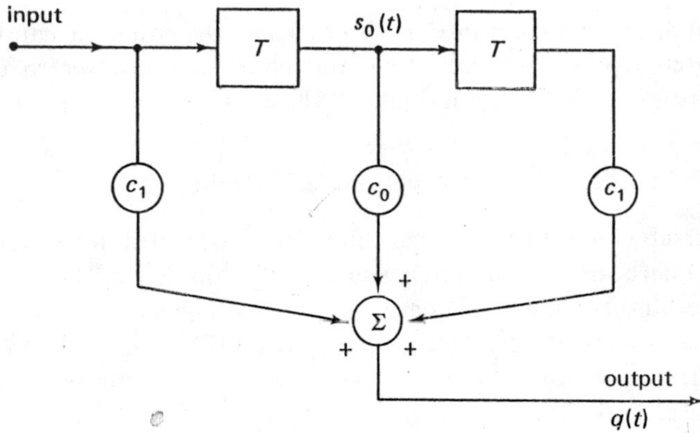

(a) amplitude equalization

set $c_{-1} = c_1 = c$

$c_0 = 1$

transfer function

$$\frac{Q(j\omega)}{S_0(j\omega)} = 1 + 2c \cos \omega T$$

(b) phase equalization

$c_{-1} = -c_1 = c$

$c_0 = 1$

transfer function

$$\frac{Q(j\omega)}{S_0(j\omega)} = 1 + 2cj \sin \omega T$$

Fig. 13-16 Elementary forms of transversal equalizers for amplitude and phase equalization. The phase equalization configuration provides a sinusoidal phase correction when $c \ll 1$.

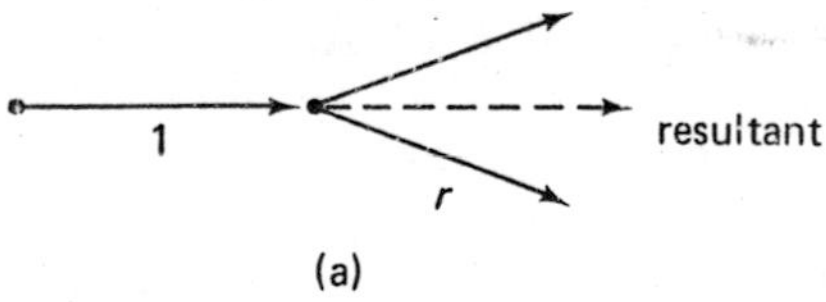

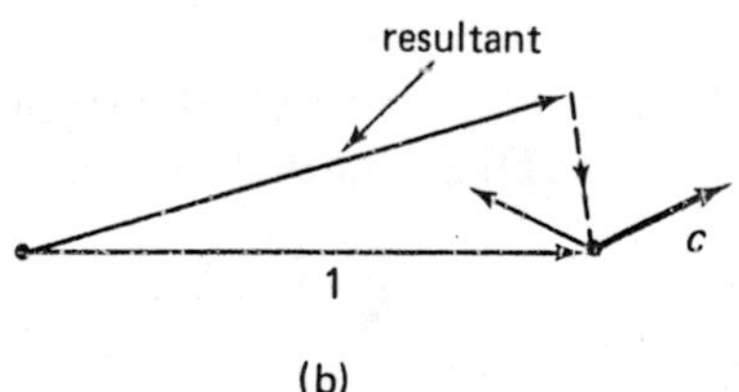

Fig. 13-17 Phasor representation of transversal equalizer outputs for (a) amplitude equalization, and (b) phase equalization

and the resultant is as shown in Fig. 13-17. Thus, the equalizer can be used either to match a prescribed amplitude and phase response, or to equalize the impulse response using a test digital PSK modulation as just described.

13-6 DECISION-FEEDBACK EQUALIZATION

The transversal equalizer described above compensates for intersymbol interference effects on the present pulse being examined by both previous pulses and yet-to-be-examined binary pulses. The effects of these "future" pulses on the present pulse *is not* a violation of the causality principle in realizable filters, but rather an acknowledgement of the group delay of the present pulse caused by the channel.

The decision-feedback equalizer operates on the principle that the intersymbol interference of the previously detected pulses can be removed entirely in the absence of bit error if the signal amplitude and the intersymbol interference components (channel impulse response) are known. This approach is in contrast to the transversal filter approach where the effects of previously detected pulses are always corrupted by noise because the operations are performed in analog form. In the equalizer of Fig. 13-12 the equalization inserts an analog correction signal even though some elements of the adaption algorithm may rely on binary decision-directed operations in the bit detector output.

Figure 13-18 illustrates the form of a simplified Decision-Feedback Equalizer (DFE) which operates on the intersymbol interference of previous pulses only. An example impulse response function is shown in the same figure. Note that we are assuming that the channel impulse response is known and varies very slowly with time, if at all. Adaptive versions of this equalizer are discussed below.

The channel output for intersymbol effects caused only by previous pulses is

$$x_i = \sum_{k=0}^{\infty} b_{i-k} h_k + n_i \tag{13-59}$$

where h_k is the channel response normalized to $h_0 = 1$, the n_i are the additive noise samples, and the b_i are the transmitted circumflex pulses. All signals are sampled at $t = iT$. If the DFE binary decisions are b_i, then the feedback signal is

$$y_i = \sum_{k=1}^{\infty} \hat{b}_{i-k} h_k \tag{13-60}$$

for a perfectly known channel. Then the residual data signal is

$$d_i \triangleq x_i - y_i = b_i + n_i + \sum_{k=1}^{\infty} (b_{i-k} - \hat{b}_{i-k}) h_i \tag{13-61}$$

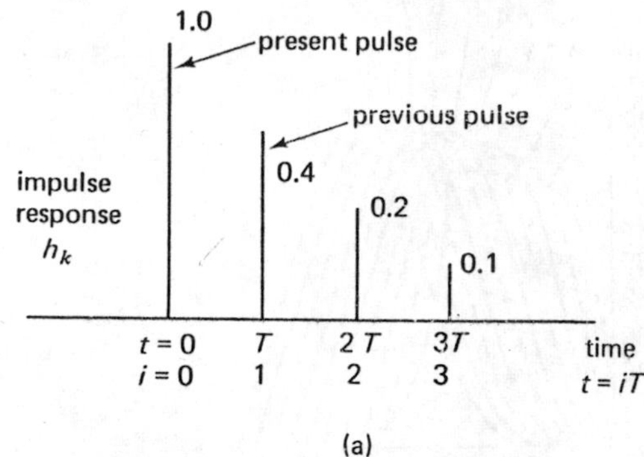

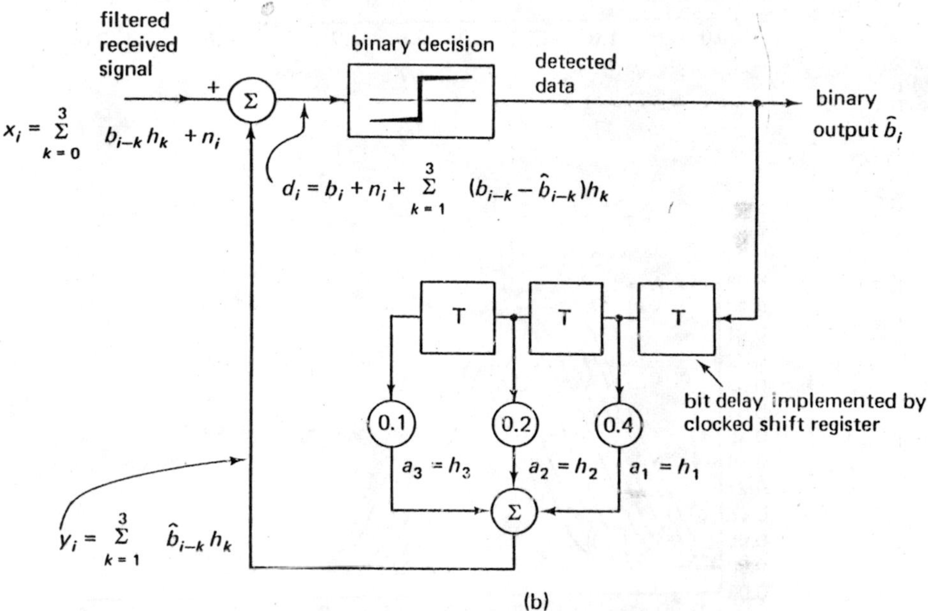

Fig. 13-18 Decision-feedback equalizer (DFE) in simplified form for an example impulse response h_k, (a) channel impulse response, (b) block diagram of the DFE for the impulse response of (a)

Thus, if there are no bit errors and $b_i = \hat{b}_i$, this intersymbol interference is removed completely. Even if the bit-error rate is relatively high ($\mathcal{P}_b = 10^{-2}$), the average intersymbol interference is still reduced to approximately 1 percent or 20 dB of its average power without equalization. If an error does occur, however, the adjacent bit is more likely to be in error. These error effects tend to occur in bursts because of error propagation.

Examples of the filter response to an NRZ binary waveform are given

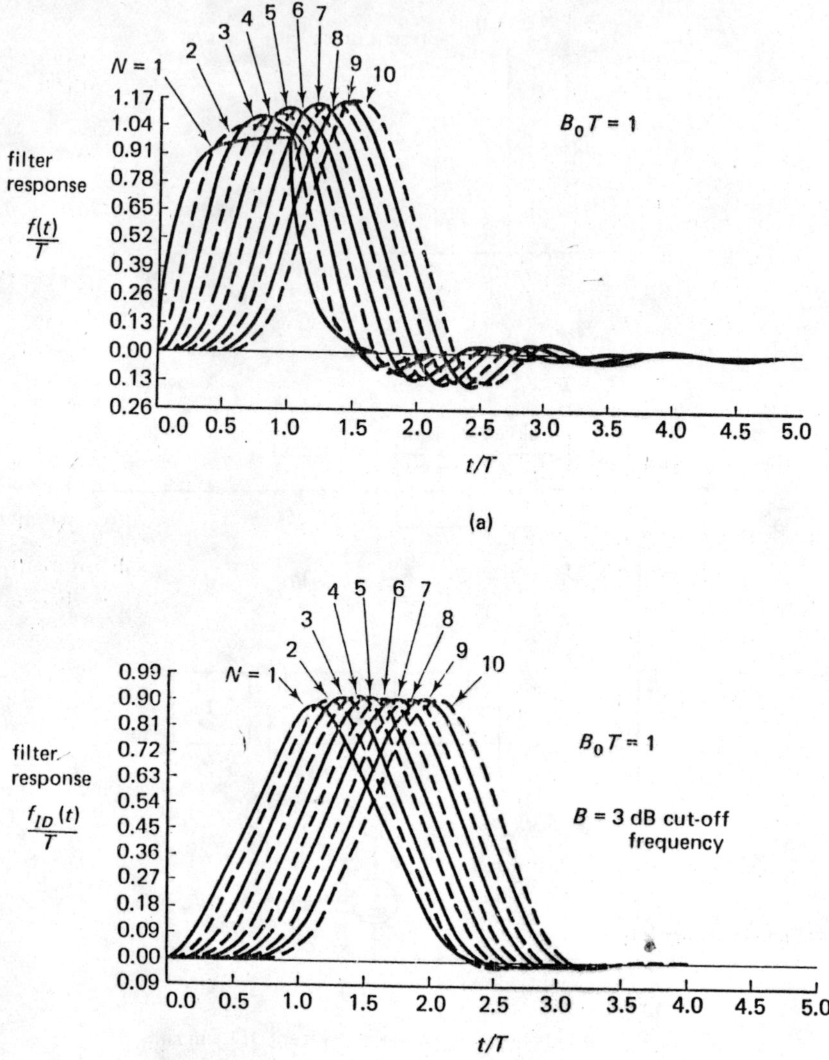

Fig. 13-19 Response to a single NRZ pulse of duration T sec by a Nth order Butterworth filter. (a) Butterworth filter output $f(t)$, (b) output of the Butterworth filter followed by finite memory (T sec) integration detection $f_{ID}(t)$ [From Korn, 1973]

in Fig. 13-19 [I. Korn, 1973, 878–890]. The filter output is shown in 13-19*a* for a T-sec duration pulse of unit amplitude. The filter is an Nth order Butterworth. In Fig. 13-19*b* the response is given for a Butterworth filter followed by an integrate-and-dump filter. The 3-dB filter bandwidth is

$B_0 = 1/T$. In either figure the sampled channel response h_k can be found by taking samples at spacing T sec. Note that for an Nth order Butterworth filter for $N = 10$, the peak output $f(t)$ occurs at $t/T \cong 1.6$, and at $t/T = 2.6$ the output is approximately -0.13.

George, Bowen, and Storey [1971] have reported on a more general version of the DFE which compensates for intersymbol interference effects of both previous and future pulses. This DFE is shown in Fig. 13-20. The effects of the future pulses are compensated in a linear transversal equalizer just as shown in the previous section. This feedforward equalizer is then followed by the decision-feedback equalizer to complete the equalization.

This more general DFE equalizer thus makes the estimate of the bit stream $\hat{b}_i$

$$\hat{b}_i = \operatorname{sgn}\left[\sum_{k=0}^{-N} a_k x_{i-k} - \sum_{k=1}^{M} a_k \hat{b}_{i-k}\right] \tag{13-62}$$

where the $k = 0$ to $-N$ terms represent the feedforward-equalizer (FFE) contribution and the $k = 1$ to M terms represent the decision-feedback equalizer (DFE). The equalizer taps can be adjusted to minimize the output mean-square error

$$\mathbf{E}(e_j^2) = \mathbf{E}[(b_j - \hat{b}_j)^2] \qquad e_j = b_j - \hat{b}_j \tag{13-63}$$

The optimum tap settings are found by setting

$$\frac{\partial \mathbf{E}(e_j^2)}{\partial a_k} = 0 = \begin{cases} 2\mathbf{E}[e_j x_{j-k}] & k = 0, -1, \ldots, -N \\ -2\mathbf{E}[e_j \hat{b}_{j-k}] & k = 1, 2, \ldots, M \end{cases} \tag{13-64}$$

These equations reduce to [George et al., 1971] solutions for a_j

$$\begin{aligned} &\sum_{k=0}^{N} a_k \psi(k-j) + \phi_n(j-k) = h_j \qquad && j = 0, -1, \ldots, -N \\ &a_j = \sum_{k=0}^{-N} a_k h_{j-k} && j = 1, 2, \ldots, M \end{aligned} \tag{13-65}$$

where $\phi_n(j-k)$ is the autocorrelation function of the noise, and $\psi(k-j)$ is the autocorrelation function of the sampled channel impulse response h_k

$$\psi(k-j) \triangleq \sum_{\ell=0}^{i} h_\ell h_{\ell+k-j} \tag{13-66}$$

The feedforward equalizer is sometimes preceded by a filter $q(-\tau)$ matched to the channel impulse response $q(t)$. If in addition the input noise

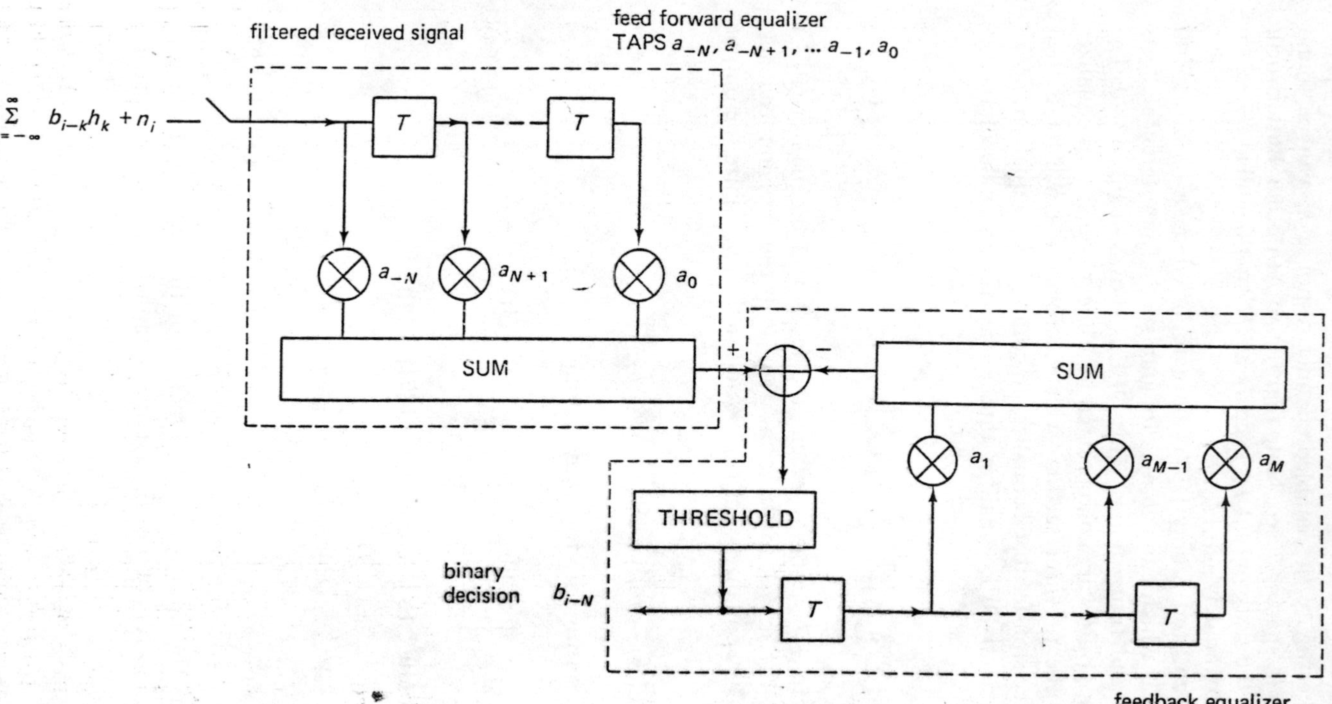

Fig. 13-20 Block diagram of a decision-feedback equalizer utilizing both feedforward and feedback equalization. The feedback equalizer can be completely digital and can use clocked shift register stages for delay elements if a bit synchronization clock is available.

is white with spectral density N_0, the feedforward-tap equation (13-65) of the equalizer reduces to

$$\sum_{j=0}^{\infty} a_j[\phi_q(j-m) + N_0\delta(j-m)] = \delta(m) \qquad m \leq 0 \tag{13-67}$$

and $\phi_q(j-m) \triangleq \int_0^{\infty} q(\tau)\, q[\tau-(j-m)T]\, d\tau$. If the past decisions are all correct, the decision-feedback output mean-square error is

$$\mathrm{E}[e^2(j)] = N_0 a_0$$

Define the z transform representation of $\phi_q(i)$ as

$$\Phi_q(z) \triangleq \sum_{i=-\infty}^{\infty} \phi_q(i) z^{-i} \qquad z \triangleq e^{-st}$$

Factor the summation of noise density and $\Phi_q(t)$ as

$$P(z)P(z-1) \triangleq N_0 + \Phi_q(z)$$

and $P(z) \triangleq \sum_{i=0}^{\infty} p_i\, z^{-i}$ where p_i is the so-called "equivalent received pulse." Define also the reciprocal transform

$$R(z) \triangleq \frac{1}{P(z)} = \sum_{i=-\infty}^{\infty} r_i z^{-i}$$

For $p_i = P \exp(-i\gamma)$ the output mean-square error is

$$\mathrm{E}[e^2(j)] = N_0 r_0^2 = PN_0$$

where P is the equivalent response at $i = 0$.

Modulator Equalization

If the filters in the communication link are constant and known ahead of time, the transversal equalization of Fig. 13-13 can sometimes be placed in the baseband inputs to the phase modulator. Equalization can be achieved in this manner without the addition of further noise components in the receiver. Such equalization can be performed more easily because the delay storage elements in the phase and quadrature channels need only store binary numbers for BPSK or QPSK and can be implemented by using shift registers. This predistortion coding bears some resemblance to quadrature amplitude shift-keying (QASK) and partial response coding (Chapters 11 and 16).

CHAPTER 14

BIT SYNCHRONIZERS FOR DIGITAL COMMUNICATIONS

14-1 INTRODUCTION

Power-efficient digital receivers generally require the existence of a digital clock synchronized to the received bit stream to control the integrate-and-dump detection filters or to control otherwise the timing of the output bit stream. Bit synchronizers are also required when a nearly-synchronous bit stream is received over a cable transmission system and must be detected and perhaps multiplexed with other parallel bit streams. Also, the received bit-transition time clock has some time jitter and frequency drift corresponding to the oscillator phase noise and path change velocity.

Bit synchronization as discussed here is restricted to self-synchronization techniques that extract clock time directly from a noisy NonReturn-to-Zero (NRZ) bit stream. This bit format has no spectral line component at the bit rate or its harmonic frequencies for random data-bit sequences with 50% transition density.

Scramblers are available to randomize the data-bit sequences, preventing a long string of "ones" or "zeros" [Savage, 1967, 449–487]. (See also Chap. 16.) These devices eliminate the potential line component in the input bit stream by producing a 50% transition density. They also improve the performance of self-synchronizing types of bit synchronizers. In other bit synchronization techniques some of the signal energy is solely dedicated to synchronization, as for example where a known bit pattern or additive sinusoidal component is transmitted along with the data.

In this chapter, we discuss four different classes of bit synchronizers:

1. *Nonlinear-filter synchronizer* This open loop type of synchronizer functions by linearly filtering the received bit stream to reduce the noise and to magnify the observability of the bit transitions. The filter output is then passed through a memoryless even-law nonlinearity to produce a spectral line at the bit rate. This synchronizer is commonly used in high-bit rate links and links which normally operate at high SNR.
2. *Inphase/Midphase (IP/MP) synchronizer* (Also termed the data transition tracking synchronizer) This synchronizer operates in a closed loop and combines the operations of bit detection and bit synchronization. The bit detector determines which bits represent a change from the previous bit and whether a (10) or a (01) transition has occurred. This transition information is then utilized to provide the correct sign to a tracking error channel. The IP/MP synchronizer can be employed even at low SNR and medium data rates. It also operates well even in the presence of relatively long times between transitions.
3. *Early-late bit synchronizer* This type of bit synchronizer is similar to the inphase/midphase technique. It too involves a closed-loop system but has a somewhat different method of obtaining the bit-timing error estimate.
4. *Optimum (maximum-likelihood) synchronizer* This type of synchronizer an optimal means for searching for the correct synchronization time cell during acquisition. This synchronizer is an open-loop system rather than a tracking technique. This approach is generally not practical; nevertheless it does represent a bound on obtainable performance.

In general, the discussion of each synchronizer includes a functional description of the principle of operation, comparisons with other types of synchronizers, and a linearized (approximate) analysis of performance in a noise environment. Performance is discussed both in terms of the rms timing error for the expected magnitude of timing error and the effects of timing error on bit error probability. References are given to more detailed nonlinear analyses. (See Lindsey and Simon [1973, Chap. 9] for a detailed analysis of synchronizer performance.) Another key effect of bit synchronizers is the possible loss of bit integrity. Bit integrity is lost whenever the recovered clock slips or gains one or more clock pulses, thereby causing the addition or deletion of one or more data bits. This loss in bit integrity can cause demultiplexers or switches in the data link to produce at least momentarily, a totally corrupted bit pattern.

14-2 NONLINEAR-FILTER BIT SYNCHRONIZERS

Three different types of nonlinear-filter bit synchronizers are shown in Fig. 14-1. These synchronizers are a cascade of a linear filter and an instantaneous nonlinearity. Other variations of these approaches are available which use different filters and nonlinearities. The first (Fig. 14-1*a*) operates on the received noisy bit stream and first passes it through a filter matched to the bit waveform. The matched filter output is then passed through an even-law nonlinearity [Wintz and Luecke, 1969, 380–389]. The second type of bit synchronizer (Fig. 14-1*b*) forms the product of the input waveform and the delayed version of the input waveform. Both of these nonlinearities produce a spectral line component at the clock frequency. The third type (Fig. 14-1*c*) operates by differentiating the input waveform after bandpass filtering to produce a sequence of positive and negative spikes at the transitions. A square-law device then converts these biphase spikes to unipolar pulses having a line component at the clock rate.

Matched Filter Bit Synchronizer

In this synchronizer the noisy input bit stream is first fed to a filter matched to the input waveshape as shown in Fig. 14-1*a*. The filter output for a . . . 000100 . . . input is then the same shape as the autocorrelation function of a single input pulse as shown in Fig. 14-2. An *RC* low-pass filter can be used as an approximation to the matched filter if the cutoff frequency is selected as $f_c = 1/T$ the bit rate. The filtering eliminates some of the noise not already removed by previous filters, and produces an output waveform that is the summation of the delayed autocorrelations of the input pulse waveform shown in Fig. 14-2. Clearly, if a sequence of bits $d_1, d_2, \ldots, d_i, \ldots$ appears at the input, the filter output is

$$x(t) = \sum_{i=1}^{\infty} d_i R(t - iT) \tag{14-1}$$

where $R(t)$ is the filtered pulse function for a single pulse.

The output of the even-law nonlinearity for three types of nonlinearities of interest is

$$y(t) = f[x(t)] = \begin{cases} x^2(t) \\ \log\cosh x(t) \\ |x(t)| \end{cases} \tag{14-2}$$

depending on the nonlinearity selected. All three types of nonlinearity produce a spectral line component at the clock frequency (and its harmonics).

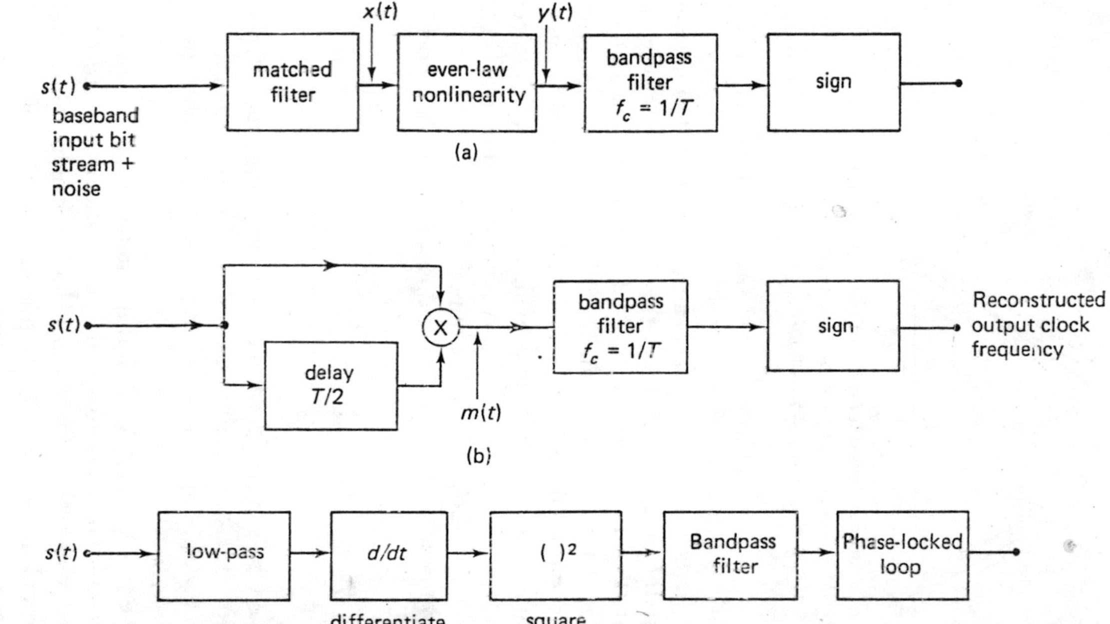

Fig. 14-1 Three types of nonlinear-filter bit synchronizers; (a) nonlinear filter using a matched filter and even-law nonlinearity, (b) delay-and-multiply synchronizer, and (c) differentiate-and-square synchronizer. A phase-locked loop can be employed in place of or in addition to the bandpass filter to provide better waveshape and narrower bandwidth.

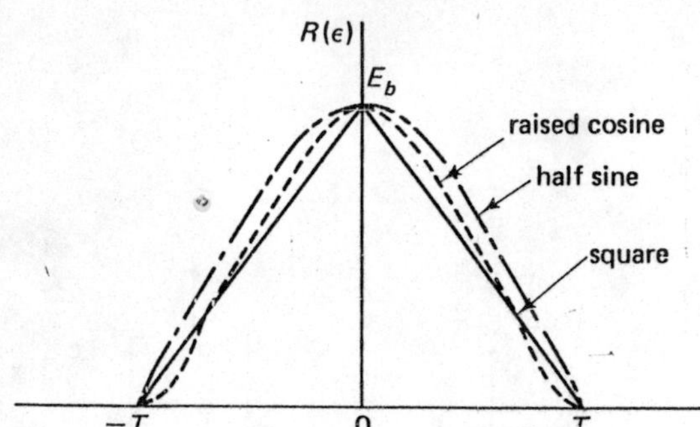

Fig. 14-2 Autocorrelation function of the square, half-sine, and raised-cosine input waveforms, each of which has the same energy E_b for a single bit.

The log cosh x nonlinearity, shown later to be the nonlinearity of an optimal type bit synchronizer, acts as a square-law device for small inputs and a magnitude device for large inputs. The square-law nonlinearity flips the sign of the pulse outputs so that at each transition either a 10 or a 01 produces an identical pulse output (Fig. 14-3). For trapezoidal-wave inputs, a triangle-pulse output is produced at each transition by a magnitude nonlinearity. This waveform produces a spectral line component at the bit rate that can be amplified by the bandpass filter amplifier to provide a sine wave at the clock rate (zero crossings at the bit-transition times). A phase-locked oscillator can then be used to recover the clock frequency component.

Define the effective memory time of the bandpass filter or phase-locked loop as [Proakis et al., 1964, 55]

$$KT \triangleq \frac{(\sum C_i)^2 T}{\sum C_i^2}$$

where C_i is the relative weighting assigned to the ith preceding time interval. If k preceding T sec intervals are equally weighted for kT sec memory, then $K = k$. If exponential weighting is used where $C_i = DC_{i-1}$, and $D < 1$, then $K = (1 + D)/(1 - D)$.

Wintz and Luecke [1969, Fig. 9] have shown results for an RC low-pass filter with a 3-dB bandwidth $B = 1/T$ followed by a square-law detector. The clock component is recovered with a phase-locked loop or bandpass filter with an equivalent bandpass memory of KT. The expected magnitude of error for a raised-cosine input waveform is approximately equal to

$$\frac{\overline{|\epsilon|}}{T} \simeq \frac{0.33}{\sqrt{KE_b/N_0}} \qquad \frac{E_b}{N_0} > 5, \qquad K \geq 18 \tag{14-3}$$

where N_0 is the received baseband noise density (one-sided).

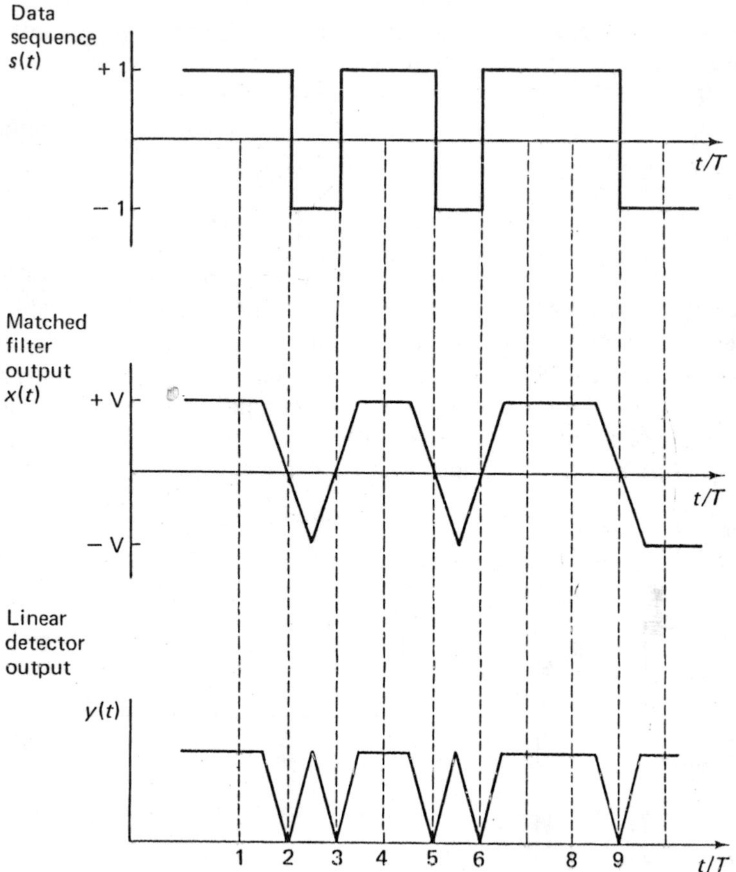

Fig. 14-3 Bit synchronizer waveform: the matched filter synchronizer of Fig. 14-1*a*. A magnitude nonlinearity is employed, and the input is noise-free. The inputs are rectangular pulses.

At high SNR the probability density of timing error ϵ is approximately Gaussian and the rms timing error is approximately

$$\frac{\sigma_\epsilon}{T} = 1.25\frac{\overline{|\epsilon|}}{T} = \frac{0.411}{\sqrt{KE_b/N_0}} \qquad E_b/N_0 \gg 1 \tag{14-4}$$

Delay-and-Multiply Bit Synchronizer

The second type of nonlinear-filter bit synchronizer operates on a rectangular waveform by forming the product $m(t) \triangleq s(t)\,s(t-\Delta)$ where $s(t)$ is the received periodically clocked random or pseudorandom sequence. The best value of delay Δ to be used is $\Delta = T/2$. This product $m(t)$ contains a periodic component at the bit rate which can be filtered by a bandpass filter at $f_c = 1/T$. In addition there is a random or pseudonoise component $r(t, \Delta)$. The expected value of $m(t, \Delta)$ is, of course, the autocorrelation function of the input sequence $s(t)$

$$\mathbf{E}[m(t, \Delta)] = R_s(\Delta)$$

To show the periodicity in $m(t)$, we can use the decomposition property for products of pseudorandom sequences [Gill and Spilker, 1963, 246–247]. As an example, consider the maximal-length linear feedback shift-register sequence (four-stage FSR) shown in Fig. 14-4. The decomposition $m(t, \Delta) = p(t, \Delta) + r(t, \Delta)$ is also shown for $\Delta = T/4$. The periodic pulse train $p(t, \Delta)$ has amplitude 0, 1 and takes on the $+1$ value for that portion of each digit interval ($T - \Delta$ sec) when the unshifted and shifted waveforms match.

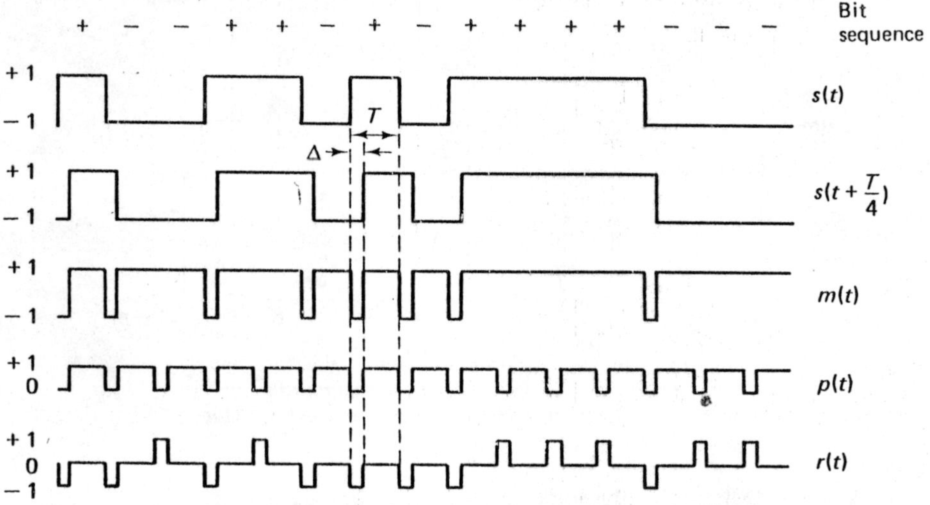

Fig. 14-4 Product of maximal-length sequence and its decomposition; $m(t) = p(t) + r(t)$, displacement $\Delta = T/4$ where $p(t)$ is periodic and $r(t)$ is pseudorandom.

The ternary random or pseudorandom $r(t)$ waveform has amplitudes $-1, 0, +1$, and can take on the ± 1 amplitude only in that fraction of each digit interval (Δ seconds long) for which adjacent digits overlap. The sequence of ± 1 digits represents a cyclically shifted and inverted version of the sequence $s(t)$. If the input sequence is random with independent bits, the sequence $r(t, \Delta)$ is also random. Because the possible transition times for both

$p(t)$ and $r(t)$ are identical, the two waveforms are not necessarily uncorrelated, and this fact must be kept in mind in manipulations of these waveforms.

For random binary sequences with time displacements $kT < \Delta < (k+1)T$, $k \geq 1$, the product can be decomposed into the sum $m(t, \Delta) = r_1(t, \Delta) + r_2(t, \Delta)$, where r_1 has amplitude $-1, 0, +1$ and takes on the ± 1 value in time intervals iT to $iT + \Delta$ in a random manner; r_2 also has amplitude $-1, 0, +1$ and takes on the ± 1 value in the intervals $iT + \Delta$ to $(i+1)T$, also in a random manner. If the original sequence was a random binary waveform, adjacent digits being independent of one another, then adjacent digits in r_1 and r_2 are independent of one another, and the random digit amplitudes in r_1 are independent of those in r_2. On the other hand, if the original sequence $s(t)$ is pseudorandom, then r_1 and r_2 are simply two different cyclic shifts of the same input sequence $s(t)$. This relationship makes use of the "cycle and add" property of maximal length PN sequence, which shows that the product of two shifted versions of a PN sequence is the same sequence shifted in time. (See Chap. 18.)

Before computing the power spectral density of $m(t, \Delta)$, we note that its autocorrelation function for $\Delta < T$ can be expressed by

$$\begin{aligned} R_m(\tau, \Delta) &= \mathbf{E}[m(t, \Delta)m(t+\tau, \Delta)] \\ &= \mathbf{E}\{[p(t, \Delta) + r(t, \Delta)][p(t+\tau, \Delta) + r(t+\tau, \Delta)]\} \\ &= R_p(\tau, \Delta) + R_r(\tau, \Delta) + R_{p\times r}(\tau, \Delta) + R_{p\times r}(-\tau, \Delta) \end{aligned} \tag{14-5}$$

where $R_p(\tau, \Delta) \triangleq \mathbf{E}[p(t, \Delta)\, p(t+\tau, \Delta)]$, $R_{p\times r}(\tau, \Delta) \triangleq \mathbf{E}[p(t, \Delta) r(t+\tau, \Delta)]$. A similar expression can be written for the autocorrelation when $p(t, \Delta) = 0$ and the cross-correlation term $R_{r_1\times r_2}(\tau, \Delta) = 0$ for all $\Delta > T$ if $s(t)$ is a random binary sequence. The cross-correlation $R_{p\times r}(\tau, \Delta) = 0$. (This zero cross-correlation is to be expected from the spectral decomposition theorem discussed in Wold [1954, appendix by Whittle].) The power spectral density of $m(t, \Delta)$ can be found by showing that all of the (nonzero) correlation functions have triangular or periodic triangular shapes. The form of the power spectral density for a random input is

$$\begin{aligned} G_m(f, \Delta) = {} & \left(1 - \frac{\Delta}{T}\right)^2 \delta(f) + \left(\frac{\Delta}{T}\right)^2 \sum_{\substack{n=-\infty \\ n\neq 0}}^{\infty} \operatorname{sinc}^2 \frac{\pi n \Delta}{T}\, \delta\left(f + \frac{n}{T}\right) \\ & + \frac{\Delta^2}{T} \operatorname{sinc}^2 \pi f \Delta \qquad \Delta < T \end{aligned} \tag{14-6}$$

where the first two terms represent the spectrum of the dc and periodic components $R_s(\Delta)$ and $p(t, \Delta)$, respectively. If $\Delta = T/2$, the spectral components at the clock rate are

$$\left(\frac{1}{2}\right)^2 \operatorname{sinc}^2 \frac{\pi}{2}\left[\delta\left(f+\frac{1}{T}\right)+\delta\left(f-\frac{1}{T}\right)\right] \tag{14-7}$$

which can be filtered and used for bit timing. Even if the input waveform $s(t)$ has finite rise and fall times, the product $m(t, T/2)$ contains a line component at $f_c = 1/T$.

The delay-and-multiply operation described here is similar in one respect to the differentiate-and-square operation bit synchronizer of Fig. 14-1*c*. For sufficiently small Δ, this bit synchronizer produces a quantity

$$\begin{aligned}\left\{\frac{ds(t)}{dt}\right\}^2 &\cong \left\{\frac{s(t)-s(t-\Delta)}{\Delta}\right\}^2 = \frac{2}{\Delta} - \frac{2s(t)s(t-\Delta)}{\Delta} \\ &\cong \frac{2}{\Delta}[1-m(t)] \qquad \Delta \ll T\end{aligned} \tag{14-8}$$

since $s^2(t) = 1$ in the absence of noise. Clearly, (14-8) contains the product $s(t) \times s(t-\Delta)$ of the delay-and-multiply operation. Thus the quantity in (14-8) also contains a periodic line component similar to that in (14-7).

14-3 IN-PHASE/MID-PHASE (IP/MP) BIT SYNCHRONIZER

An improved performance closed-loop bit synchronizer was first presented by Lindsey and Tausworthe [1968, 272–276] and also discussed by Simon [1969, 20-9–20-15] and Hurd and Anderson [1970, 141–146]. The synchronizer is also termed the data transition tracking (DTT) bit synchronizer because of the method of operation. Both an inphase channel and a midphase channel are utilized in providing a timing-error discriminator function.

An analogy can be made of the inphase × midphase bit stream product to the inphase × quadrature carrier product of the Costas type of carrier-recovery loop. Similarly, the open loop nonlinear bit synchronizers can be compared to the squaring-type carrier-recovery circuits for BPSK signals.

Figure 14-5 is a block diagram of the IP/MP bit synchronizer. The inphase branch determines the polarity of the bit transitions when and if they occur, while the midphase channel determines the magnitude of the bit-timing error. Use of both channels in the multiplier provides the correct sign of the timing error. The midphase error signal Z_k is multiplied by $I_k = \pm 1$ if a transition has been sensed or by $I_k = 0$ if no transition has been sensed. These decisions as to the proper value of I_k are, of course, subject to bit error effects.

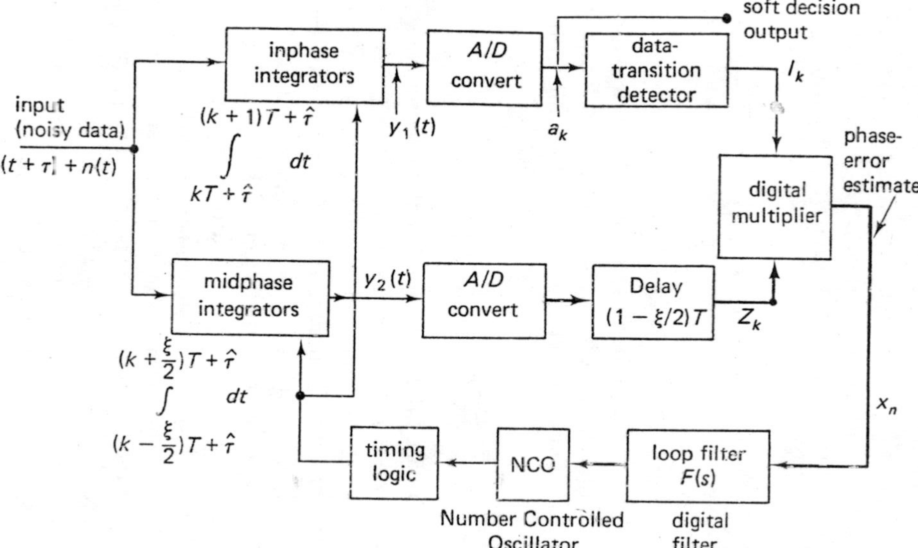

Fig. 14-5 in-phase/Mid-phase bit synchronizer with inphase and midphase channels. The input clock offset is τ, and the clock phase estimate is $\hat{\tau}$. The midphase integrator window width is ξT sec. The timing error is $\epsilon \triangleq \tau - \hat{\tau}$.

The filtered output of the multiplier is used to drive the VCO and to control the integrate-and-dump operations. It is possible to improve noise performance while the SNR is above threshold by narrowing the midphase integration window to $T/4$. The bandwidth of the closed loop must be consistent with the dynamics of the timing jitter and the width of the timing window ξT. An advantage of this inphase/midphase synchronizer is that during long periods between transitions, caused by a long sequence of 1s or 0s, where there are no data errors, the discriminator does not allow any noise to perturb the loop: it "holds" the last valid estimate. This statement assumes that there are no data errors, and hence $I_k = 0$, during this interval.

Assume that the input signal $s(t)$ is a random pulse train of rectangular pulses $p(t)$ with clock offset τ:

$$s(t) = \sum_i \sqrt{P_s}\, d_i p[t - iT + \tau(t)] + n(t) \tag{14-9}$$

$$p(t) = 1 \qquad 0 < t < T \tag{14-10}$$

where $d_i = \pm d_i = \pm 1$ and the input noise $n(t)$ is white Gaussian noise of one-sided density N_0. The bit decisions are made with a hard limiter ($a_k = \pm 1$). Adjacent decisions are used to detect transitions according to the algorithm

$$I_k = \frac{a_k - a_{k+1}}{2} = \pm 1, 0 \tag{14-11}$$

The output of the midphase filter I_k is sampled at T-sec intervals offset from the inphase intervals by a delay of $T/2$ sec, which places the two channels in coincidence. Waveforms at the input and within the loop for a typical, noise-free input bit stream are shown in Fig. 14-6. The output of the midphase integrator is more easily multiplied by $I_k = \pm 1, 0$, and filtered with a narrow bandwidth if it is first digitized by a quantizer of 3 or more bits/sample. The sign of the midphase integrator gives a "hard" bit decision, and a 3-bit quantizer gives a "soft" decision which is also useful for driving a Viterbi decoder or other soft decision decoder. The phase detector output is $a_k I_k = 0$ for this example where the bit-timing error is zero. The phase discriminator characteristic increases linearly until $\epsilon = T/2$.

Set the timing error $\epsilon \triangleq \tau - \hat{\tau}$ at a constant value. Then we define the synchronizer phase discriminator characteristic as the conditional expectation.

$$D(t) = \mathbf{E}[x(t, \epsilon) \mid \epsilon]$$

The effective gain of the loop is the slope of the input-output discriminator curve at timing error $\epsilon = 0$. This gain can be defined as

$$A = \frac{\partial}{\partial \epsilon} \mathbf{E}[x(\epsilon, R)]\bigg|_{\epsilon=0} \triangleq \frac{\partial}{\partial \epsilon} D(\epsilon, R)\bigg|_{\epsilon=0} \tag{14-12}$$

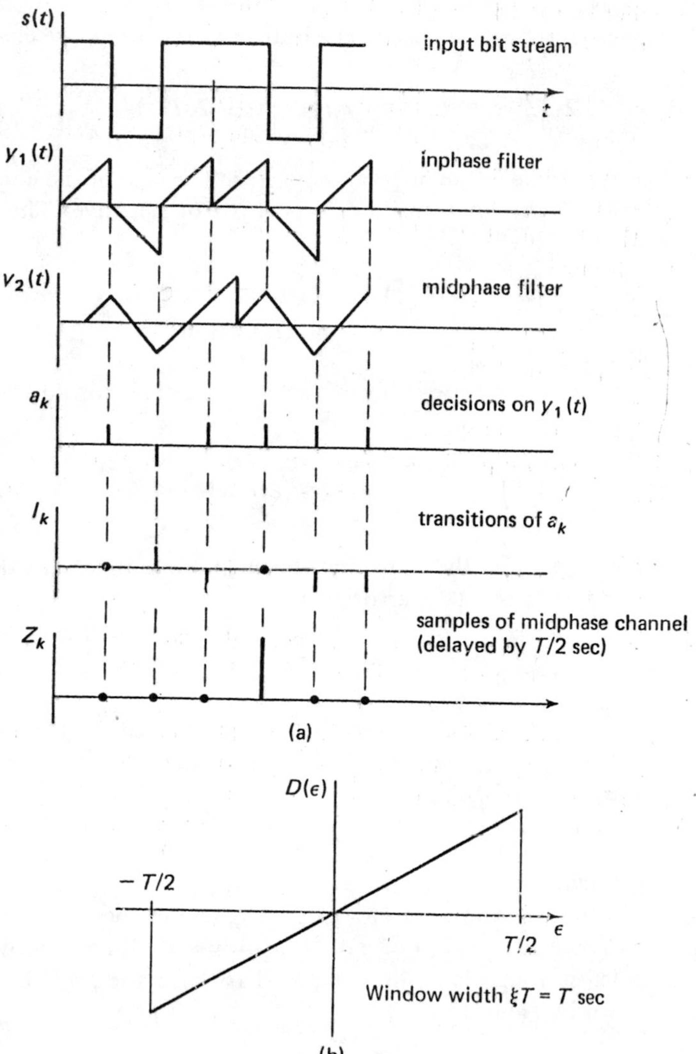

Fig. 14-6 Waveforms of the in-phase/midphase loop showing outputs of the inphase and midphase integrate-and-dump filter. The phase detector output is $a_k I_k = 0$ for this example where the bit timing error is zero. (a) Waveforms for $\xi = 1$. (b) The synchronizer phase discriminator characteristic increases linearly until $\epsilon = T/2$.

where ϵ is the actual phase error and D is the discriminator characteristic—the expected value of the multiplier output—for a given value of ϵ, R. The SNR $R \triangleq P_s T/N_0$.

When a symbol transition occurs, the output of the midphase channel is

$$Z_k = \pm\left[2\epsilon P_s^{(1/2)} + \int_{(k-1/2)T}^{(k+1/2)T} n(t)\,dt\right] = \pm\, 2\epsilon P_s^{1/2} + N_k \tag{14-13}$$

where ϵ is assumed to be in the interval $|\epsilon| < T/2$. The sign of the correction depends on whether the bit transition is positive or negative. The phase-detector output is

$$\hat{\epsilon} = \frac{Z_k I_k}{\sqrt{P_s}} = \begin{cases} Z_k(\pm 1/\sqrt{P_s}) & \pm \text{ transition} \\ 0 & \text{no transition} \end{cases} \tag{14-14}$$

The discriminator characteristic is the expected value of Z_k for no transition errors:

$$D(\epsilon) = \left(E\frac{Z_k I_k}{\sqrt{P_s}}\right) = \frac{1}{2}\left[\frac{2\epsilon\sqrt{P_s}}{\sqrt{P_s}}\right] = \epsilon \qquad |\epsilon| < \frac{T}{2} \tag{14-15}$$

where the 1/2 indicates that the transitions occur half the time. The detector output produces bit errors at an error rate

$$\mathcal{P}_b(R) = \frac{1}{2}\,\text{erfc}\sqrt{\frac{E_b}{N_0}} \tag{14-16}$$

where

$$\text{erfc}\, x = \frac{2}{\sqrt{\pi}}\int_x^{\infty} e^{-t^2}\,dt \tag{14-17}$$

is the co-error function.

An error in the transition-detector output I_k causes the estimate of the timing error ϵ to reverse in sign. For a 01 transition and bit-error probability $\mathcal{P}_b$, the phase-detector sign is easily computed to have the probabilities of occurrence shown in Table 14-1.

Table 14-1 PROBABILITY OF VARIOUS OUTPUT BIT PATTERNS FOR AN INPUT BIT PATTERN 01

True Bit Pattern	01	*Probability*	*Phase-Detector Multiplier I_k*
Detected output bits	01	$(1-\mathcal{P}_b)^2$	+1
	10	$\mathcal{P}_b^2$	−1
	00	$\mathcal{P}_b(1-\mathcal{P}_b)$	0
	11	$\mathcal{P}_b(1-\mathcal{P}_b)$.	0

Neglect the interdependence between inphase transition errors and mid-phase noise, and assume that $\mathcal{P}_b$ is constant with respect to ϵ for small $|\epsilon| \ll T$. Then we can write the discriminator gain A at $\epsilon = 0$ from (14-12) as

$$A \approx [(1 - \mathcal{P}_b)^2 - \mathcal{P}_b^2]\frac{\partial}{\partial \epsilon}\mathrm{E}[x(\epsilon, R = \infty)]$$
$$\approx 1 - 2\mathcal{P}_b(R) = \operatorname{erf}\sqrt{R} = 1 - \operatorname{erfc}\sqrt{R} \tag{14-18}$$

for a data transition density of 50%. Thus, transition errors lower the effective loop gain of the bit synchronizer. However, note that a relatively high error rate of $\mathcal{P}_b$-10^{-2} produces only a 2% decrease in loop gain.

The general expression for the synchronizer discriminator characteristic for $|\epsilon| < T/2$ is given by [Simon 1969, 20-9–20-15]

$$\frac{D(\epsilon)}{T} = \frac{\epsilon}{T}\operatorname{erf}\left[\sqrt{R}\left(1 - \frac{2\epsilon}{T}\right)\right] - \frac{1}{8}\left(1 - \frac{2\epsilon}{T}\right)$$
$$\left\{\operatorname{erf}\sqrt{R} - \operatorname{erf}\left[\sqrt{R}\left(1 - \frac{2\epsilon}{T}\right)\right]\right\} \qquad |\epsilon| \leq \frac{T}{2} \tag{14-19}$$

where the signal energy is P_sT, the one-sided noise density is N_0, and $R \triangleq P_sT/N_0 = E_b/N_0$. The performance of the synchronizer is transition-density sensitive. The discriminator characteristic of (14-19) assumes a transition density of 50%.

The noise spectral density at the multiplier output $\epsilon = 0$ for stationary inputs is

$$G_N(\omega, \epsilon)|_{\epsilon=0} = \frac{1}{T}\sum_{m=-\infty}^{\infty} R(m, \epsilon)\cos\omega mT|_{\epsilon=0}$$
$$R(m, \epsilon)|_{\epsilon=0} = \mathrm{E}(N_k N_{k+m}) \tag{14-20}$$

where

$$N_k = \int_{(k-1/2)T}^{(k+1/2)T} n(t)\,dt \tag{14-20a}$$

for values of k where a transition occurs and zero otherwise. Only the noise density in the vicinity of $\omega = 0$ is of interest because of the narrow-band-width loop filters. Successive values of N_k are independent, because the time intervals of integration are nonoverlapping. Hence, the noise density at $\omega = 0$ is

$$G_N(0, 0) = \frac{1}{T}\mathrm{E}(N_k^2) = \frac{1}{2}\left(\frac{N_0}{2}\right) = \left(\frac{N_0}{4}\right) \tag{14-21}$$

where the 1/2 factor occurs because transitions occur half the time. For an equivalent closed-loop noise bandwidth B_L Hz of the loop in Fig. 14-7, the

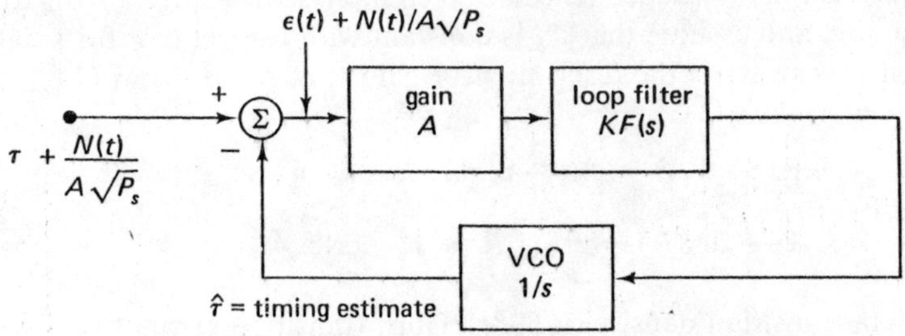

Fig. 14-7 Equivalent linearized block diagram of the inphase/midphase bit synchronizer; the timing error $\epsilon = \tau - \hat{\tau}$, and the noise $N(t)$ is the phase-detector output noise $N(t) \triangleq \sum N_k s(t - kT)$.

inverse output SNR is

$$\frac{1}{(\text{SNR})_0} \cong \frac{G_N(0,0)}{A^2P_s} B_L = \left(\frac{N_0/4}{A^2P_s}\right) B_L \cong \frac{N_0 B_L}{4P_s(\text{erf}\sqrt{R}\,)^2} = \frac{B_L T}{4R(\text{erf}\sqrt{R}\,)^2} \tag{14-22}$$

where $R = P_s T/N_0$. Thus, the mean-square timing error for large $B_L T$ is

$$\frac{\sigma_\epsilon^2}{T^2} = \frac{1}{(\text{SNR})_0} \cong \frac{B_L T \xi}{4R} \qquad \frac{R}{B_L T} \gg 1 \tag{14-23}$$

where ξT is the window width. Simon [1969] has shown that for larger values of $B_L T$ but large R, the result becomes

$$\sigma_\epsilon^2 \cong \frac{B_L T}{4R} (1 - B_L T)^{-1} \xi \qquad \sigma_\epsilon \ll \xi T \tag{14-24}$$

where the second factor accounts for the change in the noise spectrum. He also has shown that use of a midphase window of only $\xi T = T/4$ gives a 3-dB improvement. The acquisition window is only half as wide as for a window of $T/2$, however, and the B_L would have to be increased for the same acquisition time. For smaller ξ, one has to compute the equivalent gain. (See Sec. 12-3.)

For small values of $R/B_L T$, the mean-square timing error is uniformly distributed over $|\epsilon| < T/2$ and $\sigma_\epsilon^2/T^2 = 1/12$. The relationships between loop-filter parameters and closed-loop bandwidth are given in Chap. 12 (Table 12-3).

14-4 EARLY- LATE-GATE BIT SYNCHRONIZER

Another type of closed-loop bit synchronizer is the absolute-value early-late-gate bit synchronizer [Simon, 1970a] shown in Fig. 14-8. This unit uses early- and late-gate integrate-and-dump channels, having an absolute value

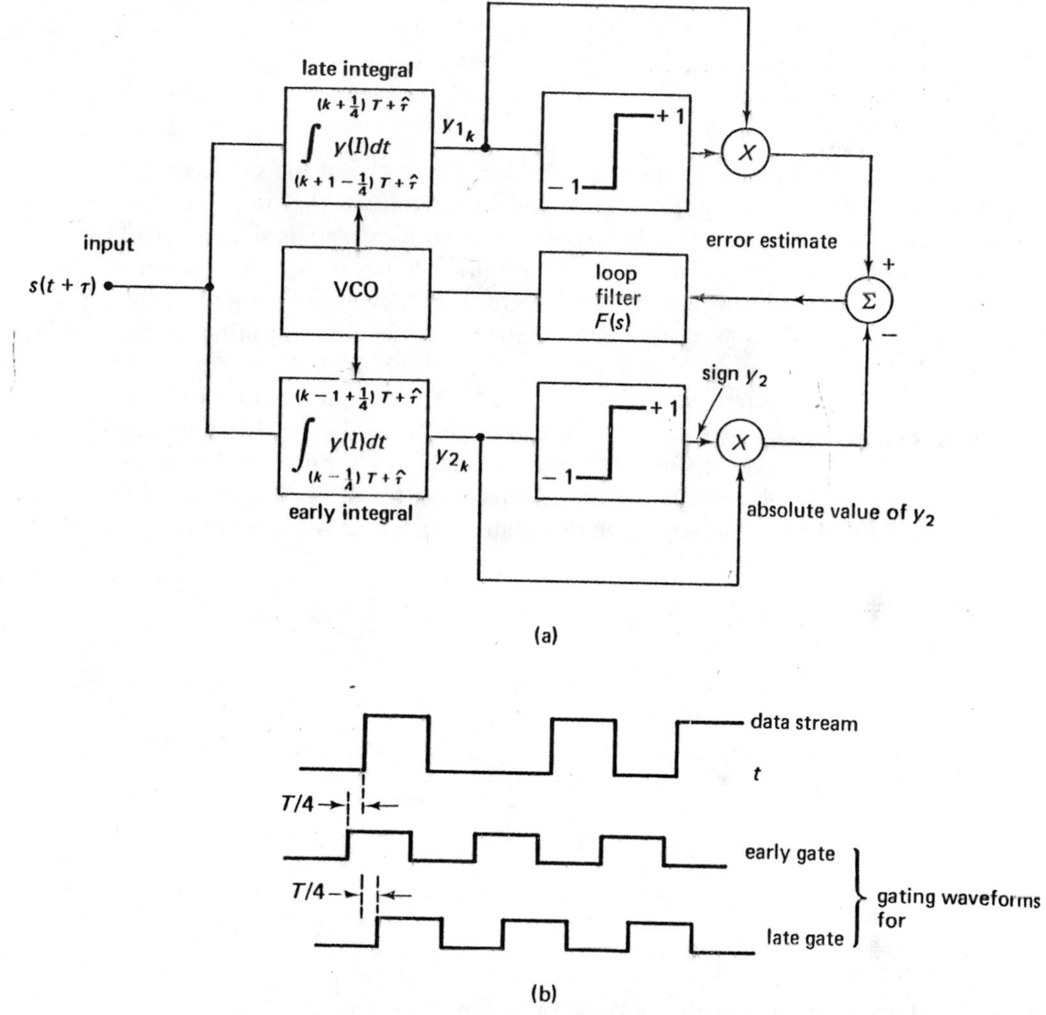

Fig. 14-8 Block diagram and waveforms for an absolute-value early- late-gate bit synchronizer. (a) Block diagram: the $T/4$ overlap used can be shown to be optimum. (b) Waveforms for $\hat{\tau} = \tau$

operation that makes it bit independent. The channels are then subtracted, and if the early channel has the greater absolute value, a correction voltage is developed to slow down the clock timing Number Controlled Oscillator NCO or VCO (after appropriate loop filtering).

Simon [1970*a*] has shown that the absolute-value bit synchronizer has a mean square timing error identical to the in-phase/mid-phase loop with an integration window of $T/4$, namely,

$$\frac{\sigma_\epsilon^2}{T^2} = \frac{1}{8}\frac{B_L T}{R} \qquad \frac{R}{B_L T} \gg 1 \tag{14-25}$$

where B_L is the closed-loop noise bandwidth. The shape of the synchronizer discriminator characteristic can be found by examining the output of the early and late integrals in Fig. 14-8 when the input bit stream is shifted ahead in time. The late integral for a $T/4$ sec shift ahead in time has maximum output. If there is a transition, the early integral has zero output; otherwise the early integral also has maximum output. Since random data transitions occur 1/2 the time, the average difference in these outputs is maximum. On the other hand, if the input bit stream is offset by $T/2$ or 0, the output of both integrations is equal and the difference averages to zero. The discriminator characteristic for large R for this early- late-gate synchronizer is shown in Fig. 14-9. Table 14-2 compares the relative features of the in-phase/mid-phase (IP/MP) and early-late gate (absolute value) type bit synchronizers.

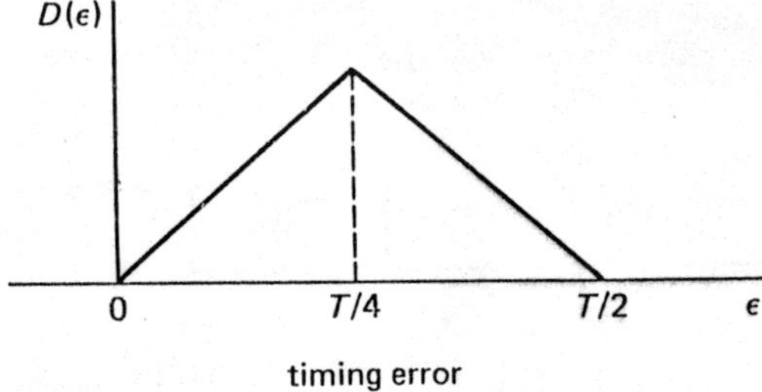

Fig. 14-9 Bit synchronizer phase discriminator characteristic for the early- late-gate bit synchronizer. $D(\epsilon)$ is an odd function of ϵ.

Table 14-2 ADVANTAGES OF DATA TRANSITION TRACKING AND EARLY- LATE-GATE SYNCHRONIZERS RELATIVE TO ONE ANOTHER

Synchronizer Type	*Advantages*
IP/MP	• Better noise tracking performance if the window width ξT is small ($\xi \leq 1/4$) • Shorter pull-in time because it has a stronger correction $D(\epsilon)$ in the vicinity of $\epsilon = (1/2)T$
Early-Late	• Simpler to implement • Less sensitive to dc offsets

14-5 OPTIMUM BIT SYNCHRONIZER

We next examine the optimum (maximum likelihood) method of searching for the correct clock time during the acquisition mode. This problem is distinguished from the closed-loop tracking techniques in the previous paragraphs (Sec. 14-4). Assume that the time jitter τ of the received data bit stream is constant over an observation interval (memory time) of KT sec, and that τ is uniformly distributed over $0 \leq \tau \leq T$. The bit stream is received with additive white Gaussian noise of variance σ^2.

Sample each T-sec duration bit at n samples/bit to develop samples x_{ij} as the ith sample from the jth bit—that is, $1 \leq i \leq n$, $1 \leq j \leq K$. Quantize the timing offset τ as $\tau_m = mT/n$. The signal samples over one bit are labeled $s_i(\tau_m)$.

The maximum likelihood (ML) estimate of τ_m for a given set of received signal samples $\mathbf{X} = \mathbf{S} + \mathbf{N} = (x_{11}, \ldots, x_{nk})$ is the value of τ_m, which maximizes

$$p(\tau_m \,|\, \mathbf{X}) = \frac{p(\tau_m)}{p(\mathbf{X})}\, p(\mathbf{X} \,|\, \tau_m) \tag{14-26}$$

Since τ_m is uniformly distributed, $p(\tau_m) = 1/n$, and $\mathbf{X}$ is given, the ML estimate is the value of τ_m that maximizes

$$p(\mathbf{X} \,|\, \tau_m) = C' \prod_{j=1}^{k} \left[p(0) \exp - \left\{ \sum \frac{[x_{ij} - s_i(\tau_m)]^2}{2\sigma^2} \right\} + p(1) \exp - \left\{ \sum \frac{[x_{ij} + s_i(\tau_m)]^2}{2\sigma^2} \right\} \right] \tag{14-27}$$

where the probability of "1" and "0" data bits are equally likely $p(1) = p(0) = 1/2$. The constant C' contains the normalizing factors and we have used the Gaussian distributions of the noise $\mathbf{X} - \mathbf{S}$. The x_{ij} are the observed samples. Thus, we can rewrite (14-27) and define Y_j as

$$p(\mathbf{X} \,|\, \tau_m) = C \prod_{j=1}^{k} \cosh \sum_i \left[\frac{x_{ij} s_i(\tau_m)}{\sigma^2} \right] = C \prod_j Y_j \tag{14-28}$$

since $\sum x_{ij}^2$ is a constant for any given j the received signal bit interval and $\sum s_i^2(\tau_m)$ is a constant independent of τ_m. This expression is maximized by maximizing the natural logarithm $\log \prod_j Y_j = \sum_j \log Y_j$ with respect to τ_m to obtain the optimum estimate [Wintz and Luecke, 1969]:

$$\tau_m = \max_{\tau_m} \sum_{j=1}^{k} \log \cosh \sum_i x_{ij} s_i(\tau_m)/\sigma^2 \tag{14-29}$$

Thus, the optimum detector consists of a bank of correlators performing the computation $\sum x_{ij}s_i(\tau_m)$ for each value of τ_m and employing log cosh weighting of the output. The optimum decision is then made by maximizing the sum over all available bits. The same result is achieved by a matched filter with log cosh weighting followed by a sampler and set of summation devices.

This computation of the optimum value of τ_m is generally too complex to be useful for a practical system. However, this detector performance provides a useful bound on performance. Computer simulation of the performance for a memory KT sec yields an expected magnitude of error (AME) of [Wintz and Luecke, 1969, 383]

$$\mathrm{E}\left[\frac{|\epsilon|}{T}\right] \cong \frac{0.2}{\sqrt{KE_b/N_0}} \qquad \frac{E_b}{N_0} > 2 \tag{14-30}$$

for raised cosine input waveforms. Notice that the nonlinear bit synchronizer performance in (14-4) is about 1.5 times as large a timing error. The error increases by approximately a factor of 2 above this value when $E_b/N_0 = 1/2$. The rms timing error is slightly larger than the AME

$$\frac{\sigma_\epsilon}{T} \cong \frac{0.25}{\sqrt{KE_b/N_0}}$$

The performance for rectangular and half-sine input waveforms is approximately the same as that given in (14-31) for SNRs where bit synchronization errors can have strong influence on error rate, $E_b/N_0 < 20$.

14-6 BIT ERROR PROBABILITY WITH CLOCK TIMING ERRORS

The primary measure of performance of the bit synchronizer is its effect on the bit error probability in the recovered data. In this section we calculate the effects of the rms clock timing error on the degradation in error rate performance. We first review the bit error probability for NRZ waveform in a Gaussian channel conditioned on a given clock timing error ϵ. Then we compute the expected bit error by averaging over the clock timing error statistics. The bit error probability for a given clock timing error ϵ is [Lindsey and Simon, 1973, 468]

$$\begin{aligned}\mathcal{P}_b(\epsilon) &= \frac{1}{4}\,\mathrm{erfc}\left\{\sqrt{\frac{E_b}{N_0}}\,\frac{[R_s(\epsilon) + R_s(T-\epsilon)]}{R_s(0)}\right\} \\ &\quad + \frac{1}{4}\,\mathrm{erfc}\left\{\sqrt{\frac{E_b}{N_0}}\,\frac{[R_s(\epsilon) - R_s(T-\epsilon)]}{R_s(0)}\right\}\end{aligned} \tag{14-31}$$

where $R_s(\epsilon) = \mathrm{E}[s(t)s(t+\epsilon)]$ is the autocorrelation function of the input waveform. For random NRZ inputs we have

$$\begin{aligned} R_s(\epsilon) &= 1 - |\epsilon|/T, \qquad |\epsilon| < T \\ R_s(\epsilon) &= 0, \qquad |\epsilon| \geq T \end{aligned}$$

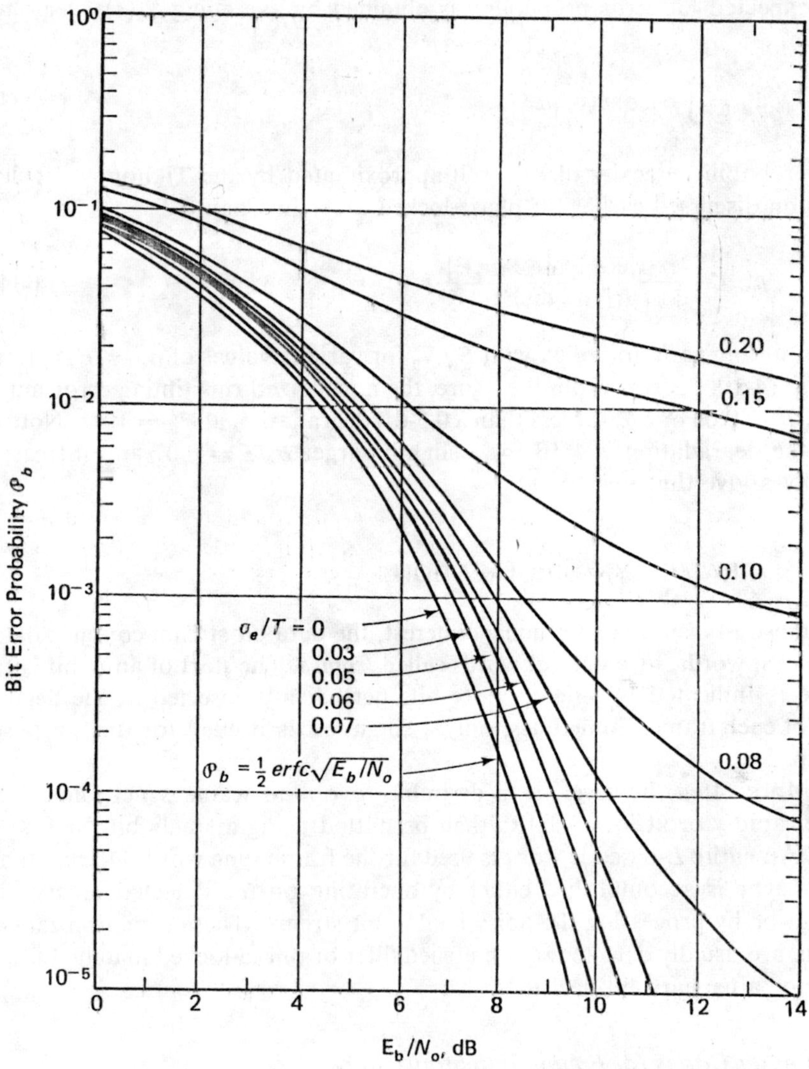

Fig. 14-10 Average probability of bit error versus E_b/N_0 with standard deviation of the symbol sync error σ_ϵ as a parameter (NRZ) [From Lindsey and Simon, 1973, 469, Fig. 9-37]

By substituting this expression for $R_s(\epsilon)$ in the expression for $\mathcal{P}_b(\epsilon)$ (14-31), we obtain

$$\mathcal{P}_b(\epsilon) = \frac{1}{4}\,\text{erfc}\left(\sqrt{\frac{E_b}{N_0}}\right) + \frac{1}{4}\,\text{erfc}\left[\sqrt{\frac{E_b}{N_0}}(1 - 2|\epsilon|)\right] \tag{14-32}$$

The expected bit error probability is obtained by averaging $\mathcal{P}_b(\epsilon)$ over ϵ to obtain

$$\mathcal{P}_b = \int \mathcal{P}_b(\epsilon)\, p(\epsilon)\, d\epsilon \tag{14-33}$$

The probability density of ϵ is well approximated by the Tikhonov density function discussed earlier for phase-locked loops in Chap. 12

$$p(\epsilon) = \frac{\exp\,[\cos 2\pi\epsilon/(2\pi\sigma_\epsilon)^2]}{I_0[1/(2\pi\sigma_\epsilon)^2]} \tag{14-34}$$

The resulting variation of $\mathcal{P}_b$ with E_b/N_0 for various values of $\sigma_\epsilon = T$ is given in Fig. 14-10. As seen from the figure, the normalized rms timing error must be $\sigma_\epsilon/T \leq 0.05$ or 5% for less than a 0.5-dB degradation at $\mathcal{P}_b = 10^{-5}$. Notice that the degradation is 4 dB for a slightly larger $\sigma_\epsilon/T = 0.07$ and increases rapidly above that point.

14-7 FRAME SYNCHRONIZATION

In almost all instances of practical interest, the data bit stream contains data in blocks, words, or blocks of words called *frames*. The start of an N-bit data frame is indicated by one or more bits periodically inserted at the beginning of each frame. Sometimes, only a single bit is needed for this purpose (Chap. 5).

More often, however, it is desirable to obtain frame synchronization more rapidly or at lower SNRs than permitted using a single bit for frame sync. An entire L-bit code word is used for the frame sync word. Detection of frame sync is accomplished either by operating on the detected binary bit stream or by processing the noisy analog bit stream. These synchronization words are usually detected by a matched filter or phase-locked matched filter detector after hard bit decisions.

Barker Codes for Frame Synchronization

Barker has devised a method of frame sync in which the sync word is located by correlating successive L-bit segments of the received bit sequence with the stored sync word. Binary Barker codes are often used for frame synchronization [Barker, 1953, 273–287; Turyn, 1963, 1256; Toerper, 1968, 278–289]. They consist of a finite number of $\pm$ symbols and have unique properties in that their autocorrelation function C_k sidelobes (adjacent data symbols are all assumed to be zero)

$$C_k = \sum_{i=1}^{N-k} x_i x_{i+k} \tag{14-35}$$

never exceed unity in magnitude for $k \neq 0$. This property allows one to determine with little ambiguity the start position for word synchronization. Even then there is always some potential ambiguity because adjacent data bits are random. Thus, they are often used as the start of a frame of data bits in a time-division multiplexing system. The only known binary Barker words are listed in Table 14-3. Williard codes, which minimize the probability of false sync with random data words in the overlap region for all values of overlap, are also shown [Williard, 1962]. For each Barker sequence the complement, reflected (mirror image), and reflected complement operations generate three other codes with good synchronization properties.

Table 14-3 BARKER WORDS AND WILLIARD CODES OF LENGTH N UP TO $N = 13$

N	*Barker Sequence*	*Neuman-Hofman Sequence*	*Williard Code*
1	+		+
2	++ or +−		+−
3	++−		++−
4	+++− and ++−+		++−−
5	+++−+		++−+−
7	+++−−+−		+++−+−−
11	+++−−−+−−+−		+++−++−+−−−
13	+++++−−++−+−+	−−−−−−++−−+−+	+++++−−+−+−−−

Table 14-4 shows a set of code words with good synchronization properties, which were found by extensive computer search by Neuman and Hofman [1971, 272–281]. This Neuman-Hofman (NH) code set includes code words up to length 24. The synchronization word of ± 1 symbols has zero value outside the word. The autocorrelation function (14-35) of these words is also given in Table 14-4. Also listed is the worst case autocorrelation sidelobe for signals with and without a sign ambiguity. [Neuman, Hofman, 1971, 281.] The list of codes in Table 14-4 includes 2 Barker words.

Optimum Frame Synchronization

The transmitted frame consists of an L-bit sync word $(s_0, s_1, \ldots, s_{L-1})$ imbedded in an $N - L$ bit data stream $(d_L, d_{L+1}, \ldots, d_{N-1})$ to form the frame $(s_0, s_1, \ldots, s_{L-1}, d_L \ldots, d_{N-1})$ where $s_i, d_i = \pm 1$. These data are received along

Table 14-4 LIST OF NEUMAN-HOFMAN SYNCHRONIZATION WORDS UP TO LENGTH 24 THE CODE WORD LENGTH IS C_0 (Courtesy NASA Ames Research Center)

Sync Word	Autocorrelation Function C_k																									Max Sidelobe	
	Shift																										
	0	1	2	3	4	5	6	7	8	9	10	11	12	13	14	15	16	17	18	19	20	21	22	23	24	C_k	$\|C_k\|$
0001101	7	0	−1	0	−1	0	−1	← Barker word																		0	1
00001101	8	1	0	1	−2	−1	0	−1																		1	2
00011101	8	1	0	−3	0	−1	0	−1																		1	3
000011101	9	2	1	−2	−1	−2	−1	0	−1																	2	2
001111101	9	2	1	0	−1	−2	1	0	−1																	2	2
0000011010	10	1	2	1	−2	−1	0	1	0	1																2	2
0000110101	10	−1	2	−1	−2	1	0	−1	0	−1																2	2
00001011001	11	0	−1	2	−1	−2	1	0	1	0	−1															2	2
00011101101	11	0	−1	0	−1	0	−1	0	−1	0	−1	← Barker word														0	1
000111101101	12	1	0	1	−4	1	0	−1	0	−1	0	−1														1	4
001100000101	12	1	0	−1	0	−1	2	1	0	1	0	−1														2	2
0000001100101	13	2	1	2	1	0	1	0	1	0	−1	0	−1													2	2
0000010110011	13	2	−1	2	1	−2	1	0	−1	0	−1	−2	−1													2	2
00001100110101	14	−1	−2	−1	2	1	0	−1	−2	1	0	−1	0	−1												2	2
00110011111010	14	1	−2	−1	2	−1	0	−1	−2	−1	0	−1	0	1												2	2
001111100110101	15	0	−1	−2	−1	−2	1	2	1	−2	1	0	1	0	−1											2	2
000011001001010	15	−2	−1	2	−1	0	1	2	1	2	−1	0	1	0	1											2	2
0000011001101011	16	1	−2	1	2	1	0	−1	−2	1	0	−1	−2	−1	−2	−1										2	2
0000111011101101	16	1	0	1	2	−1	−2	1	−2	−1	−2	1	−2	−1	0	−1										2	2

[NASA Ames Research Center/Pioneer Project]

Table 14-4 (Continued)

	Autocorrelation Function C_k																									Max Sidelobe	
	Shift																										
Sync Word	0	1	2	3	4	5	6	7	8	9	10	11	12	13	14	15	16	17	18	19	20	21	22	23	24	C_k	$\|C_k\|$
00001011001110101	17	−2	1	−2	1	0	1	0	1	−4	1	−4	1	0	−1	0	−1									1	4
00001111011011101	17	2	1	2	−1	−2	−1	2	1	0	−1	−2	−1	−2	−1	0	−1									2	2
000010101101100111	18	−1	0	1	0	1	−2	1	0	−1	−2	1	0	1	−2	−3	−2	−1								1	3
001100111110100101	18	−1	−2	1	−2	1	−2	−1	−2	1	0	−1	0	1	0	1	0	−1								1	2
0000111000100010010	19	2	−1	−2	1	0	1	2	1	2	1	2	1	2	−1	2	1	0	1							2	2
0001110111011011010	19	−2	−1	2	1	0	1	−2	1	−2	1	−2	1	−2	−1	−2	1	0	1							2	2
00000100110101001110	20	−1	0	1	2	1	−2	−1	2	1	−2	−1	2	1	−2	−1	−2	−1	0	1						2	2
00010001111100101101	20	1	0	1	−4	−1	0	−1	0	−1	−2	1	0	1	0	−3	0	−1	0	−1						1	4
000000101110100111001	21	2	1	0	1	−2	−1	0	−1	−2	−1	2	−1	−2	1	−2	−1	0	1	0	−1					2	2
001101100001000010101	21	−2	1	0	1	2	−1	0	−1	2	−1	−2	−1	2	1	2	−1	0	1	0	−1					2	2
0001000111110011011010	22	1	−2	1	0	−3	0	1	0	−3	−2	1	0	−3	−2	1	−2	1	−2	1	0	0				1	3
00000010101100110100111	23	0	1	2	1	0	1	2	−5	2	−3	2	1	−2	1	0	1	−2	−1	−2	−3	−2	−1			2	5
00000011110011001001010	23	2	−1	2	−1	−2	−3	0	−3	−2	1	−2	1	2	1	2	−1	2	1	0	1	0	1			2	3
000111111001000011001010	24	3	0	−1	−2	−1	−4	−9	−2	3	−2	1	2	3	2	−3	−2	1	0	1	−2	1	0	1		3	9
000001110011101010110110	24	−1	0	−3	0	1	0	1	0	−1	−4	1	0	−1	0	1	0	−1	−4	−1	0	−1	0	1		1	4

with additive Gaussian noise samples $(n_0, \ldots, n_{N-1})$ and observed with some cyclic shift. Because the sync word is not received by itself but is imbedded in the data bit stream, the optimum detector is not a pure correlator but generally includes a received-energy-correction term [Massey, 1972, 115–118].

When there is no ambiguity in the sign of the received sequence, the optimum* sync word detector observes the received samples ρ_i and maximizes the statistic

$$\mathcal{S}(\mu) = \sum_{i=0}^{L-1} s_i \rho_{i+\mu} - \sum_{i=0}^{L-1} \frac{N_0}{2\sqrt{E_b}} \log \cos \left(\frac{2\sqrt{E_b}\rho_{i+\mu}}{N_0} \right) \tag{14-36}$$

with respect to time-shift μ where E_b is the energy/bit, N_0 is the one-sided noise density, and ρ_i is the received sample value of the data-bit plus noise, where each received sample has value $r_i = \pm\sqrt{E_b} + n_i$, and $s_i = \pm 1$, $d_i = \pm 1$. The first term in (14-36) is simply the correlation term. The value of μ that maximizes $\mathcal{S}(\mu)$ is the Bayes estimate of the sync word epoch. At high $E_b/N_0 \gg 1$, this statistic (14-36) becomes

$$\mathcal{S}(\mu) \cong \sum_{i=0}^{L-1} s_i \rho_{i+\mu} - \sum_{i=0}^{L-1} |\rho_{i+\mu}| \tag{14-37}$$

Thus, wherever $s_i \rho_{i+\mu}$ is positive, the contribution to $\mathcal{S}$ is canceled by $|\rho_{i+\mu}|$, only negatively correlated terms contribute, and the optimum decision rule reduces to:

> Choose the location of the sync word as the point where there is least total *negative* correlation.

For small $E_b/N_0 \ll 1$, the statistic to be maximized reduces to

$$\mathcal{S}(\mu) = \sum_{i=0}^{L-1} \left[s_i \rho_{i+\mu} - \frac{\sqrt{E_b}}{N_0} \rho_{i+\mu}^2 \right] \tag{14-38}$$

The second term in (14-38) is the received energy correction—that is, the correction to the usual correlation measure.

If there is a sign ambiguity in the received sequence, as occurs with PSK, then the statistic replacing (14-36) to be maximized becomes

$$\mathcal{S}'(\mu) = \left| \sum_{i=1}^{L-1} s_i \rho_{i+\mu} \right| - \sum_{i=0}^{L-1} \frac{N_0}{2\sqrt{E_b}} \log \cos \left(\frac{2\sqrt{E_b}\rho_{i+\mu}}{N_0} \right) \tag{14-39}$$

This time the **magnitude** of the correlation term is measured, whereas the actual value is used in (14-36). Again, there is a correction term to the correlation statistic.

---

*The Bayes estimate for equally likely ± 1 data bits.

For the $L = 13$-bit sync words and relatively low values of $E_b/N_0 = 1$, 2, the results for probability of erroneous sync are given by Table 14-5 for optimum detection [Massey, 1972, 115–118]. Results are given for the optimum detectors of (14-36) and (14-39), that is, both with and without the sign ambiguity of BPSK.

Table 14-5 PROBABILITY OF ERRONEOUS SYNC FOR 13-BIT SYNC WORDS FOR PSK WITH A SIGN AMBIGUITY AND INPUTS WITH NO SIGN AMBIGUITY

Sync Word 13-Bit	$E_b/N_0 = 1$		$E_b/N_0 = 2$	
	No Ambiguity	*BPSK*	*No Ambiguity*	*BPSK*
Barker word	0.09	0.14	0.00	0.00
Neuman-Hofman word	0.07	0.14	0.00	0.00

CHAPTER 15

VITERBI DECODING OF CONVOLUTIONAL CODES

15-1 INTRODUCTION

A substantial body of literature has emerged on coding—both block and convolutional codes [Peterson and Weldon, 1972]. In this chapter, we shall concentrate on convolutional codes and a maximum-likelihood decoding algorithm for these convolutional codes, which is credited to Viterbi [1967, 260–269; 1971, 751–771]. It is one of the best techniques yet evolved for digital communications where energy efficiency dominates in importance.

Maximum-likelihood decoding using the Viterbi decoding algorithm, offers a number of advantages; it permits a major equipment simplification while obtaining the full performance benefits of maximum-likelihood decoding. In particular, the Viterbi decoding benefits are: (1) E_b/N_0 for 10^{-5} error rate is reduced by approximately 4 to 6 dB relative to uncoded BPSK or QPSK; (2) the decoder structure is relatively simple for short constraint lengths, making decoding feasible at relatively high rates of up to 100 Mbps; and (3) requirements on carrier-phase coherence, although more stringent than for uncoded BPSK, are still well within feasible limits.

The substantial improvement in performance made possible by maximum-likelihood codes requires that the transmitted bit rate and hence the bandwidth be increased by a factor of 2 for rate 1/2, or 4/3 for rate 3/4 codes. Thus, this type of coding is particularly beneficial for power-limited satellite communications channels—that is, channels in which the power limit constraint is more severe than the bandwidth constraint.

It also should be noted that satellite communications involves a substantial time delay ($\geq$0.25 sec) and often rather high data rates ($\geq$100 Mbps). This combination can make Forward Error Correction (FEC) much more desirable than automatic request for retransmission (ARQ), because of the large costs of ARQ to store data at the transmitter until a verification signal is received or a request for repeat is received for a data block. In ARQ, blocks of data are transmitted with redundancy introduced for error detection. If a data block is received in error, the receiver sends the transmitter a request for retransmission. The use of FEC and ARQ together can be advantageous.

In this chapter we review the structure of convolutional codes, describe the structure of the Viterbi decoding algorithm, and discuss the error rate performance of the decoding algorithm for PSK and QPSK signals both with and without carrier reconstruction phase noise. Many of the results described in this chapter were first derived by Viterbi and his co-workers in the cited references.

15-2 CONVOLUTIONAL CODE STRUCTURE

A convolutional encoder with constraint length K is a K-stage shift register with n linear algebraic function generators, one for each output port. A rate 1/2 code produces two output bits for every input data bit. If one of these output bits is the original data bit, the code is called systematic. Figure 15-1 shows the structure of a simple nonsystematic rate 1/2 encoder of constraint length $K = 3$.

Assuming that the encoder starts in the all-zero state, the first four bits

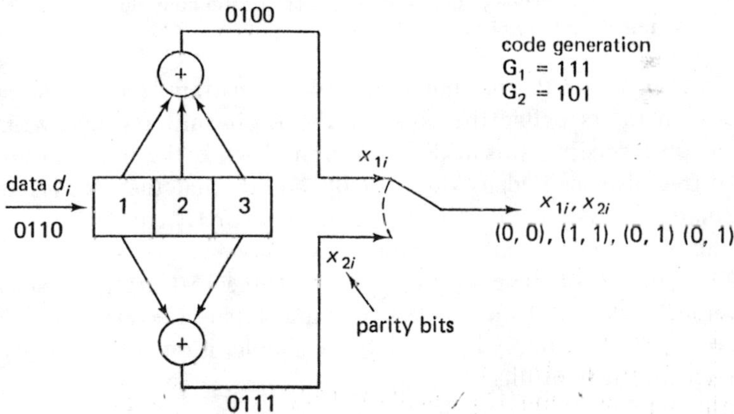

Fig. 15-1 Nonsystematic convolutional encoder with constraint length $K = 3$, and rate $1/n = 1/2$. The code generator denotes the tap positions.

0110 produce an output of 00, 11, 01, and 01, respectively, as shown. Clearly, the output of each new data bit depends on the previous bit pattern stored in stages 1, 2 of the shift register. These bit patterns can be labeled by the states defined as

$$a = 00 \qquad b = 01 \qquad c = 10 \qquad d = 11 \tag{15-1}$$

The output bits and transitions between states can be labeled by the trellis diagram of Fig. 15-2. The diagram starts in the all-zero state, node a, and makes transitions corresponding to the next data bit. These transitions are denoted by a solid line for a "0" and a dotted line for a "1." Thus, node a proceeds to node a or b with output bits 00 or 11.

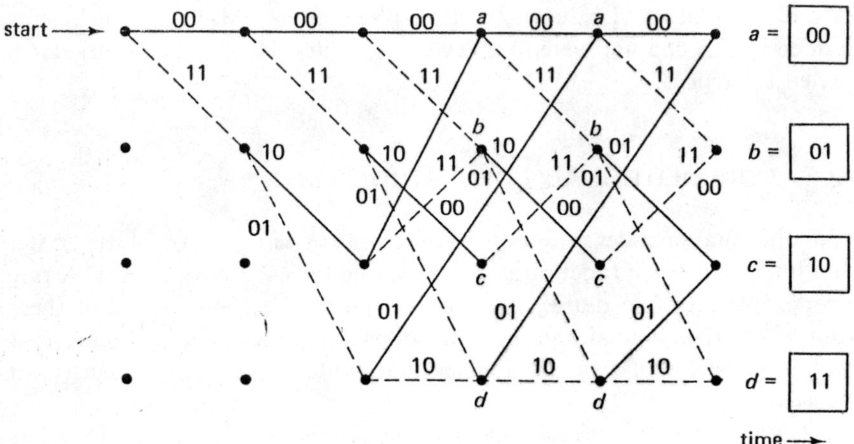

Fig. 15-2 Trellis-code representation for the convolutional encoder of Fig. 15-1

Table 15-1 shows the optimum codes for constraint lengths $K = 3$–8. The code generators define the taps for the K-bit shift register; n_e is the number of bit errors in paths at distance d_f, and d_f^* is the upper bound on minimum free distance [Odenwalder, 1970]. Notice that these codes are all nonsystematic. A systematic code would have $\mathbf{G}_1$ or $\mathbf{G}_2$ equal to 100 . . . 0; that is, one of the code generators would have only a single tap.

Note that some of the code structures in Table 15-1 are transparent to code inversion—that is, if the signs of the input bits are reversed, the coded output bit sequence is simply inverted. For example, if one of the parity bits x_{jl} is related to the data bits d_l by

$$x_{jl} = d_l \oplus d_{l-1} \oplus d_{l-2} \tag{15-2}$$

where there are an odd number of terms in the sum, reversing the sign of the d_l bits simply reverses all of these parity bits. Thus, if the numbers of "ones" or

Table 15-1 OPTIMUM RATE 1/2 CODES (MAXIMUM MINIMUM DISTANCE) [GILHOUSEN ET AL., 1971]

Constraint Length K	*Code Generators*	*Distance d_f*	*Errors n_e*	*Distance Bound d_f^**	*Code Transparent to 180° Phase Reversal*
3	$G_1 = 111$ $G_2 = 101$	5	1	5	no
4	$G_1 = 1111$ $G_2 = 1101$	6	2	6	no
5	$G_1 = 11101$ $G_2 = 10011$	7	4	8	no
6	$G_1 = 111011$ $G_2 = 110001$	8	6	9	yes
7	$G_1 = 1111001$ $G_2 = 1011011$	10	36	10	yes
8	$G_1 = 11111001$ $G_2 = 10100111$	10	2	10	no

weights of both $\mathbf{G}_1$ and $\mathbf{G}_2$ are odd, then the code is transparent to a sign inversion. That is, the decoded output bit stream has the sign ambiguity as the input. This transparency is valuable if biphase-modulated PSK is used with its ensuing sign ambiguity for it permits decoding prior to ambiguity removal. Differential decoding at the decoder output removes the sign ambiguity and simply increases the output error rate by a factor of less than 2, because decoder output errors typically occur in short bursts. Differential decoding at the decoder input would double the decoder input error rate and thus would cause a much larger increase in the bit-error rate than a factor of 2 because of the high slope in the output-versus-input-error-rate curve.

15-3 THE MAXIMUM-LIKELIHOOD DECODER FOR A BINARY SYMMETRIC CHANNEL

Maximum-likelihood decoding could be accomplished over n coded two-bit symbols for rate 1/2 codes by comparing the received $2m$ output sequences with all $4(2^m)$ possible code paths leading to each of the 4 nodes in Fig. 15-2 and selecting the code sequences with the largest cross-correlation.* This calculation is extremely difficult for large m and would result in an overly complex decoder structure.

A major simplification was made by Viterbi in the likelihood calculation by noting that each of the 4 nodes has only two predecessors, and only the path with the highest cross-correlation weight need be retained for each node. For

*The factor of 4 includes all possible initial starting states.

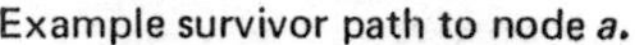

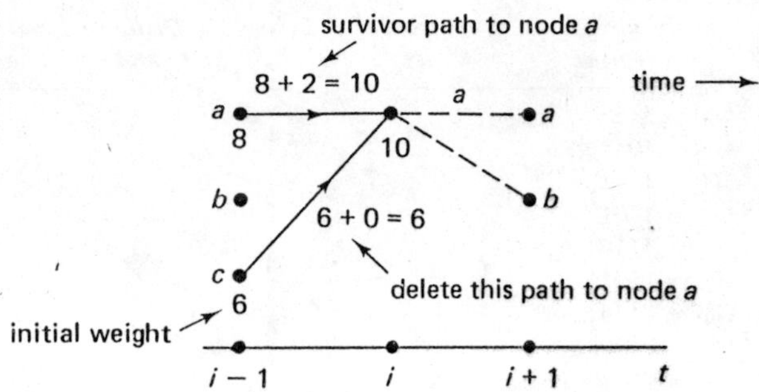

Fig. 15-3 Example of the iterative likelihood calculation showing the survivor path to node *a* at $t = i$. The alternate pattern is eliminated because of its lower weight.

example, the paths to node *a* might have weights 10 and 6 as shown in Fig. 15-3. At each node, the weight of the survivor path determines the new weight. For example, at $t = i$ the *aa* path might add a weight 2 to the weight 8 of the previous node *a*. The added weight can be computed by correlating the received code bits r_{1i}, r_{2i} with the parity bits for that transition p_{1i}, p_{2i} to produce $w_i = p_{1i}r_{1i} + p_{2i}r_{2i}$ where p, r are ± 1 for "hard" binary decisions in r_{ji}.

Figure 15-4 shows a typical path structure and weights for decoding. No errors have been introduced in the channel. Decoding has begun with no information as to the initial state (node) of the coder. Hence, all nodes at $t = 0$ have been set at zero weight. Notice that at $t = 5$ all node survivor paths began at node *a* at $t = 0$, the correct starting position. A decision on that $t = 0$ data bit, the *aa* path corresponding to a 0 data bit, can then be made. Thus in this example a correct data bit decision could be made with a 5 symbol delay. In general, with an error-producing channel, data bit decisions can be made after computing $5K$ successive nodes (a decoding delay of $5K$), where K is the coder constraint length.

If an error is made in selecting the data bit at any time instant, several data bits in succession may be decoded incorrectly before the correct path is reached. Figure 15-5 shows the surviving paths with an incorrect start position (start at node *d*) for the same data sequence as in Fig. 15-4. Note that in this example, 6 data bits, $t = 6$ (12 code bits), have been received before the correct path (correct path after $t = 2$) has produced the highest weight.

Some of the error correction and detection characteristics of the code can be established by redrawing the trellis in a state diagram (Fig. 15-6),

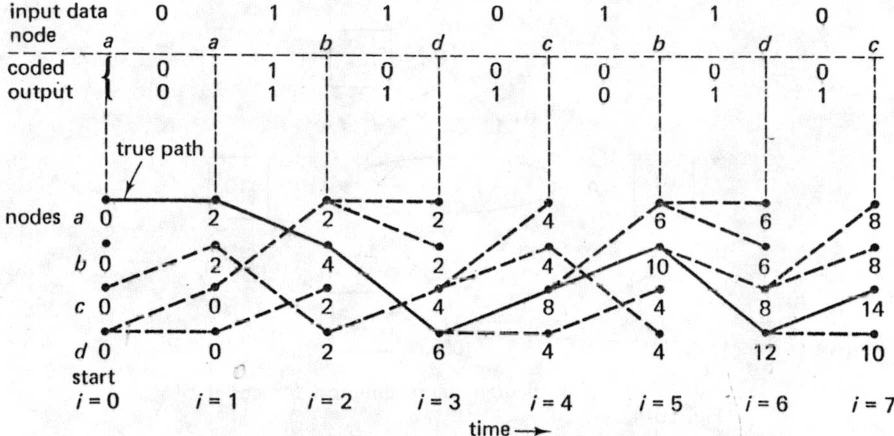

Fig. 15-4 Typical trellis pattern for decoded bit stream with code words received with no errors and unknown starting node. The true path is in solid line, and the other paths at each node are shown in dotted lines. The number adjacent to the node is its correlation metric. All nodes start at zero weight. Correlation is accomplished by replacing the codes by ±1 rather than 0, 1. Ties in the metric are broken by a random decision.

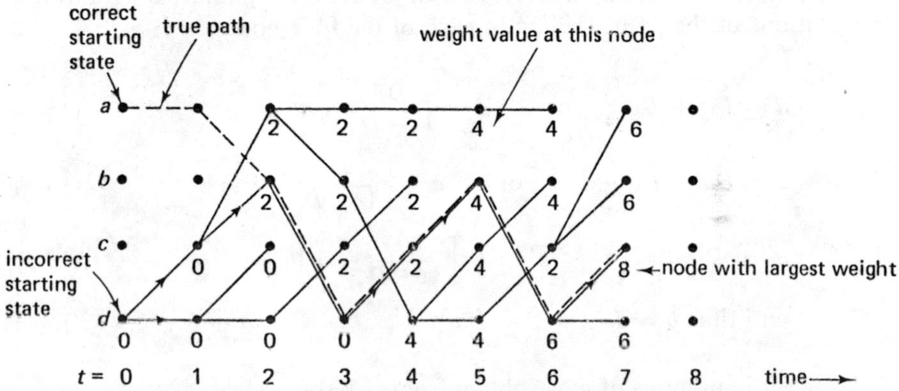

Fig. 15-5 Trellis paths after an initial starting error. The true path is shown by the dashed line.

where each path in the state diagram is labeled with the code bits corresponding to that path—for example, the *aa* path is 00. Note that the minimum-weight path from *a* to *a*, other than the direct *aa* path, is path *abca* shown by the dotted line. If *aaaa* is the correct path in three steps, the minimum alter-

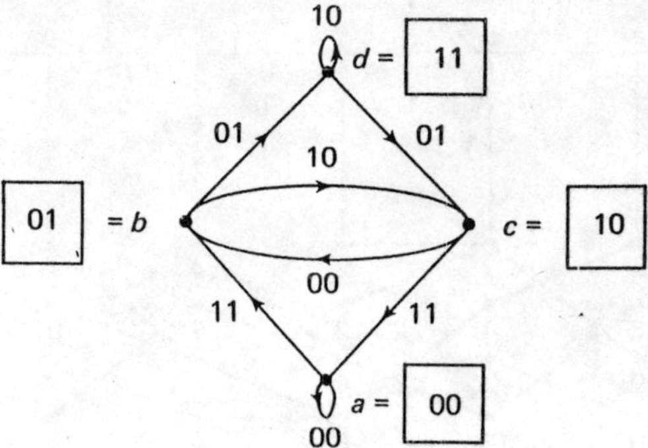

Fig. 15-6 State-diagram representation for coder of Fig. 15-4

nate-weight path *abca* corresponds to 11, 10, 11, and has weight 5, corresponding to five errors or Hamming distance $d = 5$. Thus, this code corrects any two errors over that path length and detects any three errors. Codes of minimum distance $d = 2e + 1$ can be used to correct e errors.

The error-correction properties of any code can be determined by generating the flow diagram from a to a for all paths, each labeled D^k, where k is the weight in that path. The flow diagram for the example code is given in Fig. 15-7. This diagram can be reduced to the generating function for all paths which eventually merge with the all-zeros path by the following calculations of the paths leading to each of the four nodes:

$$d = Dd + Db \qquad \text{or} \qquad d = \frac{D}{1-D}b \tag{15-3}$$

$$c = Dd + Db - b \qquad \text{or} \qquad c = \left[\frac{D}{1-D} - 1\right]b \tag{15-4}$$

$$a' = D^2c \qquad \text{or} \qquad a' = D^2\left[\frac{D}{1-D} - 1\right]b \tag{15-5}$$

$$\text{and thus } b = D^2a + c - Dd - Dc \tag{15-6}$$

Solving for a' in terms of a, we obtain from (15-3) to (15-6)

$$T(D) = \frac{a'}{a} = \frac{D^5}{1-2D} = D^5 + 2D^6 + 4D^7 + \cdots + 2^kD^{k+5} + \cdots \tag{15-7}$$

Thus, there is one path of weight 5, two of weight 6, and, in general, 2^k paths of weight $k + 5$.

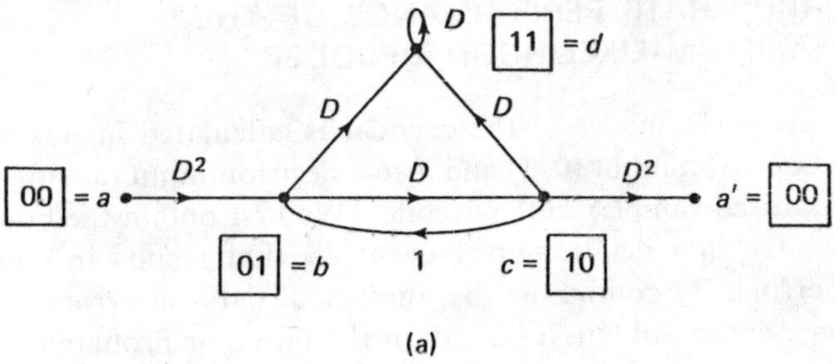

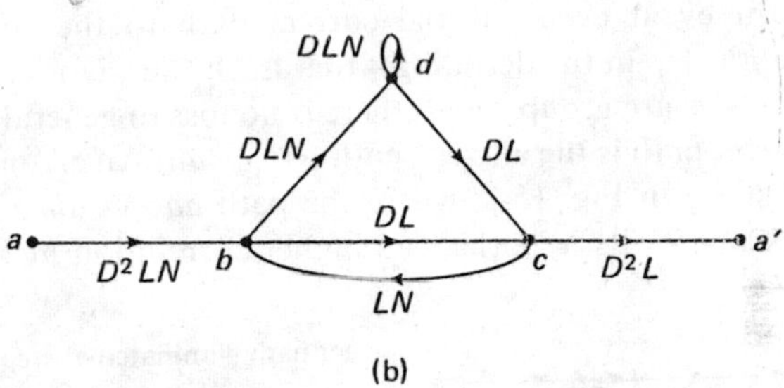

Fig. 15-7 State diagram labeled according to (a) distance from all zeros path for the coder of Fig. 15-1, (b) distance, length, and number of input "ones"

The generating function can be formed for an augmented state diagram in which each path in the state diagram contains a multiplier coefficient L with an exponent ℓ_k corresponding to the length of the path and a coefficient N^e where e is 1 for an input data 1 (a solid-line path in Fig. 15-2) and $e = 0$ for a zero data-bit path. An example augmented state diagram is given in Fig. 15-7b for the coder of Fig. 15-1. The generating function of this augmented state diagram is then

$$\begin{aligned}\frac{a'}{a} = T(D, L, N) &= \frac{D^5L^3N}{1 - DL(1 + L)N} \triangleq \sum a_k D^k N^{e_k} L^{l_k} \\ &= D^5L^3N + D^6L^4(1 + L)N^2 + \cdots \\ &\quad + D^{(5+m)}L^{(3+m)}(1 + L)^m N^{(1+m)} + \cdots \end{aligned} \tag{15-8}$$

Thus there is one distance 5 path from a to a' (the D^5 term) of length 3 corresponding to the exponent of L. There are two distance 6 paths, one of length 4, and one of length 5, both of which differ from the correct path by two data bits (the exponent of N).

15-4 ERROR-RATE PERFORMANCE OF THE MAXIMUM-LIKELIHOOD DECODER

The error-rate performance of the decoder is calculated in this section for both a hard-decision input (0, 1) and a soft-decision input (analog sample or multibit quantized sample) to the decoder. We first obtain the bound on the probability of a first decision-error event $\mathcal{P}_E$ that results in one or more output bit errors. By computing the number of data-bit errors for different error events, we then obtain a bound on the bit-error probability.

First-Error-Event Probability

A first error event occurs if the correct path to the correct node is excluded at the jth step in the decoding—that is, the survivor is an erroneous path. Since the codes are group codes, there is no loss in generality by assuming that the correct path is the all-zero path $aa \cdots aa$. An erroneous decision then occurs at step j in Fig. 15-8, where the path $aa \cdots abca$ is selected as the survivor because of its hypothesized higher correlation metric.

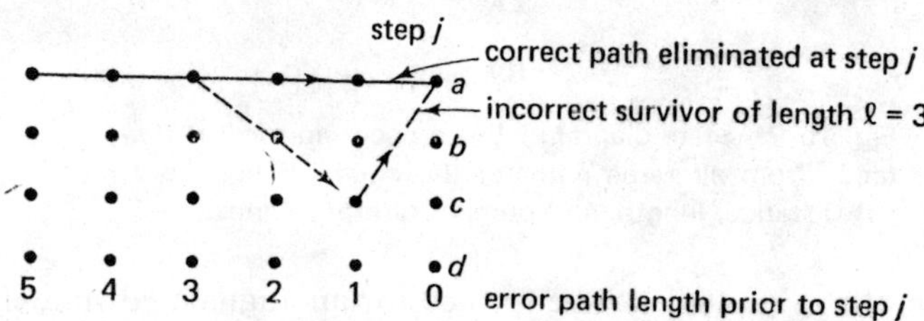

Fig. 15-8 Example of a decoding decision error for the coder of Fig. 15-2

The probability of a first error event at some arbitrary step j with length ℓ is $p(\ell)$. For hard-decision inputs, the probability of a first error event of length 3 for the one possible error path is $p(3)$ and differs from the correct path in five code bits occurring with probability $\mathcal{P}_k = \mathcal{P}_5$. In general, the total set of erroneous paths and the distance, k, from the correct path have already been described by the expansion of $T(D, L, N)$. If there are hard decisions, and we use the $T(D, L, N)$ of (15-8), the single possible length-3 error event requires three received bit errors out of the five to make the incorrect path have a higher metric than the correct path. Thus, the probability of a first error event of length $\ell = 3$ and distance $k = 5$ is expressed by

$$p(3) = \mathcal{P}_5 = \sum_{i=3}^{5} \binom{5}{i} p^i (1-p)^{5-i} \tag{15-9}$$

where p is the hard-decision received bit-error probability at the decoder input.

Note that, in general, at step j there is no simple way of computing whether the paths described in the expansion of $T(D, L, N)$ have survived previous survivor decisions at earlier nodes. However, the first-error-event probability can be overbounded by including **all the possible paths** to a decision error at step j whether they have been previously eliminated or not. Thus, since the generating function

$$T(D) = \Sigma a_k D^k \tag{15-10}$$

gives the total number of paths a_k of distance k from the correct path, the probability of a first error event at some arbitrary step j is bounded by

$$\mathcal{P}_E < \Sigma a_k \mathcal{P}_k \tag{15-11}$$

By a similar calculation using (15-8), the bit-error probability can be bounded from the expansion

$$T(D, L, N)|_{L=1} \triangleq T(D, N) = \sum_{k=1}^{\infty} a_k D^k N^{e_k} \tag{15-12}$$

since the path length ℓ_k (exponent of L) is not needed for this bound for the distance k error path. For the example encoder of Fig. 15-1, this expansion is

$$\begin{aligned} T(D, N) &= \Sigma a_k D^k N^{e_k} \\ &= D^5 N + 2D^6 N^2 + \cdots + 2^k D^{k+5} N^{k+1} + \cdots \end{aligned} \tag{15-13}$$

where the number of errors $e_k = k + 1$ in each error path. By differentiating with respect to N and setting $N = 1$, we can weight each error path by the number of output bit errors. We then have

$$\left.\frac{dT(D, N)}{dN}\right|_{N=1} = \Sigma a_k e_k D^k = \Sigma c_k D^k \qquad c_k \triangleq a_k e_k \tag{15-14}$$

Thus it can be shown that the bit-error probability is bounded by

$$\mathcal{P}_b < \Sigma c_k \mathcal{P}_k \tag{15-15}$$

in which the value of $\mathcal{P}_k$ depends on the type of input to the decoder (that is, hard or soft decision) and on the channel noise level in input bit-error rate (E_b/N_0), respectively. The code properties determine the coefficients c_k. Hence, we have a simple division between code property effects and channel effects.

Evaluation of $\mathcal{P}_k$

For a hard-decision input, the probability of a distance k survivor decision error is

$$\mathcal{P}_k = \begin{cases} \sum_{i=(k+1)/2}^{k} \binom{k}{i} p^i(1-p)^{k-i} & k \text{ odd} \\ \frac{1}{2}\binom{k}{k/2} p^{k/2}(1-p)^{k/2} + \sum_{i=k/2+1}^{k} \binom{k}{i} p^i(1-p)^{k-i} & k \text{ even} \end{cases} \tag{15-16}$$

where the factor of 1/2 in the first term for k even accounts for the random decision in case of a tie vote. Since $p < 1/2$ and $\sum_{i=1}^{k} \binom{k}{i} = 2^k$, this probability can be bounded by

$$\mathcal{P}_k < 2^k p^{k/2}(1-p)^{k/2} \tag{15-17}$$

Hence, the first-error-event probability from (15-11) and (15-17) is

$$\mathcal{P}_E < \Sigma a_k 2^k p^{k/2}(1-p)^{k/2} \tag{15-18}$$

For the example coder of Fig. 15-1 where $a_k = 2^{k-5}$ for $k \geq 5$, the probability of a first-error event $\mathcal{P}_E$ then becomes

$$\mathcal{P}_E < \sum_{k=5}^{\infty} 2^{k-5} 2^k p^{k/2}(1-p)^{k/2} = \frac{[2\sqrt{p(1-p)}]^5}{1 - 4\sqrt{p(1-p)}} \tag{15-19}$$

The bit-error probability is bounded by (15-15) using (15-17) to obtain

$$\mathcal{P}_b < \Sigma c_k 2^k p^{k/2}(1-p)^{k/2} = \left.\frac{dT(D, N)}{dN}\right|_{N=1,\, D=2\sqrt{p(1-p)}} \tag{15-20}$$

where we have also used (15-14). For the example code, $c_k = (k-4)2^{k-5}$, and the bit-error probability for hard-decision inputs is bounded by

$$\mathcal{P}_b < \sum_{k=5}^{\infty} (k-4) 2^{k-5} 2^k p^{k/2}(1-p)^{k/2} = \frac{[2\sqrt{p(1-p)}]^5}{[1 - 4\sqrt{p(1-p)}]^2} \tag{15-21}$$

For the additive Gaussian white noise channel with soft analog decision, the correlation metric is

$$\sum_{i=1}^{M} \sum_{j=1}^{n} x_{ij} y_{ij} \qquad x_{ij} = 1 \tag{15-22}$$

where j runs over the n code bits in each output code symbol, and where i runs over the number nodes in each path. For a rate 1/2 code ($n = 2$), the cross-correlation metric is evaluated over M symbols ($M = 5K$, where K is the code constraint length, is usually sufficient), and is expressed by

$$\sum_{i=1}^{M} (x_{i1} y_{i1} + x_{i2} y_{i2}) \tag{15-23}$$

If a decision error is made, an incorrect path with components x_{ij} produces a higher correlation metric than the correct path with components $x_{ij} = 1$. Thus, an erroneous decision is made if

$$\Sigma\Sigma(x'_{ij}y_{ij} - y_{ij}) > 0 \tag{15-24}$$

If the incorrect path $\mathbf{x}'$ differs from the correct path in k code bits and the probability of this occurrence is $\mathcal{P}_k$,

$$\begin{aligned}\mathcal{P}_k &= Pr\left(\sum_i \sum_j x'_{ij}y_{ij} - y_{ij} > 0\right) = Pr\left[\sum_{r=1}^{k} (x'_r - 1)y_r > 0\right] \\ &= Pr\left(\sum_{r=1}^{k} 2y_r < 0\right) = Pr\left(\sum_{r=1}^{k} y_r < 0\right)\end{aligned} \tag{15-25}$$

where $Pr\{z > 0\}$ is the probability that $z > 0$. We have used $x_r = 1$ and $x'_r = -1 \neq x_r$ by definition and we have converted the double subscripts i, j to a single subscript r, which ranges over the same set of coefficients, $r = ni + j$, and $k = n(M + 1)$.

Bound on Bit-Error Probibility

The white noise channel has noise with one-sided density N_0 and signal-energy per bit of E_b. Thus, the variables y_{ij} have mean $-\sqrt{\epsilon_s}$ where $\epsilon_s = E_b/n$ is the energy per transmitted code bit and the variance of y_{ij} is $N_0/2$. The sum $z = \Sigma y_r$ therefore is also Gaussian with mean $k\sqrt{\epsilon_s}$ and variance $kN_0/2$, and the probability of a k-bit-path error from (15-25) is

$$\begin{aligned}\mathcal{P}_k = Pr(z < 0) &= \int_{-\infty}^{0} \frac{\exp(-z - k\sqrt{\epsilon_s})^2/kN_0}{\sqrt{\pi k N_0}}\, dz \\ &= \int_{\sqrt{2k\epsilon_s/N_0}}^{\infty} \frac{\exp(-x^2/2)}{\sqrt{2\pi}}\, dx \triangleq \text{erfc}' \sqrt{\frac{2k\epsilon_s}{N_0}}\end{aligned} \tag{15-26}$$

The bit-error probability bound is then obtained from (15-15) and (15-26):

$$\mathcal{P}_b < \Sigma c_k \mathcal{P}_k = \sum_{k=d}^{\infty} c_k \,\text{erfc}' \sqrt{\frac{2k\epsilon_s}{N_0}} \tag{15-27}$$

where d is the minimum distance of the code (the smallest k where $c_k \neq 0$).

A simpler calculation involves the calculation of $[dT(D, N)]/dN$ using (15-14) rather than all the coefficients c_k. For this purpose, we want to express $\mathcal{P}_k$ as a power of k as in $\alpha\mathcal{P}^k$. First, note that the function $\text{erfc}' \sqrt{x + y}$ is bounded by

$$\text{erfc}' \sqrt{x + y} \leq \exp\left(\frac{-y}{2}\right) \text{erfc}' \sqrt{x} \tag{15-28}$$

Next, set $k = d + \ell$, $\ell = 0, 1, \ldots,$ and $x + y = 2k\epsilon_s/N_0$. Then, using (15-26) and (15-28), we obtain

$$\mathcal{P}_k = \text{erfc}' \sqrt{\frac{2k\epsilon_s}{N_0}}$$
$$= \text{erfc}' \sqrt{\frac{2(d+\ell)\epsilon_s}{N_0}} \leq \exp\left(\frac{-\ell\epsilon_s}{N_0}\right) \text{erfc}' \sqrt{\frac{2d\epsilon_s}{N_0}} \tag{15-29}$$

Thus, the bound on $\mathcal{P}_b$ in (15-27) can be slightly weakened to obtain from (15-15) and (15-29)

$$\mathcal{P}_b < \Sigma c_k \mathcal{P}_k \leq \text{erfc}' \sqrt{\frac{2d\epsilon_s}{N_0}} \sum_k c_k \exp\left[\frac{1-(k-d)\epsilon_s}{N_0}\right]$$
$$\leq \text{erfc}' \sqrt{\frac{2d\epsilon_s}{N_0}} \exp\left(\frac{d\epsilon_s}{N_0}\right) \sum_{k=d}^{\infty} c_k e^{-k\epsilon_s/N_0} \tag{15-30}$$

The expression (15-30) can be rewritten using (15-14) to replace the summation

$$\mathcal{P}_b < \text{erfc}' \sqrt{\frac{2d\epsilon_s}{N_0}} \exp\left(\frac{d\epsilon_s}{N_0}\right) \frac{dT(D, N)}{dN}\bigg|_{N=1, D=\exp(-\epsilon_s/N_0)} \tag{15-31}$$

For the example rate 1/2 code of Fig. 15-1, where $\epsilon_s = E_b/n = E_b/2$, and $d = 5$,

$$\frac{dT(D, N)}{dN}\bigg|_{N=1} = \frac{D^5}{(1-2D)^2} \tag{15-32}$$

Hence, the bound on bit-error probability from (15-31) is

$$\mathcal{P}_b < \text{erfc}' \sqrt{\frac{5E_b}{N_0}} \exp\left(\frac{5E_b}{2N_0}\right) \frac{[\exp(-E_b/2N_0)]^5}{[1 - 2\exp E_b/2N_0]^2} \tag{15-33a}$$

and (15-33a) reduces to

$$\mathcal{P}_b < \frac{\text{erfc}' \sqrt{5E_b/N_0}}{[1 - 2\exp(-E_b/2N_0)]^2} \tag{15-33b}$$

as compared to the result of no coding

$$\mathcal{P}_b = \text{erfc}' \sqrt{\frac{2E_b}{N_0}} = \frac{1}{2} \text{erfc} \sqrt{\frac{E_b}{N_0}} \tag{15-34}$$

where $\text{erfc}\, x = \frac{2}{\sqrt{\pi}} \int_x^{\infty} e^{-y^2}\, dy.$

Thus by comparing (15-33*a*),(15-34) for high E_b/N_0 this error bound shows that the coded signal is more than 10 log (2.5) = 3.98 dB better than the uncoded signal for the same output error rate.

Figures 15-9 through 15-11 show experimental curves of measured bit-error rates indicating the approximately 4- to 6-dB improvement in perform-

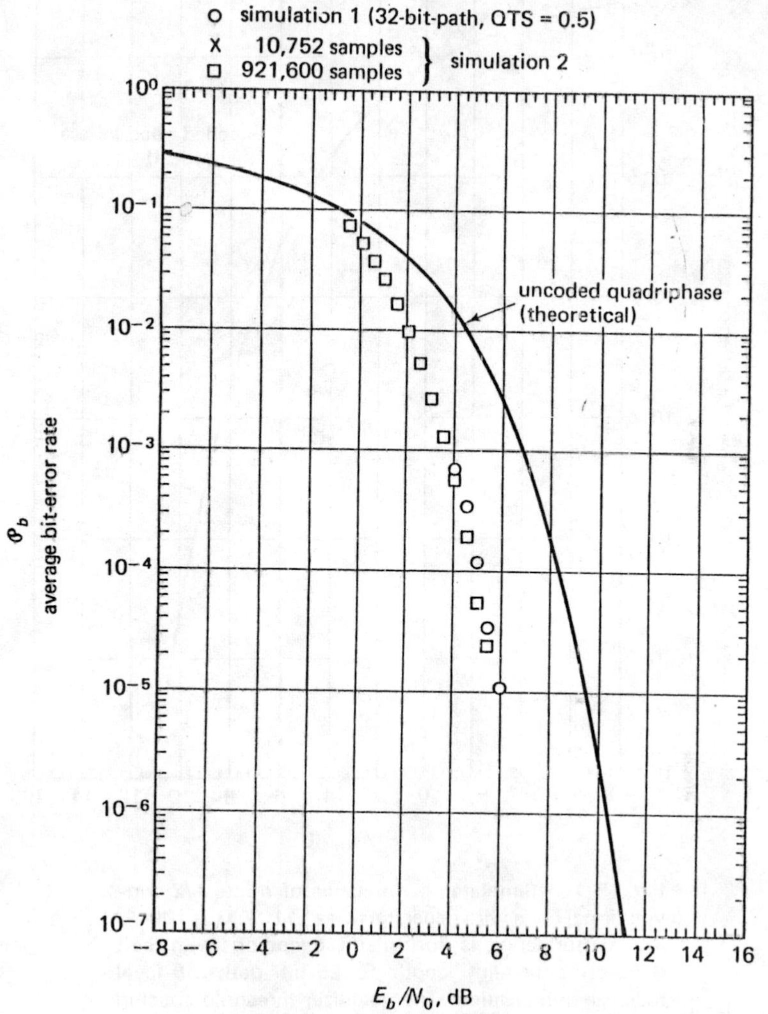

Fig. 15-9 Simulated performance of a rate 1/2 convolutional code with generators employing soft decision, $G_1 = 111$, $G_2 = 101$, constraint length 3, 15-bit paths and 32-bit paths. Soft decision 8-level receiver quantization is employed. The bit error rate is for maximum-likelihood decoding. The quantizer threshold spacing (QTS) is set equal to 0.50 σ where σ is the rms noise. [Bustamante, 1972]

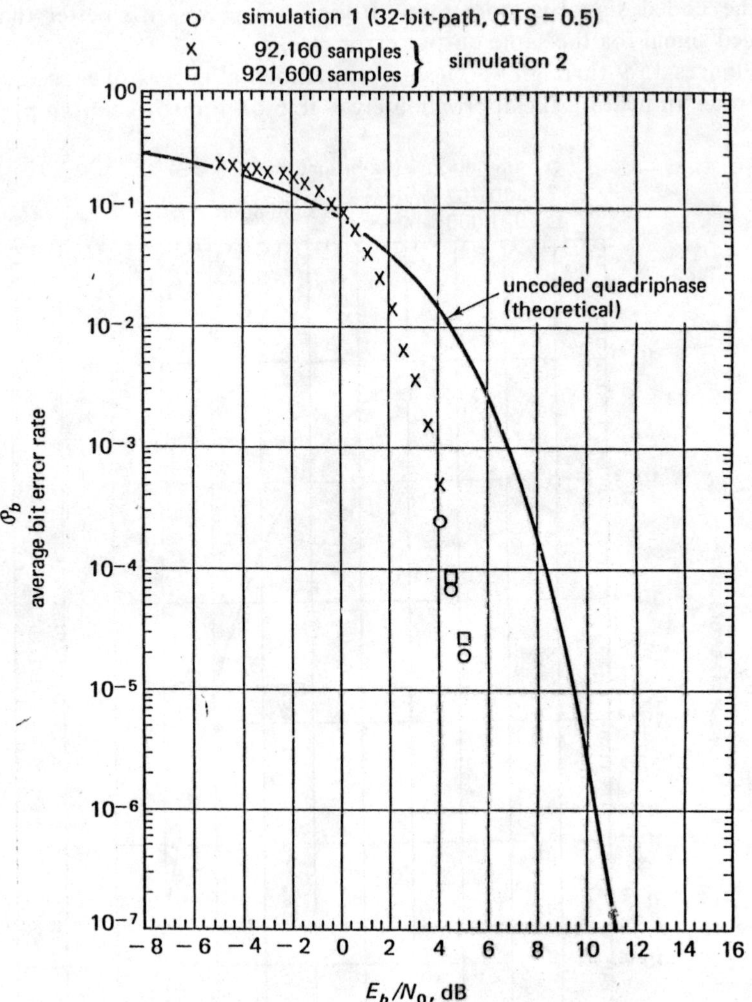

Fig. 15-10 Simulated performance of a rate 1/2 convolutional code with generators **G** = 11101, **G** = 10011. The performance is for Viterbi decoding with soft decision, constraint length 5, 25-bit paths, 8-level receiver quantization. The quantizer threshold spacing is equal to 0.50 σ.

ance at $\mathcal{P}_b = 10^{-5}$ for codes of constraint lengths 3, 5, and 7 [Heller and Jacobs, 1971, 835–848; Bustamante and Leong, 1972]. The codes are rate 1/2 codes with the parity tap positions defined by $\mathbf{G}_1$, $\mathbf{G}_2$ as indicated in the captions. Eight-level soft decision is used rather than a pure analog input; this soft decision experimentally produces a performance within 0.25 dB of the analog input sample decoder. Decoding is performed over $M = 5K$ path lengths.

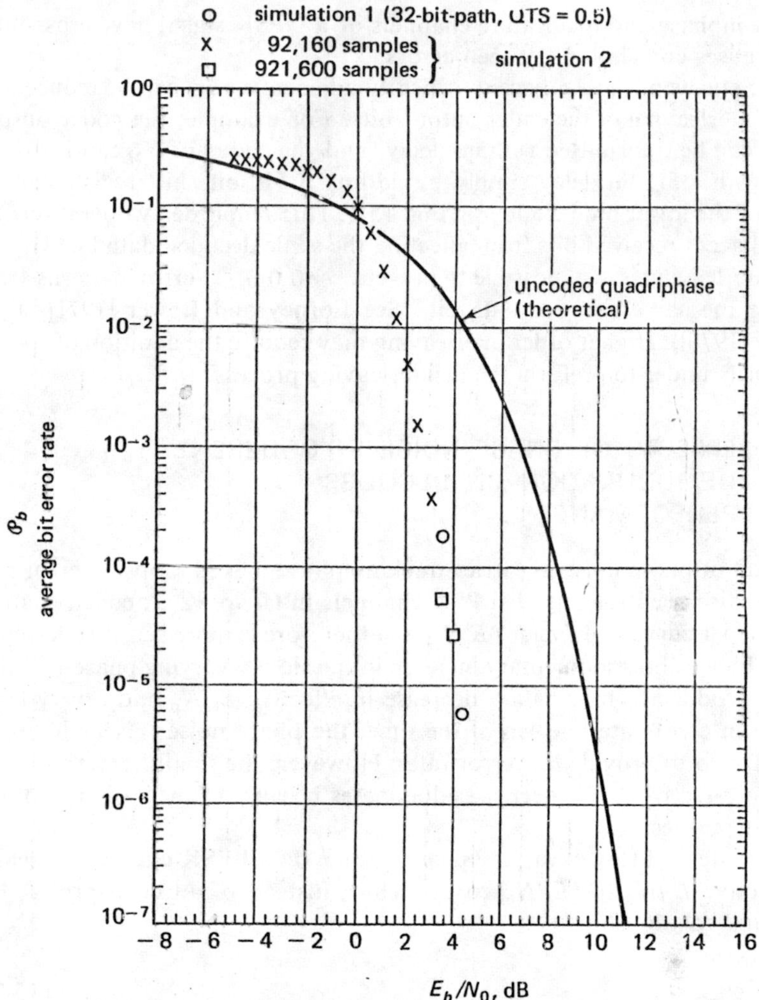

Fig. 15-11 Simulated performance of a rate 1/2 convolutional code with generators $\mathbf{G}_1 = 1111001$, $\mathbf{G}_2 = 1011011$. The performance is for Vitorbi decoding, soft decision, rate 1/2, constraint length 7, 35-bit paths, 8-level receiver quantization. The quantizer threshold spacing is set equal to 0.50 σ.

Code Interleaving

This section has assumed that the decoder input samples or bit decisions have independent noise or errors from bit to bit. This assumption may not be valid if there are significant filter distortion effects, as described in Chap. 13, or if the inphase and quadrature channels of a QPSK signal have cross-talk, which causes correlation between errors.

The effects of these adjacent or nearly adjacent errors can be reduced by proper interleaving of the coder output bits. For example, the coder output bits X_{1l} can be transmitted with no delay, and the code bits X_{2l} can be transmitted with a 5K bit delay simply by adding a 5K bit shift register in the output of the lower mod 2 adder in Fig. 15-1. This simple delay operation prevents adjacent received bits from affecting the same decoded data bit. Higher order interleaving is also possible to prevent $\cdots 0_e 0_e 0_e \cdots$ error patterns from affecting the same decoded data bit. (See Forney and Bower [1971], J. L. Ramsey [1970]). Higher order interleaving may require the addition of special sync words under to preform the deinterleaving process.

15-5 EFFECTS OF PHASE NOISE IN COHERENT DEMODULATION ON DECODER PERFORMANCE

The effect of phase noise in carrier-tracking phase-locked loops on error rate has been discussed for uncoded PSK channels in Chap. 12. A coded channel has a steeper curve of $\mathcal{P}_b$ versus E_b/N_0 and therefore is more sensitive to phase noise. Thus, a short momentary increase in the slowly varying phase-tracking error ϕ produces a momentary decrease in effective E_b/N_0 and a very large increase in error rate.* Much of the time, the phase noise is near its mean value of zero to provide low error rates. However, the small fraction of time when the effective E_b/N_0 decreases dominates because of the steepness of the error rate curves.

The effect of slowly varying phase error ϕ on a BPSK channel decreases the effective E_b/N_0 to $(E_b/N_0)\cos^2\phi$. Thus, instead of an error-probability functional relationship

$$\mathcal{P}_b = f\left(\frac{E_b}{N_0}\right) \tag{15-35}$$

the phase noise degrades performance to the error probability for a given ϕ:

$$\mathcal{P}_b(\phi) = f\left(\frac{E_b}{N_0}\cos^2\phi\right) \tag{15-36}$$

For a first-order loop the probability density of phase error ϕ for an SNR α in the closed-loop bandwidth is (see Chap. 12)

$$p(\phi) = \frac{e^{\alpha\cos\phi}}{2\pi I_0(\alpha)} \qquad \alpha \gg 1 \tag{15-37}$$

*"Slowly" refers to a small change in phase error ϕ over many ($> 10^2$) bit periods.

Thus, the average bit-error probability is

$$\bar{\mathcal{P}}_b = \int_{-\pi}^{\pi} p(\phi)\mathcal{P}_b(\phi)\, d\phi \tag{15-38}$$

The result (15-38) has been numerically integrated for a $k = 7$ rate 1/2 code to obtain the performance curve of Fig. 15-12. Note that an $\alpha \geq 16$ dB is required for small degradation with BPSK. For QPSK there is cross-channel interference in addition to E_b/N_0 attenuation, and an $\alpha \geq 25$ dB is required for the coherent demodulation [Jacobs and Heller, 1971].

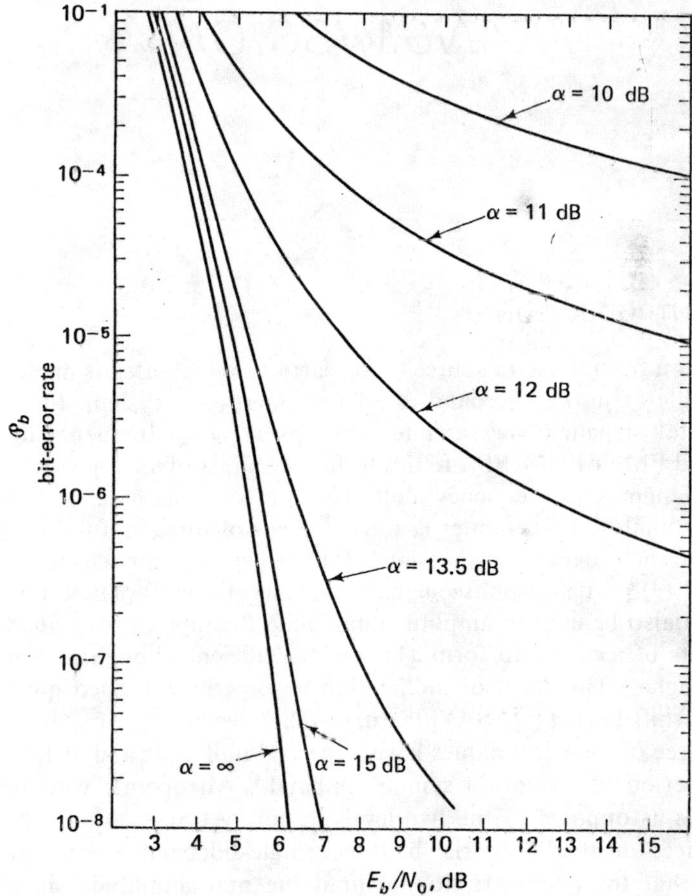

Fig. 15-12 BPSK performance curves for rate 1/2, $K = 7$ Viterbi decoder with 8-level quantization as a function of carrier phase tracking loop SNR [From Jacobs and Heller, 1971]

CHAPTER 16

BASEBAND DATA TRANSMISSION

16-1 INTRODUCTION

Transmission from the data source to the earth terminal often is made over a coaxial cable or in the baseband of an FM microwave system. In addition much digital signaling via satellite takes place using frequency-division-multiplexed FM or FDM-FM. In this technique a mix of analog voice signals and digital signals are frequency multiplexed to form the baseband input to an FM modulator. This chapter is a brief review of three useful modulation systems for such digital transmissions: (1) bipolar, (2) pair-selected ternary (PST), and (3) partial-response signaling (classes 4 and 5). These multilevel signals can also be used to amplitude modulate the inphase and quadrature components of a carrier to form a bandwidth efficient, though nonconstant, envelope signal. This form of modulation is sometimes termed quadrature amplitude shift keying (QASK) [Smith, 1975].

All three of these techniques have a spectral null at dc and at $1/T$, where T is the period of an output symbol or baud.* All operate with three or more levels as opposed to the two levels in binary transmission. They also can be placed on the RF carrier by direct single-sideband AM modulation, provided that the amplifiers do not limit the multiamplitude signal. For example, the use of partial-response signaling on both inphase and quadrature channels provides an interesting extension of QPSK to QASK.

*Note, however, that partial-response signaling consists of a broad family of techniques, only two classes of which (4 and 5) have appropriate spectral nulls [Kretzmer, 1966, 67–68].

A zero-frequency spectral null is desired in these signals to eliminate the dc wander, which occurs with nonreturn-to-zero (NRZ) bit streams for long sequences of 0s or 1s because cable repeaters and FM systems are not dc coupled. FM modulators and demodulators are subject to center frequency drift which prevents dc information transfer. If a spectral null also exists at the baud rate, a noninterfering sinusoidal component can be inserted at $f = 1/T$ to simplify the receiver clock timing.

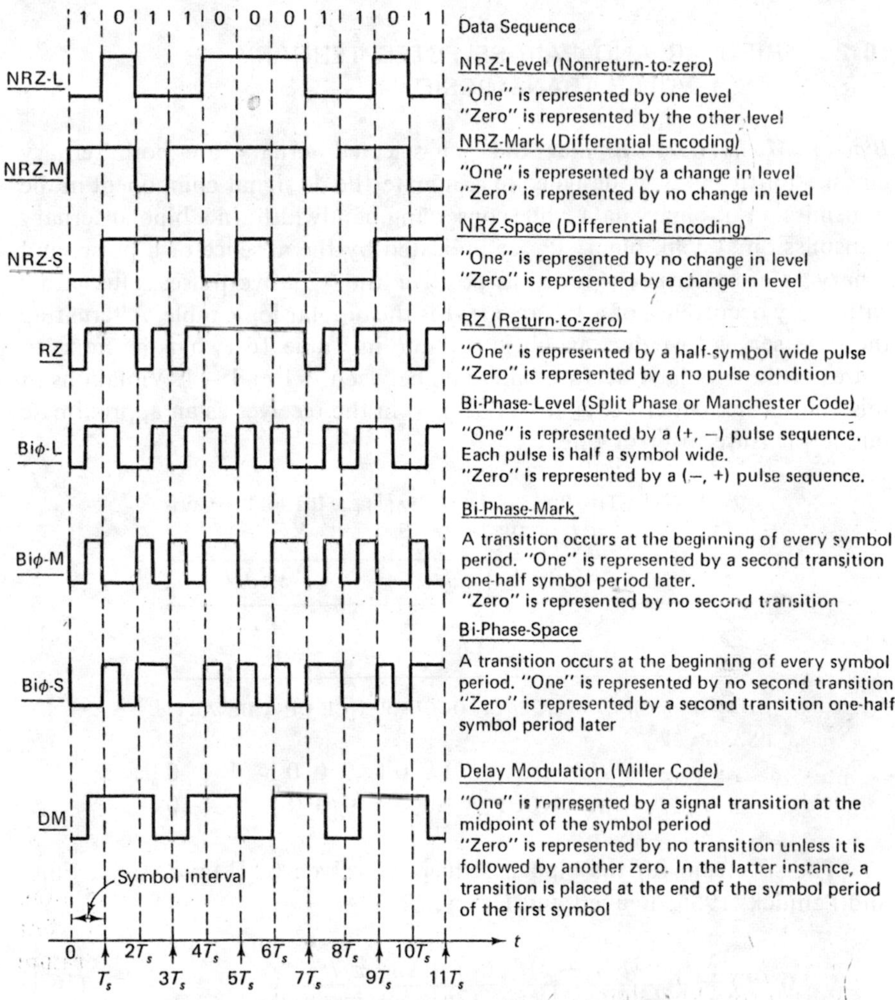

Fig. 16-1 Various binary PCM waveforms as related to a binary input data sequence [From Lindsey and Simon, 1973, Fig. 1-5]

In these waveforms, redundancy exists that can be used to provide on-line performance monitoring, which is impossible with random data in the nonreturn-to-zero (NRZ) format. It is also possible to decode some of these data codings using maximum-likelihood or Viterbi decoding algorithms. (See Magee [1975].) Although this chapter deals primarily with multilevel digital signals with spectral nulls both at dc and at the symbol rate, it is also desirable to review the standard PCM binary data formats. Figure 16-1 illustrates the output code formats for an example input data stream.

16-2 BIPOLAR AND PAIR-SELECTED TERNARY (PST) PULSE TRANSMISSION

Bipolar Modulation. Bipolar and pair-selected ternary are both ternary modulation techniques designed to eliminate the dc signal component in the transmission of binary data while conserving bandwidth. In a bipolar ternary transmission system, binary 0s are indicated by the absence of a pulse, and binary 1s are alternately coded as positive and negative pulses, alternating with every occurrence of a 1. Table 16-1 is the bipolar logic table. Alternating the ± 1s serves to reduce bandwidth. Note in Table 16-1 that the nonzero pulses in the bipolar waveform alternate between $+1$ and -1. Violations in this alternating pattern of ± 1s can be used in the receiver as an approximate online measure of error rate.

Table 16-1 THE BIPOLAR LOGIC TABLE, A BIT-BY-BIT CODING OF THE BINARY SEQUENCE

Binary	*+Mode*	*−Mode*
1	−	+
0	0	0

CHANGE MODE AFTER EACH OCCURRENCE OF A BINARY 1

EXAMPLE:

BINARY INPUT	1 0 0 1 0 0 0 1 1 1 0 1
BIPOLAR OUT	+0 0− 0 0 0+ −+ 0−

The spectrum for the bipolar sequence is given by [Mayo, 1962; Fultz and Pennick, 1965; Bennett and Davey, 1965]

$$G_b(f) = \frac{8p(1-p)f_s\,|G(f)|^2 \sin^2 \pi f/f_s}{1 + (2p-1)^2 + 2(2p-1)\cos 2\pi f/f_s} \tag{16-1}$$

where p is the probability of a binary 1 and $G(f)$ is the power spectral density of the pulse. This bipolar spectrum of (16-1) is the same as for "twinned binary" modulation which is generated from a unipolar bit train by delaying it by one bit period and subtracting it from the original. (See M. R. Aaron [1962, 99–142].) If $p = 1/2$, this density becomes

$$G_b(f) = 2f_s\,|G(f)|^2 \sin^2 \pi f/f_s \qquad f_s = 1/T \tag{16-2}$$

and is shown in Fig. 16-2.

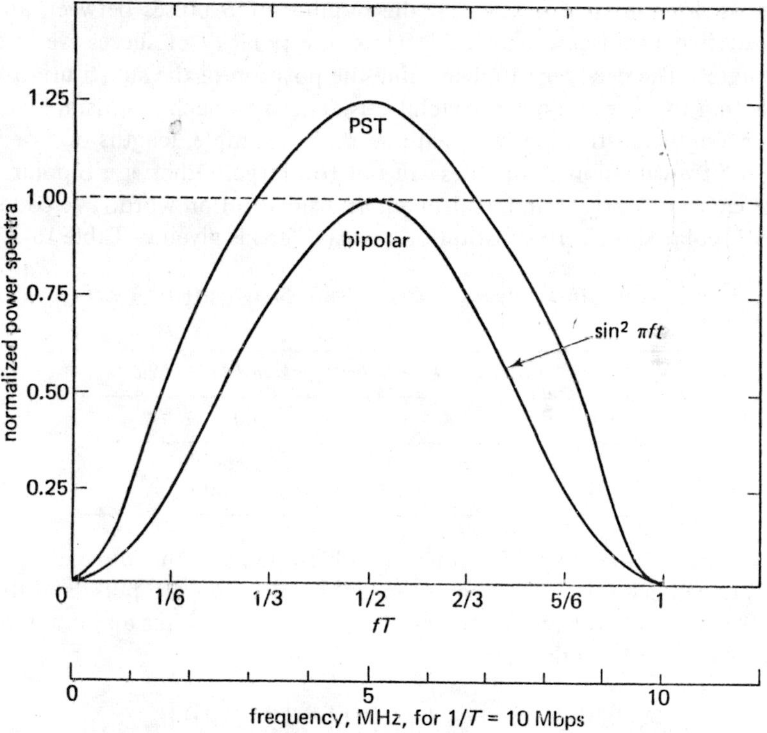

Fig. 16-2 Power spectra of bipolar and PST for random input bit streams. For balanced positive and negative pulses, probability of zero $p = 1/2$. The bit period is T = duration of binary zero or one. A bandlimited white spectrum is assumed for $G(f)$.

Timing can be recovered from this bipolar waveform by full-wave rectifying or square-law detecting the bipolar pulse train, to produce a line component at the clock rate. Notice from Table 16-1 that a long string of 0s at the input to the bipolar modulator produces a long string of 0s at the output and can cause difficulty in bit synchronization. One can generate a special "filling" sequence to avoid this difficulty, replacing, for example, a sequence of $n + 1$ zeros by an $n + 1$ bit sequence consisting of either of the two substitution words [Johannes, 1969; Croiser, 1970; Johannes, 1972]

$$00 \ldots 0B0V \quad \text{or} \quad 00 \ldots 000V \tag{16-3}$$

where B represents a normal bipolar pulse and V represents a bipolar violation—that is, a + or – "one" pulse with the same polarity as the last preceding pulse. The choice of either sequence to replace the succession of zeros is done in such a way that the number of B pulses between any two consecutive V pulses will be odd. Thus, the polarity of successive V pulses alternates. The decoder can determine the position of the substitution words by noting the inserted bipolar violations. This approach is satisfactory when larger length substitution words are used, for example, lengths of 5 or 7 bits, and the transmission error rates are not too large. Otherwise bipolar violations caused by noise can be mistaken as a substitution word. The code table for a bipolar three-zero substitution code (B3ZS) is given in Table 16-2. Thus

Table 16-2 BIPOLAR THREE-ZERO SUBSTITUTION TABLE (B3ZS)

Last Regular Data Pulse	*Last Substitute Word Pair Used*	
	00+ or +0+	00– or –0–
+	–0–	00+
–	00–	+0+

the choice of the three-bit sequence substituted for the three consecutive zeros depends on both the sign of the last regular nonzero pulse and the substitute bit pattern for the last three-zero substitutions. Thus an input sequence of zeros could be coded as

```
Binary Input   1 | 0 0 0 | 0 0 0 | 0 0 0 |
Bipolar B3ZS   + | – 0 – | 0 0 + | – 0 – | ⟶
                             ↑       ↑   Time
                             V       V
                        Bipolar   Violations
```

Similar rules can be made for five-zero substitutions.

Self-synchronizing digital scramblers can be used to randomize an input bit stream as an alternate means to avoid long sequences of 1s or 0s [Savage, 1967, 449–487]. An inverse operation can then be used at the receiver to recover the original input bit stream. These self-synchronizing data scramblers multiply the received errors by the number of shift-register taps plus one for low error rates ($K + 1$) because each error appears at a tap K times in the inverse operation. (See Sec. 16-5.)

Pair-Selected Ternary (PST) Modulation

Pair-selected ternary (PST) coding begins by breaking up the input data bit stream into pairs of data bits and coding these bit pairs, or dibits (Table 16-3). Notice that a pulse + or − occurs during each dibit, and there is a transition between the first and second symbols in each output code word. Thus there is no possibility of a long string of zeros in the coded output bit stream as there is with bipolar coding.

Table 16-3 THE PST LOGIC TABLE; THE BINARY SEQUENCE IS FRAMED INTO PAIRS OF BITS

Binary	*+Mode*	*−Mode*
11	+−	+−
10	+0	−0
01	0+	0−
00	−+	−+

CHANGE MODE AFTER EACH OCCURRENCE OF A BINARY 10 OR 01 PAIR

EXAMPLE:

BINARY INPUT	10	01	00	01	11	01
PST OUTPUT	+0	0−	−+	0+	+−	0−

The spectrum of the PST signal for equiprobable input bits is [Sipress, 1965, 366–372]

$$G_p(\omega) = \frac{1}{8T}|G(\omega)|(1 - \cos\omega T)(1 + 4\cos\omega T + 2\cos^2\omega T) \tag{16-4}$$

and is shown in Fig. 16-2 for a bandlimited white spectrum for $G(f)$. Notice that just as with bipolar transmission there is a spectral null both at dc and at $f = 1/T$. Note that there is slightly more power in the PST because the 00 output for bipolar does not occur for PST for any dibit. It is, of course, possible to get two zeros in a row at the modulation output, however; for example, 10, 01 produces +0, 0−. (Can be used to recover framing.)

16-3 PARTIAL-RESPONSE SIGNALING

A partial-response system generates an ℓ-level signal from a binary or multilevel data input using a linear weighted superposition of the input symbols A_i, the superposition memory extending over n bit periods. The original concepts for binary input data—variously known as partial response, correlative coding, duo-binary, poly-binary [Kretzmer, 1966; Lender, 1963]—permit control of the spectral properties of the signal to make efficient use of the available bandwidth and to cope better with channel distortion. Indeed, a known transmission channel with a known impulse response can serve as part or all of the linear superposition operation. In addition, known redundancy in the received signal can be used by a detector as a performance monitor.

This discussion is restricted to the so-called class 4 and 5 techniques [Kretzmer, 1966, 67–68], which provide spectral nulls at dc and at the symbol rate, which equals the maximum information bandwidth W. These quasi-band-limited signals are generated as the weighted linear superposition of impulses of form sinc $2\pi Wt \triangleq (\sin 2\pi Wt)/2\pi Wt$ spaced in time at intervals $T = 1/f_s$ $1/2W$. These impulses have a bandlimited white power spectral density $G(f)$. The modulated signal samples B_t are a superposition of the input sample values A_t using the weight coefficients k_t; that is,

$$B_t = k_1 A_t + k_2 A_{t-1} + \cdots + k_n A_{t-n+1} = \sum_{j=1}^{n} k_j A_{t-j+1} \tag{16-5}$$

where the smallest nonzero k_t is equal to unity, and the A_t can be binary or Mary. A blcck diagram of the partial-response coder is shown in Fig. 16-3a. The corresponding output power spectral density for white inputs is $|H(f)|^2$ where

$$H(f) = \int_{-\infty}^{\infty} \left[k_1 \delta(t) + k_2 \delta\left(t - \frac{1}{2W}\right) + \cdots + k_n \delta\left(t - \frac{n}{2W}\right)\right] e^{-j2\pi ft}\, dt \qquad |f| < W \tag{16-6}$$

and the coder output spectral density is zero otherwise. For a spectral null at dc, the sum of the weight coefficients must be zero. In terms of the z transform [Cadzow and Martens, 1970; Derusso et al., 1965], the transfer function of the discrete time filter of (16-5) is

$$H(z) = \sum_j k_j z^{-j+1} \tag{16-7}$$

where $z \triangleq e^{st}$ and s is the complex frequency variable. Stability of the receiver filter (Fig. 16-3b) for the partial response signal $H^{-1}(z)$ requires that the poles of $H^{-1}(z)$ lie within the unit circle in the z plane.

The weighting coefficients k_t and the amplitude spectra $|H(f)|$ for class 4 and 5 techniques are shown in Fig. 16-4. The superposition, number of input sample values summed n, and the number of output levels ℓ are given

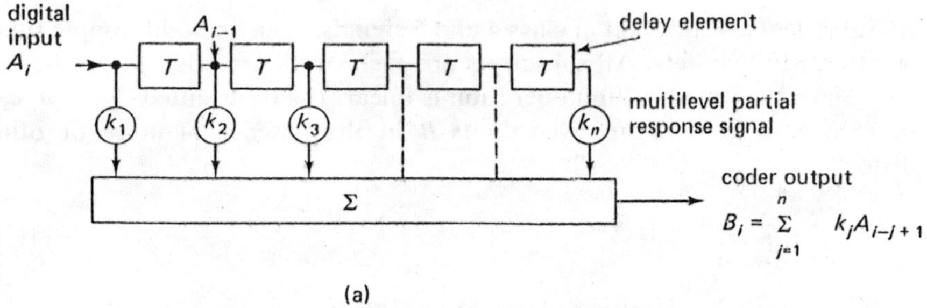

(a)

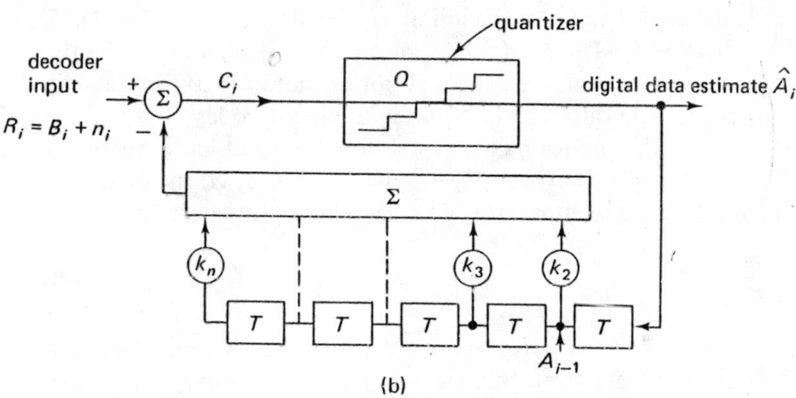

(b)

Fig. 16-3 Partial-response signal generation and detection. (a) Partial-response signal coder. The delay elements can be a shift register for binary inputs. (b) Simplified block diagram of partial-response decoder.

Table 16-4 TABLE OF PARTIAL-RESPONSE SUPERPOSITIONS, FREQUENCY RESPONSE, NUMBER OF SHIFT REGISTER STAGES $n - 1$, AND NUMBER OF LEVELS ℓ IN B_i FOR BINARY INPUT A_i

Partial Response Class	*Superposition*	*Frequency Transfer Function* $H(j\omega)$	*Number of Shift Register Stages in Code* $n - 1$	ℓ *Number of levels in* B_i
Binary	$B_i = A_i$	1	0	2
1	$B_i = A_i + A_{i-1}$	$2 \cos \pi f/2W$	1	3
2	$B_i = A_i + 2A_{i-1} + A_{i-2}$	$4 \cos^2 \pi f/2W$	2	5
3	$B_i = 2A_i + A_{i-1} - A_{i-2}$	$2 + \cos \pi f/W - \cos 2\pi f/W$ $+ j (\sin \pi f/W - \sin 2\pi f/W)$	2	5
4	$B_i = A_i - A_{i-2}$	$2 \sin \pi f/W$	3	3
5	$B_i = A_i + 2A_{i-2} - A_{i-4}$	$4 \sin^2 \pi f/W$	4	5

in Table 16-4. Notice that in class 4 and 5 signals, even and odd sample times are treated separately. All subscripts are even or all are odd.

Since the superposition operation is linear, the transmitted digits A_i can be recovered from the received digits B_i in the absence of noise or other distortion by

$$\hat{A}_i = \frac{1}{k_1}\left[B_i - \sum_{j=2}^{n} k_j \hat{A}_{i-j+1}\right] \tag{16-8}$$

if correct decisions can be made on at least the preceding $n - 1$ digits.

This inverse filter has a transfer function $H^{-1}(z)$ in the z plane. For example, for class 4 partial-response signaling, $H(z) = 1 - z^{-2} = (z^2 - 1)/z^2$, and $H^{-1}(z) = z^2/(z^2 - 1)$. The $H^{-1}(z)$ has poles at $z = \pm 1$, or equivalently, the frequency transfer function is unstable at both dc and $f = W$. Errors can propagate over an infinite duration in a purely analog decoder.

A digital decoder that performs the same operations but has a quantizer after the difference circuit is shown in Fig. 16-3*b*. For binary A_i the quantizer reduces to a simple binary comparator. The decoded outputs are then

$$\hat{A}_i = Q\left[\frac{1}{k_1}\left(B_i - \sum_{j=2}^{n} k_j \hat{A}_{i-j+1}\right)\right] \tag{16-9}$$

In this decoder a received error will tend to propagate until the input B_i reaches its top or bottom level. In a class 4 system with binary inputs, for example, an error in one output bit $\hat{A}_i$ changes the value of B_i from 0 to ± 1 and appears only at the next even time instant. If the value of B_i is ± 2 rather than 0 at that point, then the value of A_i has no effect on A_{i+2} because $B_i = 2 > \pm 1$, a $\hat{A}_i = Q(+2 \pm 1) = +1$. Hence the error propagation is halted at this point. Thus, an error in the recovery of one digit can lead to the propagation of errors over several digits in the receiver. A simplified diagram of the partial-response decoder which performs the operation of (16-8) is shown in Fig. 16-3*b*.

The probability of bit error for $M \triangleq 2^m$ input levels in A_i for a class 4 partial-response signal, transmitted over a channel with raised-cosine shaping and no intersymbol interference, is [Gunn et al., 1971, 501–520]

$$\mathcal{P}_b = \frac{2}{m}\left(1 - \frac{1}{M^2}\right) \operatorname{erfc}'\left[\sqrt{\frac{3P_s}{2(M^2 - 1)P_n}}\right] \tag{16-10}$$

where $\operatorname{erfc}' x \triangleq 1/\sqrt{2\pi} \int_x^\infty \exp(-t^2/2)\, dt$, P_s is the received signal power, and P_n = white noise power in the Nyquist frequency band 0 to W Hz.* The number of coded levels in B_i transmitted is $\ell = 2M - 1$, and the input bit rate is $m(2W)$.

---

*The error probability in a distorted channel is given in Benedetto et al. [1973, 181–190].

For the binary inputs, A_i with $M = 2$, $m = 1$, this output error probability (16-10) becomes

$$\mathcal{P}_b = \frac{3}{2}\operatorname{erfc}'\frac{P_s}{2P_n} \tag{16-10a}$$

This result shows that the class 4 partial-response signal is approximately 3 dB less efficient than the binary NRZ signal, which produces an error rate

$$\mathcal{P}_b = \operatorname{erfc}'\sqrt{\frac{2E_b}{N_0}} = \operatorname{erfc}'\sqrt{\frac{P_s}{P_n}} = \frac{1}{2}\operatorname{erfc}\sqrt{\frac{E_b}{N_0}}$$

where $P_n = N_0 W$ and $E_b = P_s/2W$.

This calculation assumes that the input signal has been Gray coded and that the errors caused by Gaussian noise produce transition errors only to levels adjacent to the transmitted level. The Gray code ensures that a single level error results in only a single bit error.

As an example, the SNR required for $\mathcal{P}_b = 2 \times 10^{-8}$ error performance for a $2M - 1 = 15$-level (-7 to $+7$) partial-response signal, with three bits per sample ($M = 2^3 = 8$ levels) is approximately 31 dB. Thus, as one would expect, the required SNR increases substantially with the number of input levels.

Error propagation can be halted or at least reduced by a precoding operation, which in effect subtracts the effect of all but one data digit from the received samples B_i [Gerrish and Howson, 1967, 186; Gunn and Lombardi, 1969, 734–737]. One can precode the partial-response code input digits A_i from an original set of Mary input data digits D_i by the inverse filter operation

$$\begin{aligned} A_i &= \frac{1}{k_1}\left[D_i - \sum_{j=2}^{n} k_j A_{i-j+1}\right] \\ &= D_i + A_{i-2} \\ &= D_i + D_{i-2} + D_{i-4} + \cdots \end{aligned} \quad \text{class 4} \tag{16-11}$$

and the coder output B_i is then identical to the original input $B_i = D_i$. Clearly this precoding makes sense only if the linear channel is ideal and the channel itself provides all of the superposition coding operation of (16-5). Note that these A_i are multilevel from $-\infty$ to ∞.

More reasonably, one can restrict the precoded digits to be Mary by interpreting the output digits A_i of the precoder in a modulo-M sense. Thus, the precoded digits are

$$A_i = \frac{1}{k_i}\left[D_i - \sum_{j=2}^{n} k_j A_{i-j+1}\right] \bmod M$$
$$= (D_i \oplus A_{i-2}) \bmod M \qquad \text{class 4} \qquad (16\text{-}12)$$

If $M > 2$, then the transmitted digits B_i take on more than the ℓ levels of Fig. 16-4 for binary inputs. As noted above, if the input data digits are three bits/digit and $M = 2^3 = 8$, then for a class 4 code the number of output levels in B_i is $\ell = 2M - 1 = 15$ since A_i ranges over 0 to 7. The decoder for this signal is simply $\hat{D}_i = B_i \bmod M$ since the modulo-M operation is commutative. The system polynomials for class 4 and 5 signals are $1 - \mathbf{D}^2$

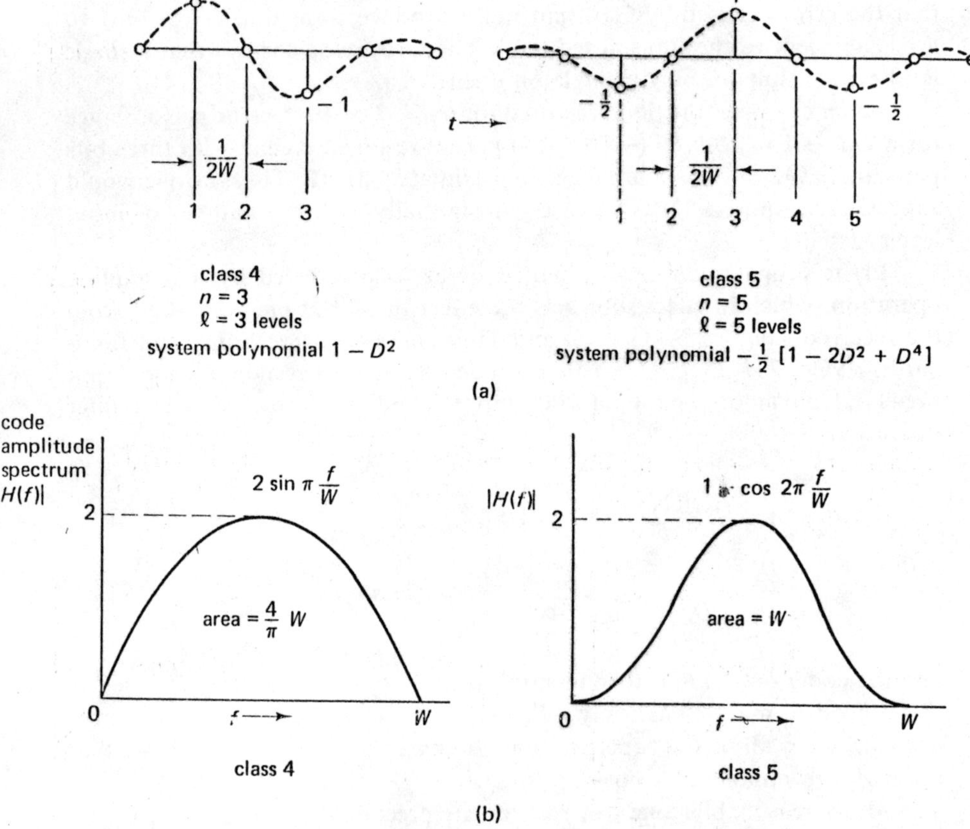

Fig. 16-4 Class 4 and Class 5 partial-response signals: (a) weighting coefficients k_i, and (b) amplitude spectra. The number of output levels ℓ refers to binary inputs.

and $(1 - 2D^2 + D^4)$, respectively. Error propagation effects increase the error probability effects in class 4 and 5 signals with binary inputs by a factor of 2 and 4, respectively, for low error rates $\mathcal{P}_e < 10^{-5}$.

Error Detection for Partial-Response Signals

Consider the class 4 signal transmitting three bits/sample where the precoder of (16-12) accepts input digits D_i and forms A_i:

$$A_i = (D_i \oplus A_{i-2}) \qquad \text{mod } 8 \tag{16-13}$$

The shaped partial-response transmitted signal B_i has 15 levels:

$$B_i = A_i - A_{i-2} \qquad -7 \text{ to } +7 \tag{16-14}$$

The initial values of B_i from (16-14) are defined as

$$\begin{aligned} B_0 &= A_0 - A_{0e} \\ B_1 &= A_1 - A_{0o} \\ B_2 &= A_2 - A_0 \\ B_3 &= A_3 - A_1 \\ &\vdots \end{aligned} \tag{16-14a}$$

where A_{0e} and A_{0o} are the initial conditions of the coder. The received quantized sequence of digits with additive noise N_i is $B_i' \triangleq Q[B_i + n_i]$. The decoder provides estimates $\hat{D}_i$ of the original input bit stream:

$$\hat{D}_i = B_i' \qquad \text{mod } 8 \tag{16-15}$$

To monitor performance, compute the following running sums of the received digits B_i separately for even and odd values of i:

$$\begin{aligned} S_i(J) &= \sum_{j=0}^{J} B'_{i-2j} = B_i' + S_{i-2}(J-2) \\ &= B_i' + B'_{i-2} + B'_{i-4} + \cdots + B_0' \qquad \text{for } i \text{ even} \\ &= B_i' + B'_{i-2} + B'_{i-4} + \cdots + B_1' \qquad \text{for } i \text{ odd} \end{aligned} \tag{16-16}$$

assuming that the signal started $2J$ samples ago. If there are no errors $B_i' =$

B_i, the sum should satisfy the relationship using (16-16) (16-14a)

$$-A_{0e} \leqq S_i(J) = \sum_{j=0}^{J} B'_{i-2j} = A_i - A_{0e} \leqq 7 - A_{0e} \tag{16-17}$$

for all even values of time i since $0 \leqq A_i \leqq 7$, and A_{0e} is the initial value of A_i for i even. For odd value of time simply replace A_{0e} with A_{0o}.

When errors occur in transmission, this constraint can be violated. Errors are assumed to occur only in adjacent levels. When we measure

$$S'_i(J) > 7 - A'_{0e} \tag{16-18}$$

the error detector indicates an error and decreases S'_i by one. Similarly, if $S'_i < -A_{0e}$, the detector indicates an error and increases S'_i by one.

EXAMPLE We build into the detector initial odd and even starting samples $A_{0o} = A_{0e} = 3$. Then as time progresses and an error e_i occurs—where we define $B'_i \triangleq B_i + e_i$, the error detector computes from (16-16) even and odd sums similar to the even running sums

$$\left.\begin{aligned} S_0 &= B'_0 = A_0 - A_{0e} \text{ (The initial value of } A_{0e} \text{ is set at 3)} \\ S_2 &= B'_2 + B'_0 = A_2 + e_2 - 3 \\ S_4 &= B'_4 + B'_2 + B'_0 = A_4 + e_4 + e_2 - 3 \\ &\dots \\ S_{2n} &= B'_{2n} + B'_{2n-1} + \cdots B'_0 \\ &= A_{2n} + e_{2n} + e_{2n-2} + \cdots + e_2 - 3 \end{aligned}\right\} \tag{16-19}$$

Thus, an error is flagged using (16-17) whenever

$$7 - 3 = 4 < S_{2n} \quad \text{or} \quad S_{2n+1} < -3 \tag{16-20}$$

Hence, these thresholds can be built into the detector, and the sum incremented appropriately if an error is detected.

16-4 QUADRATURE AMPLITUDE SHIFT KEYING (QASK)

Quadrature amplitude shift keying (QASK) [Smith, 1975] is a means for utilizing the bandwidth efficiency of a multilevel baseband channel by modulating both the inphase and quadrature components of a RF carrier. For example, one can take the outputs $B_I(t)$, $B_Q(t)$ of two partial-response coders of Fig. 16-3 and separately modulate I and Q carrier channels to form

$$B_I(t) \sin \omega_0 t + B_Q(t) \cos \omega_0 t$$

As an example, assume that the baseband modulations are formed by class 4 partial-response signals with binary inputs. The outputs B_I, B_Q are then both 3-level, $B_I, B_Q = +2, 0, -2$.

Figure 16-5 shows a phasor diagram of a simple QASK signal for this class 4 partial-response signaling. Notice that we now have an 8-phase signal with two amplitude levels, A and $\sqrt{2}A$, and an 8-phase signal with three amplitude levels, 0, A, and $\sqrt{2}A$.

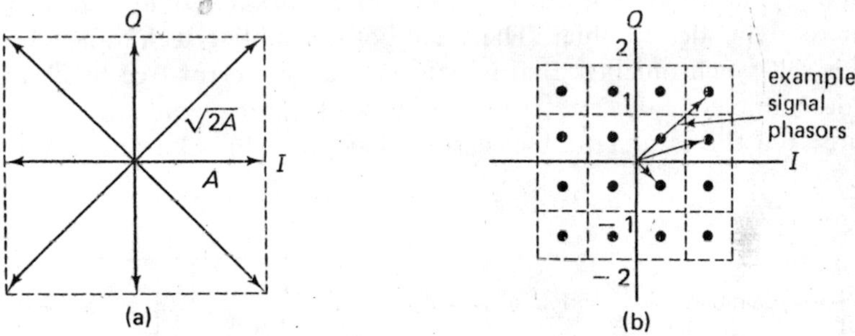

Fig. 16-5 Phasor diagram of a QASK signal (a) class 4: partial-response signaling on each of the in-phase I and quadrature Q channels; (b) phasor for 4 bits/symbol, 2 bits on each in-phase and quadrature components of QASK

The same QASK signal concept can easily be generalized to provide $2n$ bits/symbol by giving each B_I, B_Q, one of $N = 2^n$ amplitude levels. Figure 16-5*b* shows the square output signal array for a symbol carrying 4 bits—2 bits on each of the inphase and quadrature channels. There are 3 amplitude levels for this signal, $1/\sqrt{2}$, $\sqrt{5}/\sqrt{2}$, $3/\sqrt{2}$.

16-5 DIGITAL DATA SCRAMBLERS

In some instances, the digital bit streams entering any of the baseband or PSK modulators described earlier may have imbedded in them short-cycle periodic patterns. These short-cycle code patterns can sometimes cause difficulty in bit synchronization. For example, a periodic sequence of 8 bits, or long strings of all 1s or 0s—(period 1)—may be the input. These data patterns depend on the user bit streams which arrive at the modulator after multiplexing and coding if any have taken place.

Periodic patterns or strings of 1s or 0s can cause several potential problems. The principal one is that the long string of consecutive 1s might cause the bit synchronizer to lose synchronization momentarily and thereby cause a long burst of data errors in the detected output. Secondly, a periodic data pattern can create sidebands in the clock component generated by a bit-synchronizer transition-detection nonlinearity, and it is possible for the bit synchronizer to lock falsely to one of these sidebands. Periodic data patterns also cause line components in the transmitted RF spectrum that may create cochannel interference with other signals of a more serious nature than that of a purely random data pattern.

Scrambling the data is a method of substantially increasing the period of the input data while permitting recovery of the original data sequence at the other receiving descrambler. The family of scrambling techniques discussed here is self-synchronizing; that is, after a sufficient error-free block of data, the descrambler emits an error-free copy of the original input sequence. Figure 16-6 shows a scrambler/descrambler in a digital transmission link.

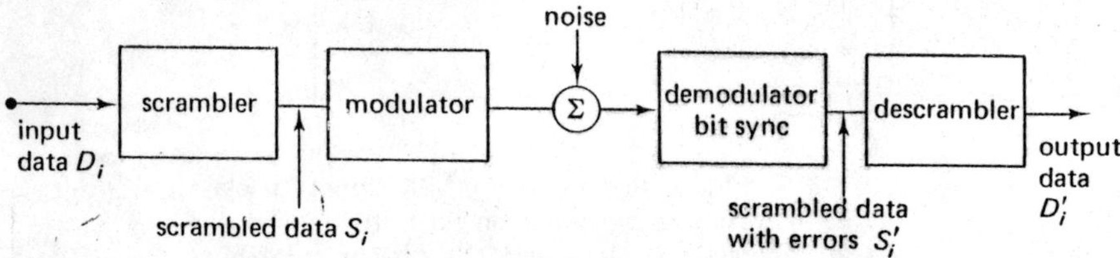

Fig. 16-6 General configuration of a data scrambler/descrambler in a digital transmission link

The basic block diagram of a scrambler and descrambler is shown in Fig. 16-7. The binary input data stream enters the scrambler feedback shift register at left where it is modulo 2 added to the pseudorandom-bit patterns R_t to produce the scrambler output $S_t = D_t \oplus R_t$. Figure 16-7*b* shows the shift register output sequence and states for a simple 3-stage feedback shift register (FSR) for an input sequence of all zeros. Clearly, a 3-stage shift register is too short to be of practical use but is useful here as an illustration.

If the tap coefficients are $C_1, C_2, \ldots, C_n$ where $C_i = 0, 1$ for an n-stage shift register, then the feedback shift register (FSR) has a tap polynomial representation

$$h(X) = X^n - C_1X^{n-1} - C_2X^{n-2} \cdots - C_n \tag{16-21}$$

Consider a set of maximal-length shift registers where the tap polynomial is primitive over the Galois field of binary elements $GF(2)$—that is, it is irreducible (has no factors except 1 and itself) and it divides $X^N - 1$ for $N \triangleq 2^n - 1$

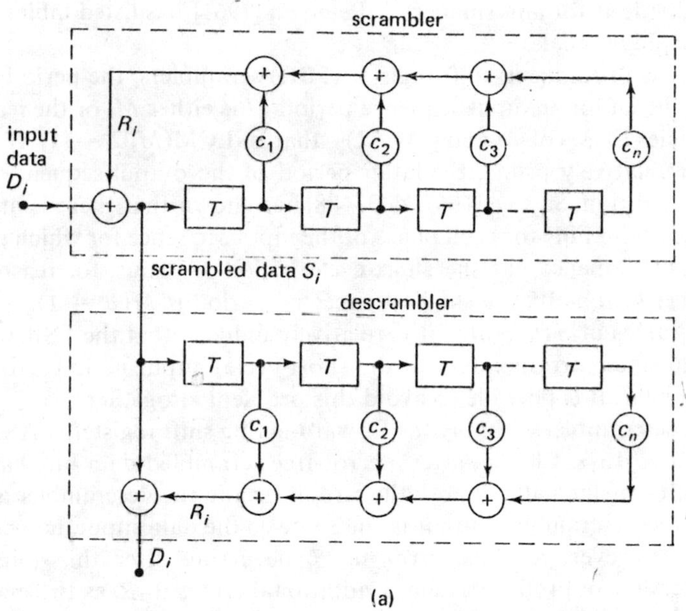

(a)

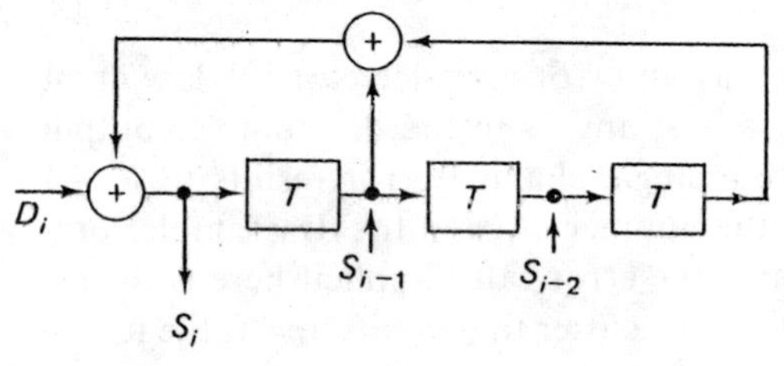

	Input	State		
Time	D_1	S_1	S_{i-1}	S_{i-2}
1	0	1	1	1
2	0	0	1	1
3	0	1	0	1
4	0	0	1	0
5	0	0	0	1
6	0	1	0	0
7	0	1	1	0
8	0	1	1	1
9	0	0	1	1

(b)

Fig. 16-7 Block diagram of a basic scrambler and descrambler; the descrambler is the filter inverse of the scrambler and is implemented as a forward acting shift register (FASR). (a) General scrambler/descrambler block diagram, (b) simple 3-bit stage scrambler output sequence and states for an all zero input. The scrambler is assumed to have an initial state (111). The scrambler output sequence has period 8.

but does not divide it for any smaller N. Peterson [1961] has listed tables of these primitive polynomials.

For this class of feedback-shift-register (FSR) scramblers, the period of the scrambler output for an input sequence period M is either M, or the least common multiple (LCM) of M and $(2^n - 1)$—that is, LCM$(M, 2^n - 1)$. If M and $2^n - 1$ are relatively prime, the latter period of the output sequence is $M(2^n - 1)$. In addition, Savage [1967, 449–488] has shown that there is only one initial state of the FSR for each phase of the input sequence for which the scrambler output sequence has the short-cycle period M. Thus, for reasonably large values of $n > 10$, the smallest expected period is $M(2^n - 1)$.

For large values of n $(n > 10)$, it is relatively unlikely that the FSR will be in the initial state corresponding to the short-cycle output period. However, as shown later, it is possible to avoid this problem altogether.

The basic descrambler is simply the forward-acting shift register (FASR) also shown in Fig. 16-7. Clearly, after n error-free scrambled data bits have arrived, the shift register states at both the scrambler and the descrambler are identical, and the descrambler output is the same as the data input, namely, $D_t = S_t \oplus R_t$. However, a single error in S_t occurring after this initial acquisition interval can produce as many additional errors in R_t as there are tap coefficients over an interval of n bits. Thus, there is an error multiplication effect by a factor of $K + 1$, where K is the number of nonzero tap coefficients and $K + 1 \geq 3$.

The scrambler can also be used as an error-rate detector for low error rates. If the scrambler is driven by all 1s, any 0s in the descrambler output correspond to channel errors. Since a single channel error results in $K + 1$ output errors, one need only count the number of 0s in the descrambler output and divide by $K + 1$ to determine the error rate. Implicit here is the assumption that no two channel errors occur closer than n bits apart, the length of the sequence generators.

The single counter scrambler (SCS), shown in Fig. 16-8, is designed to scramble periodic input sequences of periods P_1, P_2, or any submultiples thereof regardless of the state of the FSR. The SCS has two period-detecting circuits. For example, assume that a seven-stage shift-register polynomial is used $h(X) = 1 + X^4 + X^7$ and that input sequences of periods $P_1 = 7$, $P_2 = 8$ are to be scrambled. Also scrambled are inputs of period 1, 2, 4. If the scrambled sequence has period P_1 or P_2 or any submultiple of P_1, P_2, one or both of the AND gate inputs $x_1 = S_t \oplus S_{t-7}$, $x_2 = S_t \oplus S_{t-8}$ are zero, and the counter continues to count without being reset, eventually triggering the threshold. When the threshold is triggered, the state of the first stage of the FSK is changed, and the periodicity is removed.

On the other hand, if there is no periodicity in the output, the counter is randomly reset and the probability that the counter N_c reaches a threshold t

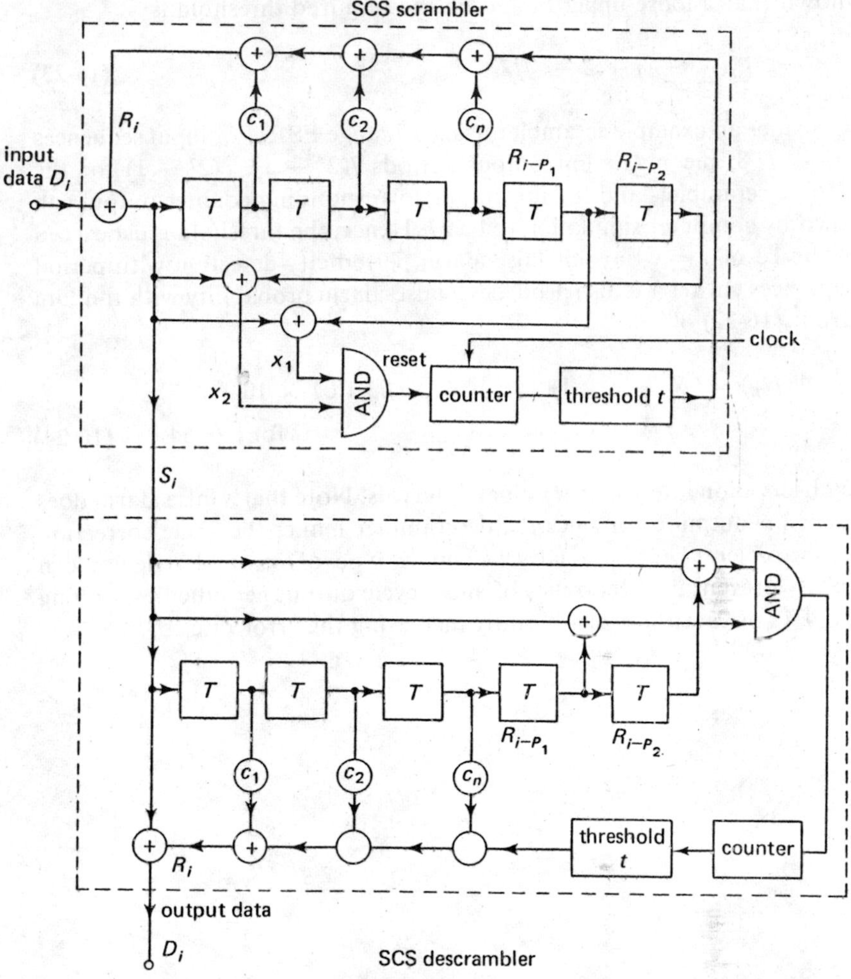

Fig. 16-8 Single counter scrambler and descrambler to prevent short period i output for input periods P_1, P_2

in N_T total clock intervals for random data is bounded by [Savage, 1967]

$$\mathcal{P}_{fa}(N_T) \leq (N_T - t + 1)(\tfrac{3}{4})^t \tag{16-22}$$

The higher the threshold is set, the lower the false-alarm probability. The threshold should not be set at too high a level, however, or it might take too long for a periodic pattern or string of 1s to be removed; and, of course, more stages of memory would be required for the threshold counter. Savage has shown that a loose upper bound on the required threshold is

$$t \leq P_2(N-1) + 2 - N/2 \tag{16-23}$$

Consider an example scrambler using a 7-stage FSR. For input sequences of period 7, 8, the maximum output periods $7(2^7 - 1)$, $8(2^7 - 1)$ for the seven-stage scrambler, and the maximum corresponding counter output can be shown by computer simulation to be 27. Hence, the threshold must exceed 27. A choice of $t = 64$ avoids false-alarm periodicity-detection with period 7, 8 sequences entirely. It also produces a false-alarm probability with random data from (16-22) of

$$\mathcal{P}(N_T) = (N_T - t + 1)(\tfrac{3}{4})^t = (N_T - 63)1.01 \times 10^{-8} \qquad \text{for } t = 64 \tag{16-24}$$

for an observation interval of N_T clock intervals. Note that a false alarm does not cause an output error since the descrambler makes the same correction as the scrambler. Thus, a relatively simple type of scrambler device can adequately prevent the occurrence of short-cycle output periodicities, or long strings of 1s or 0s without significantly increasing the error rate.

PART 4

Worldwide Timing by Satellite Relay

Distribution of accurate time becomes more important as digital communications expand to worldwide networks. The composite digital bit rates transmitted and received from each terminal are addressed to many other sites. Hence, one has an interconnected network of digital terminals with many users fed from each terminal. It would be possible to avoid the inefficiencies of pulse stuffing multiplexing if the user clocks could all be locked to the same average frequency. Elastic buffers, which require no increase in the average bit rate for pulse stuffing, could then be used if the buffer size were sufficient to accommodate the changes in propagation delay and clock-phase jitter. This approach, however, assumes the distribution of time throughout the network so that all clocks have the same average frequency.

For time-division multiple access, accurate time is even more critical. The earth terminal transmissions must be transmitted in synchronism with the satellite clock, simulated satellite clock, or master station, perhaps to accuracies of a few nanoseconds. Timing of signals arriving at a satellite from multiple earth stations has already been accomplished to an accuracy of ± 3 nsec relative to a hypothetical satellite clock.

The availability of synchronous communications satellites or subsynchronous 12-hour orbit navigation satellites, high-speed digital modulation techniques, and atomic frequency standards can lead to the distribution of time on an international scale to an accuracy never before possible. These timing-distribution techniques, which can include satellite-borne atomic frequency standards, have lead to precise means of navigation as well as efficient digital communications.

Several different time distribution techniques are discussed in Chap. 17. The emphasis is on timing for TDMA or synchronization of data sources, although these techniques are applicable to other clock-timing problems and relate closely to satellite navigation systems. Chapter 18 discusses the details of timing/ranging signals and receivers employing pseudonoise signals and delay-lock code tracking loops. The expected transient and noise performance of these loops is investigated. The properties of pseudonoise (PN) and other ranging codes are also discussed.

CHAPTER 17

SATELLITE TIMING CONCEPTS

17-1 INTRODUCTION

Satellite communications networks and satellite navigation systems often require timing systems having at least one of the following features: (1) time generation of nonoverlapping RF pulses transmitted from widely separated earth terminals in time-division multiple access (TDMA); (2) world-wide transfer of time-synchronization information so that coded bit stream transmissions from different ground terminals are in synchronism with a simulated satellite clock or system time standard; and (3) accurate tracking of the signals received from several (4 or more) satellites, each of which carries a precise clock. The one-way range to each of several satellites can then be compared and range differences can then be used to compute the user position as the intersection of several hyperboloids. Some models of (2) are given by Gersho [1966, 1689–1704] and Willard [1970, 467–483].

Desired time synchronism accuracy can range from 1 nsec to several microseconds over periods of several days or months. Even atomic standards are not presently capable of maintaining nanosecond precision without correction over such long periods. In addition, satellite orbits have nonzero eccentricities and drifts in the orbital inclination from an equatorial plane or other desired inclination that cause diurnal fluctuations in the terminal-satellite path delays. Finally, the group delay through the atmosphere/ionosphere and communication filters can vary with time, elevation angle, or location.

We begin this chapter with a brief discussion of basic time and frequency concepts, and continue with a discussion of some of the various time scales in use. Next, we discuss alternate timing and time transfer concepts and the equipment configurations used to implement this time transfer. The types of signals, side-tone, BINOR, and pseudonoise (PN) are then discussed briefly. We continue with a review of the satellite orbit geometry and the path delay fluctuations which impact this time transfer. We describe various sources of clock timing errors, clock frequency inaccuracies, and typical performance of atomic frequency standards. Finally, we discuss other sources of error in time transfer, atmospheric and ionospheric propagation effects. The effect of thermal noise on time transfer is discussed in detail in Chap. 18.

17-2 BASIC TIME AND FREQUENCY CONCEPTS

We define here four basic concepts of time: (a) **time intervals,** (b) the **date,** (c) **simultaneity,** and (d) **relativistic time dilation.** These elementary concepts are important to put into quantitative terms which can be used later in our discussions of clocks and time transfer.

The basic interval of time is the **second.** The present definition of the second has been made by the 13th Annual Conference on Weights and Measures, Paris, 1968.

> "The second is the duration of 9, 192, 631, 770 periods of the radiation corresponding to the transitions between the two hyperfine levels of the ground state of the Cesium 133 atom."

Multiple cesium frequency standards are used in averaging processes to generate the standard interval. Since the frequency of the above defined cesium radiation is 9, 192, 631, 770 Hz, one simply divides by this number to generate 1-sec time intervals.

The second concept is the **date,** or clock time t. Clock time t is a counting of the periods of the above defined cesium transitions, starting from some predetermined origin. More specifically, the ideal clock time is:

$$t = N\tau + t_0 \tag{17-1}$$

where t_0 denotes the agreed upon date of an event when the counting started, N is the number of periods that have occurred since t_0, and τ is the ideal period given by the defined cesium resonance.

Simultaneity is the third concept. Two events are simultaneous in a given coordinate frame if equivalent signals arrive in coincidence at a point in space which is geometrically equidistant from the source of each event. For example, consider the timing diagram in Fig. 17-1*a*. Points A and B are stationary in coordinate system Σ', and clocks A and B are located at each point. At the epoch of each clock, a pulse is radiated. The epochs are **simul-**

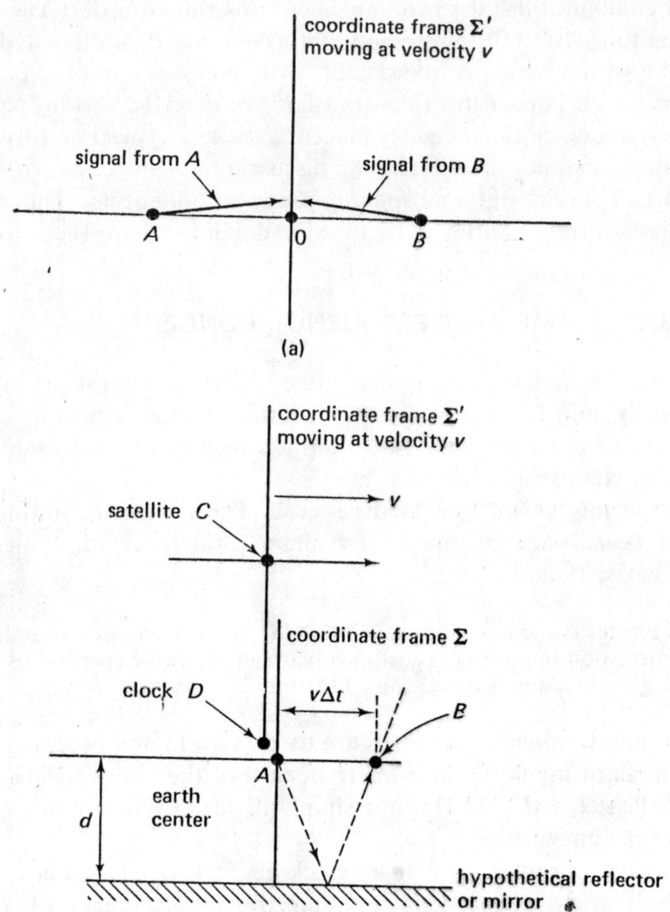

Fig. 17-1 Time concepts in moving reference frames: (a) Signals emitted from points *A*, *B* arrive simultaneously at midpoint 0 when all points are in the same coordinate frame Σ'. (b) Two coordinate frames Σ and Σ'. Coordinate frame Σ' is moving at velocity v with respect to Σ.

taneous and the clocks are synchronous if the pulses from A and B arrive at point 0—the geometric mid-point of A and B—at the same time instant.

The **time dilation** resulting from a relative motion between two coordinate systems, one moving at velocity v with respect to another, is the next concept to be discussed. This effect is a result of the special theory of relativity. Figure 17-1*b* shows two coordinate systems Σ and Σ', where Σ' is moving at velocity v with respect to Σ.

An observer at rest in Σ' measures the time interval $\Delta t'$ between the transmission of a pulse from D to an ideal reflector or mirror a distance d away and the reception as:

$$\Delta t' = 2d/c \tag{17-2}$$

Observers at rest in Σ can record the time interval Δt between the same two events if D passes over A at the time of pulse transmissions and it passes over B at the time of pulse reception. The time interval between these two events measured in the Σ coordinate frame is defined as Δt. The separation between A and B measured in Σ is then $v\Delta t$. Since the velocity of light is independent of the coordinate frame, the light travels a distance

$$c\Delta t = 2\sqrt{d^2 + \left(\frac{v\Delta t}{2}\right)^2} \tag{17-3}$$

in t sec. The pulse has traveled from A to the reflecting boundary and back to B simultaneously with reception at D. Using (17-2) and (17-3), we thus obtain:

$$\Delta t = \frac{2d}{c}\frac{1}{\sqrt{1-\beta^2}} = \frac{\Delta t'}{\sqrt{1-\beta^2}} \tag{17-4}$$

where $\beta \triangleq v/c$. If $\Delta t' = 1$ sec, then $\Delta t > 1$ sec, and the time intervals $\Delta t'$ of a clock in Σ' appear stretched out or at a lower clock rate when observed in Σ. This phenomenon is known as **time dilation.** Thus, if ideal cesium frequency standards are carried in both frames, the frequency of the clock in Σ', for example, on a satellite appears to be too low by a factor

$$\frac{\Delta f}{\Delta f'} = \frac{\Delta t'}{\Delta t} = \sqrt{1-\beta^2} \cong 1 - \frac{1}{2}\beta^2 \text{ for } \beta \ll 1 \tag{17-5}$$

Notice that when clock C in the moving frame is directly above point A there is no doppler shift to point A, yet there is still a time dilation effect. As an example, if the reference frame Σ' is moving at synchronous satellite velocities

$$v = \omega r_1 = 3.071 \text{ km/sec}$$

then $\beta = v/c = 1.024 \times 10^{-5}$ and the fractional clock frequency error is

$$\delta = \frac{\Delta f' - \Delta f}{\Delta f'} \cong 1 - \sqrt{1-\beta^2} \cong \frac{1}{2}\beta^2 = 5.24 \times 10^{-11}$$

Thus in one day the satellite clock appears to be slow by

$$\delta(3600 \times 24) = 4.53\ \mu\text{sec/day}$$

Note that although the clock frequency appears to be offset, the amount of the offset is highly predictable.

Alternative Time Scales

Before proceding with the concepts of time distribution, it is important to discuss the various types of time that can be used in a communication network or are used by others. There are at least two major categories of time which should be discussed: universal time and atomic time [Blair, 1974].

The various types of universal time are designed to relate to the earth's rotation. This type of time is, of course, valuable to the navigator. Universal Time 1 (UT1) is true navigator's time based on the earth's rotation. The UT1 second is not related to the definition of the interval of time, the second. Universal Time 2 (UT2) is a smoothed time that does not reflect the real periodic variations in the earth's angular position. Coordinated Universal Time UTC, on the other hand, has the true second as its basic time interval. Thus, there is no fluctuating frequency offset as there is with UT1, UT2. The clocks are kept in approximate step with the earth's rotation by adding or deleting seconds. Thus, there are special situations—leap seconds—where a minute contains 61 or 59 seconds instead of the conventional 60 sec. Leap seconds do not occur more than once or twice a year. However, because of the variation in the rotation rate of the earth, leap seconds are not predictable in detail. The UTC time scale is kept to within 0.7 μsec of the navigator's time scale, UT1.

Ephemeris time (ET) is determined by the orbital rotation of the earth about the sun rather than by the rotation rate of the earth about its own axes. Thus, ET should not be affected by earth movements, core mantle slippage, or geometrical changes in the shape of the earth.

Atomic time (AT) scales are based on weighted averages of multiple cesium beam frequency standards. For example, both the U.S. National Bureau of Standards (NBS) and the U.S. Naval Observatory (USNO) maintain ensembles of atomic standards to produce a reference atomic time AT.

International Atomic Time (TAI) is defined as follows:

> "International Atomic Time is the time reference coordinate established by the Bureau International de l'Heuve on the basis of the readings of clocks functioning in various establishments is accordance with the definition of the second, the S.I. Unit (International System of Units) of time."

17-3 NETWORK AND WORLDWIDE TIMING CONCEPTS

A network time standard for a limited net of earth terminals covered by one satellite can be obtained either by placing a precision clock on board the satellite, or by letting one of the earth terminals serve as a master timing station and relaying the master clock through the satellite transponder. Network timing is then established if (1) both or either station measures the round-trip path delay between satellite and itself, or (2) the total path delay from the remote terminal to the master is measured and appropriate corrections are made. Two possible configurations for different applications are shown in the next section.

Worldwide time transfer requires a number of synchronous satellites or subsynchronous satellites (for example 12-hr. orbit) in order to provide global coverage. If a system of subsynchronous satellites employs atomic frequency standards on board each satellite, the clock time of even these standards slowly drifts with respect to one another and should be corrected periodically by a master ground station which contains a reference atomic time scale. The clock time also can be transferred by transponding a master timing clock using a set of ground stations. Multi-hop satellite relay from the master timing source is required in this latter approach for worldwide coverage.

A network of satellite communications ground stations is characterized by a timing diagram as shown in Fig. 17-2. Each earth terminal has an associated clock with an apparent time $C_i(t)$, which differs from "true" system time t by $C_i(t) - t = \Delta_i(t)$.* See Fig. 17-2*b*. This clock timing error depends on the nature of the clock. A cesium standard, for example, may have a long-term ($> 10^3$ sec average) stability of better than 1 in 10^{12} whereas a high quality crystal oscillator may drift at a rate of $> 10^{-10}$/day addition to temperature effects (some of this drift is predictable). The clock time $C_i(t)$ may be represented by the state or epoch of a pseudonoise sequence or other well-structured signal as discussed in Sec 17-5.

The time delay $T_i(t)$ from a given earth terminal to the satellite varies with time depending on the satellite orbit. For a synchronous satellite, this time delay varies about its nominal 0.25-sec value because of the orbital eccentricity and inclination (see Chap. 6 and Sec. 17-6). The time delay T_s through the satellite transponder itself remains approximately constant except for thermal changes in the transponder group delay.

A signal emanating from earth terminal n at local clock time $C_n(t_0)$ and

*As already discussed, there is, in fact, only a hypothetical "true" time, and there are a number of time scales. For TDMA network timing an atomic standard at a master ground station can just as easily be assumed to be the reference time.

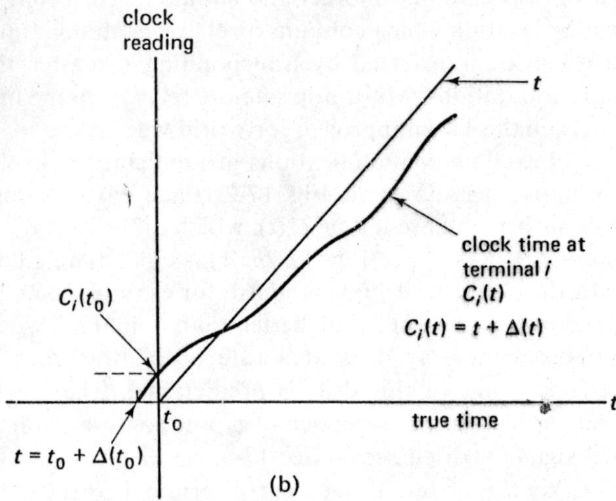

Fig. 17-2 Configuration of clocks and path delays to the satellite: (a) system configuration, (b) clock time versus real time

system time t_0 arrives at earth terminal i at system time

$$\tau_{in}(t_0) = t_0 + T_n(t_0) + T_s + T_i[t_0 + T_s + T_n(t_0)] \tag{17-6}$$

where the path delay s $T_i(t)$ is defined at the signal transmission time t rather than the reception time and account for any change in delay with time (doppler).

Note that the differential velocities are assumed to be small enough that special relativistic effects are negligible. In general, the relationship between satellite clock time τ viewed from earth and the reference coordinate time t—for example, system time at the center of the earth—is given by

$$\frac{d\tau}{dt} \approx 1 - \frac{\phi}{c^2} - \frac{1}{2}\left(\frac{v}{c}\right)^2 \tag{17-7}$$

where ϕ is the Newtonian potential at the clock, c is the velocity of light, and v is the heliocentric velocity of the clock relative to the coordinate center. The ϕ/c^2 term represents the relativistic gravitational red shift, and the $(v/c)^2$ term is the correction due to the Lorentz transformation of special relativity discussed in Sec. 17-2. For synchronous satellites the relativistic gravitational shift is approximately 50 μsec/day but is nearly constant and therefore can be compensated [Beehler, 1971]. Moyer [1971] provides a good discussion of relativistic corrections to time. Second-order gravitational effects are caused by the changing gravitational potential induced by the sun and moon.

On reception of the pulse from terminal n in Fig. 17-2 the corresponding clock at earth terminal i reads

$$C_i[\tau_{in}(t_0)] = C_i\{t_0 + T_n(t_0) + T_s + T_i[t_0 + T_s + T_n(t_0)]\} \tag{17-8}$$

At this point, of course, both the difference between $C_i(t)$ and t and the delays T_n, T_i are defined but unknown to the earth station. Hence, this single one-way transmission by itself is only a step in precise timing of the network. Several different timing approaches are possible in a network of this nature. Two alternatives are:

(1) timing slaved to satellite time, and

(2) timing slaved to the master station

As shown in Chap. 10, this second approach can require double the TDMA storage capacity in the elastic buffers relative to that required if time is slaved to the satellite. Another related technique is to slave each station clock to the average of all input clocks as discussed by Pierce [1969, 615; 636].

Throughout the rest of this discussion, we neglict relativistic effects or assume that they are already corrected.

Timing Slaved to Satellite Time

Consider a network of terminals in the coverage beam of a single communications satellite. In one timing technique, a clock correction is generated at each transmitting earth terminal so that pulses initiated at the corrected clock time $C_i'(t_0)$ all arrive at the satellite in coincidence with a hypothetical or actual satellite clock $C_s'(t_0)$. If the signals arrive at the satellite in synchronization, then they all are received by a given earth terminal in synchronism. If each earth station then transmits at suitable displacements from this clock time $C_i'(t_0 + t_i)$, the TDMA pulses arrive at the satellite in nonoverlapping time slots. This corrected clock time $C_i'(t_0)$ usually must be corrected not only as a fixed absolute time shift but also in clock frequency to account for satellite–earth-station group delay variation with time or clock frequency error.

Closed-loop timing approaches assume slowly varying doppler effects (accelerations), and slave-control-loop time constants must be long compared to the round-trip path delay (≈ 0.25 sec) to avoid loop instabilities. On the other hand, the loop bandwidths must be large enough to track the diurnal satellite–earth-station path delay changes and the clock-timing jitter and drifts. These delay and clock fluctuations are discussed in Sec. 17-7.

The satellite clock is either generated directly onboard the satellite or transmitted to a satellite transponder by a master ground station to act as a simulated satellite clock. Each individual earth terminal can measure the round-trip path delay to the satellite and back:

$$T_i(t_0) + T_i[t_0 + T_s + T_i(t_0)] + T_s \tag{17-9}$$

If the transponder delay T_s is known and $T_i(t) + T_i[t_0 + T_s + T_i(t_0)]$ is measured and tracked, the one-way path delay $T_i(t)$ can be predicted T_t sec in the future. For synchronous satellites the propagation delays vary approximately linearly with time for time increments $\lesssim 0.25$ sec; that is,

$$T_i[t_0 + T_s + T_i(t_0)] \cong T_i(t_0) + \delta(t_0)[T_i(t_0)T_s] \tag{17-10}$$

The same TDMA pulse timing can be accomplished without a direct measurement of the round-trip delay by controlling the individual earth terminal clocks in a way that forces the pulses relayed through the satellite to arrive back at the same earth terminal at time t_r in synchronism with the satellite clock. See Fig. 17-3. Thus, a given satellite clock pulse is transmitted at $t_s = t_0 + T_i(t_0) + T_s$ and received at earth terminal i at time $t_0 + T_i(t_0) + T_s + T_i[t_0 + T_i(t_0) + T_s)] = t_s + T_i(t_s)$ where $T_i(t_s) = T_i[t_0 + T_i(t_0) + T_s]$, is the down-link propagation delay measured at the time of pulse transmission. In order for the earth terminal generated pulse to be received at the correct time via the satellite transponder, it must be generated at $t_0 = t_s - [T_i(t_0) + T_s]$. Clearly, if pulses from several earth terminals are received simultaneously at terminal i, they must be in synchronism at the satellite. Hence they can be used as a TDMA frame epoch clock at each separate earth terminal.

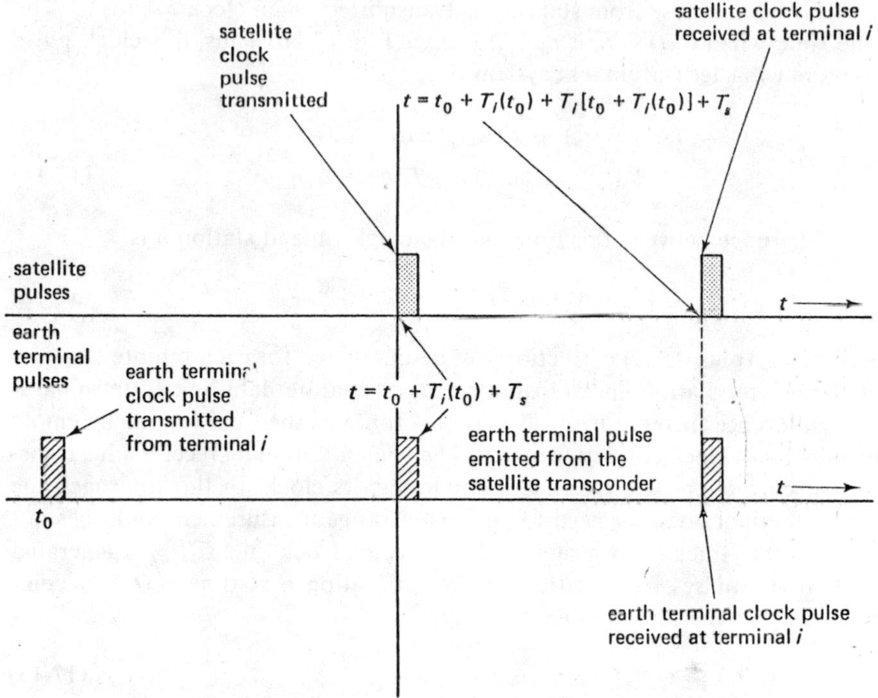

Fig. 17-3 Earth terminal time slaved to satellite time. The figure shows the pulses generated by the satellites in the upper time line and those generated by the earth terminals in the lower diagram.

Time Slaved to the Master Station

A second alternate is to slave each earth terminal's corrected clock $C_i''(t)$ so that a pulse transmitted from earth terminal i at $C_i''(t_0)$ arrives at each earth terminal j in synchronism with pulses transmitted by the master earth terminal n at $C_n''(t_0)$. The same time line diagram shown in Fig. 17-4 applies here except that the satellite clock is really generated at the master terminal rather than by the satellite itself. Since the signals are received from the satellite in synchronism, they must have arrived at the satellite in synchronism. Thus, this approach results in system timing that avoids TDMA pulse overlap. Each terminal must transmit and receive its own signal as well as receive the master timing signal.

Another method of slaving to a master station is for the master station n to compare the clock pulse received from station i with its own pulse transmission relayed by the satellite. The master station then subtracts the difference in these two arrival times.

The clock pulse from station i is transmitted when clock i reads t_0. The true time when $C_i(t) = t_0$ is $t_0 + \Delta_i(t_0)$ see Fig. 17-2b. Thus, this clock pulse arrives at master station n at system time

$$\tau_{ni}(t_0) = t_0 + \Delta_i(t_0) + T_i(t_0 + \Delta_i) + T_s + T_n[t_0 + T_s + \Delta_i + T_i(t_0 + \Delta_i)] \tag{17-11}$$

The difference between this time and the clock time at station n is

$$\tau_{ni}(t_0) - C_n[\tau_{ni}(t_0)] \triangleq \Delta_{ni}(t_0) \tag{17-12}$$

This clock (plus delay) correction can be quantized for each remote station i at the master station and transmitted on a separate data link to terminal i. The difference in these two pulse arrival times is then used at each remote terminal as a clock correction signal. Then earth terminal i receives this transmitted correction data message and corrects its clock. In this approach the slave terminal does not need to perform a range measurement and does not even need to track its own signal. The corrected clock time $C_i''(t_0)$ is generated so that the pulses from station i arrive at station n at time $\tau_{ni}(t_0)$ in coincidence with the master clock, namely,

$$\tau_{ni}''(t_0) = C_n(t_0) \tag{17-13}$$

For purposes of TDMA timing the master station does not need to correct its clock since other stations are slaving to the master time.

17-4 TIMING SYSTEM CONFIGURATIONS

Two timing system equipment configurations are described here in which remote time is slaved to a master clock. These equipment configurations are designed to implement the timing techniques described in the paragraphs in Sec. 17-3 above. In the first approach, the remote terminal itself performs the timing correction measurement and converts its clock by directly slaving its transmission to the master. In the second approach, the master station performs all the measurements and transmits the appropriate corrections by way of a return data link to the remote terminal.

Slaving to Master Time: Remote Terminal Measurement

In this approach, each remote terminal transmits and receives pulses as shown in Fig. 17-4. The timing pulses can represent either actual pulses or the reference point or epoch in a coded pulse or pseudonoise (PN) sequence

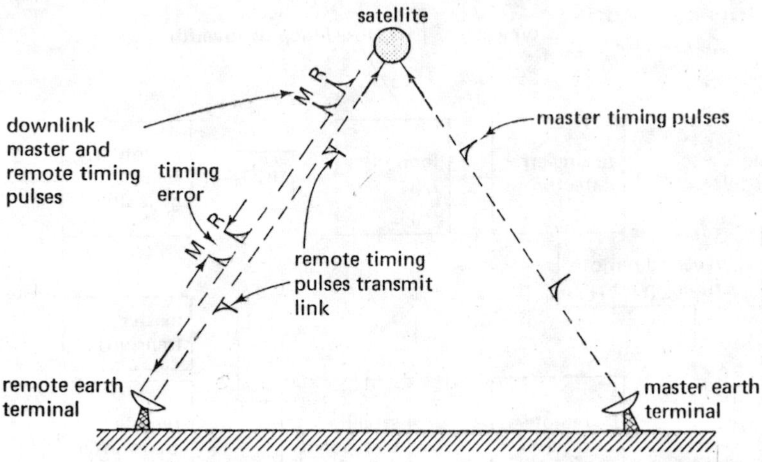

Fig. 7-4 TDMA timing system employing round-trip ranging/synchronism at the remote station

—for example, the all-1s state in a periodic PN sequence. Alternative timing/ranging signals are discussed in Sec. 17-5. The master timing pulses are transponded by the satellite as shown. A separate pulse sequence of the same period is transmitted by the remote receiver and phased by means of a closed loop so that the master and slave timing pulses arrive in synchronism (or offset in time by a prearranged amount) at the remote terminal receiver.

If the master and slave pulses are received at the remote terminal in synchronism, then they must have arrived at the satellite transponder in synchronism. Thus, any transmitted RF TDMA pulses transmitted with reference to the remote user transmit pulses can be timed to arrive at the satellite in time slots nonoverlapping those from other earth stations and remain correctly positioned. To achieve this performance, the following functions must be performed at the remote station:

1. Track the delay of the received master timing pulses using a delay-locked loop or an appropriate phase-locked loop. A delay-locked loop tracks an incoming pseudonoise or other pulse sequence as is described in detail in Chap. 18.
2. Transmit a remote timing pulse sequence with the same period as the master pulses and with a controllable phase.
3. Lock a delay-lock loop to the received remote timing pulses.
4. Measure the time delay/phase error between the received master and remote timing pulses.
5. Use the time delay/phase error to correct the timing of the transmitted remote pulse sequence, allowing for the round-trip propagation lag ($\approx$0.25 sec) as shown in Fig. 17-5.

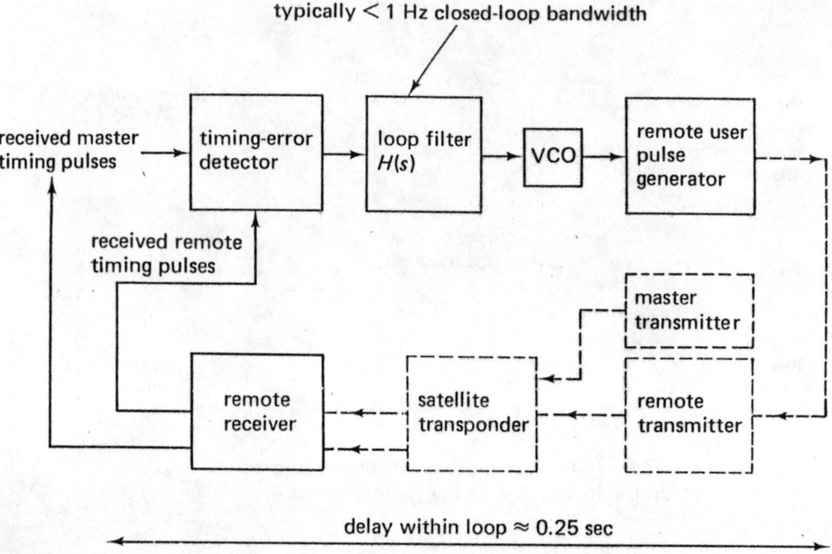

Fig. 17-5 Closed-loop remote timing pulse clock; delay within loop ≅ 0.25 sec propagation delay

Remote Terminal Slaved to Master Time: Master Terminal Measurement

Figure 17-6 shows a second type of timing network in which the remote user transmits to the master terminal by way of a phase-controlled periodic-pulse generator. In this case, the master station makes the timing-error measurement for each remote terminal and transmits the timing-error information to the remote user by way of a low-rate return data link. The round-trip delay time is ≈0.5 sec plus the delay time for the return-link path delay and modem delay, which may be substantial. Because these delays can be significantly larger than in the former system, this system is not likely to provide the same degree of timing accuracy. For networks requiring a lesser degree of timing accuracy, however, this system has the potential advantage that the remote user can be smaller in size because the only receive link is relatively narrow in bandwidth, and the remote terminal need not perform the time measurement.

Navigation Satellites

Although navigation and position location are not subjects directly addressed by this book, the use of highly accurate satellite clocks in precisely known satellite orbits can serve the function of worldwide time transfer as well as user position determination.

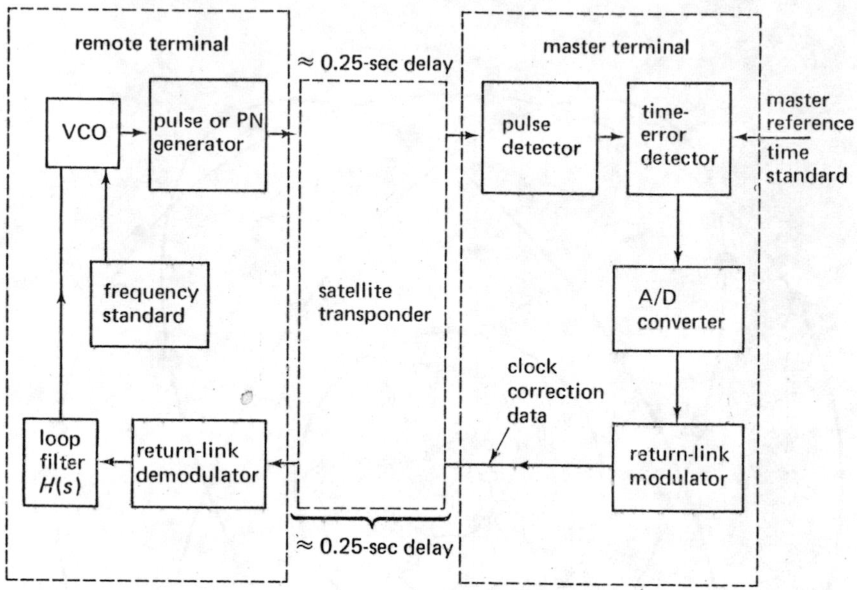

Fig. 17-6 Timing synchronizer using a return data link

The Global Positioning System employs a network of 24 NAVSTAR satellites in three orbital planes inclined at 63° relative to the equatorial planes as shown in Fig. 17-7. The ascending nodes of each orbital plane are separated 120°. Eight satellites are equally spaced in each orbital plane at an altitude of approximately 16,020 km (10,898 nm) and have 12-hour periods prograde* (sidereal time). Thus, each satellite passes over the same spot on earth every 24 hours in this subsynchronous orbit. Each satellite is three-axis stabilized (Fig. 17-8) and carries an atomic clock. The satellite down-link data stream carries clock error information relative to the on-board atomic standard. These clock error data are obtained from a ground control station. The figure shows an array of 12 helix antennas which provide an earth coverage antenna pattern (≈29° from this satellite altitude). Unlike communications satellites, the NAVSTAR satellites carry no transponder for the navigation signal. Instead, each navigating user operates in a purely passive mode, receiving and tracking the navigation signals from each of four or more satellites. Users have available 6 to 11 simultaneous satellite signals. The down-link data stream also carries satellite ephemeris information, which allows the user to know exactly where the satellite is at any given satellite clock time.

---

*Prograde means that the equatorial component of the satellite motion in its orbit is in the same direction as the earth's rotation.

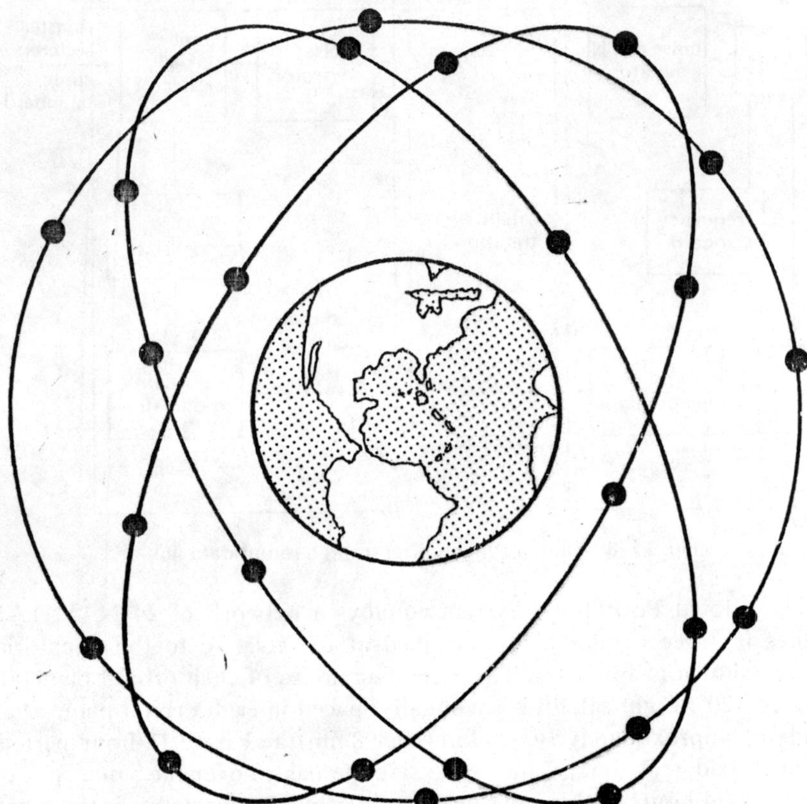

Fig. 17-7 Orbital configuration of the NAVSTAR navigation satellite in three orbital planes inclined at 63°. Each plane contains eight equally spaced satellites.

Each satellite transmits pseudonoise (PN) codes in synchronization with the stable clock at 10.23 Mbps and 1.023 Mbps. These codes PSK-modulate stable L-band carriers so that aircraft, land vehicles, and ships can perform one-way ranging to each of four satellites, thereby permitting the user to locate his position and to determine time at his location. The user has no accurate clock in general. Hence he can measure only a pseudorange (range with a fixed range offset) to each satellite. Four pseudorange measurements enable the user to solve for four unknowns: the x, y, z coordinates and time t. Two L-band carriers are transmitted at 1575.42 MHz and 1227.6 MHz so as to permit two independent measurements. This approach permits an ionospheric

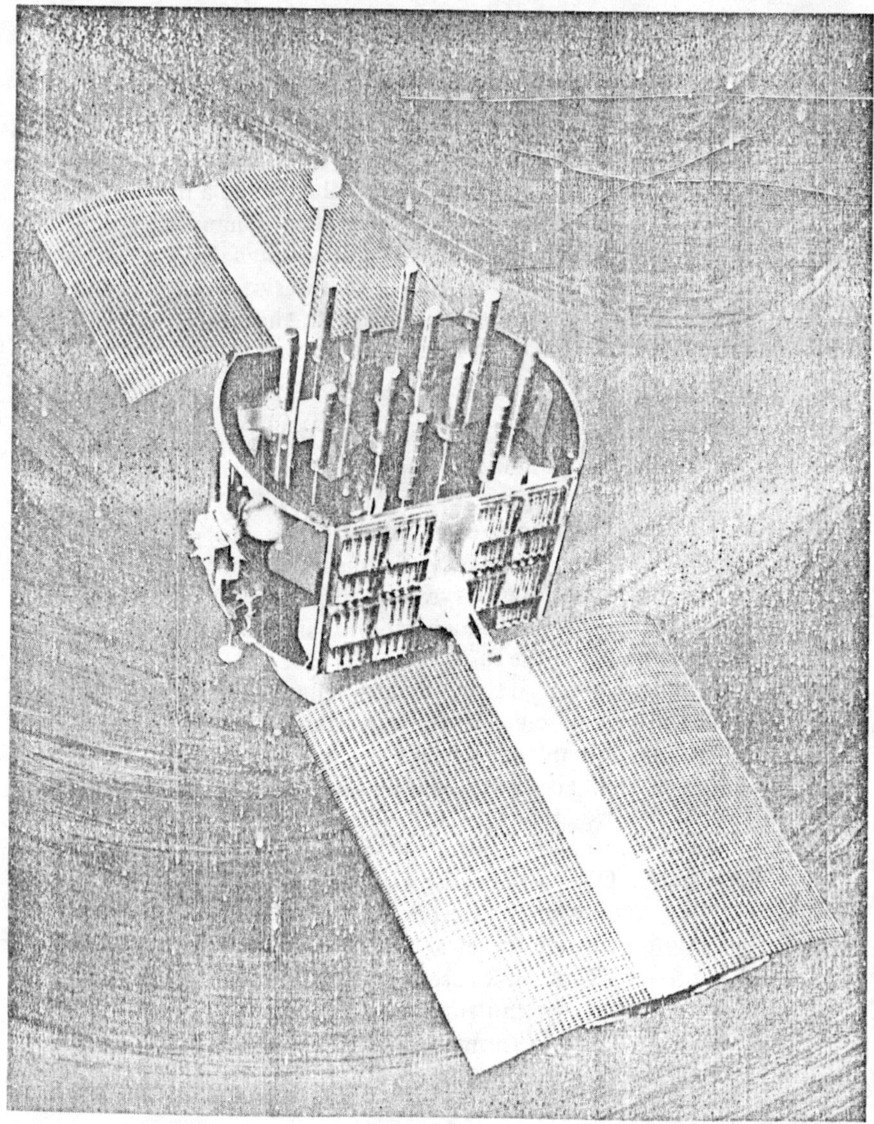

Fig. 17-8 Artist's conception of the three-axis stabilized NAVSTAR satellite with solar arrays extended on each side of the satellite. The solar array has an area of approximately 5 m² and operates with two degrees of freedom, one from the array axis and one from the physical yawing of the satellite itself.

group delay correction by making use of the group delay variation with frequency, which is approximately proportional to k/f^2. (See Sec. 17-8.)

17-5 TIMING-SYSTEM SIGNALS

Each of the timing systems discussed in Sec. 17-3 is based on the transmission of a signal $s(t + \tau_0)$ and the reception of $s(t + \tau + \tau_0)$ where the waveform $s(t)$ is known to the receiver. The receiver must then make a measurement of $\tau + \tau_0$ in the presence of thermal noise and system dynamics.

The timing system concepts can make use of any signal $s(t)$ whose period T exceeds the unambiguous range or timing of the system. For example, if the satellite round-trip range is already known to the nearest 0.05 sec, the signal period should exceed 0.05 sec. The format of that signal can take many forms, however. Widely used timing signals include:

1. *Side-tone ranging (STR) signal* A sinusoidal carrier which is PM (or AM) modulated with a sequential or simultaneous set of tones $m(t)$. This composite set of tones is $m(t) \triangleq \sum a_i \sin 2\pi f_i t$. The tone frequencies often are of the form $f_i = f_0 4^i$. The fundamental tone frequency f_0 has a period $T = 1/f_0$ where T is the unambiguous range. Figure 17-9 shows the first three tones of a side-tone ranging signal. The positive-going zero crossing of the fundamental frequency f_0 is measured first. The phase noise on this measurement should be low enough to measure unambiguously the corresponding positive-going zero covering at frequency $4f_0$. As long as no ambiguities occur, the time accuracy is determined by the phase noise on the highest frequency side tone $f_n = f_0 4^i$. These tones are sometimes transmitted in time sequence.
2. *BINOR* One can also use a square wave analogy to the side-tone ranging system. That is, one can modulate with the summation of square waves at rates $f_i = f_0 2^i$ or $f_0 4^i$ as shown in Fig. 17-10. Rather than use the multilevel summation of the five square waves, we can simply use the majority vote or sign of the summation to form the BINOR signal. One can still correlate each of the component square waves with the BINOR signal to recover time.
3. *Pseudonoise (PN) ranging signal* A sinusoidal PSK or QPSK modulated by one or more binary pseudonoise sequences, whose period of this pseudonoise sequence equals or exceeds T sec. Composite sequences, such as a majority logic combination of several shorter period sequences, can be employed to allow a more rapid acquisition time. (See Chapter 18.)
4. *Time-gated pseudonoise ranging signal* A signal having the same form as the preceding signal except that the signal is time gated and may appear periodically for only a small duty factor.

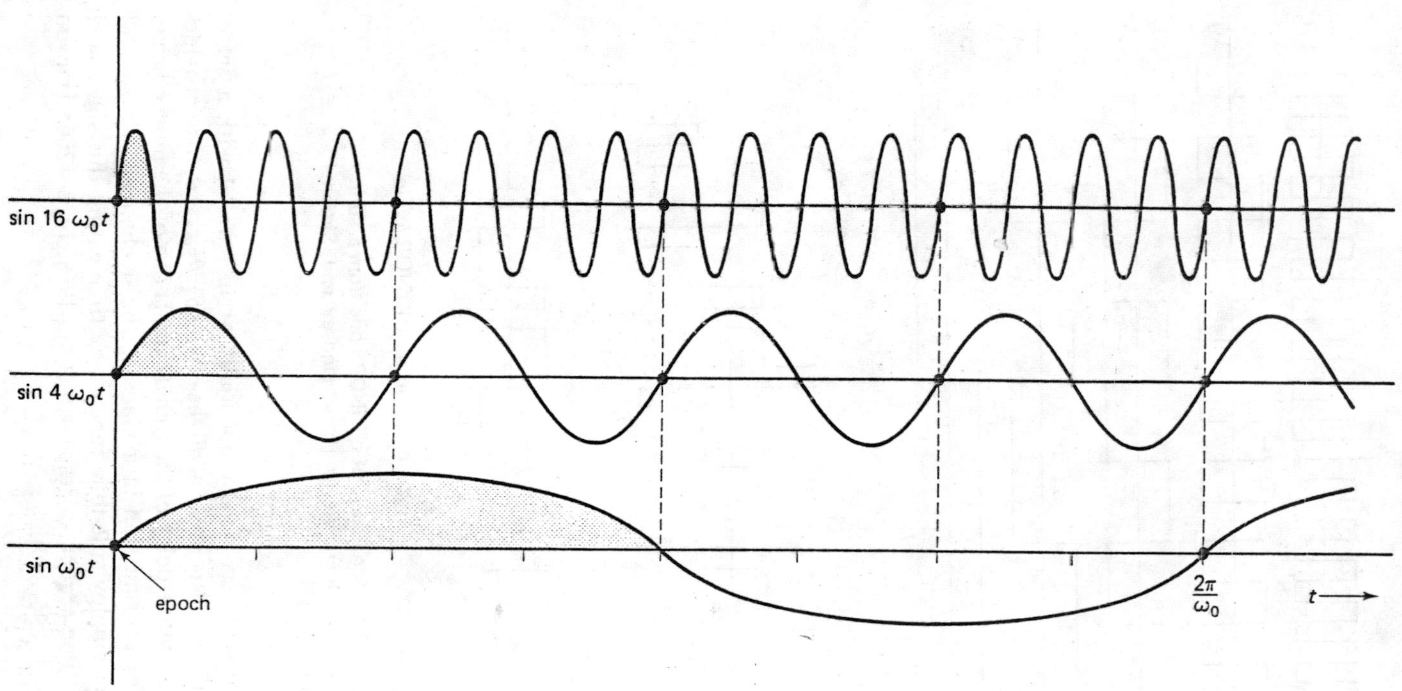

Fig. 17-9 Waveforms for the first few tones of a side-tone ranging system. The phase noise in each received carrier is assumed to be small enough that the carrier phase can be recovered unambiguously at 4 times its frequency.

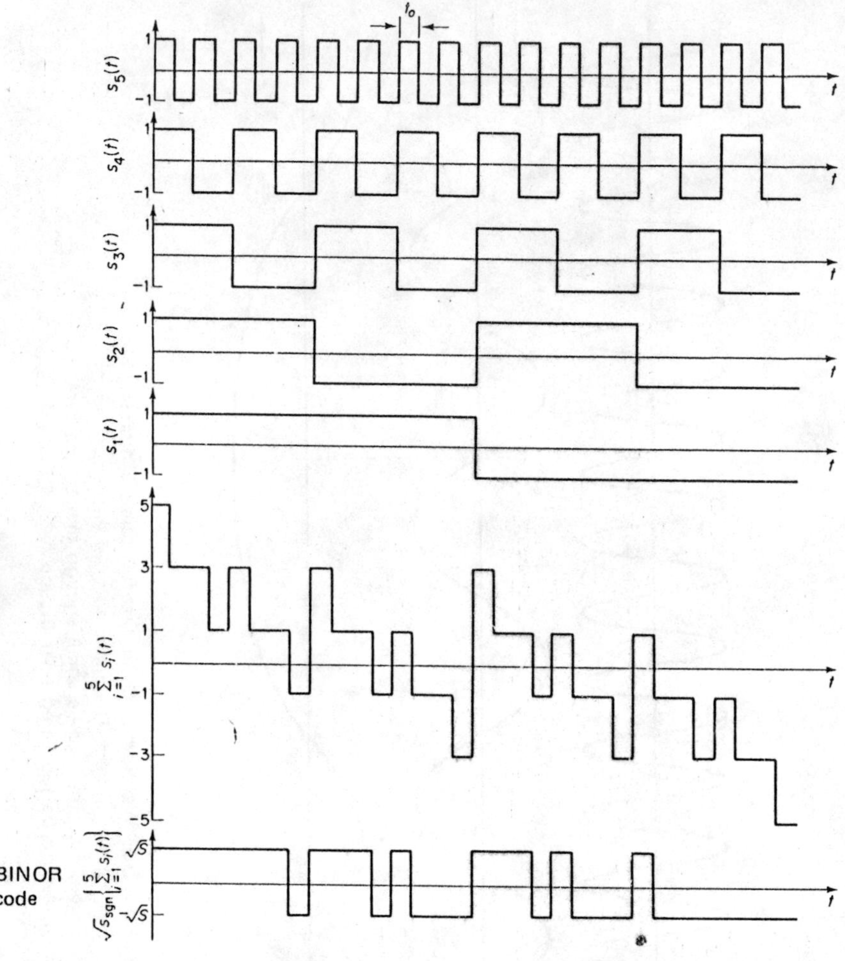

Generation of a BINOR code

Fig. 17-10 Generation of a BINOR code from a summation of square waves [From Lindsey and Simon, 1973, Fig. 4-7]

The STR signal is recovered sequentially or in parallel with a set of phase-locked oscillators. Recovery of all the phases $\theta_1, \theta_2, \ldots, \theta_n$ then provides an estimate of the relative delay of the signals and is one of the most efficient systems known for transmission in a Gaussian noise channel. A system of multiple users can employ side-tone ranging by time gating this signal and transmitting it in time sequence [Jespersen et al., 1968, 1202–1206; Hanson and Hamilton, 1971, 895–899].

Provided all tone ambiguities are resolved to the correct cycle, the accuracy is largely determined by the phase error (SNR) on the highest frequency side tone. In addition, doppler information is available from tracking the RF carrier itself. Simultaneous transmission of continuous STR signals—that is, multiple access in exactly the same overlapping frequency band—is not generally feasible, because the phase-locked oscillators exhibit a strong capture behavior whereby the strongest of these signals captures the phase-locked loop and the rest are to an extent suppressed.

Pseudonoise (PN) ranging signals can be tracked with a form of delay-lock loop discriminator or one of its variations. One advantage of this type of signal is its relative tolerance to interference generated by other multiple-access signals, intermodulation products generated by other signals, or spurious sinusoids. The RF carrier doppler of the PN signal can be tracked after bandwidth compression of the PN signal as well as the code delay. For this reason as well as our emphasis on digital signals, we have concentrated on PN ranging and time-gated PN ranging in this book. Chapter 18 describes the techniques and performance of delay-lock tracking loops for PN code tracking. The PN sequences are generated by feedback shift registers.

Figure 17-11 shows the block diagram of a simple 4-stage feedback shift register. At time $t = 1$, the shift register delay element state is (0100). Every possible 4-bit pattern shows up in the 4-bit window and every possible shift register state appears except for the all-zero state. (Four zeros in a row, the all-zero state, would permit no further change in state.) Figure 17-12 shows the output bit sequence for this shift register and the autocorrelation function $R(i)$ for the corresponding ± 1 sequence. The autocorrelation function for all maximal-length linear shift register sequences of period $2^n - 1$ has this 2-level autocorrelation property. Further details of the properties of PN codes are discussed in Chap. 18. Time-gated PN signals can be tracked using a time-gated version of the delay-lock loop (DLL); or if the time-gated signal is of sufficiently short duration, matched-filter detection can be used with a surface-wave delay line or digital matched filter. The pulse outputs of matched filter appear with each periodic time-gated PN signal, and these periodic pulse outputs can in turn be tracked using a special time-gated delay- or phase-locked loop.

Signal design to optimize range accuracy (rms) in the presence of noise with a bandwidth constraint is well known [Woodward, 1957] to yield a signal with all the energy peaked at the band edges as a pair of tones. RMS error is inversely proportional to the normalized second moment of the signal spectrum about the center frequency. Thus, side-tone ranging can yield a lower rms error than a PN signal of the same power.

Nevertheless PN ranging signals can yield important advantages over side-tone ranging. A PN system allows multiple signals to occupy the same

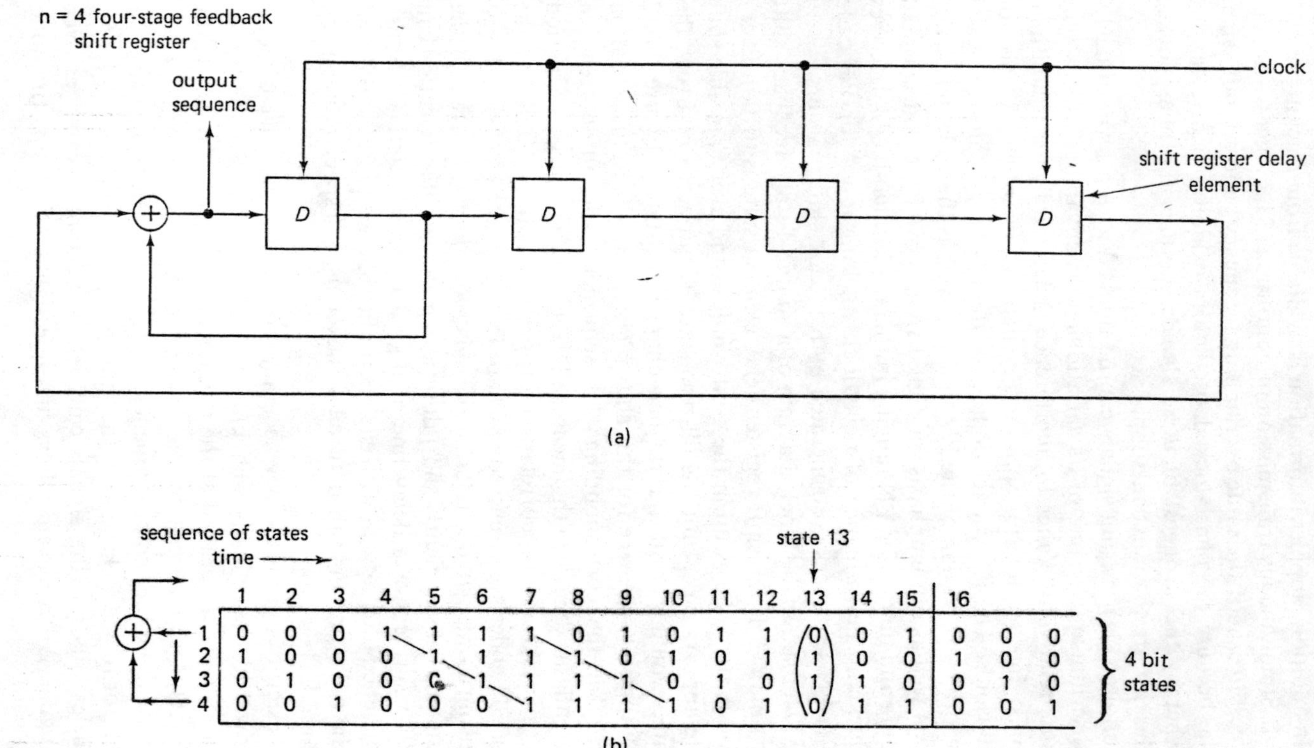

Fig. 17-11 Four-stage pseudonoise code generator for a 15-bit PN sequence: (a) Feedback shift register block diagram, (b) Sequence of shift register states

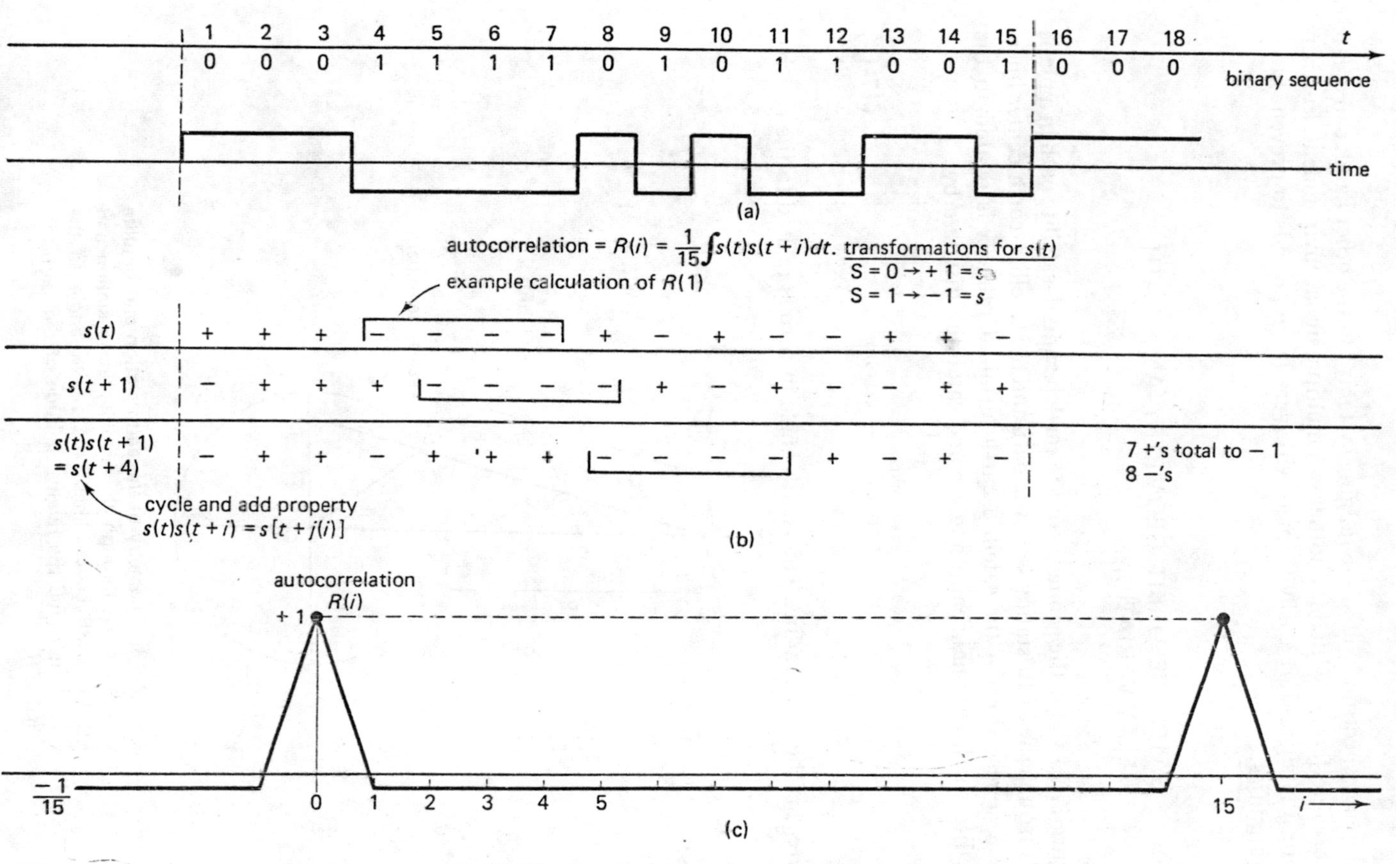

Fig. 17-12 Pseudonoise codes and their autocorrelation function: (a) 15-bit PN sequence, (b) example calculations of autocorrelation for 1-bit offset R (1), (c) autocorrelation function

frequency spectrum. Side-tone ranging systems, on the other hand, can create severe interference with one another in a multiple signal environment. Furthermore, PN signals are easily formed with very large periods to prevent range ambiguities.

17-6 SATELLITE ORBIT GEOMETRY AND PATH DELAY VARIATIONS

Figure 17-13 shows the geometry of the earth terminal–satellite path in a plane cut through the earth-station–satellite path and the satellite–earth-center lines. For a spherical earth the satellite–earth terminal range x is related to the satellite–earth terminal zenith angle η and satellite altitude d by

$$x = r_e\left\{\left[\left(1 + \frac{d}{r_e}\right)^2 - \sin^2 \eta\right]^{1/2} - \cos \eta\right\} \qquad (17\text{-}14)$$

where r_e is the earth radius. The zenith angle in turn is related to the relative

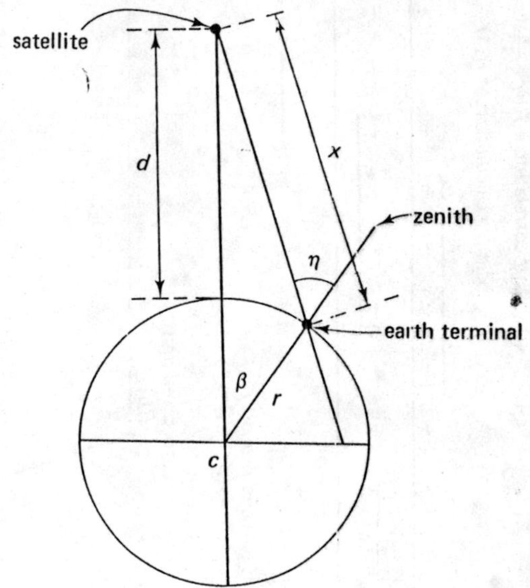

Fig. 17-13 Geometry of the earth station and satellite in a plane cut through the satellite–earth station and satellite–earth center lines; the zenith angle to the satellite is η and the relative latitude of the earth terminal is β.

latitude of the earth station (relative to the subsatellite point) by [Jacobs and Stacey, 1971]*

$$\sin(\pi + \eta) = \frac{r_e + d}{x} \sin\beta \tag{17-15}$$

or

$$\eta = \sin^{-1} \frac{(1 + d/r_e)\sin\beta}{[4(1 + d/r_e)\sin^2\beta/2 + (d/r_e)^2]^{1/2}} \tag{17-16}$$

For an inclined synchronous orbit, the satellite zenith angle undergoes diurnal changes, which cause delay variations similar to those given in Fig. 17-14. (See also Secs. 6-2 and 6-3.) The range and range-rate changes are shown for the synchronous satellite ATS 1. The range-tracking loop must have a loop bandwidth consistent with the dynamics, as small as they are, of the satellite motion. Satellite eccentricities [Royal Aircraft Establishment, 1971] for synchronous communications satellites are typically on the order of 10^{-4} and produce only a small path delay change compared to the inclination.

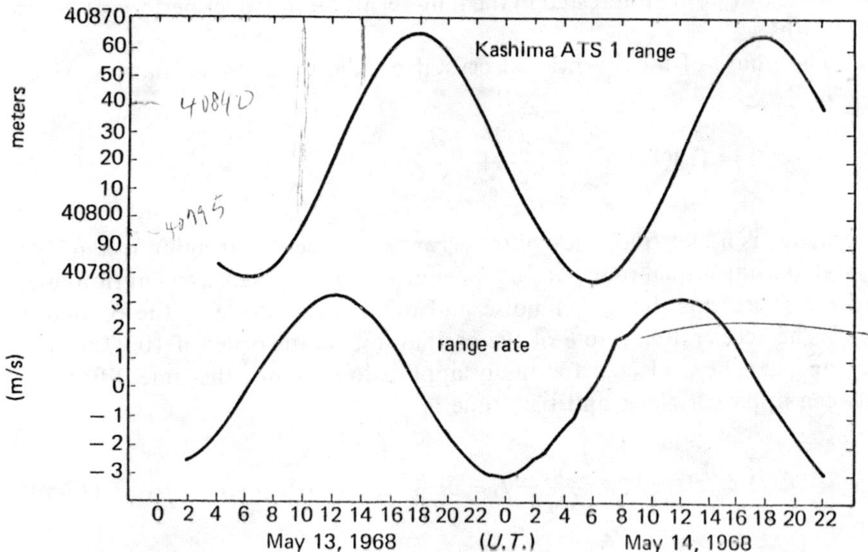

Fig. 17-14 Measured range and range rate for ATS 1 on May 13 and 14, 1968 [From Takahashi, 1970, 770–779]

---

*This equation assumes a spherical earth for simplicity. A more accurate model is obtained using a reference ellipsoid model of the earth with an equatorial radius 6378.165 km, a polar radius of 6356.785 km (an equatorial bulge of 21.4 km), and an eccentricity of 0.08181.

17-7 CLOCK-TIMING ERROR

In a worldwide timing system, it is important to know the statistics for rate of clock-time drift and the total clock-time error between the master timing source and the slave clocks. Clock phase $\phi(t)$ is the directly measurable quantity and relates to frequency $\Omega(t) \triangleq d\phi(t)/dt$. The apparent time $c_i(t)$ of the clock is expressed as the number of clock zero counts divided by the frequency, or, more precisely,

$$c_i(t) = \frac{\phi_i(t)}{\Omega_0} \tag{17-17}$$

where Ω_0 is the nominal frequency of the clock. The exact frequency $\Omega(t)$, of course, is not known at this clock.

Clock or oscillator phase and phase noise spectral density were discussed in Chap. 14 with respect to the short-term stability (a fraction of a second) and the phase noise spectral density. The same clock phase is under discussion here, but we are interested in the long-term stability over periods of hours to months.

The phase of a conventional crystal oscillator varies as [Barnes, 1966, 207–220]

$$\phi(t) = \Omega_0[(1 + \delta)t + \frac{\alpha t^2}{2} + \frac{\beta t^3}{3} + c(t)] \tag{17-18}$$

where $\delta\Omega_0$ is a bias frequency offset error, α represents the aging rate of the crystal (perhaps equal to 0.5×10^{-10} per day), β is the clock acceleration rate, and $\epsilon(t)$ represents the flicker noise and other phase noise in the oscillator itself. The acceleration rate β of the oscillator is on the order of 10^{-14} sec per sec^3 and can be neglected for many applications. Thus, the time difference between apparent clock and true time is

$$\Delta(t) \triangleq c(t) - t = \frac{\phi}{\Omega_0} - t = \delta t + \alpha t^2 + \beta t^3 + \epsilon(t) \tag{17-19}$$

If clock corrections are made at some time t_0 to correct both the bias frequency error δ and the initial time error (phase), then the residual time error at time t is

$$\Delta(t) = \frac{\alpha}{2}(t - t_0)^2 + \frac{\beta}{3}(t - t_0)^3 + \epsilon(t) - \epsilon(t_0) \tag{17-20}$$

For a crystal oscillator, and time differences $t - t_0$ on the order of several days, the first term in (17-17) often dominates, and the time error is approximately equal to

$$\Delta(t) \cong \frac{\alpha}{2}(t - t_0)^2 \tag{17-21}$$

As an example, if the aging coefficient $\alpha = 10^{-10}$/day, then (17-21) yields

$$\Delta(t) \cong \frac{10^{-10}}{\text{day}} \frac{(t - t_0)^2}{2}$$

For $t - t_0 = 10$ days, the time difference becomes

$$\Delta(t) \cong \frac{1}{2}[10 \times 24 \times 3600]\left(\frac{10^{-10}}{\text{day}} \times 10 \text{ days}\right) \text{sec} = 432\ \mu\text{sec}$$

The aging coefficient α, however, tends to be relatively predictable, and its effect can be significantly reduced.

Atomic frequency standards typically provide output frequencies directly from the quantum transitions of electrons in an atom or atomic resonances of a gas. In active atomic hydrogen masers or ammonia masers, a crystal oscillator is usually phase-locked to the maser signal. Alternatively the atomic resonances of a gas, such as cesium or rubidium vapor in other frequency standards can be used as an extremely stable, narrow-bandwidth filter in an AFC feedback loop to control the frequency of a crystal oscillator.

Table 17-1 shows some of the typical characteristics of atomic frequency standards. These standards either avoid or significantly reduce the long-term aging effects; aging effects are negligible in cesium and hydrogen units, and the aging coefficient for rubidium is often reduced to $\alpha \leq 10^{-14}$/day.

The rms frequency error σ_f of a cesium atomic frequency standard varies as a Brownian motion process for moderate values of averaging time $|\tau| < 1$ day [McCoubrey, 1966, 116–135]:*

$$\frac{\sigma_f(\tau)}{f} = \left\{\frac{E\{[\phi(t + \tau) - \phi(t)]^2\}}{\Omega_0 \tau}\right\}^{1/2} \cong \frac{1.29}{fF}\tau^{-1/2} \tag{17-22}$$

where F is the figure of merit, f is the oscillator frequency, and $\tau \triangleq t - t_0$ is the averaging time. Thus, the rms fractional frequency taken over an ensemble of measurements of averaging duration τ decreases as $\tau^{-1/2}$, and the rms error σ_t in apparent time taken over that same interval from (17-20) is

*The time error is proportional to $\int_0^\tau \epsilon(t)\,dt$ where $\epsilon(t)$ is a white frequency noise.

Table 17-1 CHARACTERISTICS OF TYPICAL ATOMIC FREQUENCY STANDARDS* [FROM MCCOUBREY, 1966, 116–135; 1967, 805–814; BLAIR, 1974]

Characteristic	*Atomic Hydrogen Maser*	*Rubidium Gas-Cell Controlled Oscillator*	*Cesium Atomic-Beam Controlled Oscillator*
Nominal resonance frequency	1420.405751 MHz	6834.682608 MHz	192.631770 MHz
Resonance width	1 Hz	200 Hz	250 Hz
Intrinsic reproducibility	$\pm 5 \times 10^{-13}$	Does not apply since sensitive to gas pressure in resonance cell and temperature of gas and cavity	$\pm 3 \times 10^{-12}$
Stability (rms deviation from the mean)			
1 sec	5×10^{-13}	5×10^{-12}	3×10^{-12}
1 min	6×10^{-14}	9×10^{-13}	8.5×10^{-13}
1 hr	1×10^{-14}	5×10^{-13}	2×10^{-13}
1 day	6×10^{-15}	5×10^{-13}	2×10^{-14}
Temperature effects	$< 1 \times 10^{-13}/°C$	$\approx 10^{-12}/°C$	$\approx 2 \times 10^{-13}/°C$
Systematic drift	None detectable $< 10^{-12}$/year	$\approx 1 \times 10^{-11}$/month	None detectable $< 3 \times 10^{-12}$/year

$$\sigma_t = \{E[C(t) - t]^2\}^{1/2} = \left\{\frac{E\{[\phi(t+\tau) - \phi(t)]^2\}}{\Omega_0}\right\}^{1/2} \cong \frac{1.29}{fF}\tau^{1/2}$$

$$\cong \frac{1.29}{F}\tau^{1/2}10^{-10} \qquad (17\text{-}23)$$

for $f = 10^{+10}$ Hz, and σ_t increases in proportion to $\tau^{1/2}$. Thus, for a cesium standard with $F = 3$, the rms time error increases as $(0.43)\tau^{1/2}\,10^{-10}$. For $\tau = 10^6$ sec and $f = 10^{+10}$, the rms time error is

$$\sigma_t(\tau) = 0.43\tau^{1/2}10^{-10} = 0.43 \times 10^{+3} \times 10^{-10} = 0.43 \times 10^{-7}$$
$$= 43 \text{ nsec}$$

As $t \longrightarrow \infty$, the stability σ_f/f does not continue indefinitely to 0 because of flicker noise ($1/f$ noise) and systematic errors [Lacy, 1966, 170–178]. Thus, sooner or later σ_f is affected by long-term drifts, flicker noise ($1/f$ noise) and fluctuations caused by aging, and small changes in ambient conditions, such as temperature magnetic field; σ_f/f then approaches some constant, and $\sigma_t \sim \tau$ for sufficiently large τ [Barnes, 1967, 822–826].

A more useful measure of the standard deviation for sampled frequency measurements has been described by Allan [1966, 221–230] and was described in Chap. 12 where we emphasized the phase noise spectral density relations. Here we use a somewhat more general definition of the Allan variance and concentrate on long-term timing errors. If T is the period of sampling, τ is the sampling time interval ($\tau \leq T$), and N is the number of samples taken in the averaging, then the Allan variance of frequency is the infinite time average

$$\langle \sigma^2(N, T, \tau) \rangle = \left\langle \frac{1}{N-1} \left\{ \sum_{n=0}^{N-1} \left[\frac{\phi(nT+\tau) - \phi(nT)}{\tau} \right] \right. \right. \\ \left. \left. - \frac{1}{N} \left[\sum_{n=0}^{N-1} \frac{\phi(nT+\tau) - \phi(nT)}{\tau} \right] \right\}^2 \right\rangle \tag{17-24}$$

where the averaging intervals are defined in Fig. 17-15. If there is no dead time between sample intervals, then $T = \tau$ and this expression reduces to

$$\langle \sigma^2(N, \tau) \rangle = \left\langle \frac{1}{N-1} \left\{ \sum_{n=0}^{N-1} \left(\frac{\Delta\phi_n}{\tau} \right) - \frac{1}{N} \left[\frac{\phi(NT) - \phi(0)}{\tau} \right] \right\}^2 \right\rangle \tag{17-25}$$

where $\phi[(n+1)T] \equiv \phi(nT+\tau)$ and we define $\Delta\phi_n \triangleq \phi[(n+1)\tau] - \phi(n\tau)$.

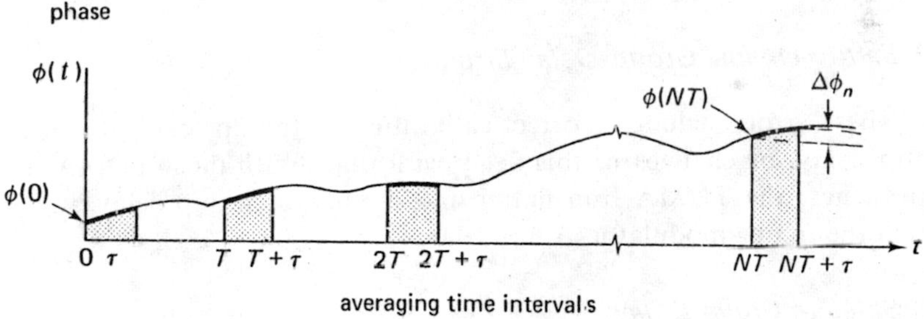

Fig. 17-15 Averaging time interval used in computing the Allan variance of frequency

The second term in the parentheses subtracts the average frequency over the NT-sec interval. The quantity

$$\tau\sigma(N, \tau)$$

is then the sample rms time error expected over a τ-sec interval.

17-8 POSSIBLE SOURCES OF TIMING ERROR

The accuracy of the timing measurements discussed in the previous section is of critical importance since it affects such performance measures as the guard time requirements and efficiency of TDMA systems, the buffer sizes and efficiency in digital communication nodes, and the accuracy of navigation and ranging measurements. Most of the contributors to timing error can be grouped into three categories:

1. Thermal noise effects affecting the timing error measurement device
2. Propagation and equipment group delay effects
3. Physical motions and dynamics of the transmitting and receiving equipment

Thermal Noise Effects

Thermal receiver and system noise enters the receivers and signal tracking devices and causes a delay error variance similar to that described for phase noise in a phase-locked loop. The noise analysis of the delay-locked tracking equipment is discussed in Chap. 18.

The effects of tracking error caused by noise or cochannel interference are also discussed in Chap. 18. Other error sources are discussed in this section.

Earth Terminal Group-Delay Errors

These errors include incorrect calibration of the up- or downconverter group-delay, or changes in this delay with time. With most TDMA timing approaches, the TDMA transmitter and modulator group delay must also match the timing modulator group delay.

Satellite Group Delay

The group-delay effects of the satellite transponder and antenna system can also change from their calibration value because of either aging or thermal effects. The group delay may not be the same from one portion of the frequency channel to another. Hence, it can be important to have all timing signals pass through the same portion of the satellite transponder passband. If the satellite itself generates the signal from its internal clock, as in the NAVSTAR satellite, then the group delay from the satellite clock to the satellite amplifier output should be held constant.

Satellite Motion

There is always some motion in the satellite relative to the earth station (see Chap. 6) that is significant where timing accuracy desired is of the order of 1 nsec. However, the doppler motion caused by satellite drifts, orbital inclination, and nonzero eccentricity is generally accurately tracked, as discussed in Chap. 18. Doppler rate effects are very small usually.

Ionospheric Delays

The ionosphere produces an apparent path range increment (group velocity effect*) of approximately

$$\Delta R \cong kQ(E)N_T(1/f^2) \text{ m} \tag{17-26}$$

where $k = 1.2 \times 10^{-6}$, $Q(E) \triangleq \sec \sqrt{(E^2) + (18°)^2}$ is the approximate effect of elevation angle, E is the elevation angle, f is the frequency in GHz, and N_T is the integrated electron density over the propagation path in electrons/m^2 [Lawrence et al., 1964, 4–27; Arendt, 1965, 268–277].

The integrated electron density is on the order of 10^{16} to 10^{18} electrons/m^2; ΔR at 1.6 GHz is $\approx$20 m. A simplified expression for electron content of the ionosphere is

$$N = N_m[2h/h_m - (h/h_m)^2] \qquad \text{electrons/m}^2 \tag{17-27}$$

where N has a peak electron density at approximately 150-km altitude and a peak electron density of 10^{12} electrons/m^2, and extends as high as 250 km or above [Poularikas and Golden, 1969, 865–867]. These errors are of particular significance if the difference in up- and downlink frequency group delays affects the timing measurement. In some timing links, this ionospheric delay is tracked and largely removed as an error source, by dual frequency measurements at f_1, f_2.
The difference in path delay at frequencies $f_1 f_2$ is

$$\Delta\tau = \frac{K}{f_1^2} - \frac{K}{f_2^2} = K\left(\frac{f_2^2 - f_1^2}{f_1^2 f_2^2}\right) \tag{17-28}$$

in the path delay measurement allows one to compute K, which is proportional to integrated electron density and, hence, to compute the total propagation delay.

Atmospheric Propagation Errors

Atmospheric effects caused by water vapor and dry air can also cause ray bending, particularly at low elevation angles. The variation in the index of refraction of the atmosphere causes a group-delay change and also causes the satellite-ground path length to differ from a straight line. Hence, the apparent increase in path length causes both the range and range rate to be in error.

The atmosphere can be modeled by a biexponential model that accounts

---

*The group velocity v_g and phase velocity v_p are related by [Johnson, 1954]

$$v_g = c\, df/d(fn) < c < v_p$$

where $n \triangleq c/v_p$ is the index of refraction.

for both dry air and water vapor effects separately. The index of refraction is $n \triangleq c/v_p$, the ratio of free space velocity to the phase velocity of the medium. For the biexponential model, the index of refraction is expressed by

$$n = 1 + N_d e^{-ah} + N_w e^{-bh}) \times 10^{-6} \tag{17-29}$$

where h is the altitude in kilometers, N_d is the dry air coefficient, and N_w is the wet air coefficient at sea level. The dry air coefficient varies from 260 to 290 with the seasons of the year, and a varies from 0.103 to 0.110 km^{-1}. The wet air coefficient varies even more widely with the seasons, ranging over 20 to 120, and b typically varies from 0.33 to 0.45 km^{-1} for microwave signals. The tropospheric range error is due mainly to a change in the propagation velocity, causing a delay error $\Delta\tau = 1/c\,[\int n\, dR - D]$ where D is the true range. The error caused by ray bending is usually substantially less than that amount.

Figures 17-16 and 17-17 show typical errors in apparent range and elevation error as a function of apparent elevation angle. Curves for both the

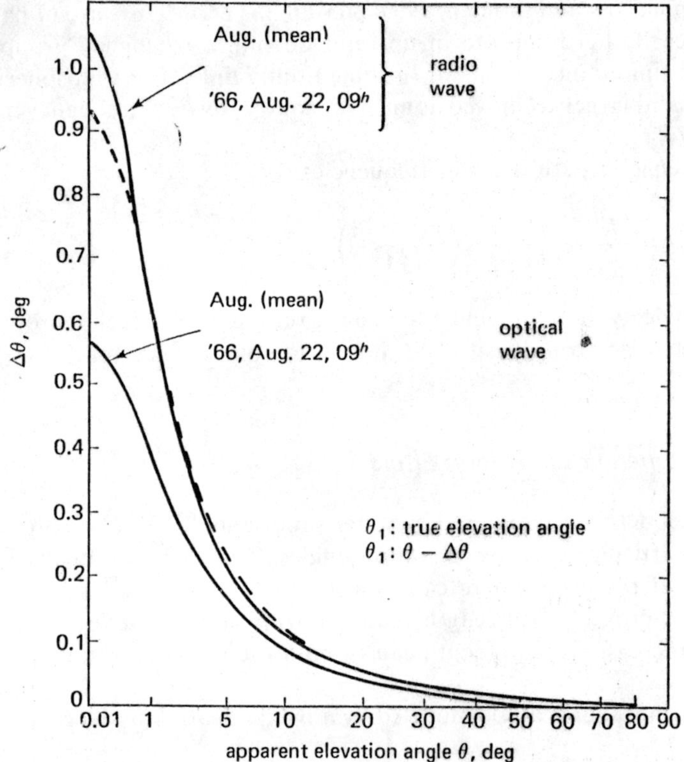

Fig. 17-16 Tropospheric elevation angle error $\Delta\theta$ versus apparent elevation angle θ [From Takahashi, 1970, 770–779]

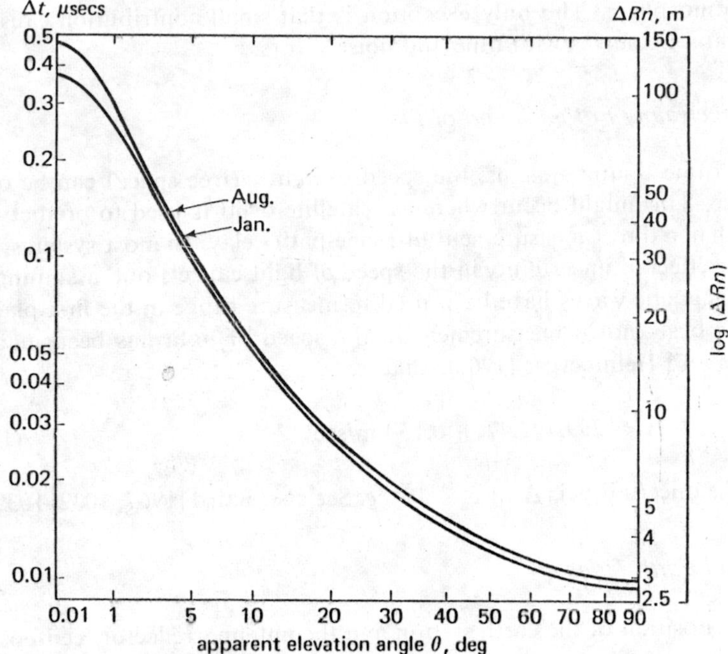

Fig. 17-17 Errors in apparent and true range due to the troposphere [After Takahashi 1970, 770–779]

mean for August and a particular time in August are shown. As can be seen, the range error can be as large as 0.5 μsec at low elevation angles at *C*-band, and the apparent elevation angle can be off as much as 1°. Fortunately, however, most of these errors either do not have much effect on TDMA timing systems because they are largely frequency independent,* and can be predicted. Above 20° elevation angles and 10 GHz the path delay deviation can be predicted to an accuracy of approximately 12 cm [Bean and Thayer, 1963, 273–285].

Rainfall has a relatively small effect on transit time fluctuation. Measurements by Gray [1970, 1059–1068] indicate that their effects are more than an order of magnitude below the fluctuations caused by the gaseous atmosphere. The fluctuations in a line-of-sight link through the atmosphere are usually less than 0.5 nsec/km.

Some of these atmospheric error sources for TDMA timing can be almost entirely removed by a timing technique, as in Fig. 17-1 where the measurement is made on two signals which have both passed through the

*The refraction coefficient does vary with frequency, however, for frequencies close to the oxygen or water vapor absorption lines, where the ratio of group velocity to phase velocity $v_g/v_p = 1 + (\lambda/n)(dn/d\lambda)$ varies with frequency.

same atmosphere. The only exception is that small contribution caused by temperature changes with time and noise errors.

Uncertainty in the Speed of Light

In some timing systems, the speed of light c (free space) can be of significance. This might occur where the satellite orbit is used to predict range rather than a direct measurement of range-path delay. In most systems, however, the effect of uncertainty in the speed of light cancels out, assuming that electromagnetic waves have been used to measure range in the first place.

The best-known measurement of the speed of light has been made by Karolus and Helmberger [1967]; that is,

$$c \pm \Delta = 299{,}792.47 \pm 0.15 \text{ km/sec} \tag{17-30}$$

Thus, the uncertainty is $\Delta = 5 \times 10^{-7}c$. See also Baird [1967, 1032–1039].

Solid Earth Tides

The position of the earth station and the antenna-reflector feed position must be accurately known for extremely precise measurements of satellite orbit and time. A second-order effect is the movement of the earth station itself with time. Among the contributors to this motion are the lunar and solar tides, which have an amplitude variation of 0.20 m at low latitudes well away from the coast [Bender, 1967, 1039–1045].

CHAPTER 18

DELAY-LOCK TRACKING OF PSEUDONOISE SIGNALS

18-1 INTRODUCTION

Each of the timing systems of Chap. 17 employs a closed-loop timing/ranging unit in order to measure the relative delay of the signal $s(t + \tau)$. When the signal $s(t)$ is a pseudonoise (PN) modulated signal, it is often convenient to employ a delay-lock tracking system to perform the measurement of τ in the presence of additive noise.

The delay-lock loop (DLL) has been described previously [Spilker and Magill, 1961, 1403–1416] as an optimal device for tracking the delay difference between two correlated waveforms. The DLL is a nonlinear feedback system, which employs a form of cross-correlation in the feedback loop. This device tracks the delay of a broadband signal much in the same manner that a phase-lock loop tracks the phase of a sinusoidal signal. In this chapter, several modified versions of the delay-lock discriminator are described which are particularly designed to track binary signals generated from feedback shift-registers [Spilker, 1963]. One of the main reasons for the interest in this type of signal is that it is easy to regenerate the signal in the discriminator for use in the cross-correlation operations with any desired amount of delay. Furthermore, certain classes of these sequences possess desirable autocorrelation properties; for example, the maximal-length linear shift-register (PN) sequences introduced in Chap. 17 have "two-level" autocorrelation functions and can be designed to have periods which are long enough so that ambiguities are of no concern.

In an actual tracking operation the binary sequence is applied to an RF carrier and either transmitted from the satellite directly or transmitted to the satellite and returned to the ground tracking equipment via a transponder. The RF signal can then be demodulated coherently. The demodulated binary sequence plus additive noise is then fed into the delay-lock loop in order to obtain the estimate of the delay between the original and received sequences. The first portion of this chapter is concerned with the delay-estimation operation after demodulation of the RF carrier.

We next discuss envelope correlation versions of the delay-lock loops which operate without coherent demodulation of the carrier. Envelope correlation delay-lock loops are particularly useful when the PN signal has been biphase modulated with data. One can then track the PN signal noncoherently without first removing the data modulation.

In this chapter, we discuss the principles of delay-lock tracking and its variations. Linearized models are developed for the loop, and the threshold noise performance is discussed. The acquisition (lock-on) performance of both coherent and envelope correlation delay-lock loops is analyzed. The acquisition and multiple access properties of these tracking systems depend heavily on the properties of the codes used. For this reason we also discuss the autocorrelation, partial correlation, and doppler offset effects of PN and several closely related codes.

The objectives of this chapter are to present the fundamental principles of PN tracking systems. In particular, we discuss the following key aspects of the delay-lock system:

- Baseband delay-lock discriminator
- Baseband delay-lock discriminator with amplitude limiting or signal quantization
- Bandpass delay-lock discriminator employing envelope correlation
- Cross-correlation reference waveforms
- Transient performance of the delay-lock discriminator
- Code search and acquisition
- Partial-correlation properties of PN codes. Doppler offset effects. Filter distortion effects on cross-correlation.
- Multiple-access codes possessing good cross-correlation properties, namely, Gold codes

18-2 DELAY-LOCK RANGING TECHNIQUES

In this section, we describe the basic principles of operation of a coherent or baseband version of the delay-lock discriminator for PN signals. The noise and transient performances of the delay-lock loop are described next. Using the results of Chap. 17, it is easy to demonstrate that if one can measure the range or time-delay difference between signals received at the satellite from

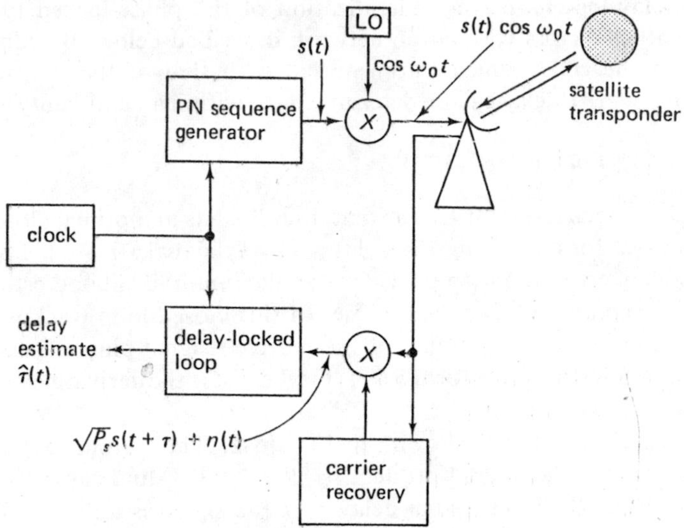

Fig. 18-1 Block diagram of the ranging system: the round-trip path delay is τ sec.

other transmitters and that from one's own transmitter, a common time base can be established relative to an actual or ground generated satellite clock.

A block diagram of a timing ranging system employing a delay-lock loop to make this measurement is shown in Fig. 18-1. The binary PN sequence is applied to an RF carrier and transmitted to the satellite, where it is transponded and returned to the earth station. The RF carrier is then coherently demodulated, and the demodulated binary sequence plus additive noise is fed into a delay-lock discriminator, which measures the delay difference between the original and received sequences. Noncoherent delay-lock loops (Sec. 18-7) can be used to avoid the requirement for coherent carrier recovery.

Pseudonoise codes are employed because they are easily generated using feedback shift registers and can be precisely clocked at high speeds ($\gtrsim$100 Mbps). Furthermore, certain classes of codes [Elspas, 1959, 45–60] possess desirable autocorrelation properties with arbitrarily long periods.

Define a PN code $s(t) = \pm 1$ which is clocked at a rate $f_s = 1/\Delta$. We take $s(t)$ to be a linear-feedback-shift-register (LFSR) sequence of period $M\Delta = (2^n - 1)\Delta$. Properties of LFSR sequences include: (1) two-level autocorrelation functions (1 or $-1/M$); (2) all 2^n patterns of n bits are observed in an n-bit window except the all-zero pattern; and (3) cycle-and-add property, $S(t) \oplus S(t + k\Delta) = S(t + j\Delta)$, where k, j are integers, and we define $S = 0$ if $s = +1$, and $S = 1$ if $s = -1$. Thus $S(t) \oplus S(t + \tau)$ is equivalent to $s(t)\, s(t + \tau)$.

The delay-lock loop is a generalization of the phase-locked loop and operates with the cross-correlation network described below. By comparing the state of the code sequence transmitted with that of the receiver shift register, the relative delay can be determined (to within an ambiguity $\pm nM\Delta$)

The Delay-Lock Loop

The delay-lock loop (DLL) was first described as an optimal closed-loop tracking device for estimating the delay τ of any signal $s(t + \tau)$. For small values of delay error, the loop provides the maximum likelihood estimate of delay in the presence of Gaussian noise. In this closed-loop tracking cross-correlator, one forms the product of the received signal plus noise $s(t + \tau) + n(t)$ with a reference waveform $s'(t + \hat{\tau})$ where s' is the derivative of s and $\hat{\tau}$ is the delay estimate.

The delay-lock loop is shown* in Fig. 18-2. The output, $x(t)$, of the multiplier contains the filtered product $s(t + \tau)\dot{s}(t + \tau)$ and causes the delay estimate to track the input signal delay τ. If the signal is a pure sine wave, namely, $s'(t) = \sin(\omega_0 t + \tau)$, then $s'(t + \hat{\tau}) = \omega_0 \cos(\omega_0 t + \hat{\tau})$, and the delay-lock loop for this special case reduces to the phase-locked loop. If the signal, on the other hand, is a broadband waveform, the differentiated waveform is also broadband. For example, if $s(t)$ is a PN sequence, then $s(t)$ is a sequence of impulses of the same sign as the transitions in the PN sequence.

The quasi-linear operation of the delay-lock loop (DLL) can be analyzed by examining the multiplier output, $x(t)$. The delay error may be defined as $\epsilon(t) \triangleq \tau(t) - \hat{\tau}(t)$. We can then write the Taylor's series expansion for the delayed received signal

$$s(t + \tau) = s(t + \hat{\tau}) + \epsilon s'(t + \hat{\tau}) + \frac{\epsilon^2}{2} s''(t + \hat{\tau}) + \cdots$$

where the primes refer to differentiation with respect to the argument, and all derivatives of $s(t)$ are assumed to exist.† Initially, the delay error $\epsilon(t)$ is assumed small so that the Taylor's series expansion of $s(t + \tau)$ about $s(t + \hat{\tau})$ converges rapidly. The multiplier output then has the series expansion

$$\frac{x(t)}{k} = A\Big[s(t + \hat{\tau})s'(t + \hat{\tau}) + \epsilon(t)[s'(t + \hat{\tau})]^2 + \frac{\epsilon^2(t)}{2!} s''(t + \hat{\tau})s'(t + \hat{\tau}) + \cdots\Big] + n(t)s'(t + \hat{\tau}) \qquad (18\text{-}1)$$

*An alternative is to use a matched filter for a pulsed PN code and then lock a phase-lock loop or delay-lock loop to the envelope-detected pulse outputs of the matched filter. Surface wave delay lines can be used as the bandpass matched filter.

†RC filtered white noise, for example, is nondifferentiable. It seems, however, that for most physical systems parasitic effects cause the signal functions to be differentiable. (See S. O. Rice [1954].)

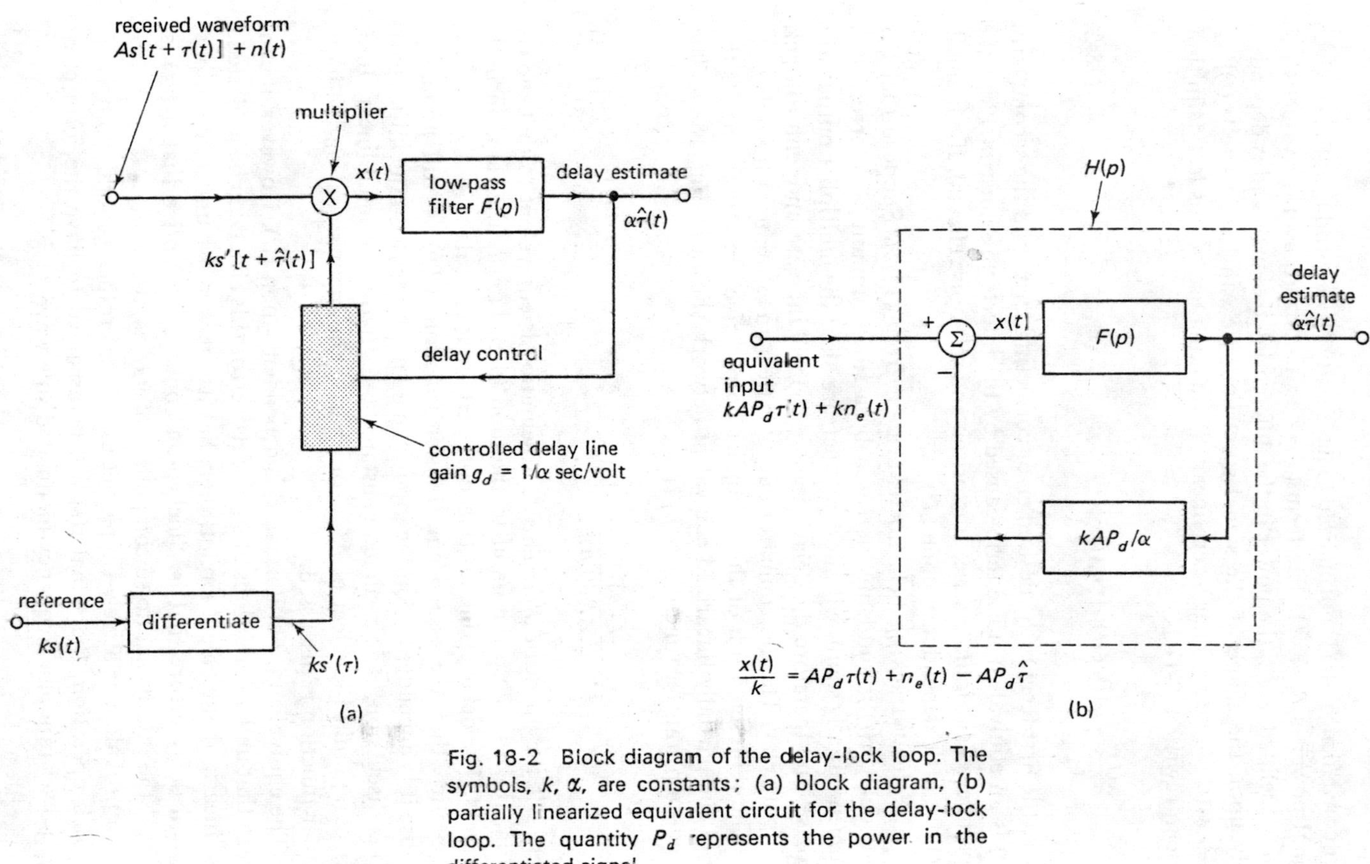

Fig. 18-2 Block diagram of the delay-lock loop. The symbols, k, α, are constants; (a) block diagram, (b) partially linearized equivalent circuit for the delay-lock loop. The quantity P_d represents the power in the differentiated signal.

For convenience, $s(t)$ is normalized to have unity power, and thus the received signal power is $P_s = A^2$. The term $(s')^2$ has a nonzero average value which will be defined as P_d, the power in the differentiated signal, and is dependent only upon the shape of the signal spectrum. We can then write $[s'(t)]^2 \triangleq P_d + s_2(t)$ where $s_2(t)$ has a zero mean. By making use of this last definition, we can rewrite (18-1) as

$$\frac{x(t)}{k} = AP_d\,\epsilon(t) + n_e(t) \tag{18-2}$$

where the first term is the desired error correcting term, and the second term $n_e(t)$ is an equivalent noise term caused by the interfering noise $n(t)$ and the remainder of the infinite series (distoriton and self-noise effects). If ϵ is small, $n_e(t)$ has little dependence upon $\epsilon(t)$.

The delay tracking behavior is evident from (18-2). Suppose that the input delay $\tau(t)$ is suddenly increased by a small amount. The error, $\epsilon(t)$, assumed initially small, will suddenly increase also; the multiplier output will increase, and therefore the delay estimate $\hat{\tau}(t)$ will increase and tend to track the input delay. The discriminator output is then an estimate of the delay.

The representation of the multiplier output given in (18-2) permits the use of the partially linearized equivalent network shown in Fig. 18-2*b*. The closed-loop transfer function $H(p)$ is

$$H(p) = \frac{F(p)}{1 + kAP_dF(p)/\alpha} \tag{18-3}$$

where p is the operator d/dt. This representation is equivalent to that shown in Fig. 18-2*a* because the input to the loop filter $F(p)$ is the same in both instances. The delay estimates thus obtained are identical.

Notice that the equivalent transfer function $H(p)$ is still nonlinear because it is dependent upon the input signal amplitude A. In the initial part of this discussion, A is assumed constant, and $H(p)$ is assumed linear. In a later paragraph, the effect of AGC or limiting the input signal on the loop transfer function is discussed.

The equivalent input noise $n_e(t)$ is dependent upon $\hat{\tau}(t)$. However, it can be seen that under conditions of small delay error this effect can be neglected. This linearized equivalent circuit, then, has its greatest use under conditions of small delay error: that is, "locked-on" operation. Notice that if $n(t)$ is "white," the interfering noise component of $n_e(t)$ is also white.

To provide a relatively simple yet useful and rather general analysis of the DLL operation, the signal $s(t)$ will be assumed to have the form of a unity power random, frequency-modulated sine wave

$$s(t) = \sqrt{2}\,\sin\,[\omega_0 t + \phi(t)] \triangleq \sqrt{2}\,\sin\left[\omega_0 t + \int_0^t \omega_i(t')dt'\right] \tag{18-4}$$

The spectrum of $s(t)$ will be taken to be rectangular with bandwidth B_s and center frequency f_0. (It is assumed that $B_s < 2f_0$ and that $\omega_i(t)$ has a zero average value.) Then we can write the expressions:

$$s'(t) = \sqrt{2}\,\omega_s(t)\cos[\omega_0 t + \phi(t)]$$

which has power

$$P_d = (2\pi)^2\left[f_0^2 + \frac{1}{3}\left(\frac{B_s}{2}\right)^2\right] \tag{18-5}$$

where we have defined $\omega_s(t) \triangleq \omega_0 + \omega_i(t)$.

Define the quantities $a_n = \mathrm{E}[s'(t)s^{(n)}(t)]$, where $s^{(n)}(t)$ is the nth derivative of $s(t)$. Note that $a_n = 0$ if n is even, and $a_1 = \mathrm{E}\{[s'(t)]^2\}$ is the power in the differentiated signed, $a_1 = P_d$.

In general, the input to the linearized equivalent circuit can be written as the sum of an equivalent input signal and three types of interference noise terms,

$$\begin{aligned} \text{signal term} &= kAP_d\epsilon(t) \\ \text{noise term} &= kn_e(t) = k[n_d(t) + n_s(t) + n_n(t)] \end{aligned} \tag{18-6}$$

where $n_d(t)$ represents a nonlinear distortion term (it is small for small ϵ), $n_s(t)$ is a self-noise term that is dependent upon the carrier characteristics, and $n_n(t)$ is an external interference term that is dependent upon external noise at the DLL input. By making use of (18-4), (18-5), these noise terms can be evaluated as

$$\begin{aligned} n_d(t) &= A\left[a_3\frac{\epsilon^3(t)}{3!} + a_5\frac{\epsilon^5(t)}{5!} + \cdots\right] \\ n_s(t) &= A\Big\{\epsilon(t)[s'(t+\hat{\tau})^2 - a_1] + \frac{\epsilon^2(t)}{2!}s'(t+\hat{\tau})s''(t+\hat{\tau}) \\ &\quad + \frac{\epsilon^3(t)}{3!}[s'(t+\hat{\tau})s'''(t+\hat{\tau}) - a_3] + \cdots\Big\} \\ n_n(t) &= n(t)s'(t+\hat{\tau}) \\ &= \sqrt{2}\,n(t)\omega_s(t+\hat{\tau})\cos[\omega_0 t + \phi(t+\hat{\tau})] \end{aligned} \tag{18-7}$$

The distortion terms are taken as those terms of the form ϵ^n for $n \neq 1$.

The terms in the multiplier output with spectra centered about $\omega = 2\omega_0$ are neglected because they are assumed to be above the passband of the loop filter. This is not possible for low-pass input spectra with $\omega_0 = 0$, of course.

Thus far, it has been indicated that the DLL tends to track the delay variations of an incoming signal provided that the delay error magnitude,

$|\epsilon| = |\tau - \hat{\tau}|$ is small. In the next paragraph we determine how small this error must be and what occurs as the error becomes larger.

Assume that $s(t)$ is a stationary (wide sense), ergodic, random variable with zero mean, and that the delays $\tau(t)$ and $\hat{\tau}(t)$ are constant or slowly varying with time. Under these conditions, the loop filter, when properly optimized, forms the average of the multiplier output to obtain

$$\begin{aligned} \mathrm{E}[x(t)] &= \mathrm{E}\{[As(t+\tau) + n(t)]ks'(t+\hat{\tau})\} \\ &= -kAR'_s(\tau - \hat{\tau}) = -kAR'_s(\epsilon) \end{aligned}$$

where $n(t)$ and $s(t)$ are assumed independent, and $R'_s(\epsilon) = (d/d\epsilon)[R_s(\epsilon)]$, the derivative of the autocorrelation function of $s(t)$. Thus the important component in the multiplier output is not always linearly dependent upon the delay error, but, more generally, is functionally dependent upon the error through the differentiated autocorrelation function, and thereby causes changes in the effective loop gain as ϵ changes.

The multiplier output can be written using (18-6), (18-7) as

$$\frac{x(t)}{k} = -AR'_s[\epsilon(t)] + n_s(t) + n_n(t) \tag{18-8}$$

where we have used the relationship $R'_s(\epsilon) = \sum_{n=1}^{\infty} a_n \epsilon^n/n!$. A further general statement can be made with respect to the effective loop gain for small $|\epsilon|$. The correction component of the multiplier output for small $|\epsilon|$ is $kA\epsilon(t)a_1$, where a_1 is, in general, given by

$$a_1 = \int_0^\infty \omega^2 G_s(f)\, df = P_d \tag{18-9}$$

and depends only on the shape of the signal spectrum. Example signal spectra and the corresponding DLL discriminator characteristics are shown in Fig. 18-3. Thus the actual nonlinear correction signal in the delay-lock loop is the derivative of the autocorrelation function of the signal, $R'_s(\epsilon)$. The differentiating delay-lock loop for the PN sequence has a delay discriminator function

$$\begin{aligned} D(\epsilon) = R'_s(\epsilon) &= +K \text{ for } 0 < \epsilon < \Delta \\ &= -K \text{ for } -\Delta < \epsilon < 0 \\ &= \quad 0 \qquad |\epsilon| < \Delta \end{aligned} \tag{18-10}$$

and represents a "bang-bang" type of servo operation. The equivalent gain is dependent on rms error (See Chapter 12).

For binary PN sequences, the impulses generated in the $s'(t)$ operation

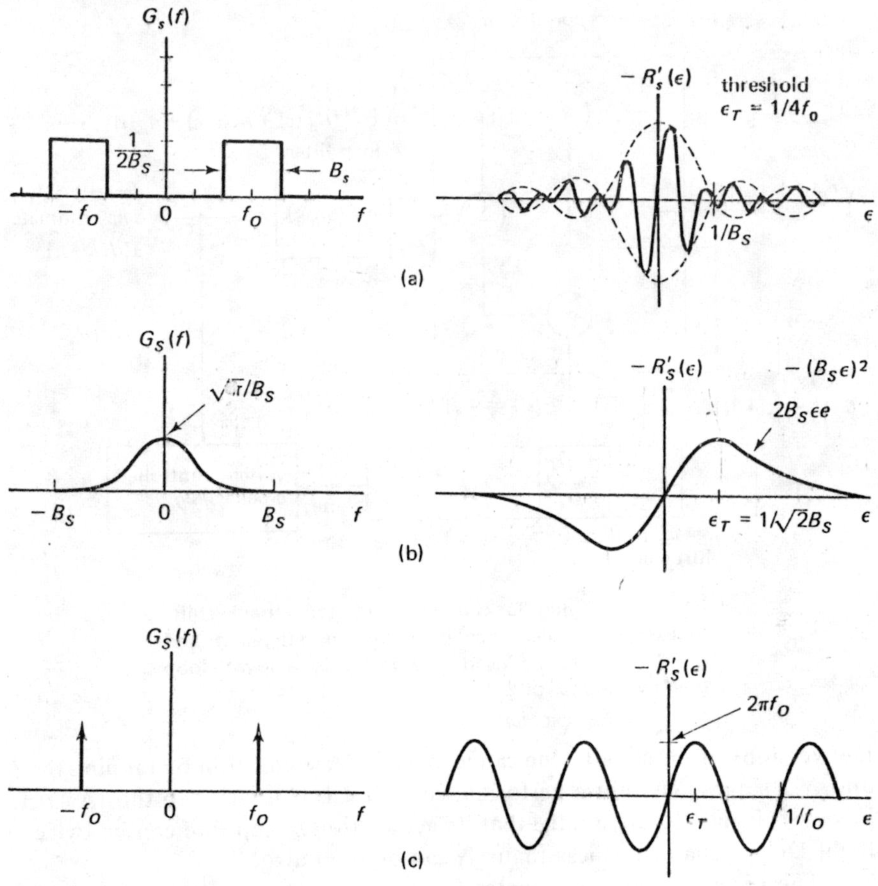

Fig. 18-3 Signal power spectral density (unity signal power) and the corresponding discriminator characteristics: (a) rectangular bandpass spectrum $\epsilon_T = 1/4f_0$, (b) Gaussian low-pass spectrum

$$G(f) = (\sqrt{\pi}/B_S) \exp - (\pi f/B_S)^S$$
$$\epsilon_T = 1/\sqrt{2}B_S$$

(c) pure sine wave signal

are not handled conveniently in practical circuits. Hence one often uses a variation of the delay-lock technique specially designed for binary signals [Spilker, 1963]. Much of the rest of this chapter deals with variations on the delay-lock tracking concept.

Figure 18-3 is a block diagram of the baseband or coherent 1Δ delay-lock loop. The reference to 1Δ refers to the separation between the early and

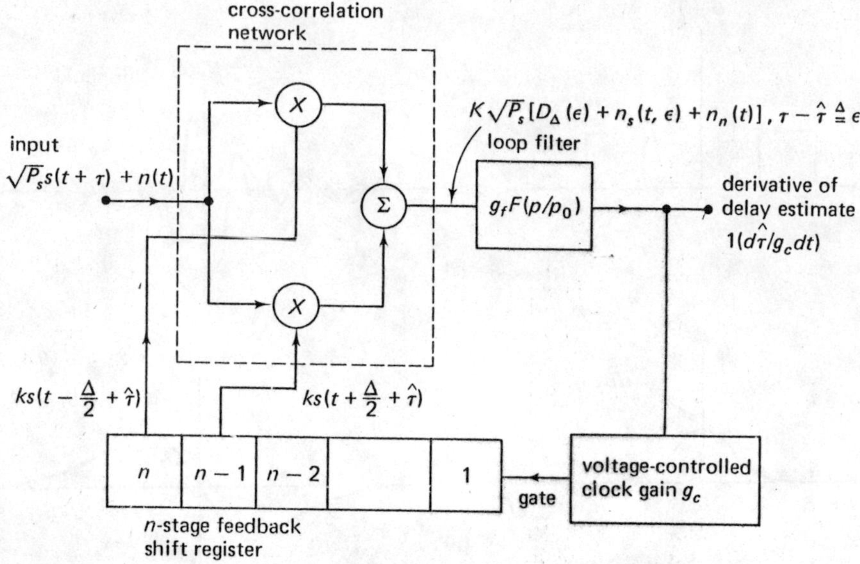

Fig. 18-4 Delay-lock loop (DLL) for binary shift-register sequences. Feedback taps at stages n and $n-2$ can also be used and result in a lower loop gain by a factor of 2.

late versions of the signal. One can also use a 2Δ separation by tapping the N and N-2 stages. The noise performance 2Δ DLL is inferior to the 1Δ DLL; however, it has the advantage that its acquisition sweep rate can be twice as high. Delay separations less than 1Δ can be used also.

The received signal originates from an n-stage maximal-length LFSR, and has a period $M\Delta = (2^n - 1)\Delta$, where Δ is the digit width. This binary signal, plus additive Gaussian white noise resulting from the coherent detection of the RF signal, is fed into the cross-correlation network, where it is compared with time-displaced versions* of the same pseudorandom binary signal as used in the transmitter. The cross-correlation network shown in Fig. 18-3 has an output

$$k\delta s(t+\hat{\tau})[\sqrt{P_s}s(t+\tau)+n(t)]$$
$$= k\left[s\left(t+\frac{\Delta}{2}+\hat{\tau}\right)-s\left(t-\frac{\Delta}{2}+\hat{\tau}\right)\right][\sqrt{P_s}s(t+\tau)+n(t)] \tag{18-11}$$

*A variation of this loop is the "τ-Jitter" loop where the correlation operations are performed in time sequence with $s(t-\Delta/2+\hat{\tau})$ and then with $s(t+\Delta/2+\hat{\tau})$. The differences are taken between the time-successive outputs of a single correlator [Hartman, 1974, 2–9].

where $\delta s(t) \triangleq s(t + \Delta/2) - s(t - \Delta/2)$, and P_s is the received signal power. The term δs is reminiscent of the expression for the differentiated signal used in the cross-correlation operation

$$\frac{ds}{dt} = \lim_{\delta \to 0} \frac{s(t + \delta) - s(t - \delta)}{2\Delta}$$

of (18-1) [Spilker and Magill 1961, 1403–1416]. In fact, decreasing the separation below Δ between the two reference signals can increase loop gain farther, but also can produce a more sensitive threshold.

As is shown below, the product in (18-3) contains a low-pass spectral component that serves to keep the DLL accurately tracking the signal delay once the system has locked-on. The purpose of the low-pass filter in Fig. 18-3 is to remove as much noise and other interference as possible. The filter is in series with the cross-correlation network and the filter provides the desired closed-loop response. The closed-loop response, in turn, is designed on the basis of the expected dynamics of signal delay. Thus the output of the loop filter is an estimate of the time derivative of delay (proportional to the transmitter's radial velocity) and is used to control the clock rate of the FSR.

The delay estimate $\hat{\tau}$ can be easily obtained by sensing the time instants when the transmitter and receiver FSRs go through a particular state—for example, the all-1s state—and then by determining the time difference between these time instants. A clock counter can be reset at each epoch and clocked with the LFSR clock so as to provide a digital clock readout at all time. In this manner the DLL clock can be subtracted digitally from the reference clock to provide a continuous digital reading of the delay separation.

18-3 SYSTEM EQUATIONS

To analyze the response of the system, we express the cross-correlation network output as

$$k\delta s(t + \hat{\tau})[\sqrt{P_s}\, s(t + \tau) + n(t)] = k\sqrt{P_s}[D_\Delta(\epsilon) + n_s(t, \epsilon) + n_n(t)/\sqrt{P_s}] \qquad (18\text{-}12)$$

where $\epsilon = \tau - \hat{\tau}$ is the delay error. The term that is not explicitly dependent on time—namely, $D_\Delta(\epsilon)$, the delay discriminator characteristic of the 1Δ DLL—is the time average of the cross-correlation network for fixed ϵ and has been separated from $n_s(t, \epsilon)$, the self-noise term. The discriminator characteristic is easily shown to be

$$D_\Delta(\epsilon) = \mathbf{E}[\delta s(t + \hat{\tau})s(t + \tau)]$$
$$= R_s\left(\epsilon - \frac{\Delta}{2}\right) - R_s\left(\epsilon + \frac{\Delta}{2}\right) \tag{18-13}$$

where $R_s(\epsilon)$ is the signal autocorrelation function, and the $\mathbf{E}$ operator represents the ensemble average for a sequence $s(t)$ with an equiprobable random time origin.

If a reference signal $s(t + \Delta + \hat{\tau}) - s(t - \Delta + \hat{\tau})$ with a 2Δ separation is used, the corresponding DLL discriminator characteristic

$$D_{2\Delta}(\epsilon) = R_s(\epsilon - \Delta) - R_s(\epsilon + \Delta) \tag{18-14}$$

has half the loop gain at $\epsilon = 0$. As shown later, the maxmum search velocity for the 2Δ DLL is double that for the 1Δ DLL. The threshold value of rms delay error, however, is reduced somewhat. This reference signal and others are discussed in Sec. 18-8. The binary signal has amplitudes $s = \pm 1$. The autocorrelation function [Golomb, 1955] and the discriminator characteristic for maximal-length sequences are plotted in Fig. 18-5. As can be seen, $D_{2\Delta}(\epsilon)$ varies linearly with ϵ for $|\epsilon| < \Delta$ and is zero for $2\Delta \leq |\epsilon| \leq (M - 2)\Delta$. Since $s(t)$ is periodic with period $M\Delta$, $D_{2\Delta}(\epsilon)$ has the same periodicity.

The self-noise term in the output of the cross-correlation multiplier network is the difference between the total output and the desired $D_{2\Delta}(\epsilon)$ and can be expressed as

$$n_s(t, \epsilon) = s(t + \Delta + \hat{\tau})s(t + \tau) - s(t - \Delta + \hat{\tau})s(t + \tau) - D_{2\Delta}(\epsilon) \tag{18-15}$$

In order to determine the effect of the self-noise on the system performance, it is necessary to determine the power spectrum of $n_s(t, \epsilon)$. This computation is performed for values of delay error $\epsilon = 0, \pm m\Delta$.

It has already been pointed out that maximal-length linear binary sequences possess the "cycle-and-add" property [Golomb, 1957; Titsworth and Welch, 1959, Sec. 3]:

$$S(t + i\Delta) \oplus S(t + j\Delta) = S(t + r\Delta) \tag{18-16}$$

In other words, if $\mathbf{S}_i$ is the state vector of the FSR at time $t + i\Delta$ and $\mathbf{T}$ is the transformation matrix of the FSR $\mathbf{S}_{i+1} = \mathbf{T}\mathbf{S}_i$, then $(\mathbf{I} + \mathbf{T}^{j-i})\mathbf{S}_i = \mathbf{T}^{r-i}\mathbf{S}_i$ where $\mathbf{I}$ is the identity matrix. The relationship between i, j, r can be found

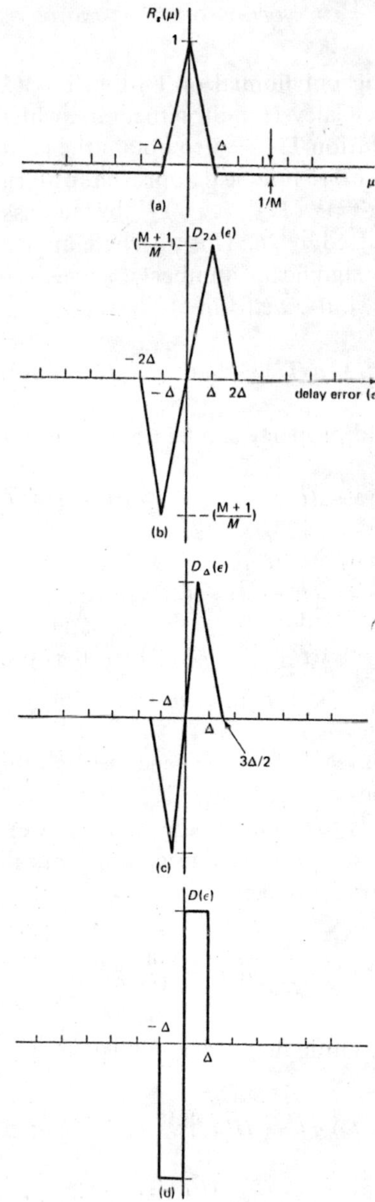

Fig. 18-5 Autocorrelation function $R_s(\mu)$ and delay-lock loop (DLL) discriminator characteristic for maximal-length binary shift-register sequences: (a) the autocorrelation function, (b) the 2Δ DLL characteristic, (c) the 1Δ DLL characteristic, (d) the DLL characteristic for a differentiated reference signal of (18-10).

using the characteristic polynomial of **T**—that is, $\phi(\lambda) = |\mathbf{T} - \lambda\mathbf{I}|$. In particular, one can use the Caley-Hamilton theorem, which shows that $\phi(\mathbf{T}) = 0$, and the periodicity relation $\mathbf{T}M = \mathbf{I}$ to obtain the relationship between these integers. An alternate more practical approach for large **M** is to compute the matrix products $\mathbf{T}^2, (\mathbf{T}^2)^2, (\mathbf{T}^2)^4, \ldots, \mathbf{T}^{2^n}$ by successive multiplication to determine the code shifted by $2n$. This alternate approach is the more useful for shift registers of a significant number of stages. Thus, in order to obtain the state vector for m shifts, we form

$$\mathbf{T}^m\mathbf{S} = [a_0\mathbf{T}^{2n} + a_1\mathbf{T}^{2n-1} + \cdots + a_{2n-1}\mathbf{T} + a_{2n}]\mathbf{S} \tag{18-17}$$

This cycle-and-add property can be used to simplify $n_s(t, \epsilon)$ in (18-15):

$$\begin{aligned} n_s(t, \epsilon = m\Delta) &= s(t + \tau + j\Delta) - s[t + \tau + (j - 1)\Delta] \\ &\qquad\qquad m = 0, M, 2M, \ldots \\ &= \pm\left[s(t + \tau + n\Delta) - \frac{1}{M}\right] \qquad m = \pm 1, M \pm 1, \ldots \\ &= s(t + \tau + r\Delta) - s(t + \tau + q\Delta) \\ &\qquad\qquad \text{for other } m \text{ where for } q \neq r, r \pm 1 \end{aligned} \tag{18-18}$$

where m, j, n, r, q are integers that are dependent on the delay error $\epsilon = m\Delta$. For no value of m does $r = q$. Hence, it is seen that for $\epsilon = \pm\Delta$, the self-noise power spectrum is the same as the signal power spectrum $G_s(f)$. The signal power spectral density has line components at frequencies that are integer multiples of $1/M\Delta$; that is,

$$G_s(f) = \frac{1}{M\Delta}\mathcal{G}(f) \sum_{\nu=-\infty}^{\infty} \delta\left(f - \frac{\nu}{M\Delta}\right) \tag{18-19}$$

where $\delta(f - f_0)$ is the Dirac delta function defined by the operator equation

$$\int H(f)\,\delta(f - f_0)\,df = H(f_0) \tag{18-20}$$

The quantity

$$\mathcal{G}(f) = \Delta\left(\frac{\sin \pi\,\Delta f}{\pi\,\Delta f}\right)^2 \triangleq \Delta \operatorname{sinc}^2(\pi\Delta f) \tag{18-21}$$

plotted in Fig. 18-6, can be considered as the "envelope" of the signal spectrum. For other integer values of delay error, the self-noise is expressed as the difference between two versions of the same sequence having different time

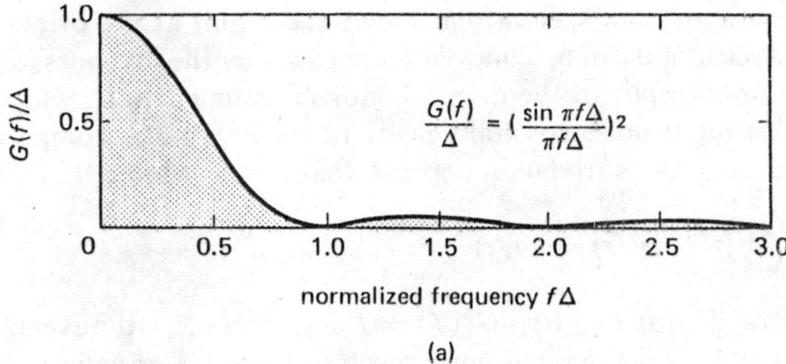

(a)

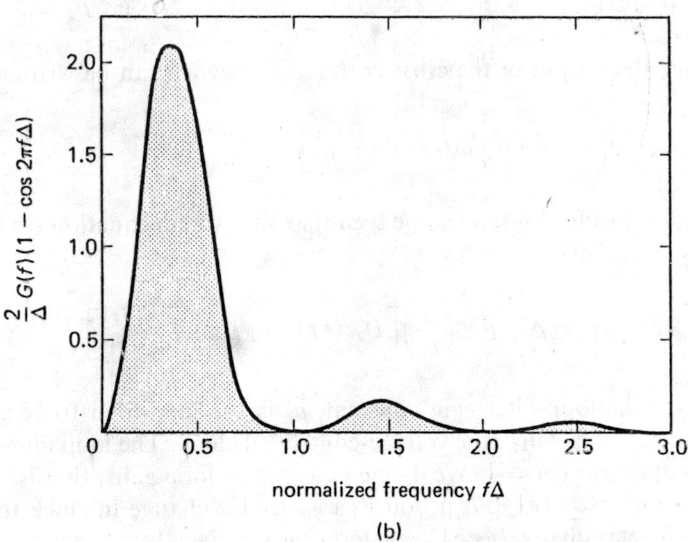

(b)

Fig. 18-6 "Envelopes" of power spectral densities: (a) binary sequence spectral density, (b) self-noise spectral density for $\epsilon = \Delta$

origins, and the self-noise spectrum is

$$\begin{aligned} G_{ns}(f, \epsilon = m\Delta) &= 2G_s(f)[1 - \cos 2\pi f\Delta] \qquad \text{if } m = 0, M, \ldots \\ &= 2G_s(f)[1 - \cos 2\pi(r - q)f\Delta] \\ &\qquad\qquad \text{if } 1 < |m| < M - 1, \ldots \end{aligned} \tag{18-22}$$

and the envelope of the self-noise density is

$$\mathcal{G}_{ns}(f, \epsilon = m\Delta) = 2\mathcal{G}(f)[1 - \cos 2\pi f\Delta]$$

The self-noise spectrum for $\epsilon = 0$ is depicted in Fig. 18-6. It is a general characteristic of such spectra that they have a null at the origin, and as a result the self-noise can be removed more easily by the low-pass loop filter.

The noise input to the discriminator is assumed to be white Gaussian noise. This input noise spectral density produces a noise component in the output of the cross-correlation network that is also white:

$$k\delta s(t + \tau)n(t) = kn_n(t) \tag{18-23}$$

The spectral density of $n_n(t)$ is $G_{n_n}(f) = P_d N_0$, where P_d is the average power of $\delta s(t + \hat{\tau})$ and N_0 is the input noise spectral density. For values of the delay estimate $\hat{\tau}$ that are fixed or vary slowly in a time interval $M\Delta$, the average power in $\delta s(t + \hat{\tau}) \triangleq s(t + \hat{\tau} + \Delta) - s(t + \hat{\tau} - \Delta)$ is $P_d = 2$. Hence, $G_{n_n}(f) = 2N_0$ w-sec.

Thus, the output of the cross-correlator network can be written

$$k\sqrt{P_s}\left[D_{2\Delta}(\epsilon) + n_s(t, \epsilon) + \frac{n_n(t)}{\sqrt{P_s}}\right] \tag{18-24}$$

By referring to Fig. 18-4, it can be seen that the system equation (in operator notation $p \triangleq d/dt$) is

$$p\hat{\tau} = kg_c g_f \Delta\sqrt{P_s}F\left(\frac{p}{p_0}\right)\left[D_{2\Delta}(\epsilon) + n_s(t, \epsilon) + \frac{n_n(t)}{\sqrt{P_s}}\right] \tag{18-25}$$

where g_f is the loop-filter gain constant, p_0 is the loop-filter frequency constant, and g_c is the gain of the voltage-controlled clock. The loop filter transfer function at dc is $F(0) = 1$. We define g_0 as the dc loop gain; that is, a steady delay error of ϵ sec, $|\epsilon| < \Delta$ produces a $g_0\epsilon/\Delta$-Hz change in clock frequency or a g_0-sec delay change/sec. The dc loop gain is given by

$$g_0 = kg_f g_c\sqrt{P_s}\,\frac{M + 1}{M} \tag{18-26}$$

The system equation (18-25) can then be written as

$$p\hat{\tau} = g_0\,\Delta F\left(\frac{p}{p_0}\right)\left(\frac{M}{M + 1}\right)\left[D_{2\Delta}(\epsilon) + n_s(t, \epsilon) + \frac{n_n(t)}{\sqrt{P_s}}\right] \tag{18-27}$$

18-4 NOISE PERFORMANCE

In this section, we determine the effects of receiver noise and self-noise on system accuracy. The DLL is assumed to be in the locked-on state—that is, $|\epsilon| < \Delta$. Thus, the system is operating in the linear region where $D_{2\Delta}(\epsilon) = [(M + 1)/M](\epsilon/\Delta)$.

Define the normalized loop gain $g \triangleq g_0/p_0$. The linearized system equation can then be obtained from (18-27) as

$$\frac{\hat{\tau}}{\Delta} = g\frac{F(p/p_0)}{(p/p_0)}\left[\frac{\epsilon}{\Delta} + \frac{n_s(t,\epsilon) + n_n(t)/\sqrt{P_s}}{(M+1)/M}\right] \tag{18-28}$$

Since $\epsilon \triangleq \tau - \hat{\tau}$, this expression can be written as

$$\frac{\hat{\tau}}{\Delta} = H\left(\frac{p}{p_0}\right)\left[\frac{\tau}{\Delta} + \frac{n_s(t,\epsilon) + n_n(t)/\sqrt{P_s}}{(M+1)/M}\right] \tag{18-29}$$

where $H(p/p_0)$ is the linearized closed-loop transfer function

$$H\left(\frac{p}{p_0}\right) = \frac{gF(p/p_0)}{p/p_0 + gF(p/p_0)} \tag{18-30}$$

The linear feedback network depicted in Fig. 18-7 can be shown to have the

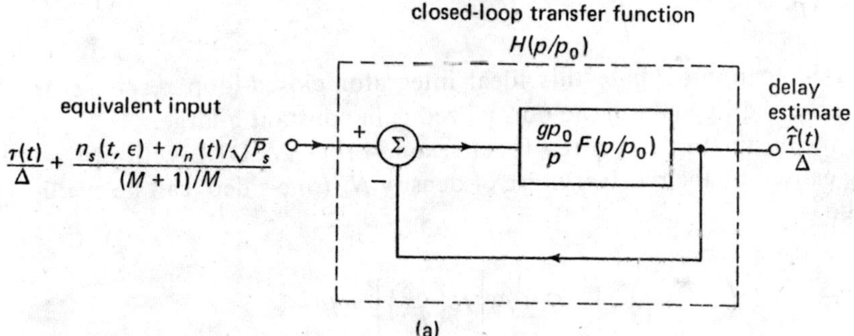

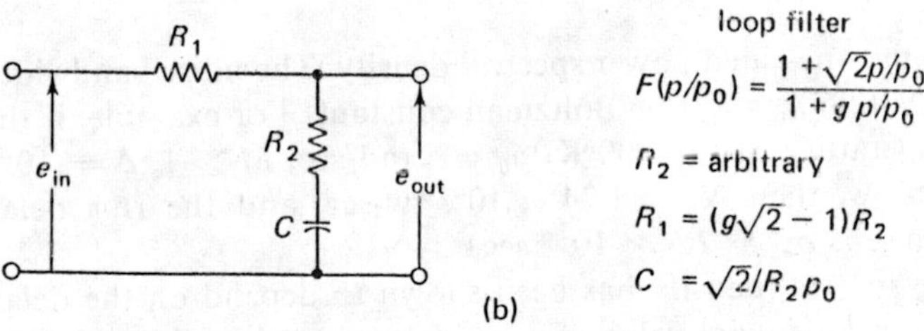

Fig. 18-7 Linearized equivalent circuit for the delay-lock loop, valid for $|\epsilon| < \Delta$: (a) block diagram, (b) loop-filter referred to in the analysis (18-32)

same system equation (18-28), and therefore it provides performance equivalent to that of Fig. 18-4 when the equivalent input

$$\frac{\tau(t)}{\Delta} + \frac{n_s(t, \epsilon) + n_n(t)/\sqrt{P_s}}{(M + 1)/M}$$

is applied and $|\epsilon| < \Delta$.

Consider that the closed-loop transfer function has the form

$$H\left(\frac{p}{p_0}\right) = \frac{1 + \sqrt{2}\,p/p_0}{1 + \sqrt{2}\,p/p_0 + (p/p_0)^2} \tag{18-31}$$

This transfer function has been shown optimum for ramp inputs of delay in the presence of white noise in that it minimizes the total squared transient error plus the mean-squared error caused by interfering noise [Jaffe and Rechtin, 1955, 66–72]. The relationship between p_0 and the transient response is discussed further in Sec. 18-9. The loop filter shown in Fig. 18-7—that is,

$$F\left(\frac{p}{p_0}\right) = \frac{1 + \sqrt{2}\,p/p_0}{1 + gp/p_0} \tag{18-32}$$

can be used to approximate this ideal integrator closed-loop response as closely as desired by making the normalized gain constant g large.

The mean-square delay error for a second-order 2Δ DLL with critical damping caused by the receiver noise of density N_0 (one-sided) can be evaluated to be

$$\sigma_{n_n}^2 = \frac{\Delta^2}{P_s}\left(\frac{M}{M+1}\right)^2 \int_0^\infty G_{n_n}(f)\left|H\left(\frac{j\omega}{p_0}\right)\right|^2 df$$

$$= 1.06\,\Delta^2\left(\frac{M}{M+1}\right)^2 \frac{N_0 p_0}{P_s} \tag{18-33}$$

where $N_0 = \kappa \mathfrak{T}$ is the one-sided power spectral density. The noise bandwidth of the DLL is $B_n = 1.06\,p_0$ and κ is Boltzman constant. For example, if the receiver noise temperature is $\mathfrak{T} = 900°\text{K}$, $p_0 = 1$ rad/sec, $M \gg 1$, $\Delta = 10^{-6}$ sec, and $P_s = 10^{-15}$ w, then $N_0 = 1.24 \times 10^{-20}$ w-sec, and the rms delay error caused by noise is $\sigma_{n_n} = 2.5 \times 10^{-9}$ sec.

The self-noise spectral density has been shown to depend on the delay error ϵ. Hence, the mean-square value of delay error caused by the self-noise also depends on ϵ. An upper bound on the mean-square delay error can be found by assuming the most pessimistic spectral density function for $|\epsilon| \leq \Delta$ the self-noise spectrum obtained for $\epsilon = \pm\Delta$. The mean-square self-noise delay error using (18-18) is then

$$\sigma_{n_s}^2 = 2\Delta^2 \int_0^\infty G_{n_s}(f) \left| H\left(\frac{j\omega}{p_0}\right) \right|^2 df$$
$$= \frac{\Delta^2}{M} \sum_{\nu=-\infty}^{\infty} \frac{1 + 2(2\pi\nu/M\,\Delta p_0)^2}{1 + (2\pi\nu/M\,\Delta p_0)^4} \left(\frac{\sin \nu\pi/M}{\nu\pi/M}\right)^2 \tag{18-34}$$

Under conditions where the number of digits in a period M is large and the loop-filter constant p_0 is small compared to the signal bandwidth ($\approx 1/2\Delta$), this expression simplifies to

$$\sigma_{n_s}^2 \cong \Delta^2 \left(\frac{p_0\,\Delta}{2}\right) \tag{18-35}$$

For the parameters just considered, $p_0 = 1$ rad/sec, $\Delta = 10^{-6}$ sec, this expression gives an rms self-noise delay error of $\sigma_{n_s} \cong 7.07 \times 10^{-10}$ sec.

When the delay-error fluctuations become too large, the discriminator encounters a threshold effect because large delay errors force the discriminator to operate, at least temporarily, in a region of negative or zero loop gain, since $D_{2\Delta}(\epsilon)$ has a negative slope for $\Delta < |\epsilon| < 2\Delta$ and zero loop gain for $|\epsilon| \geq 2\Delta$.

Experimental measurements for the 2Δ DLL indicate that the system threshold occurs when the total rms delay error is about $\sigma_n \cong 0.3\Delta$ (the 1Δ DLL has a threshold of approximately $\sigma_n \cong 0.2$). For delay errors having $\sigma_n < 0.3\Delta$, experiments showed a negligible probability of losing the locked-on state in the absense of transient errors. Threshold in a delay-lock loop (DLL) is somewhat analogous to that for a phase-locked loop (PLL). When the value of delay error exceeds the range of the discriminator characteristic—that is, $|\epsilon| > 1.5\Delta$ for a 1Δ DLL—then there is no longer a correction voltage and the DLL will no longer track the delay of the incoming signal. Loss of lock in a 1Δ DLL can be more serious than that in a PLL because there is no longer any correction voltage at all for $|\epsilon| > 1.5\Delta$ until $|\epsilon| = \pm(M - 1)\Delta$, whereas a PLL has a stable lock-on point every 360°. For the delay-error fluctuations having a Gaussian amplitude statistic and $\sigma_n = 0.3\Delta$, the probability that $|\epsilon| > \Delta$ in the linearized model is only 0.00087. This statement assumes that transient errors caused by fluctuations in τ are small compared to Δ. Assuming that the self-noise error is negligible, the system threshold 2Δ DLL can be computed and is given approximately by using (18-33) and $\sigma_{n_n} = 0.3\Delta$ to obtain

$$\frac{P_s}{p_0 N_0} = 11.1\left(\frac{M}{M+1}\right)^2 \quad \text{or} \quad \frac{P_s}{B_n N_0} = 10.5\left(\frac{M}{M+1}\right)^2 \tag{18-36}$$

It should be pointed out that certain improvements in noise performance can result if the loop filter is made adaptive. One can sense whether or not the system is locked on by means of an external correlator and reference waveform in coincidence with the received signal. When a high level of correlation is attained, the loop is considered locked. When the loop is unlocked and the discriminator is in its search mode, the loop filter can be designed to permit target acquisition with the desired search velocity. Once the target has been acquired, the closed-loop bandwidth can be decreased, and the noise performance can thereby be improved. Hysteresis (requiring several out-of-lock indications in sequence) can prevent undesired resumption of the search mode.

18-5 EFFECT OF AMPLITUDE-LIMITING OR QUANTIZING THE RECEIVED DATA

Considerable simplification in the circuitry is possible if the received signal is amplitude limited before entering the cross-correlation network. The input to the cross-correlation network then has the binary form

$$u(t) = A \operatorname{sgn}[\sqrt{P_s}\, s(t) + n(t)] \triangleq A \operatorname{sgn} x(t) \tag{18-37}$$

where A is the limiter output amplitude $s(t) = \pm 1$, and sgn x is the sign function defined as

$$\begin{aligned} \operatorname{sgn} x &= 1 \qquad x \geq 0 \\ &= -1 \qquad x < 0 \end{aligned} \tag{18-38}$$

Thus, the waveforms fed into the discriminator have $\pm A$ amplitude and, as a result, the multiplication circuits of Fig. 18-2 can be replaced by modulo-2 adders as shown in Fig. 18-8. It is apparent that this change is inconsequential for a noise-free input. We shall determine how much this limiting action degrades the noise performance when the input SNR is small.

The first step in the solution of this problem is to compute the cross-correlation between the output of the limiter $u(t + \mu)$ and the signal component $s(t)$. It is this cross-correlation that determines the useful signal component in the correlator network output through its relationship to the discriminator characteristic. Assume that band-limited white Gaussian noise enters the limiter. The noise $n(t)$ is low-pass and has a maximum frequency f_{max} and average power P_n. It can be shown that the cross-correlation $R_{us}(\mu)$ is related to the signal autocorrelation function $R_s(\mu)$ by (See Appendix C.)

$$\mathbf{E}[u(t + \mu)s(t)] \triangleq AR_{us}(\mu) = AR_s(\mu) \operatorname{erf} \sqrt{\frac{P_s}{2P_n}} \tag{18-39}$$

where $\operatorname{erf} x \triangleq \dfrac{2}{\sqrt{\pi}} \displaystyle\int_0^x e^{-z^2}\, dz$ and $R_s(0) = 1$.

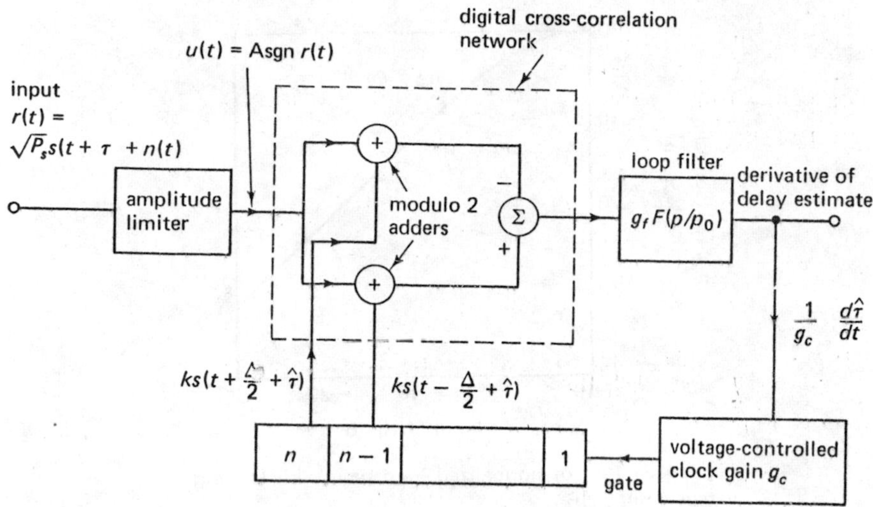

Fig. 18-8 Block diagram of the delay-lock discriminator operating on amplitude-limited inputs

is the error function. For small input SNR, $AR_{us}(\mu)$ of (18-39) becomes

$$AR_{us}(\mu) \cong \sqrt{\frac{2}{\pi}\frac{P_s}{P_n}}\,AR_s(\mu) \tag{18-40}$$

It is interesting to compare this cross-correlation function with that obtained without the limiter when the average total input power is held constant at the same level A^2. This type of operation is obtained if the discriminator is preceded by an "ideal" automatic gain control (AGC) amplifier having a constant average total output power A^2. The cross-correlation for this situation is

$$AR_s(\mu)\sqrt{\frac{P_s}{P_s + P_n}} \tag{18-41}$$

When this expression is compared with $AR_{us}(\mu)$, it is apparent that at small values of input SNR, the cross-correlation with amplitude-limited inputs is degraded by a factor of $\sqrt{2/\pi}$. Since both expressions are proportional to the square root of the input SNR for small SNR, the loop gain of the discriminator g_0 decreases in the same proportion. More generally, the loop gain with limiting g_{0L} varies with SNR as

$$g_{0L} = g_0 \operatorname{erf}\sqrt{\frac{P_s}{2P_n}} \tag{18-42}$$

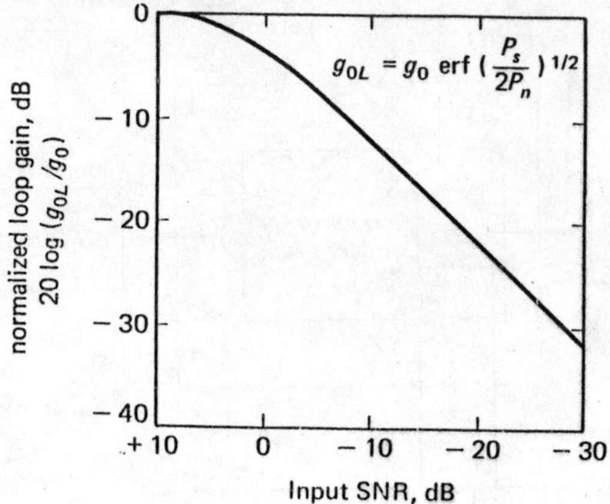

Fig. 18-9 Plot of normalized loop gain with limiting versus input SNR

where g_0 is the loop gain for noise-free inputs. This expression is plotted in Fig. 18-9.

We next determine the spectral density of the noise output of the cross-correlator network. It is known that at low SNR the autocorrelation function of the noise in the hard-limiter (lower-pass) output is [Price, 1958, 69–72]

$$R_{n0}(\mu) = \frac{2}{\pi} A^2 \sin^{-1} \frac{R_n(\mu)}{P_n} \tag{18-43}$$

Hence, the two-sided noise spectrum (primed notation) in the limiter output is

$$G'_{no}(f) = \frac{2}{\pi}\frac{A^2}{P_n} \times \left[G'_n(f) + \frac{0.167\,{}_3G'_n(f)}{P_n^2} + \frac{0.075\,{}_5G'_n(f)}{P_n^4} + \cdots \right] \tag{18-44}$$

where $G'_n(f)$ is the input noise spectrum and the notation ${}_kG'_n(f)$ represents the convolution in the frequency domain

$${}_3G'_n(f) \triangleq G'_n(f) * G'_n(f) * G'_n(f) \tag{18-45}$$

$${}_5G'_n(f) \triangleq G'_n(f) * G'_n(f) * G'_n(f) * G'_n(f) * G'_n(f) \tag{18-46}$$

Assume that the noise bandwidth is large compared to the effective correlator-filter bandwidth. Then noise effects in the cross-correlator output depend mainly upon $G'_{no}(0)$. For bandlimited white Gaussian noise of low-pass bandwidth B Hz, output noise power spectral density at $f = 0$ has the value (from 18-43)

$$G'_{no}(0) \cong 2\int_0^\infty \frac{2}{\pi} A^2 \sin^{-1}\left[\frac{R_n(\mu)}{P_n}\right] d\mu$$

$$\cong \frac{4}{\pi} A^2 \int_0^\infty \sin^{-1}\left[\frac{\sin 2\pi B\mu}{2\pi B\mu}\right] d\mu$$

$$\cong \frac{2}{\pi}\frac{A^2}{\pi B}\int_0^\infty \left[\frac{\sin x}{x} + \frac{1}{6}\left(\frac{\sin x}{x}\right)^3\right.$$

$$\left. + \frac{3}{40}\left(\frac{\sin x}{x}\right)^5 + \cdots\right] dx = \frac{A^2}{2B}\left(\frac{2.2}{\pi}\right) \quad \text{two-sided} \qquad (18\text{-}47)$$

The degradation in system performance at low SNR caused by the limiter action can now be determined by taking the ratio of the square of the signal component amplitude at the cross-correlator output to the noise spectral density $G_{n0}(0) \Delta 2G'_{n0}(0)$. This ratio is then compared with the corresponding ratio obtained with an ideal AGC amplifier having average output power A^2. For operation with the limiter, this ratio can be obtained from (18-30) and (18-37) as

$$\frac{R^2_{us}(0)}{G_{n0}(0)} = \frac{P_s}{P_n/B}\left(\frac{1}{1.1}\right) = \frac{P_s}{N_0}\frac{1}{1.1} \qquad \text{for } P_s/P_n \ll 1 \qquad (18\text{-}48)$$

where $P_n = N_0 B$

Thus, even though the hard-limiter suppresses the effective signal power by a factor of $2/\pi$ or 1.96 dB, the bandwidth spreading of the bandlimited white noise power by the limiter partially compensates for this effect because of the noise density reduction at $f = 0$. If the noise were much narrower in bandwidth, however, there would be no significant reduction in noise density after the correlator and the limiter degradation would be 1.96 dB.

Operation with the ideal AGC yields a ratio that is larger by only a factor of 1.1, or 0.4 dB. Hence, the use of an amplitude limiter is attractive since it allows the use of binary logic for the cross-correlation circuitry with only a small loss in the theoretical noise performance.

With either binary or analog inputs to the cross-correlation discriminator, if the average input power is fixed, the loop gain decreases in proportion to the square root of the SNR. Hence, the loop gain should be made large enough to satisfy the search velocity requirements at the lowest SNR likely to be encountered.

Effects of One-Bit Quantizing of the Received Signal

In a digital DLL, the received signal is first sampled and quantized before processing as shown in Fig. 18-10. In the simplest digital DLL, the quantizing is 1 bit/sample and is identical to the sampled limiter signal.

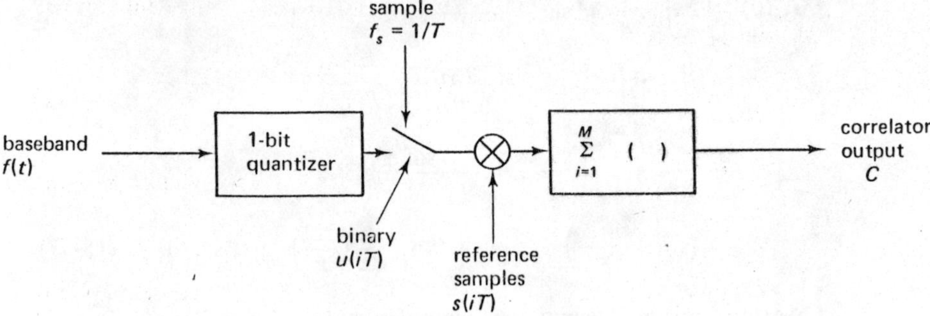

Fig. 18-10 Binary sampled-data correlator (coherent)

The received signal samples are thus equal to

$$u(iT + \mu)$$

where $u(t)$ is the limiter output as in (18-37), and the received samples are spaced T sec apart. Correlation with the reference signal samples $s(iT)$ over M samples then produces the result

$$c(\mu) \triangleq \sum_{i=1}^{M} u(iT + \mu)s(iT) \tag{18-49}$$

and the expected value of $c(\mu)$ is obtained from (18-39)

$$\bar{c}(\mu) \triangleq \mathbf{E}[c(\mu)] = \sum_{i=1}^{M} \mathbf{E}[u(iT + \mu)s(iT)] = MAR_s(\mu) \operatorname{erf} \sqrt{\frac{P_s}{2P_n}} \tag{18-50}$$

Thus the mean square signal power output for this coherent correlator is

$$M^2A^2R_s^2(\mu) \operatorname{erf}^2 \sqrt{\frac{P_s}{2P_n}} = M^2A^2 \operatorname{erf}^2 \sqrt{\frac{P_s}{2P_n}} \qquad \text{if } \mu = 0 \tag{18-51}$$

The mean square noise output at each sample is

$$\sigma^2 = A^2 - A^2R_s^2(\mu) \operatorname{erf}^2 \sqrt{\frac{P_s}{2P_n}} \tag{18-52}$$

since the mean square sample value is simply A^2. For independent noise samples the mean square noise in the correlator output is $M\sigma^2$.

The correlator output signal-to-noise power ratio for the digital correlator after M independent samples is then

$$(\text{SNR})_d = \frac{[\bar{c}(\mu)]^2}{M\sigma^2} = \frac{MR_s^2(\mu)\,\text{erf}^2\sqrt{P_s/2P_n}}{1 - \text{erf}^2\sqrt{P_s/2P_n}} \tag{18-53}$$

The corresponding output $(\text{SNR})_a$ for the analog correlator is $MR_s^2(\mu)$. Thus the performance ratio is

$$\mathcal{R} \triangleq \frac{(\text{SNR})_d}{(\text{SNR})_a} = \frac{\text{erf}^2\sqrt{P_s/2P_n}}{1 - \text{erf}^2\sqrt{P_s/2P_n}} \tag{18-54}$$

At low input, SNR $= P_s/P_n$. This result corresponds to a $2/\pi$ or 1.96 dB suppression. At high input SNR, the performance ratio increases above unity because errors in the binary decisions are rare. Clearly, however, the output noise is non-Gaussian at high P_s/P_n and the output $(\text{SNR})_d$ does not have the same meaning. Notice that the result for limiting and sampling is generally different from that with limiting alone. The noise spectrum spreading effect of the nonsampled limiter does not appear in the digital correlator.

Coherent Digital Correlation with 2-Bit Quantization

In some systems the 1.96-dB loss of the 1-bit digital correlation may be excessive, and 2 or more bits/sample may be required. This subsection derives the correlation loss for a 2-bit quantizer shown in Fig. 18-11.

The input is assumed to be a baseband PN sequence plus additive Gaussian noise of power P_n. The received signal is then

$$r(t) = \sqrt{P_s}\, s(t) + n(t) \tag{18-55}$$

where $n(t)$ is bandlimited Gaussian noise, and $s(t) = \pm 1$. The sampled received waveform is then

$$r_i = \sqrt{P_s}\, s_i + n_i \tag{18-56}$$

where n_i has rms value $\sigma = \sqrt{P_n}$, and the noise samples are assumed to be independent. The correlation over M samples from quantized samples $Q(r_i)$ is

$$c(0) = \sum_{i=1}^{M} s_i\, Q[\sqrt{P_s}\, s_i + n_i] = \sum_{i=1}^{M} Q[\sqrt{P_s} + n_i'] = \sum_{i=1}^{M} Q[y_i] \tag{18-57}$$

where $n_i' \Delta s_i n_i$ also has independent samples $y_i \triangleq \sqrt{P_s} + n_i$, and rms value $\sigma = \sqrt{P_n}$. Figure 18-11 defines the 2-bit quantizer characteristic of interest

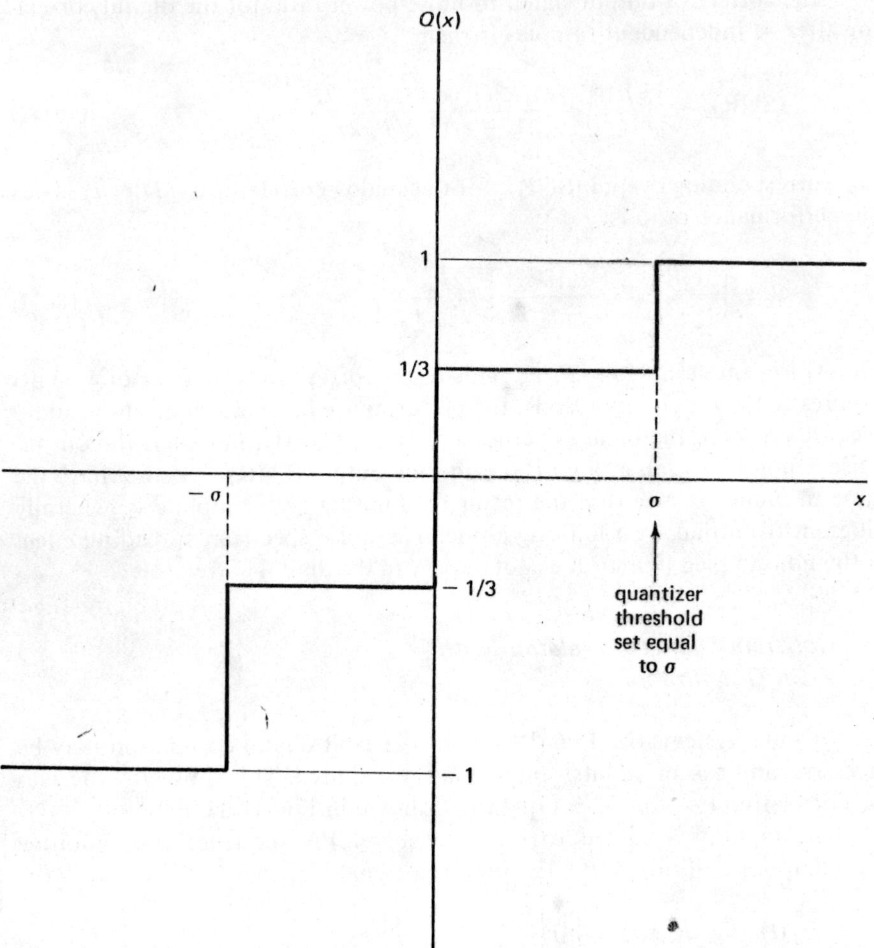

Fig. 18-11 Four-level quantizer for a digital correlator producing 2 bits per sample

here. The mean and standard deviation of the y_i are $m_y = \sqrt{P_s}$ and $\sigma_y = \sigma = \sqrt{P_n}$.

The output SNR of the quantizer is to be calculated for low input SNR $= P_s/P_n$. The first step in the calculation is to calculate the mean and variance of $Q(y)$, the quantized samples for an equivalent "all ones" input.

The mean of the quantizer output in Fig. 18-11 is

$$m_t \triangleq \mathbf{E}\,|\,Q(y_i)\,|$$

$$m_t = 1p(y_i > \sigma) - 1p(y_i < -\sigma) + (\tfrac{1}{3})p(0 < y_i < \sigma) - (\tfrac{1}{3})p(-\sigma < y_i < 0) \qquad (18\text{-}58)$$

This expression can be restated as

$$\begin{aligned} m_i &= (\tfrac{1}{3})p(y_i > 0) - (\tfrac{1}{3})p(y_i < 0) + (1 - \tfrac{1}{3})p(y_i > \sigma) \\ &\quad - (1 - \tfrac{1}{3})p(y_i < -\sigma) \\ &= (\tfrac{2}{3})[p(0 < y_i < m_y) + p(y_i > \sigma) - p(y_i < -\sigma)] \\ &= \left(\frac{1}{3}\right)\left[\operatorname{erf}\frac{\rho}{\sqrt{2}}\right) + \operatorname{erf}\left(\frac{1+\rho}{\sqrt{2}}\right) - \operatorname{erf}\left(\frac{1-\rho}{\sqrt{2}}\right)\right] \end{aligned} \tag{18-59}$$

where

$$\rho \triangleq \sqrt{\frac{P_s}{P_n}}, \quad \operatorname{erf}(x) = \frac{2}{\sqrt{\pi}} \int_0^x e^{-y^2}\, dy.$$

At low input SNR we approximate

$$\begin{aligned} &\operatorname{erf}(z) \cong \frac{2}{\sqrt{\pi}} z, \qquad z \ll 1 \\ &\operatorname{erf}(a + z) - \operatorname{erf}(a - z) \cong \frac{4}{\sqrt{\pi}} z\, e^{-a^2} \qquad z \ll 1 \end{aligned} \tag{18-60}$$

The resulting approximation for m_i from (18-59) and (18-60) is

$$\begin{aligned} m_i &\cong \left(\frac{1}{3}\right)\left(\frac{2}{\sqrt{\pi}}\right)\left[\frac{\rho}{\sqrt{2}} + \frac{2\rho}{\sqrt{2}} \exp\left(-\frac{1}{2}\right)\right] \\ &= \frac{2\rho}{3\sqrt{2\pi}}\left[1 + 2\exp\left(-\frac{1}{2}\right)\right] \end{aligned} \tag{18-61}$$

The variance of the quantizer output is

$$\begin{aligned} \sigma_i^2 &= \mathrm{E}[Q^2(y_i)] \\ &= 1p(y_i > \sigma) + 1p(y_i < -\sigma) \\ &\quad + \tfrac{1}{9}p(0 < y_i < \sigma) + \tfrac{1}{9}p(-\sigma < y_i < 0) \\ &= 1 - (1 - \tfrac{1}{9})p(-\sigma < y_i < \sigma) \end{aligned} \tag{18-62}$$

This expression can be written

$$\begin{aligned} \sigma_i^2 &= 1 - \left(\frac{8}{9}\right)\frac{1}{2}\left[\operatorname{erf}\left(\frac{1+\rho}{\sqrt{2}}\right) + \operatorname{erf}\left(\frac{1-\rho}{\sqrt{2}}\right)\right] \\ &\cong 1 - \frac{8}{9}\operatorname{erf}\left(\frac{1}{\sqrt{2}}\right) \qquad \rho \ll 1 \end{aligned} \tag{18-63}$$

for low input SNR. The output SNR can be written [Chang, 1975] using

$$(\text{SNR})_{\text{out}} \cong \frac{M^2 m_i^2}{M\sigma_i^2} = \frac{M(2/3)^2(1/2\pi)\rho^2[1 + 2\exp(-1/2)]^2}{[1 - 8/9\,\text{erf}(1/\sqrt{2})]}$$

$$\cong \frac{M\rho^2 2[1 + 2\exp(-1/2)]^2}{9\pi[1 - 8/9\,\text{erf}(1/\sqrt{2})]} = 0.88\,M\rho^2 \qquad (18\text{-}64)$$

as compared to $M\rho^2$ for an infinite level quantizer. This factor of 0.88 corresponds to 0.55 dB degradation rather than 1.96 dB degradation for the 1-bit quantizer at small input SNR. Hence there is a 1.41 dB improvement by using 2 bits rather than 1-bit/sample.

18-6 DELAY-LOCK TRACKING OF TIME-GATED SIGNALS

If time-gated PN sequences are employed for TDMA timing, the closed-loop operation of the loop can be easily computed using the sampled-data analysis of the spectrum of Chap. 2. Assume that each T_f-sec frame, a d-sec sample of the signal and noise is taken. See Fig. 18-12. Assume further that both the bandwidth of the signal-delay fluctuation and the closed-loop noise bandwidth of the delay-lock loop are small compared to $1/T_f$, the sampling rate of the loop. This latter assumption is critical; otherwise the loop will decay between successive time observations of the signal, and the performance can differ considerably from that described below [Natali, 1972, 89–107].

The signal output of the cross-correlation network is now equal to the product

$$k\left[s\left(t + \frac{\Delta}{2} + \hat{\tau}\right) - s\left(t - \frac{\Delta}{2} + \hat{\tau}\right)\right][-\sqrt{P_s}\,s(t + \tau) + n'(t)]r(t) \qquad (18\text{-}65)$$

where $r(t)$ is the sampling function of duty factor d/T_f. Figure 18-12 shows the sampled-data delay-lock loop and the sampled correlator output waveform. A finite memory integrator of NT sec integration time is employed. This equation assumes that the received input is time-gated in synchronism with the time-gated signal. Thus, the expected value of the output is the delay discriminator characteristic

$$D_\Delta^*(\epsilon) = \frac{d}{T_f}\left[R_s\left(\epsilon - \frac{\Delta}{2}\right) - R_s\left(\epsilon + \frac{\Delta}{2}\right)\right] = \frac{d}{T_f} D_\Delta(\epsilon) \qquad (18\text{-}66)$$

Thus, the loop gain of the delay-lock loop has been reduced by the duty factor d/T_f.

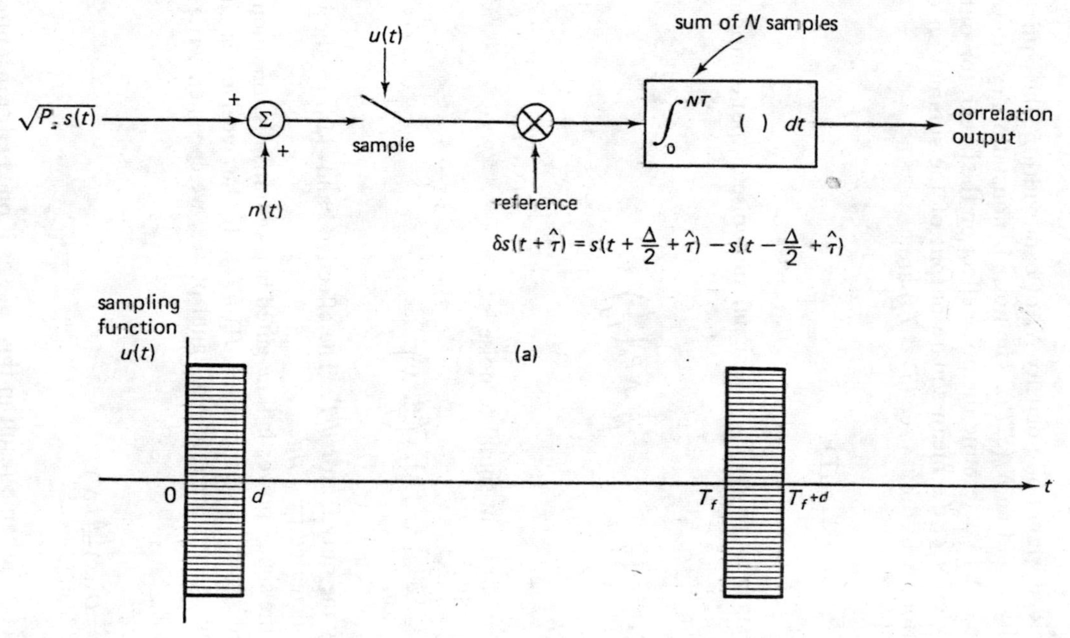

Fig. 18-12 Time-gated correlation: (a) block diagram, (b) the input signal and noise are sampled at a duty factor of d/T_f as shown.

The noise power spectral density has also been reduced by the duty factor since the equivalent input noise is now

$$n(t) = n'(t)r(t) \tag{18-67}$$

If $n'(t)$ had one-sided white noise density N_0 of bandwidth B, the equivalent input noise density would be $N_0(d/T_f)$. The noise is nonstationary, however, and one must use care to determine the noise effect on the integrator output. For a fixed delay error ϵ, the mean square output of the summation of N samples (integration) spaced by T sec ($T \ll T_f$) are

$$\frac{1}{N}\sum_{i=1}^{n} \sqrt{P_s}\, D_\Delta(\epsilon) + n(iT)$$

for a continuous signal of power P_s. The signal and noise output components have powers

$$P_{so} = P_s D_\Delta^2(\epsilon)\left(\frac{d}{T_f}\right)^2 \sigma_n^2 = \frac{1}{N^2} NP_n\left(\frac{d}{T_f}\right)$$

where $P_n \triangleq N_0 B$

Then the ratio of output signal-to-noise is

$$(\text{SNR})_a = \frac{P_{so}}{\sigma_n^2} = \frac{P_s}{P_n} D_\Delta^2(\epsilon)\left(\frac{d}{T_f} N\right) \tag{18-68}$$

and is reduced by the duty factor d/T_f. The effective bandwidth of the integrating filter has increased by T_f/d.

Note that if the signal power is time-gated before transmission, then the actual transmitted signal power is $P_{st} = P_s(d/T_f)$. If the receiver is time-gated in synchronism with the received signal pulses, we can then write the output SNR as

$$(\text{SNR})_a = D_\Delta^2(\epsilon)\frac{P_{st}}{P_n}(N)$$

Thus one has exactly the same result in this case as if one had transmitted a continuous signal of average power P_{st}, received continuous noise of density N_0, and then integrated for the entire period, NT sec. Time gating both signal and noise in this manner causes no degradation with respect to continuous transmission.

18-7 DELAY-LOCK LOOP USING ENVELOPE CORRELATION

Principle of Operation

A block diagram of the cross-correlation portion of the envelope correlation delay-lock discriminator using a square-law detector is shown in Fig. 18-13. This is often called the noncoherent delay-locked loop. The received input consists of a PSK signal, modulated by the PN sequence and possible binary data $d(t) = \pm 1$ plus noise and is given by

$$r(t) = \sqrt{2P_s}\, d(t+\tau)s(t+\tau) \sin [\omega_0(t+\tau)] + n(t) \tag{18-69}$$

where P_s = received power, $s(t)$ = pseudorandom sequence of ± 1s, τ = total time delay, and $n(t)$ = bandlimited (centered about ω_0) white Gaussian noise of one-sided density N_0 and bandwidth $W \gg 1/\tau$. We represent the noise inphase and quadrature components as

$$n(t) = N_s(t) \sin \omega_0 t + N_c(t) \cos \omega_0 t = N(t) \sin [\omega_0 t + \phi_n(t)] \tag{18-70}$$

where the power spectral density of $N(t)$ is N_0.

The incoming waveform is now multiplied by two locally generated reference signals given by the two waveforms (the $\pm$ signs indicate two separate reference waveforms)

$$\sqrt{2}\, s\left(t + \hat{\tau} \pm \frac{\Delta}{2}\right) \cos (\omega_0 + \omega_1)t$$

where Δ = bit length (chip length) and $\hat{\tau}$ = receiver estimate of the delay. The effects of using offset delays of $\pm\Delta$ instead of $\pm\Delta/2$ are examined later in this section. The output of the multiplier is filtered to remove the frequency sum term. The two waveforms at the IF input are then

$$\sqrt{P_s}\, d(t+\tau)s(t+\tau)s\left(t \pm \frac{\Delta}{2} + \hat{\tau}\right) \sin \omega_1 t + s\left(t \pm \frac{\Delta}{2} + \hat{\tau}\right) n'(t) \tag{18-71}$$

where

$$n'(t) = N_s(t) \sin \omega_1 t + N_c(t) \cos \omega_1 t = N(t) \sin [\omega_1 t + \phi_n(t)] \tag{18-72}$$

Define the baseband equivalent noise $s(t \pm \Delta/2 + \tau)n'(t)$ as $n_i'(t)$ where $i = 1, 2$ for the two cross-correlation channels 1, 2, and is bandlimited white

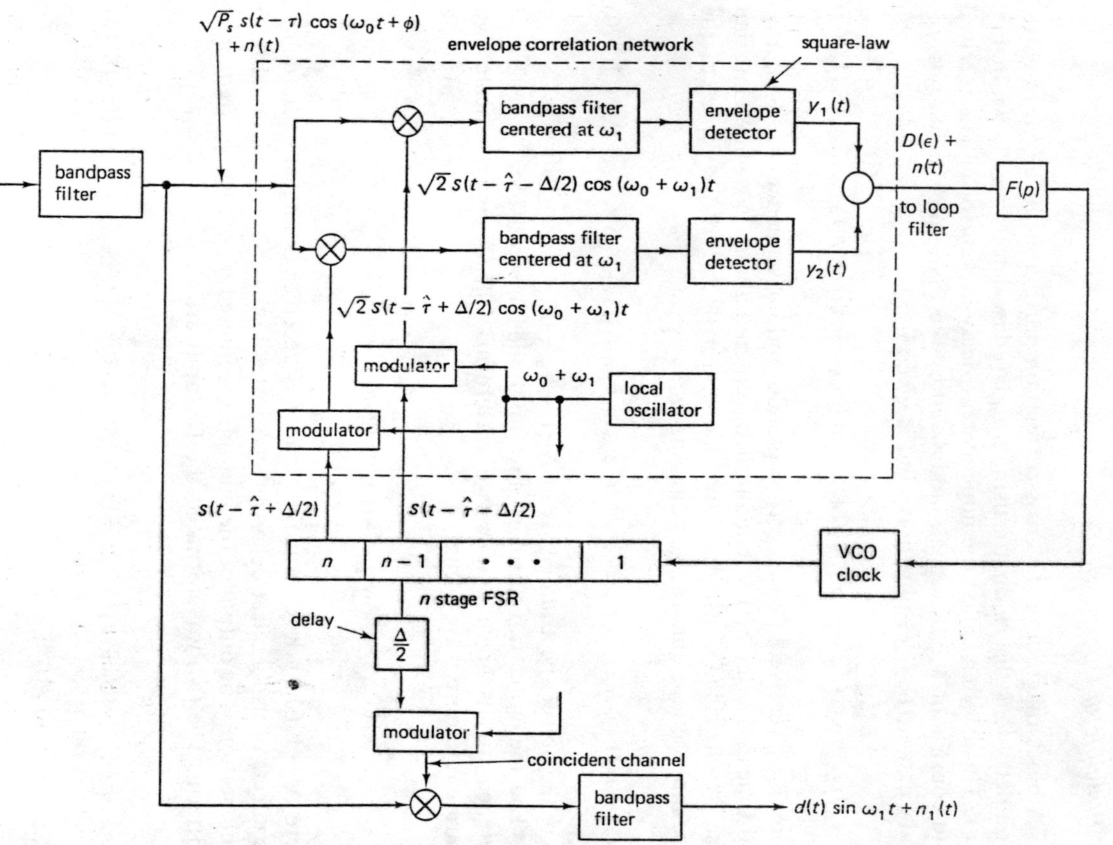

Fig. 18-13 An alternate form of the delay-locked loop using envelope correlation [Gill, 1966, 415–426]. The delay error is $\epsilon \triangleq \tau = \hat{\tau}$.

Gaussian noise of density N_0 (one-sided), since the noise is simply translated down in frequency.

The data $d(t)$ is recovered using a coincident channel as shown in Fig. 18-13. The IF filter output is a BPSK signal of the form

$$A_{IF}\, d(t) \sin \omega_{IF} t + n_{IF}(t) \tag{18-73}$$

and can be demodulated using a BPSK demodulator as shown in Chap. 11. The IF filters in the coincident channel, as well as the early and late channels, must be wide enough in bandwidth to pass the data modulation plus any doppler frequency offset.

The sequence multiplying the noise term consists of ± 1s. That is, the sequence merely inverts or does not invert the phase of the noise. Since the white noise has random phase, randomly inverting or not inverting still leaves the phase random. Hence, the noise terms are still bandlimited white noise and the IF correlator inputs can be written as

$$x_1(t) = \sqrt{P_s}\, d(t) s(t+\tau) s\left(t + \frac{\Delta}{2} + \hat{\tau}\right) \sin \omega_1 t + n_1'(t) \tag{18-74}$$

$$x_2(t) = \sqrt{P_s}\, d(t) s(t+\tau) s\left(t - \frac{\Delta}{2} + \hat{\tau}\right) \sin \omega_1 t + n_2'(t) \tag{18-75}$$

where

$$n_1'(t) = n'(t) s\left(t + \frac{\Delta}{2} + \hat{\tau}\right) \qquad n_2'(t) = n'(t) s\left(t - \frac{\Delta}{2} + \hat{\tau}\right) \tag{18-76}$$

are the uncorrelated noise terms. Thus, the signal portion of the IF output of (18-74), (18-75) contains a sinusoid with amplitude proportional to the autocorrelation function $R_s[\tau - \hat{\tau} + (\Delta/2)]$ plus a self-noise term completely analogous to the coherent delay-locked discriminator.

When either of the two reference sequences is perfectly aligned with $s(t+\tau)$—that is, when $\hat{\tau} = \tau \pm \Delta$—one product of the two sequences is unity since each individual sequence is composed of ± 1s. Therefore, the signal component in the multiplier output is a pure sinusoid (biphase modulated by the data $d(t)$ if any) of frequency f_{IF} and is passed without modification through the IF filter.

On the other hand, when the sequences are misaligned by an integer number of pulse lengths Δ, the product of the two sequences is another pseudorandom sequence. When the sequence is produced from a maximal-length LFSR, the above statement follows directly from the cycle-and-add property (Sec. 18-2). Thus, for any pseudorandom sequence, the out-of-phase correlation between waveforms goes rapidly to zero or to some small value. Since there are frequent transitions relative to a narrowband IF filter, the spectrum is wideband and most of this self-noise is removed by the IF filter if the IF bandwidth is relatively small.

More generally, one can use the decomposition property of products of random or psuedorandom sequences as shown by Gill and Spilker [1963, 246–247] and discussed in Sec. 14-2. The product of two very long-period pseudorandom sequences has a finely spaced (essentially continuous) line spectrum plus possible line components at the PN code rate $1/\Delta$ and harmonics n/Δ, and a dc component proportional to the code cross-correlation. This decomposition is valid even if the separation between sequences is not an integer number of pulse lengths. Thus, the signal output of the narrow-bandwidth IF filter is essentially zero when the sequences are misaligned by more than one pulse length.

Now, consider what happens when the displacement between input and reference sequences is equal to or less than one pulse length. In this situation the signal $\times$ signal product in one correlator output is

$$\begin{aligned} &d(t)s(t+\tau)s\left(t+\frac{\Delta}{2}+\hat{\tau}\right)\sin\omega_1 t \\ &= d(t)\left[R_s\left(\epsilon-\frac{\Delta}{2}\right)+p\left(t,\epsilon-\frac{\Delta}{2}\right)+r\left(t,\epsilon-\frac{\Delta}{2}\right)\right]\sin\omega_1 t, \\ &\qquad |\epsilon|<\Delta \end{aligned} \tag{18-77}$$

where $R_s(x)$ is the signal autocorrelation function $\epsilon \triangleq \tau - \hat{\tau}$, $p(t)$ is the periodic component at the clock rate, and $r(t)$ is the pseudorandom self-noise term discussed in Sec. 14-2. This value of correlation $R_s(x)$ represents the relative constant term multiplying the sinusoid. For example, when $R_s = 1$, the signal consists of a sinusoid only; when $R_s = 0$, it consists of a rapidly varying sequence only. Since most of the rapidly varying self-noise and clock components are filtered out for small IF bandwidth, the IF filter outputs for the two channels are approximately

$$s_1(t) + n_1(t) = R_s\left(\epsilon - \frac{\Delta}{2}\right)\sqrt{P_s}\sin\omega_1 t + n_1(t) \qquad -\frac{\Delta}{2} \leqq \epsilon \leqq \frac{3\Delta}{2} \tag{18-78}$$

and

$$s_2(t) + n_2(t) = R_s\left(\epsilon + \frac{\Delta}{2}\right)\sqrt{P_s}\sin\omega_1 t + n_2(t) \qquad -\frac{3\Delta}{2} \leqq \epsilon \leqq \frac{\Delta}{2} \tag{18-79}$$

and $R_s(\epsilon - \Delta/2)$, $R_s(\epsilon + \Delta/2)$ are constant with time if ϵ is constant.

The autocorrelation terms in this region are

$$R_s\left(\epsilon - \frac{\Delta}{2}\right) = \left[1 - \frac{|\epsilon - \Delta/2|}{\Delta}\right] \tag{18-80}$$

$$R_s\left(\epsilon + \frac{\Delta}{2}\right) = \left[1 - \frac{|\epsilon + \Delta/2|}{\Delta}\right] \tag{18-81}$$

and $n_1(t)$, $n_2(t)$ are each bandlimited Gaussian noise since they are merely $n_1'(t)$, $n_2'(t)$ after being filtered by the IF filter.

The signals are then squared in the square-law detector. These square-law detector terms are later low-pass filtered by the low-pass filter to remove double frequency terms; the resultant output is the squared envelope of the IF.

The output noise density of $n_1(t)$, $n_2(t)$ at the correlator output is $N_0/2$. This one-sided noise density has been reduced by a factor 2 because of equal translation of power to f_1 and to the sum frequency $2f_0 + f_1$. The square-law detector output then has a baseband noise power spectral density (one-sided)

$$P_s\frac{N_0}{2} + \frac{N_0^2}{2}(B_{\text{IF}} - f) \qquad 0 < f < B_{\text{IF}} \tag{18-82}$$

When the signals are subtracted, the resulting form (ignoring multiplicative constants) is the delay discriminator characteristic. Thus, for constant ϵ, the expected value of the output of the difference circuit in Fig. 18-8 is

$$D_\Delta'(\epsilon) = R_s^2\left(\epsilon - \frac{\Delta}{2}\right) - R_s^2\left(\epsilon + \frac{\Delta}{2}\right) \tag{18-83}$$

which is shown in Fig. 18-14*a*.

We can also use a reference waveform $s(t \pm \Delta + \hat{\tau})$. The only change in the analysis for this reference occurs in adding the two waveforms together. The 2Δ discriminator characteristic is then

$$D_{2\Delta}'(\epsilon) = R_s^2(\epsilon - \Delta) - R_s^2(\epsilon + \Delta) \tag{18-84}$$

which is also shown in Fig. 18-14*b*. In both cases, the shape of the discriminator characteristics produces a correction signal. However, the gain produced using $\pm\Delta/2$ is at least double that produced employing $\pm\Delta$, for a coherent DLL, and furthermore the gain at $\epsilon = 0$ goes to zero for the $\pm\Delta$ or 2Δ noncoherent DLL. Thus the 2Δ square-law DLL is nonlinear even for small ϵ and has a null zone at $\epsilon = 0$. Hence, it is inferior to the 1Δ square-law DLL for most purposes.

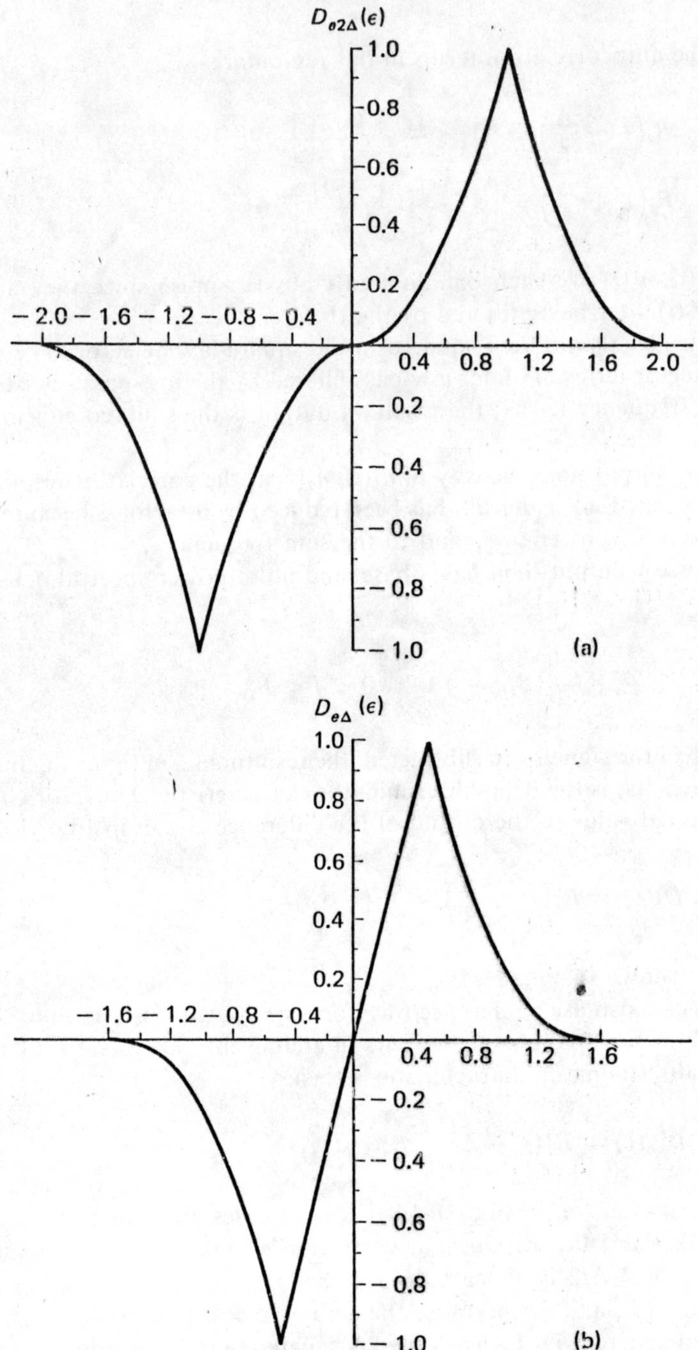

Fig. 18-14 Delay discriminator characteristic of a delay-lock loop with an envelope correlator (square-law detector) for code separations of (a) $\pm\Delta$ and (b) $\pm\Delta/2$

Noise Effects

In these paragraphs we compute the output signal and noise density components at various points within the 1Δ DLL and conclude with a calculation of the rms delay tracking error for linear operation of the delay-lock loop $|\epsilon| < \Delta/2$. All noise densities are one-sided.

The received signal plus noise $r(t)$ contains wideband white noise $n(t)$ as in (18-69) of power spectral density N_0.

$$n(t) = N(t) \sin [\omega_0 t + \phi_n(t)] \tag{18-85}$$

and the baseband noise envelope $N(t)$ also has density N_0. After multiplication by $2s(t \pm \Delta/2 + \tau) \sin (\omega_0 + \omega_1)t$, and removal of the double frequency terms, the noise component of (18-72) is

$$\begin{aligned} n_1'(t) &= s\left(t + \frac{\Delta}{2} + \hat{\tau}\right) N(t) \sin [\omega_1 t + \phi_n(t)] \\ &\triangleq N_1'(t) \sin [\omega_1 t + \phi_n(t)] \end{aligned} \tag{18-86}$$

This reference $s(t)$ has unity power. Since the noise $n(t)$ is assumed to be wide in bandwidth, multiplication by $s(t)$ does not significantly spread the bandwidth of $n(t)$, and the noise density at the correlator output remains at N_0.

Filtering at IF reduces the bandwidth of the noise to B_{IF}, but the noise density remains the same. The two filtered noise components are

$$\begin{aligned} n_1(t) &= N_{s1}(t) \sin \omega_1 t + N_{c1}(t) \cos \omega_1 t \\ n_2(t) &= N_{s2}(t) \sin \omega_1 t + N_{c2}(t) \cos \omega_1 t \end{aligned} \tag{18-87}$$

where the noise densities of n_1 at IF and $N_{s1}(t)$ at baseband are both N_0.

The outputs of the square-law envelope detector are then the baseband portions of the square of $s_1 + n_1$ and $s_2 + n_2$, respectively where $s_1 t$ is the IF filtered signal. The resulting expressions for detector output are derived from (18-78), (18-79), (18-87):

$$\begin{aligned} y_1(t) &= 2\mathfrak{F}_{LP}[s_1^2(t) + 2s_1(t)n_1(t) + n_1^2(t)] \\ &= P_s R_s^2\left(\epsilon - \frac{\Delta}{2}\right) + 2\sqrt{P_s} N_1(t) R_s\left(\epsilon - \frac{\Delta}{2}\right) + N_{s1}^2(t)\Big] \end{aligned} \tag{18-88}$$

$$\begin{aligned} y_2(t) &= 2\mathfrak{F}_{LP}]s_2^2(t) + 2s_2(t)n_2(t) + n_2^2(t)] \\ &= P_s R_s^2\left(\epsilon + \frac{\Delta}{2}\right) + 2\sqrt{P_s} N_2(t) R_s\left(\epsilon + \frac{\Delta}{2}\right) + N_{s2}^2(t)\Big] \end{aligned} \tag{18-89}$$

where $\mathfrak{F}_{LP}$ represents the low-pass filter transfer function that removes the double IF frequency terms at $2f_1$. The spectrum of $N_{s1}(t)$ is low-pass with density N_0 for $0 < f < B_{IF}$, and the spectrum of $N_{s1}^2(t)$ has a triangular spectrum

$$G_{N_{s1}^2}(f) = 2N_0^2\left[B_{IF} - \frac{f}{2}\right] + (N_0 B_{IF})^2 \qquad 0 < f < 2B_{IF} \qquad \text{(18-90)}$$

Because the two waveforms $s(t + \Delta/2)$ and $s(t - \Delta/2)$ are uncorrelated —that is, PN bits separated by Δ or more are independent in $s(t)$—and the input noise $n(t)$ is white Gaussian noise, the noise terms $n_1(t)$, $n_2(\tau)$ are uncorrelated; that is, $R_{n_1n_2}(\tau) = 0$. We evaluate $R_{n_1n_2}(\tau)$ by defining an IF filter with an impluse response $h(t)$. Thus, the filtered noise of (18-87) can be written as

$$n_1(t) = \int h(t - \sigma)n_1'(\sigma)\,d\sigma$$

$$n_2(t) = \int h(t - \sigma)n_2'(\sigma)\,d\sigma \qquad \text{(18-91)}$$

The cross-correlation between n_1 and n_2 is then obtained using (18-76), (18-91)

$$\begin{aligned} R_{n_1n_2}(\tau) &= \mathbf{E}[n_1(t)n_2(t + \tau)] = \iint d\sigma\, d\gamma h(t - \sigma)h(t + \tau - \gamma) \\ &\quad \times \mathbf{E}\left[s\left(\sigma + \frac{\Delta}{2} + \hat{\tau}\right)n'(\sigma)s\left(\gamma - \frac{\Delta}{2} + \hat{\tau}\right)n'(\gamma)\right] \\ &= 0 \end{aligned} \qquad \text{(18-92)}$$

The difference circuit output in Fig. 18-14 is then*

$$\begin{aligned} d(t) &\triangleq y_1(t) - y_2(t) \\ &= P_s D_{\epsilon\Delta}(\epsilon) + 2\sqrt{P_s}\left[N_{s1}(t)R_s\left(\epsilon - \frac{\Delta}{2}\right) - N_{s2}(t)R_s\left(\epsilon + \frac{\Delta}{2}\right)\right] \\ &\quad + [N_{s1}^2(t) - N_{s2}^2(t)] \triangleq \sqrt{P_s}^{1/2} D_{\epsilon\Delta}(\epsilon) + n_{s\times n}(t) + n_{n\times n}(t) \end{aligned} \qquad \text{(18-93)}$$

where $D_{\epsilon\Delta}(\epsilon) \triangleq R_s^2\left(\epsilon - \frac{\Delta}{2}\right) - R_s^2\left(\epsilon + \frac{\Delta}{2}\right)$

---

*One can show that if the input noise $n(t)$ is a sine wave interference rather than bandlimited Gaussian white noise, then the correlator output interference in the two channels is simply a pair of sine waves biphase modulated by $s(t + \Delta/2 + \hat{\tau})$ and $s(t - \Delta/2 + \hat{\tau})$, respectively. The bandpass filtered output of these two waveforms is nearly identical; hence, there is a substantial degree of interference rejection at the output of the difference circuit for $B_{IF} \ll f_c$.

The output noise density of the first bracketed noise term, $n_{s\times n}(t)$, is

$$G_{s\times n}(f) = N_0 4P_s\left[R_s^2\left(\epsilon - \frac{\Delta}{2}\right) + R_s^2\left(\epsilon + \frac{\Delta}{2}\right)\right] \qquad 0 < f < B_{IF}$$
$$\cong 2N_0P_s \qquad |\epsilon| \ll \Delta \tag{18-94}$$

since $R_s^2(\Delta/2) = 1/4$. The output noise density of the second bracketed noise term in (18-93) is

$$G_{n\times n} = 4N_0^2\left[B_{IF} - \frac{f}{2}\right] \qquad 0 < f < 2B_{IF} \tag{18-95}$$

where B_{IF} is the IF bandwidth. For $|\epsilon| \ll \Delta$ the difference-circuit output noise in (18-93) has two components, $s \times n$ and $n \times n$, and then has a one-sided, spectrum

$$G_{dn}(f) \triangleq G_{s\times n}(f) + G_{n\times n}(f)$$
$$G_{dn}(f) \cong 2P_sN_0 + 4N_0^2\left(B_{IF} - \frac{f}{2}\right) \qquad |\epsilon| \ll \Delta \tag{18-96}$$

where B_{IF} is the IF filter bandwidth. Loop filtering passes only frequency components in the vicinity of $f = 0$. Hence, the power spectral density of interest is

$$G_{dn}(0) \cong 2P_sN_0 + 4N_0^2B_{IF} \tag{18-97}$$

Define B_n as the closed-loop noise bandwidth of the delay-lock loop. The signal output is $P_sD_{\epsilon\Delta}(t) = P_s2\epsilon$ for $|\epsilon| < \Delta/2$. The output noise power in a closed-loop noise bandwidth B_n for $B_n \ll B_{IF}$ is

$$\sigma_n^2 = B_n[G_{s\times n}(0) + G_{n\times n}(0)] = [2N_1P_s + 4N_0^2B_{IF}]B_n \tag{18-98}$$

Thus, the variance output delay error ϵ normalized to Δ for $|\epsilon| \ll \Delta$ is

$$\frac{\sigma_\epsilon}{\Delta} = \frac{\sigma_n}{\Delta(2P_s)} \cong \frac{[2N_0P_s + 4N_0^2B_{IF}]^{1/2}B_n^{1/2}}{2P_s}$$
$$= \left[\frac{N_0^2B_n}{2P_s} + \frac{N_0^2B_{IF}B_n}{P_s^2}\right]^{1/2}$$
$$\cong \left\{\frac{P_n}{2P_s}\left[1 + 2\frac{P_{nIF}}{P_s}\right]\right\}^{1/2} \tag{18-99}$$

where $N_0B_{IF} = P_{nIF}$, $N_0B_n \triangleq P_n$, signal power. As long as the SNR in the IF is moderately high—that is, $(\text{SNR})_{IF} \triangleq P_s/P_{nIF} > 8$—the effect of the second term is relatively small. Thus, the envelope correlator operating in a noise environment produces an rms timing jitter in the delay measurement given by [Gill, 1966, 415–426]

$$\frac{\sigma_\epsilon}{\Delta} = \left[\frac{B_n}{2C/N_0}\left(1 + \frac{2B_{\mathrm{IF}}}{C/N_0}\right)\right]^{1/2} \tag{18-100}$$

where Δ is the chip duration, B_L is the loop bandwidth, B_{IF} is the post-correlator IF bandwidth $C \equiv P_s$, and C/N_0 is the carrier-to-thermal-noise density

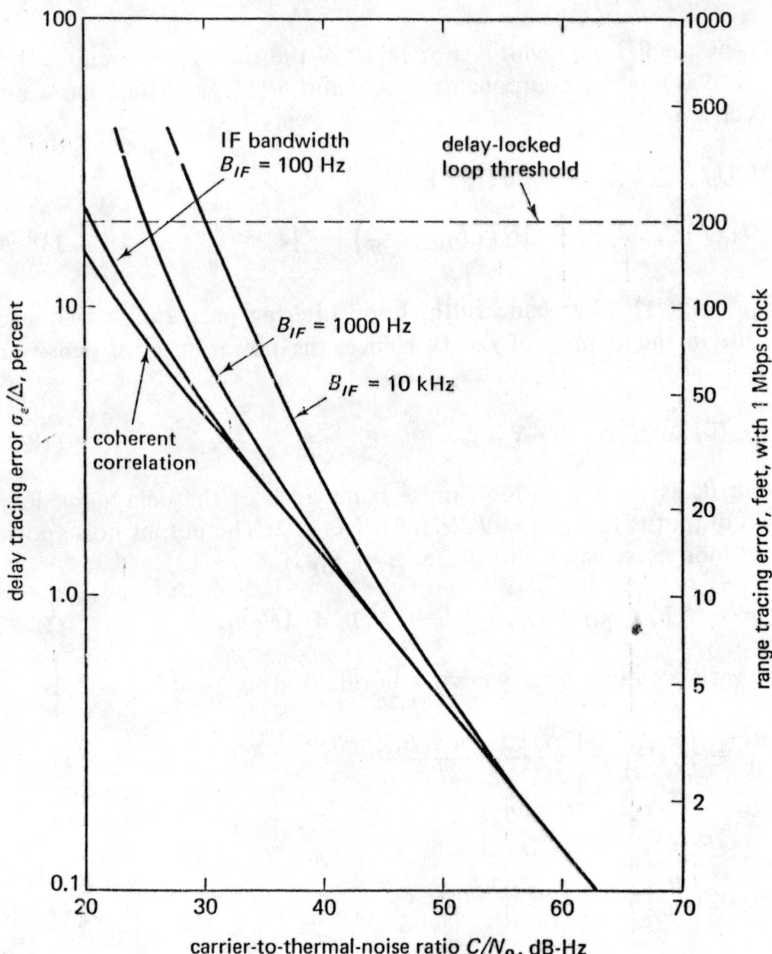

Fig. 18-15 Delay-lock loop (1Δ) range-tracking performance for 6 Hz or 8 dB-Hz closed loop bandwidth: the delay-lock loop threshold is shown here at $\sigma_t/\Delta = 0.2$ to allow for transient error [Natali, 1972].

ratio. Figure 18-15 shows a typical plot of σ_e/Δ versus C/N_0 for a 1-Mbps clock—that is, $\Delta = 1$ μsec—and a noise bandwidth $B_n = 6$ Hz. Note that for $B_{IF} \leqq 100$ Hz, the noncoherent DLL is very close to the coherent DLL for $C/N_0 > 20$ dB. The noise bandwidth must be wide enough to track the dynamics of target motion just as discussed in Chap. 12 for the phase-lock loop. The same linear transient response characteristic applies here to the delay-lock loop for small delay errors. The steady-state transient delay error τ_e for a signal-delay acceleration of $\ddot{\tau}$ and a second-order delay-lock loop is

$$\frac{\tau_e}{\Delta} = \frac{1.12}{\Delta} \frac{\ddot{\tau}}{(2B_n)^2} \tag{18-101}$$

The delay acceleration is related to the path-delay acceleration $\ddot{s}$ by

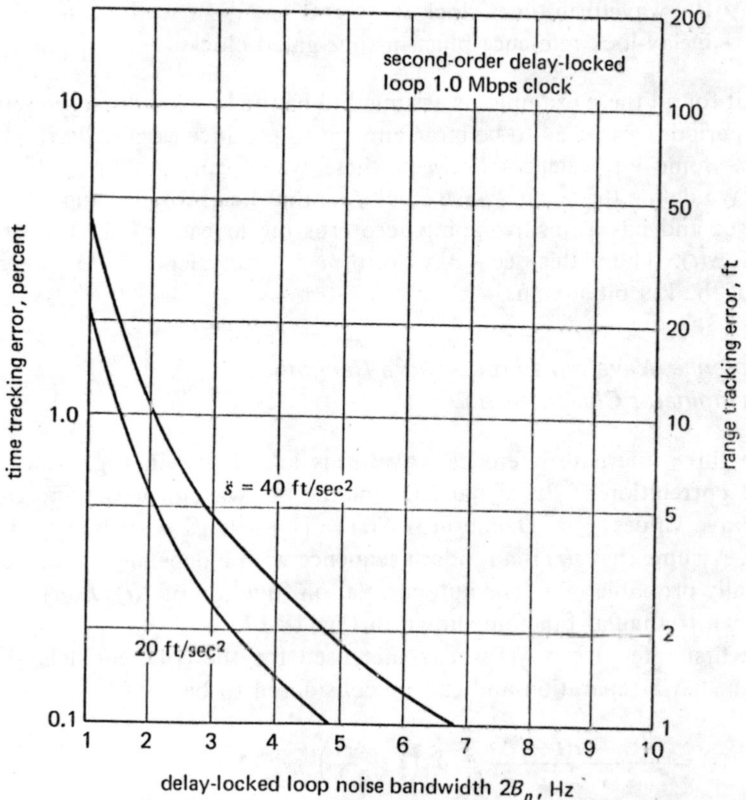

Fig. 18-16 Dynamic range-tracking error of delay-lock loop due to range acceleration $\ddot{s}$

$\ddot{\tau} = \ddot{s}/c$. Accelerations of $\ddot{s} = 20$, 40 ft/sec² are shown in Fig. 18-16 versus noise bandwidth and tracking error for a second-order loop.

18-8 COMPARISON OF DELAY-LOCK TRACKING WAVEFORMS FOR PN SIGNALS

We have described two versions of the delay-lock discriminator which employ either coherent or noncoherent correlators. However, in tracking a PN sequence $x(t)$ having bit duration Δ, several different reference waveforms can be used for delay-locked tracking and correlation in coherent and noncoherent tracking systems. The particular reference waveforms to be considered are:

- Delay-lock loop reference $\delta s(t) = \dfrac{s(t) - s(t - \Delta)}{2}$
- PN waveform times clock $s(t) f_c(t)$
- Delay-lock reference plus—a time-gated clock.

The input for all these examples is assumed to be the baseband PN sequence, and the period is assumed to be large enough to produce essentially random statistics. Some equivalences between these waveforms are given in this section. We define the square-wave clock $f_c(t)$ that has zero crossings spaced by $\Delta/2$ sec and has a positive-going zero crossing at each of the transition times of $x(t)$. Thus, this second waveform is equivelent to Manchester coding of the PN bit stream.

Reference Waveforms Producing a Ranging Discriminator Characteristic

The three alternate reference waveforms are shown in Fig. 18-17 for coherent correlation of PN signals. Define the PN waveform $x(t)$ and clock $f_c(t)$ to have values ± 1. Define also $S(t) = [1 - s(t)]/2$ as a binary (0, 1) variable. Assume that $s(t)$ is a random sequence with independent transitions and equally probable ± 1. The autocorrelation function of $s(t)$, $R_s(\tau)$, is the well-known triangular function shown in Fig. 18-17.

The first reference waveform is that used for the coherent delay-lock loop with $\pm\Delta/2$ separation and can be considered to be

$$v(t) = \frac{s(t) - s(t - \Delta)}{2} \triangleq \delta s\left(t + \frac{\Delta}{2}\right) \tag{18-102}$$

This waveform is ternary. Note that $v(t)$ is identical to $s(t)$ where $s(t) = s(t - \Delta)$ and that $v(t) = 0$ otherwise. Each condition occurs half the time.

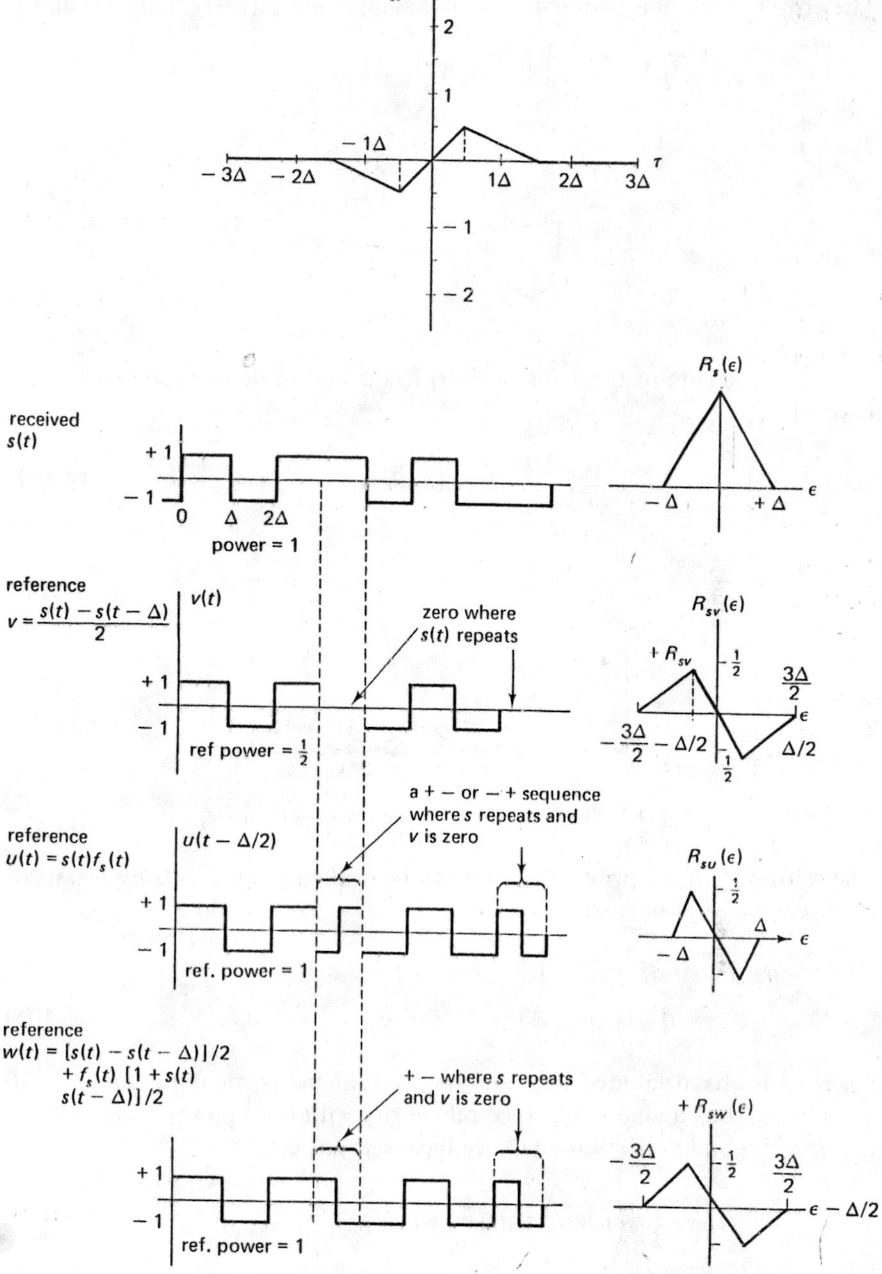

Fig. 18-17 Alternative reference waveforms for delay-lock loop tracking functions

The cross-correlation between s and v is shown in Fig. 18-17 as $R_{sv}(\tau)$ and is

$$\begin{aligned} R_{sv}(\tau) &\triangleq \mathbf{E}\left\{s(t+\tau)\left[\frac{s(t)-s(t-\Delta)}{2}\right]\right\} \\ &= \frac{1}{2} \qquad \tau = 0 \\ &= -\frac{1}{2} \qquad \tau = -\Delta \\ &= 0 \qquad \tau = \frac{-\Delta}{2} \end{aligned} \tag{18-103}$$

The spectrum of the reference $v(t)$ has a null at dc as is shown in Fig. 18-18.

$$G_v(\omega) = K\left(\frac{\sin \omega\,\Delta/2}{\omega\,\Delta/2}\right)^2 [1 - \cos \omega\,\Delta] \tag{18-104}$$

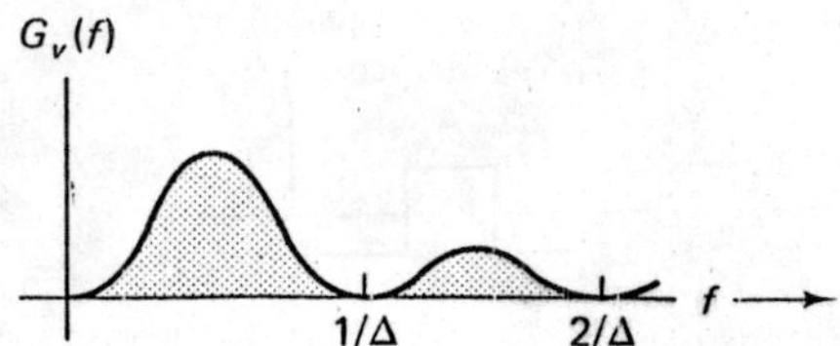

Fig. 18-18 Spectrum of reference waveform $v(t)$

The self-noise in the product $s(t)v(t)$ can be evaluated at $\tau = 0$ by means of the cycle-and-add property as

$$\begin{aligned} n_{vs0}(t) &\triangleq s(t)v(t) - \tfrac{1}{2} = v^2(t) - \tfrac{1}{2} \\ &= \tfrac{1}{2}s(t)s(t-\Delta) = \tfrac{1}{2}s(t+\Delta) \end{aligned} \tag{18-105}$$

where $j\Delta$ is a fixed value of delay and j depends on the particular code selected. Thus, $n_{vs0}(t)$ has a sinc $\omega\Delta/2$ spectral density with total power $P_{ns} = 1/4$. At the in-lock condition $\tau = -\Delta/2$, we have self-noise

$$\begin{aligned} n_{vs1}(t) &\triangleq \frac{1}{2}s\left(t - \frac{\Delta}{2}\right)[s(t) - s(t - j\Delta)] \\ &= \frac{1}{2}\left[z(t) - z\left(t - \frac{\Delta}{2}\right)\right] \end{aligned} \tag{18-106}$$

where $z(t) \triangleq s[t - (\Delta/2)]s(t)$.

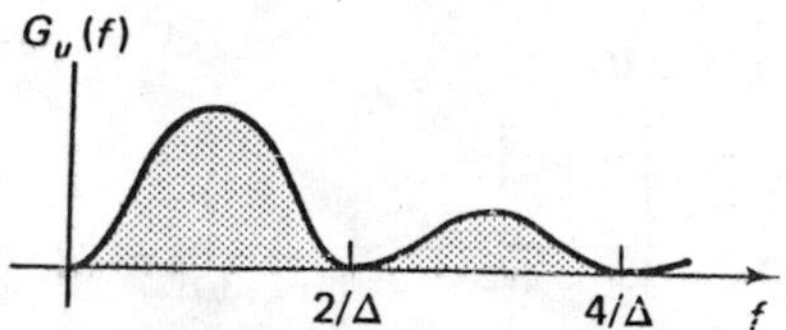

Fig. 18-19 Spectrum of reference waveform $u(t)$

The second reference waveform shown in Fig. 18-17 is $u(t) = s(t)f_c(t)$, the product of the PN sequence with the clock. This waveform has the spectrum of Fig. 18-19 and can be written

$$u(t) = s(t)f_c\left(t + \frac{\Delta}{2}\right)$$

$$u\left(t + \frac{\Delta}{2}\right) = s\left(t + \frac{\Delta}{2}\right)f_c\left(t + \frac{\Delta}{2}\right)$$

$$= v(t) + \left(\frac{1 - s(t)s(t - \Delta)}{2}\right)f_c(t)s(t) \qquad (18\text{-}107)$$

Thus, the waveform $u(t)$ is equal to $v(t)$ where $v = \pm 1$. In the gaps where $v(t)$ has a zero, $u(t + \Delta/2)$ has either a $(+1, -1)$ or a $(-1, +1)$ sequence, depending on the sign of $s(t)$. This $(+1, -1)$ clock doublet is the second term in the above expression.

The cross-correlation $R_{us}(\tau)$ between u and s is similar to $R_{sv}(\tau)$ of (18-103) except for a time shift $\Delta/2$:

$$R_{us}(\tau) \triangleq \mathbf{E}[s(t + \tau)u(t)] = \mathbf{E}\Bigg\{s(t + \tau)$$

$$\times \left[v\left(t - \frac{\Delta}{2}\right) + \frac{1 - s\left(t - \frac{\Delta}{2}\right)s\left(t - \frac{3\Delta}{2}\right)}{2} f_c\left(t - \frac{\Delta}{2}\right)s(t)\right]\Bigg\}$$

$$= R_{sv}\left(\tau + \frac{\Delta}{2}\right) \qquad (18\text{-}108)$$

and is plotted in Fig. 18-22. The self-noise in (18-107) at $\tau = 0$ is the entire product, since the correction term $R_{sv}(\tau + \Delta/2)$ is zero

$$n_{us0}(t) = s(t)u(t)$$

$$= s(t)[s(t)f_c(t)] = f_c(t) \qquad (18\text{-}109)$$

and it is simply a square wave with zero crossings spaced at $\Delta/2$. The spectrum of the self-noise has no dc component and is shown in Fig. 18-20.

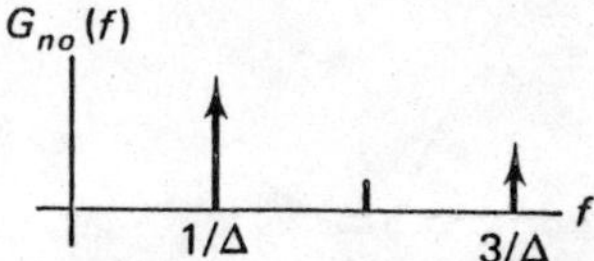

Fig. 18-20 Spectrum of self-noise of $u(t)s(t)$

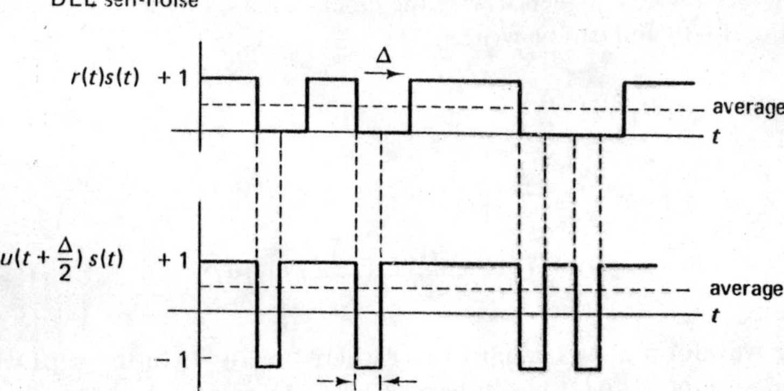

Fig. 18-21 Comparison of self-noise produced by reference waveforms $u(t)$ and $v(t)$ in the tracking loops

The difference between self-noise of u and y can be observed from the product shown in Fig. 18-21. The pulses from $+1$ values of the products occur at the same rate and at random times (multiples of Δ sec) for both references. The average power in the noise pulses of $\pm 1/2$ in sv is 1/4. The autocorrelation functions of sv and us are shown in Fig. 18-22. The shape of the self-noise spectral density $G_n(0)$ of n_{sv} is identical to that of n_{us}, although n_{us} has double the power (but twice the bandwidth) in its random components.

A third reference waveform that can be used is

$$w(t) = v(t) + f_c(t)[1 + s(t)s(t - \Delta)]/2 \tag{18-110}$$

This waveform is similar to $v(t)$, except that where $v(t) = 0$, $w(t)$ always has a $(+1, -1)$ sequence. Hence, $w(t)$ contains a periodic component that makes it correlate with an incoming square wave or other interference of the same period. Otherwise, its cross-correlation function $R_{sw}(\tau)$ is identical to that of $R_{sv}(\tau)$.

Since the cross-correlation functions of v, u, w are identical, the thermal noise performance of the system is determined by the power and spectral

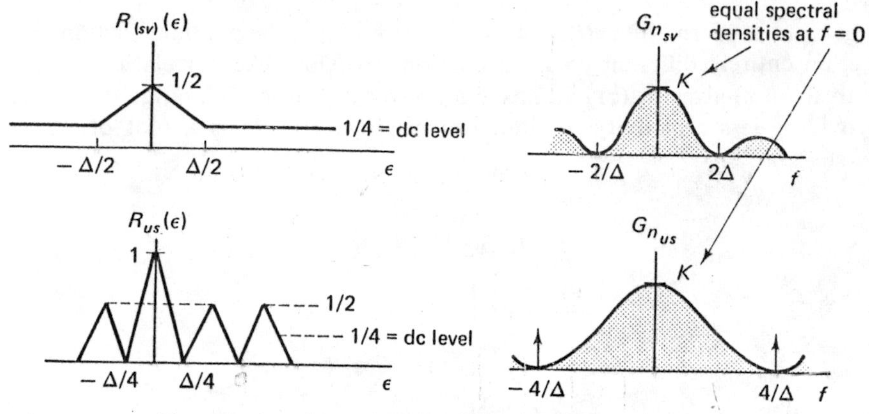

Fig. 18-22 Autocorrelations and spectra of product waveforms

density of $v(t)n(t)$. If $n(t)$ is white noise of power P_n, then $v(t)n(t)$ is time-gated white noise of power $P_n/2$, since half the time the noise is gated off. The products $w(t)n(t)$, $u(t)n(t)$ are also white but have power P_n—that is, 3 dB more noise power than the correlator output noise of $v(t)n(t)$. Thus, for white noise inputs, the use of $v(t)$ has an advantage since $v(t)n(t)$ provides a 3-dB improvement in noise performance relative to $w(t)$ or $u(t)$.

Reference Waveforms for Wide Acquisition Range

It is sometimes desirable to produce a reference waveform that will allow correlation while acquiring synchronization or imperfectly tracking—that is, a nonzero delay error. These waveforms can permit coherent carrier recovery even though perfect time synchronization has not been obtained. Noise performance is degraded, however.

For example, we can generate a reference that is the sum of time-displaced code sequences:

$$z_a(t) = \frac{1}{N}\sum_{i=1}^{N} s(t - i\Delta) \tag{18-111}$$

and for $N = 2$, we have the ternary waveform

$$z(t) = \tfrac{1}{2}[s(t) + s(t + \Delta)] \tag{18-112}$$

where $u(t)$ takes on values -1, 0, $+1$, and $s(t) = \pm 1$. Note that this is a different waveform than $v(t) = \frac{1}{2}[s(t) - s(t - \Delta)]$ of the previous section and has an entirely different cross-correlation behavior. The correlation of $z(t)$ with $s(t + \tau)$ gives $R_{sz}(\tau)$ and has a nonzero value over a 3Δ range of τ, thus providing less sensitivity to timing error. Figure 18-23 is a plot of $R_{sz}(\tau)$ versus τ.

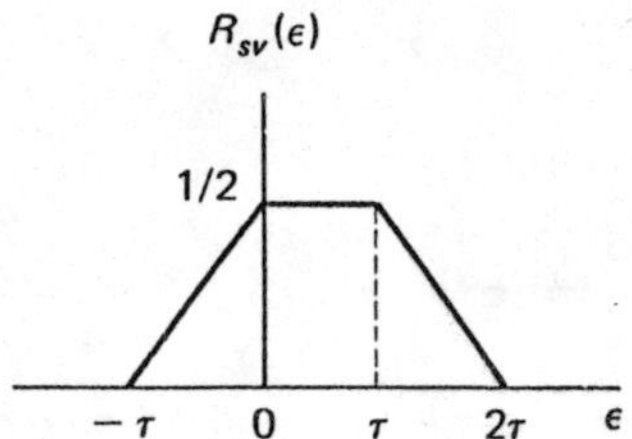

Fig. 18-23 Cross-correlation $R_{sv}(\epsilon)$

Previously we were able to generate binary waveforms u and w in (18-107) (18-110) having the same cross-correlation properties with s as for v. One can also obtain similar binary waveforms for this acquisition problem. For example, let us define

$$z_b(t) = z(t) + (1 - |z(t)|)s(t - 2\Delta)f_c(t) \tag{18-113}$$

$$z_c(t) = z(t) + (1 - |z(t)|)f_c(t) \tag{18-114}$$

where $z_c(t)$ contains a gated periodic clock and hence has line components in its spectrum as did $w(t)$. The cross-correlation functions of both z_b and z_c with s are the same as $R_{sz}(\tau)$.

Coherent Delay-Lock Loop

The reference waveform of (18-102) can be used in an alternative form of coherent delay-locked loop. In Section 18-2 we discussed coherent delay lock where the PN bit stream is recovered using a conventional BPSK carrier recovery technique ($\times 2$ or Costas type loops). In this section we show a technique for a coherent carrier DLL that does not require as high an input SNR as do these previous approaches.

Figure 18-24 shows this approach. The received waveform is first correlated with $z_1(t)$ to get an approximate correlation match in timing after an initial search and bandpass (noncoherent) correlation for acquisition. The resulting product $s(t)y(t) \sin \omega_0 t$ then contains enough pure carrier component to permit a phase-locked loop to recover the carrier from coherent detection. The coherently detected signal $s(t)$ is then tracked in a baseband coherent delay-locked loop as shown.

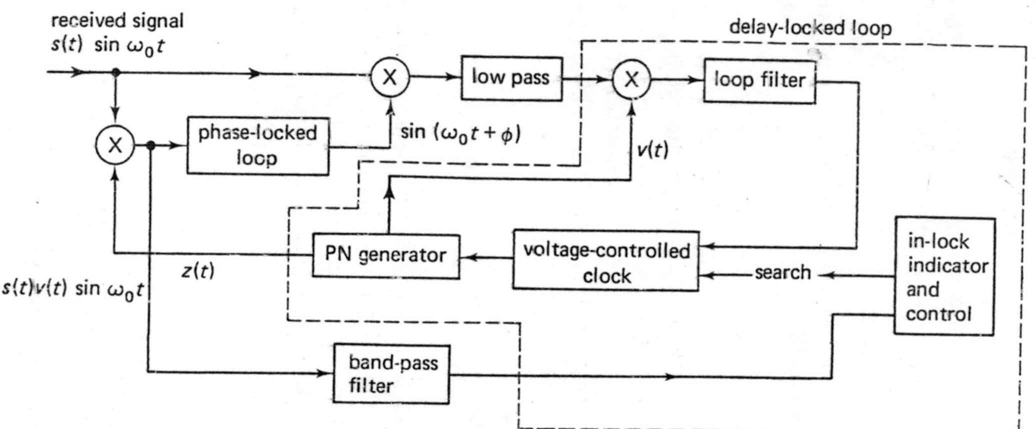

Fig. 18-24 Coherent PN tracking system for a spread spectrum transmission without bandpass correlations

18-9 TRANSIENT PERFORMANCE

In this section, two problems are investigated concerning the transient performance of the baseband and envelope correlator delay-lock discriminators in Figs. 18-4 and 18-13 in the absence of noise:

1. How rapidly can a given region be searched and the target acquired for a given closed-loop noise bandwidth, $B_n = 1.06p_0$ Hz?
2. Once the target signal has been acquired, what is the maximum change in velocity that can be tolerated without losing the locked-on condition?

These results represent bounds on the performance. In the presence of high-level noise, one generally has to decrease the search rate. Improved detection strategies are possible using sequential correlation detectors or a simplified version thereof in an incremental delay search technique. (See Sec. 18-10.)

We first rewrite the closed-loop response for the general baseband delay-lock loop, neglecting the noise effects and making use of the time normalizations $x \triangleq \epsilon/\Delta$, $y \triangleq \tau/\Delta$:

$$y - x = \frac{\hat{\tau}}{\Delta}, \quad \mathrm{s} \triangleq \frac{p}{p_0} = \frac{1}{p_0}\frac{d}{dt}, \quad g \triangleq g_0/p_0$$

Rewrite (18-27) using these newly defined terms and set $M \longrightarrow \infty$ and n_s, $n_n = 0$. Define $D(x)$ as the delay-lock loop discriminator characteristic for any type of DLL. We also normalize $D'(0) = 1.0$ so that g is the loop gain. The differential equation for this general DLL (noise free) is then

$$\mathrm{s}(y - x) = gF(\mathrm{s})D(x) \tag{18-115}$$

The integral plus proportional-control loop filter now becomes

$$gF(\mathrm{s}) = \frac{1 + \sqrt{2\mathrm{s}}}{1/g + \mathrm{s}} \tag{18-116}$$

and the operator equation for the loop becomes

$$\left(\frac{1}{g} + \mathrm{s}\right)\mathrm{s}(y - x) = (1 + \sqrt{2\mathrm{s}})D(x) \tag{18-117}$$

This system equation can be rewritten in time-derivative notation

$$\frac{\dot{y}}{g} + \ddot{y} = \frac{\dot{x}}{g} + \ddot{x} + D(x) + \sqrt{2}\,D'(x)\dot{x} \tag{18-118}$$

where $\dot{x} \triangleq 1/p_0(dx/dt)$, $D'(x) \triangleq dD/dx, \ldots$. The phase-plane method of solving this second-order differential equation is to compute the trajectories in $x, \dot{x}$ space. These trajectories describe the solution of this equation for the desired sets of initial conditions. Define a new phase plane slope variable $y \triangleq \ddot{x}/\dot{x} = d\dot{x}/dx$, which from (18-118) is given by

$$\begin{aligned}\frac{d\dot{x}}{dx} &= y[x, \dot{x}, \dot{y}, \ddot{y}] \\ &= -\frac{D(x) + [\sqrt{2}\,D'(x) + 1/g]\dot{x} - \dot{y}/g - \ddot{y}}{\dot{x}}\end{aligned} \tag{18-119}$$

If the dc loop gain $g \longrightarrow \infty$ and $\ddot{y} = 0$, then this equation simplifies to

$$\frac{d\dot{x}}{dx} = -\left[\frac{D(x)}{\dot{x}} + \sqrt{2}\,D'(x)\right] \tag{18-120}$$

Figure 18-25 shows the slope $d\dot{x}/dx$ of the phase-plane trajectories for the 2Δ DLL for the phase plane at $\dot{x} = 0.707$. Note that if the phase-plane trajectory hits ($\dot{x} = 1$, $x = 1$), the trajectory does not converge to the origin since the slope changes sign at $x = +1$. A similar statement can be made for the 1Δ DLL at $x = 0.5$. Thus, one can get an approximate feel for the conver-

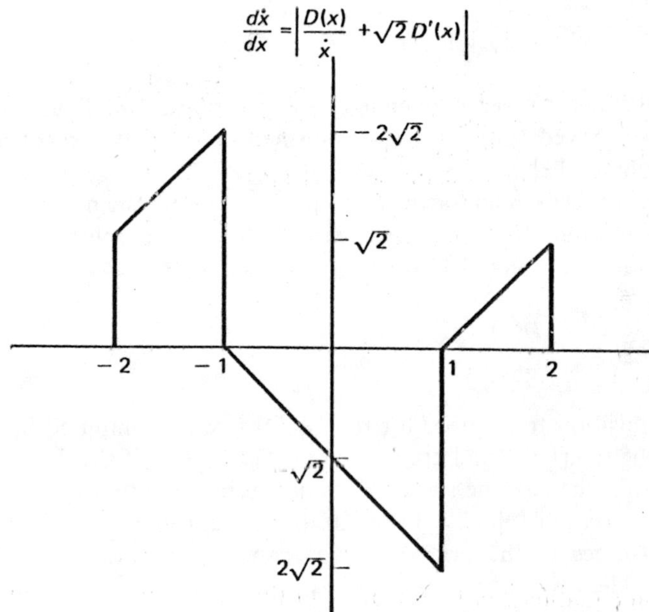

Fig. 18-25 Plot of $d\dot{x}/dx$ for $\dot{x} = 0.707$ for the 2Δ delay-lock loop

gence regions from a simple examination of $d\dot{x}/dx$ at a constant value of $\dot{x}$. Computer solutions to these trajectories can be obtained by approximating the differential equation with the difference equation

$$\begin{aligned}\dot{x}_{n+1} - \dot{x}_n &\triangleq \dot{x}\left(x_0 + \sum_{j=0}^{n} \delta_j\right) - \dot{x}\left(x_0 + \sum_{j=0}^{n-1} \delta_j\right) \\ &\cong \gamma(\dot{x}_n, x_n)\, \delta_n \end{aligned} \tag{18-121}$$

where x_0, $\dot{x}_0$ are the initial values of x, $\dot{x}$, and where δj is the jth increment in x. Satisfactory solutions can be obtained by letting the computer function as an adaptive device, so that the size of the interval is made variable. In this particular sequential computation [Bellman, 1961, Chap. 4], the interval size is taken as

$$|\delta_j| = \frac{\delta}{1 + |\gamma(\dot{x}_j, x_j)|} \tag{18-122}$$

The value δ to be used is 0.02. Notice that the distance Δr_n moved in the x, $\dot{x}$ plane in one increment is

$$\begin{aligned}\Delta r_n &= \sqrt{(x_{n+1} - x_n)^2 + (\dot{x}_{n+1} - \dot{x}_n)^2} \\ &= \delta\left(\frac{1 + \gamma_n^2}{1 + 2|\gamma_n| + \gamma_n^2}\right)^{1/2}\end{aligned} \tag{18-123}$$

Hence, the distance moved is bounded by $\delta\sqrt{2} \leq r_n \leq \delta$. Thus, the maximum distance moved is δ, yet needlessly small increments are not used for small or moderate $|\gamma|$.

Consider the search-and-acquisition problem where the normalized loop gain $g = \infty$. Assume that the target search velocity is a constant, $\dot{y}(t) = \dot{y}$, $\ddot{y} = 0$. The variable γ in (18-119) now becomes

$$\gamma(x, \dot{x}) = -\frac{D(x) + \sqrt{2}\, D'(x)\dot{x}}{\dot{x}} \tag{18-124}$$

The acquisition trajectories for the 2Δ DLL were computed by Mulloy [1962] and Nielsen [1975] and are plotted in Fig. 18-26*a*. If the delay error is decreasing from the left, the system does not respond until $x = -2$. As can be seen, if the search velocity $|\ddot{y}| \leq 2.0$, the system locks on and the state variable converges to the origin. As an example, consider $\Delta = 10^{-6}$ sec, $p_0 = 10$ rps. Then we have

$$\dot{y} = \frac{(d\tau/dt)}{p_0\,\Delta} = \frac{10^5\, d\tau}{dt} \tag{18-125}$$

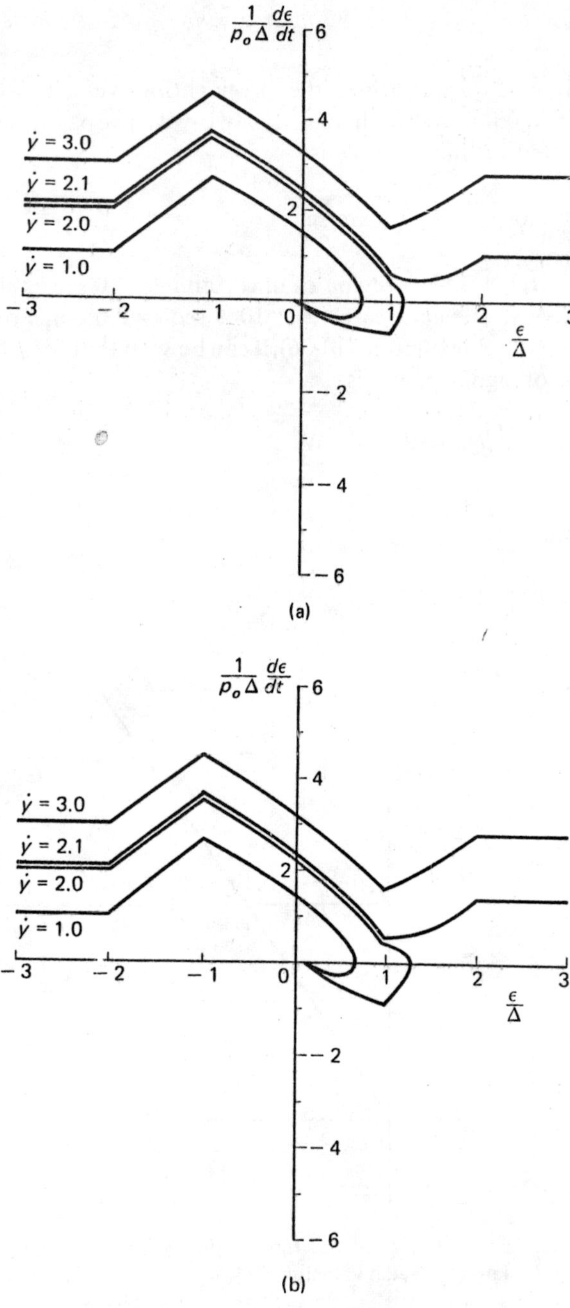

Fig. 18-26 Baseband 2Δ delay-lock loop acquisition trajectories for various values of $(1/p_0\Delta)\,(d\epsilon/dt)$ the normalized search velocity. The loop gain in (a) is $g = \infty$, and in (b) it is $g = 10$ [From Nielsen, 1975].

For electromagnetic propagation, the propagation velocity, $c = 3 \times 10^5$ km/sec and the maximum search velocity (two-way propagation time) for a 2Δ DLL permitted in this example is

$$v = (\tfrac{1}{2})2.0cp_0\,\Delta = 3.0 \text{ km/sec} \tag{18-126}$$

Thus, we can search a 30-km region of uncertainty in 10 sec as shown in Fig. 18-27. Alternatively, one can search 2×10^{-6} sec or 2.0 chips of code in 1.0 sec. From the threshold equation (18-36), it can be seen that for $p_0 = 10$ rps the threshold value of signal power is

$$P_s = 23.53p_0N_0 = 235.3N_0\text{W} \tag{18-127}$$

in this example ($M \gg 1$).

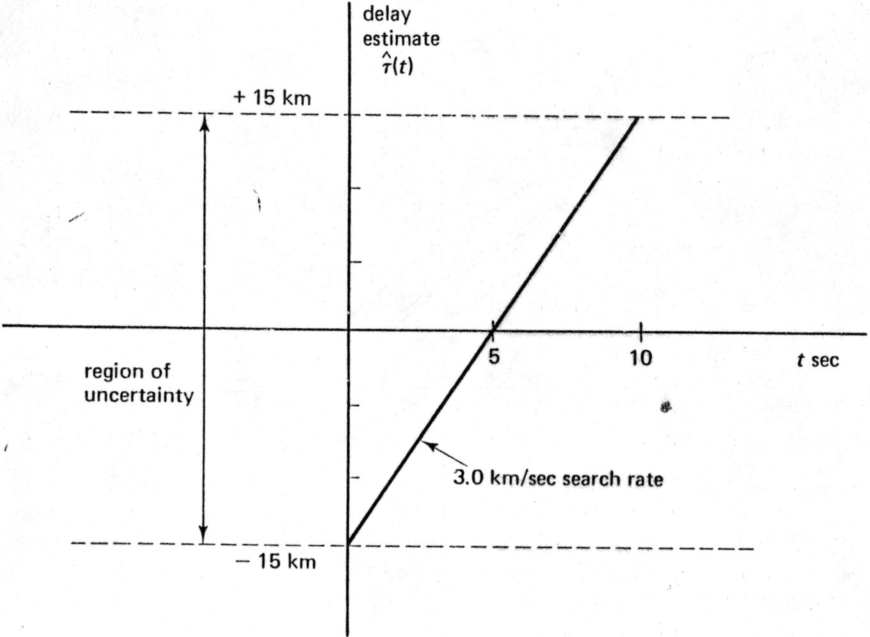

Fig. 18-27 Linear search rate of the delay-lock loop at 3.0 km/sec search velocity

Notice, however, that even if the system fails to lock on in this particular interval of x, the velocity error is less at the end of the transient than at the beginning. Since $D(x)$ is periodic every M, the system always locks on eventually, because with $g = \infty$ there is no decay in x outside of the intervals

$|x| < 2$. It may, however, pass through stable regions of x many times before locking on. In practice, this behavior might rely on unrealistic storage times (unless digital filtering is utilized) in the filter as is implicit in the assumption that $g = \infty$. Furthermore, the time required for lock-on might be intolerable for search velocities $|\dot{y}| > 2.0$, and there is an additional high sensitivity to noise effects.

The effect of target velocity transients can also be ascertained from Fig. 18-26. If the system is initially locked on, ($x = 0$, $\dot{x} = 0$) and the target suddenly changes velocity to $\dot{y}$, the system response is described by that portion of the trajectories which begins at ($x = 0$, $\dot{x} = \dot{y}$). As can be seen, normalized velocity transients of $\dot{y} = 2.3$ can be tolerated without losing the locked-on state for the 2Δ DLL.

Acquisition trajectories have also been obtained for finite-loop-gain systems with $g = 10$ (Fig. 18-26*b*). The equation for γ in this example is

$$\gamma(x, \dot{x}, \dot{y}) = -\frac{D(x) + (\sqrt{2}\,D'(x) + 0.1)\dot{x} - 0.1\dot{y}}{\dot{x}} \tag{18-128}$$

Because of the finite loop gain, the maximum steady-state clock frequency change is $g = 10p_0$. Thus, for this reason alone, the system would never lock on for $|\dot{y}| > 10$.

Note that the reduction in loop gain from ∞ to 10 has very little effect on the maximum tolerable search velocity; $|\dot{y}| \leq 2.0$ is still permitted. However, the curves that lock on converge to $x = 0.1\dot{y}$ or $\epsilon = x\Delta = 0.1\Delta\dot{y}$ rather than to $x = 0$. For $\Delta = 10^{-6}$ sec and $\dot{y} = 1$, the steady-state lock-on point is $\epsilon = 0.1$ μsec. In principle, by measuring $\dot{y}$ one could correct for this steady-state bias error since the loop gain is known if the signal level is known.

Although it is not shown on the trajectory curves, the value of $\dot{x}$, for the curves that do not lock on, decays with a time constant $g/p_0 = 10/p_0$ sec. Thus, if the period M of the sequence is large so that $M \gg 1$, the value of $\dot{x}$ will have decayed with time almost back to its original value when $x = M - 2$, the next opportunity for acquisition.

As well as knowing whether or not the system will lock on for a given search velocity, it is also important to know how long the transient lasts. The normalized time response of the system can be obtained from computer solutions to the difference equation

$$\begin{aligned}\Delta t \triangleq t_{n+1} - t_n \triangleq t\left(x_0 + \sum_{i=0}^{n} \delta_i\right) - t\left(x_0 + \sum_{i=0}^{n-1} \delta_i\right) \\ \cong \frac{\delta}{(1 + |\gamma_n|)\dot{x}(x_n)}\end{aligned} \tag{18-129}$$

where t_n is the time required for x to progress from x_0 to $x_0 + \sum_{i=0}^{n-1} \delta_i$.

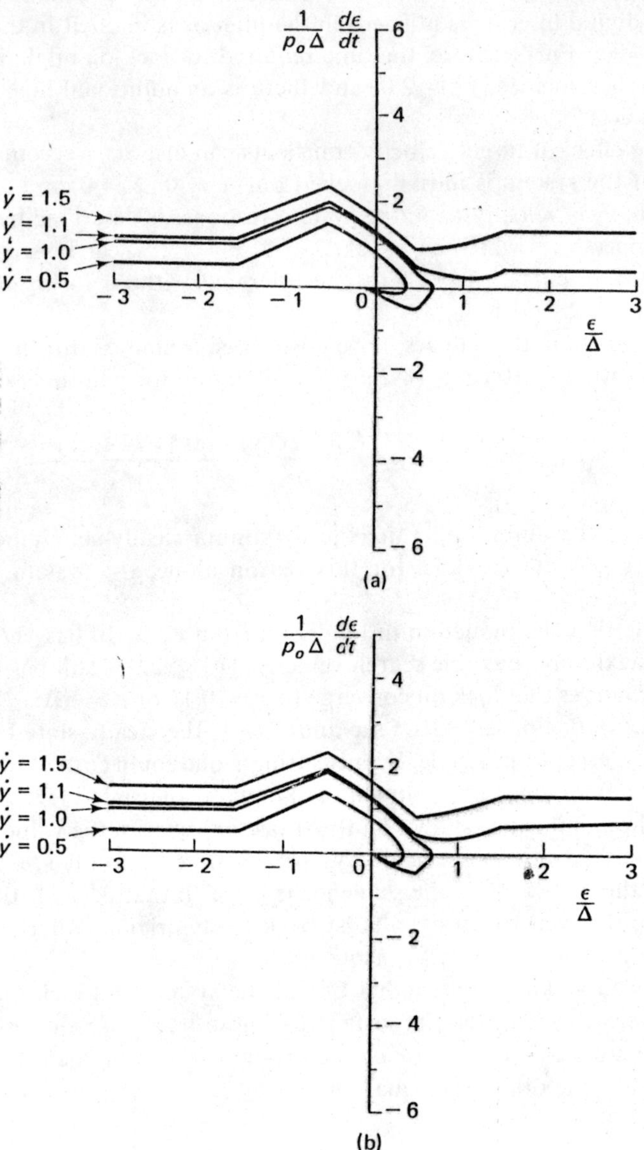

Fig. 18-28 Baseband 1Δ delay-lock loop acquisition trajectories for various values of $(1/p_0\Delta)(d\epsilon/dt)$, the normalized search velocity. The loop gain in (a) is $g = \infty$ and in (b) $g = 10$ [From Nielsen, 1975].

Transient Response for the 1Δ Baseband DLL

Figure 18-28 shows the acquisition transients for a 1Δ baseband delay-lock loop. Notice that the maximum search velocity has decreased by a factor of 2 to $\dot{y} = 1.0$. Note that this decrease in maximum search velocity is the same factor as the decrease in the interval between peaks in the discriminator characteristic, 2Δ for the 2Δ DLL and 1Δ for the 1Δ DLL. *Only during the time when the delay error is within this interval does the phase plane trajectory progress towards the stable point* (the origin for the $g = \infty$ loop).

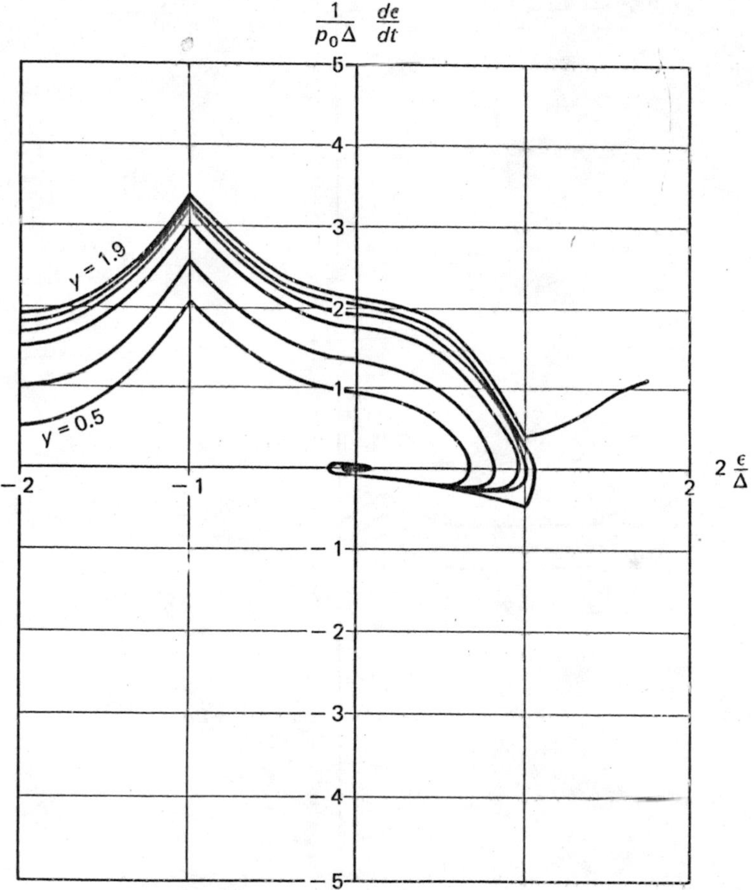

Fig. 18-29 Lock-on trajectories for the 2Δ delay-lock loop using envelope correlation; plotted for various values of the normalized search velocity $\dot{y}$

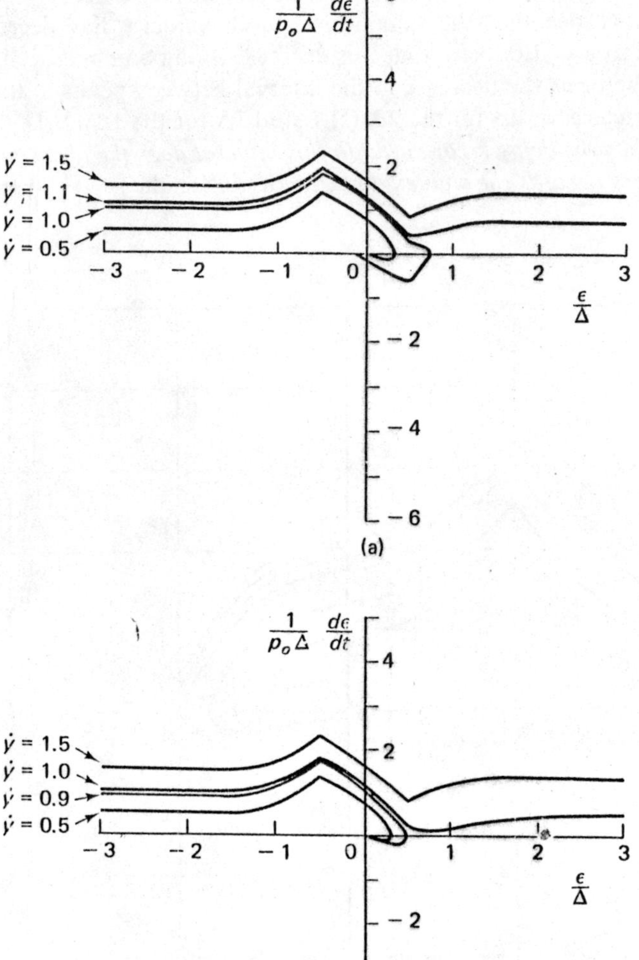

Fig. 18-30 Acquisition trajectories for the 1Δ delay-lock loop with envelope correlation. The loop gains (a) $g = \infty$, (b) $g = 10$ [From Nielsen, 1975].

Transient Response Using Envelope Correlation

The phase-plane trajectories for the 1Δ and 2Δ envelope correlation (square-law detector) delay-lock loop are plotted in Figs. 18-29 and 18-30. These can be compared with those for coherent (video) delay-lock loops; of particular interest are the normalized search velocity $\dot{y}$ and the behavior of the trajectories in the region of the origin of the phase plane.

The actual search velocity expressed in PN sequence bits or chips per second is related to the normalized search velocity by

$$\text{search velocity in chips/sec} = \dot{y}p_0$$

where p_0 is the loop frequency constant discussed earlier. For the coherent

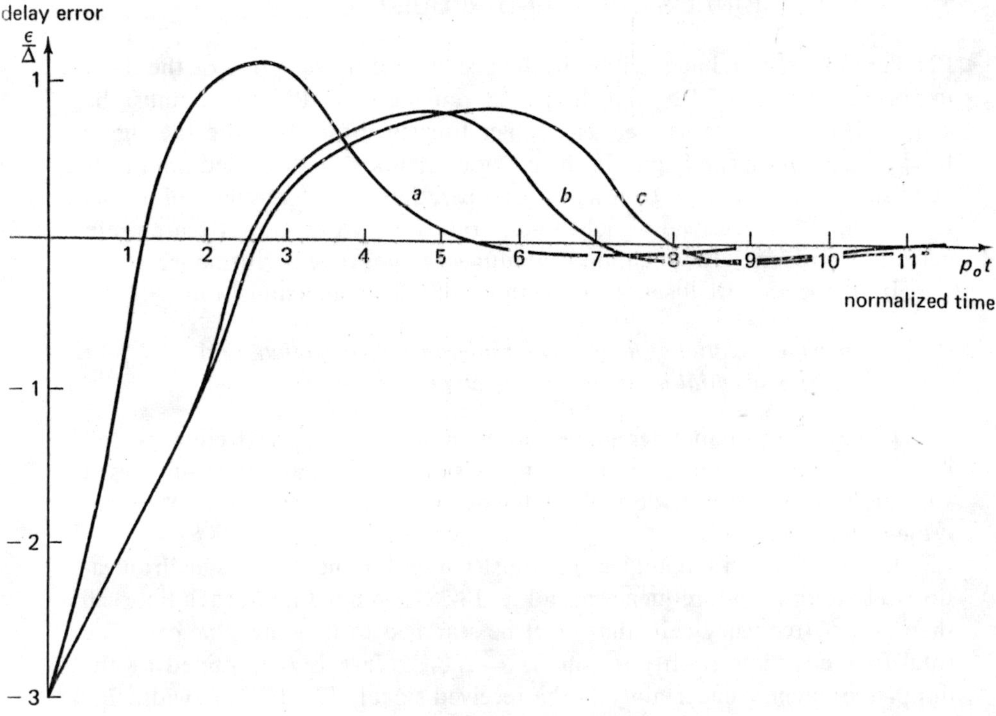

Fig. 18-31 Acquisition transients vs time for the delay-lock loop, loop gain $g = \infty$: (a) 2Δ loop with video correlation, $\dot{y} = 2.0$, (b) 1Δ loop with video correlation, $\dot{y} = 1.0$, (c) 1Δ loop with envelope correlation, $\dot{y} = 1.0$ [From Nielsen, 1975]

correlation 2Δ loop, the maximum $\dot{y}$ is 2.0. For the envelope correlation* delay-lock loop, the maximum $\dot{y}$ is slightly lower: 1.8 for the 2Δ loop. In the region of the origin, the trajectory of the 2Δ loop slows down considerably since the sensitivity of the loop with square-law detectors approaches zero as ϵ approaches zero. (See Fig. 18-25.)

The acquisition transients 1Δ for the 2Δ envelope correlation delay-lock loop are shown in Figure 18-30. The maximum search velocities from $g = \infty$, 10, are $\dot{y} = 1.0$ and 0.9, respectively. Values of $\dot{x}$ are about half those for the 2Δ DLL in Fig. 18-29; hence response time for the same loop natural frequency p_0 is slower. The 1Δ acquisition trajectories are essentially the same for the 1Δ delay-lock loop for envelope correlation as for baseband correlation. In Fig. 18-31, the lock-on transient response is plotted vs time for maximum search velocities and high loop gain.

18-10 PN SIGNAL SEARCH AND ACQUISITION

PN signal search and acquisition in the presence of noise is one of the most important aspects of PN signalling and ranging. The PN signal must be acquired in both time as well as doppler frequency offset before the signal tracking operation can begin. In the previous sections we discussed nonlinear acquisition and hold-in transients for operation in the absence of noise. Operation at low received signal-to-noise ratios requires a slower search rate generally, and special acquisition procedures can improve performance.

In this section we discuss two methods of PN signal acquisition:

a) *conventional fixed format noncoherent detection,* and
b) *sequential noncoherent detection.*

Fixed format noncoherent detection samples each time-frequency cell for a fixed time T_d and then makes a decision. Sequential detection uses a variable time to reach a desired confidence level for either decision: *signal present* or *signal absent*.

In either search tenchique, one must search for the signal synchronization in both time and frequency as in Fig. 18-32. Assume that at each time cell there are N frequency cells that must be searched to find the sync cell. The total frequency uncertainty region $\Delta F = NB_{\text{IF}}$ is largely determined by the doppler frequency uncertainty in the received signal. The IF bandwidth B_{IF} must be large enough to pass any data modulation or other spectral spreading that remains after the PN code is removed, e.g., effects of oscillator phase noise or finite observation time.

---

*Gill, 1966.

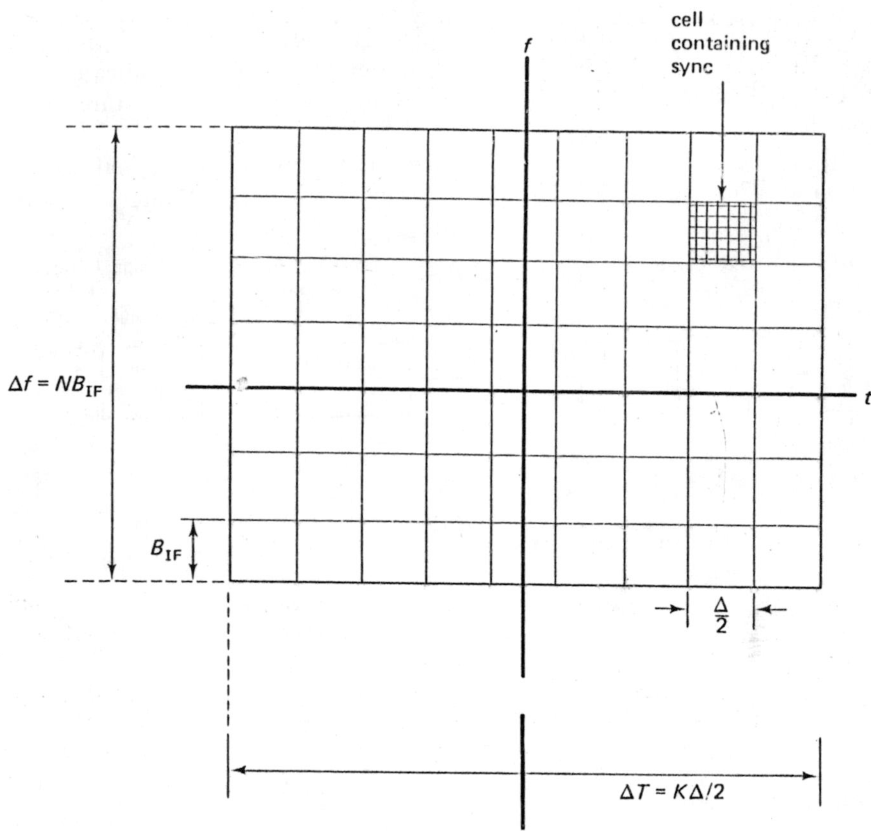

Fig. 18-32 PN signal search in time and frequency

The time uncertainty ΔT is determined either by the code period or by the initial range or clock uncertainty. Usually one steps in time increments of $\Delta/2$ rather than increments of Δ in order that at least one sample be near the correlation peak.

Fixed Format Search Techniques

In conventional fixed format search techniques, one samples each cell for T_d sec and makes a decision as to whether or not the cell contains sync. The threshold for this decision is set to produce a maximum probability of detection for a prescribed false alarm rate at the given signal-to-noise ratio.

One can generalize this technique to require that there be n detection decisions out of m trials ($n < m$) for an "in-lock" indication. Furthermore, one can utilize a form of hysteresis to prevent a false "out of lock" indication after acquisition has been achieved. For example, one can indicate out-of-lock or loss of sync only if there are four consecutive below-threshold indications.

The block diagram of Fig. 18-33 shows the fixed format PN signal search technique. The received signal is passed through a bandpass RF filter centered

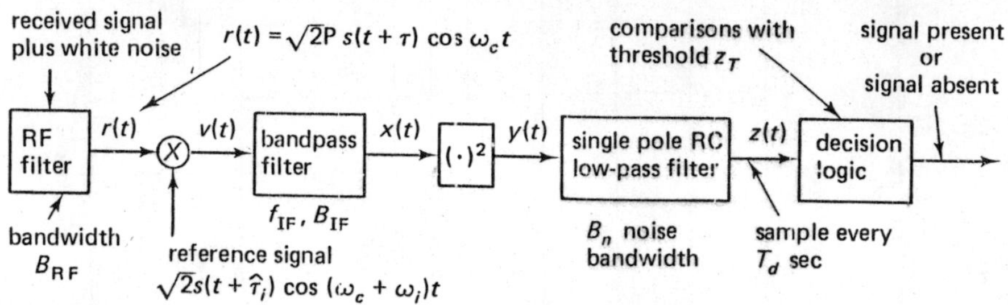

Fig. 18-33 Fixed format PN signal acquisition using noncoherent detection. The code phase estimate $\hat{\tau}_i$ or reference frequency offset ω_i is switched every T_d sec. The output of the low-pass filter is sampled every T_d sec in synchronism with the signal switching.

at f_c to prevent image frequency noise and interference from entering the correlation mixer. The RF filter bandwidth B_{RF} is assumed to be large compared to the signal bandwidth. The received signal is then mixed in the cross-correlation mixer with a reference signal

$$q(t) = \sqrt{2}\, s(t + \tau_k) \cos [(\omega_c + \omega_i)t] \tag{18-130}$$

where $s(t) = \pm 1$ is the PN sequence, $\omega_i = \omega_{IF} + i2\pi B_{IF}$, and $\tau_k = k\,\Delta/2 + \tau_0$. The value of (i, k) is changed each decision time interval T_d. The output of the mixer contains white noise plus a possible pure sinusoid (or a sinusoid modulated with narrow-band data). If the value of (i, k) is proper, the sinusoid representing the signal correlation level passes through the bandpass filter, is squared in the square-law detector, and the detector output is then filtered by a low-pass RC filter of bandwidth B_n. At decision time nT_d, filter output $z(t)$ is sampled and compared with a threshold value z_T. If $z(n\,T_d) > z_T$, then the cell is assumed to contain sync; if not, it is assumed to contain only noise, and the search is continued.

Define the input signal plus noise at the mixer input to be

$$\begin{aligned} r(t) &\triangleq \sqrt{2P_s}\, s(t) \cos (\omega_c t + \phi) + n(t) \\ n(t) &\triangleq N(t) \cos [(\omega_c t + \phi_n(t)] \end{aligned} \tag{18-131}$$

where P_s is the received signal power, $s(t)$ is the PN sequence, and $n(t)$ is bandlimited white noise of power spectral density N_0 watt/Hz (one-sided).

The output of the mixer in the IF frequency band is then the sum of difference frequency components of signal and noise

$$v(t) = \sqrt{P_s}\, s(t)s\left(t + k\frac{\Delta}{2}\right) \cos\,(\omega_{\mathrm{IF}} + i2\pi B_{\mathrm{IF}})t$$

$$+ \frac{N(t)}{\sqrt{2}} s\left(t + k\frac{\Delta}{2}\right) \cos\,[(\omega_{\mathrm{IF}} + i2\pi B_{\mathrm{IF}})t + \phi_n(t)] \qquad (18\text{-}132)$$

The noise term remains as bandlimited white noise of power spectral density $N_0/2$ w/ Hz.

The IF filter bandwidth is assumed to be narrow with respect to the signal bandwidth; that is, $B_{\mathrm{IF}} \ll 2/\Delta$ but is wide compared to the observation time. When the signal falls in the IF passband, the IF filter output is then

$$x(t) = \sqrt{P}\, R_s\left(k\frac{\Delta}{2}\right) \cos\,(\omega_{\mathrm{IF}} + i2\pi B_{\mathrm{IF}})t$$

$$+ N'(t) \cos\,[\omega_{\mathrm{IF}}t + \phi_n'(t)] \qquad (18\text{-}133)$$

where $R_s(\ )$ is the code autocorrelation function, and the noise term has spectral density $N_0/2$ and power $(N_0/2)B_{\mathrm{IF}}$. If the signal is outside of the IF passband, or if $k > 2$, the first term is zero.

The low-pass component of the square-law detector output for signal present in the IF ($i = 0,\ k = 0,\ \phi_n \ll 1$) is

$$y(t) = x^2(t)\Big|_{\text{low pass}} \cong \frac{P_s}{2} + \frac{N'^2(t)}{2} + \sqrt{P_s}\, N'(t) \qquad (18\text{-}134a)$$

The probability density of this $y(t)$ is Rician; that is, signal is absent.

If the signal component falls outside of the IF passband, or if $|k| > 2$, then

$$y(t) = \frac{N'^2(t)}{2} \qquad (18\text{-}134b)$$

This $y(t)$ has a Rayleigh probability density.

The noise components at the low-pass filter output (18-134) have power either

$$\frac{P_s N_0}{2} + \left(\frac{N_0}{2}\right) B_{\mathrm{IF}} \qquad \text{signal present}$$

$$\left(\frac{N_0}{2}\right) B_{\mathrm{IF}} \qquad \text{signal absent} \qquad (18\text{-}135)$$

Assume that the low-pass detection filter is a single pole RC with a time constant T_c and a noise bandwidth B_n, $T_c = \frac{1}{2} B_n$. Assume further that the noise bandwidth is small compared to the IF bandwidth $B_n \ll B_{IF}$ so that the low-pass filter output is approximately Gaussian. If the signal is switched into the IF for T_d sec while the noise is present continuously, then the mean output of the low-pass filter is either

$$\left.\begin{aligned} \mathbf{E}[z(T_d)] &= \frac{P_s}{2}(1-\alpha) + \frac{N_0 B_{IF}}{2} \\ &= \frac{N_0 B_{IF}}{2} \end{aligned}\right\} \begin{array}{l} \text{signal present} \\ \text{signal absent} \end{array} \tag{18-136}$$

where we define $\alpha \triangleq \exp - (T_d/T_c)$. Similarly the variance of the low-pass filter output is

$$\left.\begin{aligned} \sigma_z^2 &= P_s(1-\alpha^2)\frac{N_0 B_n}{2} + \left(\frac{N_0}{2}\right)^2 \frac{B_n B_{IF}}{2} \\ &= \left(\frac{N_0}{2}\right)^2 \frac{B_{IF} B_n}{2} \end{aligned}\right\} \begin{array}{l} \text{signal present} \\ \text{signal absent} \end{array} \tag{18-137}$$

For a threshold setting of z_T the probabilities of false dismissal for signal present and false alarm signal absent are [Huang, 1974]

$$\mathcal{P}_d = \mathcal{P}(z < z_t)\Big|_{\text{signal present}}$$

$$= \frac{1}{2}\operatorname{erfc}\left\{\frac{1 - \alpha + \dfrac{1}{\mathrm{SNR}_{IF}} - z_T}{\sqrt{\left[2(1-\alpha^2) + \dfrac{1}{(\mathrm{SNR})_{IF}}\right]\dfrac{1}{(\mathrm{SNR})_{IF}}\dfrac{B_n}{B_{IF}}}}\right\} \tag{18-138}$$

where $(\mathrm{SNR})_{IF} = \dfrac{P_s}{B_{IF} N_0}$, $\operatorname{erfc} x \triangleq \dfrac{2}{\sqrt{\pi}} \displaystyle\int_x^\infty e^{-y^2}\, dy.$

The probability of false alarm is

$$\mathcal{P}_a = \mathcal{P}(z > z_T)\Big|_{\text{signal absent}}$$

$$= \frac{1}{2}\operatorname{erfc}\left[\frac{z_T - \dfrac{1}{(\mathrm{SNR})_{IF}}}{\sqrt{\dfrac{4B_n}{B_{IF}}\dfrac{1}{(\mathrm{SNR})_{IF}}}}\right] \tag{18-139}$$

The optimum value of T_d for a $(SNR)_{IF}$ range of 0 dB to 4 dB is $T_d \cong 1.1\ T_c = 1.1/4B_n$. This value of T_d is used in the calculation below.

Figure 18-34 shows a plot of the probability of detection for a fixed setting of false alarm probability $\mathcal{P}_a = 5 \times 10^{-3}$. The results are given as a function of the IF SNR for various values of the ratio of B_n/B_{IF}. Note that at $\mathcal{P}_d = 0.9$, a factor of 2 decreases in the low-pass filter bandwidth decreases the required $(SNR)_{IF}$ by approximately 1.8 dB. At lower $(SNR)_{IF}$, a factor of 2 decreases in bandwidth results in only a (3 dB)/2 = 1.5 dB decrease in $(SNR)_{IF}$ because of square-law noise effects.

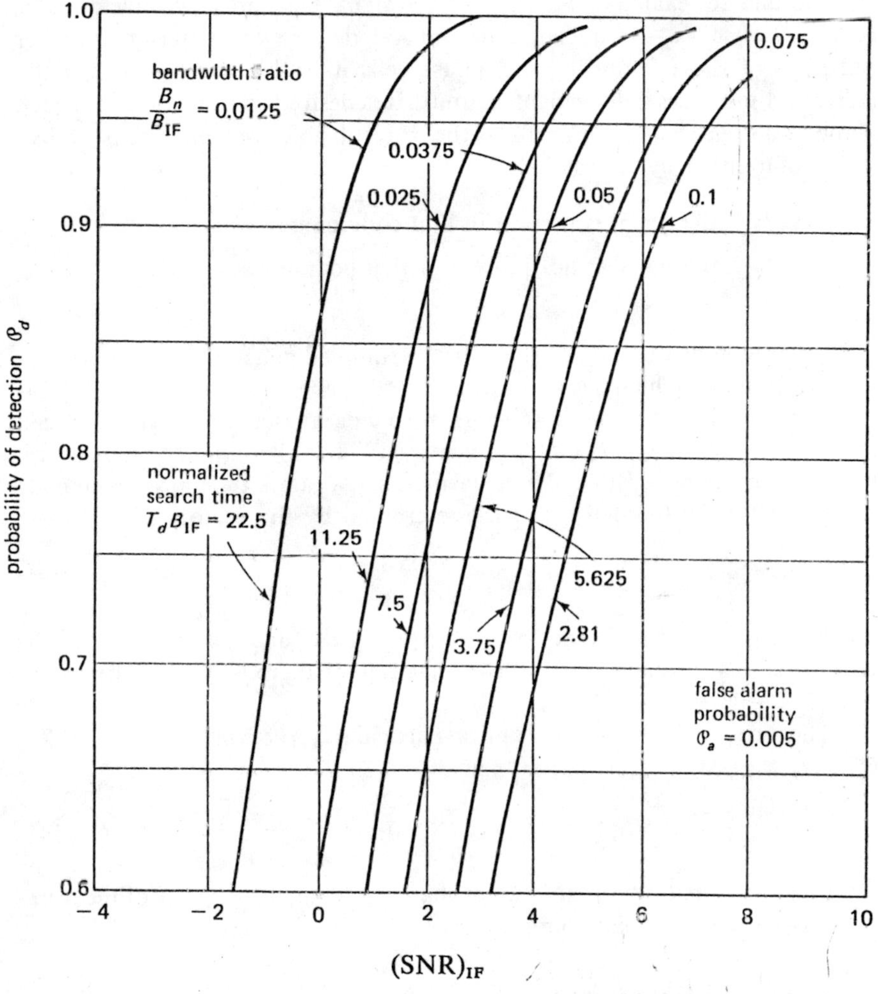

Fig. 18-34 Probability of detection (dB) vs. $(SNF)_{IF}$ for fixed format noncoherent sync detector. The time frequency cells are searched at a rate $f_s = 4\ B_n/1.125 = 1/T_d$. Thus, $T_d B_{IF} = (B_{IF}/B_n)(1.125/4)$. [After Huang, 1974]

Code Acquisition by Sequential Detection

Rather than skew the reference code with respect to the received signal in the delay-locked loop as described above, one can sequentially step the reference code phase over the region of delay uncertainty. This approach can lead to more rapid acquisition in a noise environment. Once code synchronization is obtained, the delay-lock loop can be closed, and tracking proceeds from there.

Assume, for example, that the received signal has both a possible doppler frequency offset of N frequency cells and K code offset cells. Assume further that each of these frequency and phase cells must be searched, is equally likely, and produces independent outputs. It is desired to test (correlate) each of the code phases in turn at all of the N resolvable frequency cells and to decide that either hypothesis

H_A: the signal is present in that code phase

or H_B: the signal is not present at that code phase

is correct.

A variable number, m, of sequentially sampled coherent or noncoherent correlation measurements is made at each cell. Each measurement is of duration δt sec. Additional measurements are made at that time/frequency cell until either hypothesis H_A or H_B is accepted. Thus the number of measurements is a random variable. The measured signal amplitude at a given code phase and the ith frequency cell is represented by the vector

$$\mathbf{X}_i^m \triangleq (x_{i_1}, x_{i_2}, \ldots, x_{im}) \tag{18-140}$$

Let $p(\mathbf{X}_i^m \mid a)$ be the conditional probability density function of $\mathbf{X}_i^m$ where the parameter a is the signal strength in that cell. If there is no signal present, $a = 0$.

The probabilities of observing a particular set of measurements, $\mathbf{X}_1^m$, $\mathbf{X}_2^m, \ldots, \mathbf{X}_N^m$ after m samples are then

$$p_{0m} = p(\mathbf{X}_1^m \mid 0) p(\mathbf{X}_2^m \mid 0) \cdots p(\mathbf{X}_N^m \mid 0) \tag{18-141}$$

given that no signal is present. The probability of making this set of measurements for a signal present condition is

$$\begin{aligned} p_{1m} = \frac{1}{N}[&p(\mathbf{X}_1^m \mid a) p(\mathbf{X}_2^m \mid 0) \cdots p(\mathbf{X}_N^m \mid 0) \\ &+ p(\mathbf{X}_1^m \mid 0) p(\mathbf{X}_2^m \mid a) \cdots p(\mathbf{X}_N^m \mid 0) \\ &+ \cdots + p(\mathbf{X}_1^m \mid 0) p(\mathbf{X}_2^m \mid 0) \cdots p(\mathbf{X}_N^m \mid a)] \end{aligned} \tag{18-142}$$

given that signal of amplitude a is present in one of the frequency cells. The likelihood ratio r is then computed for the observed set of measurements from (18-141) and (18-142) and compared with threshold value A, B:

$$r \triangleq \frac{p_{1m}}{p_{0m}} = \frac{1}{N} \sum_{i=1}^{N} \frac{p(\mathbf{X}_i^m \mid a)}{p(\mathbf{X}_i^m \mid 0)} \tag{18-143}$$

If $r \leq B$, then H_B is accepted, and it is decided that no signal is present. The correlator measurement then proceeds to examine the next code-phase cell. If $r \geq A$, then the code phase is accepted and search stops. If $B < r < A$, then the receiver makes neither decision, examines the next sample, and repeats the calculation until either threshold is crossed. In principle, one need never accept the signal present hypothesis H_A for signal search and acquisition. One simply never departs from that code phase. In practice, however, one usually wants a good indication that the receiver is "in-lock" and is functioning properly.

The thresholds A, B are set by the false-alarm probability $\mathcal{P}_a$ and the probability $\beta \triangleq 1 - \mathcal{P}_d$ of missing sync when the signal is present. Wald [1947] has computed these threshold levels:

$$A = \frac{1-\beta}{N\mathcal{P}_a} \qquad B = \frac{\beta}{1 - N(\mathcal{P}_a)} \tag{18-144}$$

Marcus and Swerling [1962, 237–245] have computed the performance for noncoherent sequential detection. The IF bandwidth at the correlator output is assured to provide statistically independent samples having the Rice distribution

$$p(x_{ij}) = x_{ij} \exp\left[-\frac{(x_{ij}^2 + a^2)}{2}\right] I_0(ax_{ij}) \qquad x_{ij} > 0 \tag{18-145}$$

where x_{ij} is the magnitude of the envelope correlator output. The normalized signal amplitude is a, and the SNR $R = a^2/2$.

The likelihood ratio (18-143) is then

$$r = \frac{p_{1m}}{p_{0m}} = \frac{1}{N} \exp\left(-\frac{ma^2}{2}\right) \sum_{i=1}^{N} \exp\left[\sum_{j=1}^{m} \ln I_0(ax_{ij})\right] \tag{18-146}$$

where $I_0(X)$ is the modified Bessel function of the first kind of order zero.

For $\mathcal{P}_a = 10^{-8}$ and probability of detection $\mathcal{P}_d \triangleq 1 - \beta = 0.9$, the required average number of samples per cell m is as given in Table 18-1. The total number of samples to be taken over the ambiguity region is $\bar{m}NK$. These results were obtained by Marcus and Swerling for the no-signal case using computer simulation. The result for coherent detection is also shown for

Table 18-1 EXPECTED NUMBER OF SAMPLES PER RESOLUTION ELEMENT m, FOR $\mathcal{P}_a = 10^{-8}$, AND $\mathcal{P}_d = 0.9$. THE NUMBER N IS THE NUMBER OF FREQUENCY CELLS TO BE EXAMINED FOR EACH TIME CELL.

IF Signal-to-Noise Ratio, R, dB	Expected Number of Samples $\bar{m}$: Noncoherent Detection $N = 1$	Noncoherent Detection $N = 4$	Coherent Detection $N = 1$
−3	32	40	12
0	10	13	5.5
3	4	5	3
7	2	2	1.2

$N = 1$. Coherent detection, however, assumes knowledge of RF phase, which is unrealistic for many search problems. For the case where $N = 1$ and coherent detection, Gumacos [1963, 89–99] established the expected number of samples per cell to be $\bar{m} = -2(\ln \beta)/R$ where R is the $(\text{SNR})_{\text{IF}}$. For $R = 1$ or 0 dB and $\beta = 0.1$, then $\bar{m} = 4.6$ for a uniform a priori distribution of sync. This result compares with $\bar{m} = 5.5$ obtained by Marcus and Swerling for slightly different assumptions.

Thus, if there are $NK = 4000$ cells to be searched, $N = 4$ frequency cells for each of the $K = 1000$ time cells and the IF SNR is −3 dB, then $\bar{m} = 40$ samples are required for each cell. If noncoherent detection samples are taken at 1-msec intervals, the expected search time (worst case) is

$$mNK(10^{-3}\text{ sec}) = 16{,}000 \times 10^{-3}\text{ sec} = 16\text{ sec} \tag{18-147}$$

To assure that a cell occurs near the correlation peak of the PN code, it is common to sample twice per PN code chip. Hence, K is twice the number of code chips of uncertainty, and this example corresponds to 500 code chips of uncertainty. The expected number of samples $\bar{m}$ per resolution element for sequential detection can be compared loosely to the number $T_d B_{\text{IF}}$ for the fixed-format signal detection discussed earlier in this section. Notice that at $(\text{SNR})_{\text{IF}} = 0$ dB a time bandwidth product $T_d B_{\text{IF}} \cong 25$ is required in Fig. 18-34 to give a $\mathcal{P}_d = 0.9$ as compared to $\bar{m} = 10$ for the sequential detector. Thus the sequential detector can have a significantly smaller expected acquisition time under some conditions.

In practice, a simpler configuration (Fig. 18-35) often is used rather than the optimum sequential detector. The IF correlator samples can be integrated and compared with a time-varying rejection threshold rather than computing the likelihood ratio. Acceptance of a cell as containing sync occurs when M consecutive tests fail to reject the cell. If a code cell is accepted as

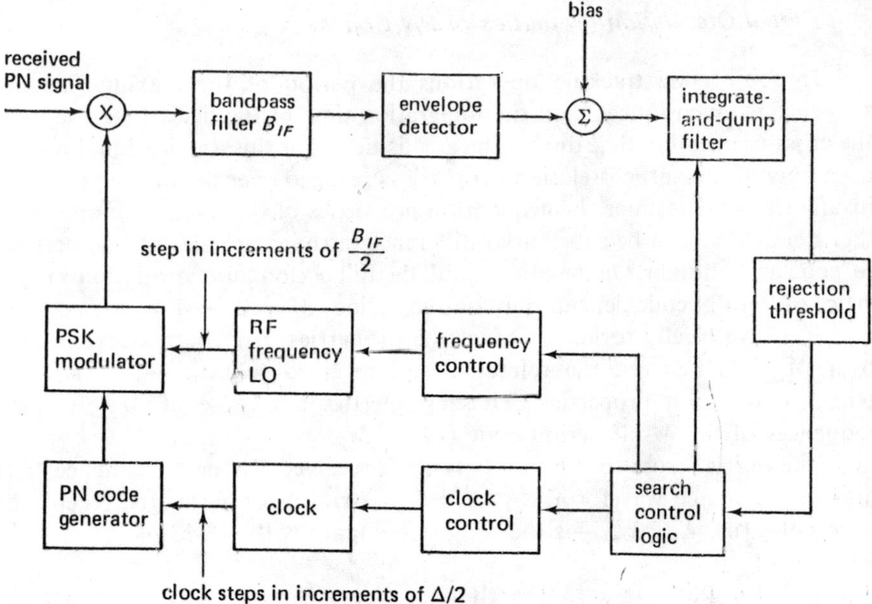

Fig. 18-35 Block diagram of a simplified sequential search-and-acquisition circuit using noncoherent detection, the detector both a frequency and a code delay uncertainty [From Cahn, 1973]

containing sync, the delay-lock loop can then be closed and closed-loop tracking continues from there.

18-11 PROPERTIES OF PN SEQUENCES AND RELATED CODES

In the previous paragraphs and in Chapter 6 we discussed some properties of PN codes as related to tracking and acquisition. In this subsection we discuss additional properties of PN codes for the short correlation times often encountered in code acquisition. We also discuss the properties of two related families of codes: in particular, two codes derived from linear maximal length shift register codes:

- Majority vote PN codes
- Gold codes

We analyze briefly some of the autocorrelation and cross-correlation properties of these code classes. Doppler effects are considered.

Partial Correlation Properties of PN Codes

In some code tracking operations the period of the maximal-length sequence is much longer than the integration time of the delay-lock loop or the cross-correlation time during the acquisition. For this application, knowledge only of the autocorrelation properties averaged over the full sequence is insufficient to determine the acquisition properties of the tracking loop. Furthermore, there can be substantial difference in the properties of one period M code and another. On the other hand, the full period autocorrelation properties of the PN code depend only on the period, $M = 2^n - 1$, of the code.

Here we briefly review some of the properties of the cross-correlation over M_0 bits between the reference and received sequence—that is, the partial correlation properties. These properties are those of M_0-bit subsequences of the M-bit period code ($M_0 < M$).

The partial cross-correlation between a received PN code sequence $\{y_i\}$ and the reference waveform $[x_{i+1}, x_{i+2}, \ldots, x_{i+M_0}]$ for error-free reception of y_i— that is, $y_i = x_{i+k}$—is the weight [compare with (18-131)]

$$W_i' = \sum_{j=1}^{M_0} y_{i+j} \oplus x_{i+j} = \sum_{j=1}^{M_0} x_{i+k+j} \oplus x_{i+j} = \sum_{j=1}^{M_0} x_{i+l+j} \tag{18-148}$$

where we assume $k \neq 0$, $x_i = (0, 1)$, and have used the cycle-and-add property of the maximal-length linear PN codes, which shows that the modulo-2 addition of two shifted versions of the same sequence is a third version of the same sequence which is shifted by l bits [Golomb, 1967]. The quantity W_i is then the weight (number of 1s) in an M_0-bit subsequence of the code. If synchronization occurs, then $y_{i+j} = x_{i+j}$. We then have $W_i = 0$.

To determine the acquisition properties of the code, we compute the first three moments of the sequences and compare them with the moments of a purely random sequence. These moments were first calculated by Lindholm [1968, 569–576] and the results were extended by Wainberg and Wolf [1970, 606–612].

The first moment of the sequence is the summation over all code offsets of the partial-correlation subsequence weight; that is,

$$\begin{aligned} S^{(1)} &= \frac{1}{M}\sum_{i=0}^{M-1} W_i \triangleq \frac{1}{M}\sum_{i=0}^{M-1}\sum_{j=1}^{M_0} b_{i+j} W_i \triangleq \sum_{j=1}^{M_0} b_{i+j} \\ &= \frac{1}{M}\sum_{j=1}^{M_0}\sum_{i=0}^{M-1} b_{i+j} = \frac{1}{M}\sum_{j=1}^{M_0}(-1) = -\frac{M_0}{M} \end{aligned} \tag{18-149}$$

where the b_i are ± 1 and $b_i \triangleq 1 - 2x_i$, and there is one more (-1) than $(+1)$ in the full period of the code. This value of $S^{(1)}$ is compared to $S^{(1)} = M_0$ when $y_{i+j} = x_{i+j}$, and the two codes are in synchronization. Thus, for code periods much larger than the subsequence length $M_0 \ll M$, the first moment $S^{(1)} \cong 0$. The second moment is defined as

$$\begin{aligned}
S^{(2)} &= \frac{1}{M}\sum_{i=0}^{M-1}{}' W_i^2 = \frac{1}{M}\sum_{i=0}^{M-1}\left(\sum_{j=0}^{M_0-1}{}' b_{i+j}\right)^2 \\
&= \frac{1}{M}\sum_{i=0}^{M-1}\left(M_0 + 2\sum_{j=0}^{M_0-2}\sum_{k=j+1}^{M_0-1} b_{i+j}b_{i+k}\right) \\
&= M_0 + \frac{2}{M}\sum_{i=0}^{M-1}\sum_{j=0}^{M_0-2}\sum_{k=j+1}^{M_0-1} \underbrace{b_{i+j}b_{i+k}}_{b_{i+l}} \qquad j \neq k
\end{aligned}$$

where we have again used the cycle-and-add property. Thus, the second moment is

$$\begin{aligned}
S^{(2)} &= M_0 + \frac{2}{M}\sum_{j=0}^{M_0-2}\sum_{k=j+1}^{M_0-1}(-1) \\
&= M_0 + \frac{2}{M}\binom{M_0}{2}(M_0-1) = M_0\left[1 - \frac{M_0-1}{M}\right] \qquad (18\text{-}150) \\
&\cong M_0 \qquad \text{and } [S^2]^{1/2} \cong \sqrt{M_0} \qquad \text{for } M_0 \ll M
\end{aligned}$$

where we have used the cycle-and-add property and the fact that there is one more (-1) than $+1$ in the M-bit sequence. Notice that both the first and second moments are independent of the particular maximal-length sequence with an M-bit period. Also note that $S^{(2)}$ decreases to 1 as $M_0 \rightarrow M$.

Similarly, one can compute the third moment as

$$\begin{aligned}
S^{(3)} = \frac{1}{M}\sum_{n=0}^{M-1}\left(\sum_{i=0}^{M_0-1}{}' b_{n+i}\right)^3 &= \frac{1}{M}\sum_{n=0}^{M-1}{}'\Big\{(3M_0-2)\sum_{i=0}^{M_0-1}{}' b_{n+i} \\
&+ (3!)\sum_{i=0}^{M_0-1}\sum_{j=0}^{M_0-2}\sum_{k=0}^{M_0-1} b_{n+i}b_{n+j}b_{n+k} \qquad (18\text{-}151)
\end{aligned}$$

where the code-dependent property is

$$b_{n+i}b_{n+j}b_{n+k} = \begin{cases} b_{n+l} & b_{n+i}b_{n+j} \neq b_{n+k} \\ 1 & b_{n+i}b_{n+j} = b_{n+k} \end{cases} \qquad (18\text{-}152)$$

Thus, the third moment is

$$\begin{aligned}
S^{(3)} &= -\frac{M_0(3M_0-2)}{M} + \frac{3!}{M}\left\{\binom{M_0}{3}(-1) + (M+1)B_3\right\} \\
&= \frac{-M_0^3}{M} + \frac{3!}{M}(M+1)B_3 \qquad (18\text{-}153)
\end{aligned}$$

where B_3 is the number of 3-tuples $(i, j, k,)$ for which $b_{n+i}\, b_{n+j} = b_{n+k}$ where $0 \leqq i < j < k \leqq M_0 - 1$; B_3 is the number of three-term polynomials of degree $M_0 - 1$ or less divisible by the sequence-generating polynomial.

The generating or characteristic polynomial of the PN sequence

$$f(X) = \sum_{n=0}^{r} C_n X^n \tag{18-154}$$

where $C_n = 0$ or 1 is related to the shift-register-feedback tap generator

$$x_i = \sum_{n=1}^{r} C_n x_{i-n} \tag{18-155}$$

Addition is modulo 2. The feedback tap arrangements can be selected from Peterson [1961] and Peterson and Weldon [1972]. This moment depends on the particular $M = 2^n - 1$ length sequence selected.

These three moments and their values for random sequences $(M \gg M_0)$ are given in Table 18-2. The third moment is a measure of the skewness of the distribution of weights. As an example of the range of values for the third moment for $M = 2^{23} - 1 \gg M_0 = 500$, four codes from Lindholm [1968, 569–576] are given in Table 18-3. Note that the last two codes are

Table 18-2 Moments of Random and Maximal-Length Sequence Subsequences for Subsequence Length M_0 and Code Period M $(M \gg M_0)$ (Code Values $b_i = \pm 1$)

Order of Moment	*Random Sequence*	*PN Sequence*
1	0	$\frac{M_0}{M} \cong 0$
2	M_0	$M_0\left[1 - \left(\frac{M_0 - 1}{M}\right)\right] \cong M_0$
3	0	$\frac{-M_0^2}{M} + 3!\left(\frac{M+1}{M}\right)B_3 \cong 3!\, B_3$

Table 18-3 Partial Correlation Properties of Four $2^{23} - 1$ Sequences. Correlation over $M_0 = 500$ bits $x_i = (0, 1)$

Code Characteristic Polynomial (Octal Notation)	*Number of Feedback Taps*	*Partial Correlation $M_0 = 500$*		
		Minimum Weight	*Maximum Weights*	*Third Moment*
40000041	2	121	293	10707
66666667	16	139	302	10335
40404041	4	179	307	−15
40435651	10	203	300	−15

much more random for subsequence cross-correlation than the first two, and hence are less susceptible to false lock-on at an erroneous time offset.

Majority Vote Composite Codes

As noted, it is possible to generate sequences having arbitrarily long periods by using enough shift-register stages. For extremely long period sequences, however, a straightforward search procedure may lead to an intolerably long acquisition time. It may sometimes be better to use a short-period acquisition signal or a composite sequence made up of several sequences each of relatively prime periods. Then each sequence can be acquired in parallel and tracked with a separate delay-lock discriminator. In this way, the period of the composite sequence is the product of the periods, whereas the minimum acquisition time is directly proportional to the period of the longest individual sequence. If N sequences are properly combined, the power available for the acquisition of each sequence is $\cong \sqrt{2/\pi N}$ of the total power.

Easterling [1962, 76–84] has described one interesting means for combining an odd number of binary sequences; the sequences are simply combined in a majority logic. For example, the composite code $c(i)$ can be the majority vote combination of a number of shorter sequences $a_k(i)$ of period P_k:

$$c(i) = \operatorname{sgn}\left[\sum_{k=l}^{N} a_k(i)\right] \tag{18-156}$$

for N odd, where $a_k(i) = \pm 1$. Composite codes can also be formed by mod-2 addition of separate PN codes. Both this code and the majority vote code are easily and quickly shifted in phase. However, the mod-2 addition composite code does not possess the rapid acquisition properties of the majority code. If all the subsequences $a_k(i)$ have periods P_k that are relatively prime, then the period of $c(i)$ is

$$P = \prod_{k=l}^{N} P_k \tag{18-157}$$

For a composite code to be useful for rapid acquisition, it must correlate well with each of the subsequences. That is, the cross-correlation

$$R_{ca_k}(j) \triangleq \frac{1}{P} \sum_{i=1}^{P} c(i) a_k(i + j) \tag{18-158}$$

should be maximized at $j = 0$ for all subsequences a_k. Tausworthe [1970, 19–30] and Braverman [1963] have shown that the majority vote composite code maximizes this cross-correlation and is the same for all of the component sequences a_k; that is,

$$\rho(N) \triangleq R_{ca_1}(0) = R_{ca_2}(0) = \cdots = R_{ca_N}(0) \tag{18-159}$$

Thus, separate delay-lock loops can be used in parallel to track each of the subsequences. The search time for each delay-lock loop is proportional to the search time τ_a per bit multiplied by the period P_k of that subsequence. Alternatively, a delay-lock loop can acquire each of the subsequences sequentially. In this case, the acquisition time is

$$T_{\text{acq}} = \tau_a \sum_{k=1}^{N} P_k \tag{18-160}$$

where τ_a is determined by the acquisition and partial-correlation properties of the subsequence and the normalized variance of the partial correlation $\sigma_{m_0}^2(k)$. As an example, consider a composite code of period

$$P = 19,418,400 = (25)(27)(29)(31)(32) \tag{18-161}$$

composed of five sequences of relatively prime periods 25, 27, 29, 31, 32. The acquisition time of the composite code is then

$$T_{\text{acq}} = \tau_a \sum_{k=1}^{N} P_k = 144\tau_a \tag{18-162}$$

for sequential composite-code acquisition rather than $19 \times 10^6 \tau_a'$ for a sequential search over the entire code. The search time τ_a' per bit for bit-by-bit correlation is slightly less than τ_a because the composite-code cross-correlation is less than the cross-correlation of $c(i)$ with itself; that is,

$$\rho(N) < 1 \qquad N > 1 \tag{18-163}$$

The value of $\rho(N)$ has been computed from (18-158) as

$$1 > \rho(N) = \binom{N-1}{\frac{N-1}{2}} \Big/ 2^{N-1} \cong \frac{(2/\pi)^{1/2}}{\sqrt{N-1}} \qquad N \gg 1 \tag{18-164}$$

Values of $\rho(N)$ for a majority vote of N code subsequence components are given in Table 18-4.

The partial-correlation properties of the composite code are more important than the full-period correlation since the whole objective of composite

Table 18-4 MAJORITY VOTE CROSS-CORRELATION PROPERTIES. THE PARTIAL CORRELATION AVERAGE IS OVER M_o BITS.

Number of Subsequences N	*Inphase Cross-Correlation* $\rho(N)$	*Normalized Partial-Correlation Variance* $M_o\sigma^2_{M_o} = 1 - \rho^2(N)$
1	1	0
3	0.500	0.75
5	0.375	0.859
7	0.312	0.903
9	0.272	0.926
11	0.230	0.947
13	0.226	0.949
15	0.209	0.956
17	0.200	0.960
19	0.188	0.965
21	0.179	0.968

codes is to permit more rapid search and acquisition. The partial correlation over M_o bits of the code starting at bit ℓ, the phase shift of the code, is given by

$$R_{ca_k}(i, \ell, M_o) \triangleq \frac{1}{M_o} \sum_{j=1}^{M_o} c(j + \ell)a_k(j + \ell + i) \tag{18-165}$$

where i is the relative phase shift of the two codes. This partial correlation for $M_o < P$ acts as a random variable with a mean value equal to the true value of the cross-correlation

$$\mathbf{E}[R_{ca_k}(i, \ell, M_o)] = R_{ca_k}(i) \tag{18-166}$$

but it also has a variance caused by the self-noise of the pseudorandom sequence product. The variance of this partial correlation is

$$\begin{aligned}\sigma^2_{M_o}(N) &\triangleq \mathbf{E}[R_{ca_k}(i, \ell, M_o) - R_{ca_k}(i)]^2 = \frac{1}{M_o}[1 - R^2_{ca_k}(i)]\\ \sigma^2_{M_o} &\cong \frac{1}{M_o} \qquad i \neq 0\\ \sigma^2_{M_o} &= \frac{1}{M_o}[1 - \rho^2(N)] \qquad i = 0\end{aligned} \tag{18-167}$$

The normalized variance $M_o\sigma^2_{M_o}$ is tabulated in Table 18-4 for $i = 0$.

Assume that we use cross-correlation measurements to test for synchronization. The Chebyshev inequality can then be used to bound the probability of false acquisition due to self-noise. The Chebyshev inequality states that

$$Pr(|x| > \epsilon) \leq \frac{\sigma_x^2}{\epsilon^2}$$

where σ_x^2 is the variance of the random variable x. Thus we obtain

$$\mathcal{P}_a = Pr[|R_{ca_k}(i, \ell, M_o)| \geq \rho, i = 0] \leq \frac{1}{[\rho(N)]^2 M_o} \qquad \text{(18-168)}$$

As an example, set the upper bound on false acquisition at $\mathcal{P}_a = 10^{-3}$. For the $N = 5$ component code $\rho(5) = 3/8$, and the value of M_o required is $M_o = 1000(8/3)^2 = 7110$ bits, or for the $P_k = 30$-bit code, about 230 subsequence periods.

One final comment should be made on the majority vote composite codes. The very property that makes them have a high cross-correlation with the subsequences can make them poor for multiple access. For example, one cannot simply use codes of an arbitrary different phase to provide code division multiple access (CDMA), because the codes have high autocorrelation sidelobes at the subsequence periods. Furthermore, the power spectral density of the codes has line components at frequencies corresponding to each of the code periods.

Gold Codes

One advantage of PN ranging over sidetone ranging is that multiple PN coded ranging signals can occupy the same frequency band without interfering heavily with one another. Thus, it is important for these codes to have low cross-correlation over all possible phase shifts of each code. This use of multiple spread-spectrum codes is termed spread-spectrum multiple access (SSMA) or code division multiple access (CDMA). A set of codes can be constructed of products of linear maximal-length sequences that has uniformly low cross-correlations of one member of the set **V** with any other member of the set. Let $f_a(t)$ represent one linear maximal-length sequence of period $M = 2^n - 1$ and $f_b(t)$ be another maximal-length sequence of the same period with a cross-correlation bounded by

$$|\theta_{a,b}(i)| = \left|\sum_{t=1}^{P} f_a(t) f_b(t+i)\right| < K \qquad \text{any } i \qquad \text{(18-169)}$$

This relationship holds only for "preferred" pairs of PN codes. A family of codes **V** with $M + 2$ members has been defined by Gold [1967, 619–621]:

$$\mathbf{V} = \begin{cases} g_i(t) = f_a(t) f_b(t+i) & i = 0, 1, 2, \ldots, M-1 \\ g_{P+1}(t) = f_a(t) \\ g_{P+2}(t) = f_b(t) \end{cases} \qquad \text{(18-170)}$$

Thus, there are $M + 2 = 2^n + 1$ codes in this set. Each code is generated by a pair of PN generators, each of n stages. An example Gold code generator is shown in Fig. 18-36. Note that if one placed all of the $2^n + 1$ Gold codes of period $2^n - 1$ end-to-end, the period of the resulting sequence of codes is $(2^n + 1)(2^n - 1) = 2^{2n} - 1$. This period is the same as that generated by a maximal-length shift register of $2n$ stages.

The cross-correlation of any code in **V** with any other can easily be evaluated using the cycle-and-add property

$$f_a(t)f_a(t + i) = f_a(t + j) \tag{18-171}$$

discussed earlier. Thus, the code cross-correlation ϕ_{ik} between any two codes $g_i(t)$ and $g_k(t)$ $(g_i \in \mathbf{V})$ is

$$\begin{aligned}\phi_{ik}(\ell) &\triangleq \sum_{t=1}^{P} g_i(t)g_k(t + \ell) \\ &= \sum_{t=1}^{P} [f_a(t)f_b(t + i)][f_a(t + \ell)f_b(t + i + \ell)] \\ &= \sum_{t=1}^{P} f_a(t + m)f_b(t + n) = \theta_{a,b}(n - m)\end{aligned} \tag{18-172}$$

Hence, using (18-169) the cross-correlation between any two Gold codes is bounded by

$$|\phi_{ik}(\ell)| < K$$

Gold [1968, 154–156] has shown that, for preferred pairs of PN codes, these cross-correlation functions are three-valued as opposed to the two-valued autocorrelation functions of the linear maximal-length sequence itself. The cross-correlation has value -1 approximately half the time or more precisely $(2^n - 2)/2 = 2^{n-1} - 1$ times, and properly selected codes f_a, f_b provide a maximum cross-correlation magnitude

$$K = \begin{cases} 2^{(n+1)/2} + 1 & n \text{ odd} \\ 2^{(n+2)/2} + 1 & n \text{ even}, n \neq 0, \text{mod } 4 \end{cases} \tag{18-173}$$

Table 18-5 lists the cross-correlation properties of Gold codes. Thus, the cross-correlation maximum at dc decreases only at every other value of n by 6 dB.

As an example, if $M = 2^{13} - 1 = 8191$, the maximum cross-correlation is

$$K = 2^{(13+1)/2} + 1 = 129 \qquad \text{and } \frac{K}{M} = 0.0157 \tag{18-174}$$

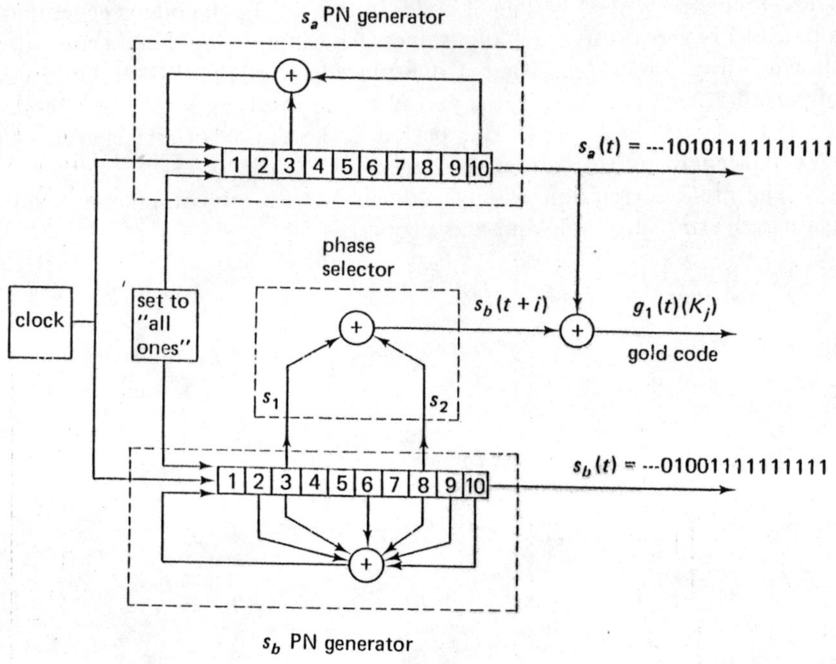

Fig. 18-36 Gold code, $g_i(t)$, generation from 2 PN sequences $s_a(t)$, $s_b(t)$. The code period is $M = 1023$ bits.

Table 18-5 CROSS-CORRELATION PROPERTIES OF GOLD CODES FOR CODES OF PERIOD $M = 2^n - 1$

Code Period	*n = Number of Shift-Register Stages*	*Normalized Cross-Correlation Level*	*Probability of Level*
$M = 2^n - 1$		$-\frac{[2^{(n+1)/2} + 1]}{M}$	0.25
	n odd	$-\frac{1}{M}$	0.50
		$\frac{[2^{(n+1)/2} - 1]}{M}$	0.25
$M = 2^n - 1$		$-\frac{[2^{(n+2)/2} + 1]}{M}$	0.125
	n even and $n \neq 4i$	$-\frac{1}{M}$	0.75
		$\frac{[2^{(2+n)/2} - 1]}{M}$	0.125

Note that for purely random sequences the expected 2σ cross-correlation value is $2\sqrt{8191} \cong 181.01$

The cross-correlation maximum for frequency offsets (of several inverse periods) as well as phase offsets is larger than that for zero frequency offset, and it decreases by 3 dB every time n is is increased by 1. This frequency offset can result from different doppler shifts from one transmitter to another. The statistics of the cross-correlation overall code phases for several different doppler frequency offsets are given in Fig. 18-37 for an $M = 1023$-bit code, and a clock rate of 1.023 Mbps. Notice that the cross-correlation sidelobes

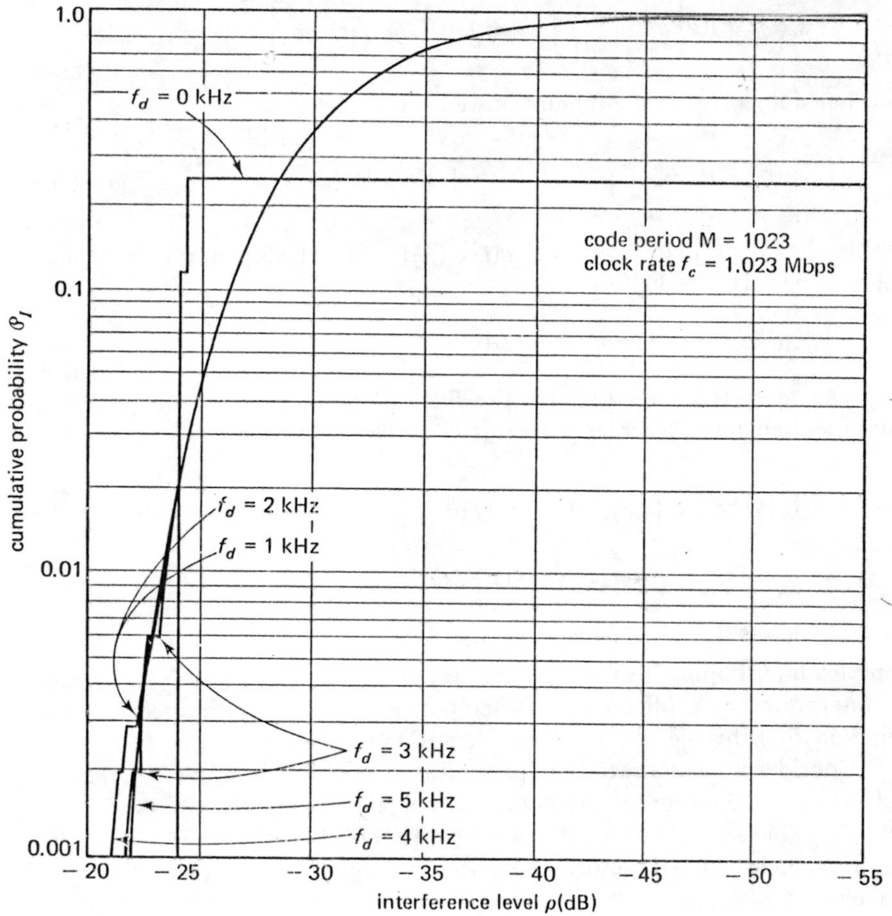

Fig. 18-37 Cumulative probability of interference level exceeding ρ(dB) for 1023-bit Gold code at clock rate f_c = 1.023 Mbps for various doppler frequency offsets f_d [After Chang, 1975]

(worst case) are higher with doppler offset by approximately 3 dB than those for zero doppler offset.

18-12 FILTER DISTORTION EFFECTS ON CROSS-CORRELATION

The PSK-PN signal, represented by Re $[a(t)e^{j\omega_0 t}]$, at center frequency ω_0 is generally filtered by a bandpass transmission filter $H(j\omega)$. Thus the transmitted signal is

$$s_0(t) = \text{Re}\left[e^{j\omega_0 t}\left(\frac{1}{2\pi}\right)\int_{-\infty}^{\infty} A(j\omega)H(j\omega)e^{j\omega t}\,d\omega\right] \tag{18-175}$$

and has autocorrelation $R(\tau)$ and power

$$R(0) \triangleq P_0 = \frac{1}{2\pi}\int_{-\infty}^{\infty} |A(j\omega)|^2\,|H(j\omega)|^2\,d\omega \tag{18-176}$$

where the ideal PN waveform, $a(t) \in (+1, -1)$ of chip duration Δ, has spectral density (2-sided)

$$|A(j\omega)|^2 = \Delta[\sin \pi f\Delta/\pi f\Delta]^2$$

The receiver cross-correlates the filtered signal with the delayed and phase shifted reference $s_r(t) = \text{Re}\,[a(t-\tau)e^{j(\omega_0 t+\phi)}]$ to form

$$\begin{aligned} R_{sr}(\tau) \triangleq \text{Re}\int_{-\infty}^{\infty} s(t)s_r^*(t-\tau)\,dt &= \left|\frac{1}{2\pi}\int_{-\infty}^{\infty} |A(j\omega)|^2 H(j\omega)e^{j\omega\tau}\,d\omega\right| \\ &= \left|\int_{-\infty}^{\infty} R(\tau)e^{j\omega(\tau-t)}h(\tau-t)\,dt\right| \end{aligned} \tag{18-177}$$

where ϕ has been adjusted for maximum cross-correlation and $h(t)$ is the complex filter impulse response. Note that the cross-correlation function can be interpreted as a filter output after passing the original autocorrelation function $R(t)$ through the transmit filter [Cahn, 1973].

Consider a bandlimited bandpass filter of bandwidth $B_{IF} = 2/\Delta$ rectangular amplitude response, and parabolic phase response $H(j\omega) = e^{jD(\omega-\omega_0)^2}$ for $|f - f_0| < 1/\Delta$. In Table 18-6 we show the effective cross-correlation loss $R_{sr}(0)/P_0$ measured with respect to transmitted signal power P_0 and the cross-correlation loss itself. The parameter used is $\theta_d = (2\pi)^2 D/\Delta^2$, the phase distortion at the first spectral null. Comparison with results in Chapter 13 shows that cross-correlation loss is much less sensitive to filter distortion than is error rate. Intersymbol interference effects make error rate degradation substantially larger for the same filter distortion because one is then trying to recover the received signal on a bit-by-bit basis. In addition, it is clear that cross-correlation loss can be reduced to 0 dB by using a reference waveform which is filtered the same as the transmitted signal.

Table 18-6 FILTER DISTORTION EFFECTS ON CROSS-CORRELATION LOSS FOR A BANDLIMITED SIGNAL $B_{IF} = 2/\Delta$, AND PARABOLIC PHASE DISTORTION WITH PHASE DISTORTION θ_d RADIANS AT BANDEDGE

Phase Distortion θ_d *radians*	*Loss* $R_{sr}(0)$ (dB)	*Effective Loss* $R_{sr}(0)/P_0$(dB)
0	0.448 dB	0 dB
[illegible]	0.482	0.034
2	0.605	0.157
3	0.788	0.340
4	1.029	0.581
5	1.308	0.860

Appendices

Appendix A

EVALUATION OF THE COVARIANCE COEFFICIENTS FOR DELTA MODULATION ERROR SAMPLES

The error covariance coefficients from Sec. 4-3 are calculated here. Define the error covariance $R_e(n) = \sigma_x^2 \rho_n - 2\phi_n + r_n$ where σ_x^2 is the input variance, ρ_n is the input autocovariance, r_n is the autocovariance of the sampled output y, and ϕ_n is the cross-covariance between input x and output y. Each of these terms is evaluated, and approximate values are given for fine quantizing $\delta/\sigma_y \ll 1$.

Assume a zero mean Gaussian input $x(t)$ of variance σ_x^2 and equiprobable quantizer parity at $t = 0$; that is, $p(Q_e) = p(Q_0) = 1/2$. As a first step, we compute the probability function of the sampled output at time $t = nT$.

$$\begin{aligned} p(y_n = k\delta) &= p(y_n = k\delta \,|\, Q_e)p(Q_e) + p(y_n = k\delta \,|Q_0)p(Q_0) \\ &= \tfrac{1}{2}[p(y_n = k\delta \,|\, Q_e) + p(y_n = k\delta \,|\, Q_0)] \end{aligned} \tag{A-1}$$

Given a time instant corresponding to the even quantizer levels Q_e, we obtain only even-level outputs with probability

$$p(y_{2m} = 2k\delta \,|\, Q_e) = \frac{1}{\sqrt{2\pi}\sigma_x} \int_{(2k-1)\delta}^{(2k+1)\delta} (e^{-x^2/2\sigma_x^2})\, dx \tag{A-2}$$

Similarly, we find that given an odd-level quantizer Q_0, the output level probability is

$$p[y_{2m+1} = (2k-1)\delta \mid Q_0] = \frac{1}{\sqrt{2\pi}\sigma_x} \int_{(2k-2)\delta}^{2k\delta} (e^{-x^2/2\sigma_x^2})\, dx \qquad \text{(A-3)}$$

Define $\beta \triangleq \delta/\sigma_x$, and set $u \triangleq x/\sigma_x$. Then, using $p(Q_e) = p(Q_0) = 1/2$, the output level probability is

$$p(y_n = k\delta) = \frac{1}{2\sqrt{2\pi}} \int_{(k-1)\beta}^{(k+1)\beta} (e^{-u^2/2})\, du \qquad \text{(A-4)}$$

The expected output level is $\mathrm{E}(y_n) = 0$ because of symmetry in the input. By also defining $u \triangleq \beta(v+k)$, the mean-square output is

$$\mathrm{E}(y_n^2) \triangleq r(0) = \frac{1}{2\sqrt{2\pi}} \sum_{k=-\infty}^{\infty} k^2\delta^2 2 \int_0^1 e^{-\beta^2/2(v+k)^2} \beta\, dv \qquad \text{(A-5)}$$

We use the Poisson sum formula [Bellman, 1961, 7–12] in the form

$$\sum_{n=-\infty}^{\infty} e^{-t(z+n)^2} = \sqrt{\frac{\pi}{t}}\left[1 + 2\sum_{k=1}^{\infty} e^{-n^2k^2/t} \cos 2nkz\right] \qquad \text{(A-6)}$$

The summation of integrals can then be written [Goodman, 1969, 1197–1218]

$$\sum_{n=-\infty}^{\infty} n^2 \int_0^1 e^{-t(y+n)2}\, dy = \frac{1}{2t}\sqrt{\frac{\pi}{t}}\left(1 + 4\sum_{k=1}^{\infty} e^{-\pi^2k^2/t}\right) + \sqrt{\frac{\pi}{t}}\left(\frac{1}{3} + \sum_{k=1}^{\infty}\left(\frac{1}{\pi k}\right)^2 e^{-\pi^2k^2/t}\right) \qquad \text{(A-7)}$$

By setting $t = \beta^2/2$, the expression (A-7) corresponds to (A-5) for the variance of output y, which can be evaluated as

$$r(0) = \frac{\beta}{\sqrt{2\pi}}\delta^2\sqrt{\frac{2\pi}{\beta^2}}\left[\frac{2}{2\beta^2}\left(1 + 4\sum_k e^{-2\pi^2k^2/\beta^2}\right) + \frac{1}{3} + \sum\left(\frac{1}{\pi k}\right)^2 e^{-2\pi^2k^2/\beta^2}\right] \qquad \text{(A-8)}$$

or in normalized form, the variance becomes

$$\frac{r(0)}{\sigma_x^2} = 1 + 4\sum_{k=1}^{\infty} e^{-2\pi^2k^2/\beta^2} + \beta^2\left[\frac{1}{3} + \sum_{k=1}^{\infty}\left(\frac{1}{\pi k}\right)^2 e^{-2\pi^2k^2/\beta^2}\right]$$

$$\simeq \frac{1+\beta^2}{3} \qquad \text{if } 2\pi^2\sigma_x^2/\delta^2 \gg 1 \qquad \text{(A-9)}$$

where the approximation in (A-9) corresponds to fine quantizing with respect to the input variance.

Autocovariance of Output y

For both n, μ even integers, the joint probability of y_n, $y_{n+\mu}$ conditioned on Q_e is

$$p(y_n = 2k\delta, y_{n+\mu} = 2\ell\delta \,|\, Q_e)$$
$$= \frac{1}{2\pi\sigma_y^2\sqrt{1-\rho_{\mu^2}}}\int_{(2k-1)\delta}^{(2k+1)\delta}\int_{(2\ell-1)\delta}^{(2\ell+1)\delta} e^{-(u^2+v^2-2\rho_\mu uv)/2\sigma_x^2(1-\rho_\mu^2)}\,du\,dv \tag{A-10}$$

Because $p(Q_e) = p(Q_0) = 1/2$, for $k+\ell+\mu$ even, we have the joint probability

$$p(y_n = k\delta, y_{n+\mu} = \ell\delta) \triangleq \frac{1}{2\pi\sqrt{1-\rho_{\mu^2}}}\int_{(k-1)\beta}^{(k+1)\beta}\int_{(\ell-1)\beta}^{(\ell+1)\beta} e^{-(u^2+v^2-2\rho_\mu uv)}\,du\,dv \tag{A-11}$$

and this probability is zero for $k+\ell+\mu$ odd. Hence, the autocovariance of y can be expressed as

$$\begin{aligned} r(\mu) &= \sum_k\sum_\ell (k\delta)p(y_n = k\delta, y_{n+\mu} = \ell\delta) \\ &= \delta^2\sum_k\sum_\ell (2k)(2\ell)\overbrace{p(2k, 2\ell, \mu)}^{\text{both even}} \\ &\quad + \delta^2\sum_k\sum_\ell (2k-1)(2\ell-1)\overbrace{p(2k-1, 2\ell-1, \mu)}^{\text{both odd}} \qquad \mu \text{ even} \\ &= 2\delta^2\sum_k\sum_\ell (2k)(2\ell-1)p(2k, 2\ell-1, \mu) \qquad \mu \text{ odd} \end{aligned} \tag{A-12}$$

Use the integral results [Goodman, 1969, 1197–1218]

$$\sum_{k=-\infty}^{\infty} 2k\int_{(2k-1)\beta}^{(2k+1)\beta} ve^{-v^2/2}\,dv = 2\sum_{k=-\infty}^{\infty} e^{-\beta^2/2(2k-1)^2} \tag{A-13}$$

and

$$\sum_{k=-\infty}^{\infty} k\int_{(2k-1)\beta}^{(2k+1)\beta} e^{-v^2/2}\sin\frac{\pi m\rho_v}{\beta}\,dv$$
$$= \sqrt{\frac{\pi}{2}}e^{-(\pi^2m^2\rho_\mu)/2\beta^2}\left[\frac{\pi m\rho_\mu}{\beta^2} + 2\sum_{k=1}^{\infty}\frac{(-1)^k}{\pi k}e^{-\pi^2k^2/2\beta^2}\sinh\frac{\pi^2km\rho_\mu}{\beta^2}\right] \tag{A-14}$$

The autocovariance for μ odd can then be evaluated using (A-13), (A-14) as

$$\frac{r(\mu)}{\sigma_x^2} = \rho_\mu\left[1 + 4\sum_{k=1}^{\infty} e^{-2k^2/\rho^2}\right]$$

$$+ 2\beta^2 \sum_{m=1}^{\infty}\sum_{k=1}^{\infty}\frac{(-1)^m}{mk}[e^{(-2/\beta^2)(k^2+m^2-2mk\rho_\mu)} - e^{-(2/\beta^2)(k^2+v^2+2mk\rho_\mu)}]$$

$$\cong \rho_\mu + 4\beta^2 \sum_{k=1}^{\infty}\frac{(-1)^{\mu k}}{k^2}e^{-k^2/\beta^2}\sinh k^2\frac{\rho_\mu}{\beta^2} \qquad \frac{\sigma_x^2}{\beta^2} \gg 1 \tag{A-15}$$

Cross-Covariance Between x and y

The joint probability of input x, and output y is

$$p(x_n = u, y_{n+\mu} = k\delta)$$

$$= \frac{1}{4\pi\sigma_x^2\sqrt{1-\rho_{\mu^2}}}\int_{(k-1)\delta}^{(k+1)\delta}\exp\left[-\frac{u^2+v^2-vu\rho_\mu}{2\sigma_x^2(1-\rho_{\mu^2})}\right]du\,dv$$

Note that x is continuous and y is discrete. Thus, the cross-covariance is

$$\phi_\mu = \sum_{k=-\infty}^{\infty} k\delta \int up(x_n = u, y_{n+\mu} = k\delta)$$

$$= \frac{\rho_\mu\delta\sigma_x}{2\sqrt{2\pi}}\sum_{k=-\infty}^{\infty} k\int_{(k-1)\beta}^{(k+1)\beta} ve^{-v^2/2}\,dv = \sigma_x^2\rho_\mu\left[1 + 2\sum_{k=1}^{\infty}e^{-2\pi^2k^2/\beta^2}\right]$$

$$\cong \rho_\mu\sigma_x^2 \qquad \beta^2 = \frac{\delta^2}{\sigma_x^2} \ll 1 \tag{A-16}$$

This approximation is used in (4-10) in Sec. 4-3.

Appendix B

FILTER CHARACTERISTICS

Requirements for filter selectivity for typical communications links have been discussed in Chap. 13. Bandpass filtering is generally required to reduce adjacent channel interference by other signals and to prevent sidebands in our transmitted signal from interfering with others (See Chap. 8.). Filtering can also reduce the thermal noise effects in certain parts of the receiver. A measure of the noise filtering effectiveness can be expressed as the equivalent noise bandwidth B_n of the filter. In the process, however, the filters distort the signal to some extent because of amplitude gain and group-delay variations over the frequency band of the signal.

Eggen and McAllister [1966, 50–55] have presented a set of filter characteristics for Chebyshev (equal-ripple), Butterworth, transitional Butterworth-Thompson, Bessel, and Legendre filters. The curves of Figs. B-1 through B-3 give the amplitude response, phase response, and group-delay response, respectively, for each of these filter types in the low-pass configuration. Pole locations are also provided in Table B-1. All filters are normalized to unity 3-dB bandwidth.

The equivalent noise bandwidth of a filter $H(j\omega)$

$$B_n = \frac{1}{|H(0)|^2} \int_0^\infty |H(j\omega)|^2 \, df \qquad \text{one-sided}$$

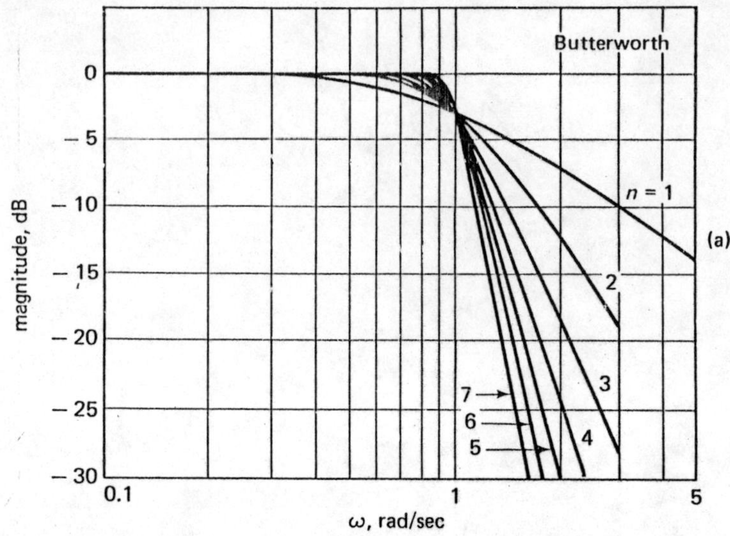

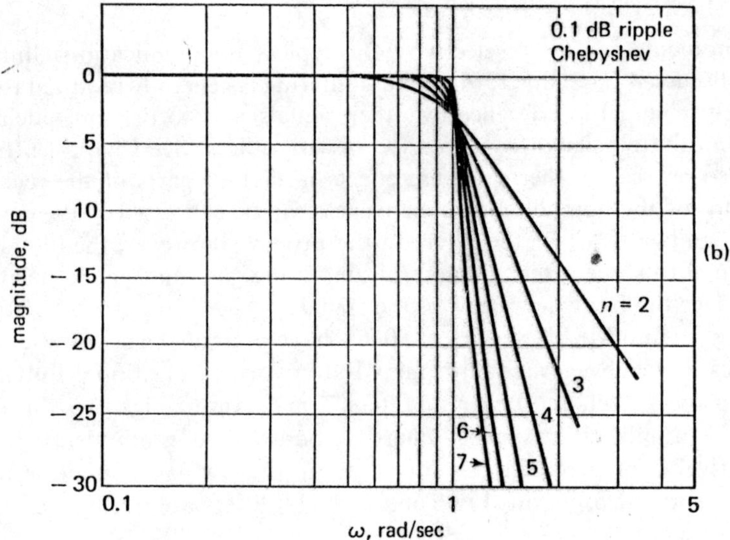

Fig. B-1 Amplitude characteristics of various filters: (a) Butterworth, (b) 0.1-dB ripple Chebyshev [Eggen and McAllister, 1966]

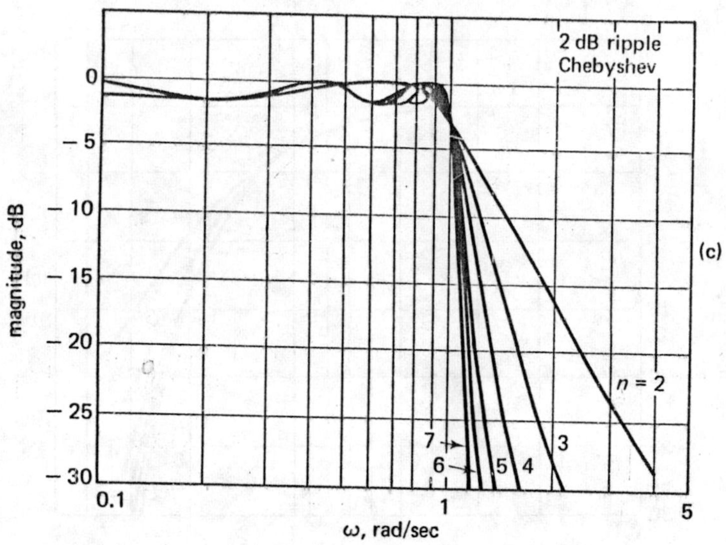

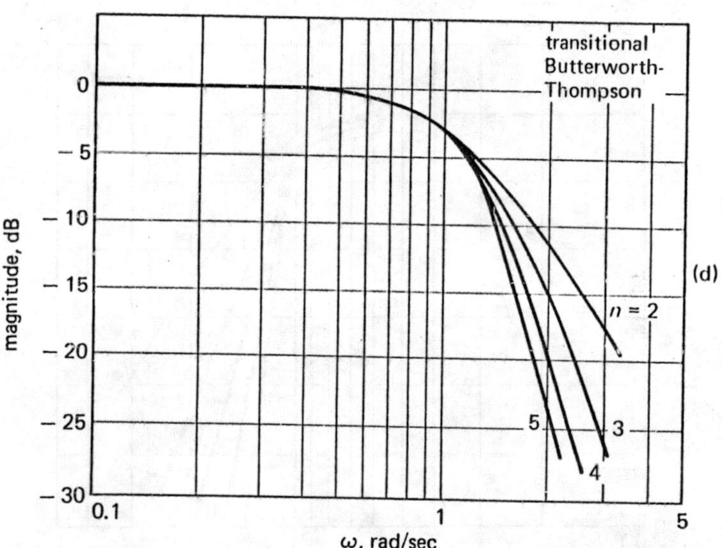

Fig. B-1 (Continued) (c) 2-dB ripple Chebyshev, (d) Transitional Butterworth-Thompson [Eggen and McAllister, 1966]

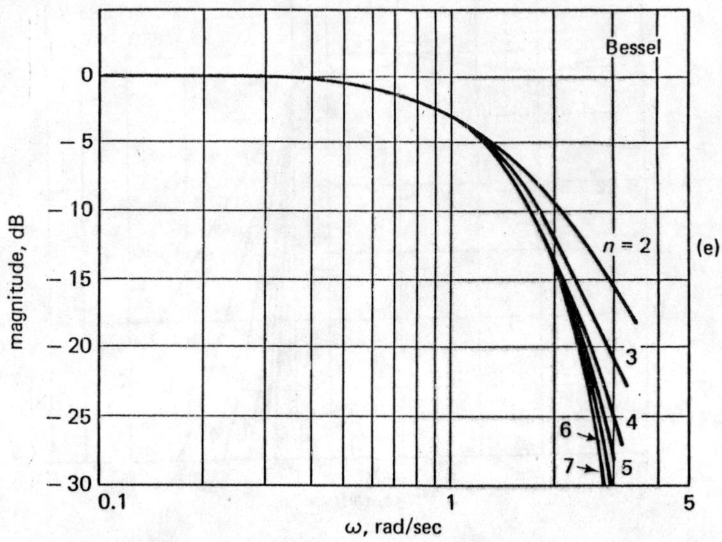

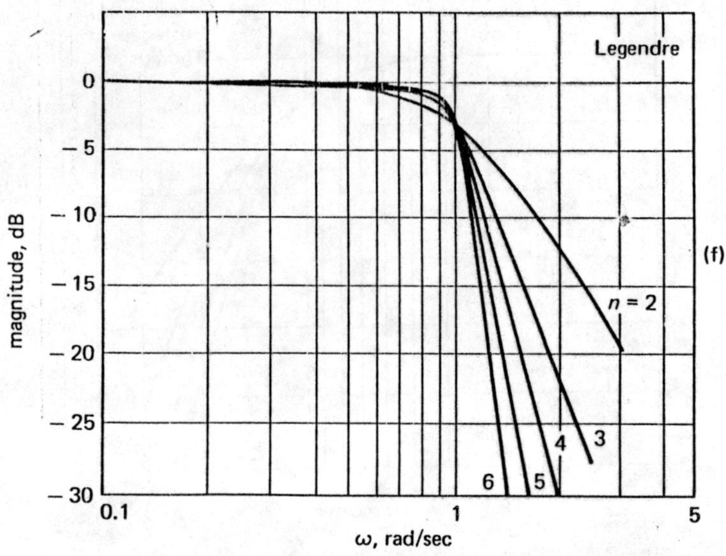

Fig. B-1 (Continued) (e) Bessel, (f) Legendre [Eggen and McAllister, 1966]

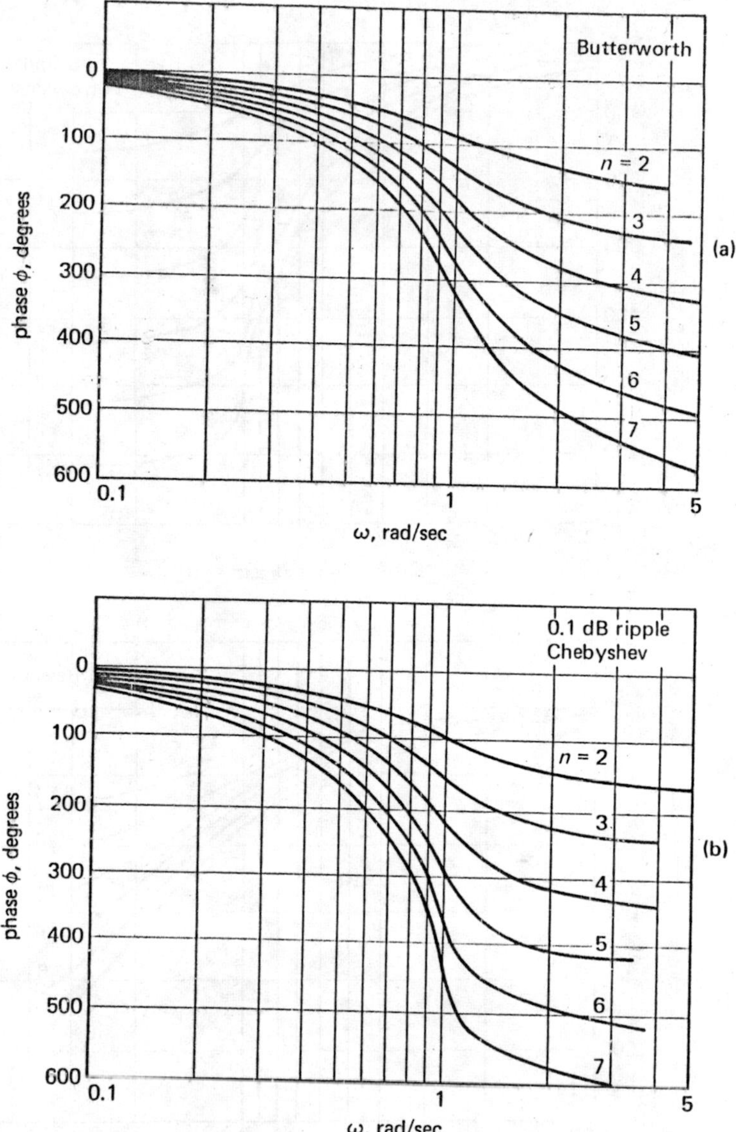

Fig. B-2 Phase characteristics of various filters: (a) Butterworth, (b) 0.1-dB ripple Chebyshev [Eggen and McAllister, 1966]

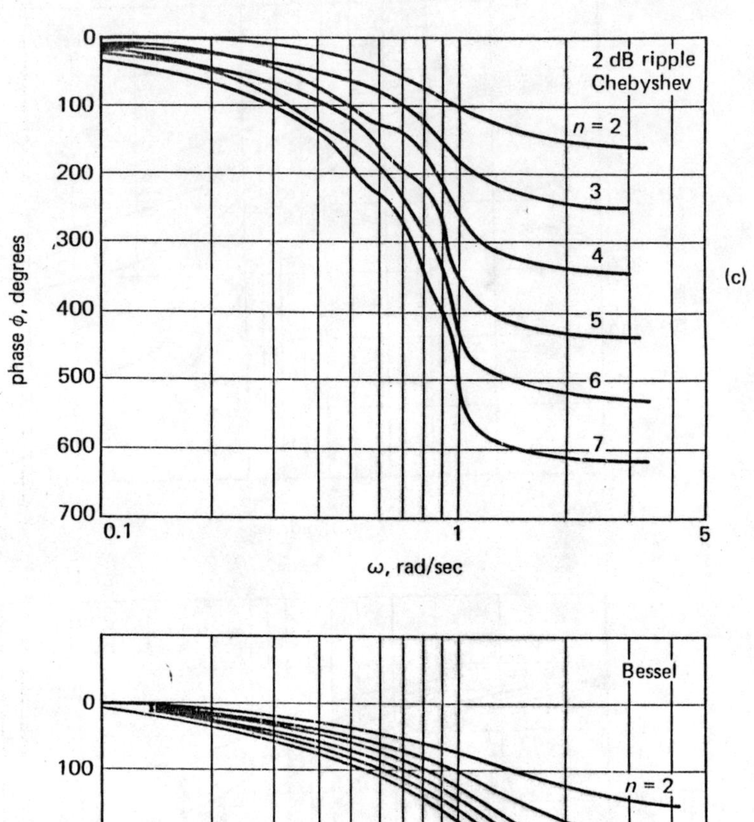

(c)

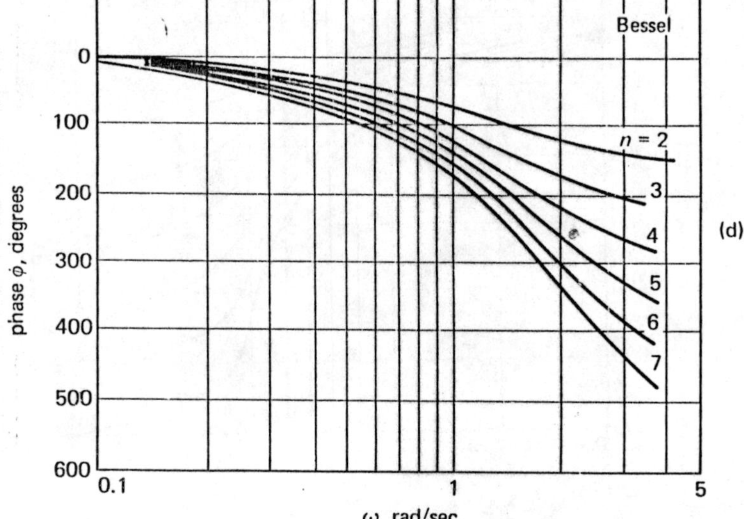

(d)

Fig. B-2 (Continued) (c) 2-dB ripple Chebyshev, (d) Bessel [Eggen and McAllister, 1966]

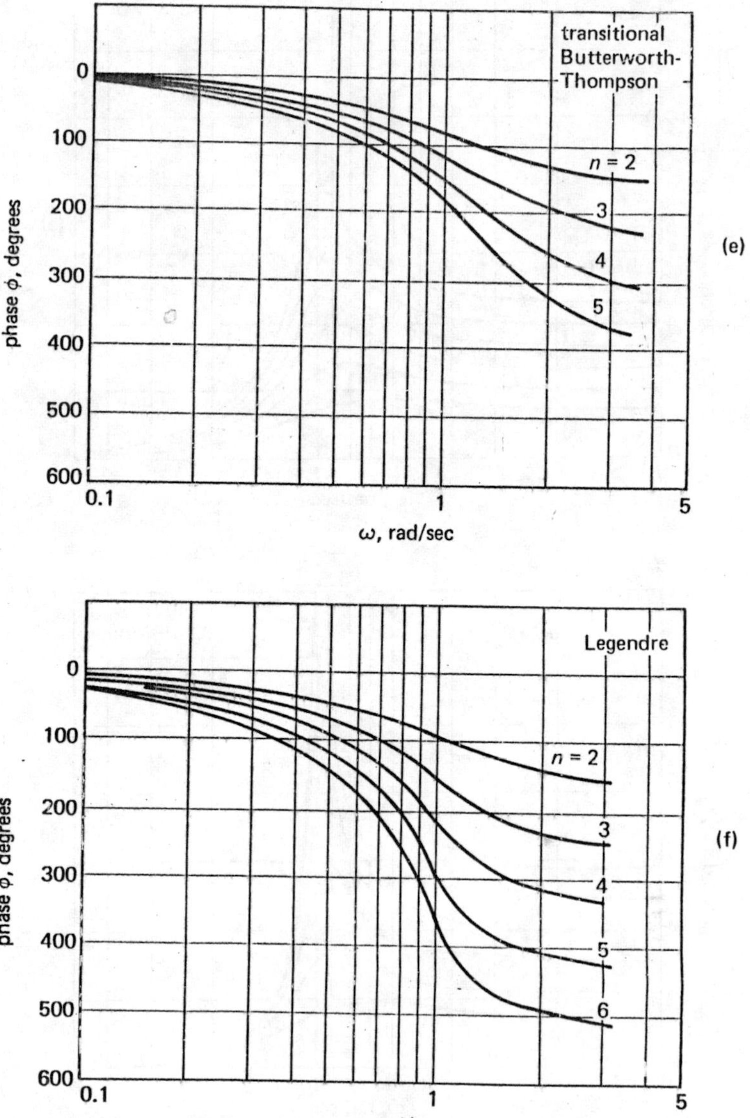

Fig. B-2 (Continued) (e) Transitional Butterworth-Thompson, (f) Legendre [Eggen and McAllister, 1966]

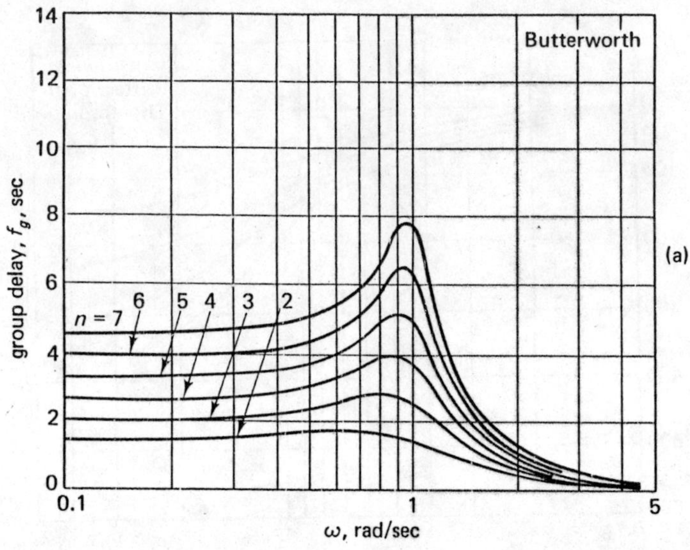

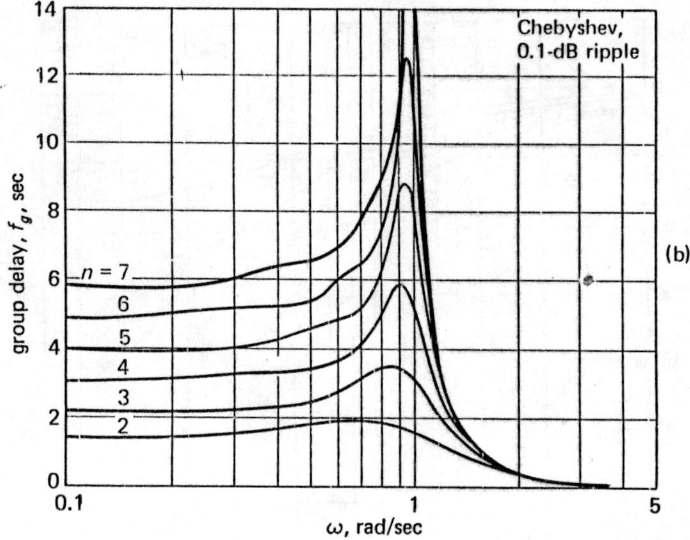

Fig. B-3 Group-delay characteristics of various filters. (a) Butterworth (b) 0.1 dB ripple

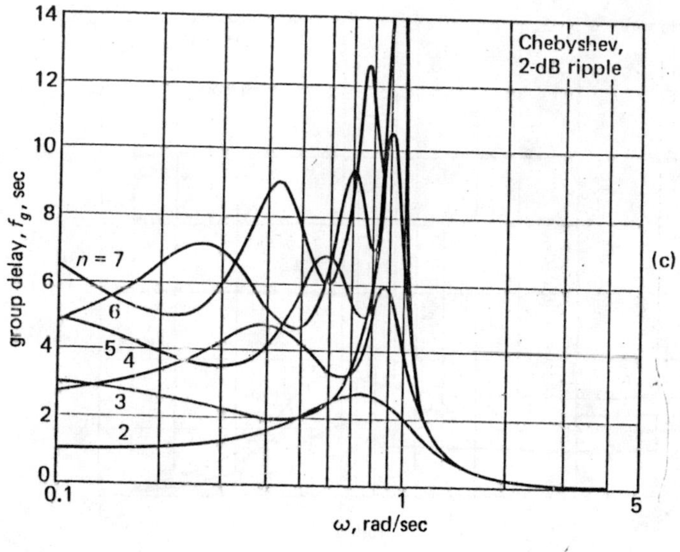

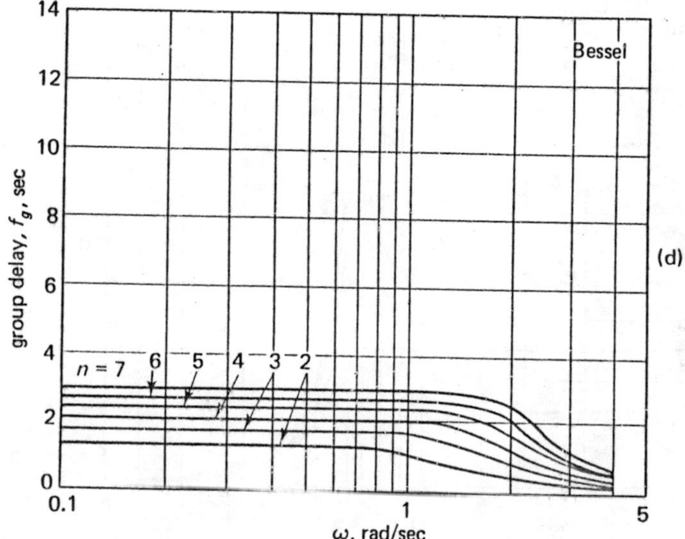

Fig. B-3 (Continued) (c) 2 dB ripple Chebyshev (d) Bessel

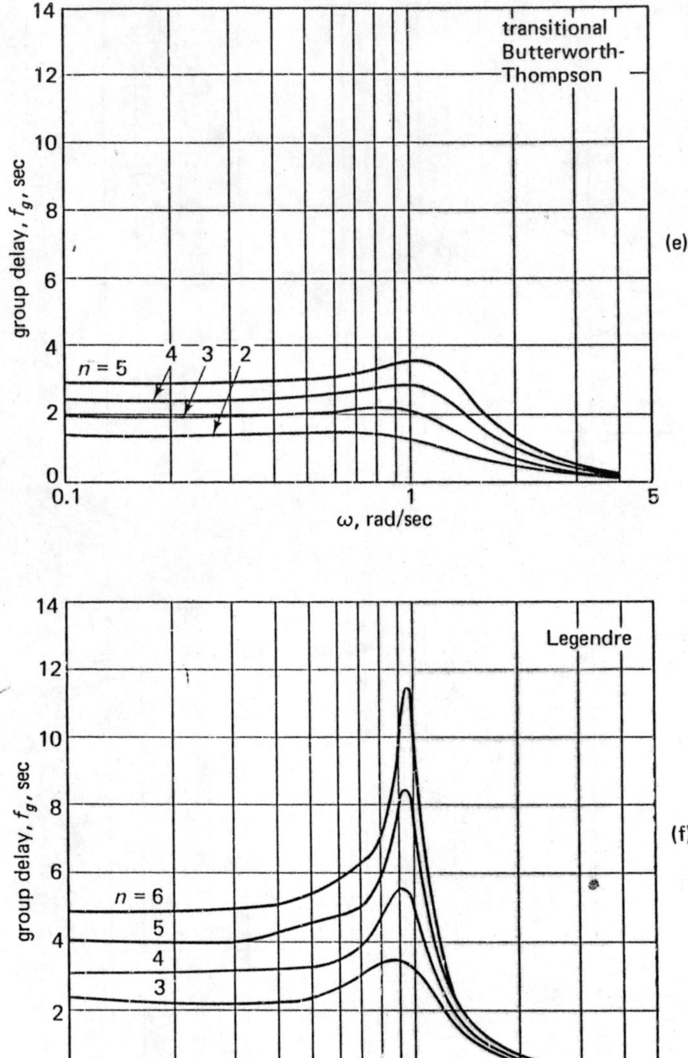

Fig. B-3 (Continued) (e) Transitional Butterworth-Thompson (f) Legendre

Table B-1 NORMALIZED POLE LOCATIONS [FROM EGGEN AND McALLISTER, 1966]

Filter Type	$n = 2$	$n = 3$	$n = 4$	$n = 5$	$n = 6$	$n = 7$
Butterworth	$s_{1,2} = -0.7071 \pm j0.7071$	$s_1 = -1.0000 \pm j0$ $s_{2,3} = -0.5000 \pm j0.8660$	$s_{1,2} = -0.9239 \pm j0.3827$ $s_{3,4} = -0.3827 \pm j0.9239$	$s_1 = -1.0000 \pm j0$ $s_{2,3} = -0.8090 \pm j0.5878$ $s_{4,5} = -0.3090 \pm j0.9511$	$s_{1,2} = -0.9659 \pm j0.2588$ $s_{3,4} = -0.7071 \pm j0.7071$ $s_{5,6} = -0.2588 \pm j0.9659$	$s_1 = -1.0000 \pm j0$ $s_{2,3} = -0.9010 \pm j0.4339$ $s_{4,5} = -0.6235 \pm j0.7818$ $s_{6,7} = -0.2225 \pm j0.9749$
Chebyshev, 0.1-dB ripple	$s_{1,2} = -0.6104 \pm j0.7106$	$s_1 = -0.6979 \pm j0$ $s_{2,3} = -0.3489 \pm j0.8683$	$s_{1,2} = -0.5257 \pm j0.3833$ $s_{3,4} = -0.2177 \pm j0.9254$	$s_1 = -0.4749 \pm j0$ $s_{2,3} = -0.3842 \pm j0.5884$ $s_{4,5} = -0.1467 \pm j0.9521$	$s_{1,2} = -0.3916 \pm j0.2590$ $s_{3,4} = -0.2867 \pm j0.7076$ $s_{5,6} = -0.1049 \pm j0.9666$	$s_1 = -0.3527 \pm j0$ $s_{2,3} = -0.3178 \pm j0.4341$ $s_{4,5} = -0.2199 \pm j0.7822$ $s_{6,7} = -0.0785 \pm j0.9754$
Chebyshev, 2-dB ripple	$s_{1,2} = -0.3741 \pm j0.7572$	$s_1 = -0.3572 \pm j0$ $s_{2,3} = -0.1786 \pm j0.8938$	$s_{1,2} = -0.2486 \pm j0.3896$ $s_{3,4} = -0.1029 \pm j0.9406$	$s_1 = -0.2157 \pm j0$ $s_{2,3} = -0.1745 \pm j0.5946$ $s_{4,5} = -0.0666 \pm j0.9621$	$s_{1,2} = -0.1738 \pm j0.2609$ $s_{3,4} = -0.1272 \pm j0.7128$ $s_{5,6} = -0.0465 \pm j0.9737$	$s_1 = -0.1544 \pm j0$ $s_{2,3} = -0.1391 \pm j0.4364$ $s_{4,5} = -0.0962 \pm j0.7865$ $s_{6,7} = -0.0343 \pm j0.9807$
Transitional Butterworth-Thompson	$s_{1,2} = -0.8615 \pm j0.6977$	$s_1 = -1.1249 \pm j0$ $s_{2,3} = -0.6942 \pm j0.9368$	$s_{1,2} = -1.0858 \pm j0.3987$ $s_{3,4} = -0.5843 \pm j1.0605$	$s_1 = -1.1771 \pm j0$ $s_{2,3} = -1.0059 \pm j0.6428$ $s_{4,5} = -0.5103 \pm j1.1442$		
Bessel	$s_{1,2} = -1.1016 \pm j0.6364$	$s_1 = -1.3226 \pm j0$ $s_{2,3} = -1.0474 \pm j0.9992$	$s_{1,2} = -1.3700 \pm j0.4102$ $s_{3,4} = -0.9952 \pm j1.2571$	$s_1 = -1.5023 \pm j0$ $s_{2,3} = -1.3808 \pm j0.7179$ $s_{4,5} = -0.9576 \pm j1.4711$	$s_{1,2} = -1.5716 \pm j0.3209$ $s_{3,4} = -1.3819 \pm j0.9715$ $s_{5,6} = -0.9307 \pm j1.6620$	$s_1 = -1.6827 \pm j0$ $s_{2,3} = -1.6104 \pm j0.5886$ $s_{4,5} = -1.3775 \pm j1.1904$ $s_{6,7} = -0.9089 \pm j1.8346$
Legendre		$s_1 = -0.6200 \pm j0$ $s_{2,3} = -0.3450 \pm j0.9010$	$s_{1,2} = -0.5500 \pm j0.3590$ $s_{3,4} = -0.2320 \pm j0.9460$	$s_1 = -0.4680 \pm j0$ $s_{2,3} = -0.3880 \pm j0.5890$ $s_{4,5} = -0.1540 \pm j0.9680$	$s_{1,2} = -0.4390 \pm j0.2400$ $s_{3,4} = -0.3090 \pm j0.6980$ $s_{5,6} = -0.1152 \pm j0.9780$	

is tabulated in Table B-2 along with the one-sided 3-dB bandwidth for Bessel, Chebyshev, and Butterworth filters for filters of order $n = 1, 2, 3, 4, 5, 6$ [Shelton, and Adkins, 1970, 828–830]. Clearly, as the order of the filter increases, the ratio of one-sided noise bandwidth to 3-dB bandwidth (one-sided) approaches 1.0 for filters of small or no ripple.

Table B-2 RATIO OF NOISE BANDWIDTHS TO 3-DB BANDWIDTH [AFTER SHELTON AND ADKINS, 1970]

Filter Type	*Order*	B_n/B_3 dB
Butterworth	1	1.57 = $\pi/2$
	2	1.11
	3	1.04
	4	1.03
	5	1.02
	6	1.01
Bessel	1	1.57
	2	1.16
	3	1.08
	4	1.04
	5	1.04
	6	1.04
Chebyshev (½ dB)	1	1.57
	2	1.15
	3	1.00
	4	1.08
	5	0.96
	6	1.07
Chebyshev (1 dB)	1	1.57
	2	1.21
	3	0.96
	4	1.15
	5	0.92
	6	1.13
Chebyshev (2 dB)	1	1.57
	2	1.33
	3	0.86
	4	1.28
	5	0.82
	6	1.27
Chebyshev (3 dB)	1	1.57
	2	1.48
	3	0.78
	4	1.43
	5	0.73
	6	1.41

Another useful filter relationship is the integral of $(\sin x/x)^2$ tabulated in Table B-3. This table enables one to determine the fraction of noise power removed from an integrate-and-dump filter output when white noise is passed through an ideal lowpass filter of bandwidth B before the integrate-and-dump filter operation. The integrate-and-dump filter has $|H(j\omega)|^2 = (\sin \pi fT/\pi fT)^2$ as its transfer function.

The same result gives the fractional amount of power truncated by a lowpass filter with a NRZ binary input spectrum $(\sin \pi f\Delta/\pi f\Delta)^2$ where Δ is the NRZ chip width. If the spectrum is cut off at the first spectral null $x = \pi$, then 0.44 dB of power is removed.

Table B-3 FRACTION OF TOTAL POWER IN A $(\sin x/x)^2$ SPECTRUM IN THE FREQUENCY RANGE 0 TO B. THE NORMALIZED ANGULAR FREQUENCY FOR A PULSE WIDTH Δ SEC IS $x \triangleq \Delta f\pi$ B/π is the band width in H_z

Normalized Video Bandwidth		*Power Ratio*	
B	B/π H_z	$P_B = \dfrac{\int_0^B \left(\dfrac{\sin x}{x}\right)^2 dx}{\pi/2}$	
0	0.0	0 or	$-\infty$ dB
0.1	.0318	0.0636	-11.97
0.2	.0637	0.1268	$-$ 8.97
0.3	.0955	0.1891	$-$ 7.23
0.4	.1273	0.2502	$-$ 6.02
0.5	.1592	0.3097	$-$ 5.09
1.0	.3183	0.5712	$-$ 2.43
1.5	.4775	0.7546	$-$ 1.22
2.0	.6366	0.8561	$-$ 0.675
2.5	.7958	0.8955	$-$ 0.479
3.0	.9549	0.9028	$-$ 0.444
3.5	1.1141	0.9037	$-$ 0.440
4.0	1.2732	0.9110	$-$ 0.405
4.5	1.4324	0.9248	$-$ 0.340
5.0	1.5915	0.9386	$-$ 0.275
6.0	1.9099	0.9498	$-$ 0.224
7.0	2.2282	0.9514	$-$ 0.216
10.0	3.1831	0.9668	$-$ 0.147
15.0	4.7746	0.9795	$-$ 0.090
∞	∞	1.0	0 dB

Appendix C

CROSS-CORRELATION BETWEEN LIMITER INPUT AND OUTPUT

The purpose of this appendix is to compute the cross-correlation between the limiter output $u(t + \mu)$ and the reference signal component $s(t)$ as used in Sec. 18-5. This cross-correlation can be written

$$\mathbf{E}\left[z(t) \triangleq \frac{u(t+\mu)s(t)}{A}\right] = R_{us}(\mu) = \Pr[z(t) = 1] - \Pr[z(t) = -1] \tag{C-1}$$

where $u(t) = A \operatorname{sgn} [\sqrt{P_s}\, s(t) + n(t)]$. To simplify the notation, define $s(t) = s_o$, $s(t + \mu) = s_\mu$, $n(t + \mu) = n_\mu$. Define the signal probabilities $\Pr(s = 1) = p = (M + 1)/2M$, and $\Pr(s = -1) = q = 1 - p$. The probability that $z = 1$ can then be written

$$\begin{aligned}
\Pr(z = 1) &= p \Pr(n_\mu > -\sqrt{P_s} s_\mu \,|\, s_o = 1) \\
&\quad + q \Pr(n_\mu < -\sqrt{P_s} s_\mu \,|\, s_o = -1) \\
&= \Pr(n_\mu > -\sqrt{P_s})[p \Pr(s_\mu = 1 \,|\, s_o = 1) \\
&\quad + q \Pr(s_\mu = -1 \,|\, s_o = -1)] \\
&\quad + \Pr(n_\mu > \sqrt{P_s})[p \Pr(s_\mu = -1 \,|\, s_o = 1) \\
&\quad + q \Pr(s_\mu = 1 \,|\, s_o = -1)]
\end{aligned} \tag{C-2}$$

where use has been made of the fact that the probability density of n is symmetric about the origin. The probability that $z = -1$ can be obtained in a similar manner

$$\begin{aligned}\Pr(z = -1) = \Pr(n_\mu < -\sqrt{P_s})[p \Pr(s_\mu = 1 \mid s_o = 1) \\ + q \Pr(s_\mu = -1 \mid s_o = -1)] \\ + \Pr(n_\mu < \sqrt{P_s})[p \Pr(s_\mu = -1 \mid s_o = 1) \\ + q \Pr(s_\mu = 1 \mid s_o = -1)]\end{aligned} \tag{C-3}$$

The autocorrelation function of the signal is expressed by

$$\begin{aligned}R_s(\mu) &= \mathbf{E}[s(t)s(t+\mu)] \\ &= p[\Pr(s_\mu = 1 \mid s_o = 1) - \Pr(s_\mu = -1 \mid s_o = 1)] \\ &\quad + q[\Pr(s_\mu = -1 \mid s_o = -1) - \Pr(s_\mu = 1 \mid s_o = -1)]\end{aligned} \tag{C-4}$$

Hence, by using (C-1) and (C-4) the cross-correlation function $R_{us}(\mu)$ can be written

$$R_{us}(\mu) = R_s(\mu) \Pr(|n_\mu| < \sqrt{P_s}) \tag{C-5}$$

Under the assumption that the noise has stationary Gaussian amplitude statistics and has a mean-square value P_n, we obtain

$$R_{us}(\mu) = R_s(\mu) \operatorname{erf} \sqrt{\frac{P_s}{2P_n}} \tag{C-6}$$

Bibliography

AARON, M. R. "PCM Transmission in the Exchange Plant," *BSTJ*, Jan. 1962: 99–142.

———, J. S. FLEISCHMAN, R. A. MCDONALD, and E. N. PROTONOTARIOS. "Response of Delta Modulation to Gaussian Signals," *BSTJ*, May–June 1969: 1167–1196.

AARONS, J. "Geophysical Aspects of Radio Star and Satellite Ionospheric Scintillations," in P. Newman, Ed., North Atlantic Treaty Organization AGARDograph 95, *Spread F and Its Effect Upon Radiowave Propagation and Communications*, 1966: 247–266.

AARONS, J., H. E. WHITNEY, R. J. ALLEN, "Global Morphology of Ionospheric Scintillation," *Proceedings of IEEE Satellite Communications*, February 1971, 159–172.

ABATE, J. E. "Linear and Adaptive Delta Modulation," *Proc. IEEE*, Mar. 1967: 298–307.

ABRAMOWITZ, M., and I. A. STEGUN. *Handbook of Mathematical Functions*, National Bureau of Standards, Applied Math. Series, 1964.

AEIN, J. M. "Multiple Access to Hard-Limiting Communications Satellite Repeater," *IEEE Trans. Space Electronics & Telemetry*, Dec. 1964: 159–167.

———, "On the Output Power Division in a Captured Hard -Limiting Repeater," *IEEE Trans. Comm. Tech.*, June 1966: 347–349.

ALLAN, D. W. "Statistics of Atomic Frequency Standards," *Proc. IEEE*, Feb. 1966: 221–230.

"Antenna Design Supplement," *Microwaves*, June 1967.

ARENDT, P. R. "Determination of the Ionospheric Electron Content Using Satellite Signals," *Proc. IEEE*, Mar. 1965: 268–277.

ARTHURS, E., and H. DYM. "On the Optimum Detection of Digital Signals in the Presence of White Gaussian Noise—A Geometric Interpretation and a Study of Three Basic Data Transmission Systems," *IRE Trans. Comm. Systems*, Dec. 1962: 336–372.

ATAL, B. S., and M. R. SCHROEDER. "Adaptive Predictive Coding of Speech Signals," *BSTJ*, Oct. 1970: 1973–1986.

BABCOCK, W. C. "Intermodulation Interference in Radio Systems," *BSTJ*, Jan. 1953: 63–73.

BAGHDADY, E. J., R. N. LINCOLN, and B. D. NELIN. "Short-Term Frequency Stability: Characterization, Theory, and Measurement," *Proc. IEEE-NASA Symp. on Short-Term Frequency Stability*, U.S. Government Printing Office, Washington, D.C., NASA SP-80: 65–87. Also published in *Proc. IEEE*, July 1965: 704–722.

BAIRD, R. C. "RF measurements of the Speed of Light," *Proc. IEEE*, June 1967: 1032–1039.

BALAKRISHNAN, A. V. *Advances in Communication Systems*, New York: Academic Press, 1965: 98–102.

———, *Communication Theory*, New York: McGraw-Hill Book Company, 1968: 97.

BARBER, R. E. "Short-Term Frequency Stability of Precision Oscillators and Frequency Generators," *BSTJ*, Mar. 1971: 881–916.

———, "Studies of the Short-Term Frequency Stability of Precision Oscillators and Frequency Generators," Raleigh, NC: North Carolina State University Press, 1970.

BARGELLINI, P. L., Ed., *Communications Satellite Technology*, Cambridge, MA: MIT Press, 1972.

BARKER, R. H. "Group Synchronization of Binary Digital Systems," in W. Jackson, Ed., *Communication Theory*, New York: Academic Press, 1953: 273–287.

BARNES, A. A., and R. L. FEY. "Synchronization of Two Remote Atomic Time Scales," *Proc. IEEE* (Correspondence), Nov. 1963: 1665.

BARNES, J. A. "Atomic Timekeeping and the Statistics of Precision Signal Generators," *Proc. IEEE*, Feb. 1966: 207–220.

———, "Tables of Bias Functions, B_1 and B_2 for Variances Based on Finite Samples of Processes with Power Law Spectral Densities," NBS Technical Note No. 275, U.S. Government Printing Office, Washington, D.C., 1969.

———, "The Development of an International Atomic Time Scale," *Proc. IEEE*, June 1967: 822–826.

——— and D. W. ALLAN. "A Statistical Model of Flicker Noise," *Proc. IEEE* Feb. 1966: 176–178.

——— et al. "Characterization of Frequency Stability," National Bureau of Standards, Tech. Note 394, Oct. 1970.

BATE, R. R. *Fundamentals of Astrodynamics*, New York: Dover Publications, 1971.

BEAN, B. R., and G. D. THAYER. "Comparison of Observed Atmospheric Radio Refraction Effects with the Values Predicted through the Use of Surface Weather Observations," *J. Res. NBS*, vol. 670, May–June 1963: 273–285.

BEEHLER, R. E. "A Historical Review of Atomic Frequency Standards," *Proc. IEEE*, June 1967: 792–805.

———, "Spaceborne Clock System," National Bureau of Standards, NBS report 10735, Nov. 1971.

BELL TELEPHONE LABORATORIES TECHINICAL STAFF. "Transmission Systems for Communications," Bell Telephone Laboratories, Inc., 1971.

BELLMAN, R. *A Brief Introduction to the Theta Functions*, New York: Holt, Rinehart, Winston, 1961: 7–12.

———, *Adaptive Control Processes: A Guided Tour*, Princeton, NJ: Princeton University Press, 1961: Ch. 4.

BENDER, P. L. "Laser Measurements of Long Distances," *Proc. IEEE*, June 1967: 1039–1045.

BENEDETTO, S., G. DEVICENTIS, and A. LUVISON. "Error Probability in the Presence of Intersymbol Interference and Additive Noise for Multilevel Digital Signals," *IEEE Trans. Comm. Tech.*, Mar. 1973: 181–190.

BENNETT, S., J. PUENTE, and R. STAMMINGER. "Multiple-Access for Satellites; Methods and Performance," *COMSAT*, Clarksburg, MD, Tech. Memo SCED-5-64, July 9, 1964.

BENNETT, W. R. "Spectra of Quantized Signals," *BSTJ*, July 1948: 446–472.

——— and J. R. DAVEY. *Data Transmission*, New York: McGraw-Hill Book Company, 1965: Ch. 15.

BENOIT, A. "Signal Attenuation Due to Neutral Oxygen and Water Vapour, Rain and Clouds," *Microwave Journal*, Nov. 1968: 73–80.

BERGLUND, C. N. "A Note on Power Law Devices and Their Effect on Signal-to-Noise Ratio," *PGIT*, Jan. 1964: 52–57.

BERKS, W. I., W. LUFT, "Photovoltaic Solar Arrays for Communications Satellites," *Proc. IEEE*, Feb. 1971: 263–271.

BERLEKAMP, E. R. *Algebraic Coding Theory*, New York: McGraw-Hill Book Company, 1968.

BERMAN, A. L., and C. E. MAHLE. "Nonlinear Phase Shift in Traveling Wave Tubes as Applied to Multiple Access Communications Satellite," *IEEE Trans. Comm. Tech.*, Feb. 1970: 37–48.

BEYER, J. P., R. D. BRISKMAN, D. W. LIPKE, and K. MANNING. "Some Orbital Spacing Considerations for Geostationary Communication Satellites Using the 4- and 6-GHz Frequency Bands," *Communications Satellite Corp.*, Tech. Memo DS-1-68, Oct. 25, 1968.

BHARUCHA-REID, A. T. *Elements of the Theory of Markov Processes and Their Applications*, New York: McGraw-Hill Book Company, 1960.

BLACHMAN, N. M. "Band-Pass Nonlinearities," *IEEE Trans. Info. Theory*, Apr. 1964: 162–164.

———, "Detectors, Band-pass Nonlinearities, and their Optimization: Inversion of the Chebyshev Transform," *IEEE Trans. Info. Theory*, July 1971: 398–404.

BLACKMAN, R. B., and J. W. TUKEY. *The Measurement of Power Spectra*, New York: Dover Publications, Inc., 1959.

BLACKWELL, L. A., and K. L. KOTZEBUE: *Semiconductor-Diode Parametric Amplifiers*, Englewood Cliffs, NJ: Prentice-Hall, Inc., 1961: Ch. 2.

BLAIR, B. E. "Time and Frequency: Theory and Fundamentals," U. S. Dept. of Commerce, National Bureau of Standards, Monograph 140, Washington, D.C., 1974.

BODILY, L. "Correlating Time from Europe to Asia with Flying Clocks," *Hewlett-Packard J.*, vol. 16, Apr. 1965: 1–8.

———, D. HARTKE, and R. C. HYATT. "World-Wide Time Synchronization," *Hewlett-Packard J.*, vol. 17, Aug. 1966: 13–20.

BODY-COMB, J. V., and A. H. HEDDAD. "Some Properties of Predictive Quantizing Systems," *IEEE Trans. Comm. Tech.*, Oct. 1970: 682–686.

BOND, F. E., and H. F. MEYER. "Intermodulation Effects in Limiter Amplifier Repeaters," *IEEE Trans. Comm. Tech.*, Apr., 1970: 127–135.

BOONTON, R. C. "Nonlinear Control System with Random Inputs," *Proc. Symposium on Nonlinear Circuit Analysis*, Polytech Inst. of Brooklyn, 1953: 369–391.

BOSWORTH, R. H., and J. C. CANDY. "A Companded One-Bit Coder for Television Transmission," *BSTJ*, May–June 1969: 1459–1480.

BOYARSKY, A., and M. FUKADA. "The Effects of Self-Noise on the Error Voltage of the Delay-Lock Discriminator," *IEEE Trans. Comm. Tech.* Aug. 1970: 443–447.

BRADLEY, W. E. "Communications Strategy of Geostationary Orbit," *Astronautics* and *Aeronautics*, vol. 6, Apr. 1968: 34–41.

BRAVERMAN, D. J. "A Discussion of Spread Spectrum Composite Codes," Aerospace Corp., Report TDK-269, Dec. 1963.

BRAYER, K. "Analysis of Quadrature Crosstalk in Bandpass Signals," *Proc. IEEE*, Feb. 1971: 292–294.

BRIGGS, B. H., and J. A. PERKIN. "On the Variation of Radio, Star and Satellite Scintillation with Zenith Angle," *J. Atmos. Terr. Physics*, vol. 25, 1963: 339–350.

BRISKMAN, R. D. "A Domestic Satellite Design," *IEEE Trans. AES*, Mar. 1971: 263–268.

———, "Domestic Communications Services via Satellite," *J. Spacecraft and Rockets*, July 1969: 835–870.

——— and R. C. BARTHLE. "Trends in Design of Communications Satellite Earth Stations," *Microwave J.*, vol. 51, Oct. 1967: 26–108.

BRITT, C. L., and D. F. PALMER. "Effects of CW Interference on Narrowband Second Order PLL," *IEEE Trans. AES*, Jan. 1967: 123–135.

BROWN, W. M. "Optimum Prefiltering of Sampled Data," *IRE Trans. Info. Theory*, Oct. 1961: 269–270.

BRUCE, J. D. "On the Optimum Quantization of Stationary Signals," *IEEE 1964 International Convention Record*, Part 1: 118–124.

BUCHNER, M. M., JR. "An Asymmetric Encoding Scheme for Word Stuffing," *BSTJ*, Mar. 1970: 379–398.

———, "A System Approach to Quantization and Transmission Error," *BSTJ*, May–June 1969a: 1219–1248.

———, "The Equivalence of Certain Harper Codes," *BSTJ*, Nov. 1969b: 3113–3130.

BURNS, A. A. "Preliminary Analysis of a Proposed High Stability Frequency Standard Distribution System," *IEEE Trans. AES*, Mar. 1971: 385–392.

BUSTAMANTE, HERMAN A., Private communication, 1972.

———, and W. K. LEONG. "Simulation and Implementation of Soft Decision Viterbi Decoding for High Efficiency Communication Links," Philco-Ford Communication Sciences TM 215, Sept. 1972.

BUTMAN, S. "Synchronization of PCM Channels by the Method of Word Stuffing," *IEEE Trans. Comm. Tech.*, COM-16, no. 2, Apr. 1968: 252–254.

BYKOV, V. L., V. A. BOROKOV, and S. M. KHOMUTOV. "Improving the Channel Capacity of Satellite Communication Systems with Multistation Access and Frequency Multiplexing," *Telecommun. Radio Eng.*, vol. 22, Jan. 1, 1968.

CADZOW, J. A., and H. R. MORTENS. "Discrete-Time and Computer Control Systems," Englewood Cliffs, NJ: Prentice-Hall, Inc., 1970.

CAHN, CHARLES, Magnavox Research Report, Magnavox Research Laboratory, Torrence, California, 1973.

———, "A Note on Signal-to-Noise Ratio in Band-Pass Limiters," *IRE Trans. IT*, Jan. 1961: 39–43.

———, "Cross-Talk Due to Finite Limiting of Frequency Multiplexed Signals," *Proc. IRE*, Jan. 1960: 53–59.

———, "Performance of Digital Phase Modulation Communication System," *IRE Trans. Comm. Systems*, May 1959: 3–6.

CASSELMAN, C. J., and M. L. TIBBALS. "The Radux-Omega Long Range Navigation System," *Proc. 2nd Nat'l Conf. on Military Electronics*, June 1958: 385–389.

CCIR "Technical factors influencing the efficiency of use of the geostationary satellite orbit by communication satellites sharing the same frequency band—General summary," presented at the International Radio Consultative Committee (CCIR) XII Plenary Assembly, New Delhi, India, Doc. IV/1058-E, 1970.

CCIR "Techniques for calculating interference noise in terrestrial radio-relay systems and communication satellite systems carrying multichannel telephony," presented at the CCIR XII Plenary Assembly, New Delhi, India, Doc. IV/1050-E, 1970.

Cesium Beam Frequency Standard, Model 5061A, Palo Alto, CA: Hewlett-Packard Company, Jan. 1973.

CHAN, D., and R. W. DONALDSON. "Optimum Pre- and Post-Filtering of Sampled Signals with Applications to Pulse Modulation and Data Compression Systems," *IEEE Trans. Comm. Tech.*, Apr. 1971: 141–157.

CHANDRASEKAR, S. "Stochastic Problems in Physics and Astronomy," *Rev. Mod. Physics*, vol. 15, no. 1, 1943.

CHANG, H. "Analysis of Quantization Effects on Digital Matched Filters," Private Communications, 1975.

CHARLES, F. J., and W. C. LINDSEY. "Some Analytical and Experimental Results for Low SNR", *Proc. IEEE*, Sept. 1966: 1152–1166.

CHI, A., and S. N. WITT. "Time Synchronization of Remote Clocks Using Dual VLF Transmissions," *Proc. 30th Annual Symp. on Frequency Control*, 1966.

CHIU, H. C., and R. S. SIMPSON. "Effect of Quadrature and Demodulation Phase Errors in a Quadrature PSK System," *IEEE Trans. Comm. Tech.*, Aug. 1973: 945–948.

CACCIAMANI, E. R. "The SPADE System as Applied to Data Communications and Small Earth Terminal Operation," *Comsat Tech. Rev.*, Fall 1971: 121–182.

COOLEY, W. C., and M. J. BARRETT. "Handbook of Space Environmental Effects on Solar Cell Power Systems," Final Report, NASA Contract NASW-1345, Jan. 1968.

COSTAS, J. P. "Synchronous Communications," *Proc. IRE*, Dec. 1956: 1713–1718.

COURANT, R., and D. HILBERT. *Methods of Mathematical Physics*, vol. I, New York: Wiley Interscience, 1953: 184–206.

CRAFT, H. D., JR., and L. H. WESTERLUND. "Scintillation at 4 and 6 GHz Caused by the Ionosphere, *Proc. AIAA 10th Aerospace Sciences Meeting*, San Diego, CA, Jan. 1972.

CRAMER, H. *Mathematical Methods of Statistics*, Princeton, NJ: Princeton University Press, 1945.

CRANE, R. K. "Propagation Phenomena Affecting Satellite Communication Systems," *Proc. IEEE*, Feb. 1971, p. 178.

CRAWFORD, A. B., and R. H. TURRIN. "A Packaged Antenna for Short Hop Microwave Radio Systems," *BSTJ*, July 1969: 1605–1622.

CROISER, A. "Compatible High Density Bipolar Coder," *IEEE Trans. Comm. Tech.*, June 1970: 265–268.

CROOK, J. S., and A. J. GIGER. "All-Weather Earth Station Satellite Communications Antennas," *BSTJ*, Sept. 1965: 1225–1228.

CUMMISKEY, P. "Adaptive Differential Pulse-Code Modulation for Speech Processing," Ph.D. Dissertation, Newark College of Engineering, Newark, NJ, 1973.

CUTLER, C. C., Ed., Special Issue on Redundancy Reduction, *Proc. IEEE*, Mar. 1967.

CUTLER, L. S., and C. L. SEARLE. "Some Aspects of the Theory and Measurement of Frequency Fluctuations in Frequency Standards," *Proc. IEEE*, Feb. 1966: 136–154.

DAMMANN, C. L., L. D. MCDANIEL, and C. L. MADDOX. "D2 Channel Bank—Multiplexing and Coding," *BSTJ*, Oct. 1972: 1675–1700.

DAVENPORT, W. B., JR. *Probability and Random Processes*, New York: McGraw-Hill Book Company, 1970.

———, "Signal-to-Noise Ratios in Band-Pass Limiters," *J. Appl. Phys.*, 24, June 1953: 720–727.

——— and W. L. ROOT. *An Introduction to the Theory of Random Signals and Noise*, New York: McGraw-Hill Book Company, 1958: 277–311.

DAVIES, T. L., and R. H. DOHERTY. "Widely Separated Clocks with Microsecond Synchronization and Independent Distribution System," *IRE Trans. SET*, Sept.–Dec. 1960: 138–146.

DAVIS, R. C., and D. D. MCRAE. "A More General Approach to the Interpolation of Sampled Data by Realizable Filters," *IEEE Trans. Comm. Tech.*, Feb. 1969: 65–73.

DEBUDA, R. "Coherent Demodulation of Frequency-Shift Keying with Low Direction Ratio," *IEEE Trans. Comm. Tech.*, June 1972: 429–435.

——— "The Fast FSK Modulation System," Inter. Conf. on Comm.: 41-25–41-27.

DE JAGER, F. "Delta Modulation—A Method of PCM Transmission Using the 1-Unit Code," Philips Research Report, Dec. 1962: 442–466.

DERUSSO, P. M., R. J. ROY, and C. M. CLOSE. *State Variables for Engineers*, New York: John Wiley & Sons, Inc., 1965.

DEUTSCH, R., *Estimation Theory*, Prentice-Hall, Englewood Cliffs, NJ., 1966.

DEVELET, J. A. "A Threshold Criterion for Phase-Lock Demodulation," *Proc. IEEE*, Feb. 1963: 349–356; *Proc. IEEE*, Apr. 1963: 580.

———, "The Influence of Time Delay on Second-Order Phase-Locked Loop Acquisition Range," *Proc. Int. Telemetering Conf.*, London, Eng., Sept. 1963: 432–437.

DILL, G., and N. SHIMASAKI. "Signaling and Switching for Demand Assigned Satellite Communications," *Proc. 1969 INTELSAT/IEE Conf. Digital Satellite Communication*, 297–307.

DION, A. R., and L. J. RICARDI. "A Variable Coverage Satellite Antenna System," *Proc. IEEE*, Feb. 1971: 252–262.

DONALDSON, R. W., and D. CHAN. "Analysis and Subjective Evaluation of Differential Pulse-Code Modulation Voice Communications Systems," *IEEE Trans. Comm. Tech.*, Feb. 1969: 10–19.

——— and R. J. DOUVILLE. "Analysis, Subjective Evaluation, Optimization, and Comparison of PCM, DPCM, DM, AM, PM Voice Communication Systems," *IEEE Trans. Comm. Tech.*, Aug. 1965: 421–431.

DUNCAN, J. W., et. al., "Polarization Isolation Characteristics of a Dual Beam Reflector Antenna," in *Communications Satellite Technology* edited by P. L. Bargellini, MIT Press, Cambridge, Mass., 1974.

DUNN, J. G. "An Experimental 9600 bit/sec Voice Digitizer Employing Adaptive Prediction," *IEEE Trans. Comm. Tech.*, Dec. 1971: 1021–1032.

EASTERLING, M. F. "A Skin-Tracking Radar Experiment Involving the COURIER Satellite," *IRE Trans. SET*, June 1962: 76–84.

EDELSON, B. I., and A. M. WERTH. "SPADE System Progress and Application," *COMSAT Tech Rev.*, Spring 1972: 221–242.

EDRICH, J. "Parametric Amplification of Millimeter and Submillimeter Waves, Results, Potential, and Limitation," *Digest of 1970 Int. Microwave Symposium*, May 1970.

EGGEN, C. P., M. S. MCALLISTER, "Modern Low-Pass Filter Characteristics," *Electro Technology*, Aug. 1966: 50–55.

ELIAS, P. "Bounds on Performance of Optimum Quantizers," *IEEE Trans. Info. Theory*, Mar. 1970: 172–184.

———, *A Radar System Based on Statistical Estimation and Resolution Considerations*, Stanford Electronics Laboratory, Stanford University, Stanford, CA., No. 361–1, App. A, 1955.

ELSPAS, B. "The Theory of Autonomous Linear Sequential Switching Networks," *IRE Trans. Circuit Theory*, vol. CT-6, Mar. 1959: 45–60.

1970 Proc. 5th Intersoc. Energy Convers. Eng. Conf., Las Vegas, NV, Sept. 1970; Box 9123, Albuquerque, NM 87119.

ESSEN, L., and J. MCA. STEELE. "The international comparison of atomic standards of time and frequency," *Proc. IEE* (London), vol. 109B, Jan. 1962: 41–47.

ESSMAN, J. E., and P. A. WINTZ. "The Effects of Channel Errors in DPCM Systems and Comparison with PCM Systems," *IEEE Trans. Comm. Tech.*, Aug. 1973: 867–877.

FANO, R. M. *Transmission of Information*, New York: John Wiley & Sons, Inc., 1961.

FARRELL, J. L., and J. C. MURTHA. "Statistical Bit Synchronization in Digital Communications," *IEEE Trans. Comm. Tech.*, Aug. 1971: 487–491.

FARRIELLO, E. "A Novel Digital Speech Detector for Improving Effective Satellite Capacity," *IEEE Trans. Comm. Tech.*, Feb. 1972: 55–60.

FELDMAN, N. E., and C. M. KELLY, eds., *Communications Satellites for the 1970s, Technology*, Progress in Astronautics and Aeronautics Series, vols. 25, 26, M.I.T. Press, Cambridge, MA., 1971.

FELLER, W., *An Introduction to Probability Theory and Its Applications*, vol. II, John Wiley, New York, NY., 1976.

FENERSTEIN, E. "Intermodulation Products for the *m*th Law-Biased Wave Rectifier," *Quarterly App. Math*, Apr.–Jan. 1956–57: 184–192.

FILIPOWSKI, R. F., and E. I. MUEHLDORF. *Space Communications Systems*, Englewood Cliffs, NJ: Prentice-Hall, Inc., 1965: Ch. 3.

FISZ, M. *Probability Theory and Mathematical Statistics*, New York: John Wiley & Sons, Inc., 1963.

FORNEY, G. D., JR. *Coding System Design for Advanced Solar Missions*, Codex Corp., Watertown, MA., Final Rept. Contract NAS2-3637 (NASA-Ames Research Center), Dec. 1967.

———, "Maximum Likelihood Sequence Estimation of Digital Sequences in the Presence of Intersymbol Interference," *IEEE Trans. Info. Theory*, May 1972: 363–377.

———, "The Viterbi Algorithm," *Proc. IEEE*, Mar. 1973: 268–278.

——— and E. K. BOWER. "A High Speed Sequential Decoder: Prototype Design and Test," *IEEE Trans. Comm. Tech.*, Oct. 1971: 821–834.

FRANKLIN, G. F. "The Optimum Synthesis of Sampled-Data Systems," Columbia University, Elec. Res. Lab. Tech. Rpt. T-6B, May 1955.

FRASER, J. M., D. B. BULLOCK, and N. G. LONG. "Overall Characteristics of a TASI System," *BSTJ*, July 1962: 1439–1454.

FRAZIER, J. P., and J. PAGE. "Phase-Lock Loop Frequency Acquisition Study," *IRE Trans. SET*, Sept. 1962: 210–227.

"Frequency and Time Standards," *Application Note No. 52*, Hewlett-Packard Company, Palo Alto, CA, 1965.

FUENZALIDA, J. C. "A Comparative Study of the Utilization of the Geostationary Orbit," *Proc. Int. Conf. Digital Satellite Communication, Intelsat/IEE*, 1969: 213–225.

FULTZ, K. E., and D. B. PENNICK. "The T1 Carrier System," *BSTJ*, Sept. 1965: 1405–1452.

GALLAGER, R. G. *Information Theory and Reliable Communication*, John Wiley, New York, NY., 1968.

GARDNER, F. M. *Phase Lock Techniques*, New York: John Wiley & Sons, Inc., 1966.

GATTERER, L. E., et al. "World-Wide Clock Synchronization Using a Synchronous Satellite," *IEEE Trans. Inst. and Meas.*, Dec. 1968: 372–378.

GEORGE, D. A., R. R. BOWEN, JR., and R. STOREY. "An Adaptive Decision Feedback Equalizer," *IEEE Trans. Comm.*, June 1971: 281–293.

GERBER, E. A., and R. A. SYKES. "State of the Art-Quartz Crystal Units and Oscillators," *Proc. IEEE*, 54, Feb. 1966: 103–116.

GERRISH, A. M., and R. D. HOWSON. "Multilevel Partial-Response Signaling," *Proc. Int. Conference on Comm.*, 1967: 186.

GERSHO, A. "Mutual Synchronization of Geographically Separated Oscillators," *BSTJ*, Dec. 1966: 1689–1704.

GIBSON, R. C. "On the *N* Equal Amplitude Sinusoidal Intermodulation Problem," *Int. Comm. Convention Record*, vol. II, June 1973: 31.1–31.6.

GIGER, A. J., and R. H. TURRIN. "The Triply-Folded Horn Reflector; A Compact Ground Station Antenna Design for Satellite Communications," *BSTJ*, Sept. 1965: 1229–1254.

GILHOUSEN, K. S., S. A. HELLER, J. M. JACOBS, and A. J. VITERBI. "Coding Study for High Data Rate Telemetry Links," Linkabit Corp. Report, Jan. 1971: 30.

GILL, W. J. "A Comparison of Delay-Lock Loop Implementations," *IEEE Trans. AES*, July 1966: 415–426.

———, "Effect of Synchronization Error in Pseudo-Random Carrier Communications," 1st IEEE Ann. Communications Conv., Boulder, CO, June 7–9, 1965.

——— and J. J. SPILKER, JR. "An Interesting Decomposition Property for the Self-Products of Random or Pseudorandom Binary Sequences," *IEEE Trans. on Communications Systems*, June 1963: 246–247.

GITLIN, R. D., E. Y. HO, and J. E. MAZO. "Pass-band Equalizer of Differentially Phase-Modulated Data Signals," *BSTJ*, Feb. 1973: 219–238.

GLANCE, B. "Power Spectra of Multilevel Digital Phase-Modulated Signals," *BSTJ*, Sept. 1971: 2857–2878.

GOLD, R. "Maximal Recursive Sequences with 3-Valued Recursive Cross-Correlation Functions," *IEEE Trans. Info. Theory*, Jan. 1968: 154–156.

———, "Optimal Binary Sequences for Spread Spectrum Multiplexing," *IEEE Trans. Info. Theory*, Oct. 1967: 619–621.

GOLDEN, T. S. "A Note on Equatorial Scintillation at 136 MHz and 1550 MHz, GFSC Report X-520, Oct. 1970: 70–397.

GOLDING, L. S., and J. E. D. BALL. "Satellite Television Covers the World," *IEEE Spectrum*, Aug. 1973: 24–31.

GOLDMAN, S. L. "Second Order Phase-Lock Loop Acquisition Time in the Presence of Narrow Band Gaussian Noise," *IEEE Trans. Comm. Tech.*, Apr. 1973: 297–300.

GOLDSTEIN, A. J., and C. J. BYRNE. "Pull-In Frequency of the Phase-Controlled Oscillator," *Proc. IRE Letters*, vol. 49, July 1961: 1209.

GOLOMB, S. W. "Sequences with the Cycle and Add Property," Jet Propulsion Lab., Calif. Inst. of Tech., Pasadena, CA, Section Rept. No. 8–573, 1957.

———, "Sequences with Randomness Properties," Glenn Martin Co., Baltimore, MD, Final Rept., Contract No. SC-54-36611, 1955.

———, *Shift Register Sequence*, San Francisco: Holden-Day, 1967.

———, L. D. BANMERT, M. F. EASTERLING, J. J. STIFFLER, and A. J. VITERBI. *Digital Communications with Space Applications*, Englewood Cliffs, NJ: Prentice-Hall, Inc., 1964.

GOODING D. J. "Performance of a Digital Matched Filter," Private Communication Aug. 1966.

GOODMAN, D. J. "Delta Modulation Granular Quantizing Noise," *BSTJ*, May–June, 1969a: 1197–1218.

———, "The Application of Delta Modulation to Analog to PCM Encoding," *BSTJ*, Feb. 1969b: 321–343.

GOODMAN, L. M., and P. R. DROUILET, JR. "Asymptotically Optimum Pre-emphasis and De-emphasis Network, Pre-sampling and Quantizing," *Proc. IEEE*, May 1966: 795–796.

GRAY, D. A. "Transit Time Variations in Line of Sight Tropospheric Propagation Paths," *BSTJ*, July–Aug. 1970: 1059–1068.

GRAY, R. M., and R. C. TAUSWORTHE. "Frequency Counted Measurements and Phase Locking to Noisy Oscillators," *IEEE Trans. Comm. Tech.*, Feb. 1971: 21–30.

GREEFKES, J. A., and F. DE JAGER. "Continuous Delta Modulation," Philips Res. Rep., vol. 23/2, 1968: 233–246.

GREENSTEIN, L. J. "Slope Overload Noise in Linear Delta Modulators with Gaussian Inputs," *BSTJ*, Mar. 1973: 387–422.

GRUEN, W. J. "Theory of AFC Synchronization," *Proc. IRE*, Aug. 1953: 1043–1048.

GRUMIN, J. K., and W. F. MCGEE. "Microwave System Intermodulation Simulation," *Proc. 1968 Int. Conf. on Comm.*, 403–406.

GUIDACE, D. A., and J. P. CASTELLI. "The Use of Extraterrestrial Radio Sources in the Measurement of Antenna Parameters," *IEEE Trans. AES*, Mar. 1971: 226–234.

GUMACOS, C. "Analysis of an Optimum Sync Search Procedure," *IEEE Trans. Comm. Systems*, Mar. 1963: 87–99.

GUNN, J. F., J. S. RONNE, and D. C. WELLER. "Mastergroup Digital Transmission in Modern Coaxial Systems," *BSTJ*, Feb. 1971: 501–520.

——— and J. A. LOMBARDI. "Error Detection for Partial Response Systems," *IEEE Trans. Comm. Tech.*, Dec. 1969: 734–737.

MAUNG GYI. "Some Topics on Limiters and FM Demodulators," Stanford TR #7053-3, July 1965.

HALL, H. M. "Power Spectrum of Hard-Limited Gaussian Processes," *BSTJ*, Nov. 1969: 3031–3058.

HANSEN, R. C. *Microwave Scanning Antennas*, New York: Academic Press, 1966.

HANSON, D. W., and W. P. HAMILTON. "Clock Synchronization from Satellite Tracking," *IEEE Trans. AES*, Sept. 1971: 895–899.

HARTMAN, H. P. "Analysis of a Dithering Loop for PN Code Tracking," *IEEE Trans. AES*, Jan. 1974: 2–9.

HASTINGS, H. F. "Transmission to Naval Observatory of Reference Frequency Derived from Hydrogen Masers at NRL," *NRL Progress Rept.*, Mar. 1965: 46–47.

HATCH, G. M. "Communication Subsystem Design Trends for the Defense Satellite Communications Program," *IEEE Trans. AES*, Sept. 1969: 724–730.

HAYES, J. F. "Optimum Prefiltering and Postfiltering of Sampled Waveforms," *IEEE Trans. Comm. Tech.*, Aug. 1971: 532–535.

HELLER, J. A. "Improved Performance of Short Constraint Length Convolutional Codes," *Space Program Summary 37–56*, vol. III, Jet Propulsion Laboratory, Feb.–Mar. 1969: 83–84.

———, "Short Constraint Length Convolutional Codes," *Space Program Summary 37–54*, vol. III, Jet Propulsion Laboratory, Oct.–Nov. 1968: 171–177.

——— and I. M. JACOBS. "Viterbi Decoding for Satellite and Space Communications," *IEEE Trans. Comm. Tech.*, Oct. 1971: 835–848.

HELSTROM, C. W., *Statistical Theory of Signal Detection*, Pergamon Press, New York, NY., 1960.

HENRY, J. C. "DPSK versus FSK with Frequency Uncertainty," *IEEE Trans. Comm. Systems*, Dec. 1970: 814–816.

HENRY, J. C. III, "DPSK Versus FSK with Frequency Uncertainty," *IEEE Transactions Communications Systems*, December 1970, 814–816.

HILDEBRAND, R. B. *Advanced Calculus for Engineers*, Englewood Cliffs, NJ: Prentice-Hall, Inc., 1949: 2–10.

HOGG, D. C. "Ground Station Antennas for Space Communications," in L. Young, Ed., *Advance in Microwaves*, New York: Academic Press, 1968.

HOLZAR, W. "Atmospheric Attenuation in Satellite Communications," *Microwave Journal*, Mar. 1965: 119–125.

HUANG, J. Y. Private Communication, 1974.

HUANG, R. Y., and P. HOOTEN. "Communication Satellite Processing Repeaters," *Proc. IEEE*, Feb. 1971: 238–251.

HUDSON, G. E. "Some Characteristics of Commonly Used Time Scales," *Proc. IEEE*, June 1967: 815–821.

HULT, J. L., and E. E. REINHARDT. "Satellite Spacing and Frequency Sharing for Communication and Broadcast Services," *Proc. IEEE*, Feb. 1971: 118–128.

HURD, W. J., and T. O. ANDERSON. "Digital Transition Tracking Symbol Synchronizer for Low SNR Coded Systems," *IEEE Trans. Comm. Tech.*, Apr. 1970: 141–146.

IEEE Special Issue on Noise, *IEEE Trans. Microwave Theory and Techniques*, Sept. 1968.

INORE, H., Y. YASUDA, Y. KAWAI, and M. TAKAGI. "Subscriber Line Circuits for an Experimental TDM Exchange Featuring Delta Modulation Techniques," *J. Elec. Comm. Eng. of Japan*, vol. 44, 1961: 1322–1328.

Proc. INTELSAT/IEE, Conf. Digital Satellite Communication, No. 59, London, Eng., 1969.

IPPOLITO, L. J. "Effects of Precipitation on 15.3, 31.65 GHz Earth Space Transmissions with the ATS V Satellite," *Proc. IEEE*, Feb. 1971: 189–205.

ISLEY, W. C., and K. I. DUCK. "Propulsion Requirements for Communications Satellites," in P. L. Bargellini, Ed., *Communications Satellite Technology*, Cambridge, MA: MIT Press, 1974: 165–190.

JACOBS, E., and J. M. STACEY. "Earth Footprints of Satellite Antennas," *IEEE Trans. AES*, Mar. 1971: 235–242.

JACOBS, I. M., and J. A. HELLER. "Performance Study of Viterbi Decoding as Related to Space Communications," Linkabit Corp. Report, Aug. 1971.

JAFFE, R., and E. RECHTIN. "Design and Performance of Phase-Locked Circuits Capable of Near-Optimum Performance Over a Wide Range of Input Signals and Noise Levels," *IRE Trans. IT*, Mar. 1955: 66–72.

JAIN, P. C. "Limiting of Signals in Random Noise," *IEEE Trans. Info. Theory*, May 1972: 332–340.

JANSKY, D. M., and M. C. JERUCHIM. "A Technical Basis for Communication Satellites to Share the Geostationary Orbit," presented at the AIAA 3rd Communications Satellite Systems Conf., Los Angeles, CA, Apr. 1970, Paper 70–441.

JASIK, H. *Antenna Engineering Handbook*, New York: McGraw Hill Book Company, 1961.

JAYANT, N. S. "Adaptive Quantization with a One Word Memory," *BSTJ*, Sept. 1973: 1119–1144.

———, "Digital Coding of Speech Waveforms: PCM, DPCM, and DM Quantizers," *Proc. IEEE*, May 1974: 611–632.

——— and L. R. RABINER. "The Application of Dither to the Quantization of Speech Signals," *BSTJ*, July–Aug. 1972: 1293–1304.

JERUCHIM, M. C., and F. E. LILLEY. "Spacing Limitations of Geostationary Using Multi-level Coherent PSK Signals," *IEEE Trans. Comm. Tech.*, Oct. 1972: 1021–1025.

——— and T. C. SAYER. "Orbit/Spectrum Utilization Study," General Electric, Valley Forge, PA, Doc. No. 69SD4270, May 15, 1969.

JESPERSEN, J. L., G. KAMAS, L. E. GATTERER, and P. F. MACDORAN. "Satellite VHF Transponder Time Synchronization," *Proc. IEEE*, July 1968: 1202–1206.

JOHANNES, V. I. "Comments on Compatible High Density Bipolar Codes," *IEEE Trans. Comm. Tech.*, Feb. 1972: 78–79.

——— and R. H. MCCULLOUGH. "Multiplexing of Asynchronous Digital Signals Using Pulse Stuffing with Added-Bit Signaling," *IEEE Trans. Comm. Tech.*, Oct. 1966: 562–568.

———, et al. "Bipolar Pulse Transmission with Zero Extraction," *IEEE Trans. Comm. Tech.*, Apr. 1969, 303–310.

JOHNSON, J. C., *Physical Meteorology*, John Wiley, New York, NY., 1954.

JONES, J. J. "Hard-Limiting of Two Signals in Random Noise," *IEEE Trans. Info. Theory*, Jan. 1963: 34–42.

———, "Filter Distortion and Intersymbol Interference Effects on PSK Signals," *IEEE Trans. Comm. Tech.*, Apr. 1971: 120–132.

———, "Filter Distortion and Intersymbol Interference Effects on QPSK Signal Transmission," System Sciences Conference, Hawaii, 1970.

———, Private Communication, 1976.

JOWETT, J.K.S., and A. K. JEFFERIS. "Ultimate Capacity of the Geostationary-Satellite Orbit," *Proc. IEEE*, vol. 116, Aug. 1969: 1304–1310.

KAPLAN, M. H. "Active Attitudes and Orbit Control of Body-Oriented Geostationary Communications Satellites," in P. L. Bargellini, Ed., *Communications Satellite Technology*, Cambridge, MA: MIT Press, 1974: 29–56.

KAROLUS, A., and J. HELMBERGER. "Messung der Lichtgeschwindkeit, auf der 48 m Basis des DGFI in München," Bayerischen Akademie der Wissenschoften, Munich, Germany, 1967.

KATZENELSON, J. "On Errors Introduced by Combined Sampling and Quantization," *IRE Trans. Automat. Control*, Apr. 1962: 58–68.

KAY, A. R., D. A. GEORGE, and M. J. ERIC. "Analysis and Compensation of Bandpass Nonlinearities for Communication," *IEEE Trans. Comm. Tech.*, Oct. 1972.

KEMENY, J. G., and J. L. SNELL. *Finite Markov Chains*, New York: D. Van Nostrand Co., 1963.

KENT, G. S., and J. R. KOSTER. "Some Studies of Night-time F-layer Irregularities at the Equator Using Very High Frequency Signals Radiated from Earth Satellites," *Annels de Geophysique*, vol. 22, no. 3, 1966: 405–417.

KIESLING, J. D., R. R. ELBERT, W. B. GARNER, and W. L. MORGAN. "A Technique for Modeling Communications Satellites," *COMSAT Tech. Rev.*, Spring 1972: 73–101.

KINSEL, G. V. "Multibeam Satellite Repeater Tradeoffs Applied to a Multi-functional System," *Proc. Int. Conf. on Comm.*, 1973: 12-19–12-26.

KLAPPER, J., and J. T. FRANKLE. *Phase-Locked and Frequency-Feedback Systems*, New York: Academic Press, 1972.

KOELLE, H. H. *Handbook of Astronautical Engineering*, New York: McGraw-Hill Book Company, 1961.

KORN, I., "Probability of Error in Binary Communications with Causal Band-Limiting Filters, Part 1, Non Return-to-Zero Signal," *IEEE Trans. Comm. Tech.*, Aug. 1973: 878–890.

KOSTER, J. R. "Ionospheric Studies Using the Tracking Beacon on Early-Bird Synchronous Satellite," *Annels de Geophysique*, vol. 22, no. 3, 1966: 435–439.

KRESSNER, G. N. *Introduction to Space Communications Systems*, New York: McGraw-Hill Book Company, 1964: Ch. 4.

KRETSMER, E. G. "Generalization of a Technique of Binary Data Communication," *IEEE Trans. Comm. Tech.*, Feb. 1966: 67–68.

KUNO, H., and Y. HATA. "The Concept of Position Fixing by a Passive Mode Navigation Satellite System," *IEEE Trans. AES*, Nov. 1970: 780–789.

LACY, R. F. "Short Term Stability of Passive Atomic Frequency Standards," *Proc. IEEE*, Feb. 1966: 170–178.

LANDAU, H. J. "Sampling, Data Transmission and the Nyquist Rate," *Proc. IEEE*, Oct. 1967: 1701–1706.

——— and H. D. POLLAK. "Prolate Spheroidal Wave Functions, Fourier Analysis and Uncertainty, III: The Dimension of Essentially Time- and Band-Limited Signals," *BSTJ*, July 1962: 1295–1336.

LAWRENCE, et al. "Ionospheric Effects in Earth-Space Satellites," *Proc. IEEE*, Jan. 1964: 4–27.

LEE, G. M., and J. J. KOMO. "PCM Bit Synchronization and Detection by Nonlinear Filter Theory," *IEEE Trans. Comm. Tech.*, Dec. 1970: 757–762.

LEE, Y. W. *Statistical Theory of Communication*, Englewood Cliffs, NJ: Prentice-Hall, Inc., 1971.

LENDER, A. "The Duo-binary Technique for High Speed Data Transmission," *IEEE Trans. Comm. Tech.* May 1963: 214–218.

LENEMAN, O.A.Z. "On Finite Pulse-Width Sampling," *IEEE Trans. Comm. Tech.*, Oct. 1967: 718–720.

LIM, J., and J. O. SCANLON. "Group Delay Characteristics of Chebyshev Filters," *IEEE Trans. Circuit Theory*, Sept. 1964: 427–430.

LIN, S. H. *Error-Correcting Codes*, Englewood Cliffs, NJ: Prentice-Hall, Inc., 1970.

———, "Statistical Behavior of Rain Attenuation," *BSTJ*, Apr. 1973: 557–582.

LINDHOLM, J. H. "An Analysis of the Pseudo-Randomness Properties of Subsequences of Long *M* Sequences," *IEEE Trans. Info. Theory*, July 1968: 569–576.

LINDSEY, W. C. "Phase-Shift-Keyed Signal Detection with Noisy Reference Signals," *IEEE Trans. AES*, July 1966: 393–401.

———, *Synchronization Systems in Communications*, Englewood Cliffs, NJ: Prentice-Hall, Inc., 1972.

——— and T. O. ANDERSON. "Digital Data-Transition Tracking Loops," Int. Telemetry Conference, Oct. 1968.

——— and F. J. CHARLES. "A Model Distribution for the Phase Error in Second-Order PLL's," *IEEE Trans. Comm. Tech.*, Oct. 1966: 662–664.

——— and M. K. SIMON. "Carrier Synchronization and Detection of Polyphase Signals," *IEEE Trans. Comm. Tech.*, June 1972a: 441–454.

———, "Data-Aided Carrier-Tracking Loops," *IEEE Trans. Comm. Tech.*, Apr. 1971: 152–169.

———, "On the Detection of Differentially Encoded Polyphase Signals," *IEEE Trans. Comm. Tech.*, Dec. 1972b: 1121–1127.

——— and M. K. SIMON. *Telecommunications Systems Engineering*, Englewood Cliffs, NJ: Prentice-Hall, Inc., 1973.

——— and R. C. TAUSWORTHE. "Digital Data-Transmitter Tracking Loops," JPL, SPS 37-50, vol. III, Apr. 1968: 272–276.

LLOYD, S. P., and B. MCMILLAN. "Linear Least Squares Filtering and Prediction of Sampled Signals," *Proc. Symposium on Modern Network Synthesis*, Polytech Inst. of Brooklyn, 1955: 241–247.

LOO, C. "Calculation of Intermodulation Noise Due to Hard and Soft Limiting of Multiple Carriers," Int. Conf. on Comm., 1974: 3A-1–3A-4.

LOONEY, C. H., JR. "VLF Utilization at NASA Satellite Tracking Stations," *J. Res. NBS* (Radio Sci.), vol. 68D, 1964: 43–45.

LOTHALLER, W. E. "System Considerations for European Communications Satellites," 1972 Int. Comm. Conf., Philadelphia, PA, June 1972.

LUNDGREN, C. W. "A Satellite System for Avoiding Serial Suntransit Outages and Eclipses," *BSTJ*, Oct. 1970: 1943–1972.

——— and A. S. MAY. "Radio Relay Antenna Pointing for Controlled Interference with Geostationary Satellites," *BSTJ*, Dec. 1969: 3387–3422.

LUNDQUIST, L. "Digital PM Spectra by Transform Techniques," *BSTJ*, Feb. 1969: 1164–1189.

LUTZ, S. G. "Orbit Utilization—from Both Sides," presented at the AIAA 3rd Communications Satellite Systems Conf., Los Angeles, CA, Apr. 1970.

MAGEE, E. R., JR. "A Comparison of Compromise Viterbi Algorithm and Standard Equalization Techniques over Bandlimited Channels," *IEEE Trans. Comm. Systems.*, Mar. 1975: 361–366.

MAGILL, D. T. "Comparative Acquisition Performance of Second- and Third-Order Phase Locked Loops," Philco-Ford, Comm. Sciences Dept., June 1967.

———, "Multiple-Access Modulation Techniques," in *Communications Satellite System Technology*, New York: Academic Press, 1966: 667–680.

———, "Noise Theory of Tracking Cross-Correlators," 1965 IEEE International Conv., New York, Mar. 22–26, 1965.

———, Private Communication, 1968.

MALLONY, P. "A Maximum Likelihood Bit Synchronizer," *Proc. 1968 Int. Telemetry Conference*: 1–16.

MANASSE, R., R. PRICE, and R. M. LERNER. "Loss of Signal Detectability in Band-Pass Limiters," *IRE Trans. IT*, Mar. 1958: 34–38.

MANN, H., H. M. STRAUBE, and C. P. VILLARS. "A Commanded Coder for an Experimental PCM Terminal," *BSTJ*, Jan. 1962: 173–226.

MARCUS, M. B., and P. SWERLING. "Sequential Detection in Radar with Multiple Resolution Elements," *IEEE Trans. Info. Theory*, Apr. 1962: 237–245.

MARCUSE, D. "Losses and Impulse Response in Parabolic Index Fibers with Square Cross-Section," *BSTJ*, Feb. 1974: 195–216.

MARK, J. W., and P. S. BUDIHARDY. "Performance of Jointly Optimized Prefilter-Equalizer Receivers," *IEEE Trans. Comm. Tech.*, Aug. 1973: 941–944.

MARKOWITZ, W. "International Frequency and Clock Synchronization," *Frequency*, vol. 2, no. 4, 1964: 30–31.

———, C. A. LIDBACK, H. UYEDA, and K. MURAMATSU. "Clock Synchronization via Relay II satellite," *IEEE Trans. Inst. and Meas.*, Dec. 1966: 177–184.

MARSTEN, R. B., ed., *Communication Satellite System Technology*, Progress in Astronautics and Aeronautics Series, vol. 19, M.I.T. Press, Cambridge, MA., 1966.

MASSEY, J. L. "Optimum Frame Synchronization," *IEEE Trans. Comm. Tech.*, Apr. 1972: 115–118.

———, *Threshold Decoding*, Cambridge, MA: MIT Press, 1963.

MAX, JOEL. "Quantizing for Minimum Distortion," *IRE Trans. IT*, Mar. 1960: 7–12.

MAY, A. S., and J. J. PAGONES. "Model for Computation of Interference to Radio-Relay Systems from Geostationary Satellites," *BSTJ*, Jan. 1971: 81–102.

MAYO, J. S. "A Bipolar Repeater for Pulse Code Signals," *BSTJ*, Jan. 1962: 25–98.

———, "An Approach to Digital System Networks," *IEEE Trans. Comm. Tech.*, Apr. 1967: 307–310.

———, "Experimental 224 Mb/s PCM Terminals," *BSTJ*, Nov. 1965: 1813–1841.

MCBRIDE, A. L., and A. P. SAGE. "On Discrete Sequential Estimation of Bit Synchronizations," *IEEE Trans. Comm. Tech.*, Feb. 1970: 48–58.

———, "Optimum Estimation of Bit Synchronization," *IEEE Trans. AES*, May 1969: 525–536.

MCCLURE, R. B. "Analysis of Intermodulation Distortion in a FDMA Satellite Communication System with a Bandwidth Constraint," *IEEE Int. Conf. on Comm.*, 1970.

MCCOUBREY, A. O. "A Survey of Atomic Frequency Standards," *Proc. IEEE*, June 1967: 805–814.

———, "The Relative Merits of Atomic Frequency Standards," *Proc. IEEE*, Feb. 1966: 116–135.

MESSERSCHMIDT, D. B. "Design of a Finite Impulse Response for the Viterbi Algorithm and Decision-Feedback Equalizer," *Int. Conf. Communications*, 1974: 370-1–370-5.

METHWICK, H. R., J. F. BALCEWICZ, and M. HECT. "The Effect of Tandem Band and Amplitude Limiting on the E_{b/N_0} Performance of Minimal (Frequency) Shift Keying (MSK)," *IEEE Trans. Comm. Tech.*, Oct. 1974: 1525–1537.

MIDDLETON, D., *An Introduction to Statistical Communication Theory*, McGraw-Hill, New York, NJ., 1960.

MIEDEMA, H., and M. G. SCHACHTMAN, "TASI Quality—Effect of Speech Detectors and Interpolation," *BSTJ*, July 1962: 1455–1476.

MILLER, S. E. "Delay Distortion in Generalized Lens-Like Media," *BSTJ*, Feb. 1974: 174–194.

MITCHELL, A.M.J. "Frequency Comparison of Atomic Standard by Radio Signals," *Nature*, vol. 198, June 1963: 1155–1158.

MONSEN, P. "Adaptive Equalization of Finite Angle Modulated Signals," *Proc. IEEE Int. Conf. on Comm.*, June 1972: 44-6–44-11.

MORGAN, A. H. "Distribution of Standard Frequency and Time Signals," *Proc. IEEE*, June 1967: 827–836.

———, "Precise Time Synchronization of Widely Separated Clocks," National Bureau of Standards, Boulder, CO, Tech. Note 22, July 1959.

MORIKAMI, H., et al. "A Newly Developed Intelsat Earth Station," *Int. Conf. on Comm.*, 1973: 8-30–8-35.

MORROW, W. E., JR., ed. *Proc., IEEE*, Special Issue on Satellite Communication, Feb., 1971.

MORSE, P. M., and H. FESHBECK. *Methods of Mathematical Physics*, New York: McGraw-Hill Book Company, 1953: 786–787.

MOYER, T. D. "Mathematical Formulation of the Double-Precision Orbit Determination Program," TR32-1527, Jet Propulsion Laboratory, California Inst. of Tech., Pasadena, CA, May 1971.

MULLOY, C. S. "Digital Analysis of the Delay-Lock Discriminator," M.S.E.E. Thesis, U.S. Naval Postgraduate School, Monterey, CA, 1962.

NATALI, F. D. "All Digital Coherent Demodulator Techniques," *Proc. Int. Telemetry Conf.*, 1972: 89–107.

———, Private Communication, 1973.

——— and W. J. WALBESSEN. "Phase-Locked Loop Detection of Binary PSK Signals Utilizing Decision Feedback," *IEEE Trans. AES*, Jan. 1969: 83–90.

NEUMAN, F., and L. HOFMAN. "New Pulse Sequences with Desirable Correlation Properties," *Proc. Nat. Telemetry Conf.*, 1971: 272–282.

NIELSEN, P. T. "On the Acquisition Behavior of Binary Delay-Lock Loops," *IEEE Trans. Aerospace and Elec. Syst.*, May 1975: 415–418.

NIKITIN, N. P. "Cut-Off in Tracking in a Network of Phase Automatic Frequency Control," *Automatica and Telemekhanika*, vol. 26, Apr. 1965: 669–675.

NOLL, P. "Non-adaptive and Adaptive DPCM of Speech Signals," Overdruk uit Polytech. Tijdschr. Ed. Elektrotech./Electron. (the Netherlands), no. 19, 1972.

NOSAKI, K. "PSK Demodulator with Delay Line for PCM—TDMA Systems," *IEEE Trans. Comm. Tech.*, Aug. 1970: 425–434.

NUTTALL, A. H. "Theory and Application of the Separable Class of Random Processes," Technical Report 343, Research Laboratory of Electronics, Massachusetts Institute of Technology, Cambridge, MA, May 1958.

———, "Error Probabilities for Equicorrelated *M*-ary Signals under Phase-Coherent and Phase-Incoherent Reception," *IRE Trans. IT,* July 1962: 305–314.

OBERST, J. F., and D. L. SCHILLING. "Performance of Self-Synchronized Phase-Shift Keyed Systems," *IEEE Trans. Comm. Tech.*, Dec. 1970: 664–669.

ODENWALDER, J. P. "Optimum Decoding of Convolutional Codes," Ph.D. Thesis, System Sciences Department, UC Los Angeles, 1970.

OGATA, M. "Station Configuration of Malaysia Earth Station," IEEE Conf. on Earth Station Technology, Oct. 1970.

OLIVER, B. M., J. R. PIERCE, and C. E. SHANNON. "The Philosophy of PCM," *Proc. IRE,* Nov. 1948: 1324–1331.

OMURA, J. K. "On the Viterbi Decoding Algorithm," *IEEE Trans. Info. Theory,* Jan. 1969: 177–179.

O'NEAL, J. B. "Delta Modulation Quantization Noise Analytical and Computer Simulation Results for Gaussian and Television Signals," *BSTJ,* Jan. 1966: 117–141.

———, "Predictive Quantization (DPCM) for the Transmission of Television Signals," *BSTJ,* May–June 1966: 689–721.

OSBORNE, W. P., and M. B. LUTZ, "Coherent and Noncoherent Detection of CPFSK," *IEEE Trans. Comm.*, Aug. 1974: 1023–1036.

O'SULLIVAN, M. R. "Tracking Systems Employing the Delay-Lock Discriminator," *IRE Trans. SET,* Mar. 1972: 1–7.

PAEZ, M. D., and T. H. GLISSOM. "Minimum Mean Square Error Quantization in Speech, PCM, and DPCM Systems," *IEEE Trans. Comm. Tech.*, Apr. 1972: 225–230.

PALMER, L. C., and S. A. KLEIN. "Phase-Slipping in Phase-Locked Loop Configurations that Track Biphase or Quadriphase Modulated Carriers," *IEEE Trans. Comm. Tech.*, Oct. 1972: 984–991.

PANTER, P. F. *Modulation, Noise, and Spectral Analysis,* New York: McGraw-Hill Book Company, 1965: 495–503.

——— and W. DITE. "Quantization Distortion in Pulse-Count Modulations with Non-Uniform Spacing of Levels," *Proc. IRE,* Jan. 1951: 44–48.

PAPOULIS, A. *Probability, Random Variables, and Stochastic Processes,* New York: McGraw-Hill Book Company, 1965.

PARADYSZ, R. E., and W. L. SMITH. "Measurement of FM Noise Spectra of Low-Noise VHF Crystal Controlled Oscillators," *IEEE Trans. Inst. and Meas.*, Dec. 1966: 202–211.

PARK, J. H. "Effect of Band Limiting on the Detection of Binary Signals," *IEEE Trans. AES,* Sept. 1969: 867–870.

PETERSON, W. W. *Error-Correcting Codes,* New York: John Wiley & Sons, Inc., 1961.

——— and E. J. WELDON, JR. *Error-Correcting Codes,* 2d ed., Cambridge, MA: MIT Press, 1972.

PIERCE, J. A. "Intercontinental Frequency Comparison by Very Low-Frequency Radio Transmission," *Proc. IRE*, June 1957: 794–803.

———, H. T. MITCHELL, and L. ESSEN. "World-Wide Frequency and Time Comparisons by Means of Radio Transmissions," *Nature*, vol. 174, Nov. 1954: 922–923.

———, G.M.R. WINKLER, and R. L. CORKE. "The GBR experiment: A Trans-Atlantic Frequency Comparison between Cesium Controlled Oscillators," *Nature*, vol. 187, no. 4741, 1960: 914–916.

PIERCE, J. R. "Synchronizing Digital Networks," *BSTJ*, Mar. 1969: 615–636.

PODRACZKY, E., and J. KIESLING, "Intelsat V System Concepts and Technology," *Proc. AIAA Comm. Satellite Systems Conference*, 1972: 72–536.

POULARIKAS, A. D., and T. S. GOLDEN. "A Note on the Effects of the Ionosphere on Satellite Orbital Correction in Near Real Time," *IEEE Trans. AES*, Oct. 1969: 865–867.

PRABHU, V. K. "Error Probability of *M*-ary CPSK Systems with Inter-symbol Interference, *IEEE Trans. Comm. Tech.*, Feb. 1973: 97–109.

———, "Error Rate Considerations for Coherent Phase-Shift Keyed Systems with Co-Channel Interferences," *BSTJ*, Mar. 1969: 743–767.

PRESTON, C. N., and J. C. TELLIER. "The Lock-In Performance of an AFC Circuit," *Proc. IRE*, vol. 41, Feb. 1953: 249–251.

PRICE, R. "A Fixed Reflector, Steerable Beam Earth Station Antenna," *Proc. National Telecomm. Conf.*, 1972: 25B–1.

———, "A Note on the Envelope and Phase-Modulated Components of Narrow-Band Gaussian Noise," *IRE Trans. IT*, Sept. 1955: 9–13.

———, "A Useful Theorem for Nonlinear Devices Having Gaussian Inputs," *IRE Trans. IT*, June 1958: 69–72.

——— and P. E. GREEN, JR. "Signal Processing in Radar Astronomy Communication via Multipath Media," M.I.T. Lincoln Lab., Lexington, MA, Rept. No. 234, Oct. 1960.

PRITCHARD, W. L., and P. L. BANGELLINI. "Trends in Technology for Communications Satellites," *Astronautics and Aeronautics*, Apr. 1972: 36.

PROAKIS, J. G., and J. H. MILLER. "An Adaptive Receiver for Digital Signalling through Channels with Intersymbol Interference," *IEEE Trans. Info. Theory*, July 1969: 484–497.

———, P. R. DROUILHET, JR., and R. PRICE, "Performance of Coherent Detection Systems Using Decision Directed Channel Measurement," *IEEE Trans. Comm. Systems*, Mar. 1964: 54–63.

PROTONARIOUS, E. N. "Slope Overload Noise in Differential PCM," *BSTJ*, Nov. 1967: 2119–2162.

PUENTE, J. G. "A Practical and Efficient Multiple Access System," COMSAT Labs., Clarksburg, MD, Tech. Memo ED-9-64, Oct. 30, 1964.

———, W. G. SCHMIDT, and A. M. WERTH. "Multiple Access Techniques for Commercial Satellites," *Proc. IEEE*, Feb. 1972: 218–229.

PURTON, R. F. "A Survey of Telephone Speech-Signal Statistics and Their Significance in the Choice of a P.C.M. Companding Law," *Proc. IEE* (London), 109B, Jan. 1962: 60–66.

PUSTARBI, H. S. "An Improved 5-MHz Reference Oscillator for Time and Frequency Standard Applications," *IEEE Trans. Inst. and Meas.*, Dec. 1966: 196–202.

RAMJO, S., and P. SAVITZ. "Orbital Design Strategy for Domestic Communication Satellite Systems," *Proc. Int. Conf. on Comm.*, 1973: 36-1–36-6.

RAMON, S., and J. R. WHINNERY. *Fields and Waves in Modern Radio*, New York: John Wiley & Sons, Inc., 1953: 368.

RAMSEY, J. L. "Realization of Optimum Interleavers," *IEEE Trans. Info. Theory*, May 1970: 338–345.

REDER, F. H., P. BROWN, G.M.R. WINKLER, and C. BICKART. "Final Results of a Worldwide Clock Synchronization Experiment," *Proc. 15th Annual Symp. on Frequency Control*, 1961.

——— and G.M.R. WINKLER. "World-Wide Clock Synchronization," *IRE Trans. on Military Electronics*, vol. MIL-4, Apr.–July 1960: 366–376.

———, G.M.R. WINKLER, and C. BICKART. "Results of a Long-Range Clock Synchronization Experiment," *Proc. IRE*, vol. 49, June 1961: 1028–1032.

REIFFEN, B. "A System for Minimizing Intermodulation Interference in Communications Systems Based on Frequency Selection," MIT. Lincoln Lab., TR 311-5, 1955.

REY, T. J. "Further on the Phase-Locked Loops in the Presence of Noise," *Proc. IEEE*, May 1965: 494–495.

RHODES, S. A. "Performance of Offset-QPSK Communications with Partially Coherent Detection," National Telecomm. Conf., 1973.

RICE, D. L., et al. "Transmission Loss Prediction for Tropospheric Loss Communication Circuits," Tech. Note 101, National Bureau of Standards, U. S. Government Printing Office, Jan. 1967.

RICE, S. O. "Mathematical Analysis of Random Noise," *BSTJ*, vol. 23, no. 3, July 1944: 282–332. Also published in N. Wax, Ed., *Selected Papers on Noise and Stochastic Processes*, New York: Dover Publications, Inc., 1958: 133–294.

———, "Noise in FM Receivers, Time Series Analysis," in *Time Series Analysis*, M. Rosenblatt, Ed., New York: John Wiley & Sons, Inc., 1963: Ch. 25.

RICHARDSON, J. M. "Progress in the Distribution of Standard Time and Frequency," presented at the URSI XVth General Assembly, Munich, Sept. 1966.

RICHMAN, D. "The DC Quadricorrelator: A Two-Mode Synchronization System," *Proc. IRE*, vol. 42, Jan. 1954: 288–299.

ROBERTS, L. G. "Picture Coding Using Pseudo-Random Noise," *IRE Trans. IT*, Feb. 1962: 145–154.

———, "PCM Television Bandwidth Reduction Using Pseudo-Random Noise," S.M. Thesis, Massachusetts Institute of Technology, Cambridge, MA, Feb. 1961.

ROBINSON, G., et al. "PSK Signal Power Spectrum Spread Produced by Memoryless Nonlinear TWTs," *COMSAT Technical Review*, Fall 1973: 227–256.

ROE, G. M. "Quantizing for Minimum Distortion," *IRE Trans. IT*, Oct. 1964: 384–385.

ROSENBAUM, A. S. "PSK Error Performance with Gaussian Noise and Intereference," *BSTJ*, Feb. 1969: 413–442.

———, "Error Performance of Multiphase DPSK with Noise and Interference," *IEEE Trans. Comm. Tech.*, Dec. 1970: 821–824.

RUZE, J. "Antenna Tolerance Theory—A Review," *Proc. IEEE*, Apr. 1966: 633–640.

RYAN, C. R., and R. F. ZIEMER. "Adaptive Equalizers for High Data Rate QPSK Transmission in Two-Component Multipath," *Proc. Int. Conf. on Comm.*, June 1973: 18-4–18-9.

SABELHAUS, A. B. "Applications Technology F and G Communication Subsystem," *Proc. IEEE*, Feb. 1971: 206–212.

SAGE, G. F. "Serial Synchronization of Pseudo-Noise Systems," *IEEE Trans. Comm. Tech.*, Dec. 1964: 123–127.

SAKRISAN, D. J. *Notes on Analog Communications*, Princeton, NJ: Van Nostrand, 1970: Ch. 5, Sec. 2.

———, "Satellite Communications Earth Stations," *IEEE Spectrum*, 1972.

SAVAGE, J. E. "Some Simple Self-Synchronizing Digital Data Scramblers," *BSTJ*, Feb. 1967: 449–487.

SCHILLING, D. L., and J. BILLIG. "On the Threshold Extension Capability of the PLL and FMFB," *Proc. IRE*, May 1964: 621.

SCHMIDT, W. "An On-Board Switched Multiple-Access System for Millimeter-Wave Satellites," *Proc. 1969 INTELSAT/IEE Int. Conf. Digital Satellite Communication*, 388–407.

SCHROEDER, K. G. "Characteristics and Applications of Multibeam Satellite Antennas," in P.L. Bargellini, Ed., *Communications Satellite Technology*, Cambridge, MA: MIT Press, 1974: 503–532.

SCHUCHMAN, L. "Dither Signals and Their Effect on Quantizing Noise," *IEEE PG Comm. Technology*, Dec. 1964: 162–165.

SCHWARTZ, J. W., et al. "Modulation Techniques for Multiple Access to a Hard-Limiting Satellite Repeater," *Proc. IEEE*, May 1966: 763–777.

SCHWARTZ, M. *Information Transmission, Modulation, and Noise*, New York: McGraw-Hill Book Company, 1970.

———, W. R. BENNETT, and S. STEIN. *Communication Systems and Techniques*, New York: McGraw-Hill Book Company, 1966: 161–163.

SELTZER, D. E. "Computed Transmission Through Rain at Microwave and Visible Frequencies," *BSTJ*, Oct. 1970: 1873–1892.

SEMPLAK, R. A. "Simultaneous Measurements of Depolarization by Rain Using Linear and Circular Polarizations at 18 GHz," *BSTJ*, Feb. 1974: 400–404.

——— and H. E. KELLER. "A Dense Network for Rapid Measurement of Rainfall Rate," *BSTJ*, July–Aug. 1969: 1745–1756.

SENGO, J. R., JR., and J. F. HAYES. "Analysis and Simulation of a PN Synchronization System," *IEEE Trans. Comm. Tech.*, Oct. 1970: 676–679.

SEVY, J. L. "The Effect of Multiple CW and FM Signals Passed through a Hard Limiter or TWT," *IEEE Trans. Comm. Tech.*, Oct. 1966: 568–578.

———, "Interference Due to Limiting and Demodulation of Two Angle Modulated Signals, *IEEE Trans. AES*, July 1968: 580–587.

SHAFT, P. D. "Intermodulation Interference in Frequency Division Multiple Access," Int. Conf. on Communications, June 1967.

———, "Limiting of Several Signals and Its Effect on Communication System Performance," *IEEE Trans. Comm. Tech.*, Dec. 1965: 504.

———, "Signals through Nonlinearities and Suppression of Undesired Intermodulation Terms," *IEEE Trans. Info. Theory*, Sept. 1972: 657–659.

SHANNON, C. E., and W. WEAVER, *The Mathematical Theory of Communication*, University of Illinois Press, Urbana, IL., 1959.

SHAVER, H. N. "Topics in Statistical Quantization," EE Thesis, Stanford University, Stanford, CA, 1965.

SHELTON, R. D., and A. F. ADKINS. "Noise Bandwidth of Common Filters," *IEEE Trans. Comm. Tech.*, Dec. 1970: 828–830.

SHERMAN, R. J. "Quadraphase Shift-Keyed Detection with a Noisy Reference Signal," *IEEE Eascon Convention Record*, 1969: 45–52.

SILVER, S. *Microwave Antenna Theory and Design*, Radiation Lab. Series, New York: McGraw-Hill Book Company, 1949.

SIMON, M. K. "An Analysis of the Steady-State Phase-Noise Performance of a Digital-Data-Transition Tracking Loop," *1969 Int. Commun. Conference Record:* 20-9–20-15.

———, "Nonlinear Analysis of an Absolute Value Type of an Early-Late Gate Bit Synchronizer," *IEEE Trans. Comm. Tech.*, Oct. 1970a: 589–596.

———, "Optimization of the Performance of a Digital Data Transition Tracking Loop," *IEEE Trans. Comm. Tech.*, Oct. 1970b: 686–689.

SIPRESS, J. M. "A New Class of Selected Ternary Pulse Transmission Plans for Digital Transmission Lines," *IEEE Trans. Comm. Tech.*, Sept. 1965: 366–372.

SLABINSKI, V. J. "Variation in Range, Range-Rate, Propagation Time Delay, and Doppler Shift for a Nearly Geostationary Satellite," in P. L. Bargellini, Ed., *Communications Satellite Technology*, Cambridge, MA: MIT Press, 1974: 3–28.

SMITH, B. M. "A Semi-Empirical Approach to the PLL Threshold," *IEEE Trans. AES*, July 1966: 463–468.

SMITH, BERNARD. "Instantaneous Companding of Quantized Signals," *BSTJ*, vol. 36, May 1957: 653–710.

SMITH, F. L. "A Nomogram for Look Angles to Geostationary Satellites," *IEEE Trans. Aerospace & Electronic Systems*, May 1972: 394.

SMITH, J. G. "Odd-Bit Quadrature Amplitude Shift Keying," *IEEE Trans. Comm., Tech.*, March 1975: 385–389.

SONES, W. K., and L. F. GRAY. "Design Features of an Unattended Earth Terminal for Satellite Communication," *Proc. Int. Conf. on Comm.*, 1973: 3-12–3-17.

SPANG, H. A., III and P. M. SCHULTHEISS. "Reduction of Quantizing Noise by Use of Feedback," *IRE Trans. Comm. Systems*, Dec. 1962: 373–380.

SPILKER, J. J., JR. "Delay-Lock Tracking of Binary Signals," *IRE Trans. SET*, Mar. 1963: 1–8.

———, "Theoretical Bounds on the Performance of Sampled Data Communications Systems," *IRE Trans. Circuit Theory*, Sept. 1960: 335–341.

——— and D. T. MAGILL. "The Delay-Lock Discriminator—An Optimum Tracking Device," *Proc. IRE*, vol. 49, Sept. 1961: 1403–1416.

STAMMINGER, R., and A. K. JEFFERIS. "Transmission System Planning for INTELSAT IV," presented at the 1970 IEE Conf. Earth Station Technology, London, England.

STEELE, J. MCA., W. MARKOWITZ, and C. A. LIDBACK. "Telstar Time Synchronization," *IEEE Trans. Inst. and Meas.*, Dec. 1964: 164–170.

STEIN, S., and J. J. JONES. *Modern Communication Principles*, New York: McGraw-Hill Book Company, 1967.

STEWART, R. M. "Statistical Design and Evaluation of Filters for the Restoration of Sampled Data," *Proc. IRE*, Feb. 1956: 253–258.

STIFFLER, J. J. *Theory of Synchronous Communications*, Englewood Cliffs, NJ: Prentice-Hall, Inc., 1971.

STILTZ, H. E., ed., *Aerospace Telemetry*, vols. I, II, Prentice-Hall, Englewood Cliffs, NJ., 1961.

STONE, R. R., JR., W. MARKOWITZ, and R. G. HALL. "Time and Frequency Synchronization of Navy VLF Transmissions," *IRE Trans. on Instrumentation*, Sept. 1960: 155–161.

STRATONOVICH, R. L., *Topics in the Theory of Random Noise*, Gordon and Breach, New York, NY., 1967.

SULLIVAN, D. P. "Future Trends in Military Communications Satellite Repeaters," *IEEE Trans. AES*, Mar. 1970: 466–470.

SULLIVAN, W. A. "High Capacity Microwave System for Digital Data Transmissions," *IEEE Trans. Comm. Tech.*, June 1972: 129–136.

SUNDE, E. D. *Communications System Engineering Theory*, New York: John Wiley & Sons, Inc., 1968.

———, "Intermodulation Distortion in Multi-Carrier FM Systems," *1965 IEEE Int. Conv. Record*, Part 2: 130–146.

SYDNOR, R., and J. J. CALDWELL. "Frequency Stability Requirements for Space Communications and Tracking Systems," *Proc. IEEE*, Feb. 1966: 231–236.

"Table of Earth Satellites," Royal Aircraft Establishment, Tech Report 71082, April 1971.

TAKAHASHI, K. "Atmospheric Error in Range and Range Rate Measurements between a Ground Station and an Artificial Satellite," *IEEE Trans. AES*, Nov. 1970: 770–779.

TAUB, H., and D. L. SCHILLING, *Principles of Communications Systems*, McGraw-Hill, New York, NY., 1972.

TAUSWORTHE, R. C. "Theory and Practical Design of Phase-Locked Receivers," Jet Propulsion Lab., Calif. Inst. of Tech., Feb. 1966.

——— and R. B. CROW. "Practical Design of Third-Order Phase-Locked Loops," Jet Propulsion Lab., Calif. Inst. of Tech., (Internal Document) TR 900–450, April 1971.

TEUPSER, R. "The Raisting Satellite Earth Station and Its System Noise Temperature," *IEEE Trans. Comm. Tech.*, Dec. 1967: 848–854.

TIKHONOV, V. I. "Phase-Lock Automatic Frequency Control Applications in the Presence of Noise," *Automatica and Telemekhanika*, vol. 23, no. 3, 1960.

TITSWORTH, R. C. "Asymptotic Results for Optimum Equally Spaced Quantization of Gaussian Data," JPL Space Programs Summary 37–29, IV, California Institute of Technology, Pasadena, CA, Oct. 1964: 242–244.

———, "Optimal Threshold and Level Selection for Quantizing Data," JPL Space Programs Summary 37–23, IV, California Institute of Technology, Pasadena, CA, Oct. 1963: 196–200.

———, "Optimal Ranging Codes," *IEEE Trans. SET*, Mar. 1964: 19–30.

——— and L. R. WELCH. "Modulation by Random and Pseudo-Random Sequences," Jet Propulsion Lab., Progress Rept. No. 20-387, California Institute of Technology, Pasadena, CA, June 1959: Sec. 3.

TOERPER, K. E. "Biphase Barker-Coded Data Transmission," *IEEE Trans. AES*, Mar. 1968: 278–289.

TOTTY, R. E., and G. C. CLARK. "Reconstruction Error in Waveform Transmission," *IEEE Trans. Info. Theory*, Apr. 1967: 336.

TURYN, R. "On Barker Codes of Even Length," *Proc. IEEE*, Sept. 1963: 1256.

VAN DE WEG, H. "Quantizing Noise of a Single Integration Delta Modulation System with an *N*-Digit Code," Philips Research Report, Oct. 1953: 367–385.

VAN HORN, J. H. "A Theoretical Synchronization System for Use with Noisy Digital Signals," *IEEE Trans. Comm. Tech.*, Sept. 1961: 82–90.

VAN LINT, J. H. *Coding Theory*, Berlin: Springer Verlag, 1971.

VAN TREES, H. L. "A Threshold for Phase Locked Loops," MIT Lincoln Lab., TR 246, Aug. 1961.

———, *Detections, Estimation, and Modulation Theory*, New York: John Wiley & Sons, Inc., Part I, 1968; Part II, 1971.

———, "Functional Techniques for the Analysis of the Nonlinear Behavior of Phase-Locked Loops," *Proc. IEEE*, vol. 52, Aug. 1964: 894–911.

VICTOR, W. K., R. STEVENS, and S. W. GOLOMB. "Radar Exploration of Venus," Goldstone Observatory Report for March–May 1961, Jet Propulsion Lab., Rept. No. 32–132, California Institute of Technology, Pasadena, CA, Aug. 1961.

VITERBI, A. J. "Acquisition and Tracking Behavior of Phase-locked Loops," *Proc. Symposium on Active Networks*, New York: Polytechnic Press, 1960: 583–619.

———, "Convolutional Codes and Their Performance in Communication Systems," *IEEE Trans. Comm. Tech.*, Oct. 1971: 751–771.

———, "Error Bounds for Convolutional Codes and an Asymptotically Optimum Decoding Algorithm," *IEEE Trans. Info. Theory*, Apr. 1967: 260–269.

———, "Phase-Locked Loop Dynamics in the Presence of Noise by Fokker-Planck Techniques," *Proc. IEEE*, vol. 51, Dec. 1963: 1737–1753.

———, *Principles of Coherent Demodulation*, New York: McGraw-Hill Book Company, 1966: Chs. 7, 10.

——— and J. P. ODENWALDER. "Further Results on Optimal Decoding of Convolutional Codes," *IEEE Trans. Info. Theory*, Nov. 1969: 732–734.

"Voice Processing for Narrow Band Channels," *Convention Record*, Session 20, National Telecommunications Conference, 1973.

WAINBERG, S., and J. K. WOLF. "Subsequences of Pseudo-random Sequences," *IEEE Trans. Comm. Tech.*, Oct. 1970: 606–612.

WALD, A. *Sequential Analysis*, New York: John Wiley & Sons, Inc., 1947.

WALL, V. W. "Military Communication Satellites," *Astronautics and Aeronautics*, vol. 6, 1968: 52–58.

WANG, M. C., and G. E. UHLENBECK. "On the Theory of Brownian Motion," *Rev. Mod. Physics*, vol. 17, nos. 2, 3, Apr.–July 1945.

WARD, R. B. "Acquisition of Pseudo-Noise Signals by Sequential Estimation," *IEEE Trans. Comm. Tech.*, Dec. 1965: 475–484.

———, "Application of Delay-Lock Radar Techniques to Deep-Space Tasks," *IEEE Trans. SET*, June 1964: 49–65.

WATT, A. D., R. W. PLUSH, W. W. BROWN, and A. H. MORGAN. "Worldwide VLF Standard Frequency and Time Broadcasting," *J. Res. NBS* (Radio Prop.) vol. 65D, no. 6, 1961: 617–627.

WATTS, D. G. "A General Theory of Amplitude Quantization with Applications to Correlation Determination," *Proc. IEE* (London), 109C, Monograph No. 481 M, Nov. 1961: 209–218.

WEIS, W. G., and M. EVANS. "Application of the Delay-Lock Discriminator to the Satellite Rendezvous Problem," presented at the Aerospace and Navigation Electronics Conf., Baltimore, MD, October 1962.

WHITNEY, H. E., J. AARONS, and C. MALIK. "A Proposed Index for Measuring Ionospheric Scintillations," *Planet. Space Science*, vol. 17, 1969: 1069–1073.

WIDROW, B. "A Study of Rough Amplitude Quantization By Means of Nyquist Sampling Theory," *IRE Trans. Circuit Theory*, Dec. 1956: 266–276.

———, "Statistical Analysis of Amplitude-Quantized Sampled-Data Systems," Technical Report 2103–1, Stanford Electronics Laboratories, Stanford University, Stanford, CA, May 10, 1960.

WIENER, N. *Interpolation, Extrapolation, and Smoothing of Stationary Time Series*, New York: John Wiley & Sons, Inc., 1949.

WILKINS, L. C., and P. A. WINTZ. "Bibliography on Data Compression, Picture Properties, and Picture Coding," *IEEE Trans. Info. Theory*, Mar. 1971: 180–197.

WILKINSON, E. J. "A Dual Polarized Cylindrical-Reflector Antenna for Communication Satellites," *Microwave Journal*, Dec. 1973: 27–62.

WILLARD, M. W. "Analysis of a System of Mutually Synchronized Oscillators," *IEEE Trans. Comm. Tech.*, Oct. 1970: 467–483.

———, "Optimum Code Patterns for PCM Synchronization," 1962 National Telemetry Conference, paper 5–5.

WINKLER, M. R. "Pictorial Transmission with HIDM," *IEEE Intel. Convention Record*, Part 1, 1965: 285–290.

WINTZ, P. A., and E. J. LUECKE. "Performance of Optimum and Suboptimum Synchronizers," *IEEE Trans. Comm. Tech.*, June 1969: 380–389.

WITT, F. J. "An Experimental 224 Mb/s Digital Multiplexer-Demultiplexer Using Pulse Stuffing Synchronization," *BSTJ*, Nov. 1965: 1843–1885.

WOLD, H. *A Study in the Analysis of Stationary Time Series*, Stockholm: Almquist and Wiksell, 1954: Appendix by P. Whittle.

WOLF, J. K. "A Survey of Coding Theory," *IEEE Trans. Info. Theory*, July 1973: 381–388.

———, "Effect of Channel Errors on Delta Modulation," *IEEE Trans. Comm. Tech.*, Feb. 1966: 2–7.

WOOD, R. C. "On Optimum Quantization," *IEEE Trans. Info. Theory*, Mar. 1969: 248–252.

WOODWARD, P. M. *Probability and Information Theory with Applications to Radar*, New York: McGraw-Hill Book Company, 1957.

WORDFORD, J. B., and R. L. DUTCHER. "A Satellite System to Support an Advanced Air Traffic Control Concept," *Proc. IEEE*, Mar. 1970: 438–447.

WOZENCRAFT, J. M., and I. M. JACOBS. *Principles of Communication Engineering*, New York: John Wiley & Sons, Inc., 1965.

YAMAMOTO, H., K. HIRADE, and Y. WATANABE. "Carrier Synchronization for Coherent Detection of High-Speed Four-Phase-Shift-Keyed Signals," *IEEE Trans. Comm. Tech.*, Aug. 1972: 803–807.

YAN, J., and R. W. DONALDSON. "Subjective Effects of Channel Transmission Errors on PCM and DPCM Voice Communications Systems," *IEEE Trans. Comm. Systems*, June 1972: 281–290.

YEH, L. P. "Geostationary Satellites Orbital Geometry," *IEEE Trans. Comm. Tech.*, Apr. 1972: 252.

ZIMMERER, R. W., and M. MIZUSHIMA. "Precise Measurement of the Microwave Absorption Frequencies of the Oxygen Molecule and the Velocity of Light," *Phys. Rev.*, vol. 121, 1961: 152–155.

ZUCKER, H. "Gain of Antennas with Random Surface Deviation," *BSTJ*, Oct. 1968: 1637–1651.

Glossary of Notation

Glossary of Notation

Symbol	*Definition*	*Typical Equation or Section Reference*
a_i	Binary values (0, 1)	(3-72)
a_i	Filter coefficient	(4-65)
$A(i)$	Integer representing the quantizer output level	(3-73)
A	Signal amplitude	Sec. 9-4
$A(t)$	Envelope function at time t	(9-1)
A/D	Analog/Digital Converter	Sec. 1-3
AFC	Automatic Frequency Control	Sec. 9-3
AME	Average Magnitude of Error	Sec. 3-6
b_i	Binary values (0, 1)	(3-72)
B_i	Bandwidth measure $B_i = \int_0^\infty (2\pi f)^{i+2} G_y(f)\, df$ rad/sec	(4-46)
B	Signal Amplitude	Sec. 9-4
B	Information bandwidth Hz (one-sided)	(3-11)
B_n	Noise bandwidth Hz (one-sided)	(12-30)
BSC	Binary Symmetric Channel	Sec. 3-2
BPSK	BiPhase Shift Keying	Sec. 3-2
c	Velocity of light	(17-2)
$\mathcal{C}$	Channel Capacity	(3-16)
C/N_0	Carrier power/noise density (one-sided)	(6-16)

Symbol	*Definition*	*Typical Equation or Section Reference*
$C_i(t)$	Clock time at terminal i at system time t	(17-1)
$C(jv_1, jv_2)$	Joint characteristic function	(9-39)
$\text{Ci}(x)$	Cosine integral $\text{Ci}(x) \triangleq \int_{\infty}^{x} \left(\frac{\cos t}{t}\right) dt$	(12-7)
CSDM	Continuous Slope Delta modulation	Sec. 4-7
d_m	Minimum distance between codes	Sec. 5-4
D	Frequency sweep rate	(12-68)
$D_\Delta(\epsilon)$	Delay-lock-loop (1Δ DLL) characteristics as a function of delay error	(18-4)
$\mathfrak{D}$	Distortion metric	(3-24)
dB	Decibel 10 $\log_{10} P_2/P_1$	(6-19)
dBm	Decibel relative to 1 milliwatt	(6-24)
DFE	Decision Feedback Equalizer	Sec. 13-7
DLL	Delay-lock loop	Sec. 18-2
DPCM	Differential Pulse Code Modulation	Sec. 4-1
e	Number 2.71828	(2-1)
$e(t)$	Error in message estimate	(3-5)
e_i	Binary (0, 1) error value at time i	(3-74)
$\mathbf{E}$	Expected value operator	(3-6)
E	Elevation angle (degrees)	Fig. 6-19
E_b	Energy per bit	(3-11)
erfc x	Co-error function $\frac{2}{\sqrt{\pi}} \int_x^{\infty} e^{-y^2} dy \leq 2$	(3-12)
erfc′ x	Alternate Co-error function $\frac{1}{\sqrt{2\pi}} \int_x^{\infty} e^{-y^2/2} dy \leq 1$	(3-12)
erf x	Error function $(2/\pi) \int_0^x e^{-y^2} dy$, erf x + erfc $x = 2$	(4-102)
f	Frequency in Hz	(2-1)
f_s	Sampling rate $f_s \triangleq 1/T$	(2-1)
f_i	Fibonacci sequence	Sec. 4-7
$f(k)$	Waveform at time $t = kT$	(4-85)
$f(\omega)$	Magnitude of $F(j\omega)$	(2-2)
$F(j\omega)$	Filter transfer function	(2-2)
$\mathbf{F}(u)$	Amplitude nonlinearity	(9-2)
FP	Fokker-Planck equation	Sec. 12-10
$g(A)$	Envelope of nonlinearity in fundamental frequency zone	(9-5)
G	Antenna gain	Sec. 6-7

Symbol	*Definition*	*Typical Equation or Section Reference*
$G_s(\omega)$	Power spectral density of $s(t)$	Sec. 2-2
$\mathbf{G}_i$	Code generator	Sec. 15-2
$\mathbf{G}_k(m)$	Gray code representation of integer m	Table 3-3
$h(t)$	Impulse response of smoothing filter	(4-89)
$H(j\omega)$	Smoothing filter transfer function	(4-15)
$\mathcal{H}_n(x)$	Hermite polynomials	(9-45)
i	Integer	(3-7)
I	Integral	(2-8)
$I_0(x)$	Modified Bessel function of the first kind and order zero	(12-133)
$\mathcal{I}$	Satellite inclination	
IM	Intermodulation product power	9-82
IF	Intermediate Frequency	Sec. 8-1
j	$\sqrt{-1}$ Imaginary value, or a summing integer	(2-1)
$J_m(x)$	Bessel function of order m and argument x	(9-66)
k	Number of information bits in a block of length n for a (n, k) code	Sec. 3-6
K	Constant	
l	Number of bits/sample	Sec. 3-2
$\log x$	Natural logarithm of x	(3-14)
$m(t)$	Analog message	Sec. 2-2
M	Size of ensemble of signal sources	(4-53)
M	Link Margin	(6-23)
MUX	Multiplexer	Sec. 1-3
MSK	Minimal Shift Keying	Sec. 11-4
$M = 2^n$	Number of phases in *M*ary PSK (MPSK)	Sec. 11-5
$M = 2^n - 1$	Period of n stage maximal length shift register sequence	Sec. 18-2
n	Number of bits/symbol	Sec. 11-5
nm	Nautical mile 1 nm = 6076.1 ft = 1.852 km	Sec. 6-2
$n(t)$	Additive noise	Sec. 2-2
N	Number of quantizing levels $N - 2^l$	(3-10)
N_0	Noise power spectral density (one-sided)	(3-11)
p	Heaviside operator $p = d/dt$ or probability of error	(12-25)
$p(x)$	Probability density of variable x	(3-30)
P_i	Power in the ith signal	(3-11)
$\mathcal{P}_b$	Probability of bit error	(3-11)

Symbol	*Definition*	*Typical Equation or Section Reference*
PCM	Pulse Code Modulation	Sec. 3-1
PST	Pair Selected Ternary	Sec. 16-1
q	Probability of correct code bit, $q = 1 - p$	(3-84)
$Q(S_i)$	Quantizer output for input S_i	(3-1)
QPSK	Quadriphase Shift Keying	Sec. 11-1
r	Satellite range from earth station	Sec. 6-3
$r(t)$	Sampling function	Sec. 2-2
$\mathcal{R}$	Information bit rate $\mathcal{R} = 1/T = \ell f_s$	(3-11)
$R_y(nT)$	Autocorrelation function of $y(T)$ for offset nT	(4-24)
$\mathcal{R}_s$	Transmission symbol rate $\mathcal{R}_s = \mathcal{R}/n$, n bits per symbol	Sec. 13-3
R_s	Normalized sampling rate $R_s = f_s/2f_m$	(4-27)
R_b	Normalized bit rate $\mathcal{R}_b \triangleq \ell f_s/2f_m$	(4-8)
R_i	Signal autocorrelation $R_i \triangleq R_y(iT)$	(4-69)
re	Earth radius re = 3444 nm = 6373 km	Sec. 6-2
RC	Resistor-Capacitor Filter	Sec. 4-4
RF	Radio Frequency	Sec. 7-3
Re( )	Real part of complex quantity	(13-2)
s	Complex frequency $s = \sigma + j\omega$	(12-25)
$s(t)$	Input signal	Sec. 2-2
$s^*(t)$	Sampled version of $s(t)$, $s^*(t) \triangleq s(t)r(t)$	Sec. 2-2
$S(j\omega)$	Fourier transform of $s(t)$	Sec. 2-2
Si(x)	Sine integral function $\mathrm{Si}(x) \triangleq \int_0^x (\sin y/y)\, dy$	(4-38)
S_{max}	Maximum ΔM or DPCM Slope	(4-2)
S_i	Input signal samples at time iT	Sec. 4-6
$\mathbf{S}$	Stuff word vector used in multiplexing	Sec. 5-4
sinc x	Function $\sin x/x$	Sec. 4-4
SNR	Signal-to-Noise Power Ratio (usually in dB)	Sec. 2-4
SQPSK	Staggered (or offset) QPSK modulation	Sec. 11-1
t	Time variable	Sec. 2-2
T	Sampling period $T \triangleq 1/f_s$	Sec. 2-2
T	Time delay	(18-1)
$\hat{T}$	Estimate of time delay	(18-1)
$\mathfrak{T}$	Noise Temperature	(6-19)
T_f	Frame duration for TDMA	Sec. 10-3
TDMA	Time Division Multiple Access	Sec. 10-1

Symbol	*Definition*	*Typical Equation or Section Reference*
TWT	Traveling Wave Tube	Sec. 9-9
u	Normalized frequency $u \triangleq \omega/\omega_n$	(12-35)
$u(t)$	Unit step function	(13-2)
UHF	Ultra-High Frequency	Sec. 6-3
$v(x)$	Compressor characteristic	(3-60)
V	Maximum quantizer level	(3-42)
$\mathcal{V}$	Quantizer output variance	Sec. 3-7
VCO	Voltage Controlled Oscillator	Sec. 12-4
VSDM	Variable Slope Delta Modulation	Sec. 4-7
$W(A)$	Weight of code word A	(5-27)
W	RF transmission bandwidth	Sec. 3-2
x_i	Time samples $x(iT)$	(3-24)
$y'(t)$	Time derivative of $y(t)$, $y' \triangleq dy/dt$	(4-31)
y_i	Time samples $y(iT)$	(3-24)
$Y(j\omega)$	Smoothing filter transfer function	(2-1)
$\mathbf{z}$	Channel vector	(12-39)
α_{ij}	jth quantizer output bit at time i, ± 1 values	(3-6)
α	Attenuation coefficient in dB/km	(6-8)
β_{ij}	jth quantizer bit, including error effects at time i	(3-6)
γ	Constant = 0.577216	(12-8)
$\Gamma(x)$	Gamma function	(4-54)
$\delta(t)$	Dirac delta function	(12-118)
δ	Quantizer interval (uniform)	(3-31)
δ_{ij}	Kronecker delta function, $\delta_{ij} = 1$ if $i = j$	(3-6)
Δ	Duration of burst noise event	(4-41)
Δ_i	Nonuniform quantizing interval	(3-34)
Δx	Effective input step size	(3-62)
ΔM	Delta Modulation	Sec. 4-1
ϵ_o	Dielectric constant	Sec. 8-3
ϵ	Phase error or delay error	(12-23)
ε_m	Neumann factor	(9-61)
ζ	Damping coefficient in second-order loop	(12-29)
η	Second derivative of $y(t)$, $\eta \triangleq y''(t)$	(4-46)
η	Efficiency	(10-8)
η_o	$\eta_o \triangleq \sqrt{\mu_o/\epsilon_o}$	Sec. 8-3
$\theta(\omega)$	Phase angle of filter transfer function	(2-2)
$\theta(A)$	AM/PM transfer function	(9-90)
κ	Boltzmann's constant 198.6 dBm/°K Hz	(6-19)
λ	Signal wavelength	Sec. 6-3

Symbol	Definition	Typical Equation or Section Reference
μ_0	Permeability	
$\boldsymbol{\mu}_0$	Earth gravitational parameter	(6-1)
$\boldsymbol{\mu}$	Compressor characteristic constant	(3-60)
μ	Time derivative of $y(t)$, $\mu \triangleq y'(t)$	(4-46)
ν	Integer used in summations	(2-1)
ξ	Third derivative of $y(t)$, $\xi \triangleq y''(t)$	(4-46)
ρ_n	Sample values of autocovariance $\rho_n \triangleq \rho(nT)$	(4-9)
$\sum$	Summation	(2-1)
σ_s, σ_e	RMS values of s, e	(4-74)
τ	Time duration of one bit or propagation delay	(3-10) or Sec. 18-2
$\hat{\tau}$	Estimate of time delay τ	(18-7)
ϕ_μ	Cross-covariance	(4-9)
$\phi(\omega)$	Phase angle of filter transfer function	(2-2)
$\psi(t)$	Carrier phase jitter	(11-12)
ω	Frequency in radians/sec, $\omega = 2\pi f$	(2-1)
$\hat{\omega}$	Measured average frequency in radians/sec	(12-3)
ω_n	Natural frequency of a linear transfer function	(12-45)
$\nabla\epsilon$	Gradient of ϵ	(13-42)

Index

Index